AF335543

Shapes and Geometries

Advances in Design and Control

SIAM's Advances in Design and Control series consists of texts and monographs dealing with all areas of design and control and their applications. Topics of interest include shape optimization, multidisciplinary design, trajectory optimization, feedback, and optimal control. The series focuses on the mathematical and computational aspects of engineering design and control that are usable in a wide variety of scientific and engineering disciplines.

Editor-in-Chief
John A. Burns, Virginia Polytechnic Institute and State University

Editorial Board
H. Thomas Banks, North Carolina State University
Stephen L. Campbell, North Carolina State University
Eugene M. Cliff, Virginia Polytechnic Institute and State University
Ruth Curtain, University of Groningen
Michel C. Delfour, University of Montreal
John Doyle, California Institute of Technology
Max D. Gunzburger, Iowa State University
Rafael Haftka, University of Florida
Jaroslav Haslinger, Charles University
J. William Helton, University of California at San Diego
Art Krener, University of California at Davis
Alan Laub, University of California at Davis
Steven I. Marcus, University of Maryland
Harris McClamroch, University of Michigan
Richard Murray, California Institute of Technology
Anthony Patera, Massachusetts Institute of Technology
H. Mete Soner, Carnegie Mellon University
Jason Speyer, University of California at Los Angeles
Hector Sussmann, Rutgers University
Allen Tannenbaum, University of Minnesota
Virginia Torczon, William and Mary University

Series Volumes
Delfour, M. C. and Zolésio, J.-P., *Shapes and Geometries: Analysis, Differential Calculus, and Optimization*

Betts, John T., *Practical Methods for Optimal Control Using Nonlinear Programming*

El Ghaoui, Laurent and Niculescu, Silviu-Iulian, eds., *Advances in Linear Matrix Inequality Methods in Control*

Helton, J. William and James, Matthew R., *Extending H^∞ Control to Nonlinear Systems: Control of Nonlinear Systems to Achieve Performance Objectives*

Shapes and Geometries

Analysis, Differential Calculus, and Optimization

M. C. Delfour

Université de Montréal
Montréal, Canada

J.-P. Zolésio

CNRS & INRIA
Sophia Antipolis, France

Society for Industrial and Applied Mathematics
Philadelphia

Library of Congress Cataloging-in-Publication Data
Delfour, Michel C., 1943-
    Shapes and geometries : analysis, differential calculus, and optimization / M.C. Delfour, J.-P. Zolésio.
      p. cm. – (Advances in design and control)
    Includes bibliographical references and index.
    ISBN 0-89871-489-3
      1. Shape theory (Topology) I. Zolésio, J.-P. II. Title. III. Series.

QA612.7 .D45 2001
514'.24--dc21

                              2001020303

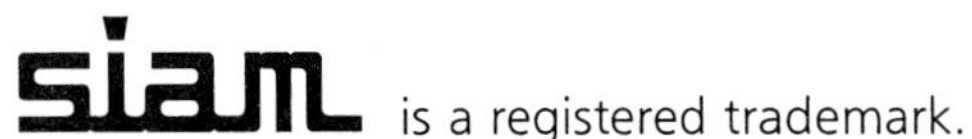 is a registered trademark.

This book is dedicated to
Alice, Jeanne, Jean, and Roger

Contents

List of Figures

Preface

Objectives and Scope of the Book

The objective of this book is to give a comprehensive presentation of mathematical constructions and tools that can be used to study problems where the modeling, optimization, or control variable is no longer a set of parameters or functions but the shape or the structure of a geometric object. In that context, a good analytical framework and good modeling techniques must be able to handle the occurrence of singular behaviors whenever they are compatible with the mechanics or the physics of the problems at hand. In some optimization problems the natural intuitive notion of a geometric domain undergoes *mutations* into relaxed entities such as microstructures. So the objects under consideration need not be smooth open domains, or even sets, as long as they still make sense mathematically.

This book covers the basic mathematical ideas and methods, which often come from very different areas of applications and fields of mathematical activities and which have traditionally evolved in parallel directions. The field of research is frighteningly broad because it touches on areas that include classical geometry, modern partial differential equations, geometric measure theory, topological groups, and constrained optimization, with applications to classical mechanics of continuous media such as fluid mechanics, elasticity theory, fracture theory, modern theories of optimal design, optimal location and shape of geometric objects, free and moving boundary problems, and image processing. New issues raised in some applications force researchers to take a fresh view of the fundamentals of well-established mathematical areas such as boundary value problems, to find suitable relaxation of solutions, or geometry, to relax basic notions of volume, perimeter, and curvature. In that context Henri Lebesgue was certainly a pioneer when in 1907 he relaxed the intuitive notion of volume to the one of measure on an equivalence class of sets. He was followed in that spirit in the early 1950s by the celebrated work of E. De Giorgi, who used the relaxed notion of *finite perimeter* defined on the class of Caccioppoli sets to solve Plateau's problem of minimal surfaces.

Thus the material that is pertinent to the study of geometric objects and the quantities and functions that are defined on them is so broad that it would necessitate an encyclopedic investment to bring together the basic theories and their fields of applications. This objective is obviously beyond the scope of a single book and two authors. The coverage of this book is more modest but critical at this stage of evolution of the field. Even if shape analysis and optimization have undergone considerable and important developments on the theoretical and numerical fronts, there are still cultural barriers between areas of applications and between theories. The whole field is extremely active, and the best is yet to come with fundamental structures and tools beginning to emerge. It is hoped that this book will help to build new bridges and stimulate cross-fertilization of ideas and methods.

Structure of the Book

The book is informally divided into two parts. The first part (Chapters 1–6) presents a global classification of domains according to their level of smoothness. With each level is associated a set parametrized function: the characteristic function for measurable sets, the distance function for Hausdorff-type metric topologies, the oriented distance function for finer metric topologies involving curvatures and their derivatives, and the Courant metric for families of diffeomorphisms. The oriented distance function appears as a very promising tool for describing in a single framework the whole range of sets, from those with a nonempty boundary to C^∞-domains. The second part (Chapters 7–9) is devoted to shape derivatives, their definitions, their structure, the latest version of the shape calculus, a new modern approach to the tangential differential calculus, and finally the shape calculus under state equation constraints. Chapter 7 is also a transition between transformation and velocity methods.

Intended Audience

The targeted audience is mainly applied mathematicians and advanced engineers, but the book is also suitable for a broader audience of mathematicians as a relatively well-structured initiation to shape analysis and calculus techniques. Some of the chapters are fairly self-contained and of independent interest. They can be used as lecture notes for a mini-course. The material at the beginning of each chapter is accessible to a broad audience, while the latter sections may sometimes require more mathematical maturity. Thus, the book can be used as a graduate text as well as a reference book. It complements existing books which emphasize specific mechanical or engineering applications or numerical methods. It can be considered a companion to J. Sokołowski and J. P. Zolésio's *Introduction to Shape Optimization* [9], published in 1992.

Earlier versions of parts of this book have been used as lecture notes in graduate courses at the Université de Montréal in 1986–1987, 1993–1994, 1995–1996, and 1997–1998, and at international meetings, workshops, and schools: Séminaire de Mathématiques Supérieures on *Shape Optimization and Free Boundaries* (Montréal, Canada, June 25 to July 13, 1990); short course on Shape Sensitivity Analysis (Kénitra, Morocco, December 1993); course of the COMETT MATARI European Program on *Shape Optimization and Mutational Equations* (Sophia-Antipolis, France, September 27 to October 1, 1993); CRM Summer School on *Boundaries, Interfaces and Transitions* (Banff, Canada, August 6–18, 1995); and the CIME course on Optimal Design (Troia, Portugal, June 1998).

Acknowledgments

The first author is pleased to acknowledge the support of the Canada Council, which initiated the work presented in this book through a Killam Fellowship; the constant support of the National Sciences and Engineering Research Council of Canada; and the FCAR program of the Ministère de l'Éducation du Québec. Many thanks also to Louise Letendre and André Montpetit of the Centre de Recherches Mathématiques, who provided their technical support, experience, and talent over the extended period of gestation of this book.

Michel Delfour
Jean-Paul Zolésio

August 13, 2000

Chapter 1
Introduction

1 Geometry as a Variable

The central object of this book[1] is the *geometry* as a variable. As in the theory of functions of real variables, we want to retain the concepts of differential calculus, spaces of geometries, evolution equations, and other familiar concepts in analysis when the variable is no longer a scalar, a vector, or a function, but a geometric domain. This is motivated by many important problems in physics and engineering which involve the geometry as a modeling, design, or control variable. In general the geometric objects we shall consider will not be parametrized or structured. Yet we are not starting from scratch, and several building blocks are already available from many fields: geometric measure theory, physics of continuous media, free boundary problems, the parametrization of sets by functions, the set derivative as the inverse of the integral, the parametrization of functions by sets, the Hausdorff metric, and so on.

As is often the case in mathematics, spaces of geometries and notions of derivatives with respect to the geometry are built on well-established elements of functional analysis and differential calculus. There are many ways to structure families of geometries. For instance, a domain can be made variable by considering the images of a fixed domain by a family of diffeomorphisms that belong to a function space

---

[1] The numbering of equations, theorems, lemmas, corollaries, definitions, examples, and remarks is by chapter. When a reference to another chapter is necessary, it is always followed by the words *in Chapter* and the number of the chapter. For instance, "equation (2.14) in Chapter 8." The text of theorems, lemmas, and corollaries is italic. The text of definitions, examples, and remarks is roman and is ended by a square ($\square$). This makes it possible to aesthetically emphasize certain words, particularly in definitions. The bibliography is by author in alphabetical order. For each author or group of coauthors, there is a number in square brackets, beginning with one. A reference to an item by a single author is of the form J. Dieudonné [1], and a reference to an item by several coauthors is of the form S. Agmon, A. Douglis, and L. Nirenberg [2]. Boxed formulae or statements are used in some chapters for two distinct purposes: first, to emphasize certain important definitions, results, or identities; second, in long proofs of some theorems, lemmas, or corollaries, to isolate key intermediary results that are necessary for the reader to follow the subsequent steps of the proof.

over a fixed domain. This often occurs in physics and mechanics, where the deformations of a continuous body or medium are smooth, or in the numerical analysis of optimal design problems for working on a fixed grid. This construction naturally leads to a group structure induced by the composition of the diffeomorphisms. The underlying spaces are no longer topological vector spaces, but topological groups that can be endowed with a nice complete metric space structure by introducing the *Courant metric*. The practitioner might or might not want to use the underlying mathematical structure associated with his or her constructions, but it is there and it contains information that might guide the theory and influence the choice of the numerical methods used in the solution of the problem at hand.

The parametrization of a fixed domain by a fixed family of diffeomorphisms obviously limits the family of variable domains. The topology of the images is similar to the topology of the fixed domain. Singularities that were not already present there cannot be created in the images. Other constructions make it possible to considerably enlarge the family of variable geometries and possibly open the doors to pathological sets that are no longer open sets with a nice boundary. Instead of parametrizing the domains by functions or diffeomorphisms, certain families of functions can be parametrized by sets. A single function completely specifies a set, or at least an equivalence class of sets. This includes the distance functions and the characteristic function, but also the support function from convex analysis. Perhaps the best-known example of that construction is the Hausdorff metric topology. This is a very weak topology that does not preserve the volume of a set. When the volume, the perimeter, or the curvatures are important, such functions must be able to yield relaxed definitions of volume, perimeter, or curvatures. The characteristic function that preserves the volume has many applications. It played a fundamental role in the integration theory of Henri Lebesgue at the beginning of the 20th century. It was also used in the 1950s by E. De Giorgi to define a relaxed notion of perimeter in the theory of minimal surfaces.

Another technique that has been successfully used in *free* or *moving boundary problems*, such as motion by mean curvature, shock waves, or detonation theory, are the level sets of a function to describe a free or moving boundary. Such functions are often the solution of a system of partial differential equations. (A good introduction to level set techniques and their applications can be found in the book of Sethian [1].) This is another way to build new tools from functional analysis. The choice of set parametrizing or parametrized functions or other appropriate constructions is obviously problem dependent, much like the choice of function spaces of solutions in the theory of partial differential equations or optimization problems. This is one aspect of the geometry as a variable. Another aspect is to build the equivalent of a differential calculus and other computational and analytical tools that are essential in the characterization and computation of geometries. Again, we are not starting from scratch and many building blocks are already available, but there are still many open questions.

This book aims to cover a small but fundamental part of that program. This means that we had to make difficult choices and refer the reader to appropriate books and references for background material such as geometric measure theory and specialized topics such as homogenization theory and microstructures, which are now available in excellent books in English. It was unfortunately not possible to include references to the considerable literature on numerical methods, free and moving boundary problems, and optimization.

2 Shape Analysis

The terminology *shape analysis* has been introduced independently in at least two different contexts: continuum mechanics and the mathematical theory of partial differential equations (see, for instance, Haug and Céa [1]).

In continuum mechanics, shape analysis encompasses many contributions to structural mechanics of elastic bodies such as beams, plates, shells, arches, and trusses. There the objective is to optimize the *compliance*, e.g., the work of the applied loadings, by choosing the design parameters of the structure. Many of the early contributions are in dimension 2, and complete analytical solutions are often provided. Yet it is not always easy to distinguish between a *shape optimization problem*, such as the shape of a two-dimensional plate and a *distributed parameters problem*, such as the optimal thickness of that plate. The thickness that is often considered as a *shape parameter* is really a distributed parameter over the two-dimensional domain, which also specifies the plate. When the thickness goes to zero in some parts of the plate, holes are created and induce changes in the topology and the shape of the associated two-dimensional domain. This naturally leads to *topological optimization*, which deals with the connectivity of a domain, the number of holes, the fractal dimension of the boundary, and ultimately the appearance of a microstructure. This was exemplified by Cheng and Olhoff's [2] celebrated optimization of the compliance of the circular plate with respect to its thickness under prescribed loading and a constraint on the volume of material. Such questions have received a lot of attention in specific cases and have been analyzed by homogenization methods or Γ convergence (see, for instance, Murat and Tartar [1], the conference proceedings edited by Bendsøe and Mota [1], the book by Bendsøe [1], and the book edited by Cherkaev and Kohn [1], which contains a selection of translations of key papers written in French or Russian). However, many fundamental questions still remain open. For instance, how does such an analysis affect the validity of the underlying mechanical or physical models?

For convenience we shall refer to this viewpoint as the *compliance analysis*, which generally involves the extremum of the minimum of an energy or work function with respect to some design parameters.

In the mathematical theory of partial differential equations, the analysis dealt with the sensitivity of the solution of boundary value problems with respect to the shape of the geometric domain on which the partial differential equation is defined. This was done for different applications including free boundary problems, noncylindric problems,[2] and shape identification problems. This *shape sensitivity analysis* was simultaneously developed for the solution of the partial differential equation and for shape functions depending on that solution.

---

[2] A cylindrical problem is a partial differential equation in which the geometric domain is fixed and independent of the time variable. A noncylindric problem is a partial differential equation in which the geometric domain changes with time.

In that context, the compliance analysis becomes a special case, which, quite remarkably, does not require the shape sensitivity analysis of the solution of the associated partial differential equation. This important simplification arises from the fact that the compliance is the minimum of an energy or work function. An historical example that also benefits from this property is the shape derivative of the first eigenvalue of the plate studied at the beginning of the 20th century by Hadamard [1]. As for the compliance, the shape sensitivity analysis of the first eigenfunction is not required even when the first eigenvalue is repeated. This follows from the fact that the first eigenvalue can be expressed as a minimum through Rayleigh's quotient or Auchmuty's [1] dual variational principle.

Shape sensitivity analysis deals with a larger class of shape functions (e.g., minimal drag and noise reduction) and partial differential equations (e.g., the wave equation, viscous or non-Newtonian fluids), where variational energy functions are usually not available and for which the compliance analysis is no longer applicable. Yet the shape sensitivity analysis of the solution of the partial differential equation can again be avoided by incorporating the partial differential equation into a Hamiltonian or Lagrangian formulation. As in control theory, this yields an *adjoint state* partial differential equation, which is coupled with the initial partial differential equation or *state equation*. A precise mathematical justification of this approach can be given when the Lagrangian has saddle points and the shape derivative can be obtained from theorems on the differentiability of saddle points with respect to a parameter even when the saddle point solution is not unique.

The shape sensitivity analysis through a family of diffeomorphisms which preserve the smoothness of the images of a fixed domain is primarily a local analysis. It is used to establish continuity, define derivatives, or optimize in a narrow class of domains with fixed regularities and topologies: it cannot create holes or singularities that were not present in the initial domain. Consider the problem of finding the best location and shape of a hole of given volume in a homogeneous elastic plate to optimize the compliance or some other criterion under a given loading. The expectation is that the presence of the hole would improve the compliance over a homogeneous plate without holes. This problem has its analogue in control theory, for instance, the optimal placement of sensors and actuators for the control and stabilization of large flexible space structures or flexible arms of robots. Another important example is the localization of sensors and actuators to achieve noise reduction in structures. The optimal placement is usually an integral part of the control synthesis.

In the early 1970s Céa, Gioan, and Michel [1] proposed to introduce relaxed problems in which the optimal domain (here an optimal hole, the optimal location of the support of the optimal control, etc.) was systematically replaced by a density function ranging between zero and one and that would hopefully be a characteristic function (a bang bang control) in the optimal regime. Furthermore, in order to work on a fixed domain D without holes, they replaced the holes by a very weak elastic material. This changed the original topological optimization problem into the identification of distributed coefficients over the fixed domain D. Equivalently, this identification problem reduces to finding the optimal distribution of two materials for a transmission equation in D under a volume constraint on one of the two materials. Under appropriate conditions, the solution of this problem is a characteristic function, even when the space of distributed parameters is relaxed to the closed convex hull of the set of characteristic functions (cf. Chapter 3). This general technique can also be used to study the continuity of the solution of the homogeneous Dirichlet boundary value problems with respect to the domain Ω by introducing a transmission problem over a fixed domain D and letting the distributed coefficient go to infinity in the complement of Ω with respect to D. Of course, the coefficient over the complement of Ω could be replaced by a Lagrange multiplier, thus providing a formulation over the fixed domain D. This is one of the many ways to make the domain *fictitious* and avoid dealing directly with the geometry.

3 Geometric Measure Theory

The main advantage of the relaxation to a characteristic function in the formulation of problems involving a domain integral and/or a volume constraint on Lebesgue measurable sets is that the unknown set is completely described by a single function instead of a family of local diffeomorphisms.

Yet such measurable sets can be quite unstructured. For example, in problems involving a surface tension along the free boundary of a fluid or across the interface between two fluids, the domain must have a locally finite boundary measure. Another example is when the objective function is a function of the normal derivative of the state variable along the boundary (e.g., a flow or a thermal power flux through the boundary). In both cases the use of characteristic functions is limited by the fact that they are not differentiable across the boundary of the set and cannot be readily used to describe very smooth domains.

Fortunately there is enough room to make sense of a locally finite boundary measure for sets for which the characteristic function is a function of bounded variation, that is, its gradient is a vector of bounded measures. Such sets are known as *Caccioppoli* or *finite perimeter sets*. They were a key ingredient in the contribution of De Giorgi to the *theory of minimal surfaces* in the 1950s, since the norm of that vector measure (which is the total variation of the gradient of the characteristic function) turns out to be a relaxation of the boundary measure or "perimeter" of the boundary of the set. To get a compactness result for a family of Caccioppoli sets it is sufficient to put a uniform bound on their perimeter. This field of activities is known as *geometric measure theory* and its tools have been very successfully used in the theory of free and moving boundary problems. Even though it came rather late to *shape analysis*, this material is both important and fundamental.

4 Distance Functions, Smoothness, Curvatures

Another set-parametrized function that can play a role similar to the characteristic function is the *distance function* for a family of closed subsets of a fixed subset[3] D

---

[3] This set D will play several roles in this book. It is the *universe* in which a family of subsets lives. It will often be referred to as the underlying *hold-all*. In other circumstances it will have a purely technical role much like the control volume in fluid dynamics.

of $\mathbf{R}^N$. The Hausdorff metric between two sets corresponds to the uniform norm on continuous functions of the difference of their distance functions. For most applications it is not a very interesting topology since the volume is not continuous for that topology. Yet the characteristic function of the closure of the set can be expressed in terms of the gradient of the distance function. The continuity of the characteristic function and the volume can be restored by replacing the uniform norm of continuous functions by a $W^{1,p}$ norm.

In 1951, Federer [1] introduced the family of *sets of positive reach* and gave a first access to curvatures from the distance function in much the same spirit as the perimeter from the characteristic function. They are sets for which the projection onto the set is unique in a neighborhood of the set. Since the projection can be expressed in terms of the gradient of the square of the distance function, it can be shown that this is equivalent to saying that the square of the distance function is $C^{1,1}$ in a neighborhood of the set. Since the gradient of the distance function has a jump discontinuity at the boundary of the set, he managed to recover from the distance function the *curvature measures* of the boundary and made sense of the Steiner formula.

The jump discontinuity of the gradient across the boundary can be removed for domains for which the boundary is a submanifold of codimension 1, by going to the *algebraic distance function*. For smooth domains this function is quite remarkable since it inherits the same degree of smoothness in a neighborhood of the boundary as the boundary itself. Then the gradient, the Hessian matrix, and the higher order derivatives in the neighborhood of the boundary can be used to characterize and compute curvatures and derivatives of curvatures along the boundary. This correspondence remains true for domains of class $C^{1,1}$. As defined in this book the gradient of the algebraic distance function coincides with the outward unit normal on the boundary. Because of this implicit orientation, it will also be called the *oriented distance function*. The restriction of the Hessian matrix to the boundary coincides with the second fundamental form of differential geometry. Its eigenvalues are zero and the principal curvatures of the boundary.

A nice relaxation of the curvatures is obtained by considering sets for which the elements of the Hessian matrix of the oriented distance function are bounded measures. They are called *sets of locally bounded curvature* (cf. Delfour and Zolésio [17, 32]). To get a compactness result, it is sufficient to put a uniform bound on the total variation of the Hessian matrix.

The oriented distance function also provides a framework to compare the relative smoothness of sets ranging from arbitrary sets with a nonempty boundary to sets of class C^∞ via the smoothness of the oriented distance function in a neighborhood of the boundary of the set. As a result Sobolev spaces can be used to introduce the notion of a *Sobolev domain*, which becomes intertwined with the classical notion of C^k-domains.

5 Shape Optimization

As in the vector space case, optimization and control problems with respect to geometry present various degrees of difficulty. When the objective function does not depend on the solution of a state equation or variational inequality defined on the varying domain, it is sufficient to invoke compactness and continuity arguments. In special cases such as the optimization of the compliance or of the first eigenvalue the problem can be transformed into the optimization of an objective function which is itself the minimum of some appropriate function defined on a fixed function space. There a direct study of the dependence of the solution of the state equation with respect to the underlying domain can be bypassed. In the general case a state equation constraint has to be very carefully handled from both the mathematical and the application viewpoints. When the analysis can be restricted to families of Lipschitzian or convex domains, it is usually possible to give a meaning and prove the continuity of the solution of the state equation with respect to the underlying varying domain.

When families of arbitrary bounded open domains are considered new phenomena can occur. As the domains converge in some sense the corresponding solutions may converge in a weak sense to the solution of a different type of state equation over the limit domain: boundary conditions may no longer be satisfied, strange terms may occur on the right-hand side of the equation, etc. In such cases it is often a matter of modeling of the physical or technological phenomenon. Is it natural to accept a generalized solution or a relaxed formulation of the state equation, or should the family of domains be sufficiently restricted to preserve the form of the original state equation and maintain the continuity of the solution with respect to the domain? The most suitable relaxation is not necessarily the most general mathematical relaxation. It must always be compatible with the physical or technological problem at hand. Too general a relaxation of the problem or too restrictive conditions on the varying domains can yield completely unsatisfactory solutions however nice the underlying mathematics may be. A good balance of mathematical, physical, and engineering intuition is essential.

Of course, the study of the shape continuity of the solution of the partial differential equation is also of independent mathematical interest. In the literature this issue has been addressed in various ways. Some authors simply introduce a *stability assumption*, which essentially says that the limiting domain is such that continuity occurs. Others introduce a set of more technical assumptions that correspond to a minimal set of conditions to make the crucial steps of the proof of the continuity work. On the constructive side the really challenging issue is to characterize the families of domains for which the continuity holds.

Many authors have constructed (published and unpublished) topologies and compact families for that purpose. For instance, the *Courant metric topology* was used by Micheletti [1] in 1972 for C^k-domains, the *uniform cone condition* by Chenais [1] in 1973 for uniformly Lipschitzian domains, and other metric topologies by Murat and Simon [1] in 1976 for Lipschitzian domains.

More general capacity conditions were introduced after 1994 by Bucur and Zolésio [5] in order to obtain compact subfamilies of domains with respect to the complementary Hausdorff topology and to control the curvature of the boundaries of the domains. In dimension 2 they recover the nice result of Šverák [2] in 1993, which involves a bound on the number of connected components of the complement of the sets. Intuitively the capacity conditions are such that, locally, the complement of the domains in the chosen family has "enough capacity" to preserve the homogeneous Dirichlet boundary condition in the limit. Yet the capacity conditions are not easy to use in a practical example. So Bucur and Zolésio introduced a simpler geometric constraint, called the *flat cone condition*, under which the continuity and compactness results still hold. This generalizes the *uniform cone property* to a much larger class of open domains.

6 Shape Derivatives

For functions defined on a family of domains, it is important to distinguish between a function with values in a topological vector space over a fixed domain and a function such as the solution of a partial differential equation, which takes its values in a Sobolev space defined on the varying domain. In the latter case special techniques have to be used to transport the solution onto a fixed domain or to first embed the varying domains into a fixed *hold-all* D, extend the solutions to D, and enlarge the Sobolev space to a large enough space of functions defined over the fixed hold-all D. In both cases the function is defined over a family of domains or sets belonging to some *shape space* which is generally nonlinear and nonconvex. Thus defining derivatives on those spaces is more related to defining derivatives on differentiable manifolds than in vector spaces.

It is perhaps for topological groups of diffeomorphisms that the most complete theory of shape derivatives is available. This material will be covered in detail in Chapter 2, but it is useful to briefly introduce the main ideas and definitions here. Given a Banach space Θ of mappings from a fixed open hold-all D into $\mathbf{R}^N$, first consider the group of diffeomorphisms

$$\mathcal{F}(\Theta) \stackrel{\text{def}}{=} \left\{ F : D \to D \ : \ F - I \in \Theta \text{ and } F^{-1} - I \in \Theta \right\}.$$

Then consider the images of a fixed open domain Ω_0 in D by $\mathcal{F}(\Theta)$:

$$\mathcal{X}(\Omega_0) \stackrel{\text{def}}{=} \{F(\Omega_0) \ : \ \forall F \in \mathcal{F}(\Theta)\}.$$

They can be identified with the quotient group

$$\mathcal{F}(\Theta)/\mathcal{G}(\Omega_0), \quad \mathcal{G}(\Omega_0) \stackrel{\text{def}}{=} \{F \in \mathcal{F}(\Theta) \ : \ F(\Omega_0) = \Omega_0\}$$

of diffeomorphisms of D. For the specific choices of D, Θ, and Ω_0 that are of interest, this quotient group can be endowed with the so-called *Courant metric* to make it a complete metric space. This metric space is neither linear nor convex.

For unconstrained domains ($D = \mathbf{R}^N$) and "sufficiently small" elements θ of Θ, transformations of the form $F = I + \theta$ belong to $\mathcal{F}(\Theta)$ and hence *perturbations of the identity* (i.e., of Ω_0) can be chosen in the vector space Θ. This makes it possible to define directional derivatives and speak of Gâteaux and Fréchet differentiability with respect to Θ as in the classical case of functions defined on vector spaces. Unfortunately, this approach does not extend to submanifolds D of $\mathbf{R}^N$ or to domains that are constrained in one way or another (e.g., constant volume, perimeter, etc.).

If, in the general case, we had an infinite-dimensional differentiable manifold structure on the quotient group $\mathcal{F}(\Theta)/\mathcal{G}(\Omega_0)$, we could directly apply the techniques of *differential geometry*. But, even in the absence of adequate structures, we can at least use some of its fundamental ideas. For instance, the continuity in I (i.e., at Ω_0) can be characterized by the continuity along one-dimensional flows of velocity fields in $\mathcal{F}(\Theta)$ through the point I (that is, $I(\Omega_0) = \Omega_0$). This suggests defining a notion of directional derivative along one-dimensional flows associated with "velocity fields" V which keep the flows inside D. When D is a smooth submanifold of $\mathbf{R}^N$ such velocities are tangent to D and the set of all such velocities has a vector space structure. This key property generalizes to other types of hold-all Ds. So it is possible to define directional derivatives and speak of shape gradient and shape Hessian with respect to the associated vector space of velocities. This second approach has been known in the literature as the *velocity method*. Another very nice property of that method is that perturbations of the identity, as previously defined in the unconstrained case, can be recovered by a special choice of velocity field, thus creating a certain unity in the methodology.

A new concept of derivative, the *topological derivative*, was recently introduced by Sokołowski and Zochowski [1] for problems where the knowledge of the optimal topology of the domain is important. It gives some sensitivity of a shape function to the presence of a small hole at a point of the domain as the size of that hole goes to zero. For a domain integral, that is, the integral of a locally Lebesgue integrable function, this derivative is the negative of the classical *set-derivative* which is equal to the function at almost every point. This is a direct consequence of the Lebesgue differentiation theorem. So, at least in its simplest form, this approach aims to extend the classical concept of *set-differentiation* as the inverse of integration over sets.

This type of derivative becomes more intricate as we look at the sensitivity of the solution of a boundary value problem or at shape functions which are functions of that solution. For functions that are the integrals of an integrable function with respect to a Radon measure, the Lebesgue differentiation extends in the form of the Lebesgue–Besicovitch differentiation theorem, which says that the set-derivative of that integral is again equal to the function almost everywhere with respect to the Radon measure. So one could try to determine the class of shape functions that can be expressed as an integral with respect to some Radon measure.

We felt that, since the work is relatively new, it would be premature to include a chapter on that topic in this book, but the approach is certainly a worthwhile addition to the global arsenal!

7 Shape Calculus and Tangential Differential Calculus

The velocity method has been used in various applications and contexts, and a very complete *shape calculus* is now available (for instance, the reader is referred to the book of Sokołowski and Zolesio [9]). In the computations, and especially in that of second-order derivatives, an intensive use is made of the *tangential differential calculus*. In order to avoid parametrizations and local bases, tangential derivatives are defined through extensions of functions from the boundary to some small neighborhood. Their importance should not be underestimated from either the theoretical or computational point of view. The use of an intrinsic tangential gradient, divergence, or Laplacian can considerably simplify the computation and the final form of the expressions, making more apparent their fine structure. Too many computations using local coordinates, Christoffel symbols, or intricate parametrizations are often difficult to decipher or effectively use.

This book will give the latest and most important developments of that calculus. They mainly result from the use of the oriented distance function in the theory of thin and asymptotic shells. In that context, it has been realized that extending functions defined on the boundary Γ of a domain Ω by composition with the projection onto Γ results in sweeping simplifications to the tangential calculus. This is due to the fact that the projection can be expressed in terms of the gradient of the oriented distance function. Computing derivatives on Γ becomes as easy as computing derivatives in the Euclidean space. Curvature terms, when they occur, appear in the right place and in the right form through the Hessian matrix of the oriented distance function which coincides with the second fundamental form of Γ. Chapter 8 will provide a self-contained introduction to these new techniques and show how the combined strengths of the shape calculus and the tangential differential calculus considerably simplify computations and expand our capability to tackle complex and challenging problems.

8   Shape Analysis in This Book

Problems in which the design, control, or optimization variable is no longer a vector of parameters or functions but the shape of a geometric domain, a set, or even a "fuzzy entity" cover a much broader range of applications than those for which the compliance or the shape sensitivity analysis have been used. Yet their analysis makes use of common mathematical techniques: partial differential equations, functional analysis, geometry, modern optimization and control theories, finite element analysis, large-scale constrained numerical optimization, etc.

In this book the terminology *shape* will be used for domains ranging from unstructured sets to C^∞-domains. Relaxations of their geometric characteristics such as the volume, perimeter, connectivity, curvatures, and their derivatives will be considered. Shape spaces (often metric and complete) corresponding to different levels of smoothness or degrees of relaxation of the geometry will be systematically constructed. For smooth domains we will emphasize the use of topological groups of diffeomorphisms and the Courant metric. For more general domains we will use a generic construction based on the use of set-parametrized functions. The characteristic function will be associated with metric spaces of equivalence classes of Lebesgue measurable sets. The distance function will be associated with the uniform Hausdorff metric topology in the space of continuous functions, but also with $W^{1,p}$ topologies for which the characteristic function and hence the volume function are continuous. The oriented (algebraic) distance function will be used in the same way to generate new metric topologies. In each case, the metric is constructed from the norm of one of the set-parametrized functions in an appropriate function space. The construction is generic and applies to other choices of set-parametrized functions and function spaces. For instance, the support function of convex analysis can be used to generate a complete metric topology on equivalence classes of sets with the same closed convex hull (cf. Delfour and Zolésio [17]).

The nice property of the characteristic and distance functions over classical local diffeomorphisms is that the set is globally described in term of the analytical properties of a single function. For instance, the gradient of the characteristic function yields a relaxed definition of the perimeter, and the Hessian of the distance function, the boundary measure and curvature terms. But the characteristic function and the gradient of the distance function are both discontinuous at the boundary of the set. This seriously limits their use in the description of smooth domains.

In contrast, the oriented (algebraic) distance function can describe a broad spectrum of sets ranging from arbitrary sets with nonempty boundary to C^∞ open domains according to its degree of smoothness in a neighborhood of the boundary of the set. It readily combines the advantages of local diffeomorphisms and the characteristic and distance functions that are readily obtained from it. This provides, as in functional analysis, a common framework for the classification and comparison of domains according to their relative degree or lack of smoothness. As in geometric measure theory, compact families of sets will be introduced based on the degree of differentiability. One interesting family is made up of the sets with *locally bounded curvature*, which provide a sufficient degree of relaxation for most applications and for which nice compactness theorems are available. This family includes Federer's sets of positive reach and hence closed convex and semiconvex sets.

9 Overview of the Book

The book can be divided into two parts. The first part (Chapters 2 to 6) deals with the analysis of domains and sets ranging from classical smooth domains to Lebesgue measurable sets, and beyond, to equivalence classes of sets associated with the distance and the oriented distance functions. It concentrates on their basic properties, the construction of spaces of domains and topologies, compact families, characterization of the continuity, problem formulations, and some generic examples of optimization problems. The second part (Chapters 7 to 9) concentrates on the choice of perturbations of sets to define shape derivatives. It presents a modern version of the *shape calculus*, an introduction to the latest developments of the *tangential differential calculus*, and the shape derivatives under a state equation constraint. In this partition of the book, Chapter 7 is a transition chapter.

9.1 Analysis, Shape Spaces, and Optimization

Chapter 2 starts with the classical characterizations and properties of open sets or *domains*. It emphasizes the three points of view: local C^k- or Hölderian diffeomorphisms, local epigraph of C^k- or Hölderian functions (in particular, Lipschitzian epigraphs), and level sets of C^k-functions. A section is devoted to the *uniform cone property* which provides an equivalent geometric characterization of domains that are locally the epigraph of Lipschitzian functions. The last two sections of the chapter present several examples of metric spaces of domains using quotient spaces of diffeomorphisms. They use the generic construction of the so-called Courant metric as introduced by Micheletti [1]. Those spaces correspond to quotient groups of diffeomorphisms. It is the nonlinear and nonconvex character of *shape spaces* that will make the differential calculus and the analysis of shape optimization problems more challenging than their counterparts in topological vector spaces.

Chapter 3 relaxes the family of classical domains to the larger class of Lebesgue measurable sets. Using the *characteristic function* associated with a set, *metric spaces* of equivalence classes of characteristic functions are constructed. They are also nonlinear and nonconvex. On one hand this type of relaxation is desirable in optimization problems where the topology of the optimal set is not a priori specified; on the other hand, it necessitates the relaxation of the theory of partial differential equations on a smooth open domain to measurable sets that have no smoothness and may not even be open. Furthermore, some optimization problems yield optimal solutions where the characteristic function is naturally relaxed to a function between zero and one. Such solutions can be interpreted as *microstructures, fuzzy sets*, probability measures, etc. As a first illustration of the use of characteristic functions in optimization, the solution of the original problem of Céa and Malanowski [1] for the optimization of the compliance with respect to the distribution of two materials is given with complete details. A second example deals with the buckling of column, which is one of the very early optimal design problems formulated by Lagrange in 1770. This is followed by the construction of the *nice representative* of an equivalence class of measurable functions: the *measure theoretic representative*. The last section is devoted to the *Caccioppoli or finite perimeter sets*, which have been introduced to solve the Plateau problem of minimal surfaces (Plateau [1]). Even if, by nature, a characteristic function is discontinuous at the boundary of the associated set, the characteristic function of Caccioppoli sets has some smoothness: it belongs to $W^{\epsilon,p}(D)$, $1 \le \epsilon < 1/p$, $p \ge 1$. This is sufficient to obtain compact families of sets by putting a uniform bound on the perimeter. One such family is the set of (locally) Lipschitzian (epigraph) domains contained in a fixed bounded *hold-all* and satisfying a uniform cone property. This property puts a uniform bound on the perimeter of the sets. The use of the theory of finite perimeter sets is illustrated by an application to a free boundary problem in fluid mechanics: the modeling of the Bernoulli wave, where the surface tension of the water enters via the perimeter of the free boundary. The chapter concludes with an approximation of the Dirichlet problem by transmission problems over a fixed larger space in order to study the continuity of its solution with respect to the underlying moving domains.

Chapter 4 moves on to the classical Hausdorff metric topology which is associated with the space of equivalence classes of distance functions of sets with the same closure. As in Chapter 2, the construction of the metric is generic. By going to the distance function of the complement of the set in the uniform topology of the continuous functions, we get the *complementary Hausdorff topology*. These uniform topologies are often too coarse for applications to physical or technological systems. But, since the distance function is uniformly Lipschitzian, it can also be embedded into $W^{1,p}$-Sobolev spaces, and finer metric topologies can be generated. They offer definite advantages over the uniform Hausdorff topology in the sense that they preserve the volume of sets since the characteristic function is continuous with respect to $W^{1,p}$-topologies. Yet we lose the compactness of the family of subsets of a fixed bounded hold-all of $\mathbf{R}^N$. Compact families are recovered by imposing some smoothness on the Hessian matrix as in the case of the characteristic functions of Chapter 3. Sets for which the gradient of the distance function is a vector of functions of bounded variation are said to be of *bounded curvature* since their Hessian matrix is intimately connected with the curvatures of the boundary. This class of sets is sufficiently large for applications and at the same time sufficiently structured to obtain interesting theoretical results. For instance, such sets turn out to be Caccioppoli sets. Closed convex sets that are completely characterized by the convexity of their distance function are also of locally bounded curvature. To complete the list of families of sets that are associated with the distance function we introduce Federer's *sets of positive reach* and a first compactness theorem. The chapter is complemented with a general compactness theorem for families of sets of global or local bounded curvature in a tubular neighborhood of their boundary.

Chapter 5 further extends the use of the distance function in the characterization of families of sets. Its use to characterize the smoothness of sets is limited by the fact that its gradient presents a jump discontinuity at the boundary. This is similar to the jump discontinuity of the characteristic function. To get around this difficulty it is natural to subtract from the distance function to the set the distance function to the complement of the set. This is the *algebraic distance function* which gives a level set description of the set. One remarkable property of this new function is the fact that a set is of class $C^{1,1}$ (resp., C^k, $k \geq 2$) if and only if its algebraic distance function is locally $C^{1,1}$ (resp., C^k) in a neighborhood of its boundary. It also provides an orientation of the boundary since its gradient coincides with the unit outward normal. For this reason we use the terminology *oriented distance function*. As in Chapter 4 Hausdorff and $W^{1,p}$-topologies and sets of global or locally bounded curvature can be introduced. We now have a continuous classification of sets ranging from sets with a nonempty boundary to C^∞-sets, much as in the theory of functions. Closed convex sets are characterized by the convexity of their oriented distance function. Convex sets and semiconvex sets are of locally bounded curvature. This property extends to Federer's sets of positive reach. Natural compactness theorems are also available for sets of bounded curvature in tubular neighborhoods of their boundary. The chapter concludes with an expanded version of the compactness of the family of Lipschitzian subsets of a bounded hold-all satisfying a uniform cone property.

Chapter 6 illustrates the use of distance functions for a generic minimization problem under a state equation constraint and the optimization of the first eigenvalue of an elliptic operator with or without constraint on the volume of the domain. Existence of solution is first obtained for a family of Lipschitzian open subsets of a bounded hold-all verifying the uniform cone property. The second part extends some of the results to a larger family of domains satisfying *capacity conditions*, which turn out to be important to obtain the continuity of solutions of partial differential equations with respect to their underlying domain of definition. From this continuity the existence of optimal domains can be obtained for shape functions that depend on the solution of a boundary value problem over the variable domain. One special case of a capacity condition is the *flat cone condition* which generalizes the condition of Šverák [2] involving a bound on the number of connected components of the complement of the sets.

9.2   Continuity, Derivatives, Shape Calculus, and Tangential Differential Calculus

The second part of the book builds on the constructions, structures, and characterizations developed in the first part. A first observation is that all the shape spaces considered are nonlinear and nonconvex spaces. This means that the development of a shape calculus is more related to the differential calculus on a manifold than the differential calculus in vector spaces.

Chapter 7 is a pivot chapter. It clarifies the longstanding issue of the equivalence of the continuity of shape functions with respect to the Courant metrics and along the flow of velocity fields. A general equivalence between transformations and velocities is given for unconstrained and constrained families of domains. This is specialized to specific families of transformations associated with the Courant metrics studied in Chapter 2.

This chapter also prepares the ground for and motivates the definition of shape semiderivatives which will be given in Chapter 8. As in the case of the continuity the equivalent characterizations via transformations and flows of velocities is very much in the background and at the origin of the many seemingly different definitions which can be found in the literature. Preliminary considerations are first given to the definition of a shape function and to two candidates for the definition of a directional shape semiderivative. They respectively correspond to *perturbations of the identity* associated with any one of the metric spaces constructed in Chapter 2 and to the *velocity method* associated with the flow of a generally nonautonomous vector field. The first one seems to be limited to domains in $\mathbf{R}^N$, while the second one naturally extends to domains living in a fixed smooth submanifold of $\mathbf{R}^N$. Moreover, the shape directional derivative obtained by perturbations of the identity can be recovered by a special choice of velocity field. Most definitions of shape derivatives which can be found in the literature can be brought down to one of the two approaches. Specific constructions of perturbations of the identity and of velocities are given for the classical examples of C^∞-domains, C^k-domains, Cartesian graphs, polar coordinates, and level sets.

Chapter 8 is central to the second part of the book. After a self-contained review of differentiation in topological vector spaces, it introduces basic definitions of first- and second-order Eulerian shape semiderivatives and derivatives by the *velocity method*. General structure theorems are given for Eulerian semiderivatives of a shape function. They arise from the fact that shape functions are usually defined over equivalence classes of sets, and hence only the normal part of the velocity along the boundary really affects the shape function. Bridges are provided with the *method of perturbations of the identity* which is later exploited to isolate the symmetrical part of the shape Hessian. A section is devoted to a modern version of the shape calculus. It gives the general formulae for the shape derivative of domain and boundary integrals. From these formulae several examples are worked out, including the semiderivative of the boundary integral of the square of the normal derivative. For the computation of a broader range of shape derivatives the reader is referred to the book by Sokołowski and Zolésio [9]. In most cases, both domain and boundary expressions are available for derivatives. The boundary expression usually contains more information on the structure of the derivative than its domain counterpart. Finally, to effectively deal with the differential calculus in boundary integrals, we provide the latest version of the *tangential calculus* on C^2-submanifolds of codimension 1, which has been developed in the context of the theory of shells (cf. Delfour and Zolésio [28, 33] and Delfour [3, 7]). This calculus has been significantly simplified by using the projection associated with the oriented distance function studied in Chapter 5. This powerful tool combined with the shape calculus makes it possible to obtain clean, explicit expressions of second-order shape derivatives of domain integrals along with a better understanding of their fine structure.

Chapter 9, the final chapter, completes the shape calculus by introducing the basic theoretical results and computational tools for the shape derivative of functions that depend on a state variable that is usually the solution of a partial differential equation or inequality defined over the varying underlying domain. A first section concentrates on shape functions that are of the *compliance type*; that is, they are the minimum of an energy function associated with the state equation or inequality. Such functions are very nice in the sense that they do not generate an adjoint state equation, and their derivative can be obtained by theorems on the differentiability of a minimum with respect to a parameter even when the minimizers are not unique. A detailed generic example is provided to illustrate how to use the *function space parametrization* to transport the functions in Sobolev spaces over variable domains to a Sobolev space over the fixed larger hold-all. These techniques extend to more complex situations. For instance, Sobolev spaces of vector functions with zero divergence can be transported by the so-called Piola transformation. Domain and boundary expressions are provided. The main theorem is applied to the example of the buckling of columns. An explicit expression of the semiderivative of *Euler's buckling load* with respect to the cross-sectional area is obtained from the main theorem and a necessary and sufficient analytical condition is given to characterize the maximum Euler's buckling load with respect to a family of cross-sectional areas. The theory is further illustrated by providing the semiderivative of the first eigenvalue of several boundary problems over a bounded open domain: Laplace equation, bi-Laplace equation, linear elasticity. In general, the first eigenvalue is not simple over an arbitrary bounded open domain and the eigenvalue is not differentiable; yet the main theorem provides explicit domain and boundary expressions of the semiderivatives.

For general shape functions, a Lagrangian formulation is used to incorporate the state equation and to avoid the study of the derivative of the state equation with respect to the domain. The computation of the shape derivative of a state-constrained function reduces to the computation of the derivative of a saddle point with respect to a parameter even when the saddle point solution is not unique. It yields an expression that depends on the associated *adjoint state equation* much like in control theory, but here the domains play the role of the controls. The technique is illustrated on both the homogeneous Dirichlet and Neumann boundary value problems by function space parametrization. An alternative to this method is the *function space embedding* combined with the use of Lagrange multipliers. It consists in extending solutions of the boundary value problems over the variable domains to a larger fixed hold-all, rather than transporting them. This approach offers many technical advantages over the other one. The computations are easier and they apply to larger classes of problems. This is illustrated on the nonhomogeneous Dirichlet boundary value problem. Again domain and boundary expressions for the shape gradient are obtained. Yet the relative advantages of one method over the other are very much problem and objective dependent. Finally, it is important to acknowledge that the above techniques seem quite robust and are systematically used for nonlinear state equations and in contexts where optimization or saddle point formulations are not available.

Chapter 2
Classical Descriptions and Properties of Domains

1 Introduction

In this chapter we consider the classical families of subsets of the finite-dimensional Euclidean space with nonempty boundary and interior. Their smoothness, which is characterized by the smoothness of their boundary, can be defined in several ways. A first approach is to assume that we can associate with each point of the boundary a diffeomorphism from a neighborhood of that point that locally flattens the boundary. Another way is to assume that the set is the union of the positive level sets of a continuous function and that the zero level is its boundary. A third way is to assume that in each point of the boundary the set is the epigraph of a function. The smoothness of the set is characterized by the smoothness of the corresponding diffeomorphism, function, or graph. Those definitions are equivalent for families of sufficiently smooth sets. The classical domains with C^k and Hölderian boundaries belong to the first category, while Lipschitzian domains that are fundamental in the theory of partial differential equations and the associated Sobolev spaces belong to the third category.

The basic definitions and constructions for the first two categories are given in section 3, and for the third one, in section 5. Section 6 gives the *uniform cone property*, which is an equivalent geometric way to characterize Lipschitzian domains. This property has provided one of the early examples of a compact family of Lipschitzian domains (cf. section 5.4 in Chapter 3). The material provided in the first five sections is relatively standard in geometry. Section 8 gives the construction of what may be the first complete metric topology on a family of domains of class C^k, which are the images of a fixed domain of class C^k through a family of C^k-diffeomorphisms of $\mathbf{R}^N$. This was introduced by Micheletti [1] in 1972 under the name of *Courant metric* (metrica di Courant). What is especially nice about this work is that the constructions readily extend to other families of domains by changing the space of transformations and relaxing the conditions on the fixed domain to an open or closed domain. Section 9 extends the generic framework of Micheletti to other spaces of transformations associated with the spaces $\mathcal{B}^k(\mathbf{R}^N, \mathbf{R}^N)$, $\mathcal{B}^k(D, \mathbf{R}^N)$,

$C_0^k(D, \mathbf{R}^N)$, $W^{k,\infty}(\mathbf{R}^N, \mathbf{R}^N)$, and $W^{k,\bar{c}}(\mathbf{R}^N, \mathbf{R}^N)$. The last two spaces were used by Murat and Simon [1] in 1976 to construct similar metrics by a different method.

2   Notation and Definitions

2.1   Basic Notation

$\mathbb{N}$ is the set of integers $\{1, 2, \dots\}$, and $\mathbf{R}$, the field of real numbers. The *interior* and the *closure* of a subset A in $\mathbf{R}^N$ will be denoted, respectively, int A and $\overline{A}$. The *relative complement* of A in B will be written

$$\complement_B A \ (\text{or } B \backslash A) \ \stackrel{\text{def}}{=} \ \{x \in B : x \notin A\}.$$

When $B = \mathbf{R}^N$, simply write $\complement A$, or A^c. The *boundary* ∂A of A is defined as $\overline{A} \cap \overline{\complement A}$.

Since the superscript t will often appear in the book, the transpose of a vector v and a matrix A will be denoted, respectively, $^*v$ and $^*A$. The inverse of A will be denoted by A^{-1}. The inner product and norm in $\mathbf{R}^N$ will be written

$$x \cdot y \stackrel{\text{def}}{=} \sum_{i=1}^{N} x_i\, y_i, \quad |x| = \sqrt{x \cdot x}.$$

For a linear transformation $A : \mathbf{R}^N \to \mathbf{R}^K$,

$$|A| \stackrel{\text{def}}{=} \max_{|x|_{\mathbf{R}^N} \leq 1} |Ax|_{\mathbf{R}^K},$$

and if A_{ij} and B_{ij} are the matrix representations of A and B with respect to some bases $\{a_1, \dots, a_N\}$ and $\{b_1, \dots, b_K\}$ of, respectively, $\mathbf{R}^N$ and $\mathbf{R}^K$,

$$A \cdot\cdot B \stackrel{\text{def}}{=} \sum_{i=1}^{N} \sum_{j=1}^{K} A_{ij} B_{ij}.$$

The norm $|A|$ of A is equivalent to the norm $\sqrt{A \cdot\cdot A}$.

2.2   Continuous and C^k Functions

Let Ω be an open subset of $\mathbf{R}^N$. Denote by $C(\Omega)$ or $C^0(\Omega)$ the space of continuous functions from Ω to $\mathbf{R}$, and, for an integer $k \geq 1$,

$$C^k(\Omega) \stackrel{\text{def}}{=} \left\{ f \in C^{k-1}(\Omega) : \partial^\alpha f \in C(\Omega),\ \forall \alpha,\ |\alpha| = k \right\},$$

where $\alpha = (\alpha_1, \dots, \alpha_N) \in \mathbb{N}^N$ is a multi-index, $|\alpha| = \alpha_1 + \cdots + \alpha_N$ is the order of the derivative, and

$$\partial^\alpha f \stackrel{\text{def}}{=} \frac{\partial^{|\alpha|} f}{\partial x_1^{\alpha_1} \cdots \partial x_N^{\alpha_N}}. \tag{2.1}$$

By convention $\partial^0 f$ will be the function f in order to make sense of the case $\alpha = 0$. When $|\alpha| = 1$, we also use the standard notation $\partial_i f$ or $\partial f / \partial x_i$. $\mathcal{D}^k(\Omega)$ or $C_c^k(\Omega)$ (resp., $\mathcal{D}(\Omega)$ or $C_c^\infty(\Omega)$) will denote the space of all k-times (resp., infinitely) continuously differentiable functions with compact support contained in the open set Ω.

Denote by $\mathcal{B}^0(\Omega)$ the space of bounded continuous functions from Ω to $\mathbf{R}$, and, for an integer $k \geq 1$, the space

$$\mathcal{B}^k(\Omega) \overset{\text{def}}{=} \left\{ f \in \mathcal{B}^{k-1}(\Omega) : \partial^\alpha f \in \mathcal{B}^0(\Omega), \forall \alpha, |\alpha| = k \right\},$$

that is, the space of all functions in $\mathcal{B}^0(\Omega)$ whose derivatives of order less than or equal to k are continuous and bounded in Ω. Endowed with the norm

$$\|f\|_{C^k(\Omega)} \overset{\text{def}}{=} \max_{0 \leq |\alpha| \leq k} \sup_{x \in \Omega} |\partial^\alpha f(x)|, \tag{2.2}$$

$\mathcal{B}^k(\Omega)$ is a Banach space.

If a function f is bounded and uniformly continuous[1] on Ω, it possesses a unique, continuous extension to the closure $\overline{\Omega}$ of Ω. Denote by $C^k(\overline{\Omega})$ the space of functions f in $C^k(\Omega)$ for which $\partial^\alpha f$ is bounded and uniformly continuous on Ω for all α, $0 \leq |\alpha| \leq k$. A function f in $C^k(\Omega)$ is said to *vanish at the boundary* of Ω if for every α, $0 \leq |\alpha| \leq k$, and $\varepsilon > 0$ there exists a compact subset K of Ω such that, for all $x \in \Omega \cap \complement K$, $|\partial^\alpha f(x)| \leq \varepsilon$. Denote by $C_0^k(\Omega)$ the space of all such functions. Clearly $C_0^k(\Omega) \subset C^k(\overline{\Omega}) \subset \mathcal{B}^k(\Omega) \subset C^k(\Omega)$. Endowed with the norm (2.2), $C_0^k(\Omega)$, $C^k(\overline{\Omega})$, and $\mathcal{B}^k(\Omega)$ are Banach spaces. Finally

$$C^\infty(\Omega) \overset{\text{def}}{=} \bigcap_{k \geq 0} C^k(\Omega), \quad C_0^\infty(\Omega) \overset{\text{def}}{=} \bigcap_{k \geq 0} C_0^k(\Omega), \quad \text{and} \quad \mathcal{B}(\Omega) \overset{\text{def}}{=} \bigcap_{k \geq 0} \mathcal{B}^k(\Omega).$$

When f is a vector function from Ω to $\mathbf{R}^m$, the corresponding spaces will be denoted $C_0^k(\Omega)^m$ or $C_0^k(\Omega, \mathbf{R}^m)$, $C^k(\overline{\Omega})^m$ or $C^k(\overline{\Omega}, \mathbf{R}^m)$, $\mathcal{B}^k(\Omega)^m$ or $\mathcal{B}^k(\Omega, \mathbf{R}^m)$, $C^k(\Omega)^m$ or $C^k(\Omega, \mathbf{R}^m)$, etc.

We quote the following classical compactness theorem.

Theorem 2.1 (Ascoli–Arzelà theorem). *Let Ω be a bounded open subset of $\mathbf{R}^N$. A subset $\mathcal{K}$ of $C(\overline{\Omega})$ is precompact in $C(\overline{\Omega})$ provided the following two conditions hold:*

(i) *There exists a constant M such that for all $f \in \mathcal{K}$ and $x \in \Omega$, $|f(x)| \leq M$;*

(ii) *for every $\varepsilon > 0$ there exists $\delta > 0$ such that if $f \in \mathcal{K}$, x, y in Ω, and $|x - y| < \delta$, then $|f(x) - f(y)| < \varepsilon$.*

2.3 Hölder and Lipschitz Continuous Functions

Given λ, $0 < \lambda \leq 1$, a function f is $(0, \lambda)$-*Hölder continuous* in Ω if

$$\exists c > 0, \forall x, y \in \Omega, \quad |f(y) - f(x)| \leq c\,|x - y|^\lambda.$$

---

[1] A function $f : \Omega \to \mathbf{R}$ is uniformly continuous if for each $\varepsilon > 0$ there exists $\delta > 0$ such that for all x and y in Ω such that $|x - y| < \delta$ we have $|f(x) - f(y)| < \varepsilon$.

When $\lambda = 1$, we also say that f is *Lipschitz* or *Lipschitz continuous*. Similarly for $k \geq 1$, f is (k, λ)-*Hölder continuous* in Ω if

$$\forall \alpha, \; 0 \leq |\alpha| \leq k, \; \exists c > 0, \; \forall x, y \in \Omega, \quad |\partial^\alpha f(y) - \partial^\alpha f(x)| \leq c \, |x - y|^\lambda.$$

Denote by $C^{k,\lambda}(\Omega)$ the space of all (k, λ)-Hölder continuous functions on Ω. Define for $k \geq 0$ the subspaces[2]

$$C^{k,\lambda}(\overline{\Omega}) \stackrel{\text{def}}{=} \left\{ f \in C^k(\overline{\Omega}) : \begin{array}{c} \forall \alpha, \; 0 \leq |\alpha| \leq k, \; \exists c > 0, \; \forall x, y \in \Omega \\ |\partial^\alpha f(y) - \partial^\alpha f(x)| \leq c \, |x - y|^\lambda \end{array} \right\} \qquad (2.3)$$

of $C^k(\overline{\Omega})$. By definition for each α, $0 \leq |\alpha| \leq k$, $\partial^\alpha f$ has a unique, bounded, continuous extension to $\overline{\Omega}$. Endowed with the norm

$$\|f\|_{C^{k,\lambda}(\Omega)} \stackrel{\text{def}}{=} \|f\|_{C^k(\Omega)} + \max_{0 \leq |\alpha| \leq k} \; \sup_{\substack{x,y \in \Omega \\ x \neq y}} \frac{|\partial^\alpha f(y) - \partial^\alpha f(x)|}{|x - y|^\lambda}, \qquad (2.4)$$

$C^{k,\lambda}(\overline{\Omega})$ is a Banach space.

In general $C^{k+1}(\overline{\Omega}) \not\subset C^{k,1}(\overline{\Omega})$, but the inclusion is true for a large class of domains including convex domains.[3] We quote the following embedding theorem.

Theorem 2.2 (Adams [1]). *Let $k \geq 0$ be an integer and $0 < \nu < \lambda \leq 1$ be real numbers. Then the following embeddings exist:*

$$C^{k+1}(\overline{\Omega}) \to C^k(\overline{\Omega}), \qquad (2.5)$$

$$C^{k,\lambda}(\overline{\Omega}) \to C^k(\overline{\Omega}), \qquad (2.6)$$

$$C^{k,\lambda}(\overline{\Omega}) \to C^{k,\nu}(\overline{\Omega}). \qquad (2.7)$$

If Ω is bounded, then the embeddings (2.6) and (2.7) are compact. If Ω is convex, we have the further embeddings

$$C^{k+1}(\overline{\Omega}) \to C^{k,1}(\overline{\Omega}), \qquad (2.8)$$

$$C^{k+1}(\overline{\Omega}) \to C^{k,\nu}(\overline{\Omega}). \qquad (2.9)$$

If Ω is convex and bounded, then embedding (2.5) and (2.9) are compact.

As a consequence of the second part of the theorem, the definition of $C^{k,\lambda}(\overline{\Omega})$ simplifies when Ω is convex:

$$C^{k,\lambda}(\overline{\Omega}) \stackrel{\text{def}}{=} \{ f \in \mathcal{B}^k(\Omega) : \forall \alpha, \; |\alpha| = k, \; \exists c > 0, \; \forall x, y \in \Omega,$$
$$|\partial^\alpha f(y) - \partial^\alpha f(x)| \leq c \, |x - y|^\lambda \}, \qquad (2.10)$$

---

[2]The notation $C^{k,\lambda}(\overline{\Omega})$ should not be confused with the notation $C^{k,\lambda}(\Omega)$ for (k, λ)-*Hölder continuous* functions in Ω *without* the uniform boundedness assumption in Ω. In particular, $C^{k,\lambda}(\mathbf{R^N})$ is contained in but not equal to $C^{k,\lambda}(\mathbf{R}^N)$.

[3]For instance, the convexity can be relaxed to the more general condition: there exists $M > 0$ such that for all x and y in Ω, there exists a path $\gamma_{x,y}$ in Ω (that is, a C^1 injective map from the interval $[0, 1]$ into Ω with $\gamma(0) = x$, $\gamma(1) = y$, and $\int_0^1 \|\gamma'(t)\| \, dt \leq M \, |x - y|$).

and its norm is equivalent to the norm

$$\|f\|_{C^{k,\lambda}(\Omega)} \stackrel{\text{def}}{=} \|f\|_{C^k(\Omega)} + \max_{|\alpha|=k} \sup_{\substack{x,y\in\Omega \\ x\neq y}} \frac{|\partial^\alpha f(y) - \partial^\alpha f(x)|}{|x-y|^\lambda}. \tag{2.11}$$

Given an integer $s \geq 0$ and a real number $p \geq 1$, the *Sobolev spaces* are defined as

$$W^{s,p}(\Omega) \stackrel{\text{def}}{=} \{f \in L^p(\Omega) : \partial^\alpha f \in L^p(\Omega), \forall\alpha, 0 \leq |\alpha| \leq s\}.$$

The reader is referred to Adams [1] for details and extension of this definition to the case where s is not an integer. The notation $H^s(\Omega)$ will also be used when $p = 2$. By Rademacher's theorem (cf., for instance, Evans and Gariepy [1])

$$C^{k,1}(\overline{\Omega}) \subset W_{\text{loc}}^{k+1,\infty}(\Omega)$$

and, when Ω is convex,

$$C^{k,1}(\overline{\Omega}) = W^{k+1,\infty}(\Omega).$$

When f is a vector function from Ω to $\mathbf{R}^m$, the corresponding spaces will be denoted $C^{k,\lambda}(\overline{\Omega})^m$ or $C^{k,\lambda}(\overline{\Omega}, \mathbf{R}^m)$.

3 Smoothness of Domains, Boundary Integral, Boundary Curvatures

The smoothness of the boundary of a set is classically characterized in section 3.1 by introducing at each point of its boundary a local diffeomorphism (that is, defined in a neighborhood of the point) that locally *flattens* the boundary. This family of diffeomorphisms is then used in section 3.2 to define the associated boundary integral and, more generally, the integral over smooth submanifolds of arbitrary dimension in $\mathbf{R}^N$. Those definitions can be generalized by introducing the Hausdorff measures which extend the integration theory on smooth submanifolds to arbitrary subsets of $\mathbf{R}^N$, thus making the writing of the integral completely independent of the choice of the local diffeomorphisms. Section 3.3 completes the section by defining the fundamental forms and curvatures for a smooth submanifold of $\mathbf{R}^N$ of codimension 1.

3.1 C^k and Hölderian Sets

Let $\{e_1, \ldots, e_N\}$ be the standard unit orthonormal basis in $\mathbf{R}^N$. We use the notation $\zeta = (\zeta', \zeta_N)$ for a point $\zeta = (\zeta_1, \ldots, \zeta_N)$ in $\mathbf{R}^N$, where $\zeta' = (\zeta_1, \ldots, \zeta_{N-1})$. Denote by B the open unit ball in $\mathbf{R}^N$ and define the sets

$$B_0 \stackrel{\text{def}}{=} \{\zeta \in B : \zeta_N = 0\}, \tag{3.1}$$

$$B_+ \stackrel{\text{def}}{=} \{\zeta \in B : \zeta_N > 0\}, \quad B_- \stackrel{\text{def}}{=} \{\zeta \in B : \zeta_N < 0\}. \tag{3.2}$$

The main elements of Definition 3.1 are shown in Figure 2.1 for $N = 2$.

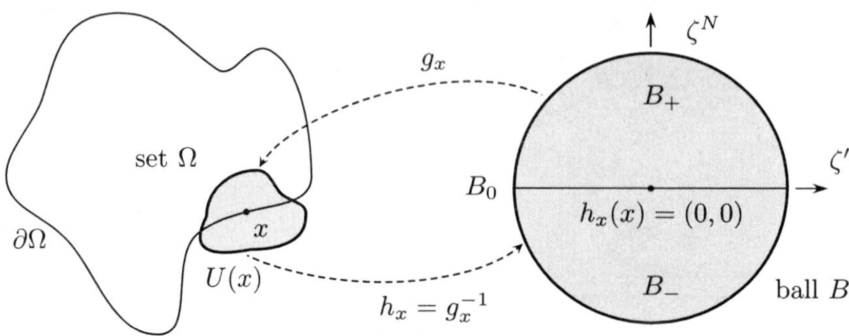

Figure 2.1. *Diffeomorphism h_x from $U(x)$ to B.*

Definition 3.1.

Let Ω be a subset of $\mathbf{R}^N$ such that $\partial\Omega \neq \varnothing$.

  (a) Ω is said to be of *class C^k*, $0 \leq k \leq \infty$, if for each $x \in \partial\Omega$ there exist

    (i) a neighborhood $U(x)$ of x, and

    (ii) a bijective map $g_x : U(x) \to B$ with the following properties:

$$g_x \in C^k\big(U(x), B\big), \quad h_x \stackrel{\text{def}}{=} g_x^{-1} \in C^k\big(B, U(x)\big), \tag{3.3}$$

$$\operatorname{int}\Omega \cap U(x) = h_x(B_+), \tag{3.4}$$

$$\Gamma_x \stackrel{\text{def}}{=} \partial\Omega \cap U(x) = h_x(B_0), \quad B_0 = g_x(\Gamma_x). \tag{3.5}$$

  (b) Ω is said to be *(k,ℓ)-Hölderian* or of *class $C^{k,\ell}$*, $0 \leq k$, $0 < \ell \leq 1$, if the conditions of part (a) are satisfied with a map $g_x \in C^{k,\ell}\big(U(x), B\big)$ with inverse $h_x = g_x^{-1} \in C^{k,\ell}\big(B, U(x)\big)$. $\square$

Remark 3.1.

The above definitions are usually given for an open set Ω called a *domain*. This terminology naturally arises in partial differential equations, where this open set is indeed the domain over which the solution of the equation is defined.[4] At this point it is not necessary to assume that the set Ω is open. The family of diffeomorphisms $\{g_x : x \in \partial\Omega\}$ characterizes the equivalence class of sets

$$[\Omega] = \{A \subset \mathbf{R}^N : \operatorname{int} A = \operatorname{int}\Omega \text{ and } \partial A = \partial\Omega\}$$

with the same interior and boundary. The sets $\operatorname{int}\Omega$ and $\partial\Omega$ are invariants for the equivalence class $[\Omega]$ of sets of class $C^{k,\ell}$. The notation Γ for $\partial\Omega$ and the standard

---

[4] Classically for $k \geq 1$ a bounded open domain Ω in $\mathbf{R}^N$ is said to be of class C^k if its boundary is a C^k-submanifold of $\mathbf{R}^N$ of codimension 1 and Ω is located on one side of its boundary $\Gamma = \partial\Omega$ (cf. Agmon [1, Def. 9.2, p. 178]).

terminology *domain* for the unique (open) set $\operatorname{int}\Omega$ associated with the class $[\Omega]$ will be used. In what follows a $C^{k,\ell}$ mapping g_x with $C^{k,\ell}$ inverse will be called a $C^{k,\ell}$-*diffeomorphism*. $\square$

Consider the solution $y = y(\Omega)$ of the Dirichlet problem on the bounded open domain Ω

$$-\Delta y = f \text{ in } \Omega, \quad y = g \text{ on } \partial\Omega. \tag{3.6}$$

When Ω is of class C^∞, the solution $y(\Omega)$ can be sought in any Sobolev space $H^m(\Omega)$, $m \geq 1$, by choosing sufficiently smooth data f and g and appropriate compatibility conditions. However, when Ω is only of class C^k, $1 \leq k < \infty$, m cannot be made arbitrarily large by choosing smoother data f and g. The reader is referred to Dautray and Lions [1, Chap. VII, section 3, pp. 1271–1304] for classical smoothness results of solutions to elliptic problems in domains of class C^k. It is important to understand that we shall consider shape problems involving the solution of a boundary value problem over domains with minimal smoothness. Since we know that, at least for elliptic problems, the smoothness of the solution not only depends on the smoothness of the data but also on the smoothness of the domain. This issue will be of paramount importance to ensure that the shape problems are well posed.

Remark 3.2.
We shall see in section 5 that domains that are locally the epigraph of a $C^{k,\ell}$, $k \geq 1$, (resp., Lipschitzian) function are of class $C^{k,\ell}$ (resp., $C^{0,1}$), but domains that are of class $C^{0,1}$ are not necessarily locally the epigraph of a Lipschitzian function. Nevertheless, globally $C^{0,1}$-mappings with a $C^{0,1}$ inverse are important since they transport L^p-functions onto L^p-functions and $W^{1,p}$-functions onto $W^{1,p}$-functions (cf., for instance, Nečas [1, Lems. 3.1 and 3.2, pp. 65–66]). $\square$

For sets of class C^1, the unit exterior normal to the boundary $\Gamma = \partial\Omega$ can be characterized through the Jacobian matrices of g_x and h_x. By definition of B_0, $\{e_1, \ldots, e_{N-1}\} \subset B_0$ and the tangent space $T_y\Gamma$, $\Gamma = \partial\Omega$, at y to Γ_x is the vector space spanned by the $N-1$ vectors

$$\{Dh_x(\zeta',0)e_i : 1 \leq i \leq N-1\}, \quad (\zeta',0) = g_x(y) \in B_0, \tag{3.7}$$

where $Dh_x(\zeta)$ is the Jacobian matrix of h_x at the point ζ:

$$(Dh_x)_{\ell m} \stackrel{\text{def}}{=} \partial_m(h_x)_\ell.$$

So from (3.7) a normal vector field to Γ_x at $y \in \Gamma_x$ is given by

$$m_x(y) = -\,^*(Dh_x)^{-1}(\zeta',0)\,e_N = -\,^*Dg_x(y)\,e_N, \quad h_x(\zeta',0) = y, \tag{3.8}$$

since

$$-m_x(y) \cdot Dh_x(\zeta',0)\,e_i = e_N \cdot e_i = \delta_{iN}, \quad 1 \leq i \leq N.$$

Thus the *outward unit normal field* $n(y)$ at $y \in \Gamma_x$ is given by

$$
\begin{aligned}
n(y) &= -\frac{{}^*(Dh_x)^{-1}(\zeta', 0)\, e_N}{|\,{}^*(Dh_x)^{-1}(\zeta', 0)\, e_N|} \quad \forall h_x(\zeta', 0) = y \in \Gamma_x, \\
n(y) &= -\frac{{}^*(Dh_x)^{-1}(h_x^{-1}(y))\, e_N}{|\,{}^*(Dh_x)^{-1}(h_x^{-1}(y))\, e_N|} \quad \forall y \in \Gamma_x.
\end{aligned}
\tag{3.9}
$$

It can be verified that n is uniquely defined on Γ by checking that for $y \in \Gamma_x \cap \Gamma_{x'}$, n is uniquely defined by (3.9).

3.2 Boundary Integral, Canonical Density, and Hausdorff Measures

The family of neighborhoods $U(x)$ associated with all the points x of Γ is an open cover of Γ. If $\Gamma = \partial\Omega$ is assumed to be compact, then there exists a finite open subcover; that is, there exists a finite sequence of points $\{x_j : 1 \le j \le m\}$ of Γ such that $\Gamma \subset U_1 \cup \cdots \cup U_m$, where $U_j = U(x_j)$. For simplicity, index all the previous symbols by j instead of x_j.

The boundary integration on Γ is obtained by using a partition of unity $\{r_j : 1 \le j \le m\}$ for the family of open neighborhoods $\{U_j : 1 \le j \le m\}$ of Γ:

$$
\begin{cases}
r_j \in \mathcal{D}(U_j), \quad 0 \le r_j(x) \le 1, \\
\displaystyle\sum_{j=1}^{m} r_j(x) = 1 \text{ in a neighborhood } U \text{ of } \Gamma,
\end{cases}
\tag{3.10}
$$

such that $\overline{U} \subset \cup_{j=1}^{m} U_j$, where $\mathcal{D}(U_j)$ is the set of all infinitely continuously differentiable functions with compact support in U_j. If $f \in C(\Gamma)$, then

$$
(fr_j) \circ h_j \in C(B_0), \quad 1 \le j \le m.
\tag{3.11}
$$

Define the boundary integral of fr_j on Γ_j as

$$
\int_{\Gamma_j} fr_j \, d\Gamma \stackrel{\text{def}}{=} \int_{B_0} (fr_j) \circ h_j(\zeta', 0)\, \omega_j(\zeta')\, d\zeta', \quad \Gamma_j = U(x_j) \cap \Gamma,
\tag{3.12}
$$

where $\omega_j = \omega_{x_{y}}$ and ω_x is the *density term*

$$
\begin{aligned}
\omega_x(\zeta') &= |m_x(h_x(\zeta', 0))|\, |\det Dh_x(\zeta', 0)|, \\
m_x(y) &= -\,{}^*(Dh_x)^{-1}(h_x^{-1}(y))\, e_N = -\,{}^*Dg_x(y)\, e_N.
\end{aligned}
\tag{3.13}
$$

From this define the boundary integral of f on Γ as

$$
\int_{\Gamma} f \, d\Gamma \stackrel{\text{def}}{=} \sum_{j=1}^{m} \int_{\Gamma_j} fr_j \, d\Gamma.
\tag{3.14}
$$

The expression on the right-hand side of (3.12) results from the parametrization of the boundary Γ_j by B_0 through the diffeomorphism h_j.

In differential geometry there is a general procedure to define the canonical density for a d-dimensional submanifold V in $\mathbf{R}^N$ parametrized by a C^k-mapping (cf., for instance, Berger and Gostiaux [1, Def. 2.1.1, p. 48 (p. 56 in French edition) and Prop. 6.62, p. 214 (p. 239 in French edition)]).

Definition 3.2.
Fix integers $k \geq 1$ and $1 \leq d < N$ and a real number $0 \leq \ell \leq 1$. A subset S of $\mathbf{R}^N$ is said to be a d-*dimensional submanifold* in $\mathbf{R}^N$ of class C^k (resp., $C^{k,\ell}$) if for each $x \in S$, there exists an open subset $U(x)$ of $\mathbf{R}^N$ containing x, a diffeomorphism g_x of class C^k (resp., $C^{k,\ell}$) from $U(x)$ onto its open image $g_x(U(x))$, such that

$$g_x(U(x) \cap S) = g_x(U(x)) \cap R^d,$$

where $R^d = \left\{ (x_1, \ldots, x_d, 0, \ldots, 0) \in \mathbf{R}^N \colon \forall (x_1, \ldots, x_d) \in \mathbf{R}^d \right\}.$ $\qquad\square$

For $k \geq 1$ the canonical density ω_x on the submanifold S at a point $y \in U(x) \cap S$ is given by

$$\omega_x = \sqrt{|\det B|},$$

where the $d \times d$ matrix B is given by

$$(B)_{ij} = (Dh_x\, e_i) \cdot (Dh_x\, e_j), \ 1 \leq i, j \leq d, \quad h_x = g_x^{-1} \text{ on } g_x(U(x)).$$

In the case of interest, $d = N - 1$ and it is easy to verify that

$$^*Dh_x\, Dh_x = \begin{bmatrix} ^*C\,C & ^*C\,c \\ ^*c\,C & ^*c\,c \end{bmatrix}, \quad B = {}^*C\,C,$$

where C is the $(N \times (N-1))$-matrix and c is the N-vector defined by

$$C_{ij} = \{Dh_x\}_{ij}, \quad 1 \leq i \leq N, 1 \leq j \leq N - 1, \quad c = Dh_x\, e_N.$$

Denote by $M(A)$ the matrix of cofactors associated with a matrix A: $M(A)_{ij}$ is equal to the determinant of the matrix obtained after deleting the ith row and the jth column times $(-1)^{i+j}$. Then $M(A) = (\det A)\, {}^*A^{-1}$, $M(^*A) = {}^*M(A)$, and for two invertible matrices A_1 and A_2, $M(A_1 A_2) = M(A_1)M(A_2)$. As a result

$$\det B = M(^*Dh_x\, Dh_x)_{NN} = e_N \cdot M(^*Dh_x\, Dh_x)e_N,$$

where $M(^*Dh_x\, Dh_x)_{NN}$ is the NN-cofactor of the matrix $^*Dh_x\, Dh_x$. Then

$$M(^*Dh_x\, Dh_x)_{NN} = e_N \cdot M(^*Dh_x\, Dh_x)e_N$$
$$= e_N \cdot M(^*Dh_x)\, M(Dh_x)e_N = |M(Dh_x)e_N|^2.$$

In view of the previous considerations and (3.13)

$$\sqrt{\det B} = |M(Dh_x)e_N| = |\det Dh_x|\, |^*(Dh_x)^{-1} e_N| = \omega_x.$$

Definition 3.2 gives the classical construction of a d-dimensional *surface measure* on the boundary of a C^k-domain. In 1918, F. Hausdorff [1] introduced a d-dimensional measure in $\mathbf{R}^N$ which gives the same surface measure for smooth submanifolds but is defined on all subsets of $\mathbf{R}^N$. When $d = N$, it is equal to the Lebesgue measure. To complete the discussion we quote the definition from Morgan [1, p. 8] and Evans and Gariepy [1, p. 60].

Definition 3.3.
For any subset S of $\mathbf{R}^N$, define the *diameter* of S as

$$\operatorname{diam}(S) \stackrel{\text{def}}{=} \sup\{|x - y| : x, y \in S\}.$$

Let α_d denote the Lebesgue measure of the unit ball in $\mathbf{R}^d$. The d-dimensional Hausdorff measure $H_d(A)$ of a subset A of $\mathbf{R}^N$ is defined by the following process. For δ small, cover A efficiently by countably many sets S_j with $\operatorname{diam}(S_j) \le \delta$, add up all the terms

$$\alpha_d \left(\operatorname{diam}(S_j)/2\right)^d,$$

and take the limit as $\delta \to 0$:

$$H_d(A) \stackrel{\text{def}}{=} \lim_{\delta \searrow 0} \inf_{\substack{A \subset \cup S_j \\ \operatorname{diam}(S_j) \le \delta}} \sum_j \alpha_d \left(\frac{\operatorname{diam}(S_j)}{2}\right)^d,$$

where the infimum is taken over all countable covers $\{S_j\}$ of A whose members have diameter at most δ. $\qquad\qquad\square$

For $0 \le d < \infty$, H_d is a Borel regular measure. The *Hausdorff dimension* of a set $A \subset \mathbf{R}^N$ is defined as

$$H_{dim}(A) \stackrel{\text{def}}{=} \inf\{0 \le s < \infty : H_s(A) = 0\}.$$

By definition $H_{dim}(A) \le N$ and

$$\forall k > H_{dim}(A), \quad H_k(A) = 0.$$

If a submanifold S of dimension d, $1 \le d < N$, of Definition 3.3 is characterized by a single C^1-diffeomorphism g, that is,

$$g(S) = \mathbf{R}^d \text{ and } S = h(\mathbf{R}^d), \quad h = g^{-1},$$

then for any Lebesgue-measurable set $E \subset \mathbf{R}^d$

$$\int_E \omega \, dx = H_d(h(E)).$$

This is a generalization to submanifolds of codimension greater than 1 of formula (3.12).

3.3 Fundamental Forms and Principal Curvatures

Consider a set Ω locally of class C^2 in $\mathbf{R}^N$. Its boundary $\Gamma = \partial\Omega$ is an $(N-1)$-dimensional submanifold of $\mathbf{R}^N$ of class C^2. At each point $x \in \Gamma$ there is a C^2-diffeomorphism g_x from a neighborhood $U(x)$ of x onto B. Denoting its inverse by $h_x = g_x^{-1}$, the *covariant basis* at a point $y \in U(x) \cap \Gamma$ is defined as

$$a_\alpha(y) \overset{\text{def}}{=} \frac{\partial h_x}{\partial \zeta_\alpha}(\zeta', 0), \quad \alpha = 1, \dots, N-1, \quad h_x(\zeta', 0) = y,$$

and a_N is chosen as the *inward unit normal*

$$a_N(y) \overset{\text{def}}{=} \frac{{}^*(Dh_x(h_x^{-1}(y)))^{-1}e_N}{|\,{}^*(Dh_x(h_x^{-1}(y)))^{-1}e_N|}.$$

The standard convention that the Greek indices range from 1 to $N-1$ and the Roman indices from 1 to N will be followed together with Einstein's rule of summation over repeated indices. The associated *contravariant basis* $\{a^i\} = \{a^i(y)\}$ is defined from the covariant one $\{a_i\} = \{a_i(y)\}$ as

$$a^i \cdot a_j = \delta_{ij},$$

where δ_{ij} is the Kronecker index function. The *first, second,* and *third fundamental forms a, b,* and c are defined as

$$a_{\alpha\beta} \overset{\text{def}}{=} a_\alpha \cdot a_\beta, \quad b_{\alpha\beta} \overset{\text{def}}{=} -a_\alpha \cdot a_{N,\beta}, \quad c_{\alpha\beta} \overset{\text{def}}{=} b_\alpha^\lambda \, b_{\lambda\beta},$$

where

$$a_{N,\beta} = \frac{\partial a_N}{\partial \zeta_\beta}, \quad b_\alpha^\lambda = a^\lambda \cdot a^\mu \, b_{\mu\alpha}.$$

The above definitions extend to sets of class $C^{1,1}$ for which $h_x \in C^{1,1}(B)$ and hence $h_x \in C^{1,1}(B_0)$. So by Rademacher's theorem in dimension $N-1$ (cf., for instance, Evans and Gariepy [1]), $h_x \in W^{2,\infty}(B_0)$ and the definitions of the second and third fundamental forms still make sense H_{N-1}-almost everywhere on Γ. The eigenvalues of $b_{\alpha\beta}$ are the $(N-1)$ *principal curvatures* κ_i, $1 \leq i \leq N-1$, of the submanifold Γ. The *mean curvature H* and the *Gauss curvature K* are defined as

$$H \overset{\text{def}}{=} \frac{1}{N-1} \sum_{\alpha=1}^{N-1} \kappa_\alpha \quad \text{and} \quad K \overset{\text{def}}{=} \prod_{\alpha=1}^{N-1} \kappa_\alpha.$$

The choice of the *inner normal* for a_N is necessary to make the principal curvatures of the sphere (boundary of the ball) positive. The factor $1/(N-1)$ is used to make the mean curvature of the unit sphere equal to 1 in all dimensions. The reader should keep in mind the long-standing differences in usage between geometry and partial differential equations, where the outer unit normal is used in the integration-by-parts formulae for the Euclidean space $\mathbf{R}^N$. For integration by parts on submanifolds of $\mathbf{R}^N$ the sum of the principal curvatures will naturally occur rather than the mean curvature. It will be convenient to *redefine* H as the sum of the principal curvatures and introduce the notation $\bar{H}$ for the classical mean curvature

$$H \stackrel{\text{def}}{=} \sum_{\alpha=1}^{N-1} \kappa_\alpha, \quad \bar{H} \stackrel{\text{def}}{=} \frac{1}{N-1} \sum_{\alpha=1}^{N-1} \kappa_\alpha. \tag{3.15}$$

4　Domains as Level Sets of a Function

From the definition of a set of class $C^{k,\ell}$, $k \geq 1$ and $0 \leq \ell \leq 1$, the set Ω can also be locally described by the level sets of the $C^{k,\ell}$-function

$$f_x(y) \stackrel{\text{def}}{=} g_x(y) \cdot e_N, \tag{4.1}$$

since by definition

$$\operatorname{int} \Omega \cap U(x) = \{y \in U(x) : f_x(y) > 0\},$$
$$\partial\Omega \cap U(x) = \{y \in U(x) : f_x(y) = 0\}.$$

The boundary $\partial\Omega$ is the zero level set of f_x and the gradient

$$\nabla f_x(y) = {}^*Dg_x(y)e_N \neq 0$$

is normal to that level set. Thus the *exterior normal* to Ω is given by

$$n(y) = -\frac{\nabla f_x(y)}{|\nabla f_x(y)|} = -\frac{{}^*Dg_x(y)e_N}{|\,{}^*Dg_x(y)e_N|} = -\frac{{}^*(Dh_x(g_x(y)))^{-1}e_N}{|\,{}^*(Dh_x(g_x(y)))^{-1}e_N|}.$$

From this local construction of the functions $\{f_x : x \in \partial\Omega\}$, a global function on $\mathbf{R}^N$ can be constructed to characterize Ω.

Theorem 4.1. *Given $k \geq 1$, $0 \leq \ell \leq 1$, and a set Ω in $\mathbf{R}^N$ of class $C^{k,\ell}$ with compact boundary, there exists a Lipschitz continuous function $f : \mathbf{R}^N \to \mathbf{R}$ such that*

$$\operatorname{int} \Omega = \{y \in \mathbf{R}^N : f(y) > 0\} \text{ and } \partial\Omega = \{y \in \mathbf{R}^N : f(y) = 0\} \tag{4.2}$$

and a neighborhood W of $\partial\Omega$ such that

$$f \in C^{k,\ell}(W) \text{ and } \nabla f \neq 0 \text{ on } W \text{ and } n = -\frac{\nabla f}{|\nabla f|}, \tag{4.3}$$

where n is the outward unit normal to Ω on $\partial\Omega$.

Proof. Construction of the function f. Fix $x \in \partial\Omega$ and consider the function f_x defined by (4.1). By continuity of ∇f_x, let $V(x) \subset U(x)$ be a neighborhood of x such that

$$\forall y \in V(x), \quad |\nabla f_x(y) - \nabla f_x(x)| \leq \frac{1}{2}|\nabla f_x(x)|.$$

Furthermore, let $W(x)$ be a bounded open neighborhood of x such that $\overline{W(x)} \subset V(x)$. For $\partial\Omega$ compact there exists a finite subcover $\{W_j = W(x_j) : 1 \leq j \leq m\}$ of $\partial\Omega$ for a finite sequence $\{x_j : 1 \leq j \leq m\} \subset \partial\Omega$. Index all the previous symbols by j instead of x_j and define $W = \cup_{j=1}^{m} W_j$. Let $\{r_j : 1 \leq j \leq m\}$ be a partition of unity for $\{V_j = V(x_j)\}$ such that

$$\begin{cases} r_j \in \mathcal{D}(V_j), \quad 0 \leq r_j(x) \leq 1, \quad 1 \leq j \leq m, \\ \displaystyle\sum_{j=1}^{m} r_j(x) = 1 \text{ in } \overline{W}, \end{cases} \tag{4.4}$$

where, by definition, W is an open neighborhood of $\partial\Omega$ such that $\overline{W} \subset V = \cup_{j=1}^{m} V_j$ and $\mathcal{D}(V_j)$ is the set of all infinitely continuously differentiable functions with compact support in V_j. Define the function $f : \mathbf{R}^N \to \mathbf{R}$ as

$$f(y) \overset{\text{def}}{=} d_{W\cup\Omega^c} - d_{W\cup\Omega} + \sum_{j=1}^{m} r_j(y)\frac{f_j(y)}{|\nabla f_j(x_j)|},$$

where $d_A(x) = \inf\{|y - x| : y \in A\}$ is the distance function from a point x to a nonempty set A of $\mathbf{R}^N$. The function d_A is Lipschitz continuous with Lipschitz constant equal to 1. Since for all j, $f_j \in C^{k,\ell}(V_j)$, $r_j \in \mathcal{D}(V_j)$, and $\overline{W}_j \subset V_j$, $r_j f_j \in C^{k,\ell}(V_j)$ with compact support in V_j. Therefore, since $k \geq 1$, $r_j f_j$ is Lipschitz continuous in $\mathbf{R}^N$. So, by definition, f is Lipschitz continuous on $\mathbf{R}^N$ as the finite sum of Lipschitz continuous functions on $\mathbf{R}^N$.

Properties (4.2). Introduce the index set

$$J(y) \overset{\text{def}}{=} \{j : 1 \leq j \leq m, \, r_j(y) > 0\}$$

for $y \in V$. For all $y \in \overline{W}$, $J(y) \neq \varnothing$, $d_{W\cup\Omega^c} = 0 = d_{W\cup\Omega}$, and

$$f(y) = \sum_{j\in J(y)} r_j(y)\frac{f_j(y)}{|\nabla f_j(x_j)|}. \tag{4.5}$$

For $y \in \partial\Omega = \overline{W} \cap \partial\Omega$, $f_j(y) = 0$ for all $j \in J(y)$ and hence $f(y) = 0$; for $y \in \text{int}\,\Omega \cap \overline{W}$, $f_j(y) > 0$ and $r_j(y) > 0$ for all $j \in J(y)$ and hence $f(y) > 0$; for $y \in \text{int}\,\Omega^c \cap \overline{W}$, $f_j(y) < 0$ and $r_j(y) > 0$ for all $j \in J(y)$ and hence $f(y) < 0$. For $y \in \Omega\backslash\overline{W}$, $f_j \geq 0$, $r_j \geq 0$, $\sum_{j=1}^{m} r_j f_j \geq 0$, $d_{W\cup\Omega^c} > 0$, $d_{W\cup\Omega} = 0$, and $f > 0$; for $y \in \Omega^c\backslash\overline{W}$, $f_j \leq 0$, $r_j \geq 0$, $\sum_{j=1}^{m} r_j f_j \leq 0$, $d_{W\cup\Omega^c} = 0$, $d_{W\cup\Omega} > 0$, and $f < 0$. So we have proved that $\partial\Omega \subset f^{-1}(0)$, $\text{int}\,\Omega \subset \{f > 0\}$, $\text{int}\,\Omega^c \subset \{f < 0\}$. Hence f has the properties (4.2).

Properties (4.3). Recall that on W the function f is given by expression (4.5). It belongs to $C^{k,\ell}(W)$, and a fortiori to $C^1(W)$, as the finite sum of $C^{k,\ell}$-functions on W since $k \geq 1$. The gradient of f in W is given by

$$\nabla f(y) = \sum_{j\in J(y)} \nabla r_j(y)\frac{f_j(y)}{|\nabla f_j(x_j)|} + r_j(y)\frac{\nabla f_j(y)}{|\nabla f_j(x_j)|}.$$

It remains to show that it is nonzero on $\partial\Omega$. On $\partial\Omega \cap V_j$, $f_j = 0$ and

$$\forall j \in J(y), \quad \frac{\nabla f_j(y)}{|\nabla f_j(y)|} = -n(y).$$

Therefore, in W, ∇f can be rewritten in the form

$$\nabla f(y) = -n(y) \sum_{j \in J(y)} r_j(y) \frac{|\nabla f_j(y)|}{|\nabla f_j(x_j)|}$$

and

$$\nabla f(y) + n(y) = n(y) \sum_{j \in J(y)} r_j(y) \left[1 - \frac{|\nabla f_j(y)|}{|\nabla f_j(x_j)|} \right].$$

Finally, by construction of W,

$$|\nabla f(y)| \geq |n(y)| - |n(y)| \sum_{j \in J(y)} r_j(y) \left| 1 - \frac{|\nabla f_j(y)|}{|\nabla f_j(x_j)|} \right|$$

$$\geq 1 - \sum_{j \in J(y)} r_j(y) \frac{1}{2} = \frac{1}{2}$$

and $\nabla f \neq 0$ in W. This proves the properties (4.3). $\square$

This theorem has a converse.

Theorem 4.2. *Associate with a continuous function* $f : \mathbf{R}^N \to \mathbf{R}$ *the set*

$$\Omega \stackrel{\text{def}}{=} \{y \in \mathbf{R}^N : f(y) > 0\}. \tag{4.6}$$

Assume that

$$f^{-1}(0) \stackrel{\text{def}}{=} \{y \in \mathbf{R}^N : f(y) = 0\} \neq \varnothing \tag{4.7}$$

and that there exists a neighborhood V *of* $f^{-1}(0)$ *such that* $f \in C^{k,\ell}(V)$ *for some* $k \geq 1$ *and* $0 \leq \ell \leq 1$ *and that* $\nabla f \neq 0$ *in* $f^{-1}(0)$. *Then* Ω *is a set of class* $C^{k,\ell}$,

$$\operatorname{int} \Omega = \Omega \quad \text{and} \quad \partial\Omega = f^{-1}(0). \tag{4.8}$$

Proof. By continuity of f, Ω is open, $\operatorname{int} \Omega = \Omega$, $\complement\Omega$ is closed,

$$\overline{\Omega} \subset \{y \in \mathbf{R}^N : f(y) \geq 0\}, \quad \overline{\complement\Omega} = \complement\Omega = \{y \in \mathbf{R}^N : f(y) \leq 0\},$$

and $\partial\Omega \subset f^{-1}(0)$. Conversely, for each $x \in \partial\Omega$, define the function

$$g(t) \stackrel{\text{def}}{=} f(x + t\nabla f(x)).$$

There exists $\delta > 0$ such that for all $|t| < \delta$, $x + t\nabla f(x) \in V$ and the function g is C^1 in $(-\delta, \delta)$. Hence since $g(0) = 0$ and $g'(t) = \nabla f(x + t\nabla f(x)) \cdot \nabla f(x)$,

$$f(x + t\nabla f(x)) = \int_0^t \nabla f(x + s\nabla f(x)) \cdot \nabla f(x)\, ds.$$

By continuity of ∇f in V and the fact that $\nabla f(x) \neq 0$, there exists δ', $0 < \delta' < \delta$, such that

$$\forall s,\ 0 \leq |s| \leq \delta',\quad \nabla f(x + s\nabla f(x)) \cdot \nabla f(x) \geq \frac{1}{2}|\nabla f(x)|^2 > 0.$$

Hence for all t, $0 < t \leq \delta'$, $f(x + t\nabla f(x)) > 0$. So any point in $f^{-1}(0)$ can be approximated by a sequence $\{x_n = x + t_n\nabla f(x) : n \geq 1, 0 < t_n \leq \delta'\}$ in Ω, $t_n \to 0$, and $f^{-1}(0) \subset \overline{\Omega}$. Similarly, using a sequence of negative t_n's, $f^{-1}(0) \subset \overline{\complement\Omega}$ and hence $f^{-1}(0) \subset \partial\Omega$. This proves (4.8).

Fix $x \in \partial\Omega = f^{-1}(0)$. Since $\nabla f(x) \neq 0$, define the unit vector $e_N(x) = \nabla f(x)/|\nabla f(x)|$. Associate with $e_N(x)$ unit vectors $e_1(x), \ldots, e_{N-1}(x)$, which form an orthonormal basis in $\mathbf{R}^N$ with $e_N(x)$. Define the map $g_x : V \to \mathbf{R}^N$ as

$$g_x(y) \stackrel{\text{def}}{=} \left(\{(y - x) \cdot e_\alpha(x)\}_{\alpha=1}^{N-1}, \frac{f(y)}{|\nabla f(x)|} \right) \quad \Rightarrow g_x \in C^{k,\ell}(V; \mathbf{R}^N).$$

The transpose of the Jacobian matrix of g_x is given by

$$^*Dg_x(y) = (e_1(x), \ldots, e_{N-1}(x), \nabla f(y)/|\nabla f(x)|)$$

and $Dg_x(x) = I$, the identity matrix in the $\{e_i(x)\}$ reference system. By the inverse mapping theorem g_x has a $C^{k,\ell}$ inverse h_x in some neighborhood $U(x)$ of x in V. Therefore,

$$\Omega \cap U(x) = \{y \in U(x) : f(x) > 0\} = \operatorname{int}\Omega \cap U(x),$$
$$\{y \in U(x) : f(x) = 0\} = \partial\Omega \cap U(x),$$

and the set Ω is of class $C^{k,\ell}$. $\qquad\square$

We complete this section with the important theorem of Sard.

Theorem 4.3 (J. Dieudonné [1, section 16.23, p. 167]). *Let X and Y be two differential manifolds, $f : X \to Y$ a C^∞-mapping, and E the set of critical points of f. Then $f(E)$ is negligible in Y, and $Y - f(E)$ is dense in Y.*

Combining this theorem with Theorem 4.1, this means that, for almost all t in the range of a C^∞-function $f : \mathbf{R}^N \to \mathbf{R}$, the set

$$\{y \in \mathbf{R}^N : f(x) > t\}$$

is of class C^∞ in the sense of Definition 3.1. The following theorem extends and completes Sard's theorem.

Theorem 4.4 (H. Federer [3, Thm. 3.4.3, p. 316]). *If $m > \nu \geq 0$ and $k \geq 1$ are integers, A is an open subset of $\mathbf{R}^m$, $B \subset A$, Y is a normed vector space, and*

$$f : A \to Y \text{ is a map of class } k, \quad \dim \operatorname{Im} Df(x) \leq \nu \text{ for } x \in B,$$

then

$$H_{\nu + (m-\nu)/k}(f(B)) = 0.$$

5 Domains as Local Epigraphs

After diffeomorphisms and level sets, the local epigraph description provides a third point of view and another way to characterize the smoothness of a set or a domain. Lipschitzian domains, which are characterized by the property that their boundary is locally the epigraph of a Lipschitzian function, play a central role in the theory of Sobolev spaces and partial differential equations. They can equivalently be characterized by the geometric uniform cone property. As for sets that are locally the epigraph of $C^{k,\ell}$-functions, they are completely equivalent to sets of class $C^{k,\ell}$ for $k \geq 1$.

5.1 Sets That Are Locally Lipschitzian Epigraphs

Lipschitzian domains enjoy most of the properties of smooth domains. For instance, their boundary coincides with the boundary of their interior and their boundary has zero *volume* and locally finite *boundary* measure (cf. Theorem 5.3 of section 5.4 or Theorem 6.3 of section 6). Moreover, we shall see later in Chapter 3 that a natural compactness will result when a uniform upper bound is specified on the Lipschitz constants of the local graphs.

The main elements of Definition 5.1 are illustrated in Figure 2.2 for $N = 2$.

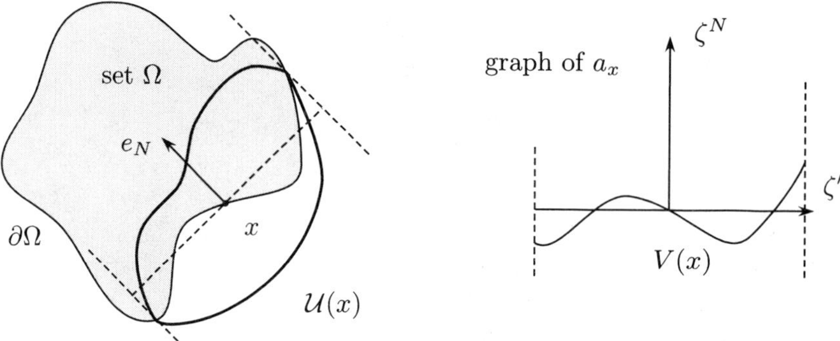

Figure 2.2. *Local epigraph representation ($N = 2$).*

Definition 5.1.
Let Ω be a subset of $\mathbf{R}^N$ such that $\partial\Omega \neq \varnothing$.

(i) Ω is said to be *locally Lipschitzian* if for each $x \in \partial\Omega$ there exist

    (a) an open neighborhood $\mathcal{U}(x)$ of x;

    (b) an orthonormal basis $\{e_1(x), \ldots, e_N(x)\}$;

    (c) a bounded open neighborhood $V(x)$ of 0 in the hyperplane $H(x) = \operatorname{span}\{e_1(x), \ldots, e_{N-1}(x)\}$ through 0 such that

$$\mathcal{U}(x) \subset \{y \in \mathbf{R}^N : P_{H(x)}(y - x) \in V(x)\}, \tag{5.1}$$

    where $P_{H(x)}$ is the orthogonal projection onto $H(x)$; and

    (d) a Lipschitzian mapping $a_x : V(x) \to \mathbf{R}$ such that

$$\mathcal{U}(x) \cap \partial\Omega = \left\{ x + \sum_{i=1}^{N} \zeta_i e_i(x) : \begin{array}{l} \zeta' \in V(x) \\ \zeta_N = a_x(\zeta') \end{array} \right\}, \tag{5.2}$$

$$\mathcal{U}(x) \cap \operatorname{int}\Omega = \mathcal{U}(x) \cap \left\{ x + \sum_{i=1}^{N} \zeta_i e_i(x) : \begin{array}{l} \zeta' \in V(x) \\ \zeta_N > a_x(\zeta') \end{array} \right\}, \tag{5.3}$$

    where $\zeta' = (\zeta_1, \ldots, \zeta_{N-1}) \in \mathbf{R}^{N-1}$.

(ii) Ω is said to be *Lipschitzian* if it is locally Lipschitzian and

$$\exists c > 0, \forall x \in \partial\Omega, \forall y, \forall z \in V(x), \quad |a_x(z) - a_x(y)| \leq c|z - y|. \tag{5.4}$$

(iii) Ω is said to be *uniformly Lipschitzian* if it is Lipschitzian and the neighborhoods $\mathcal{U}(x)$ and $V(x)$ can be chosen in such a way that $\mathcal{U}(x) - x$ and $V(x)$ are independent of x up to a rotation around 0. $\qquad\square$

Notice that it is always possible to redefine $\mathcal{U}(x)$ such that $V(x)$ is an open ball or an open hypercube centered in 0. Obviously domains that are locally the epigraph of a C^1- or smoother function are locally Lipschitzian.

According to this definition the whole space $\mathbf{R}^N$ and the closed unit ball with its center or a small crack removed are not locally Lipschitzian since in the first case the boundary is empty and in the second case the conditions cannot be satisfied at the center or along the crack, which is a part of the boundary. Similarly the set

$$\Omega \stackrel{\text{def}}{=} \bigcup_{n=1}^{\infty} \Omega_n, \quad \Omega_n \stackrel{\text{def}}{=} \left\{ y \in \mathbf{R}^N : \left| y - \frac{1}{2^n} \right| < \frac{1}{2^{n+2}} \right\}$$

is not locally Lipschitzian since the conditions of Definition 5.1 (i) are not satisfied in $0 \in \partial\Omega$. However, the set

$$\Omega \stackrel{\text{def}}{=} \bigcup_{n=1}^{\infty} \Omega_n, \quad \Omega_n \stackrel{\text{def}}{=} \left\{ y \in \mathbf{R}^N : |y - n| < \frac{1}{2^{n+2}} \right\}$$

is Lipschitzian, but not uniformly Lipschitzian.

The three cases considered in Definition 5.1 only differ when the boundary $\partial\Omega$ is unbounded. But we first introduce some notation.

Notation 5.1.

(i) Given an orthonormal basis $\{e_1(x), \ldots, e_N(x)\}$ at $x \in \partial\Omega$, a point y in $\mathbf{R}^N$ will be represented by (ζ', ζ_N), where

$$\zeta' \stackrel{\text{def}}{=} P_{H(x)}(y - x), \quad \zeta_N \stackrel{\text{def}}{=} (y - x) \cdot e_N(x),$$

and $P_{H(x)}(y - x)$ is identified with an element of $\mathbf{R}^{N-1}$.

(ii) The graph, epigraph, and hypograph of $a_x \colon V(x) \to \mathbf{R}$ will be denoted as follows:

$$A^0 \stackrel{\text{def}}{=} \left\{ x + \sum_{i=1}^{N} \zeta_i e_i(x) : \forall \zeta' \in V(x), \zeta_N = a_x(\zeta') \right\},$$

$$A^+ \stackrel{\text{def}}{=} \left\{ x + \sum_{i=1}^{N} \zeta_i e_i(x) : \forall \zeta' \in V(x), \zeta_N > a_x(\zeta') \right\},$$

$$A^- \stackrel{\text{def}}{=} \left\{ x + \sum_{i=1}^{N} \zeta_i e_i(x) : \forall \zeta' \in V(x), \zeta_N < a_x(\zeta') \right\}. \qquad \square$$

Theorem 5.1. *When the boundary $\partial\Omega$ is compact, the three types of Lipschitzian sets of Definition 5.1 coincide.*

Proof. It is sufficient to show that a locally Lipschitzian set is uniformly Lipschitzian. For each $x \in \partial\Omega$, there exists $r_x > 0$ such that $B(x, r_x) \subset \mathcal{U}(x)$, and from condition (5.1)

$$B(x, r_x) \subset \mathcal{U}(x) \subset \left\{ y \in \mathbf{R}^N : P_{H(x)}(y - x) \in V(x) \right\}.$$

Since $\partial\Omega$ is compact, there exists a finite sequence $\{x_i\}_{i=1}^{m}$ of points of $\partial\Omega$ and a finite subcover $\{B_i\}_{i=1}^{m}$, $B_i = B(x_i, R_i)$, $R_i = r_{x_i}$.

(i) We first claim that

$$\boxed{\exists R > 0, \forall x \in \partial\Omega, \exists i, 1 \leq i \leq m, \text{ such that } B(x, R) \subset B_i.}$$

We proceed by contradiction. If this is not true, then for each $n \geq 1$,

$$\exists x_n \in \partial\Omega, \ \forall i, 1 \leq i \leq m, \quad B\left(x_n, 1/n\right) \not\subset B_i,$$

$$\Rightarrow \forall i, 1 \leq i \leq m, \quad \exists y_{in} \in B\left(x_n, 1/n\right) \text{ such that } y_{in} \notin B_i.$$

Since $\partial\Omega$ is compact, there exists a subsequence of $\{x_n\}$ and $x \in \partial\Omega$ such that $x_{n_k} \to x$ as k goes to infinity. Hence for each $i, 1 \leq i \leq m$,

$$|y_{in_k} - x| \leq |y_{in_k} - x_{n_k}| + |x_{n_k} - x| \leq \frac{1}{n_k} + |x_{n_k} - x| \to 0$$

and $y_{in_k} \to x$ as k goes to infinity. For each i the set $\complement B_i$ is closed. Then $x \in \complement B_i$ and

$$y_{in_k} \in \complement B_i \to x \in \complement B_i \quad \Rightarrow \quad \exists\, x \in \partial\Omega \text{ such that } x \notin \bigcup_{i=1}^{m} B_i,$$

and this contradicts the fact that $\{B_i\}_{i=1}^{m}$ is an open cover of $\partial\Omega$. This property is obviously also satisfied for all neighborhoods N of 0 such that $\varnothing \neq N \subset B(0, R)$.

(ii) *Construction of $V'(x)$, $\mathcal{U}'(x)$, and a_x.* By construction for each $x \in \partial\Omega$, $\exists i,\ 1 \le i \le m$, such that

$$B(x, R) \subset B_i = B(x_i, R_i)$$

and $R + |x - x_i| < R_i$. From (5.1)

$$B(x_i, R_i) \subset \mathcal{U}(x_i) \subset \{y : P_{H(x_i)}(y - x_i) \in V(x_i)\}.$$

Therefore

$$B_{H(x_i)}(0, R) \subset B_{H(x_i)}(0, R_i) \subset V(x_i).$$

Choose

$$\boxed{e_j(x) = e_j(x_i), \quad 1 \le j \le N} \quad \Rightarrow \quad \boxed{H(x) = H(x_i).}$$

Furthermore, choose

$$\boxed{r \le \frac{R}{\sqrt{1 + c^2}}, \quad c = \max_{1 \le i \le m} c_{x_i},}$$

where $c_{x_i} > 0$ is the Lipschitz constant associated with a_{x_i} in $V(x_{x_i})$, $1 \le i \le m$. Define the new neighborhoods

$$\boxed{V'(x) = B_{H(x)}(0, r), \quad \mathcal{U}'(x) = B(x, R) \cap \{y : P_{H(x)}(y - x) \in V'(x)\}.}$$

They satisfy condition (5.1). Hence

$$V'(x) = B_{H(x)}(0, r) \subset B_{H(x_i)}(0, R) \subset B_{H(x_i)}(0, R_i) \subset V(x_i),$$
$$\mathcal{U}'(x) \subset B(x, R) \subset B(x_i, R_i) \subset \mathcal{U}_i = \mathcal{U}(x_i).$$

For $\zeta' \in V'(x)$ define

$$\boxed{a_x(\zeta') = a_{x_i}\big(\zeta' + P_{H(x_i)}(x - x_i)\big) - a_{x_i}\big(P_{H(x_i)}(x - x_i)\big).}$$

This map is well defined as a restriction of a_{x_i}. Indeed by construction

$$x \in B(x, R) \subset B(x_i, R_i) \subset \mathcal{U}(x_i) \subset \{y \in \mathbf{R}^N : P_{H(x_i)}(y - x_i) \in V(x_i)\}.$$

Therefore,

$$P_{H(x_i)}(x - x_i) \in V(x_i).$$

Moreover, since $R + |x - x_i| < R_i$ and $r < R$

$$|\zeta' + P_{H(x_i)}(x - x_i)| \le |\zeta'| + |P_{H(x_i)}(x - x_i)|$$
$$< r + |x - x_i| < R + |x - x_i| < R_i$$
$$\Rightarrow \zeta' + P_{H(x_i)}(x - x_i) \in B_{H(x_i)}(0, R_i) \subset V(x_i).$$

Finally, a_x is Lipschitzian since for all ζ' and ξ' in $V'(x)$

$$|a_x(\xi') - a_x(\zeta')| \le c_{x_i}|\xi' - \zeta'| \le c|\xi' - \zeta'|,$$

and the constant c is independent of x.

(iii) *Conditions* (5.1) *to* (5.3). We denote by A_i^0, A_i^+, A_i^- the sets defined in Notation 5.1 associated with $a_{x_i} : V(x_i) \to \mathbf{R}$, and by A^0, A^+, A^- those associated with the new $a_x : V'(x) \to \mathbf{R}$. By construction of $\mathcal{U}'(x)$, (5.1) is satisfied. From (5.2) at x_i

$$A_i^0 = \mathcal{U}(x_i) \cap \partial\Omega, \quad \mathcal{U}(x_i) \cap A_i^+ = \mathcal{U}(x_i) \cap \operatorname{int}\Omega,$$

and since $\mathcal{U}'(x) \subset \mathcal{U}(x_i)$

$$\mathcal{U}'(x) \cap A_i^0 = \mathcal{U}'(x) \cap \partial\Omega, \quad \mathcal{U}'(x) \cap A_i^+ = \mathcal{U}'(x) \cap \operatorname{int}\Omega.$$

To get conditions (5.2) and (5.3), it remains to show that

$$A^0 = \mathcal{U}'(x) \cap A_i^0 \quad \text{and} \quad A^+ \cap \mathcal{U}'(x) = A_i^+ \cap \mathcal{U}'(x).$$

We start with the second identity. In general for $y \in \mathbf{R}^N$, we use the following notation for the local coordinates with respect to x and x_i:

$$\zeta' = P_{H(x)}(y - x), \qquad \zeta_N = (y - x) \cdot e_N,$$
$$(\zeta^i)' = P_{H(x_i)}(y - x_i), \qquad \zeta_N^i = (y - x_i) \cdot e_N^i.$$

Then since $e_N = e_N^i$, $H(x) = H(x_i)$,

$$\zeta' = (\zeta^i)' + P_{H(x_i)}(x_i - x)$$

and

$$\zeta_N - a_x(\zeta') = \zeta_N^i + (x_i - x) \cdot e_N - a_{x_i}\big(\zeta' + P_{H(x_i)}(x - x_i)\big)$$
$$+ a_{x_i}\big(P_{H(x_i)}(x - x_i)\big)$$
$$= \zeta_N^i - a_{x_i}\big((\zeta^i)'\big),$$

since for $x \in \partial\Omega \cap B_i$

$$(x - x_i) \cdot e_N = a_{x_i}\big(P_{H(x_i)}(x - x_i)\big).$$

For $y \in A^+ \cap \mathcal{U}'(x)$

$$0 < \zeta_N - a_x(\zeta') = \zeta_N^i - a_{x_i}\big((\zeta^i)'\big),$$

and by construction of x and R

$$|y - x_i| \le |y - x| + |x - x_i| < R + |x - x_i| < R_i.$$

Thus $y \in A_i^+$ and $A^+ \cap \mathcal{U}'(x) \subset A_i^+ \cap \mathcal{U}'(x)$. Conversely by definition for $y \in A_i^+ \cap \mathcal{U}'(x)$, $(\zeta^i)' \in V(x_i)$, $\zeta' \in V'(x)$, and

$$0 < \zeta_N^i - a_{x_i}\big((\zeta^i)'\big) = \zeta_N - a_x(\zeta') \quad \Rightarrow \quad y \in A^+ \cap \mathcal{U}'(x).$$

Finally, we show that $A^0 = A_i^0 \cap \mathcal{U}'(x)$. Let $y \in A^0$; then $|\zeta'| < r$ and by construction of x and R

$$|(\zeta^i)'| \le |\zeta'| + |x - x_i| < R + |x - x_i| < R_i \quad \Rightarrow \quad (\zeta^i)' \in V(x_i),$$
$$\zeta_N^i - a_{x_i}\big((\zeta^i)'\big) = \zeta_N - a_x(\zeta') = 0 \quad \Rightarrow \quad y \in A_i^0.$$

To show that $y \in \mathcal{U}'(x)$ it is sufficient to show that

$$|y - x| < R$$

since we already know that $y \in A^0$ implies that $\zeta' \in V'(x)$. But

$$
\begin{aligned}
|y - x|^2 &= |\zeta'|^2 + |\zeta_N|^2 = |\zeta'|^2 + |a_x(\zeta')|^2 \\
&= |\zeta'|^2 + |a_{x_i}\big(\zeta' + P_{H(x_i)}(x - x_i)\big) - a_{x_i}\big(P_{H(x_i)}(x - x_i)\big)|^2 \\
&< |\zeta'|^2 + c^2|\zeta'|^2 = (1 + c^2)r^2 \le R^2
\end{aligned}
$$

by choice of r. Conversely, if $y \in A_i^0 \cap \mathcal{U}'(x)$, then $\zeta' \in V'(x) \subset V(x_i)$ and

$$\zeta_N - a_x(\zeta') = \zeta_N^i - a_{x_i}\big((\zeta^i)'\big) = 0 \quad \Rightarrow \quad y \in A^0.$$

This concludes the proof. $\qquad\qquad\qquad\qquad\qquad\qquad\qquad\qquad\qquad\square$

One of the important properties of a bounded (open) Lipschitzian domain Ω is the existence of a *continuous linear extension* of functions of the Sobolev space $H^k(\Omega)$ defined on Ω to functions of the Sobolev space $H^k(\mathbf{R}^N)$ defined on $\mathbf{R}^N$, that is,

$$E \colon H^k(\Omega) \to H^k(\mathbf{R}^N) \quad \text{and} \quad \forall \varphi \in H^k(\Omega), \quad (E\varphi)\big|_\Omega = \varphi \tag{5.5}$$

(cf. the Calderón extension theorem in Adams [1, p. 83, Thm. 4.32, p. 91] and Nečas [1, Thm. 3.10, p. 80]). This property is also important in the existence of optimal domains when it is uniform for a given family. For instance, this will occur for the family of domains satisfying the uniform cone property of section 6.

We extend the definition of locally Lipschitzian sets to locally $C^{k,\ell}$ sets.

Definition 5.2.

Let Ω be a subset of $\mathbf{R}^N$ such that $\partial\Omega \neq \varnothing$. The set Ω is said to be *locally a $C^{k,\ell}$-epigraph*, $k \geq 0$, $0 \leq \ell \leq 1$, if for each $x \in \partial\Omega$ there exist

(a) an open neighborhood $\mathcal{U}(x)$ of x;

(b) an orthonormal basis $\{e_1(x), \dots, e_N(x)\}$;

(c) a bounded open neighborhood $V(x)$ of 0 in the hyperplane $H(x) = \mathrm{span}\,\{e_1(x),$
$\dots, e_{N-1}(x)\}$ through 0 such that

$$\mathcal{U}(x) \subset \{y \in \mathbf{R}^N \ : \ P_{H(x)}(y-x) \in V(x)\}, \tag{5.6}$$

where $P_{H(x)}$ is the orthogonal projection onto $H(x)$; and

(d) a $C^{k,\ell}$-mapping $a_x : V(x) \to \mathbf{R}$ such that

$$\mathcal{U}(x) \cap \partial\Omega = \left\{ x + \sum_{i=1}^{N} \zeta_i e_i(x) \ : \ \begin{array}{l} \zeta' \in V(x) \\ \zeta_N = a_x(\zeta') \end{array} \right\}, \tag{5.7}$$

$$\mathcal{U}(x) \cap \mathrm{int}\,\Omega = \mathcal{U}(x) \cap \left\{ x + \sum_{i=1}^{N} \zeta_i e_i(x) \ : \ \begin{array}{l} \zeta' \in V(x) \\ \zeta_N > a_x(\zeta') \end{array} \right\}, \tag{5.8}$$

where $\zeta' = (\zeta_1, \dots, \zeta_{N-1}) \in \mathbf{R}^{N-1}$. $\qquad\qquad\square$

5.2   Local Epigraphs and Sets of Class $C^{k,\ell}$

Sets or domains that are locally the epigraph of a $C^{k,\ell}$-function, $k \geq 1$, $0 \leq \ell \leq 1$, (resp., locally Lipschitzian) are sets of class $C^{k,\ell}$ (resp., $C^{0,1}$). However, we shall see from Examples 5.1 and 5.2 that a domain of class $C^{0,1}$ is generally not locally the epigraph of a Lipschitzian function.

Theorem 5.2.

(i) *If Ω is locally a $C^{k,\ell}$-epigraph, $k \geq 0$, $0 \leq \ell \leq 1$, then it is of class $C^{k,\ell}$.*

(ii) *If Ω is of class $C^{k,\ell}$, $k \geq 1$, $0 \leq \ell \leq 1$, then it is locally a $C^{k,\ell}$-epigraph.*

Proof. (i) For each point $x \in \partial\Omega$ let $\mathcal{U}(x)$, $V(x)$, and a_x be the associated neighborhoods and the $C^{k,\ell}$-function. Define the $C^{k,\ell}$-mappings

$$h_x(\xi) \stackrel{\mathrm{def}}{=} x + \sum_{i=1}^{N-1} \xi_i e_i(x) + [\xi_N + a_x(\xi')]\, e_N(x), \tag{5.9}$$

$$g_x(y) \stackrel{\mathrm{def}}{=} (\{(y-x) \cdot e_i(x)\}_{i=1}^{N-1}, (y-x) \cdot e_N(x) - a_x(\{(y-x) \cdot e_i(x)\}_{i=1}^{N-1})). \tag{5.10}$$

It is easy to verify that h_x is the inverse of g_x:

$$h_x(g_x(y)) = y \text{ in } \mathcal{U}(x) \text{ and } g_x(h_x(\xi)) = \xi \text{ in } g_x(\mathcal{U}(x)),$$

and that

$$g_x(\mathcal{U}(x) \cap \Gamma) = g(\mathcal{U}(x)) \cap R^{N-1}, \quad R^{N-1} \overset{\text{def}}{=} \{(\xi', \xi_N) \in \mathbf{R}^N : \xi_N = 0\},$$
$$g_x(\mathcal{U}(x) \cap \operatorname{int}\Omega) = g_x(\mathcal{U}(x)) \cap \{(\xi', \xi_N) \in \mathbf{R}^N : \xi_N > 0\}.$$

The set Ω is of class $C^{k,\ell}$.

(ii) In the discussion preceding Theorem 4.1, we showed that a set of class $C^{k,\ell}$ is locally the level set of a $C^{k,\ell}$-function. From the definition of a set of class $C^{k,\ell}$, $k \geq 1$ and $0 \leq \ell \leq 1$, the set Ω can be locally described by the level sets of the $C^{k,\ell}$-function

$$\boxed{f_x(y) \overset{\text{def}}{=} g_x(y) \cdot e_N,}$$

since by definition

$$\operatorname{int}\Omega \cap U(x) = \{y \in U(x) : f_x(y) > 0\},$$
$$\partial\Omega \cap U(x) = \{y \in U(x) : f_x(y) = 0\}.$$

The boundary $\partial\Omega$ is the zero level set of f_x and the gradient

$$\nabla f_x(y) = {}^*Dg_x(y)e_N \neq 0$$

is normal to that level set. The exterior normal to Ω is given by

$$n(y) = -\frac{\nabla f_x(y)}{|\nabla f_x(y)|} = -\frac{{}^*Dg_x(y)e_N}{|{}^*Dg_x(y)e_N|} = -\frac{{}^*(Dh_x(g_x(y)))^{-1}e_N}{|{}^*(Dh_x(g_x(y)))^{-1}e_N|}.$$

Since f_x is C^1 and $\nabla f_x(x) \neq 0$, choose a smaller neighborhood of x, still denoted $U(x)$, such that

$$\forall y \in U(x), \quad \nabla f_x(y) \cdot \nabla f_x(x) > 0.$$

To construct the graph around the point $x \in \Gamma = \partial\Omega$, choose $e_N(x) = -n(x)$, the hyperplane $H(x)$ orthogonal to $e_N(x)$, and the orthonormal basis $\{e_1(x), \ldots, e_{N-1}(x)\}$ in $H(x)$. Consider the $C^{k,\ell}$- $(k \geq 1)$ function

$$\lambda(\zeta', \zeta_N) \overset{\text{def}}{=} f_x\left(x + \sum_{i=1}^{N} \zeta_i e_i(x)\right).$$

By construction, $\lambda(0,0) = 0$ and

$$0 = \nabla\lambda(0,0) \cdot (0, \xi_N) = \nabla f_x(x) \cdot (0, \xi_N) = \xi_N |\nabla f_x(x)| \quad \Rightarrow \xi_N = 0.$$

Therefore, by the implicit function theorem, there exists a neighborhood $V(x) \subset B_0$ of $(0,0)$ and a mapping $a_x \in C^{k,\ell}(W)$ such that $\lambda(0,0) = 0$ and

$$\lambda(\zeta', a_x(\zeta')) = f_x\left(x + \sum_{i=1}^{N-1} \zeta_i e_i(x) + a_x(\zeta')e_N(x)\right) = 0$$

for all $\zeta' \in V(x)$. By construction, $x + \sum_{i=1}^{N-1} \zeta_i e_i(x) + a_x(\zeta') e_N(x) \in \Gamma \cap U(x)$. Choose

$$\mathcal{U}(x) = U(x) \cap \{x + (\zeta', \zeta_N) : \zeta' \in V(x) \text{ and } \zeta_N \in \mathbf{R}\}.$$

By construction, $V(x)$ and $\mathcal{U}(x)$ satisfy condition (5.6). Moreover, from (3.4) and (3.5)

$$\text{int } \Omega \cap U(x) = h_x(B_+) \text{ and } \partial\Omega \cap U(x) = h_x(B_0)$$

$$\Rightarrow \left|\begin{array}{l} \text{int } \Omega \cap U(x) = \{y \in U(x) : f_x(y) > 0\}, \\ \partial\Omega \cap U(x) = \{y \in U(x) : f_x(y) = 0\}, \end{array}\right.$$

and since $\mathcal{U}(x) \subset U(x)$ we get conditions (5.7) and (5.8),

$$\text{int } \Omega \cap \mathcal{U}(x) = \{y \in \mathcal{U}(x) : f_x(y) > 0\} = A^+ \cap \mathcal{U}(x),$$
$$\partial\Omega \cap \mathcal{U}(x) = \{y \in \mathcal{U}(x) : f_x(y) = 0\} = A^0.$$

Recall that by construction of $\mathcal{U}(x)$ for all $(\zeta', \zeta_N) \in \mathcal{U}(x) - \{x\}$,

$$\zeta_N > a_x(\zeta') \iff \lambda(\zeta', \zeta_N) = f_x(x + \zeta' + \zeta_N e_N(x)) > 0$$

since

$$\frac{\partial\lambda}{\partial\zeta_N}(\zeta) = \nabla f_x(x + \zeta) \cdot e_N(x) = \nabla f_x(x + \zeta) \cdot \frac{\nabla f_x(x)}{|\nabla f_x(x)|} > 0$$

in the neighborhood $\mathcal{U}(x) - \{x\}$ of $\zeta = 0$. $\qquad\qquad\square$

Example 5.1 (Adams, Aronszajn, and Smith [1]).

Consider the open convex set $\Omega_0 = \{\rho e^{i\theta} : 0 < \rho < 1, \, 0 < \theta < \pi/2\}$ and its image $\Omega = T(\Omega_0)$ by the $C^{0,1}$-homeomorphism

$$T(\rho e^{i\theta}) = \rho e^{i(\theta - \log\rho)}, \quad T^{-1}(\rho e^{i\theta}) = \rho e^{i(\theta + \log\rho)}.$$

It is readily seen that as ρ goes to zero the image of the two pieces of the boundary of Ω_0 corresponding to $\theta = 0$ and $\theta = \pi/2$ begin to spiral around the origin. As a result Ω is not locally the epigraph of a function at the origin. $\qquad\square$

Example 5.2.

This second example can be found in Murat and Simon [1], where it is attributed to Zerner. Consider the Lipschitzian function λ defined on $[0, 1]$ as follows: $\lambda(0) = 0$, and on each interval $[1/3^{n+1}, 1/3^n]$

$$\lambda(s) \stackrel{\text{def}}{=} \begin{cases} 2\left(s - \dfrac{1}{3^{n+1}}\right), & \dfrac{1}{3^{n+1}} \le s \le \dfrac{2}{3^{n+1}}, \\[4mm] -2\left(s - \dfrac{1}{3^n}\right), & \dfrac{2}{3^{n+1}} \le s \le \dfrac{1}{3^n}, \end{cases}$$

where n ranges over all integers $n \geq 0$. Associate with λ and a real $\delta > 0$ the set

$$\Omega \stackrel{\text{def}}{=} \{(x_1, x_2) \in \mathbf{R}^2 : 0 < x_1 < 1, \; |x_2 - \lambda(x_1)| < \delta \, x_1\}.$$

The set Ω is the image of the triangle

$$\Omega_0 \stackrel{\text{def}}{=} \{(x_1, x_2) \in \mathbf{R}^2 : 0 < x_1 < 1, \; |x_2| < \delta \, x_1\}$$

through the $C^{0,1}$-homeomorphism

$$T(x_1, x_2) \stackrel{\text{def}}{=} (x_1, x_2 + \lambda(x_1)), \quad T^{-1}(y_1, y_2) = (y_1, y_2 - \lambda(y_1)).$$

Since the triangle Ω_0 is Lipschitzian, its image is a set of class $C^{0,1}$. But Ω is not locally Lipschitzian in $(0,0)$ since Ω zigzags like lightning as it gets closer to the origin. Thus, however small the neighborhood around $(0,0)$ is, a direction $e_N(0,0)$ cannot be found to make the domain locally the epigraph of a function. $\qquad\square$

5.3   Boundary Integral from the Graph of a Function

The *boundary measure* on $\partial\Omega$, which was defined in section 3.2 from the local diffeomorphism, can also be defined from the local graph representation of the boundary. Assuming that $\partial\Omega$ is compact, there is a finite subcover of open neighborhoods $\mathcal{U}(x)$ for $\partial\Omega$ that can be represented by a finite family of Lipschitzian graphs. Specifically, let $\{\mathcal{U}_j\}_{j=1}^m$, $\mathcal{U}_j = \mathcal{U}(x_j)$, be a finite open cover of $\partial\Omega$ corresponding to some sequence $\{x_j\}_{j=1}^m$ of points of $\partial\Omega$. Denote by $\{e_1^j, \ldots, e_N^j\}$, $\{V_j\}_{j=1}^m$, $V_j = V(x_j)$, and $\{a_j\}$, $a_j = a_{x_j}$, the associated elements of Definition 5.1. Introduce the notation

$$\Gamma = \partial\Omega, \qquad \Gamma_j = \Gamma \cap \mathcal{U}_j, \; 1 \leq j \leq m, \tag{5.11}$$

and the map $h_j = h_{x_j}$ and its inverse $g_j = g_{x_j}$ as defined in (5.9). By Rademacher's theorem the Lipschitzian mapping a_x is differentiable almost everywhere in $V(x)$ and belongs to $W^{1,\infty}(V(x))$ (cf., for instance, Evans and Gariepy [1]). Since h_j is defined from the Lipschitzian mapping a_j, Dh_j is defined almost everywhere in $V(x) \times \mathbf{R}$, but also is H_{N-1} almost everywhere on $V(x) \times \{0\}$. Thus the canonical density for $C^{k,\ell}$ domains, $k \geq 1$, still makes sense and is given by the same formula (3.13):

$$\omega_j(\zeta') \stackrel{\text{def}}{=} \omega_{x_j}(\zeta') = |\det Dh_j(\zeta', 0)| \, |\, {}^*(Dh_j)^{-1}(\zeta', 0) \, e_N^j|. \tag{5.12}$$

It is easy to verify that for almost all ξ

$$Dh_j = \begin{bmatrix} 1 & 0 & \cdots & \cdots & 0 \\ 0 & \ddots & & & \vdots \\ \vdots & & \ddots & & \vdots \\ 0 & & & \ddots & 0 \\ \partial_1 a_j & \partial_2 a_j & \cdots & \partial_{N-1} a_j & 1 \end{bmatrix}, \quad \det Dh_j = 1,$$

where $\partial_i a_j$ is the partial derivative of a_j with respect to the ith component of $\zeta' = (\zeta_1, \ldots, \zeta_{N-1})$. The matrix Dh_j is invertible and its coefficients belong to L^∞. By direct computation

$$^*(Dh_j)^{-1} = \begin{bmatrix} 1 & 0 & \cdots & 0 & -\partial_1 a_j \\ 0 & \ddots & & \vdots & -\partial_2 a_j \\ \vdots & & \ddots & 0 & \vdots \\ \vdots & & & \ddots & -\partial_{N-1} a_j \\ 0 & 0 & \cdots & 0 & 1 \end{bmatrix}, \tag{5.13}$$

and finally

$$\omega_j(\zeta') = |\,{}^*(Dh_j)^{-1}(\zeta', 0)\, e_N| = \sqrt{1 + |\nabla a_j(\zeta')|^2}. \tag{5.14}$$

Let $\{r_1, \ldots, r_m\}$ be a partition of unity for the $\mathcal{U}_j$'s, that is,

$$\begin{cases} r_j \in \mathcal{D}(\mathcal{U}_j), \quad 0 \le r_j(x) \le 1, \\ \displaystyle\sum_{j=1}^m r_j(x) = 1, \ \text{in a neighborhood } \mathcal{U} \text{ of } \Gamma \end{cases} \tag{5.15}$$

such that $\overline{\mathcal{U}} \subset \cup_{j=1}^m \mathcal{U}_j$. For any function f in $C^0(\Gamma)$ the functions f_j and $f_j \circ h_j$ defined as

$$\begin{cases} f_j(y) = f(y) r_j(y), & y \in \Gamma_j, \\ f_j \circ h_j(\zeta', 0) = f_j(x + \sum_{i=1}^{N-1} \zeta_i e_i^j + a_j(\zeta') e_N^j), & \zeta' \in V_j, \end{cases} \tag{5.16}$$

respectively, belong to $C^0(\Gamma_j)$ and $C^0(V_j)$. The integral of f on Γ is then defined as

$$\int_\Gamma f \, d\Gamma \stackrel{\text{def}}{=} \sum_{j=1}^m \int_{\Gamma_j} f_j \, d\Gamma_j, \qquad \int_{\Gamma_j} f_j \, d\Gamma \stackrel{\text{def}}{=} \int_{V_j} f_j(h_j(\zeta', 0)) \omega_j(\zeta') \, d\zeta'. \tag{5.17}$$

Since $\omega_j \in L^\infty(V_j)$, $1 \le j \le m$, this integral is also well defined for all f in $L^1(\Gamma)$; that is, the function $f_j \circ h_j \, \omega_j$ belongs to $L^1(V_j)$ for all j, $1 \le j \le m$.

As in section 2, the tangent plane to Γ_j at $y = h_j(\zeta', 0)$ is defined by the tangent vectors

$$\tau_i = Dh_j(\zeta', 0) \, e_i(x_j), \quad 1 \le i \le N - 1.$$

An outward normal field to Γ_j is given by

$$m_x(\zeta) = -\,{}^*(Dh_j)^{-1}(\zeta', \zeta_N) e_N^j = \,{}^*(\partial_1 a_j(\zeta'), \ldots, \partial_{N-1} a_j(\zeta'), -1),$$

and the unit outward normal to Γ_j at $y = h_j(\zeta', 0) \in \Gamma_j$ is given by

$$n(y) = n(h_j(\zeta', 0)) = \frac{{}^*(\partial_1 a_j(\zeta'), \ldots, \partial_{N-1} a_j(\zeta'), -1)}{\sqrt{1 + |\nabla a_j(\zeta')|^2}}. \tag{5.18}$$

The above constructions also apply to domains that are locally the epigraph of a $C^{k,\ell}$-function, $k \ge 1$.

5.4 Volume and Surface Measures of the Boundary

The three types of *Lipschitzian sets* are defined through a local property. Conditions (5.1) to (5.3) mean that every point of $\mathcal{U}(x)$ has a representation in terms of the epigraph, the graph, or the hypograph of the map a_x defined in $V(x)$ (cf. Lemma 5.1). Such sets and domains have nice properties: their boundary has zero "volume" and locally finite "surface measure."

Theorem 5.3. *Let $\Omega \subset \mathbf{R}^N$ be locally Lipschitzian. Then the complement of Ω is also locally Lipschitzian; the interior of Ω, the interior of its complement, and its boundary $\partial\Omega$ are not empty; and*

$$\partial\Omega = \partial(\operatorname{int}\Omega) = \partial(\operatorname{int}\complement\Omega),$$
$$\overline{\operatorname{int}\Omega} = \overline{\Omega}, \quad and \quad \overline{\operatorname{int}\complement\Omega} = \overline{\complement\Omega}. \tag{5.19}$$

Moreover, for all $x \in \partial\Omega$

$$m\big(\mathcal{U}(x)\cap\partial\Omega\big) = 0 \quad \Rightarrow \quad m\big(\partial\Omega\big) = 0 \quad and$$
$$H_{N-1}\big(\mathcal{U}(x)\cap\partial\Omega\big) \le \sqrt{1+c_x^2}\,\, m_{N-1}\big(V(x)\big), \tag{5.20}$$

where m is the N-dimensional Lebesgue measure, m_{N-1} and H_{N-1} are the respective $(N-1)$-dimensional Lebesgue measure and $(N-1)$-dimensional Hausdorff measure, and $c_x > 0$ is the Lipschitz constant of a_x in $V(x)$. If, in addition, $\partial\Omega$ is compact, then

$$H_{N-1}(\partial\Omega) < \infty. \tag{5.21}$$

Proof. The fact that the complement of Ω is Lipschitzian follows from the equivalence of conditions (b) and (c) of the next lemma.

Lemma 5.1. *Given $x \in \partial\Omega$, $\mathcal{U} = \mathcal{U}(x)$, and $V = V(x)$, the following sets of conditions are equivalent:*

(a) $\mathcal{U}\cap\partial\Omega = A^0, \mathcal{U}\cap\operatorname{int}\Omega \subset A^+, \mathcal{U}\cap\operatorname{int}\complement\Omega \subset A^-,$

(b) $\mathcal{U}\cap\partial\Omega = A^0, \mathcal{U}\cap\operatorname{int}\Omega = \mathcal{U}\cap A^+, \mathcal{U} \subset \{y : P_{H(x)}(y-x) \in V\},$

(c) $\mathcal{U}\cap\partial\Omega = A^0, \mathcal{U}\cap\operatorname{int}\complement\Omega = \mathcal{U}\cap A^-, \mathcal{U} \subset \{y : P_{H(x)}(y-x) \in V\}.$

Proof of Lemma 5.1. (a) $\Rightarrow$ (b). From the union of the three properties of (a),

$$\mathcal{U} \subset A \overset{\text{def}}{=} A^0 \cup A^+ \cup A^- = \{y : P_{H(x)}(y-x) \in V\},$$

we get the last property of (b). From the second property of (a),

$$\mathcal{U}\cap\operatorname{int}\Omega \subset A^+ \quad \Rightarrow \quad \mathcal{U}\cap\operatorname{int}\Omega \subset \mathcal{U}\cap A^+.$$

In the other direction take the complement of the first and last properties of (a) with respect to $\mathcal{U}$

$$\mathcal{U}\cap\overline{\complement\Omega} \subset A^0 \cup A^-$$
$$\Rightarrow \mathcal{U}\cap\operatorname{int}\Omega = \complement_{\mathcal{U}}(\mathcal{U}\cap\overline{\complement\Omega}) \supset \complement_{\mathcal{U}}(A^0 \cup A^-) = \mathcal{U}\cap A^+.$$

This yields the second identity in (b) and (a) $\Rightarrow$ (b).

(b) $\Rightarrow$ (c). From the first two properties of (b),

$$\mathcal{U} \cap \overline{\Omega} = \mathcal{U} \cap (A^0 \cup A^+)$$
$$\Rightarrow \mathcal{U} \cap \operatorname{int} \complement\Omega = \complement_{\mathcal{U}}(\mathcal{U} \cap \overline{\Omega}) = \complement_{\mathcal{U}}(\mathcal{U} \cap (A^0 \cup A^+)) = \mathcal{U} \cap A^-.$$

(c) $\Rightarrow$ (a). From the first two properties of (c),

$$\mathcal{U} \cap \overline{\complement\Omega} = \mathcal{U} \cap (A^0 \cup A^-)$$
$$\Rightarrow \mathcal{U} \cap \operatorname{int}\Omega = \complement_{\mathcal{U}}(\mathcal{U} \cap \overline{\complement\Omega}) = \complement_{\mathcal{U}}(\mathcal{U} \cap (A^0 \cup A^-)) = \mathcal{U} \cap A^+ \subset A^+$$

and, from the second property of (c), $\mathcal{U} \cap \operatorname{int} \complement\Omega = \mathcal{U} \cap A^- \subset A^-$. $\qquad\square$

The other properties follow from the fact that Ω is locally the epigraph of a Lipschitzian continuous map and identities (5.19) readily follow from the following simple lemma. Another proof will be given in (6.6) of section 6 by using the uniform cone property.

Lemma 5.2.

(i) $\partial(\operatorname{int}\Omega) = \partial\Omega \quad\Longleftrightarrow\quad \overline{\operatorname{int}\Omega} = \overline{\Omega}.$

(ii) $\partial(\operatorname{int}\complement\Omega) = \partial\Omega \quad\Longleftrightarrow\quad \overline{\operatorname{int}\complement\Omega} = \overline{\complement\Omega}.$

Proof. It is sufficient to prove (i). ($\Rightarrow$) By definition of the closure,

$$\overline{\Omega} = \operatorname{int}\Omega \cup \partial\Omega = \operatorname{int}\Omega \cup \partial(\operatorname{int}\Omega) = \overline{\operatorname{int}\Omega}.$$

Conversely ($\Leftarrow$), by definition $\partial\Omega = \overline{\Omega} \cap \overline{\complement\Omega}$,

$$\overline{\Omega} \cap \overline{\complement\Omega} = \overline{\operatorname{int}\Omega} \cap \overline{\complement\Omega} = \overline{\operatorname{int}\Omega} \cap \overline{\complement(\operatorname{int}\Omega)} = \overline{\operatorname{int}\Omega} \cap \overline{\complement(\operatorname{int}\Omega)} = \partial(\operatorname{int}\Omega). \qquad\square$$

From formula (5.14)

$$H_{N-1}\big(\mathcal{U}(x) \cap \partial\Omega\big) = \int_{V(x)} \big[1 + |\nabla a_x(\zeta')|^2\big]^{1/2}\, d\zeta' \le \big[1 + c_x^2\big]^{1/2} m_{N-1}\big(V(x)\big).$$

When the boundary of Ω is compact, then we can find a finite subcover of open neighborhoods $\{\mathcal{U}(x_j)\}_{j=1}^m$. From the previous estimate, $H_{N-1}(\partial\Omega)$ is bounded by a finite sum of bounded terms. $\qquad\square$

5.5   Convex Sets

Theorem 5.4. *Any convex subset Ω of $\mathbf{R}^N$ such that $\Omega \ne \mathbf{R}^N$ and $\operatorname{int}\Omega \ne \varnothing$ is a locally Lipschitzian set, and for each $x \in \partial\Omega$ the neighborhood $V(x)$ and the mapping a_x can be chosen convex.*

Proof. (i) *Construction of the mapping a_x.* Pick any point x_0 in the interior of Ω and choose $\varepsilon > 0$ such that the open ball $B(x_0, \varepsilon) \subset \mathrm{int}\,\Omega$. Since $\Omega \neq \mathbf{R}^N$, $\partial\Omega \neq \varnothing$. Associate with each $x \in \partial\Omega$ the direction

$$
e_N(x) = \frac{x_0 - x}{|x_0 - x|},
$$

the hyperplane $H(x)$ through 0 orthogonal to $e_N(x)$, the orthogonal projection operator $P_H(x)$ onto $H(x)$, and the neighborhoods

$$
\mathcal{U}(x) = \left\{ y \in \mathbf{R}^N : |P_{H(x)}(y - x)| < \varepsilon, |(y - x) \cdot e_N(x)| < |x - x_0| \right\},
$$
$$
V(x) = \{ z \in H(x) : |z| < \varepsilon \}.
$$

Then condition (5.1) is satisfied. For each $\zeta' \in V(x)$, define the line

$$
L_{\zeta'} = \{ x + (\zeta', 0) + \beta\, e_N(x) \ : \ \forall \beta \in \mathbf{R} \}
$$

and the function

$$
a_x(\zeta') = \inf_{y \in \overline{\Omega} \cap L_{\zeta'}} e_N(x) \cdot (y - x).
$$

This function is finite and there exists a unique $\hat{y} \in \partial\Omega \cap L_{\zeta'}$ such that

$$
a_x(\zeta') = e_N(x) \cdot (\hat{y} - x).
$$

We first show that

$$
|a_x(\zeta')| < |x - x_0|.
$$

To see this, notice that for $|\zeta'| < \varepsilon$, $x_0 + (\zeta', 0) \in \mathrm{int}\,\Omega$ and

$$
y \stackrel{\mathrm{def}}{=} x_0 + (\zeta', 0) - \sqrt{\varepsilon^2 - |\zeta'|^2}\, e_N(x) \in \overline{B(x_0, \varepsilon)} \subset \overline{\Omega} \ \Rightarrow \ y \in L_{\zeta'} \cap \overline{\Omega}
$$

and

$$
a_x(\zeta') \leq e_N(x) \cdot (x_0 - x) - \sqrt{\varepsilon^2 - |\zeta'|^2} \leq |x_0 - x| - \varepsilon < |x_0 - x|.
$$

In the other direction assume that $a_x(\zeta') \leq -|x - x_0|$; so there exists $\eta \geq 0$ such that

$$
y = x + (\zeta', 0) - (|x_0 - x| + \eta)\, e_N(x) \in \overline{\Omega}.
$$

Then

$$
y_0 = x_0 - \frac{|x_0 - x|}{|x_0 - x| + \eta}(\zeta', 0) \in B(x_0, \varepsilon) \subset \mathrm{int}\,\Omega
$$

and observe that

$$x = \alpha_0 \, y + (1 - \alpha_0) \, y_0 \quad \text{for } \alpha_0 = \frac{|x_0 - x|}{2|x_0 - x| + \eta}.$$

For all $z \in B(y_0, (1 - \alpha_0)\varepsilon)$

$$|z - x_0| \leq |y_0 - x_0| + (1 - \alpha_0)\varepsilon < \frac{|x_0 - x|}{2|x_0 - x| + \eta}\varepsilon + (1 - \alpha_0)\varepsilon = \varepsilon$$

$$\Rightarrow \ B(y_0, (1 - \alpha_0)\varepsilon) \subset B(x_0, \varepsilon).$$

Note that since $|x - x_0| > 0$, $(1 - \alpha_0) > 0$ for any $\eta \geq 0$. Consider the ball $B(x, (1 - \alpha_0)^2 \varepsilon)$ around x. Then

$$\begin{aligned} B(x, (1 - \alpha_0)^2 \varepsilon) &= x + (1 - \alpha_0) B(0, (1 - \alpha_0)\varepsilon) \\ &= \alpha_0 y_0 + (1 - \alpha_0) B(y_0, (1 - \alpha_0)\varepsilon) \\ &\subset \alpha_0 y_0 + (1 - \alpha_0) B(y_0, \varepsilon) \subset \text{int } \Omega \end{aligned}$$

and $x \in \text{int } \Omega$, which contradicts the assumption that $x \in \partial\Omega$. Therefore $|a_x(\zeta')| < |x - x_0|$ and there exist minimizing points $\hat{y}$ in the compact convex set $L^0_{\zeta'} \cap \overline{\Omega} \cap \overline{\mathcal{U}}$ such that $\hat{y} \in \mathcal{U}$. Moreover, $\hat{y} \in \partial\Omega$. Otherwise we could construct a small neighborhood $B(\hat{y}, \rho) \subset \text{int } \Omega$ around $\hat{y}$ and $y = \hat{y} - \rho\, e_N(x) \in \overline{\Omega} \cap L_{\zeta'}$ such that

$$e_N(x) \cdot (y - x) = e_N \cdot (\hat{y} - x) - \rho < e_N \cdot (\hat{y} - x) = \inf_{y \in \overline{\Omega} \cap L_{\zeta'}} e_N \cdot (y - x),$$

and this would contradict the minimality.

If $\hat{y}_1$ and $\hat{y}_2$ are two distinct minimizing points, then

$$e_N(x) \cdot (\hat{y}_1 - x) = e_N(x) \cdot (\hat{y}_2 - x) \text{ and } P_{H(x)}(\hat{y}_1 - x) = P_{H(x)}(\hat{y}_2 - x)$$

and $\hat{y}_1 = \hat{y}_2$. Therefore,

$$\exists! \hat{y} \in L_{\zeta'} \cap \overline{\Omega}, \quad a_x(\zeta') = e_N(x) \cdot (\hat{y} - x),$$

and $\hat{y} \in L^0_{\zeta'} \cap \partial\Omega \subset \mathcal{U}(x)$. Finally it is easy to verify that the mapping a_x satisfies conditions (5.2) and (5.3). For all $\zeta' \in V(x)$,

$$x + \zeta' + a_x(\zeta')e_N(x) \in \partial\Omega \cap \mathcal{U}(x) \quad \Rightarrow A^0 \subset \partial\Omega \cap \mathcal{U}(x),$$

and for all $\zeta' \in V(x)$ and $\zeta_N > a_x(\zeta')$ such that

$$y = x + \zeta' + \zeta_N e_N(x) \in \mathcal{U}(x),$$

$y \in \text{int } \Omega \cap \mathcal{U}(x)$ and $A^+ \cap \mathcal{U}(x) \subset \text{int } \Omega \cap \mathcal{U}(x)$. Conversely, for all $y \in \overline{\Omega} \cap \mathcal{U}(x)$,

$$y = x + P_{H(x)}(y - x) + e_N(x) \cdot (y - x)\, e_N(x) = x + \zeta' + \zeta_N e_N(x)$$

$$\Rightarrow y \in L_{\zeta'} \cap \overline{\Omega}$$

$$\Rightarrow a_x(P_{H(x)}(y - x)) \leq e_N(x) \cdot (y - x) = \zeta_N.$$

For $y \in \operatorname{int} \Omega$, clearly

$$a_x(P_{H(x)}(y - x)) < e_N(x) \cdot (y - x) = \zeta_N \quad \Rightarrow y \in A^+.$$

For $y \in \partial\Omega$, $a_x(P_{H(x)}(y - x)) = \zeta_N$.

(ii) *Convexity of the mapping* a_x. By construction the neighborhoods $V(x)$ and $\mathcal{U}(x)$ are convex. The set $\mathcal{U}(x) \cap \overline{\Omega}$ is convex. Then recall from (5.2)–(5.3) that

$$\mathcal{U}(x) \cap \left\{ x + \sum_{i=1}^{N} \zeta_i e_i(x) : \zeta' \in V(x),\ \zeta_N \geq a_x(\zeta') \right\} = \mathcal{U}(x) \cap \overline{\Omega}.$$

Thus the set on the left-hand side is convex. In particular,

$$\forall y^1, y^2 \in \overline{\Omega} \cap \mathcal{U}(x),\ \forall \alpha \in [0, 1], \quad \alpha y^1 + (1 - \alpha)y^2 \in \overline{\Omega} \cap \mathcal{U}(x).$$

In particular, for any two y^1 and y^2 in $\partial\Omega \cap \mathcal{U}(x)$,

$$y^j = x + \sum_{i=1}^{N} \zeta_i^j e_i(x), \quad \zeta_N^j = a_x(\zeta^{j'}),\ j = 1, 2,$$

and

$$y^\alpha = \alpha y^1 + (1 - \alpha)y^2 = x + \sum_{i=1}^{N} [\alpha \zeta_i^1 + (1 - \alpha)\zeta_i^2] e_i(x) \in \overline{\Omega} \cap \mathcal{U}(x)$$

$$\Rightarrow \alpha \zeta_N^1 + (1 - \alpha)\zeta_N^2 \geq a_x\left(\left(\alpha \zeta^1 + (1 - \alpha)\zeta^2 \right)' \right)$$

$$\Rightarrow \alpha a_x(\zeta^{1'}) + (1 - \alpha)a_x(\zeta^{2'}) \geq a_x\left(\alpha \zeta^{1'} + (1 - \alpha)\zeta^{2'} \right)$$

and a_x is convex.

(iii) *The mapping* a_x *is Lipschitzian.* It is sufficient to show that a_x is bounded in $V(x)$. Then we conclude from Ekeland and Temam [1, Lem. 3.1, p. 11] that the mapping is continuous and convex and hence Lipschitz continuous in $V(x)$. By construction the minimizing point $\hat{y}$ belongs to $\mathcal{U}(x)$, which is bounded. $\qquad\square$

5.6 Nonhomogeneous Neumann and Dirichlet Problems

With the help of the previous definition of boundary measure we can now make sense of the nonhomogeneous Neumann and Dirichlet problems (for the Laplace equation) in Lipschitzian domains. Assuming that the functions a_j belong to $W^{1,\infty}(V_j)$, it can be shown that the classical Stokes divergence theorem holds for such domains. Given a bounded smooth domain D in $\mathbf{R}^N$ and a locally Lipschitzian domain Ω in D, then

$$\forall \vec{\varphi} \in C^1(\overline{D}, \mathbf{R}^N), \quad \int_{\Omega} \operatorname{div} \vec{\varphi}\, dx = \int_{\Gamma} \vec{\varphi} \cdot n\, d\Gamma, \tag{5.22}$$

where the outward unit normal field n is defined by (5.18) for almost all $\in \Gamma_j$ (with $y = h_x(\zeta', 0)$, $\zeta' \in V_j$) as

$$n(y) = \frac{{}^*(\partial_1 a_j(\zeta'), \ldots, \partial_{N-1} a_j(\zeta'), -1)}{\sqrt{1 + |\nabla a_j(\zeta')|^2}}.$$

The trace of a function g in $W^{1,1}(D)$ on Γ is defined through a $W^{1,\infty}(D)^N$-extension N of the normal n as

$$\forall \varphi \in \mathcal{D}(\mathbf{R}^N), \quad \int_\Gamma g\varphi \, d\Gamma \stackrel{\text{def}}{=} \int_\Omega \operatorname{div}(g\varphi N) \, dx \tag{5.23}$$

(cf. Agmon, Douglis, and Nirenberg [1, 2]). The right-hand side is well defined in the usual sense so that by the Stokes divergence theorem we obtain the trace $g\big|_\Gamma$ defined on Γ. This trace is uniquely defined, denoted by $\gamma_\Gamma g$, and

$$\gamma_\Gamma \in \mathcal{L}\big(W^{1,1}(\Omega), L^1(\Gamma)\big). \tag{5.24}$$

Let Ω be a Lipschitzian domain in D. Given $f \in L^2(D)$ and $g \in H^1(D)$, consider the following problem:

$$\exists y \in H^1(\Omega) \text{ such that } \forall \varphi \in H^1(\Omega)$$
$$\int_\Omega \nabla y \cdot \nabla \varphi \, dx = \int_\Omega f\varphi \, dx + \int_\Gamma g\,\varphi \, d\Gamma. \tag{5.25}$$

If the condition

$$\int_\Omega f \, dx + \int_\Gamma g \, d\Gamma = 0 \tag{5.26}$$

is satisfied, then by the Lax–Milgram theorem problem (5.25)–(5.26) has a unique solution in $H^1(\Omega)/\mathbf{R}$. Any two solutions of problem (5.25) can only differ by a constant. For instance, we can associate with the solution $y = y(\Omega)$ in $H^1(\Omega)/\mathbf{R}$ of (5.25) the following objective function:

$$J(\Omega) \stackrel{\text{def}}{=} c\big(\Omega, y(\Omega)\big), \quad c(\Omega, \varphi) \stackrel{\text{def}}{=} \int_\Omega \big(|\nabla \varphi(x)| - G\big)^2 \, dx \tag{5.27}$$

for some G in $L^2(D)$.

For the nonhomogeneous Dirichlet problem let Ω be a Lipschitzian domain in D and consider the following weak form of the problem:

$$\exists y \in H^1(\Omega) \text{ such that } \forall \varphi \in H^2(\Omega) \cap H_0^1(\Omega),$$
$$-\int_\Omega y \triangle \varphi \, dx = -\int_\Gamma g \frac{\partial \varphi}{\partial n} \, d\Gamma + \int_\Omega f \, \varphi \, dx \tag{5.28}$$

for data (f, g) in $L^2(D) \times H^1(D)$. If this problem has a solution y, then by Green's theorem

$$\int_\Omega \nabla y \cdot \nabla \varphi \, dx - \int_\Gamma y \, \frac{\partial \varphi}{\partial n} \, d\Gamma = - \int_\Gamma g \, \frac{\partial \varphi}{\partial n} \, d\Gamma + \int_\Omega f \, \varphi \, dx \qquad (5.29)$$

and

$$-\triangle y = f \in L^2(\Omega), \quad y|_\Gamma = g \in H^{\frac{1}{2}}(\Gamma). \qquad (5.30)$$

Here, for instance, we can associate with y the objective function

$$J(\Omega) \stackrel{\text{def}}{=} c(\Omega, y(\Omega)), \quad c(\Omega, \varphi) \stackrel{\text{def}}{=} \frac{1}{2} \int_\Omega |\varphi - G|^2 \, dx, \quad G \in L^2(D). \qquad (5.31)$$

6   Uniform Cone Property for Lipschitzian Domains

Lipschitzian domains can also be equivalently characterized by a purely geometric
uniform cone property which seems to have been originally introduced by Agmon [1].
In this section we establish the equivalence of the two sets of definitions and express
the previous properties and formulae of Definition 5.1 in terms of the parameters
of the cone. One of the important properties of a family of open domains satisfying
the uniform cone property is that the extension operator from the domains to $\mathbf{R}^N$
is uniformly continuous.

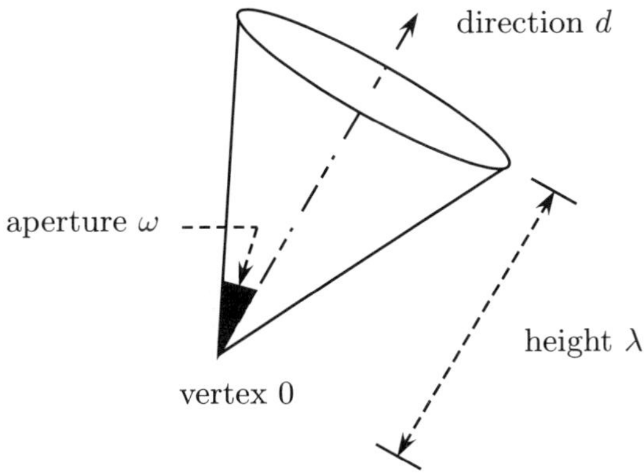

Figure 2.3. *The cone $C(\lambda, \omega, d)$.*

Notation 6.1.
Given $\lambda > 0$, $0 < \omega \le \pi/2$, and a direction $d \in \mathbf{R}^N$, $|d| = 1$, denote by $C(\lambda, \omega, d)$
the open cone

$$C(\lambda, \omega, d) \stackrel{\text{def}}{=} \left\{ y \in \mathbf{R}^N : \frac{1}{\tan \omega} |P_H(y)| < y \cdot d < \lambda \right\},$$

where P_H is the orthogonal projection onto the hyperplane H through the origin
0 and orthogonal to the direction d (cf. Figure 2.3). Further denote by $C_x(\lambda, \omega, d)$
the translated cone $x + C(\lambda, \omega, d)$ for an arbitrary $x \in \mathbf{R}^N$. $\qquad \square$

Definition 6.1.

Let Ω be a subset of $\mathbf{R}^N$ such that $\partial\Omega \neq \varnothing$.

(i) Ω is said to satisfy the *local uniform cone property* if

$$\forall x \in \partial\Omega, \ \exists \lambda > 0, \ \exists \omega > 0, \ \exists r > 0, \ \exists d \in \mathbf{R}^N, |d| = 1,$$

such that

$$\forall y \in B(x,r) \cap \overline{\Omega}, \quad C_y(\lambda, \omega, d) \subset \operatorname{int}\Omega.$$

(ii) Ω is said to satisfy the *ω-local uniform cone property* if

$$\exists \omega > 0, \ \forall x \in \partial\Omega, \ \exists \lambda > 0, \ \exists r > 0, \ \exists d \in \mathbf{R}^N, |d| = 1,$$

such that

$$\forall y \in B(x,r) \cap \overline{\Omega}, \quad C_y(\lambda, \omega, d) \subset \operatorname{int}\Omega.$$

(iii) Ω is said to satisfy the *uniform cone property* if

$$\exists \lambda > 0, \ \exists \omega > 0, \ \exists r > 0, \ \forall x \in \partial\Omega, \ \exists d \in \mathbf{R}^N, |d| = 1,$$

such that
$$\forall y \in B(x,r) \cap \overline{\Omega}, \quad C_y(\lambda, \omega, d) \subset \operatorname{int}\Omega. \qquad \square$$

As in Definition 3.1, the above definitions only involve $\partial\Omega$, $\overline{\Omega}$, and $\operatorname{int}\Omega$. Even if, classically, a Lipschitzian set is an open set, the property remains true for all sets in the class

$$[\Omega]_b = \left\{ A \subset \mathbf{R}^N \ : \ \overline{A} = \overline{\Omega} \text{ and } \partial A = \partial\Omega \right\}.$$

Moreover, they enjoy the following properties.

Theorem 6.1. *Given $\Omega \subset \mathbf{R}^N$ such that $\partial\Omega \neq \varnothing$ and*

$$\forall x \in \partial\Omega, \ \exists \lambda > 0, \ \exists d \in \mathbf{R}^N, |d| = 1, \quad \forall \alpha \in \,]0,1[, \ x + \alpha\lambda d \in \operatorname{int}\Omega, \qquad (6.1)$$

then $\operatorname{int}\Omega \neq \varnothing$ *and*

$$\overline{\operatorname{int}\Omega} = \overline{\Omega} \ \text{ and } \ \partial(\operatorname{int}\Omega) = \partial\Omega.$$

In particular, if Ω satisfies the local uniform cone property, then

$$\overline{\operatorname{int}\Omega} = \overline{\Omega}, \ \partial(\operatorname{int}\Omega) = \partial\Omega \quad \text{and} \quad \overline{\operatorname{int}\complement\Omega} = \overline{\complement\Omega}, \ \partial(\operatorname{int}\complement\Omega) = \partial\Omega.$$

Proof. (i) Pick any point $x \in \partial\Omega$. Then for all $\mu \in \,]0,1[$

$$x_\mu = x + \mu\,\lambda\,d \in \operatorname{int}\Omega$$

and $\operatorname{int}\Omega \neq \varnothing$ since $\partial\Omega \neq \varnothing$. As a result there exists a sequence $\mu_n > 0 \to 0$ such that

$$\operatorname{int}\Omega \ni x_{\mu_n} \to x \quad \Rightarrow \quad \partial\Omega \subset \overline{\operatorname{int}\Omega} \subset \overline{\Omega} \quad \Rightarrow \quad \overline{\Omega} \subset \overline{\operatorname{int}\Omega} \subset \overline{\Omega}.$$

The second identity follows from Lemma 5.2 (i). (ii) When Ω satisfies the local uniform cone property the condition of part (i) is trivially verified and the first two identities are true. For the second ones fix $x \in \partial\Omega$. Hence $x + C(\lambda, \omega, d) \subset \operatorname{int}\Omega$. Choose $x_0 = x - \tilde{r}\, d$, $\tilde{r} = \min\{r, \lambda/2\}$. Then

$$\forall \alpha \in \,]0, 1[, \quad x_\alpha = \alpha\, x_0 + (1 - \alpha)\, x \in \operatorname{int}\complement\Omega.$$

If there was an $\alpha \in \,]0, 1[$ such that $x_\alpha \in \overline{\Omega}$, then $x_\alpha \in B(x, r)$ since $|x_\alpha - x| = \alpha\,|x_0 - x| < \min\{r, \lambda/2\} \leq r$ and hence

$$x_\alpha + C(\lambda, \omega, d) \subset \operatorname{int}\Omega \quad \Rightarrow \quad \boxed{x = x_\alpha + \alpha\,\tilde{r}\,d \in \operatorname{int}\Omega,}$$

which is a contradiction. So pick $\alpha_n \to 0$ and $\operatorname{int}\complement\Omega \ni x_{\alpha_n} \to x$. The second identity follows from Lemma 5.2 (ii). $\qquad\square$

As for Lipschitzian sets the three cases of Definition 6.1 only differ when $\partial\Omega$ is unbounded.

Theorem 6.2. *If $\partial\Omega$ is compact, then the three uniform cone properties of Definition 6.1 coincide.*

Proof. It is sufficient to show that the local uniform cone property implies the uniform cone property. Since $\partial\Omega$ is compact there exists a finite open subcover $\{B_i\}_{i=1}^m$, $B_i = B(x_i, r_{x_i})$, of $\partial\Omega$ for some finite sequence $\{x_i\}_{i=1}^m$ of points of $\partial\Omega$. We now claim that

$$\exists r > 0, \forall x \in \partial\Omega, \quad \exists i, 1 \leq i \leq m, \ \text{such that}\ B(x, r) \subset B_i.$$

The proof is the same as the one in part (i) of the proof of Theorem 5.1. Define

$$\omega = \min_{1 \leq i \leq m} \omega_i \quad \text{and} \quad \lambda = \min_{1 \leq i \leq m} \lambda_i.$$

Therefore, for each $x \in \partial\Omega$, there exists i, $1 \leq i \leq m$, such that $B(x, r) \subset B_i$ and hence there exists $d_i = d_{x_i}$, $|d_i| = 1$, such that

$$\forall y \in \overline{\Omega} \cap B_i, \quad C_y(\lambda_i, \omega_i, d_i) \subset \operatorname{int}\Omega.$$

Since $B(x, r) \subset B_i$,

$$\forall y \in \overline{\Omega} \cap B(x, r), \quad C_y(\lambda, \omega, d_i) \subset C_y(\lambda_i, \omega_i, d_i) \subset \operatorname{int}\Omega.$$

With $d_x = d_i$ we conclude that Ω has the uniform cone property. $\qquad\square$

Theorem 6.3.

(i) *If Ω satisfies the local uniform cone property, then Ω is locally Lipschitzian. If for $x \in \partial\Omega$, $B(x,r)$ and $C_x(\lambda,\omega,d)$ are the open ball and the open cone and $H(x)$ is the hyperplane through 0 orthogonal to d, the neighborhoods of Definition 5.1 can be chosen as*

$$\boxed{\begin{aligned} V(x) &= B_{H(x)}(0,\rho), \\ \mathcal{U}(x) &= B(x,r_\lambda) \cap \left\{ y : P_{H(x)}(y-x) \in V(x) \right\}, \end{aligned}} \tag{6.2}$$

where $B_{H(x)}(0,\rho)$ is the open ball of radius ρ in $H(x)$ with center 0,

$$r_\lambda = \min\left\{ r, \frac{\lambda}{2} \right\}, \quad \rho = r_\lambda \sqrt{\frac{\tan^2 \omega}{1 + \tan^2 \omega}}. \tag{6.3}$$

Moreover, the associated function a_x defined by

$$\boxed{a_x(\zeta') = (y-x) \cdot d, \quad \zeta' = P_{H(x)}(y-x) \in V(x)}$$

satisfies the following Lipschitz condition in $V(x)$

$$\forall \zeta', \xi' \in V(x), \quad |a_x(\xi') - a_x(\zeta')| \le (\tan\omega)^{-1} |\xi' - \zeta'|, \tag{6.4}$$

and

$$\mathrm{m}(\partial\Omega \cap \mathcal{U}(x)) = 0, \tag{6.5}$$

$$H_{N-1}(\partial\Omega \cap \mathcal{U}(x)) \le \rho^{N-1} \alpha_{N-1} \sqrt{1 + (\tan\omega)^{-2}}, \tag{6.6}$$

where m and H_{N-1} are the respective N-dimensional Lebesgue measure and the $(N-1)$-dimensional Hausdorff measure, and α_{N-1} is the volume of the unit ball in $\mathbf{R}^{N-1}$.

(ii) *Conversely, if Ω is locally Lipschitzian, then Ω satisfies the local uniform cone property.*

Proof. (i) By assumption for each $x \in \partial\Omega$

$$\exists r > 0, \exists \omega > 0, \exists \lambda > 0, \exists d \in \mathbf{R}^N, |d| = 1,$$

such that

$$\forall y \in B(x,r) \cap \overline{\Omega}, \quad C_y(\lambda,\omega,d) \subset \operatorname{int}\Omega.$$

Let H be the hyperplane through 0 orthogonal to the direction d. Let $e_N(x) = d$ and let $\{e_i(x) : 1 \le i < N\}$ be an orthonormal basis in $H(x)$. Denote by P_H the orthogonal projection onto H and define the set

$$U = \{P_H(y-x) : \forall y \in B(x,r) \cap \partial\Omega\}.$$

Therefore,

$$\forall \zeta \in U, \quad \exists y \in B(x,r) \cap \partial\Omega, \quad P_H(y - x) = \zeta,$$

and we can associate with ζ and y the quantity

$$\boxed{\gamma(y) \stackrel{\text{def}}{=} (y - x) \cdot d.}$$

The mapping a is well defined. We wish to show that there exists an open neighborhood of 0 in U for which the mapping

$$\boxed{\zeta' \mapsto a(\zeta') \stackrel{\text{def}}{=} \gamma(y)}$$

is well defined, that is, for all y_1 and y_2 in $B(x,r) \cap \partial\Omega$

$$P_H(y_2 - x) = \zeta' = P_H(y_1 - x) \quad \Rightarrow \quad \gamma(y_2) = \gamma(y_1).$$

To do this, first reduce the radius of the ball $B(x,r)$ from r to

$$\boxed{r_\lambda = \min\{r, \lambda/2\}}$$

and define

$$\boxed{U_\lambda = \{P_H(y - x) : y \in B(x, r_\lambda) \cap \partial\Omega\}.}$$

By contradiction assume that y_1 and y_2 are two points in U_λ such that

$$P_H(y_2 - x) = \zeta' = P_H(y_1 - x) \text{ and } \gamma(y_2) > \gamma(y_1).$$

By the definition of r_λ

$$|(y_2 - y_1) \cdot d| \le |y_2 - y_1| \le |y_2 - x| + |y_1 - x| < 2r_\lambda, \le \lambda,$$

and by construction y_2 and y_1 are on the same line through $x + \zeta'$ in the direction d. In particular, since $y_1 \in \partial\Omega \cap B(x,r)$, then necessarily

$$y_2 \in C_{y_1}(\lambda, \omega, d) \subset \text{int } \Omega,$$

and this contradicts the fact that $y_2 \in \partial\Omega$. Therefore the above constructions induce a well-defined mapping

$$\boxed{\zeta' \mapsto a(\zeta') = \gamma(y) : U_\lambda \to \mathbf{R}}$$

for some unique $y \in B(x, r_\lambda) \cap \partial\Omega$ such that $P_H(y - x) = \zeta'$.

The set U_λ is a neighborhood of 0. We next show that U_λ contains an open ball $B_H(0, \rho)$ of radius $\rho > 0$. This is true if we pick

$$0 < \rho = r_\lambda \sqrt{(\tan\omega)^2/(1 + (\tan\omega)^2)}.$$

Notice that for all $\zeta' \in B_H(0, \rho)$

$$|\zeta| \frac{1}{\tan \omega} < \frac{\rho}{\tan \omega} = \sqrt{r_\lambda^2 - \rho^2} < \sqrt{r_\lambda^2 - |\zeta|^2}. \tag{6.7}$$

Associate with each $\zeta' \in B_H(0, \rho)$ the line

$$L(\zeta') = \{x + \zeta' + \alpha d : \alpha \in \mathbf{R}\} \cap B(x, r_\lambda)$$
$$= \left\{ x + \zeta' + \alpha d : |\alpha| < \sqrt{r_\lambda^2 - |\zeta'|^2} \right\}.$$

In particular, in view of (6.7)

$$L(\zeta) \cap C_x(\lambda, \omega, d) = \left\{ x + \zeta' + \alpha d : |\zeta'| \frac{1}{\tan \omega} < \alpha < \sqrt{r_\lambda^2 - |\zeta'|^2} \right\}$$

is not empty, and since

$$C_x(\lambda, \omega, d) \subset \operatorname{int} \Omega \quad \Rightarrow \quad L(\zeta') \cap C_x(\lambda, \omega, d) \subset \operatorname{int} \Omega,$$

then $L(\zeta') \cap \operatorname{int} \Omega \neq \varnothing$. Now we show by contradiction that $B_H(0, \rho) \subset U_\lambda$. If not, then there exists $\zeta \in B_H(0, \rho)$,

$$\forall \alpha, \quad |\alpha| < \sqrt{r_\lambda^2 - |\zeta'|^2}, \quad x + \zeta' + \alpha d \notin \partial\Omega,$$

and since $\operatorname{int} \Omega \cap L(\zeta') \neq \varnothing$, then $L(\zeta') \subset \operatorname{int} \Omega$ by definition of U. Thus the point

$$y = x + \zeta' - \sqrt{r_\lambda^2 - |\zeta'|^2}\, d \in \overline{\Omega} \cap B(x, r_\lambda) \subset \overline{\Omega} \cap B(x, r),$$

and by assumption

$$C_y(\lambda, \omega, d) \subset \operatorname{int} \Omega.$$

We claim that $x \in C_y(\lambda, \omega, d) \subset \operatorname{int} \Omega$, which is in contradiction with the fact that $x \in \partial\Omega$. To check this we must show that

$$\frac{1}{\tan \omega} |P_H(x - y)| < (x - y) \cdot d < \lambda.$$

But $y \in \overline{B(x, r_\lambda)}$,

$$(x - y) \cdot d \leq |x - y| \leq r_\lambda \leq \frac{\lambda}{2} < \lambda,$$

and

$$P_H(x - y) = \zeta', \quad (x - y) \cdot d = \sqrt{r_\lambda^2 - |\zeta'|^2}.$$

But

$$|\zeta'| < \rho = r_\lambda \sqrt{\tan^2 \omega / (1 + \tan^2 \omega)}$$

and finally

$$\frac{|P_H(x-y)|}{(x-y)\cdot d} < \frac{\rho}{\sqrt{r_\lambda^2 - \rho^2}} = \tan\omega.$$

This contradiction proves that $B_H(0,\rho) \subset U_\lambda$.

The neighborhoods of Definition 5.1 can now be chosen as

$$\boxed{V(x) = B_H(0,\rho),} \quad \boxed{\mathcal{U}(x) = B(x,r_\lambda) \cap \{y : P_H(y-x) \in V(x)\}.}$$

By construction, condition (5.1) in Definition 5.1 is satisfied. As for condition (5.2), by construction $V(x) \subset U_\lambda$ and for each $\zeta' \in V(x)$

$$\exists \text{ a unique } y \in \partial\Omega \cap B(x,r_\lambda) \text{ and } a(\zeta') \stackrel{\text{def}}{=} (y-x)\cdot d.$$

As a result (using Notation 5.1)

$$\begin{aligned}
A^0 &= \left\{ x + \sum_{i=1}^{N-1} \zeta_i e_i + a(\zeta')e_N : \zeta' \in V(x) \right\} \\
&= \{y : P_H(y-x) = \zeta' \in V(x) \text{ and } a(\zeta') = (y-x)\cdot d\} \\
&= \{y : P_H(y-x) \in V(x), y \in \partial\Omega \cap B(x,r_\lambda)\} \\
&= \{y : P_H(y-x) \in V(x), y \in B(x,r_\lambda)\} \cap \partial\Omega = \mathcal{U}(x) \cap \partial\Omega
\end{aligned}$$

and identity (5.2) is satisfied. It remains to check (5.3). For each $y \in \mathcal{U} \cap A^+$,

$$\zeta' = P_H(y-x) \in V(x), \quad \zeta_N = (y-x)\cdot d > a(\zeta'), \quad y \in B(x,r_\lambda),$$

and the point

$$y_{\zeta'} = x + \sum_{i=1}^{N-1} \zeta_i e_i(x) + a(\zeta')e_N(x) \in A^0 = \mathcal{U} \cap \partial\Omega.$$

Hence

$$y \in C_{y_{\zeta'}}(\lambda,\omega,d) \subset \operatorname{int}\Omega \quad \Rightarrow \quad y \in \operatorname{int}\Omega \quad \Rightarrow \quad y \in \operatorname{int}\Omega \cap \mathcal{U}.$$

Conversely, for each $y \in \operatorname{int}\Omega \cap \mathcal{U}$,

$$\zeta' = P_H(y-x) \in V(x), \quad y \in B(x,r_\lambda) \cap \operatorname{int}\Omega.$$

By definition of $A^0, y \notin A^0$. If $y \in A^- \cap \mathcal{U}$, then

$$y \in \operatorname{int}\Omega \cap \mathcal{U} \text{ and } (y-x)\cdot d < a(\zeta'),$$

and the point

$$y_{\zeta'} = x + \sum_{i=1}^{N-1} \zeta_i e_i(x) + a(\zeta')e_N(x) \in \partial\Omega$$

is above the point y on the line through $x + \zeta'$ parallel to d. Since $y \in \mathcal{U} \subset B(x, r_\lambda)$, then $y \in \overline{\Omega} \cap B(x, r_\lambda)$ and

$$C_y(\lambda, \omega, d) \subset \operatorname{int} \Omega \quad \Rightarrow \quad y_{\zeta'} \in \operatorname{int} \Omega,$$

which is a contradiction. Therefore, $\operatorname{int} \Omega \cap \mathcal{U} \subset A^+ \cap \mathcal{U}$ and condition (5.3) of Definition 5.1 is satisfied.

The mapping a is Lipschitzian. Fix ζ' and ξ' in $V(x)$ such that $a(\xi') < a(\zeta')$. By construction, the point

$$y_{\zeta'} = x + \sum_{i=1}^{N-1} \zeta_i e_i(x) + a(\zeta') e_N(x) \in \partial \Omega,$$

and by assumption, $C_{y_{\zeta'}}(\lambda, \omega, d) \subset \operatorname{int} \Omega$. But

$$y_{\xi'} = x + \sum_{i=1}^{N-1} \xi_i e_i(x) + a(\xi') e_N(x) \in \partial \Omega$$

and necessarily $y_{\xi'} \notin C_{y_{\zeta'}}(\lambda, \omega, e_N(x))$. This means that either

$$\frac{1}{\tan \omega} |P_H(y_{\xi'} - y_{\zeta'})| \geq (y_{\xi'} - y_{\zeta'}) \cdot e_N(x)$$

or

$$0 < \lambda \leq (y_{\xi'} - y_{\zeta'}) \cdot e_N(x).$$

But

$$(y_{\xi'} - y_{\zeta'}) \cdot e_N(x) = a(\xi') - a(\zeta') \leq 0,$$

which is a contradiction. Therefore, since $P_H(y_{\xi'} - y_{\zeta'}) = \xi' - \zeta'$,

$$\frac{1}{\tan \omega} |\xi' - \zeta'| \geq a(\xi') - a(\zeta').$$

A similar inequality is obtained for $a(\xi') \geq a(\zeta')$ by interchanging the role of ζ' and ξ' in the above proof. Therefore, a is uniformly Lipschitzian in $V(x)$ with constant $(\tan \omega)^{-1}$. In view of the Lipschitz continuity of a in $V(x)$, the "length" of $\partial \Omega \cap \mathcal{U}(x)$ is now given by the usual formula

$$H_{N-1}\big(\partial \Omega \cap \mathcal{U}(x)\big) = \int_{B_H(0,\rho)} \big[1 + |\nabla a(\zeta)|^2\big]^{1/2} \, d\zeta$$

$$\leq \big[1 + (\tan \omega)^{-2}\big]^{1/2} \rho^{N-1} v_{N-1}.$$

Since $\overline{\Omega}$ is the epigraph of a Lipschitzian mapping in $B(x, \rho)$, then $\operatorname{m}\big(\partial \Omega \cap \mathcal{U}(x)\big) = 0$.

(ii) Pick a point $x \in \partial \Omega$. By Definition 5.1 there exist neighborhoods $\mathcal{U}(x)$ and $V(x)$, an orthonormal basis $\{e_1, \ldots, e_N\}$, and a Lipschitzian function $a_x : V(x) \to \mathbf{R}$

with the appropriate properties. Let $c > 0$ be the Lipschitz constant associated with a_x and introduce the angle ω defined as

$$\frac{1}{\tan \omega} = c, \quad 0 < \omega \leq \frac{\pi}{2}.$$

Since $\mathcal{U}(x)$ is a neighborhood of x, $\exists R > 0$, $B(x, 2R) \subset \mathcal{U}(x)$. Pick

$$r = R \text{ and } \lambda, \quad 0 < \lambda \leq \frac{R}{\sqrt{1 + \tan^2 \omega}}.$$

For all $y \in \overline{\Omega} \cap B(x, r)$, $y \in \mathcal{U}(x)$, since

$$B(x, r) \subset B(x, 2R) \subset \mathcal{U}(x) \subset \{y : P_{H(x)}(y - x) \in V(x)\}.$$

Let $\xi_N = (y - x) \cdot e_N$ and

$$\xi' \stackrel{\text{def}}{=} P_{H(x)}(y - x) \in B_{H(x)}(0, r) \subset V(x).$$

Moreover, $\xi_N \geq a(\xi')$. For all $z \in C_y(\lambda, \omega, e_N)$, let $\zeta' \stackrel{\text{def}}{=} P_{H(x)}(z - x)$ and $\zeta_N = (z - x) \cdot e_N$ and

$$\lambda > \zeta_N - \xi_N > \frac{1}{\tan \omega} |\zeta' - \xi'| = c|\zeta' - \xi'|$$

and

$$|z - x| \leq |z - y| + |y - x| < \lambda \sqrt{1 + \tan^2 \omega} + r \leq 2R$$

and $C_y(\lambda, \omega, e_N) \subset B(x, 2R) \subset \mathcal{U}(x)$. To complete the proof, it remains to show that

$$C_y(\lambda, \omega, e_N) \subset \text{int } \Omega.$$

However, $\text{int } \Omega \cap \mathcal{U}(x) = A^+ \cap \mathcal{U}(x)$ and it is sufficient to show that for all $z \in C_y(\lambda, \omega, e_N)$, $\zeta_N > a(\zeta')$. But

$$\begin{aligned} \zeta_N - a(\zeta') &= \zeta_N - \xi_N + (\xi_N - a(\xi')) + (a(\xi') - a(\zeta')) \\ &> c|\zeta' - \xi'| + 0 - c|\zeta' - \xi'| = 0. \end{aligned}$$

Thus $\partial \Omega$ has the local uniform cone property. $\qquad \square$

Theorem 6.4.

    (i) *A set Ω satisfies the uniform (resp., ω-local uniform) cone property if and only if Ω is uniformly Lipschitzian (resp., Lipschitzian).*

    (ii) *If the boundary $\partial \Omega$ of Ω is compact and Ω satisfies the uniform cone property, then*

$$m(\partial \Omega) = 0 \quad \text{and} \quad H_{N-1}(\partial \Omega) < \infty.$$

Proof. (i) In the proof of Theorem 6.3 (i) we have shown that a set Ω which satisfies the local uniform cone property is locally Lipschitzian. We have also constructed the neighborhoods $V(x)$ and $\mathcal{U}(x)$ given by (6.2)–(6.3) and the function a_x satisfying the Lipschitz property (6.4). If Ω satisfies the ω-local uniform cone property, the aperture ω is independent of $x \in \partial\Omega$ and the Lipschitz constant $c = \tan^{-1}\omega$ is also independent of $x \in \partial\Omega$. Therefore Ω is a Lipschitzian domain. Finally, if Ω satisfies the uniform cone property, the radius r and the height λ of the cone are also independent of $x \in \partial\Omega$. Then ρ and r_λ are constant and the neighborhoods $V(x)$ and $\mathcal{U}(x) - x$ are equal up to a rotation around 0. Hence Ω is a uniformly Lipschitzian domain in the sense of Definition 5.1 (iii).

Conversely, if Ω is a Lipschitzian domain, it satisfies the local uniform cone property by Theorem 6.3 (ii). Since the Lipschitz constant $c > 0$ is independent of x, the aperture ω can be chosen as

$$\cot\mathrm{an}\,\omega = c, \quad 0 < \omega \le \pi/2,$$

and Ω satisfies the ω-local uniform cone property. If, in addition, Ω is uniformly Lipschitzian, then the neighborhood $\mathcal{U}(x) - x$ and $V(x)$ are independent of x up to a rotation around 0. In particular,

$$\exists R > 0 \text{ such that } \forall x \in \partial\Omega, \quad B(x, 2R) \subset \mathcal{U}(x).$$

Then choose

$$r = R \quad \text{and} \quad \lambda, 0 < \lambda \le R/\sqrt{1 + (\tan\omega)^2}$$

and proceed as in the proof of Theorem 6.3 (ii).

(ii) From Theorem 6.3 (i), Ω is a locally Lipschitzian domain. Moreover, since $\partial\Omega$ is compact, Ω is uniformly Lipschitzian by Theorem 5.1. So the neighborhoods $\mathcal{U}(x) - x$ and $V(x)$ are independent of x up to a rotation around 0. Since $\partial\Omega$ is compact, there exists a finite open subcover $\{\mathcal{U}_i\}_{i=1}^M$, $\mathcal{U}_i = \mathcal{U}(x_i)$, of $\partial\Omega$ for a finite sequence $\{x_i\}_{i=1}^M$ in $\partial\Omega$. Then from (5.20) in Theorem 5.3

$$\mathrm{m}(\mathcal{U}_i \cap \partial\Omega) = 0,$$

$$H_{N-1}(\mathcal{U}_i \cap \partial\Omega) \le \sqrt{1 + c^2}\, \mathrm{m}_{N-1}\big(V(x_i)\big),$$

where $c > 0$ is the uniform Lipschitz constant. Since M is finite

$$\mathrm{m}(\partial\Omega) = 0,$$

$$H_{N-1}(\partial\Omega) \le \sum_{i=1}^M H_{N-1}(\mathcal{U}_i \cap \partial\Omega) \le \sqrt{1 + c^2}\, \sum_{i=1}^M \mathrm{m}_{N-1}\big(V(x_i)\big).$$

But since the $V(x)$'s are equal up to a rotation

$$\exists v, 0 \le v < \infty, \quad \forall x \in \partial\Omega, \quad \mathrm{m}_{N-1}\big(V(x)\big) = v,$$

and $H_{N-1}(\partial\Omega)$ is finite. $\qquad\qquad\qquad\qquad\qquad\qquad\qquad\qquad\quad\square$

Remark 6.1.
Part (i) of Theorem 6.4 is an extension of Chenais [4, pp. 202–206]. Her definition of Lip (k, δ) corresponds to the one of uniformly Lipschitzian domain in Definition 5.1 (iii) for the following special choice of neighborhoods:

$$V(x) = \{\zeta \in H(x) : \max_{1 \leq i \leq N-1} |a(\zeta_i)| < \delta\},$$

$$\mathcal{U}(x) = \left\{ y \in \mathbf{R}^N : \begin{array}{l} P_{H(x)}(y - x) \in V(x) \quad \text{and} \\ |(y - x) \cdot e_N(x)| < k\,\delta(N - 1)^{1/2} \end{array} \right\},$$

where k is the Lipschitz constant of a_x,

$$|a_x(\xi) - a_x(\zeta)| \leq k \left\{ \sum_{i=1}^{N-1} |\xi_i - \zeta_i|^2 \right\}^{1/2},$$

since $\mathcal{U}(x)$ contains all the variations of a_x in $V(x)$. $\square$

7 Segment Property and C^0-Epigraphs

A larger class of domains Ω which can be characterized by a geometric property are the domains which satisfy a *segment property*. This property is sufficient for obtaining the density of $C^k(\overline{\Omega})$ in the Sobolev space $W^{m,p}(\Omega)$ for any $m \geq 1$ and $k \geq m$. As in the case of domains satisfying a *uniform cone property*, where this property is equivalent to being locally a Lipschitzian epigraph, satisfying the *segment property* is equivalent to being locally a C^0-epigraph.

An *open segment* between two distinct points x and y of $\mathbf{R}^N$ will be denoted

$$(x, y) \stackrel{\text{def}}{=} \{x + t(y - x) : \forall t, 0 < t < 1\}.$$

Definition 7.1.
Let Ω be a subset of $\mathbf{R}^N$ such that $\partial\Omega \neq \varnothing$.

(i) Ω is said to satisfy the *segment property* if

$$\forall x \in \partial\Omega, \quad \exists r > 0, \quad \exists \lambda > 0, \quad \exists d \in \mathbf{R}^N, \quad |d| = 1$$

such that

$$\forall y \in B(x.r) \cap \overline{\Omega}, \quad (y, y + \lambda d) \subset \text{int}\,\Omega.$$

(ii) Ω is said to satify the *uniform segment property* if

$$\exists r > 0, \quad \exists \lambda > 0 \quad \text{such that} \quad \forall x \in \partial\Omega, \quad \exists d \in \mathbf{R}^N, \quad |d| = 1$$

and

$$\forall y \in B(x.r) \cap \overline{\Omega}, \quad (y, y + \lambda d) \subset \text{int}\,\Omega.$$ $\square$

Of course, when Ω is a domain, Ω is open, $\Omega = \operatorname{int}\Omega$, and we are back to the standard definition. Yet, as in Definition 3.1, those definitions involve only $\partial\Omega$, $\overline{\Omega}$, and $\operatorname{int}\Omega$, and the properties remain true for all sets in the class

$$[\Omega]_b = \left\{ A \subset \mathbf{R}^N : \overline{A} = \overline{\Omega} \text{ and } \partial A = \partial\Omega \right\}.$$

A set having the segment property must have an $(N-1)$-dimensional boundary and cannot simultaneously lie on both sides of any given part of its boundary. In fact, a domain satisfying the segment property is locally a C^0-epigraph in the sense of Definition 5.2, as we shall see below. We first give a few general properties.

Theorem 7.1. *Given* $\Omega \subset \mathbf{R}^N$ *which satisfies the segment property*, $\operatorname{int}\Omega \neq \varnothing$,

$$\overline{\operatorname{int}\Omega} = \overline{\Omega} \quad \text{and} \quad \partial(\operatorname{int}\Omega) = \partial\Omega.$$

Moreover, $\operatorname{int}\complement\Omega \neq \varnothing$,

$$\overline{\operatorname{int}\complement\Omega} = \overline{\complement\Omega} \quad \text{and} \quad \partial(\operatorname{int}\complement\Omega) = \partial\Omega.$$

Proof. The first set of properties follows from Theorem 6.1. By assumption $\partial\Omega \neq \varnothing$ and for any $x \in \partial\Omega$, $(x, x + \lambda d) \subset \operatorname{int}\Omega$. Choose $x_0 = x - r_\lambda d$, $r_\lambda = \min\{r, \lambda/2\}$. Then

$$\forall \alpha \in \,]0,1[, \quad x_\alpha = \alpha\, x_0 + (1-\alpha)\, x \in \operatorname{int}\complement\Omega$$

and $\operatorname{int}\complement\Omega \neq \varnothing$. Indeed, if there was an $\alpha \in \,]0,1[$ such that $x_\alpha \in \overline{\Omega}$, then $x_\alpha \in B(x, r)$ since $|x_\alpha - x| = \alpha\,|x_0 - x| < \min\{r, \lambda/2\} \leq r$, and hence

$$(x_\alpha, x_\alpha + \lambda d) \subset \operatorname{int}\Omega \quad \Rightarrow \quad x = x_\alpha + \alpha\, r_\lambda\, d \in \operatorname{int}\Omega$$

which would yield a contradiction. Thus any point $x \in \partial\Omega$ can be approximated by a sequence $x_{\alpha_n} \in \operatorname{int}\complement\Omega \to x$, where $\alpha_n \to 0$. Therefore $\partial\Omega \subset \overline{\operatorname{int}\complement\Omega}$ and

$$\overline{\complement\Omega} = \overline{\operatorname{int}\complement\Omega \cup \partial\Omega} \subset \overline{\operatorname{int}\complement\Omega \cup \overline{\operatorname{int}\complement\Omega}} = \overline{\operatorname{int}\complement\Omega} \quad \Rightarrow \quad \overline{\complement\Omega} = \overline{\operatorname{int}\complement\Omega}.$$

The second identity follows from Lemma 5.2(ii). $\qquad\qquad\square$

As for Lipschitzian sets, the two cases of Definition 7.1 differ only when $\partial\Omega$ is unbounded.

Theorem 7.2. *If* $\partial\Omega$ *is compact, then the segment property and the uniform segment property of Definition 7.1 coincide.*

Proof. The proof is the same as that of Theorem 6.2. It is sufficient to show that the segment property implies the uniform segment property. Since $\partial\Omega$ is compact there exists a finite open subcover $\{B_i\}_{i=1}^m$, $B_i = B(x_i, r_{x_i})$ of $\partial\Omega$ for some finite sequence $\{x_i\}_{i=1}^m$ of points of $\partial\Omega$. From part (i) of the proof of Theorem 5.1

$$\exists r > 0 \quad \forall x \in \partial\Omega, \quad \exists i, 1 \leq i \leq m \quad \text{such that} \quad B(x, r) \subset B_i.$$

Define $\lambda = \min_{1 \le i \le m} \lambda_i > 0$. Therefore for each $x \in \partial\Omega$, there exists i, $1 \le i \le m$, such that $B(x, r) \subset B_i$, and hence there exists $d_i = d_{x_i}$, $|d_i| = 1$ such that

$$\forall y \in \overline{\Omega} \cap B_i, \quad (y, y + \lambda_i d_i) \subset \text{int } \Omega.$$

Since $B(x, r) \subset B_i$ and $\lambda \le \lambda_i$

$$\forall y \in \overline{\Omega} \cap B(x, r), \quad (y, y + \lambda d_i) \subset (y, y + \lambda_i d_i) \subset \text{int } \Omega.$$

Choosing $d_x = d_i$, we conclude that Ω has the uniform segment property. $\square$

Theorem 7.3. (i) *If Ω satisfies the segment property, then Ω is locally a C^0-epigraph. For $x \in \partial\Omega$, let $B(x, r_x)$, λ_x, d_x, and $H(x)$ be the associated open ball, the length and direction of the open segment, and the hyperplane through 0 orthogonal to d_x. Then there exists ρ_x,*

$$0 < \rho_x \le r_{x\lambda} \stackrel{\text{def}}{=} \min\left\{r_x, \lambda_x/2\right\}, \tag{7.1}$$

which is the largest radius such that

$$B_{H(x)}(0, \rho_x) \subset \left\{P_{H(x)}(y - x) : \forall y \in B(x, r_{x\lambda}) \cap \partial\Omega\right\}.$$

The neighborhoods of Definition 5.2 can be chosen as

$$\boxed{\begin{aligned} V(x) &\stackrel{\text{def}}{=} B_{H(x)}(0, \rho_x) \text{ and} \\ \mathcal{U}(x) &\stackrel{\text{def}}{=} B(x, r_{x\lambda}) \cap \left\{y : P_{H(x)}(y - x) \in V(x)\right\}, \end{aligned}} \tag{7.2}$$

where $B_{H(x)}(0, \rho_x)$ is the open ball of radius ρ_x in the hyperplane $H(x)$. For each $\zeta' \in V(x)$, there exists a unique $y_{\zeta'} \in \partial\Omega \cap \mathcal{U}(x)$ such that $P_{H(x)}(y_{\zeta'} - x) = \zeta'$, and the function

$$\zeta' \mapsto a_x(\zeta') \stackrel{\text{def}}{=} (y_{\zeta'} - x) \cdot d_x : V(x) \to \mathbf{R}$$

is well defined, bounded,

$$\forall \zeta' \in V(x), \quad |a_x(\zeta')| \le r_{\lambda x}, \tag{7.3}$$

and uniformly continuous in $V(x)$, that is, $a_x \in C(\overline{V(x)})$. Finally

$$\mathrm{m}(\partial\Omega \cap \mathcal{U}(x)) = 0, \tag{7.4}$$

where m is the N-dimensional Lebesgue measure.

    (ii) *Conversely, if Ω is locally a C^0-epigraph, Ω and $\complement\Omega$ satisfy the segment property.*

Proof. (i) By assumption, for each $x \in \partial\Omega$

$$\exists r > 0, \quad \exists \omega > 0, \quad \exists \lambda > 0, \quad \exists d \in \mathbf{R}^{\mathrm{N}}, \quad |d| = 1,$$

such that

$$\forall y \in B(x,r) \cap \overline{\Omega}, \quad (y, y + \lambda d) \subset \operatorname{int} \Omega.$$

Let $H = H(x)$ be the hyperplane through 0 orthogonal to the direction d. Let $e_N(x) = d$ and let $\{e_i(x) : 1 \leq i < N\}$ be an orthonormal basis in H. Denote by P_H the orthogonal projection onto H and consider the set

$$U \stackrel{\text{def}}{=} \{P_H(y - x) : \forall y \in B(x,r) \cap \partial\Omega\}.$$

Therefore

$$\forall \zeta' \in U, \quad \exists y \in B(x,r) \cap \partial\Omega, \quad P_H(y - x) = \zeta'$$

and we can associate with ζ' and y the quantity

$$\boxed{\gamma(y) \stackrel{\text{def}}{=} (y - x) \cdot d.}$$

The function a. We show that there exists a subset of U where the map

$$\boxed{\zeta' \mapsto a(\zeta') \stackrel{\text{def}}{=} \gamma(y)}$$

is well defined; that is, for all y_1 and y_2 in $B(x,r) \cap \partial\Omega$

$$P_H(y_2 - x) = \zeta' = P_H(y_1 - x) \quad \Rightarrow \quad \gamma(y_2) = \gamma(y_1).$$

To do this, first reduce the radius of the ball $B(x,r)$ from r to

$$r_\lambda = \min\{r, \lambda/2\}$$

and define the set

$$\boxed{U_\lambda \stackrel{\text{def}}{=} \{P_H(y - x) : y \in B(x, r_\lambda) \cap \partial\Omega\}.}$$

By contradiction, assume that y_1 and y_2 are two points in U_λ such that

$$P_H(y_2 - x) = \zeta' = P_H(y_1 - x) \quad \text{and} \quad \gamma(y_2) > \gamma(y_1).$$

By definition of r_λ

$$|(y_2 - y_1) \cdot d| \leq |y_2 - y_1| \leq |y_2 - x| + |y_1 - x| < 2r_\lambda \leq \lambda,$$

and by construction y_2 and y_1 are on the same line $L_{\zeta'}$,

$$L_{\zeta'} \stackrel{\text{def}}{=} \left\{ x + \sum_{i=1}^{N-1} \zeta_i e_i(x) + \zeta_N d \ : \ \forall \zeta_N \in \mathbf{R} \right\}, \quad x_{\zeta'} \stackrel{\text{def}}{=} x + \sum_{i=1}^{N-1} \zeta_i e_i(x) \qquad (7.5)$$

in the direction d through $x_{\zeta'}$. In particular, since $y_1 \in \partial\Omega \cap B(x, r)$, then necessarily

$$y_2 \in (y_1, y_1 + \lambda d) \subset \text{int } \Omega$$

and this contradicts the fact that $y_2 \in \partial\Omega$. Therefore the above constructions induce a well-defined map

$$\boxed{\zeta' \mapsto a(\zeta') = \gamma(y) : U_\lambda \to \mathbf{R}}$$

for some unique $y \in B(x, r_\lambda) \cap \partial\Omega$ such that $P_H(y - x) = \zeta'$.

U_λ *is open.* By definition, $U_\lambda \subset B_H(0, r_\lambda)$. At any point $\xi' \in U_\lambda$, the segment property is satisfied at the point

$$y_{\xi'} \stackrel{\text{def}}{=} x + \sum_{i=1}^{N-1} \xi_i e_i(x) + a(\xi') e_N(x) \in \partial\Omega \cap B(x, r_\lambda)$$

and $(y_{\xi'}, y_{\xi'} + \lambda d) \subset \text{int } \Omega$. Therefore, there exists a neighborhood $B_H(\xi', R) \subset B_H(x, r_\lambda)$ of ξ' such that, for all $\zeta' \in B_H(\xi', R)$, $L_{\zeta'} \cap \text{int } \Omega \neq \varnothing$. If ξ' is not an interior point of U_λ, there exists a sequence $\{\zeta'_n\} \subset B_H(\xi', R)$, $\zeta'_n \to \xi'$, such that for all n, $(L_{\zeta'_n} \cap B(x, r_\lambda)) \cap \partial\Omega = \varnothing$. Therefore $L_{\zeta'_n} \cap B(x, r_\lambda) \subset \text{int } \Omega$. Choose

$$\alpha \stackrel{\text{def}}{=} \frac{1}{2}\left(a(\xi') - \sqrt{r_\lambda^2 - |\xi'|^2}\right).$$

Consider the sequence

$$z_n \stackrel{\text{def}}{=} x + \sum_{i=1}^{N-1} (\zeta_n)_i e_i(x) + \alpha e_N(x) \to z \stackrel{\text{def}}{=} y_{\xi'} - \frac{1}{2}\left(a(\xi') + \sqrt{r_\lambda^2 - |\xi'|^2}\right).$$

There exists $\bar{N}$ such that, for all $n \geq \bar{N}$, $|z_n - x| < r_\lambda$ and $z_n \in \text{int } \Omega \cap B(x, r_\lambda)$. Necessarily, the limit point z belongs to $\overline{\Omega} \cap B(x, r_\lambda)$, since $P_H(z - x) = 0$ and

$$(z - x) \cdot d = \frac{1}{2}\left(a(\xi') - \sqrt{r_\lambda^2 - |\xi'|^2}\right) \quad \Rightarrow \quad -\sqrt{r_\lambda^2 - |\xi'|^2} < (z - x) \cdot d < a(\xi'),$$

since, by construction of U_λ and a, $|a(\xi')| < \sqrt{r_\lambda^2 - |\xi'|^2}$. In particular

$$(y_{\xi'} - z) \cdot d = \frac{1}{2}\left(a(\xi') + \sqrt{r_\lambda^2 - |\xi'|^2}\right)$$

$$\Rightarrow 0 < (y_{\xi'} - z) \cdot d = \frac{1}{2}\left(a(\xi') + \sqrt{r_\lambda^2 - |\xi'|^2}\right) < r_\lambda \leq \lambda \quad \Rightarrow \quad y_{\xi'} \in (z, z + \lambda d).$$

Therefore, by the segment property in z,

$$y_{\xi'} \in (z, z + \lambda d) \subset \text{int } \Omega$$

and this contradicts the fact that $y_{\xi'} \in \partial\Omega$. This proves that all points of U_λ are interior points and, hence, that U_λ is open.

In particular, $0 \in U_\lambda$ and there exists ρ, $0 < \rho \le r_\lambda$ such that ρ is the largest radius for which

$$B_H(0, \rho) \subset U_\lambda = \{P_H(y - x) : \forall y \in B(x, r_\lambda) \cap \partial\Omega\}.$$

Therefore, the neighborhoods of Definition 5.2 can be chosen as

$$V(x) \overset{\text{def}}{=} B_H(0, \rho) \quad \text{and} \quad \mathcal{U}(x) \overset{\text{def}}{=} B(x, r_\lambda) \cap \{y : P_H(y - x) \in V(x)\}.$$

By construction, condition (5.1) in Definition 5.1 is satisfied. As for condition (5.2), by construction $V(x) \subset U_\lambda$, and for each $\zeta' \in V(x)$

$$\exists \text{ a unique } y \in \partial\Omega \cap B(x, r_\lambda) \text{ and } a_x(\zeta') \overset{\text{def}}{=} (y - x) \cdot d.$$

As a result (using Notation 5.1)

$$
\begin{aligned}
A^0 &= \left\{ x + \sum_{i=1}^{N-1} \zeta_i e_i + a_x(\zeta') e_N : \forall \zeta' \in V(x) \right\} \\
&= \{y : \zeta' = P_H(y - x) \in V(x) \text{ and } (y - x) \cdot d = a_x(\zeta')\} \\
&= \{y : P_H(y - x) \in V(x), \ y \in \partial\Omega \cap B(x, r_\lambda)\} \\
&= \{y : P_H(y - x) \in V(x), \ y \in B(x, r_\lambda)\} \cap \partial\Omega = \mathcal{U}(x) \cap \partial\Omega
\end{aligned}
$$

and identity (5.2) is satisfied. It remains to check (5.3). For each $y \in \mathcal{U}(x) \cap A^+$,

$$\zeta' = P_H(y - x) \in V(x), \quad \zeta_N = (y - x) \cdot d > a_x(\zeta'), \quad y \in B(x, r_\lambda),$$

and the point

$$y_{\zeta'} = x + \sum_{i=1}^{N-1} \zeta_i' e_i(x) + a_x(\zeta') e_N(x) \in A^0 = \mathcal{U}(x) \cap \partial\Omega.$$

Hence

$$y \in (y_{\zeta'}, y_{\zeta'} + \lambda d) \subset \operatorname{int} \Omega \quad \Rightarrow \quad y \in \operatorname{int} \Omega \quad \Rightarrow \quad y \in \operatorname{int} \Omega \cap \mathcal{U}(x).$$

Conversely, for each $y \in \operatorname{int} \Omega \cap \mathcal{U}(x)$,

$$\zeta' = P_H(y - x) \in V(x), \quad y \in B(x, r_\lambda) \cap \operatorname{int} \Omega.$$

By definition of A^0, $y \notin A^0$. If $y \in A^- \cap \mathcal{U}(x)$, then

$$y \in \operatorname{int} \Omega \cap \mathcal{U}(x) \text{ and } \zeta_N = (y - x) \cdot d < a(\zeta'),$$

and the point

$$y_{\zeta'} = x + \sum_{i=1}^{N-1} \zeta_i e_i(x) + a_x(\zeta') e_N(x) \in \partial\Omega$$

is above the point y along the line $L_{\zeta'}$ through $x_{\zeta'}$ parallel to d, as defined by (7.5). Since $y \in \mathcal{U}(x) \subset B(x, r_\lambda)$, then $y \in \overline{\Omega} \cap B(x, r_\lambda)$ and

$$(y, y + \lambda d) \subset \operatorname{int} \Omega \quad \Rightarrow \quad y_{\zeta'} \in \operatorname{int} \Omega$$

which is a contradiction. Therefore $\operatorname{int} \Omega \cap \mathcal{U}(x) \subset A^+ \cap \mathcal{U}(x)$ and condition (5.3) of Definition 5.1 is satisfied.

The function a is bounded and uniformly continuous in U_λ. Consider a point $\xi' \in U_\lambda$ and let $y_{\xi'} \in \partial\Omega \cap B(x, r_\lambda)$ be the associated unique point such that

$$\xi' = P_H(y_{\xi'} - x) \quad \text{and} \quad a(\xi') = d \cdot (y_{\xi'} - x).$$

By construction, $|a(\xi')| \le |y_{\xi'} - x| \le r_\lambda$. Hence

$$\ell \overset{\text{def}}{=} \liminf_{\zeta' \to \xi'} a(\zeta') \quad \text{and} \quad L \overset{\text{def}}{=} \limsup_{\zeta' \to \xi'} a(\zeta')$$

are finite and the points

$$y_\ell \overset{\text{def}}{=} x + \sum_{i=1}^{N-1} \xi_i e_i(x) + \ell d \in \partial\Omega \cap B(x, r_\lambda),$$

$$y_L \overset{\text{def}}{=} x + \sum_{i=1}^{N-1} \xi_i e_i(x) + L d \in \partial\Omega \cap B(x, r_\lambda)$$

belong to $\partial\Omega$ as limits of points of $\partial\Omega$. If $y_\ell \ne y_{\xi'}$, then $|y_\ell - y_{\xi'}| \le |y_\ell - x| + |y_{\xi'} - x| < 2r_\lambda \le \lambda$ and either $y_{\xi'} \in (y_\ell, y_\ell + \lambda d)$ or $y_\ell \in (y_{\xi'}, y_{\xi'} + \lambda d)$. By the segment property, this means that either $y_{\xi'}$ or y_ℓ belongs to $\operatorname{int} \Omega$ which contradicts the fact that both points belong to $\partial\Omega$. Therefore $y_\ell = y_{\xi'}$, and by the same argument $y_L = y_{\xi'}$. Hence

$$\liminf_{\zeta' \to \xi'} a(\zeta') = \ell = (y_\ell - x) \cdot d = (y_{\xi'} - x) \cdot d = a(\xi'),$$

$$a(\xi') = (y_{\xi'} - x) \cdot d = (y_L - x) \cdot d = L = \limsup_{\zeta' \to \xi'} a(\zeta'),$$

and a is continuous in $\xi' \in U_\lambda$. In view of the fact that a is bounded, then a is uniformly continuous in U_λ, that is, $a \in C(\overline{U_\lambda})$.

Finally, since $\partial\Omega$ is the graph of a continuous function in $B(x, \rho)$, $\mathrm{m}\big(\partial\Omega \cap \mathcal{U}(x)\big) = 0$.

(ii) Pick a point $x \in \partial\Omega$. By Definition 5.1 there exist neighborhoods $\mathcal{U}(x)$ and $V(x)$, an orthonormal basis $\{e_1(x), \ldots, e_N(x)\}$, and a continuous function $a_x \colon V(x) \to \mathbf{R}$ with the appropriate properties. Choose $r_x > 0$ such that $B(x, 2r_x) \subset \mathcal{U}(x)$. We want to show that, for $\lambda_x = r_x$ and $d_x = e_N(x)$,

$$\forall y \in B(x, r_x) \cap \overline{\Omega}, \quad (y, y + \lambda_x d_x) \subset \operatorname{int} \Omega.$$

By Definition 5.1, $\mathcal{U}(x) \subset \{y : P_{H(x)}(y - x) \in V(x)\}$, and hence $\zeta' \overset{\text{def}}{=} P_{H(x)}(y - x) \in V(x)$. Moreover, $y \in B(x, r_x) \cap \overline{\Omega} \subset \mathcal{U}(x) \cap \overline{\Omega}$ implies that

$$(y - x) \cdot e_N(x) \ge a_x(\zeta').$$

Consider the segment $(y, y + \lambda_x d_x) = \{y + t\lambda_x d_x : 0 < t < 1\}$. For each $0 < t < 1$

$$(y + t\lambda_x d_x - x) \cdot e_N(x) = (y - x) \cdot e_N(x) + t\lambda_x \geq a_x(\zeta') + t\lambda_x > a_x(\zeta')$$
$$\Rightarrow y + t\lambda_x d_x \in A^+,$$
$$|y + t\lambda_x d_x - x| \leq |y - x| + \lambda_x < r_x + \lambda_x < 2r_x$$
$$\Rightarrow y + t\lambda_x d_x \in B(x, 2r_x).$$

Therefore

$$(y, y + \lambda_x d_x) \subset B(x, 2r_x) \cap A^+ \subset \mathcal{U}(x) \cap A^+ = \mathcal{U}(x) \cap \operatorname{int} \Omega$$
$$\Rightarrow (y, y + \lambda_x d_x) \subset \operatorname{int} \Omega$$

and Ω satisfies the segment property.

Furthermore, since Ω is locally a C^0-epigraph, so is $\complement\Omega$ by changing $e_N(x)$ into $-e_N(x)$. Hence $\complement\Omega$ also satisfies the segment property. $\qquad\square$

Theorem 7.4. *Let Ω be a subset of $\mathbf{R}^N$ with a nonempty boundary.*

(i) *If $\partial\Omega$ is compact and Ω has the segment property, Ω is locally a C^0-epigraph, where, in each $x \in \partial\Omega$, the associated neighborhoods $V(x)$ and $\mathcal{U}(x)$ and the function a_x can be chosen with the following properties: the neighborhoods $V(x)$ and $\mathcal{U}(x) - x$ are, up to a rotation around 0, independent of $x \in \partial\Omega$; the family of bounded uniformly continuous functions $a_x : V(x) \to \mathbf{R}$ can, after a rotation around 0, be chosen equicontinuous with respect to all $x \in \partial\Omega$.*

(ii) *If Ω is locally a C^0-epigraph and the neighborhoods $\mathcal{U}(x) - x$ are, up to a rotation around 0, independent of $x \in \partial\Omega$, then Ω satisfies the uniform segment property.*

Proof. (i) By assumption, from Theorem 7.2, Ω satisfies a uniform segment property for some $r > 0$ and $\lambda > 0$ and we can now repeat the constructions in the proof of part (i) of Theorem 7.3. To each $x \in \partial\Omega$ we associate a radius ρ_x, $0 < \rho_x \leq r_\lambda = \min\{r, \lambda/2\}$ and a bounded uniformly continuous function a_x defined on $V(x) = B_{H(x)}(0, \rho_x)$ such that $\partial\Omega$ is locally the graph of a_x in the neighborhood

$$\mathcal{U}(x) = B(x, r_\lambda) \cap \{y : P_{H(x)}(y - x) \in B_{H(x)}(0, \rho_x)\}$$

of x. Note that, by construction, $B(x, \rho_x) \subset \mathcal{U}(x)$, and

$$\mathcal{U}(x) - x = B(0, r_\lambda) \cap \{z : P_{H(x)}(z) \in B_{H(x)}(0, \rho_x)\}$$

depends only on r_λ and ρ_x up to a rotation around the origin. We now construct new neighborhoods $V'(x)$ and $\mathcal{U}'(x)$ which will be independent of x up to a rotation around the origin.

The family $\{B(x, \rho_x) : x \in \partial\Omega\}$ is an open cover of $\partial\Omega$. Since $\partial\Omega$ is compact, there exists a finite sequence $\{x_i\}_{i=1}^m$ of points of $\partial\Omega$ such that $\{B_i = B(x_i, \rho_{x_i}) : 1 \leq i \leq m\}$ is a finite subcover of $\partial\Omega$. At each point x_i, $\partial\Omega$ is locally the graph of the function $a_i = a_{x_i} : B_{H_i}(0, \rho_{x_i}) \to \mathbf{R}$ in the neighborhood $\mathcal{U}_i = \mathcal{U}(x_i)$ of x_i. By the same technique as in the proof of Theorem 7.2,

$$\exists \rho > 0 \quad \forall x \in \partial\Omega, \quad \exists i, 1 \le i \le m \quad \text{such that} \quad B(x, \rho) \subset B_i.$$

Consider a point $x \in \partial\Omega$ and let i be the index such that $B(x, \rho) \subset B_i$. Choose $e_N(x) = e_N(x_i)$, $H(x) = H_i$, $H_i = H(x_i)$. By construction

$$B(x, \rho) \subset B(x_i, \rho_{x_i}) \quad \Rightarrow \quad B_{H_i}(P_{H_i}(x - x_i), \rho) \subset B_{H_i}(0, \rho_{x_i}).$$

Define the new neighborhoods

$$V'(x) = B_{H_i}(0, \rho) \text{ and } \mathcal{U}'(x) = B(x, \rho) \cap \{y : P_{H_i}(y - x) \in B_{H_i}(0, \rho)\}.$$

It is readily seen that $V'(x)$ and $\mathcal{U}'(x) - x$ are now, up to a rotation around 0, independent of $x \in \partial\Omega$.

Again by construction,

$$P_{H_i}(x - x_i) + V'(x) \subset B_{H_i}(0, \rho_{x_i}) = V(x_i)$$
$$\mathcal{U}'(x) \subset B(x_i, \rho_{x_i}) \cap \{y : P_{H_i}(y - x + x - x_i) \in B_{H_i}(P_{H_i}(x - x_i), \rho)\}$$
$$\subset B(x_i, \rho_{x_i}) \cap \{y : P_{H_i}(y - x_i) \in B_{H_i}(0, \rho_{x_i})\} = \mathcal{U}(x_i).$$

In view of the above properties we can use the function a_i to construct a new function a'_x on $V'(x)$. First restrict the function a_i from $B_{H_i}(0, \rho_i)$ to $B_{H_i}(P_{H_i}(x - x_i), \rho)$. Denote by a'_x this restriction of a_i redefined (by translation) on $B_{H(x)}(0, \rho) = B_{H_i}(0, \rho)$. Since a'_x is a restriction of a_i it has the same modulus of uniform continuity as a_i for all $x \in \partial\Omega$ such that $B(x, \rho) \subset B_i$. Each function a'_x is defined on $V(x) = B_{H(x)}(0, \rho)$ and, after a rotation, can be redefined as a function $\bar{a}_x : V = B_{N-1}(0, \rho) \to \mathbf{R}$, where $B_{N-1}(0, \rho)$ is the open ball of radius ρ with center in 0 in $\mathbf{R}^{N-1}$. Similarly, each $a_i : B_{H_i}(0, \rho_i) \to \mathbf{R}$ can be redefined as a function $\bar{a}_i : V = B_{N-1}(0, \rho_i) \to \mathbf{R}$. Each $\bar{a}_x$ is the restriction of a translation in $\mathbf{R}^{N-1}$ of one of the members of the finite family of uniformly bounded and continuous functions $\{\bar{a}_i\}_{i=1}^m$ on $B_{N-1}(0, \rho)$. Therefore the family of uniformly bounded continuous functions $\bar{a}_x : V(x) \to \mathbf{R}$ is equicontinuous with respect to $x \in \partial\Omega$.

(ii) The proof is the same as the proof of part (ii) of Theorem 7.3. The uniform segment property follows from the fact that the neighborhoods $\mathcal{U}(x) - x$ of 0 are, up to a rotation around 0, independent of $x \in \partial\Omega$. Therefore, there exists $r > 0$ such that, for all $x \in \partial\Omega$, $B(0, 2r) \subset \mathcal{U}(x) - x$, and hence $B(x, 2r) \subset \mathcal{U}(x)$. As a result we can repeat the proof with r in place of r_x and $\lambda = r$ which are both independent of x. $\qquad\square$

We quote the main density results from R.A. Adams [1, Thm. 3.18, p. 54].

Theorem 7.5. *If Ω has the segment property, then the set*

$$\left\{ f|_{\text{int}\,\Omega} : \forall f \in C_0^\infty(\mathbf{R}^N) \right\}$$

of restrictions of functions of $C_0^\infty(\mathbf{R}^N)$ to $\text{int}\,\Omega$ is dense in $W^{m,p}(\text{int}\,\Omega)$ for $1 \le p < \infty$ and $m \ge 1$. In particular, $C^k(\overline{\Omega})$ is dense in $W^{m,p}(\text{int}\,\Omega)$ for any $m \ge 1$ and $k \ge m$.

8   Courant Metric Topology on Images of a Domain

One way to construct a family of variable domains is to consider the images of a fixed subset of $\mathbf{R}^N$ by some family of transformations of $\mathbf{R}^N$. The structure and the topology of the images can then be specified via the natural algebraic and topological structures of transformations or equivalence classes of transformations for which the full power of function analytic methods is available. There are many ways to do that since specific constructions and choices of structures and topologies are very much problem dependent.

In 1972 Micheletti [1] gave what may be one of the first complete metric topologies on a family of domains of class C^k which are the images of a fixed domain of class C^k through a family of C^k-diffeomorphisms of $\mathbf{R}^N$. Thus the natural underlying algebraic structure is not the *vector space structure* but the *group structure* with respect to the composition of transformations with the identity transformation as the neutral element. Her analysis culminates with the construction of a metric that she called the *Courant metric*. Her constructions are generic and naturally extend to other families of domains in $\mathbf{R}^N$ or in a fixed hold-all D associated with different spaces of transformations. In view of the pertinence of that paper, originally written in Italian, we now provide an adapted translation.

8.1   The Complete Metric Group of Diffeomorphisms $\mathcal{F}_0^k$

Following the notation of section 2 define for an integer $k \geq 1$ the spaces

$$C^k(\mathbf{R}^N) \stackrel{\text{def}}{=} C^k(\mathbf{R}^N, \mathbf{R}^N), \quad \mathcal{B}^k \stackrel{\text{def}}{=} \mathcal{B}^k(\mathbf{R}^N, \mathbf{R}^N), \quad C^k(\overline{\mathbf{R}^N}) \stackrel{\text{def}}{=} C^k(\overline{\mathbf{R}^N}, \mathbf{R}^N)$$

and $C_0^k(\mathbf{R}^N) \stackrel{\text{def}}{=} C_0^k(\mathbf{R}^N, \mathbf{R}^N)$ the subspace of maps f in $C^k(\mathbf{R}^N)$ for which f and all its partial derivatives up to order k vanish at infinity.[5] By definition, $C_0^k(\mathbf{R}^N) \subset C^k(\overline{\mathbf{R}^N}) \subset \mathcal{B}^k \subset C^k(\mathbf{R}^N)$. When endowed with the norm

$$\|f\|_{C^k} \stackrel{\text{def}}{=} \max_{0 \leq |\alpha| \leq k} \|\partial^\alpha f\|_C$$

$C_0^k(\mathbf{R}^N)$, $C^k(\overline{\mathbf{R}^N})$, and $\mathcal{B}^k$ are Banach spaces. Associate with $C_0^k(\mathbf{R}^N)$ the (nonlinear and nonconvex) space

$$\boxed{\mathcal{F}_0^k \stackrel{\text{def}}{=} \left\{ F : \mathbf{R}^N \to \mathbf{R}^N : F - I \in C_0^k(\mathbf{R}^N) \text{ and } F^{-1} \in C^k(\mathbf{R}^N) \right\}.}$$

Associate with a bounded open connected domain Ω_0 of class C^k, the family

$$\mathcal{X}(\Omega_0) \stackrel{\text{def}}{=} \left\{ F(\Omega_0) \subset \mathbf{R}^N : \forall F \in \mathcal{F}_0^k \right\} \tag{8.1}$$

of images of Ω_0 by the elements of $\mathcal{F}_0^k$. Since F is a C^k-diffeomorphism, $F(\Omega_0)$ is also a bounded open connected domain of class C^k. This induces a bijection

$$[F] \mapsto F(\Omega_0) : \mathcal{F}_0^k / \mathcal{G}(\Omega_0) \to \mathcal{X}(\Omega_0) \tag{8.2}$$

---

[5]Here $\mathbf{R}^N$ is endowed with the norm $|x| = \{\sum_{i=1}^N x_i^2\}^{1/2}$.

between $\mathcal{X}(\Omega_0)$ and the quotient space of $\mathcal{F}_0^k$ by the subset

$$\mathcal{G}(\Omega_0) \stackrel{\text{def}}{=} \left\{ F \in \mathcal{F}_0^k \; : \; F(\Omega_0) = \Omega_0 \right\} \tag{8.3}$$

of transformations which map Ω_0 onto Ω_0. The topological structure of $\mathcal{X}(\Omega_0)$ will be identified with the topological structure of the quotient space. A complete metric space topology will be introduced on $\mathcal{F}_0^k$, which will in turn induce a complete metric space topology on the quotient space. This *quotient metric* is called the *Courant metric* by Micheletti.[6] This construction is not as straightforward as it might appear at first sight. The obvious candidates for the metric do not usually satisfy the *triangle inequality* and only yield a *pseudometric*.

The first theorem gives the algebraic structure of $\mathcal{F}_0^k$.

Theorem 8.1. *For $k \geq 1$ the space $\mathcal{F}_0^k$ is a group with respect to the composition $\circ$ of transformations.*

The proof of the theorem necessitates two basic lemmas. Denote by

$$S(x)[y_1][y_2], \dots, [y_k]$$

at the point $x \in \mathbf{R}^N$ a k-linear form with arguments $y_1, \dots, y_k$.

Lemma 8.1. *Assume that F and G are two mappings from $\mathbf{R}^N$ to $\mathbf{R}^N$ such that F is k-times differentiable in an open neighborhood of x and G is k-times differentiable in an open neighborhood of $F(x)$. Then the kth derivative of $G \circ F$ in x is the sum of a finite number of k-linear applications on $\mathbf{R}^N$ of the form*

$$(h_1, h_2, \dots, h_k) \mapsto G^{(\ell)}(F(x)) \left[F^{(\lambda_1)}(x)[h_1][h_1] \dots [h_{\lambda_1}] \right]$$
$$\dots \left[F^{(\lambda_\ell)}(x)[h_{k-\lambda_\ell+1}] \dots [h_k] \right], \tag{8.4}$$

where $\ell = 1, \dots, k$ and $\lambda_1 + \dots + \lambda_\ell = k$.

Proof. We proceed by induction on k. The result is trivially true for $k = 1$. Assuming that it is true for $k - 1$, we prove that it is also true for k. This is an obvious consequence of the observation that for a mapping from $\mathbf{R}^N$ into $\mathcal{L}((\mathbf{R}^N)^{k-1}, \mathbf{R}^N)$ of the form

$$x \mapsto G^{(\ell)}(F(x))[F^{(\lambda_1)}(x)] \dots [F^{(\lambda_\ell)}(x)],$$

where $\ell = 1, \dots, k-1$ and $\lambda_1 + \dots + \lambda_\ell = k - 1$, the differential in x is

$$G^{(\ell+1)}(F(x))[F^{(1)}][F^{(\lambda_1)}(x)] \dots [F^{(\lambda_\ell)}(x)]$$
$$+ G^{(\ell)}(F(x))[F^{(\lambda_1+1)}(x)] \dots [F^{(\lambda_\ell)}(x)]$$
$$+ \dots + G^{(\ell)}(F(x))[F^{(\lambda_1)}(x)] \dots [F^{(\lambda_\ell+1)}(x)], \tag{8.5}$$

and the result is true for k. $\qquad\square$

---

[6]It turns out that the construction of the Courant metric only requires that Ω_0 be closed or open. The additional assumption that Ω_0 be a bounded, connected open domain of class C^k is to make each element of the family a bounded connected open domain of class C^k, but it is not necessary in the construction of the Courant metric.

Lemma 8.2. *Given f and g in $C^k(\mathbf{R}^N)$, let $\psi = f \circ (I + g)$. Then for each $x \in \mathbf{R}^N$*

$$|\psi(x)| = |f(x + g(x))|,$$

$$|\psi^{(1)}(x)| \le |f^{(1)}(x + g(x))|\,[1 + |g^{(1)}(x)|],$$

$$|\psi^{(i)}(x)| \le |f^{(1)}(x + g(x))|\,|g^{(i)}(x)|$$

$$+ \sum_{j=2}^{i} |f^{(j)}(x + g(x))|\, a_j(|g^{(1)}(x)|, \ldots, |g^{(i-1)}(x)|)$$

for $i = 2, \ldots, k$, where a_j is a polynomial.

Proof. This is an obvious consequence of Lemma 8.1. $\qquad\square$

Proof of Theorem 8.1. From Lemma 8.2 it readily follows that $F \circ G \in \mathcal{F}_0^k$ for all F and G in $\mathcal{F}_0^k$. For all $F = I + f \in \mathcal{F}_0^k$, $f \in C_0^k(\mathbf{R}^N)$, $F^{-1} \in C^k(\mathbf{R}^N)$, and it remains to show that $g = F^{-1} - I \in \mathcal{F}_0^k$. This amounts to showing that $|g(y)|$ and $|g^{(i)}(y)|$ go to zero as $|y| \to \infty$, $i = 1, \ldots, k$. We again proceed by induction on k. Set $y = F(x)$. By definition $g(y) = F^{-1}(y) - y = x - F(x) = -f(x)$ and $\|g\|_{C^0} = \|f\|_{C^0}$. But $y + g(y) = x$ yields $|y| - |g(y)| \le |x|$. So as $|y| \to \infty$, $x \to \infty$, and it follows from the identity $g(y) = -f(x)$ that $|g(y)| \to 0$ as $|y| \to \infty$. The theorem is true for $\mathcal{F}_0^0$.

Always, from Lemma 8.2 and the identity $g(y) = -f(x) = -f(F^{-1}(y)) = -f(y + g(y))$,

$$|g^{(1)}(y)| \le |f^{(1)}(x)|\,[1 + |g^{(1)}(y)|] \quad \Rightarrow \quad |g^{(1)}(y)|\,[1 - |f^{(1)}(x)|] \le |f^{(1)}(x)|.$$

For $|y|$ sufficiently large, $|f^{(1)}(x)|$ can be made sufficiently small and

$$0 \le |g^{(1)}(y)| \le \frac{|f^{(1)}(x)|}{1 - |f^{(1)}(x)|}. \tag{8.6}$$

Hence $|g^{(1)}(y)| \to 0$ as $|y| \to \infty$ and $\mathcal{F}_0^1$ is a group. Now show that if $\mathcal{F}_0^{k-1}$ is a group, then $\mathcal{F}_0^k$ is a group. Again, from Lemma 8.2 for $|y|$ arbitrarily large we have

$$0 \le |g^{(k)}(y)| \le \frac{\sum_{j=2}^{k} |f^{(j)}(x)|\, a_j(|g^{(1)}(y)|, \ldots, |g^{(k-1)}(y)|)}{1 - |f^{(1)}(x)|}, \tag{8.7}$$

which goes to zero. Therefore $\mathcal{F}_0^k$ is a group for every $k \ge 1$. $\qquad\square$

The next step is the definition of a metric on $\mathcal{F}_0^k$ which will subsequently define a metric on the quotient group. An obvious choice would be

$$\rho(F_2, F_1) \overset{\text{def}}{=} \|F_2 - F_1\|_{C^k} + \|F_2^{-1} - F_1^{-1}\|_{C^k}. \tag{8.8}$$

Unfortunately the natural candidate

$$\rho_{\Omega_0}(F_2, F_1) \overset{\text{def}}{=} \inf_{G_1, G_2 \in \mathcal{G}(\Omega_0)} \rho(F_2 \circ G_2, F_1 \circ G_1) \tag{8.9}$$

for a metric on the quotient space will generally not satisfy the *triangle inequality*. Micheletti's construction of the *quotient metric* on $\mathcal{F}_0^k$ is more subtle and requires a few preliminary technical lemmas.

Lemma 8.3. *Given integers $r \geq 0$ and $s > 0$, there exists a constant $c(r,s) > 0$ with the following property: if the sequence $f_1, \ldots, f_n$ in $C^r(\mathbf{R}^N)$ is such that*

$$\sum_{i=1}^n \|f_i\|_{C^r} < \alpha, \quad 0 < \alpha < s, \tag{8.10}$$

then for the map $F = (I + f_n) \circ \cdots \circ (I + f_1)$,

$$\|F - I\|_{C^r} \leq \alpha\, c(r,s). \tag{8.11}$$

Proof. We again proceed by induction on r. For simplicity define

$$F_i \overset{\text{def}}{=} (I + f_i) \circ \cdots \circ (I + f_1), \quad F_n = F.$$

By the definition of F we have

$$\begin{aligned}
F - I &= f_1 + f_2 \circ (I + f_1) + \cdots + f_n \circ (I + f_{n-1}) \circ \cdots \circ (I + f_1) \\
&= f_1 + f_2 \circ F_1 + \cdots + f_n \circ F_{n-1}.
\end{aligned} \tag{8.12}$$

Then $\|F - I\|_{C^0} \leq \sum_{i=1}^n \|f_i\|_{C^0} \leq \alpha$ and $c(0,s) = 1$ for all integers $s > 0$. From (8.12) and Lemma 8.2 we further get

$$\sup_x |(F - I)^{(1)}(x)| \leq \|f_1\|_{C^1} + \|f_2\|_{C^1}[1 + \|f_1\|_{C^1}] + \cdots$$

$$+ \|f_n\|_{C^1}[1 + \|f_{n-1}\|_{C^1}] \ldots [1 + \|f_1\|_{C^1}]$$

$$\leq \sum_{i=1}^n \|f_i\|_{C^1} \prod_{i=1}^n (1 + \|f_i\|_{C^1}) \leq \left(\sum_{i=1}^n \|f_i\|_{C^1}\right) e^{(\sum_{i=1}^n \|f_i\|_{C^1})} \leq \alpha\, e^\alpha \leq \alpha\, e^s \tag{8.13}$$

and $\|F - I\|_{C^1} \leq \alpha\, e^\alpha$. Hence $c(1,s) = e^s$ for all integers $s > 0$. We now show that if the result is true for $r - 1$, it is true for r. We have to evaluate $|(F_n - I)^{(r)}(x)|$. It is obvious that $F_n - I = (I + f_n) \circ F_{n-1} - I = F_{n-1} - I + f_n \circ F_{n-1}$ and hence

$$(F_n - I)^{(r)}(x) = (F_{n-1} - I)^{(r)}(x) + (f_n \circ F_{n-1})^{(r)}(x). \tag{8.14}$$

For $r \geq 2$ from Lemma 8.2 we have

$$|(f_n \circ F_{n-1})^{(r)}(x)| \leq |f_n^{(1)}(F_{n-1}(x))|\,|(F_{n-1} - I)^{(r)}(x)|$$

$$+ \sum_{j=2}^r |f_n^{(j)}(F_{n-1}(x))|\, a_j(|(F_{n-1} - I)^{(1)}(x)|, \ldots, |(F_{n-1} - I)^{(r-1)}(x)|).$$

By the induction assumption $|(F_{n-1} - I)^{(i)}(x)| \leq \|(F_{n-1} - I)\|_{C^{r-1}} \leq \alpha\, c(r-1,s)$, $i = 1, \ldots, r-1$, and the fact that a_j is a polynomial dependent on r for $j = 2, \ldots, r$, there exists a constant $L(r,s)$ such that

$$|(f_n \circ F_{n-1})^{(r)}(x)|$$

$$\leq |f_n^{(1)}(F_{n-1}(x))| \, |(F_{n-1} - I)^{(r)}(x)| + L(r,s) \sum_{j=2}^{r} |f_n^{(j)}(F_{n-1}(x))| \qquad (8.15)$$

$$\leq \|f_n\|_{C^r} |(F_{n-1} - I)^{(r)}(x)| + (r-1)L(r,s)\|f_n\|_{C^r}.$$

Define $M(r,s) = (r-1)L(r,s)$. From (8.14) and (8.15), we get

$$|(F_n - I)^{(r)}(x)| \leq [1 + \|f_n\|_{C^r}] \, |(F_{n-1} - I)^{(r)}(x)| + M(r,s)\|f_n\|_{C^r}.$$

After repeating this procedure $n-1$ times we have

$$|(F_n - I)^{(r)}(x)| \leq [1 + \|f_n\|_{C^r}] \dots [1 + \|f_2\|_{C^r}]\|f_1\|_{C^r}$$
$$+ [1 + \|f_n\|_{C^r}] \dots [1 + \|f_2\|_{C^r}]M(r,s)\|f_2\|_{C^r} + \dots + M(r,s)\|f_n\|_{C^r}$$
$$\leq \max[M(r,s),1]e^\alpha \alpha.$$

Henceforth $c(r,s) = \max[M(r,s),1]e^s$. $\qquad \square$

Lemma 8.4. *Given $F \in \mathcal{F}_0^k$, define $f = F - I$ and $f_\tau = f \circ \tau$, where τ denotes the pointwise translation in $\mathbf{R}^N$ by a vector τ. Then, as $|\tau|$ goes to zero, $\|f_\tau - f\|_{C^k} \to 0$.*[7]

Proof. Define the function

$$\rho(x,\tau) \stackrel{\text{def}}{=} \max \big\{ |f(x+\tau) - f(x)|, |f^{(1)}(x+\tau) - f^{(1)}(x)|,$$
$$\dots, |f^{(k)}(x+\tau) - f^{(k)}(x)| \big\}.$$

The lemma is now a simple consequence of the following properties of ρ:

    (i) $\rho(x,\tau)$ goes to zero as $|x| \to \infty$, uniformly with respect to the variable τ in a bounded neighborhood of the origin;

    (ii) ρ is continuous and $\rho(x,0) = 0$ for each $x \in \mathbf{R}^N$. $\qquad \square$

Lemma 8.5. *Given F and G in $\mathcal{F}_0^k$ and γ in $C^k(\mathbf{R}^N)$, if $\|\gamma\|_{C^k} \to 0$, then $\|G \circ (F + \gamma) - G \circ F\|_{C^k} \to 0$.*

Proof. Again proceed by induction on k. For $k = 0$ the result is obvious. Then check that if it is true for $k-1$ it is true for k. From Lemma 8.1 the kth derivative of $G \circ (F + \gamma) - G \circ F$ is the sum of a finite number of k-linear mappings of the form

$$G^{(r)}(F(x) + \gamma(x))[F^{(\lambda_1)}(x) + \gamma^{(\lambda_1)}(x)] \dots [F^{(\lambda_r)}(x) + \gamma^{(\lambda_r)}(x)]$$
$$- G^{(r)}(F(x))[F^{(\lambda_1)}(x)] \dots [F^{(\lambda_r)}(x)],$$

---

[7]The conclusions of the lemma still hold for $f \in C^k(\overline{\mathbf{R}^N})$ since the pointwise translation of a bounded uniformly continuous function is continuous in the C^0-norm.

where $r = 1, \ldots, k$ and $\lambda_1 + \cdots + \lambda_r = k$. As a result

$$|F^{(\lambda_1)}(x)|\,|F^{(\lambda_2)}(x)| \ldots |F^{(\lambda_r)}(x)| \leq (\|F - I\|_{C^k} + 1)^r$$

since the norm of this type of map can be bounded above by the following expression:

$$\sup_x |G^{(r)}(F(x) + \gamma(x)) - G^{(r)}(F(x))|$$

$$(\|F - I\|_{C^k} + 1)^r + \|\gamma\|_{C^k}\, p(\|F - I\|_{C^k}, \|G - I\|_{C^k}, \|\gamma\|_{C^k}),$$

where p is a polynomial. From this upper bound and Lemma 8.4 the result of the lemma is true for k. $\qquad\square$

We are now ready to define a metric on $\mathcal{F}_0^k$. Given $F \in \mathcal{F}_0^k$, consider finite factorizations of F and F^{-1} of the form

$$F = (I + f_n) \circ \cdots \circ (I + f_1) \quad \text{and} \quad F^{-1} = (I + g_m) \circ \cdots \circ (I + g_1).$$

First define

$$\boxed{d(I, F) \stackrel{\text{def}}{=} \inf_{(f_1, \ldots, f_n)} \sum_{i=1}^{n} \|f_i\|_{C^k} + \inf_{(g_1, \ldots, g_m)} \sum_{i=1}^{m} \|g_i\|_{C^k},} \qquad (8.16)$$

where the infima are taken with respect to all finite factorizations of F and F^{-1} in $\mathcal{F}_0^k$. It readily follows that

$$d(I, F) = d(I, F^{-1}). \qquad (8.17)$$

Extend this definition to all pairs F and G in $\mathcal{F}_0^k$,

$$\boxed{d(F, G) \stackrel{\text{def}}{=} d(I, G \circ F^{-1}).} \qquad (8.18)$$

By definition, d is right-invariant since for all F, G, and H in $\mathcal{F}_0^k$

$$\boxed{d(F, G) = d(F \circ H, G \circ H).}$$

We now check that the three axioms that define a metric on $\mathcal{F}_0^k$ are satisfied.
(i) $d(F, G) = 0 \iff F = G$. If $d(F, G) = 0$, then there exists a sequence of factorizations of $G \circ F^{-1}$ in $\mathcal{F}_0^k$,

$$G \circ F^{-1} = (I + g_{n, s_n}) \circ \cdots \circ (I + g_{n, 1})$$

such that

$$\mathcal{S}_{(n)} = \sum_{i=1}^{s_n} \|g_{n, i}\|_{C^k} \to 0 \text{ as } n \to \infty.$$

From Lemma 8.3 for n arbitrarily large (as long as $\mathcal{S}_{(n)} < 1$)

$$\|G \circ F^{-1} - I\|_{C^k} \leq c(k,1)\mathcal{S}_{(n)}$$

and hence $\|G \circ F^{-1} - I\|_{C^k} = 0$. Hence $F = G$.

(ii) $d(F,G) = d(G,F)$ from the definition and property (8.17).

(iii) $d(F,G) \leq d(F,H) + d(H,G)$. This property follows from the simpler property

$$\forall M, N \in \mathcal{F}_0^k, \quad d(I, M \circ N) \leq d(I,M) + d(I,N)$$

since $I - G \circ F^{-1} = I - (G \circ H^{-1}) \circ (H \circ F^{-1})$. But this is trivially true by the definition of $d(I, M \circ N)$ as an infimum over all factorizations of $M \circ N$ in $\mathcal{F}_0^k$, which is a bigger set than the factorizations of $M \circ N$ obtained by the compositions of the respective factorizations of M and N in $\mathcal{F}_0^k$.

Theorem 8.2. *$\mathcal{F}_0^k$ is a metric group.*

Proof. We have already shown that d is a metric on $\mathcal{F}_0^k$. Furthermore, $\mathcal{F}_0^k$ is a topological group.[8] From the results of the theory of topological groups (cf., for instance, Kelley [1]), it is sufficient to check the following condition: for each F in $\mathcal{F}_0^k$, if $d(I,H) \to 0$, then $d(I, F^{-1} \circ H \circ F) \to 0$. To get $d(I, F^{-1} \circ H \circ F) \to 0$, it is sufficient to show that

$$\|F^{-1} \circ H \circ F - I\|_{C^k} + \|F^{-1} \circ H^{-1} \circ F - I\|_{C^k}$$
$$= \|F^{-1} \circ (F + h \circ F) - I\|_{C^k} + \|F^{-1} \circ (F + k \circ F) - I\|_{C^k} \to 0,$$

where $h = H - I$ and $k = H^{-1} - I$. If $d(I,H) \to 0$, it follows from Lemma 8.3 that $\|h\|_{C^k}$ and $\|k\|_{C^k}$ go to zero. From Lemma 8.2 $\|h \circ F\|_{C^k}$ and $\|k \circ F\|_{C^k}$ go to zero and our condition follows from Lemma 8.5. $\qquad\square$

Corollary 1. *The topology[9] induced by the metric d on the topological group $\mathcal{F}_0^k$ coincides with the topology that has as a basis of neighborhoods of the identity in $\mathcal{F}_0^k$ the sets*

$$E(\varepsilon) \overset{\text{def}}{=} \left\{ F \in \mathcal{F}_0^k : \|F - I\|_{C^k} + \|F^{-1} - I\|_{C^k} < \varepsilon \right\}.$$

Proof. In fact, if

$$S(\varepsilon) \overset{\text{def}}{=} \left\{ F \in \mathcal{F}_0^k : d(I,F) < \varepsilon \right\},$$

then from the definition of the metric d and Lemma 8.3 we have for $\varepsilon < 1$

$$E(\varepsilon) \subset S(\varepsilon) \subset E(2c(k,1)\varepsilon). \qquad\square$$

---

[8]To assert that $\mathcal{F}_0^k$ is a topological group, we need the full power of Lemma 8.5 and Lemma 8.4, which says that the translation of a bounded uniformly continuous function is continuous for the C^0 norm.

[9]Note that the chain of identities in the proof does not depend on Lemmas 8.5 and 8.4.

Theorem 8.3. *The metric group $\mathcal{F}_0^k$ is complete.*

Proof. If $\{H_n\}$ is a Cauchy sequence in $\mathcal{F}_0^k$, then $d(I, H_n)$ is bounded, and hence, by Lemma 8.3, there exists $L > 0$ such that for each n

$$\|I - H_n\|_{\mathcal{C}^k} + \|I - H_n^{-1}\|_{\mathcal{C}^k} \leq L.$$

From Lemma 8.2 and the boundedness of the sequence $\{\|I - H_n\|_{\mathcal{C}^k}\}$, it follows that there exists $M > 0$ such that for all m and n

$$\|H_m - H_n\|_{\mathcal{C}^k} = \|(H_m \circ H_n^{-1} - I) \circ H_n\|_{\mathcal{C}^k} \leq M \|H_m \circ H_n^{-1} - I\|_{\mathcal{C}^k}.$$

Hence, since $\{H_n\}$ is a Cauchy sequence in $\mathcal{F}_0^k$, from Lemma 8.3 it follows that $\{H_n - I\}$ is a Cauchy sequence in $\mathcal{C}_0^k(\mathbf{R}^N)$, which converges to an element[10] denoted $H - I$. It is readily seen that $|H(x) - x|$ and $|(H - I)^{(i)}(x)|$ go to zero as $|x| \to \infty$ for $i = 1, \ldots, k$. In fact, we have the estimate

$$|H(x) - x| \leq |H(x) - H_n(x)| + |H_n(x) - x| \leq \|H - H_n\|_{\mathcal{C}^k} + |H_n(x) - x|.$$

Proceed in the same way for the derivatives. Similarly, we show that $H_n^{-1} - I$ converges in $\mathcal{C}_0^k(\mathbf{R}^N)$ to an element denoted $G - I$. It remains to check that $G^{-1} = H$. It is sufficient to show that $\|G \circ H - I\|_{\mathcal{C}^k} = 0$ and $\|H \circ G - I\|_{\mathcal{C}^k} = 0$.[11] Indeed, from Lemma 8.2

$$\|G \circ H - I\|_{\mathcal{C}^k}$$
$$\leq \|(G - H_n^{-1}) \circ H\|_{\mathcal{C}^k} + \|H_n^{-1} \circ H - I\|_{\mathcal{C}^k}$$
$$\leq c\|(G - H_n^{-1})\|_{\mathcal{C}^k} + \|H_n^{-1} \circ (H_n + H - H_n) - H_n^{-1} \circ H_n\|_{\mathcal{C}^k},$$

where c is a constant that depends on $\|H - I\|_{\mathcal{C}^k}$. From Lemma 8.5 and the convergence of $\{H_n^{-1} - I\}$ to $G - I$ and of $\{H_n - I\}$ to $H - I$ in $\mathcal{C}_0^k(\mathbf{R}^N)$, it follows that the right-hand side of the above inequality goes to zero. By the same technique we get $\|H \circ G - I\|_{\mathcal{C}^k} = 0$ and hence $H \in \mathcal{F}_0^k$.

It is now obvious that $\{H_n\}$ converges to H in $\mathcal{F}_0^k$. Indeed we have

$$d(H_n, H) \leq \|H \circ H_n^{-1} - I\|_{\mathcal{C}^k} + \|H_n \circ H^{-1} - I\|_{\mathcal{C}^k}$$
$$\leq \|(H - H_n) \circ H_n^{-1}\|_{\mathcal{C}^k} + \|(H_n - H) \circ H^{-1}\|_{\mathcal{C}^k}.$$

From the last inequality, Lemma 8.2, and the boundedness of the sequence $\{I - H_n^{-1}\}$, we get the theorem. $\qquad\square$

---

[10] Since $\mathcal{C}_0^k(\mathbf{R}^N)$ is a Banach space $H - I \in \mathcal{C}_0^k(\mathbf{R}^N)$ and it is not necessary to show that $H - I$ and all its derivatives go to zero as $|x| \to \infty$.

[11] Here we don't need Lemmas 8.5 and 8.4 since it is sufficient to use the norm C^0 rather than C^k to prove that $G = H^{-1}$ since $C^1(\overline{\mathbf{R}^N}, \mathbf{R}^N) \subset C^{0,1}(\overline{\mathbf{R}^N}, \mathbf{R}^N)$ by the embedding (2.8) of Theorem 2.2, as will be illustrated in the proof of Theorem 9.5.

8.2 The Courant Metric on the Images

Associate with a subset Ω_0 of $\mathbf{R}^N$ the family of images

$$\mathcal{X}(\Omega_0) \stackrel{\text{def}}{=} \left\{ F(\Omega_0) \subset \mathbf{R}^N : \forall F \in \mathcal{F}_0^k \right\}.$$

We have the following property.

Lemma 8.6. *Given a closed or an open subset Ω_0 of $\mathbf{R}^N$, the family*

$$\mathcal{G}(\Omega_0) \stackrel{\text{def}}{=} \left\{ F \in \mathcal{F}_0^k : F(\Omega_0) = \Omega_0 \right\}$$

is a closed subgroup of $\mathcal{F}_0^k$.

Proof. $\mathcal{G}(\Omega_0)$ is clearly a subgroup of $\mathcal{F}_0^k$. It is sufficient to show that for any sequence $\{F_n\}$ in $\mathcal{G}(\Omega_0)$ which converges to F in $\mathcal{F}_0^k$, then $F \in \mathcal{G}(\Omega_0)$. The other properties are straightforward. If $\{F_n\}$ converges to F in $\mathcal{F}_0^k$, then by Lemma 8.3 we have

$$\|F_n \circ F^{-1} - I\|_{C^k} + \|F \circ F_n^{-1} - I\|_{C^k} \to 0.$$

From $\|F \circ F_n^{-1} - I\|_{C^k} \to 0$ for each $x \in \Omega_0$ the sequence $\{F(F_n^{-1}(x))\} \subset F(\Omega_0)$ converges to $x \in \Omega_0$ and hence $\Omega_0 \subset \overline{F(\Omega_0)}$. From the convergence $\|F_n \circ F^{-1} - I\|_{C^k} \to 0$ for each $x \in F(\Omega_0)$, $F^{-1}(x) \in \Omega_0$ and the sequence $\{F_n(F^{-1}(x))\} \subset \Omega_0$ converges to $x \in F(\Omega_0)$, and hence $F(\Omega_0) \subset \overline{\Omega_0}$. Since F is a homeomorphism, $F(\Omega_0)$ is open if Ω_0 is open and closed if Ω_0 is closed. Thence, if Ω_0 is open, $\Omega_0 \subset \text{int } \overline{F(\Omega_0)} = F(\Omega_0) \subset \text{int } \overline{\Omega_0} = \Omega_0$; if Ω_0 is closed, $\overline{\Omega_0} = \Omega_0 \subset \overline{F(\Omega_0)} = F(\Omega_0) \subset \overline{\Omega_0} = \Omega_0$. Then $F(\Omega_0) = \Omega_0$. $\square$

Remark 8.1.

In her paper Micheletti [1] assumes that Ω_0 is a bounded connected open domain of class C^k in order to make all the images $F(\Omega_0)$ bounded connected open domains of class C^k. However, it is sufficient to assume that Ω_0 is either open or closed in order to prove that $\mathcal{G}(\Omega_0)$ is a closed subgroup of $\mathcal{F}_0^k$. $\square$

For simplicity we now introduce the notation $\mathcal{G}$ for $\mathcal{G}(\Omega_0)$ and $\mathcal{X} = \mathcal{X}(\Omega_0)$. By definition, for each $\Omega \in \mathcal{X}$ there exists $F \in \mathcal{F}_0^k$ such that $\Omega = F(\Omega_0)$. Therefore the map

$$\Omega \mapsto \chi(\Omega) = F \circ \mathcal{G} : \mathcal{X} \to \mathcal{F}_0^k/\mathcal{G}$$

is well defined and bijective. This induces a complete metric on $\mathcal{X}$.

Lemma 8.7. *Given an open or a closed subset Ω_0 of $\mathbf{R}^N$, the function*

$$\delta(F \circ \mathcal{G}, H \circ \mathcal{G}) \stackrel{\text{def}}{=} \inf_{G, \tilde{G} \in \mathcal{G}(\Omega_0)} d(F \circ G, H \circ \tilde{G}) \tag{8.19}$$

is a metric on $\mathcal{F}_0^k/\mathcal{G}(\Omega_0)$. The topology induced by δ coincides with the quotient topology of $\mathcal{F}_0^k/\mathcal{G}(\Omega_0)$, and the space $\mathcal{F}_0^k/\mathcal{G}(\Omega_0)$ is complete.

Proof. The key result of the lemma is the completeness. The other properties are straightforward. Consider a Cauchy sequence $\{F_n \circ \mathcal{G}\}$ in $\mathcal{F}_0^k/\mathcal{G}$. It is sufficient to show that there exists a subsequence which converges. Construct the subsequence $\{F_\nu \circ \mathcal{G}\}$ such that

$$\delta(F_\nu \circ \mathcal{G}, F_{\nu+1} \circ \mathcal{G}) < 1/(2^\nu).$$

Then there exists a sequence $\{H_\nu\}$ in $\mathcal{F}_0^k$ such that

(i) $H_\nu \in F_\nu \circ \mathcal{G}$,

(ii) $d(H_\nu, H_{\nu+1}) < \frac{1}{2^\nu}$.

We proceed by induction. By the definition of δ, if $\delta(F_1 \circ \mathcal{G}, F_2 \circ \mathcal{G}) < 1/2$, there exist $H_1 \in F_1 \circ \mathcal{G}$ and $H_2 \in F_2 \circ \mathcal{G}$ such that $d(H_1, H_2) < 1/2$. Similarly, if $H_\nu \in F_\nu \circ \mathcal{G}$, it follows that there exists $H_{\nu+1}$ with the required property. Because $\delta(F_\nu \circ \mathcal{G}, F_{\nu+1} \circ \mathcal{G}) < 1/2^\nu$, there exist G_1 and G_2 in $\mathcal{G}$ such that $d(F_\nu \circ G_1, F_{\nu+1} \circ G_2) < 1/2^\nu$. If G_3 is such that $F_\nu \circ G_1 = H_\nu \circ G_3$, then we can assume that $H_{\nu+1} = F_{\nu+1} \circ G_2 \circ G_3^{-1}$ since

$$d(H_\nu, F_{\nu+1} \circ G_2 \circ G_3^{-1}) = d(H_\nu \circ G_3, F_{\nu+1} \circ G_2) = d(F_\nu \circ G_1, F_{\nu+1} \circ G_2).$$

It is easy to check that $\{H_\nu\}$ is a Cauchy sequence in $\mathcal{F}_0^k$. Indeed

$$\forall i < j, \quad d(H_i, H_j) \le \sum_{n=i}^{j-1} d(H_n, H_{n+1}) \le \frac{1}{2^i} + \cdots + \frac{1}{2^j} \le \frac{1}{2^{i-1}}.$$

By completeness of $\mathcal{F}_0^k$, $\{H_\nu\}$ converges in $\mathcal{F}_0^k$. Since the canonical map $\pi : \mathcal{F}_0^k \to \mathcal{F}_0^k/\mathcal{G}$ is continuous, the sequence $\{\pi(H_\nu) = F_\nu \circ \mathcal{G}\}$ converges in $\mathcal{F}_0^k/\mathcal{G}$. $\square$

Finally it is natural to consider on $\mathcal{X}(\Omega_0)$ the metric[12]

$$\boxed{\rho(\Omega_1, \Omega_2) \stackrel{\text{def}}{=} \delta(\chi(\Omega_1), \chi(\Omega_2))} \tag{8.20}$$

on the images of Ω_0 by $\mathcal{F}_0^k$ induced by the bijection χ on $\mathcal{X}(\Omega_0)$. With that metric, $\mathcal{X}(\Omega_0)$ is a complete metric space. This metric is called the *Courant metric* (metrica di Courant) in Micheletti's paper. She used that metric to prove the following theorem.

Theorem 8.4. *Fix $k = 3$ and a bounded open connected domain Ω_0 of class C^k. Consider the complete metric space $\mathcal{X}(\Omega_0)$ endowed with the Courant metric d. The subset of all bounded open domains Ω in $\mathcal{X}(\Omega_0)$ such that the spectrum of the Laplace operator $-\Delta$ on Ω with homogeneous Dirichlet conditions on the boundary $\partial\Omega$ does not have all its eigenvalues simple is of the first category.[13]*

In practice this theorem says that, up to an arbitrarily small perturbation of the domain, the eigenvalues of the Laplace operator of a C^k-domain can be made simple.

---

[12]Equivalently $\rho(F_1(\Omega_0), F_2(\Omega_0)) = \delta(F_1 \circ \mathcal{G}, F_1 \circ \mathcal{G})$ for all F_1 and F_2 in $\mathcal{F}_0^k$.

[13]A set B is said to be *nowhere dense* if its closure has no interior or, alternatively, if $\complement B$ is dense. A set is said to be of the *first category* if it is the countable union of nowhere dense sets (cf. J. Dugundji [1, Def. 10.4, p. 250]).

8.3   Perturbations of the Identity

One important subfamily of $\mathcal{F}_0^k$ are the *perturbations of the identity*:

$$F = I + f, \quad f \in \mathcal{C}_0^k(\mathbf{R}^N).$$

For $k \geq 1$ and f sufficiently small F is bijective and has a unique inverse. Given $y \in \mathbf{R}^N$, consider

$$S(x) \overset{\text{def}}{=} y - f(x), \quad x \in \mathbf{R}^N.$$

Then for any x_1 and x_2,

$$S(x_2) - S(x_1) = -[f(x_2) - f(x_1)]$$
$$\Rightarrow |S(x_2) - S(x_1)| \leq |f(x_2) - f(x_1)| \leq \sqrt{N}\,\|f^{(1)}\|_{\mathcal{C}^1}\,|x_2 - x_1|.$$

For $\|f^{(1)}\|_{\mathcal{C}^1} < 1/\sqrt{N}$, S is a contraction and, for each y, there exists a unique x such that $y - f(x) = S(x) = x$, $[I + f](x) = y$. Therefore $I + f$ is bijective. But in order to show that $F \in \mathcal{F}_0^k$ we also need to prove that $[I + f]^{-1} \in \mathcal{C}^k(\mathbf{R}^N)$. The Jacobian matrix of $F(x)$ is equal to $I + f^{(1)}(x)$ and by uniform continuity with respect to f

$$\forall x \in \mathbf{R}^N, \quad |F^{(1)}(x) - I| \leq \|F^{(1)} - I\|_{\mathcal{C}^0} = \|f^{(1)}\|_{\mathcal{C}^0}.$$

If there exists $x \in \mathbf{R}^N$ such that $F^{(1)}(x)$ is not invertible, then for some $0 \neq y \in \mathbf{R}^N$, $F^{(1)}(x)y = 0$ and

$$|y| = |F^{(1)}(x)y - y| \leq |F^{(1)}(x) - I||y| \leq \|f^{(1)}\|_{\mathcal{C}^0}\,|y|$$
$$\Rightarrow 1 \leq \|f^{(1)}\|_{\mathcal{C}^0} < \frac{1}{\sqrt{N}},$$

which is a contradiction for $N \geq 1$. Therefore, for all $x \in \mathbf{R}^N$, $F^{(1)}(x)$ is invertible. As a result the conditions of the implicit function theorem are met and from Schwartz [3, Vol. 1, p. 294, Thm. 29, p. 299, Thm. 31] we have that $F^{-1} = [I + f]^{-1} \in \mathcal{C}^k(\mathbf{R}^N)$. So for $k \geq 1$

$$\forall f \in \mathcal{C}_0^k(\mathbf{R}^N), \|f^{(1)}\|_{\mathcal{C}^0} < 1/\sqrt{N}, \quad F = I + f \in \mathcal{F}_0^k.$$

Note that the condition is on $\|f^{(1)}\|_{\mathcal{C}^0}$ and not $\|f\|_{\mathcal{C}^k}$, but by definition $\|f^{(1)}\|_{\mathcal{C}^0} \leq \|f\|_{\mathcal{C}^1} \leq \|f\|_{\mathcal{C}^k}$ and the condition $\|f\|_{\mathcal{C}^k} < 1/\sqrt{N}$ yields the same result.

Note further that from the proof of Theorem 8.1

$$\|F^{-1} - I\|_{\mathcal{C}^k} \leq c\,\|F - I\|_{\mathcal{C}^k} = c\|f\|_{\mathcal{C}^k},$$

and the map $f \mapsto I + f : B(0, 1/\sqrt{N}) \subset \mathcal{C}_0^k \to \mathcal{F}_0^k$ is continuous.

9 The Generic Framework of Micheletti

The construction of a complete metric on the group of transformations and a Courant metric on the associated quotient group naturally extends to other spaces of transformations and to Banach spaces other than $\mathcal{C}_0^k(\mathbf{R}^N) = C_0^k(\mathbf{R}^N, \mathbf{R}^N)$. In this section we consider spaces of transformations of an open subset D of $\mathbf{R}^N$ onto itself and replace the space $\mathcal{C}_0^k(\mathbf{R}^N)$ by a Banach space Θ of transformations from D to $\mathbf{R}^N$. We then specialize to specific examples of Banach spaces Θ.

Given a Banach space Θ of transformations from D to $\mathbf{R}^N$, define the space

$$\mathcal{F}(\Theta) \stackrel{\text{def}}{=} \left\{ F : D \to D : F - I \in \Theta \text{ and } F^{-1} - I \in \Theta \right\}.$$

Associate with $F \in \mathcal{F}(\Theta)$ the distance

$$d(I, F) \stackrel{\text{def}}{=} \inf_{(f_1,\ldots,f_n)} \sum_{i=1}^{n} \|f_i\|_\Theta + \inf_{(g_1,\ldots,g_m)} \sum_{i=1}^{m} \|g_i\|_\Theta, \tag{9.1}$$

where the infima are taken over all finite factorizations in $\mathcal{F}(\Theta)$ of the form

$$F = (I + f_n) \circ \cdots \circ (I + f_1) \text{ and } F^{-1} = (I + g_m) \circ \cdots \circ (I + g_1)$$

for f_i, $g_i \in \Theta$. It readily follows that

$$d(I, F) = d(I, F^{-1}). \tag{9.2}$$

Extend this definition to all pairs F and G in $\mathcal{F}(\Theta)$:

$$d(F, G) \stackrel{\text{def}}{=} d(I, G \circ F^{-1}). \tag{9.3}$$

By definition, d is right-invariant since for all F, G, and H in $\mathcal{F}(\Theta)$,

$$d(F, G) = d(F \circ H, G \circ H).$$

Define for some open or closed subset Ω_0 of D the subgroup

$$\mathcal{G}(\Omega_0) \stackrel{\text{def}}{=} \{F \in \mathcal{F}(\Theta) : F(\Omega) = \Omega_0\}$$

and the following metric on $\mathcal{F}(\Theta)/\mathcal{G}(\Omega_0)$:

$$\forall F, G \in \mathcal{F}(\Theta), \quad \delta([F], [H]) \stackrel{\text{def}}{=} \inf_{G, \tilde{G} \in \mathcal{G}(\Omega_0)} d(F \circ G, H \circ \tilde{G}) \tag{9.4}$$

$$\forall F_1, F_2 \in \mathcal{F}(\Theta), \quad \rho(F_1(\Omega_0), F_2(\Omega_0)) \stackrel{\text{def}}{=} \delta([F_1], [F_2]), \tag{9.5}$$

where $[F] = F \circ \mathcal{G}$ denotes the equivalence class of F. Of course some appropriate assumptions must be incorporated to make $\mathcal{F}(\Theta)$ a well-defined complete metric space.

9.1 $\mathcal{B}^k(\mathbf{R}^N, \mathbf{R}^N)$-Mappings

The constructions of Micheletti extend from the space $\mathcal{C}_0^k(\mathbf{R}^N)$ to the Banach space $\mathcal{B}^k = \mathcal{B}^k(\mathbf{R}^N, \mathbf{R}^N)$, $k \geq 1$, of all k-times bounded, continuously differentiable transformations f of $\mathbf{R}^N$ onto itself endowed with the norm $\|f\|_{\mathcal{C}^k}$. Here $\Theta = \mathcal{B}^k$. Define

$$\boxed{\mathcal{F}^k \overset{\text{def}}{=} \left\{ F : \mathbf{R}^N \to \mathbf{R}^N \ : \ F - I \in \mathcal{B}^k \text{ and } F^{-1} - I \in \mathcal{B}^k \right\}.}$$

Clearly, for all F and G in $\mathcal{F}^k$, $F \circ G$ and $(F \circ G)^{-1} = G^{-1} \circ F^{-1}$ map $\mathbf{R}^N$ onto $\mathbf{R}^N$. So it is sufficient to show that $F \circ G - I \in \mathcal{B}^k$. This follows by linearity of the space $\mathcal{B}^k$ and Lemma 8.2 with $f = F - I$ and $g = G - I$,

$$F \circ G - I = F \circ G - G + G - I = f \circ (I + g) + g \in \mathcal{B}^k.$$

From this point it is easy to check that all the properties of $\mathcal{F}_0^k$ remain true for $\mathcal{F}^k$, except the continuity of the composition since Lemma 8.4 does not hold.

Theorem 9.1. *Let $k \geq 1$ be an integer.*

(i) *The topology induced by the metric d on $\mathcal{F}^k$ is complete.[14] Moreover, around the identity I for all $0 < \varepsilon < 1$,*

$$E(\varepsilon) \subset S(\varepsilon) \subset E(2c(k, 1)\varepsilon),$$

$$E(\varepsilon) \overset{\text{def}}{=} \left\{ F \in \mathcal{F}^k \ : \ \|F - I\|_{\mathcal{C}^k} + \|F^{-1} - I\|_{\mathcal{C}^k} < \varepsilon \right\},$$

$$S(\varepsilon) \overset{\text{def}}{=} \left\{ F \in \mathcal{F}^k \ : \ d(I, F) < \varepsilon \right\}.$$

(ii) *Given an open or closed subset Ω_0 of $\mathbf{R}^N$,*

$$\delta(F \circ \mathcal{G}, H \circ \mathcal{G}) \overset{\text{def}}{=} \inf_{G, \tilde{G} \in \mathcal{G}(\Omega_0)} d(F \circ G, H \circ \tilde{G}) \tag{9.6}$$

is a metric on $\mathcal{F}^k / \mathcal{G}(\Omega_0)$ and the topology induced by δ is complete.

As in the case of $\mathcal{F}_0^k$, $k \geq 1$, we have for the perturbations of the identity $F = I + f$,

$$\forall f \in \mathcal{B}^k(\mathbf{R}^N, \mathbf{R}^N) \text{ such that } \|f^{(1)}\|_{\mathcal{C}^0} < 1/\sqrt{N}, \quad F = I + f \in \mathcal{F}^k.$$

So in view of the inequality $\|f^{(1)}\|_{\mathcal{C}^0} \leq \|f\|_{\mathcal{C}^1} \leq \|f\|_{\mathcal{C}^k}$, the condition $\|f\|_{\mathcal{C}^k} < 1/\sqrt{N}$ yields the same result. Moreover, the map $f \mapsto I + f : B(0, 1/\sqrt{N}) \subset \mathcal{B}^k \to \mathcal{F}^k$ is continuous.

---

[14]By analogy with footnote 10 in the proof of Theorem 8.3, we use the embedding $\mathcal{B}^1(\mathbf{R}^N, \mathbf{R}^N) \subset C^{0,1}(\overline{\mathbf{R}^N}, \mathbf{R}^N)$ to prove that $G = H^{-1}$.

9.2 Lipschitzian Mappings

As a second illustration consider the spaces of transformations of $\mathbf{R}^N$ introduced by Murat and Simon [1] in 1976 in the construction of metric spaces of domains. For integers $k \geq 0$, they correspond to the following choices of the space Θ:

$$W^{k+1,\bar{c}}(\mathbf{R}^N, \mathbf{R}^N)$$
$$\stackrel{\text{def}}{=} \left\{ f \in W^{k+1,\infty}(\mathbf{R}^N, \mathbf{R}^N) : \forall 0 \leq |\alpha| \leq k+1, \partial^\alpha f \in C(\overline{\mathbf{R}^N}, \mathbf{R}^N) \right\}$$

and $W^{k+1,\infty}(\mathbf{R}^N, \mathbf{R}^N)$. The first space $W^{k+1,\bar{c}}(\mathbf{R}^N, \mathbf{R}^N)$ coincides with the space $\mathcal{C}^{k+1}(\overline{\mathbf{R}^N}) = C^{k+1}(\overline{\mathbf{R}^N}, \mathbf{R}^N)$ algebraically and topologically. The corresponding space of transformations

$$\boxed{\begin{aligned} \mathcal{F}^{k+1}(\overline{\mathbf{R}^N}) \stackrel{\text{def}}{=} \{ F : \mathbf{R}^N \to \mathbf{R}^N : \\ F - I \in \mathcal{C}^{k+1}(\overline{\mathbf{R}^N}) \text{ and } F^{-1} - I \in \mathcal{C}^{k+1}(\overline{\mathbf{R}^N}) \} \end{aligned}}$$

is a topological group for the metric d as in the case of $\mathcal{F}_0^{k+1}$ in section 8. This follows from the fact that the pointwise translation of a bounded uniformly continuous function is continuous in the C^0-norm (cf. Lemma 8.4). The second space $W^{k+1,\infty}(\mathbf{R}^N, \mathbf{R}^N)$ coincides with $\mathcal{C}^{k,1}(\overline{\mathbf{R}^N}) \stackrel{\text{def}}{=} C^{k,1}(\overline{\mathbf{R}^N}, \mathbf{R}^N)$, and

$$\boxed{\begin{aligned} \mathcal{F}^{k,1}(\overline{\mathbf{R}^N}) \stackrel{\text{def}}{=} \{ F : \mathbf{R}^N \to \mathbf{R}^N : \\ F - I \in \mathcal{C}^{k,1}(\overline{\mathbf{R}^N}) \text{ and } F^{-1} - I \in \mathcal{C}^{k,1}(\overline{\mathbf{R}^N}) \} \end{aligned}} \tag{9.7}$$

is also complete for the topology induced by the metric d, but is not a topological group. In both cases the Courant metric defines a complete metric topology on the corresponding quotient space. Also recall that $\mathcal{F}^1(\overline{\mathbf{R}^N})$ transports locally Lipschitzian (epigraph) domains onto locally Lipschitzian (epigraph) domains (cf. Bendali [1] and Djadane [1]), but that $\mathcal{F}^{0,1}(\overline{\mathbf{R}^N})$ does not, as shown in Examples 5.1 and 5.2 of section 5.2.

9.2.1 The Murat–Simon Approach

The constructions of Murat and Simon [1] to obtain a complete metric topology on the quotient spaces are different from those of Micheletti [1], which were seemingly not known to them. They worked with the *pseudometric*

$$d_p(F_2, F_1) \stackrel{\text{def}}{=} \|F_2 \circ F_1^{-1} - I\|_{W^{k+1,\infty}} + \|F_1 \circ F_2^{-1} - I\|_{W^{k+1,\infty}}$$

rather than the metric defined by the infima over finite factorizations of $F_2 \circ F_1^{-1}$ and $F_1 \circ F_2^{-1}$. They recovered a metric from the pseudometric by using an auxiliary construction which depends on a third property of a pseudometric. We briefly recall the definition and the result.

Definition 9.1.

A *pseudometric* on a space E is a function $\delta : E \times E \to \mathbf{R}^+$ with the following properties:

(i) $\delta(F_2, F_1) = 0 \iff F_2 = F_1$,

(ii) $\delta(F_2, F_1) = \delta(F_1, F_2)$ for all F_1 and F_2,

(iii) $\delta(F_1, F_3) \le \delta(F_1, F_2) + \delta(F_2, F_3) + \delta(F_1, F_2)\,\delta(F_2, F_3)\,P(\delta(F_1, F_2) + \delta(F_2, F_3))$
for all F_i's, where $P : \mathbf{R}^+ \to \mathbf{R}^+$ is a continuous increasing function. $\qquad\square$

Proposition 9.1. *Let δ be a pseudometric on E. For all α, $0 < \alpha < 1$, there exists a constant $\eta_\alpha > 0$ such that the function $\delta^{(\alpha)} : E \times E \to \mathbf{R}^+$ defined as*

$$\delta^{(\alpha)}(F_1, F_2) \overset{\text{def}}{=} \inf\{\delta(F_1, F_2), \eta_\alpha\}^\alpha$$

is a metric on E.

9.2.2 The Micheletti Approach

The pseudometric can be completely bypassed. By combining the construction of Micheletti [1] with the properties established in Murat and Simon [1], we readily get the completeness of the (Micheletti) metric for the group of transformations and of the Courant metric for the quotient space. In both cases the results can also be obtained directly by adapting, with obvious technical changes in the case of $C^{k,1}(\overline{\mathbf{R}^N})$, the sequence of lemmas and theorems of section 8. The first case is technically analogous to $\mathcal{F}_0^k$ by choosing $\Theta = C^k(\overline{\mathbf{R}^N})$.

Theorem 9.2. *Let $k \ge 1$ be an integer.*

(i) *The topology induced by d on $\mathcal{F}^k(\overline{\mathbf{R}^N})$ makes it a complete[15] topological group. Moreover, around the identity I for all $0 < \varepsilon < 1$*

$$E(\varepsilon) \subset S(\varepsilon) \subset E(2c(k,1)\varepsilon),$$

$$E(\varepsilon) \overset{\text{def}}{=} \left\{ F \in \mathcal{F}^k(\overline{\mathbf{R}^N}) : \|F - I\|_{C^k} + \|F^{-1} - I\|_{C^k} < \varepsilon \right\},$$

$$S(\varepsilon) \overset{\text{def}}{=} \left\{ F \in \mathcal{F}^k(\overline{\mathbf{R}^N}) : d(I, F) < \varepsilon \right\},$$

and the topology coincides with the topology that has as a basis of neighborhoods of the identity in $\mathcal{F}^k(\overline{\mathbf{R}^N})$ the sets $E(\varepsilon)$.

(ii) *Given an open or closed subset Ω_0 of $\mathbf{R}^N$,*

$$\delta(F \circ \mathcal{G}, H \circ \mathcal{G}) \overset{\text{def}}{=} \inf_{G, \tilde{G} \in \mathcal{G}(\Omega_0)} d(F \circ G, H \circ \tilde{G}) \tag{9.8}$$

is a metric on $\mathcal{F}^k(\overline{\mathbf{R}^N})/\mathcal{G}(\Omega_0)$. The topology induced by δ is complete and coincides with the quotient topology of $\mathcal{F}^k(\overline{\mathbf{R}^N})/\mathcal{G}(\Omega_0)$.

---

[15] By analogy with footnote 10 in the proof of Theorem 8.3, we use the embedding $C^k(\overline{\mathbf{R}^N}, \mathbf{R}^N) \subset C^{0,1}(\overline{\mathbf{R}^N}, \mathbf{R}^N)$ to prove that $G = H^{-1}$. The continuity of the composition $\circ$ follows from the extension of Lemma 8.4 to bounded uniformly continuous functions.

For the case of $\mathcal{F}^{k,1}(\overline{\mathbf{R}^N})$ with the space $W^{k+1,\infty}(\mathbf{R}^N, \mathbf{R}^N)$ we get a similar result except that $\mathcal{F}^{k,1}(\overline{\mathbf{R}^N})$ is not a topological group. Recall that the space $W^{1,\infty}(\mathbf{R}^N, \mathbf{R}^N)$ coincides with the space $C^{0,1}(\overline{\mathbf{R}^N}) \stackrel{\text{def}}{=} C^{0,1}(\overline{\mathbf{R}^N}, \mathbf{R}^N)$. By convexity of the whole space $\mathbf{R}^N$ from (2.10) in section 2.3, for $k \geq 1$ the space $W^{k+1,\infty}(\mathbf{R}^N, \mathbf{R}^N)$ coincides with $C^{k,1}(\overline{\mathbf{R}^N})$. Furthermore, from (2.11), for all $k \geq 0$ $C^{k,1}(\overline{\mathbf{R}^N})$ is a Banach space when endowed with the norm

$$\|f\|_{C^{k,1}} \stackrel{\text{def}}{=} \|f\|_{C^k} + c_k(f), \tag{9.9}$$

$$c(f) \stackrel{\text{def}}{=} \sup_{x \neq y} \frac{|f(y) - f(x)|}{|y - x|} \qquad c_k(f) \stackrel{\text{def}}{=} \max_{|\alpha|=k} \sup_{x \neq y} \frac{|\partial^\alpha f(y) - \partial^\alpha f(x)|}{|x - y|}. \tag{9.10}$$

Associate with $\Theta = C^{k,1}(\overline{\mathbf{R}^N})$ the space $\mathcal{F}^{k,1}(\overline{\mathbf{R}^N})$ defined in (9.7). In support of the claim that the results can also be obtained directly by adapting the sequence of lemmas and theorems of section 8, we provide the proof of the completeness of the Courant metric for $k = 0$. The generalization to $k \geq 1$ is lengthy but straightforward. For simplicity we shall use the notation $C^{k,1}$ and $\mathcal{F}^{k,1}$ in the proofs.

Theorem 9.3. *The space $\mathcal{F}^{k,1}(\overline{\mathbf{R}^N})$ is a group under the composition $\circ$.*

Proof. By definition, for any $F \in \mathcal{F}^{0,1}$, $F - I \in C^{0,1}$, $c(F) \leq 1 + c(F - I) < \infty$, and $F \in C^{0,1}(\mathbf{R}^N, \mathbf{R}^N)$. Therefore, for all F and G in $\mathcal{F}^{0,1}$, $c(F \circ G) \leq c(F) c(G) < \infty$, $c(F \circ G - I) \leq 1 + c(F) c(G) < \infty$, and

$$\begin{aligned}
\|F \circ G - I\|_{C^0} &\leq \|(F - I) \circ G\|_{C^0} + \|G - I\|_{C^0} \\
&\leq \|(F - I)\|_{C^0} + \|G - I\|_{C^0} < \infty.
\end{aligned}$$

Therefore, $F \circ G \in \mathcal{F}^{0,1}$. Clearly $I \in \mathcal{F}^{0,1}$ is the neutral element. By definition of an element $F \in \mathcal{F}^{0,1}$, $F^{-1} \in \mathcal{F}^{0,1}$ and $\mathcal{F}^{0,1}$ is a group under composition. $\square$

As in the previous constructions, associate with $F \in \mathcal{F}^{0,1}(\overline{\mathbf{R}^N})$

$$d(I, F) \stackrel{\text{def}}{=} \inf_{(f_1,\ldots,f_n)} \sum_{i=1}^{n} \|f_i\|_{C^{0,1}} + \inf_{(g_1,\ldots,g_m)} \sum_{i=1}^{m} \|g_i\|_{C^{0,1}}, \tag{9.11}$$

where the infima are taken over all finite factorizations in $\mathcal{F}^{0,1}(\overline{\mathbf{R}^N})$ of the form

$$F = (I + f_n) \circ \cdots \circ (I + f_1) \quad \text{and} \quad F^{-1} = (I + g_m) \circ \cdots \circ (I + g_1).$$

The other definitions are as specified at the beginning of section 9.

The next theorem requires the equivalent of Lemma 8.3 for $\mathcal{F}_0^k$.

Lemma 9.1. *Given integers $r \geq 0$ and $s > 0$ there exists a constant $c(r, s) > 0$ with the following property: for all finite sequences $f_1, \ldots, f_n$ in $C^{r,1}(\overline{\mathbf{R}^N})$ such that*

$$\sum_{i=1}^{n} \|f_i\|_{C^{r,1}} < \alpha, \quad 0 < \alpha < s, \tag{9.12}$$

then for the transformation $F = (I + f_n) \circ \cdots \circ (I + f_1)$,

$$\|F - I\|_{C^{r,1}} \le \alpha\, c(r, s). \tag{9.13}$$

Proof. For simplicity define

$$F_i \stackrel{\text{def}}{=} (I + f_i) \circ \cdots \circ (I + f_1), \quad F_n = F.$$

By the definition of F we have

$$\begin{aligned}
F - I &= f_1 + f_2 \circ (I + f_1) + \cdots + f_n \circ (I + f_{n-1}) \circ \cdots \circ (I + f_1) \\
&= f_1 + f_2 \circ F_1 + \cdots + f_n \circ F_{n-1}.
\end{aligned} \tag{9.14}$$

Then $\|F - I\|_{C^0} \le \sum_{i=1}^{n} \|f_i\|_{C^0} \le \alpha$. From (9.14) we further get for all x and y

$$|(F - I)(y) - (F - I)(x)| \le c(f_1)|y - x| + c(f_2)\left[1 + c(f_1)\right]|y - x| + \cdots$$
$$+ c(f_n)\left[1 + c(f_{n-1})\right]\ldots\left[1 + c(f_1)\right]|y - x|,$$

$$c(F - I) \le \left[\sum_{i=1}^{n} c(f_i)\right] \prod_{i=1}^{n}[1 + c(f_i)] \le \left[\sum_{i=1}^{n} c(f_i)\right] e^{\sum_{i=1}^{n} c(f_i)} \le \alpha e^{\alpha} \le \alpha e^{s},$$

and $\|F - I\|_{C^{0,1}} \le 2\alpha\, e^{\alpha}$. Hence $c(0, s) = 2e^{s}$ for all $s \ge 1$. $\qquad\square$

Theorem 9.4. *The function d is a metric on $\mathcal{F}^{k,1}(\overline{\mathbf{R}^N})$. Moreover, around the identity I for all $0 < \varepsilon < 1$*

$$E(\varepsilon) \subset S(\varepsilon) \subset E(2c(k, 1)\varepsilon),$$

$$E(\varepsilon) \stackrel{\text{def}}{=} \left\{ F \in \mathcal{F}^{k,1}(\overline{\mathbf{R}^N}) : \|F - I\|_{C^{k,1}} + \|F^{-1} - I\|_{C^{k,1}} < \varepsilon \right\},$$

$$S(\varepsilon) \stackrel{\text{def}}{=} \left\{ F \in \mathcal{F}^{k,1}(\overline{\mathbf{R}^N}) : d(I, F) < \varepsilon \right\}.$$

Proof. (a) We check that the three axioms that define a metric on $\mathcal{F}^{0,1}$ are satisfied.
(i) $d(F, G) = 0 \iff F = G$. If $d(F, G) = 0$, then there exists a sequence of factorizations of $G \circ F^{-1}$ in $\mathcal{F}^{0,1}$,

$$G \circ F^{-1} = (I + g_{n,s_n}) \circ \cdots \circ (I + g_{n,1})$$

such that

$$\mathcal{S}_{(n)} = \sum_{i=1}^{s_n} \|g_{n,i}\|_{C^{0,1}} \to 0 \text{ as } n \to \infty.$$

From Lemma 9.1 for n arbitrarily large (as long as $\mathcal{S}_{(n)} < 1$)

$$\|G \circ F^{-1} - I\|_{C^{0,1}} \le c(0, 1)\mathcal{S}_{(n)},$$

and hence $\|G \circ F^{-1} - I\|_{C^{0,1}} = 0$. Hence $F = G$.
(ii) $d(F, G) = d(G, F)$ from the definition and property (9.2).
(iii) $d(F, G) \le d(F, H) + d(H, G)$ by definition of d.
(b) This is the same proof as for Corollary 1 to Theorem 8.2 by using Lemma 9.1 instead of Lemma 8.3. $\qquad\square$

Theorem 9.5. *The metric topology defined by d on $\mathcal{F}^{k,1}(\overline{\mathbf{R}^N})$ is complete.*

Proof. If $\{H_n\}$ is a Cauchy sequence in $\mathcal{F}^{0,1}$, then

$$\forall \varepsilon > 0, \ \exists N > 0, \ \forall n, m \geq N, \quad d(H_m, H_n) < \varepsilon.$$

By Lemma 9.1, for $0 < \varepsilon < 1$

$$\|H_m \circ H_n^{-1} - I\|_{\mathcal{C}^{0,1}} + \|H_n \circ H_m^{-1} - I\|_{\mathcal{C}^{0,1}} \leq c(0,1)\,\varepsilon. \tag{9.15}$$

In particular,

$$|d(I, H_n) - d(I, H_m)| \leq d(H_m, H_n) < \varepsilon.$$

Therefore, $\{d(I, H_n)\}$ is bounded and, again by Lemma 9.1,

$$\exists L > 0, \ \forall n \geq N, \quad \|I - H_n\|_{\mathcal{C}^{0,1}} + \|I - H_n^{-1}\|_{\mathcal{C}^{0,1}} \leq L.$$

From Lemma 8.2, (9.15), and the boundedness of the sequence $\{\|I - H_n\|_{\mathcal{C}^{0,1}}\}$, it follows that for all m and n greater or equal to N

$$\|H_m - H_n\|_{\mathcal{C}^0} = \|(H_m \circ H_n^{-1} - I) \circ H_n\|_{\mathcal{C}^0} = \|H_m \circ H_n^{-1} - I\|_{\mathcal{C}^0},$$
$$c(H_m - H_n) = c((H_m \circ H_n^{-1} - I) \circ H_n)$$
$$\leq c(H_m \circ H_n^{-1} - I)\, c(H_n) \leq c(H_m \circ H_n^{-1} - I)\,(1 + L),$$

and it follows from (9.15) that for all $m, \, n > N$,

$$\|H_m - I - (H_n - I)\|_{\mathcal{C}^{0,1}} = \|H_m - H_n\|_{\mathcal{C}^{0,1}} \leq (L + 1)\, c(0,1)\,\varepsilon$$

and $\{H_n - I\}$ is a Cauchy sequence in $\mathcal{C}^{0,1}$ which converges to an element denoted $H - I$ in $\mathcal{C}^{0,1}$. Similarly, we can show that $H_n^{-1} - I$ converges to an element denoted $G - I$ in $\mathcal{C}^{0,1}$. It remains to check that $G^{-1} = H$. It is sufficient to show that $\|G \circ H - I\|_{\mathcal{C}^0} = 0$ and $\|H \circ G - I\|_{\mathcal{C}^0} = 0$. Indeed, from Lemma 8.2

$$\|G \circ H - I\|_{\mathcal{C}^0} \leq \|(G - H_n^{-1}) \circ H\|_{\mathcal{C}^0} + \|H_n^{-1} \circ H - I\|_{\mathcal{C}^0}$$
$$\leq \|(G - H_n^{-1})\|_{\mathcal{C}^0} + \|H_n^{-1} \circ (H_n + H - H_n) - H_n^{-1} \circ H_n\|_{\mathcal{C}^0}$$
$$\leq \|(G - H_n^{-1})\|_{\mathcal{C}^0} + c(H_n^{-1})\,\|H - H_n\|_{\mathcal{C}^0}$$
$$\leq \|(G - H_n^{-1})\|_{\mathcal{C}^0} + (1 + c(H_n^{-1} - I))\,\|H - H_n\|_{\mathcal{C}^0}$$
$$\leq \|(G - H_n^{-1})\|_{\mathcal{C}^0} + (1 + L)\,\|H - H_n\|_{\mathcal{C}^0} \to 0.$$

By the same technique we get $\|H \circ G - I\|_{\mathcal{C}^0} = 0$ and hence $G = H^{-1}$ and $H \in \mathcal{F}^{0,1}$. It is now obvious that $\{H_n\}$ converges to H in $\mathcal{F}^{0,1}$. Indeed we have

$$d(H_n, H) \leq \|H \circ H_n^{-1} - I\|_{\mathcal{C}^{0,1}} + \|H_n \circ H^{-1} - I\|_{\mathcal{C}^{0,1}}$$
$$\leq \|(H - H_n) \circ H_n^{-1}\|_{\mathcal{C}^{0,1}} + \|(H_n - H) \circ H^{-1}\|_{\mathcal{C}^{0,1}}$$

and

$$\|(H - H_n) \circ H_n^{-1}\|_{C^0} = \|H - H_n\|_{C^0} \to 0,$$
$$c((H - H_n) \circ H_n^{-1}) \le c(H - H_n)\, c(H_n^{-1}) \le c(H - H_n)\, (1 + L) \to 0,$$
$$\|(H_n - H) \circ H^{-1}\|_{C^0} = \|H_n - H\|_{C^0} \to 0,$$
$$c((H_n - H) \circ H^{-1}) \le c(H - H_n)\, c(H^{-1}) \le c(H - H_n)\, (1 + L) \to 0,$$

and we get the theorem. $\qquad\square$

We have the following property.

Lemma 9.2. *Given a closed or open subset Ω_0 of $\mathbf{R}^N$, the family*

$$\boxed{\mathcal{G}(\Omega_0) \stackrel{\text{def}}{=} \left\{ F \in \mathcal{F}^{k,1}(\overline{\mathbf{R}^N}) \ : \ F(\Omega_0) = \Omega_0 \right\}}$$

is a closed subgroup of $\mathcal{F}^{k,1}(\overline{\mathbf{R}^N})$.

Proof. This is the same proof as the one of Lemma 8.6 with $\mathcal{F}^{k,1}$ in place of $\mathcal{F}_0^k$. $\quad\square$

Theorem 9.6. *Given an open or closed subset Ω_0 of $\mathbf{R}^N$, the function*

$$\delta(F \circ \mathcal{G}, H \circ \mathcal{G}) \stackrel{\text{def}}{=} \inf_{G, \tilde{G} \in \mathcal{G}(\Omega_0)} d(F \circ G, H \circ \tilde{G}) \qquad (9.16)$$

is a metric on $\mathcal{F}^{k,1}(\overline{\mathbf{R}^N})/\mathcal{G}(\Omega_0)$ and the topology induced by δ is complete.

Proof. This is the same proof as the one of Lemma 8.7 with $\mathcal{F}^{k,1}$ in place of $\mathcal{F}_0^k$. $\quad\square$

Consider perturbations of the identity of the form $F = I + f$, $f \in C^{0,1}(\overline{\mathbf{R}^N})$. Consider for any $y \in \mathbf{R}^N$ the map

$$x \mapsto S(x) \stackrel{\text{def}}{=} y - f(x) \ : \ \mathbf{R}^N \to \mathbf{R}^N.$$

For all x_1 and x_2

$$S(x_2) - S(x_1) = -[f(x_2) - f(x_1)],$$
$$|S(x_2) - S(x_1)| = |f(x_2) - f(x_1)| \le c(f)\, |x_2 - x_1|,$$

and f is contracting for $c(f) < 1$. Hence for each $y \in \mathbf{R}^N$ there exists a unique $x \in \mathbf{R}^N$ such that

$$S(x) = x \quad \Rightarrow \quad y - f(x) = x \quad \Rightarrow \quad F(x) = y$$

and F is bijective. It remains to prove that $F^{-1} - I \in C^{0,1}(\overline{\mathbf{R}^{\mathrm{N}}}, \mathbf{R}^{\mathrm{N}})$ to conclude that $I + f \in \mathcal{F}^{0,1}(\overline{\mathbf{R}^{\mathrm{N}}})$. For all y_1 and y_2 and $x_i = F^{-1}(y_i)$,

$$|F^{-1}(y_2) - F^{-1}(y_1)| = |x_2 - x_1|$$
$$\leq |x_2 + f(x_2) - x_1 - f(x_1)| + |f(x_2) - f(x_1)|$$
$$\leq |y_2 - y_1| + |f(x_2) - f(x_1)|$$
$$\leq |y_2 - y_1| + c(f)|x_2 - x_1|$$
$$\leq |y_2 - y_1| + c(f)|F^{-1}(y_2) - F^{-1}(y_1)|$$
$$\Rightarrow |F^{-1}(y_2) - F^{-1}(y_1)| \leq \frac{1}{1 - c(f)} |y_2 - y_1|$$

and $c(F^{-1} - I) \leq 1 + c(F^{-1}) < \infty$. For any $y \in \mathbf{R}^{\mathrm{N}}$ and $x = F^{-1}(y)$

$$F^{-1}(y) - y = x - F(x),$$
$$\|F^{-1} - I\|_{C^0} = \sup_{y \in \mathbf{R}^{\mathrm{N}}} |F^{-1}(y) - y| = \sup_{y \in \mathbf{R}^{\mathrm{N}}} |(I - F)(F^{-1}(y))|,$$
$$= \sup_{x \in \mathbf{R}^{\mathrm{N}}} |(I - F)(x)| = \|F - I\|_{C^0} < \infty$$

since $F : \mathbf{R}^{\mathrm{N}} \to \mathbf{R}^{\mathrm{N}}$ is bijective. Therefore

$$\forall f \in \mathcal{C}^{0,1}(\overline{\mathbf{R}^{\mathrm{N}}}) \text{ such that } c(f) < 1, \quad F = I + f \in \mathcal{F}^{0,1}(\overline{\mathbf{R}^{\mathrm{N}}}).$$

Moreover, since $c(f) \leq \|f\|_{C^{0,1}}$, the condition $\|f\|_{C^{0,1}} < 1$ yields $c(f) < 1$ and $I + f \in \mathcal{F}^{0,1}(\overline{\mathbf{R}^{\mathrm{N}}})$. Furthermore, the map $f \mapsto I + f : B(0, 1/2) \subset \mathcal{C}^{0,1}(\overline{\mathbf{R}^{\mathrm{N}}}) \to \mathcal{F}^{0,1}(\overline{\mathbf{R}^{\mathrm{N}}})$ is continuous.

9.3 $\mathcal{B}^k(D, \mathbf{R}^{\mathrm{N}})$- and $C_0^k(D, \mathbf{R}^{\mathrm{N}})$-Mappings

In some applications it is interesting to consider only domains Ω that are contained in a fixed open convex subset D of $\mathbf{R}^{\mathrm{N}}$, which will be referred to as the underlying *hold-all* or *universe*. In most cases D will be a sufficiently large open ball. Given an integer $k \geq 1$, the underlying space Φ will be

$$\mathcal{C}_0^k(D) \stackrel{\mathrm{def}}{=} C_0^k(D, \mathbf{R}^{\mathrm{N}}), \quad \mathcal{B}_D^k \stackrel{\mathrm{def}}{=} \mathcal{B}^k(D, \mathbf{R}^{\mathrm{N}}), \quad \mathcal{C}^k(\overline{D}) \stackrel{\mathrm{def}}{=} C^k(\overline{D}, \mathbf{R}^{\mathrm{N}}).$$

When endowed with the norm

$$\|f\|_{C^k(D)} = \max_{|\alpha| \leq k} \|\partial^\alpha f\|_{C(D)},$$

$\mathcal{C}_0^k(D)$ and $\mathcal{B}_D^k$ are Banach spaces. Define

$$\mathcal{F}_0^k(D) \stackrel{\mathrm{def}}{=} \left\{ F : D \to D : F - I \in \mathcal{C}_0^k(D) \text{ and } F^{-1} - I \in \mathcal{C}_0^k(D) \right\},$$

$$\mathcal{F}(\mathcal{B}_D^k) \stackrel{\mathrm{def}}{=} \left\{ F : D \to D : F - I \in \mathcal{B}_D^k \text{ and } F^{-1} - I \in \mathcal{B}_D^k \right\},$$

$$\mathcal{F}^k(D) \stackrel{\mathrm{def}}{=} \left\{ F : D \to D : F - I \in \mathcal{C}^k(\overline{D}) \text{ and } F^{-1} - I \in \mathcal{C}^k(\overline{D}) \right\}.$$

For an integer $k \geq 0$, Φ can also be chosen as

$$\mathcal{C}^{k,1}(\overline{D}) \overset{\text{def}}{=} C^{k,1}(\overline{D}, \mathbf{R}^N),$$

$$\mathcal{F}^{k,1}(\overline{D}) \overset{\text{def}}{=} \left\{ F : D \to D : F - I \in \mathcal{C}^{k,1}(\overline{D}) \text{ and } F^{-1} - I \in \mathcal{C}^{k,1}(\overline{D}) \right\}.$$

When D is convex in $\mathbf{R}^N$, $\mathcal{B}^{k+1}(D, \mathbf{R}^N) \subset C^{k,1}(\overline{D}, \mathbf{R}^N) \subset \mathcal{B}^k(D, \mathbf{R}^N)$, and from the Ascoli–Arzelà Theorem 2.1 we have a sequential compactness theorem.

Theorem 9.7. *Let D be a bounded open convex subset of $\mathbf{R}^N$ and $k \geq 0$ be an integer. Given a sequence $\{F_n\}$ in $\mathcal{B}^{k+1}(D, \mathbf{R}^N)$ such that*

$$\exists c > 0, \ \forall n, \quad \|F_n\|_{C^{k+1}(D)} \leq c, \tag{9.17}$$

there exist F in $C^{k,1}(\overline{D}, \mathbf{R}^N)$ and a subsequence $\{F_{n_k}\}$ such that F_{n_k} converges to F in $\mathcal{B}^k(D, \mathbf{R}^N)$.

10　Domains in a Closed Submanifold of $\mathbf{R}^N$

The basic results of sections 8 and 9 extend to (relatively) open subsets ω_0 of a fixed closed subset Γ of $\mathbf{R}^N$ such as closed submanifolds of arbitrary codimension. For the choices of spaces Θ in the previous sections consider the closed[16] subgroup $\mathcal{G}(\Gamma)$ of $\mathcal{F}(\Theta)$ and define

$$\mathcal{X}(\omega_0) \overset{\text{def}}{=} \left\{ F(\omega_0) : \forall F \in \mathcal{G}(\Gamma) \right\},$$

which is algebraically equivalent to the quotient group

$$\mathcal{G}(\Gamma)/\mathcal{G}(\omega_0).$$

The only thing to prove is that $\mathcal{G}(\omega_0)$ is closed in the subgroup $\mathcal{G}(\Gamma)$ or $\mathcal{F}(\Theta)$ since $\mathcal{G}(\Gamma)$ is already closed in $\mathcal{F}(\Theta)$.

Lemma 10.1. *Let Γ be a nonempty closed subset of $\mathbf{R}^N$ endowed with the relative topology[17] induced by $\mathbf{R}^N$ on Γ. Assume that ω_0 is a nonempty (relatively) open subset of Γ and that Θ is one of the spaces for which we have constructed a complete Courant metric for $D = \mathbf{R}^N$. Then $\mathcal{G}(\omega_0)$ is a closed subgroup of $\mathcal{G}(\Gamma)$.*

Proof. $\mathcal{G}(\omega_0)$ is clearly a subgroup. From the proof of Lemmas 8.6 and 9.2 we can start with the following identities for the limit element F of the Cauchy sequence $\{F_n\}$:

$$\omega_0 \subset \overline{F(\omega_0)} \text{ and } F(\omega_0) \subset \overline{\omega_0}.$$

---

[16] By Lemmas 8.6 and 9.2.

[17] A subset ω of Γ is *relatively open* if there exists an open set Ω of $\mathbf{R}^N$ such that $\omega = \Omega \cap \Gamma$. The relative interior (rel. int.) of ω is the largest relatively open subset of ω.

It remains to prove that $F(\omega_0) = \omega_0$. By the definition of a relatively open subset ω_0 of Γ, there exists an open subset Ω_0 of $\mathbf{R}^N$ such that $\omega_0 = \Omega_0 \cap \Gamma$. Therefore,

$$F(\omega_0) = F(\Omega_0 \cap \Gamma) = F(\Omega_0) \cap F(\Gamma) = F(\Omega_0) \cap \Gamma.$$

Since F is a homeomorphism, $F(\Omega_0)$ is open in $\mathbf{R}^N$ and $F(\omega_0)$ is relatively open in Γ. Moreover,

$$\omega_0 \subset \overline{F(\omega_0)} \;\Rightarrow\; \omega_0 \subset \text{rel. int. } \overline{F(\omega_0)},$$
$$\text{rel. int. } \overline{F(\omega_0)} = \text{rel. int. } \overline{F(\Omega_0) \cap \Gamma} = F(\Omega_0) \cap \Gamma = F(\omega_0).$$

Similarly

$$F(\omega_0) \subset \overline{\omega_0} \;\Rightarrow\; F(\omega_0) \subset \text{rel. int. } \overline{\omega_0} = \text{rel. int. } \overline{\Omega_0 \cap \Gamma} = \Omega_0 \cap \Gamma = \omega_0$$

and $F(\omega_0) = \omega_0$. $\qquad\qquad\square$

As can be seen, the result is true for any relatively open subset of a closed subset Γ of $\mathbf{R}^N$, but the interesting cases are really when Γ is a closed sufficiently smooth submanifold of $\mathbf{R}^N$ of codimension greater than or equal to 1. This includes families of one-dimensional curves and two-dimensional surfaces. If Γ is not sufficiently smooth the subgroup $\mathcal{G}(\Gamma)$ might reduce to the identity element. Appropriate choices of Θ and Γ are essential to avoid trivial families of domains. For instance, if Γ is a closed submanifold of class C^k, $k \geq 1$, in $\mathbf{R}^N$, it is natural to choose Θ equal to $C^k(\mathbf{R}^N)$ or $C_0^k(\mathbf{R}^N)$ so that the images $F(\Gamma)$ are also of class C^k.

Chapter 3
Relaxation to Measurable Domains

1 Introduction

The constructions of the metric topologies of Chapter 2 are limited to families of
sets that are the image of a fixed domain. If it is connected, bounded, or C^k,
then the images will have the same properties under C^k-transformations of $\mathbf{R}^N$.
In this chapter we considerably enlarge the family of available sets by relaxing
the smoothness assumption to the mere Lebesgue measurability. This is done by
associating with a subset Ω its *characteristic function*

$$\chi_\Omega(x) \overset{\text{def}}{=} 1 \text{ if } x \in \Omega \text{ and } 0 \text{ if } x \notin \Omega.$$

In section 2 of this chapter the equivalence class of a measurable domain Ω is identified with the equivalence class of χ_Ω in some L^p-space for some p, $1 \le p \le \infty$.
We consider the family of such equivalence classes and its convex hull in L^p. In
section 2.1 a complete metric topology, called the *strong topology*, is induced on
that family by using as a metric the L^p-norm of the difference of two characteristic functions. In section 2.2 we consider the *weak L^p-topology* on the family of
characteristic functions. Weak limits of sequences of characteristic functions are
functions with values in $[0, 1]$ which belong to the closed convex hull. This occurs in
optimization problems where the function to be optimized depends on the solution
of a partial differential equation on the variable domain, as was shown, for instance,
by Murat [1] in 1971. It usually corresponds to the appearance of a *microstructure*
or a *composite material* in mechanics. One important example in that category was
the analysis of the optimal thickness of a circular plate by Cheng and Olhoff [2,
1] in 1981. The reader is referred to the work of Murat and Tartar [3, 1], initially
published in 1985, for a comprehensive treatment of the calculus of variations and
homogenization. Section 2.3 deals with the question of finding a *nice representative*
in the equivalence class of sets. Can it be chosen open? We introduce the *measure
theoretic representative* and characterize its interior, exterior, and boundary. It will
be used later in section 5.4 to prove the compactness of the family of Lipschitzian
domains verifying a uniform cone property. Section 2.4 shows that the family of

91

convex subsets of a fixed bounded hold-all is closed in the strong topology.

The use of the L^p-topologies is illustrated in section 3 by revisiting the optimal design problem studied by Céa and Malanowski [1] in 1970. By relaxing the family of characteristic functions to functions with values in the interval $[0,1]$ a saddle point formulation is obtained. Functions with values in the interval $[0,1]$ can also be found in the modeling of a dam by Alt [1] and Alt and Gilardi [1], where the "liquid saturation" in the soil is a function with values in $[0,1]$. Another problem amenable to that formulation is the buckling of columns, as will be illustrated in section 4 using the work of Cox and Overton [1]. It is one of the very early optimal design problems, formulated by Lagrange in 1770.

The *Caccioppoli* or *finite perimeter sets* of the celebrated Plateau problem are revisited in section 5. Their characteristic function is a function of bounded variation. They provide the first example of compact families of characteristic functions in $L^p(D)$-strong, $1 \le p < \infty$. This was mainly developed by Caccioppoli [1] and De Giorgi [1] in the context of Plateau's problem of minimal surfaces. In section 5.4 we show that the family of Lipschitzian domains in a fixed bounded hold-all verifying the uniform cone property of section 6 in Chapter 2 is also compact. This condition naturally yields a uniform bound on the perimeter of the sets in the family and hence can be viewed as a special case of the first compactness theorem. Section 6 gives an example of the use of the perimeter in the Bernoulli free boundary problem and in particular for the water wave. There the energy associated with the surface tension of the water is proportional to the perimeter, that is, the surface area of the free boundary. Section 7 illustrates the use of characteristic functions to approximate the generic Dirichlet boundary value problem by transmission problems. This technique is subsequently used to study the continuity of the initial Dirichlet boundary value problem.

2   Characteristic Functions in L^p-Topologies

2.1   Strong Topologies and C^∞-Approximations

Consider the equivalence classes $[\Omega]$ of Lebesgue measurable subsets Ω of $\mathbf{R}^N$ and the corresponding family of characteristic functions

$$\boxed{\mathrm{X}(\mathbf{R}^N) \stackrel{\text{def}}{=} \left\{ \chi_\Omega \,:\, \forall \Omega \text{ measurable in } \mathbf{R}^N \right\}.} \tag{2.1}$$

Clearly $\mathrm{X}(\mathbf{R}^N) \subset L^\infty(\mathbf{R}^N)$, and for all $p \ge 1$, $\mathrm{X}(\mathbf{R}^N) \subset L^p_{\text{loc}}(\mathbf{R}^N)$, where $L^p_{\text{loc}}(\mathbf{R}^N)$ is the set of all Lebesgue measurable functions on $\mathbf{R}^N$ which belong to $L^p(D)$ for every bounded measurable subset D of $\mathbf{R}^N$. Also associate with a nonempty measurable subset D of $\mathbf{R}^N$ the set

$$\boxed{\mathrm{X}(D) \stackrel{\text{def}}{=} \left\{ \chi_\Omega \,:\, \forall \Omega \text{ measurable in } D \right\}.} \tag{2.2}$$

This induces a complete metric space structure on $\mathrm{X}(D)$ via the bijection

$$[\Omega] \mapsto \chi_\Omega \in \mathrm{X}(D) \subset L^p(D) \,(\text{resp.}, \, \mathrm{X}(\mathbf{R}^N) \subset L^p_{\text{loc}}(\mathbf{R}^N)).$$

Theorem 2.1. *Let $1 \leq p < \infty$ be an integer.*

(i) *For a nonempty bounded measurable subset D of $\mathbf{R}^N$, $\mathrm{X}(D)$ is closed and bounded in $L^p(D)$ and*

$$\rho_D([\Omega_2],[\Omega_1]) \stackrel{\text{def}}{=} \|\chi_{\Omega_2} - \chi_{\Omega_1}\|_{L^p(D)}$$

defines a complete metric structure on the set of equivalence classes of measurable subsets of D.

(ii) *$\mathrm{X}(\mathbf{R}^N)$ is closed in $L^p_{\mathrm{loc}}(\mathbf{R}^N)$ and*

$$\rho([\Omega_2],[\Omega_1]) \stackrel{\text{def}}{=} \sum_{n=1}^{\infty} \frac{1}{2^n} \frac{\|\chi_{\Omega_2} - \chi_{\Omega_1}\|_{L^p(B(0,n))}}{1 + \|\chi_{\Omega_2} - \chi_{\Omega_1}\|_{L^p(B(0,n))}}$$

defines a complete metric structure on the set of equivalence classes of measurable subsets of $\mathbf{R}^N$, where $B(0,n)$ is the open ball of radius n in 0.

Proof. (i) Let $\{\Omega_n\}$ be a sequence of Lebesgue measurable subsets of D such that $\{\chi_{\Omega_n}\}$ is Cauchy in $L^p(D)$. It converges to some f in $L^p(D)$ and there exists a subsequence such that $\chi_{\Omega_n}(x) \to f(x)$ in D except for a subset Z of zero measure. Hence $0 = \chi_{\Omega_n}(x)\,(1 - \chi_{\Omega_n}(x)) \to f(x)\,(1 - f(x))$. Define the set

$$\Omega \stackrel{\text{def}}{=} \{x \in D\backslash Z \; : \; f(x) = 1\}.$$

Clearly Ω is measurable and $\chi_\Omega = f$ on $D\backslash Z$ since $f(x)\,(1 - f(x)) = 0$ on $D\backslash Z$. Hence $f = \chi_\Omega$ almost everywhere on D, $\chi_{\Omega_n} \to \chi_\Omega$ in $L^p(D)$ and $\chi_\Omega \in \mathrm{X}(D)$.

(ii) From part (i) using the Fréchet topology associated with the family of seminorms $q_n(f) = \|f\|_{L^p(B(0,n))}$, $n \geq 1$. $\hspace{5cm}\square$

For $p = 1$ and Ω_1 and Ω_2 in D, the metric $\rho_D([\Omega_2],[\Omega_1])$ is the measure of the symmetric difference

$$\Omega_1 \triangle \Omega_2 = (\complement_{\Omega_1}\Omega_2) \cup (\complement_{\Omega_2}\Omega_1).$$

It turns out that for $1 \leq p < \infty$ all the topologies on $\mathrm{X}(D)$ and $\mathrm{X}(\mathbf{R}^N)$ are equivalent.

Theorem 2.2. *Let D be a nonempty bounded measurable subset of $\mathbf{R}^N$. The topologies induced by $L^p(D)$ (resp., $L^p_{\mathrm{loc}}(\mathbf{R}^N)$) on $\mathrm{X}(D)$ (resp., $\mathrm{X}(\mathbf{R}^N)$) are all equivalent for $1 \leq p < \infty$.*

Proof. For D bounded, $p > 1$, $1/p + 1/q = 1$, and any χ and $\overline{\chi}$ in $\mathrm{X}(D)$

$$\|\overline{\chi} - \chi\|_{L^1(D)} = \int_D |\overline{\chi}(x) - \chi(x)|\,dx \leq \|\overline{\chi} - \chi\|_{L^p(D)}\,\mathrm{m}(D)^{1/q}$$

since $\chi(x)$ and $\overline{\chi}(x)$ are either 0 or 1 almost everywhere in D. Conversely

$$\|\overline{\chi} - \chi\|^p_{L^p(D)} = \int_D |\overline{\chi} - \chi(x)|^p\,dx = \int_D |\overline{\chi} - \chi(x)|\,dx = \|\overline{\chi} - \chi\|_{L^1(D)}.$$

Therefore, for any $\varepsilon > 0$, pick $\delta = \varepsilon^p$ and

$$\forall \, \overline{\chi} \text{ such that } \|\overline{\chi} - \chi\|_{L^1(D)} < \delta, \quad \|\overline{\chi} - \chi\|_{L^p(D)} < \varepsilon,$$

and we have the equivalence of metrics on $\mathrm{X}(D)$ for $L^1(D)$ and $L^p(D)$. $\qquad\square$

In view of this equivalence we introduce the notion of strong convergence.

Definition 2.1.

(i) If D is a bounded measurable subset of $\mathbf{R}^N$, a sequence $\{\chi_n\}$ in $\mathrm{X}(D)$ is said to be *strongly convergent* in D if it converges in $L^p(D)$-strong for some p, $1 \le p < \infty$.

(ii) A sequence $\{\chi_n\}$ in $\mathrm{X}(\mathbf{R}^N)$ is said to be *locally strongly convergent* if it converges in $L^p_{\mathrm{loc}}(\mathbf{R}^N)$-strong for some p, $1 \le p < \infty$. $\qquad\square$

The following approximation theorem will also be useful.

Theorem 2.3. *Let Ω be an arbitrary Lebesgue measurable subset of $\mathbf{R}^N$. There exists a sequence $\{\Omega_n\}$ of open C^∞-domains in $\mathbf{R}^N$ such that*

$$\chi_{\Omega_n} \to \chi_\Omega \quad in \quad L^1_{\mathrm{loc}}(\mathbf{R}^N).$$

Proof. The construction of the family of C^∞-domains $\{\Omega_n\}$ can be found in many places (cf., for instance, E. Giusti [1, section 1.14, p. 10 and Lem. 1.25, p. 23]). Associate with $\chi = \chi_\Omega$ the sequence of convolutions $f_n = \chi * \rho_n$ for a sequence $\{\rho_n\}$ of symmetric mollifiers. By construction, $0 \le f_n \le 1$. For t, $0 < t < 1$, define the sets

$$F_{nt} \stackrel{\mathrm{def}}{=} \left\{ x \in \mathbf{R}^N : f_n(x) > t \right\}.$$

By definition

$$f_n - \chi > t \text{ in } F_{nt} \backslash \Omega, \quad \chi - f_n > 1 - t \text{ in } \Omega \backslash F_{nt},$$

and

$$|\chi - \chi_{F_{nt}}| \le \frac{1}{\min\{t, 1 - t\}} |\chi - f_n| \text{ a.e. in } \mathbf{R}^N.$$

By Sard's theorem (Theorem 4.3 of Chapter 2), the set F_{nt} is a C^∞-domain for almost all t in $(0, 1)$. Fix α, $0 < \alpha < 1/2$, and choose a sequence $\{t_n\}$ such that, for all n, $\alpha \le t_n \le 1 - \alpha$, and define

$$\Omega_n \stackrel{\mathrm{def}}{=} F_{nt_n}.$$

Therefore

$$|\chi_\Omega - \chi_{\Omega_n}| \le \frac{1}{\min\{t_n, 1 - t_n\}} |\chi_\Omega - f_n| \le \frac{1}{\alpha} |\chi_\Omega - f_n| \text{ a.e. in } \mathbf{R}^N$$

and for any bounded measurable subset D of $\mathbf{R}^N$

$$\|\chi_\Omega - \chi_{\Omega_n}\|_{L^1(D)} \le \alpha^{-1} \|\chi_\Omega - f_n\|_{L^1(D)},$$

which goes to zero as n goes to infinity. $\qquad\square$

2.2 Weak Topologies and Microstructures

Some *shape optimization* problems lead to apparent paradoxes. Their solution is no longer a geometric domain associated with a characteristic function, but a *fuzzy domain* associated with the relaxation of a characteristic function to a function with values ranging in $[0, 1]$. The intuitive notion of a geometric domain is relaxed to the notion of a *probability distribution* of the presence of points of the set. When the underlying problem involves two different materials characterized by two constants, $k_1 \neq k_2$, the occurrence of such a solution can be interpreted as the *mixing* or *homogenization* of the two materials at the microscale. This is also referred to as a *composite material* or a *microstructure*. Somehow this is related to the fact that the strong convergence needs to be relaxed to the weak L^p-convergence and the space $X(D)$ needs to be suitably enlarged.

Even if $X(D)$ is strongly closed and bounded in $L^p(D)$, it is not strongly compact. However, for $1 < p < \infty$ its closed convex hull $\overline{co}\, X(D)$ is weakly compact in the reflexive Banach space $L^p(D)$. In fact,

$$\boxed{\overline{co}\, X(D) = \{\chi \in L^p(D) : \chi(x) \in [0, 1] \text{ a.e. in } D\}.} \tag{2.3}$$

Indeed, by definition $\overline{co}\, X(D) \subset \{\chi \in L^p(D) : \chi(x) \in [0, 1] \text{ a.e. in } D\}$. Conversely any χ which belongs to the right-hand side of (2.3) can be approximated by a sequence of convex combinations of elements of $X(D)$. Choose

$$\chi_n = \sum_{m=1}^{n} \frac{1}{n} \chi_{B_{nm}}, \quad B_{nm} = \left\{x : \chi(x) \geq \frac{m}{n}\right\}$$

for which $|\chi_n(x) - \chi(x)| < 1/n$. The elements of $\overline{co}\, X(D)$ are not necessarily characteristic functions of a domain; that is, the identity

$$\chi(x)\,(1 - \chi(x)) = 0 \text{ a.e. in } D$$

is not necessarily satisfied.

We first give a few basic results and then consider a classical example from the *theory of homogenization* of differential equations.

Lemma 2.1. *Let D be a bounded open subset of $\mathbf{R}^N$, K a bounded subset of $\mathbf{R}$, and*

$$\mathcal{K} \stackrel{\text{def}}{=} \{k \colon D \to \mathbf{R} : k \text{ is measurable and } k(x) \in K \text{ a.e. in } D\}.$$

(i) *For any p, $1 \leq p < \infty$, and any sequence $\{k_n\} \subset \mathcal{K}$ the following statements are equivalent:*

    (a) *$\{k_n\}$ converges in $L^\infty(D)$-weak$\star$,*

    (b) *$\{k_n\}$ converges in $L^p(D)$-weak,*

    (c) *$\{k_n\}$ converges in $\mathcal{D}(D)'$,*

    where $\mathcal{D}(D)'$ is the space of scalar distributions on D.

(ii) *If K is bounded, closed, and convex, then $\mathcal{K}$ is convex and compact in $L^\infty(D)$-weak$\star$, $L^p(D)$-weak, and $\mathcal{D}(D)'$.*

The above results remain true in the vectorial case when $\mathcal{K}$ is the set of mappings $k\colon D \to K$ for some bounded subset $K \subset \mathbf{R}^p$ and a finite integer $p \geq 1$.

Proof. (i) It is clear that (a) $\Rightarrow$ (b) $\Rightarrow$ (c). To prove that (c) $\Rightarrow$ (a) recall that since K is bounded, there exists a constant $c > 0$ such that $K \subset cB$, where B denotes the unit ball in $\mathbf{R}^N$. By density of $\mathcal{D}(D)$ in $L^1(D)$, any φ in $L^1(D)$ can be approximated by a sequence $\{\varphi_m\} \subset \mathcal{D}(D)$ such that $\varphi_m \to \varphi$ in $L^1(D)$. So for each $\varepsilon > 0$ there exists $M > 0$ such that

$$\forall m \geq M, \quad \|\varphi_m - \varphi\|_{L^1(D)} \leq \frac{\varepsilon}{4c}.$$

Moreover, there exists $N > 0$ such that

$$\forall n \geq N, \forall \ell \geq N, \quad \left| \int_D \varphi_M(k_n - k_\ell)\,dx \right| \leq \frac{\varepsilon}{2}.$$

Hence for each $\varepsilon > 0$, there exists N such that for all $n \geq N$ and $\ell \geq N$

$$\left| \int_D \varphi(k_n - k_\ell)\,dx \right| \leq \left| \int_D \varphi_M(k_n - k_\ell)\,dx \right| + \left| \int_D (\varphi - \varphi_M)(k_n - k_\ell)\,dx \right|$$
$$\leq \frac{\varepsilon}{2} + 2c\|\varphi - \varphi_M\|_{L^1(D)} \leq \varepsilon.$$

(ii) When K is bounded, closed, and convex, $\mathcal{K}$ is also bounded, closed, and convex. Since $L^2(D)$ is a Hilbert space, $\mathcal{K}$ is weakly compact. In view of the equivalences of part (i), $\mathcal{K}$ is also sequentially compact in all the other weak topologies.[1] $\qquad \square$

In view of this equivalence we adopt the following terminology.

Definition 2.2.
Let D be a bounded open subset of $\mathbf{R}^N$. A sequence $\{\chi_n\}$ in $X(D)$ is said to be *weakly convergent* if it converges for some topology between $L^\infty(D)$-weak$\star$ and $\mathcal{D}(D)'$. $\qquad \square$

It is interesting to observe that working with the weak convergence only makes sense when the limit element is not a characteristic function.

Theorem 2.4. *Let the assumptions of the theorem be satisfied. Let $\{\chi_n\}$ and χ be elements of $X(\mathbf{R}^N)$ (resp., $X(D)$) such that $\chi_n \rightharpoonup \chi$ weakly in $L^2(D)$ (resp., $L^2_{\mathrm{loc}}(\mathbf{R}^N)$). Then for all p, $1 \leq p < \infty$,*

$$\chi_n \to \chi \text{ strongly in } L^p(D)\big(resp.,\ L^p_{\mathrm{loc}}(\mathbf{R}^N)\big).$$

---

[1]In a metric space the compactness is equivalent to the sequential compactness. For the weak topology we use the fact that if E is a separable normed space, then, in its topological dual E', any closed ball is a compact metrizable space for the weak topology. Since $\mathcal{K}$ is a bounded subset of the normed reflexive separable Banach space $L^p(D)$, $1 \leq p < \infty$, the weak compactness of $\mathcal{K}$ coincides with its weak sequential compactness (cf. Dieudonne [1, Chap. XII, section 12.15.9, p. 75]).

Proof. It is sufficient to prove the result for D. The strong $L^2(D)$-convergence follows from the property

$$\int_D |\chi_n|^2 \, dx = \int_D \chi_n 1 \, dx \to \int_D \chi 1 \, dx = \int_D |\chi|^2 \, dx$$

$$\Rightarrow \int_D |\chi_n - \chi|^2 dx = \int_D |\chi_n|^2 - 2\chi\chi_n + |\chi|^2 dx \to \int_D |\chi|^2 - 2\chi\chi + |\chi|^2 dx = 0.$$

The convergence for any $p \geq 1$ now follows from Theorem 2.1. $\qquad\qquad\square$

Working with the weak convergence creates new phenomena and difficulties. For instance, when a characteristic function is present in the coefficient of the higher order term of a differential equation the weak convergence of a sequence of characteristic functions $\{\chi_n\}$ to some element χ of $\overline{co}\, X(D)$,

$$\chi_n \rightharpoonup \chi \text{ in } L^2(D)\text{-weak} \quad \Rightarrow \chi_n \rightharpoonup \chi \text{ in } L^\infty(D)\text{-weak}\star,$$

does not imply the weak convergence in $H^1(D)$ of the sequence $\{y(\chi_n)\}$ of solutions to the solution of the differential equation corresponding to $y(\chi)$,

$$y(\chi_n) \not\rightharpoonup y(\chi) \text{ in } H^1(D)\text{-weak.}$$

By compactness[2] of the injection of $H^1(D)$ into $L^2(D)$ this would have implied convergence in $L^2(D)$-strong:

$$y(\chi_n) \to y(\chi) \text{ in } L^2(D)\text{-strong.}$$

This fact was pointed out in 1971 by Murat [1] in the following example, which will be rewritten to emphasize the role of the characteristic function.

Example 2.1.

$$\begin{cases} -\dfrac{d}{dx}\left(k\dfrac{dy}{dx}\right) + ky = 0 \text{ in } D =\,]0,1[, \\ y(0) = 1 \text{ and } y(1) = 2, \end{cases} \tag{2.4}$$

where

$$\mathcal{K} = \{k = k_1(x)\chi + k_2(x)(1 - \chi) : \chi \in X(D)\} \tag{2.5}$$

with

$$k_1(x) = 1 - \sqrt{\frac{1}{2} - \frac{x^2}{6}}, \qquad k_2(x) = 1 + \sqrt{\frac{1}{2} - \frac{x^2}{6}}. \tag{2.6}$$

This is equivalent to

$$\mathcal{K} = \{k \in L^\infty(0,1) : k(x) \in \{k_1(x), k_2(x)\} \quad \text{a.e. in } [0,1]\}. \tag{2.7}$$

---

[2]This is true since D is a smooth bounded domain. Examples of domains between two spirals can be constructed where the injection is not compact.

Associate with each integer $p \geq 1$ the following functions k^p:

$$k^p(x) = \begin{cases} 1 - \sqrt{\dfrac{1}{2} - \dfrac{x^2}{6}}, & \dfrac{m}{p} < x \leq \dfrac{2m+1}{2p}, \\[2ex] 1 + \sqrt{\dfrac{1}{2} - \dfrac{x^2}{6}}, & \dfrac{2m+1}{2p} < x \leq \dfrac{m+1}{p}, \end{cases} \qquad 0 \leq m \leq p-1, \qquad (2.8)$$

and the corresponding function χ_p $(k^p = k_1 \, \chi_p + k_2 \, (1 - \chi_p))$

$$\chi_p(x) = \begin{cases} 1, & \dfrac{m}{p} < x \leq \dfrac{2m+1}{2p}, \\[2ex] 0, & \dfrac{2m+1}{2p} < x \leq \dfrac{m+1}{p}, \end{cases} \qquad 0 \leq m \leq p-1. \qquad (2.9)$$

He shows that for each p, $k^p \in \mathcal{K}$ (resp., $\chi_p \in X(D)$) and that

$$k^p \rightharpoonup k_\infty = 1 \quad \left(\text{resp., } \chi_p \rightharpoonup \frac{1}{2}\right) \text{ in } L^\infty(0,1)\text{-weak}\star, \qquad (2.10)$$

$$\frac{1}{k^p} \rightharpoonup \frac{1}{2}\left[\frac{1}{k_1} + \frac{1}{k_2}\right] = \frac{1}{1/2 + x^2/6} \text{ in } L^\infty(0,1)\text{-weak}\star. \qquad (2.11)$$

Moreover,

$$y_p \rightharpoonup y \text{ in } H^1(0,1)\text{-weak}, \qquad (2.12)$$

where y_p denotes the solution of (2.4) corresponding to $k = k^p$ and y the solution of the boundary value problem

$$\begin{cases} -\dfrac{d}{dx}\left[\left(\dfrac{1}{2} + \dfrac{x^2}{6}\right)\dfrac{dy}{dx}\right] + y = 0 \text{ in }]0,1[, \\[2ex] y(0) = 1, \quad y(1) = 2. \end{cases} \qquad (2.13)$$

Define the function

$$k_H(x) = \frac{1}{2} + \frac{x^2}{6}, \qquad (2.14)$$

which corresponds to

$$\chi_H(x) = \frac{1}{2}\left[1 + \sqrt{\frac{1}{2} - \frac{x^2}{6}}\right] \in \overline{\text{co}}\, X(D). \qquad (2.15)$$

Notice that k_H appears in the second-order term and k_∞ in the zeroth-order term in equation (2.13):

$$-\frac{d}{dx}\left[k_H \frac{dy}{dx}\right] + k_\infty y = 0 \text{ in }]0,1[. \qquad (2.16)$$

It is easy to check that

$$y(x) = 1 + x^2 \text{ in } [0,1], \tag{2.17}$$

which is not equal to the solution y_∞ of (2.4) for the weak limit $k_\infty = 1$:

$$y_\infty(x) = \frac{2(e^x - e^{-x}) + e^{1-x} - e^{-(1-x)}}{e - e^{-1}}. \tag{2.18}$$

$\square$

To our knowledge this was the beginning of the theory of homogenization in France. This is only one part of Murat's [1] example. He also constructs an objective function for which the lower bound is not achieved by an element of $\mathcal{K}$. This is a nonexistence result.

The above example uses space varying coefficients $k_1(x)$ and $k_2(x)$. However, it is still valid for two positive constants $k_1 > 0$ and $k_2 > 0$. In that case it is easy to show that

$$k_\infty = \frac{k_1 + k_2}{2}, \qquad\qquad \frac{1}{k_H} = \frac{1}{2}\left[\frac{1}{k_1} + \frac{1}{k_2}\right],$$

$$\chi_\infty = \frac{1}{2}, \qquad\qquad \chi_H = \frac{k_2}{k_1 + k_2},$$

and the solution y of the boundary value problem (2.16) is given by

$$y(x) = \frac{2\sinh cx + \sinh c(1-x)}{\sinh c},$$

where

$$c = \frac{k_1 + k_2}{2\sqrt{k_1 k_2}} \geq 1,$$

and the solution y_∞, by

$$y_\infty(x) = \frac{2\sinh x + \sinh(1-x)}{\sinh 1}.$$

Thus for $k_1 \neq k_2$, or equivalently $c > 1$, $y \neq y_\infty$.

In fact, using the same sequence of χ_p's, the sequence $y_p = y(\chi_p)$ of solutions of (3.5) weakly converges to $y_H = y(\chi_H)$, which is different from the solution $y_\infty = y(\chi_\infty)$ for $k_1 \neq k_2$. This is readily seen by noticing that since χ_∞ and χ_H are constant

$$-\Delta(k_H \, y_H) = \chi_\infty \, f = -\Delta(k_\infty \, y_\infty)$$

$$\Rightarrow k_H \, y_H = k_\infty \, y_\infty \quad \Rightarrow y_H = \frac{(k_1 + k_2)^2}{4 k_1 \, k_2} \, y_\infty \neq y_\infty, \text{ for } k_1 \neq k_2.$$

Despite this evidence to the contrary we shall see in the next section that in some cases we obtain the existence (and uniqueness) of a maximizer in $\overline{co}\, X(D)$ which belongs to $X(D)$.

2.3 Nice or Measure Theoretic Representative

Since the strong and weak L^p-topologies are defined on equivalence classes $[\Omega]$ of (Lebesgue) measurable subsets Ω of $\mathbf{R}^N$, it is natural to ask if there is a *nice representative* that is generic of the class $[\Omega]$. For instance, we have seen in Chapter 2 that within the equivalence class of a set of class C^k there is a unique open and a unique closed representative and that all elements of the class have the same interior, boundary, and exterior. The same question will again arise for finite perimeter sets. As an illustration of what is meant by a nice representative, consider the smiling and the expressionless suns in Figure 3.1. The expressionless sun is obtained by adding missing points and lines "inside" Ω and removing the rays "outside" Ω. This "restoring/cleaning" operation can be formalized as follows.

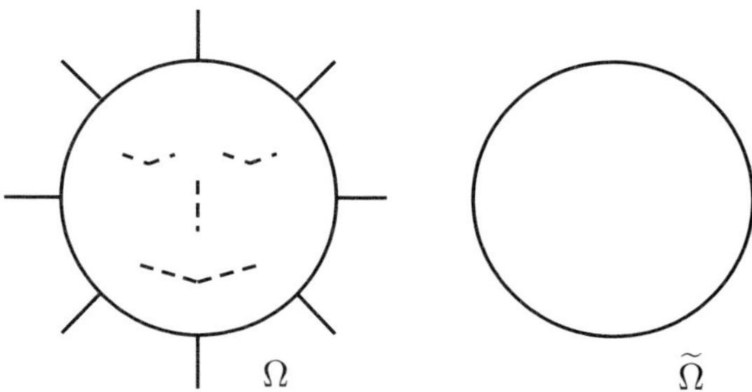

Figure 3.1. *Smiling sun Ω and expressionless sun $\widetilde{\Omega}$.*

Definition 2.3.
Associate with each Lebesgue measurable set Ω in $\mathbf{R}^N$, the sets

$$\Omega_0 = \left\{ x \in \mathbf{R}^N : \exists \rho > 0 \text{ such that } \mathrm{m}(\Omega \cap B(x,\rho)) = 0 \right\},$$
$$\Omega_1 = \left\{ x \in \mathbf{R}^N : \exists \rho > 0 \text{ such that } \mathrm{m}(\Omega \cap B(x,\rho)) = \mathrm{m}(B(x,\rho)) \right\},$$
$$\Omega_\bullet = \left\{ x \in \mathbf{R}^N : \forall \rho > 0 \text{ such that } 0 < \mathrm{m}(\Omega \cap B(x,\rho)) < \mathrm{m}(B(x,\rho)) \right\}$$

and the *measure theoretic exterior* O, *interior* I, and *boundary* $\partial_* \Omega$,

$$\mathrm{O} = \left\{ x \in \mathbf{R}^N : \lim_{r \searrow 0} \frac{\mathrm{m}(B(x,r) \cap \Omega)}{\mathrm{m}(B(x,r))} = 0 \right\},$$
$$\mathrm{I} = \left\{ x \in \mathbf{R}^N : \lim_{r \searrow 0} \frac{\mathrm{m}(B(x,r) \cap \Omega)}{\mathrm{m}(B(x,r))} = 1 \right\},$$
$$\partial_* \Omega = \left\{ x \in \mathbf{R}^N : \begin{array}{l} \liminf_{r \searrow 0} \frac{\mathrm{m}(B(x,r) \cap \Omega)}{\mathrm{m}(B(x,r))} < 1 \\ \limsup_{r \searrow 0} \frac{\mathrm{m}(B(x,r) \cap \Omega)}{\mathrm{m}(B(x,r))} > 0 \end{array} \right\}.$$

We shall say that I is the *nice* or *measure theoretic representative* of Ω.   □

The six sets Ω_0, Ω_1, $\Omega_\bullet$, O, I, and $\partial_*\Omega$ are *invariant* for all sets in the equivalence class $[\Omega]$ of Ω. They define two different partitions of $\mathbf{R}^N$,

$$\Omega_0 \cup \Omega_1 \cup \Omega_\bullet = \mathbf{R}^N \quad \text{and} \quad O \cup I \cup \partial_*\Omega = \mathbf{R}^N, \tag{2.19}$$

much like int $\complement\Omega$, int Ω, and $\partial\Omega$. We shall see from (2.20) in Theorem 2.6 that

$$\partial I = \Omega_\bullet \supset \partial_*\Omega, \quad \text{int } I = \Omega_1, \quad \text{int } \complement I = \Omega_0.$$

In general $\partial I \neq \partial_*\Omega$.

From the Lebesgue–Besicovitch differentiation theorem

$$\lim_{r\searrow 0} \frac{\mathrm{m}(B(x,r) \cap \Omega)}{\mathrm{m}(B(x,r))} = \chi_\Omega(x) \text{ almost everywhere;}$$

that is, for almost all x the limit exists and is equal to $\chi_\Omega(x)$ (cf. Evans and Gariepy [1, sections 1.7.1 and 5.8]). We have the following theorem.

Theorem 2.5. *Given a measurable subset Ω of $\mathbf{R}^N$, the sets* I, O, *and* $\partial_*\Omega$ *are Borel measurable,*

$$\chi_I = \chi_\Omega, \quad \chi_{\complement I} = \chi_O = \chi_{\complement\Omega}, \quad \chi_{\partial_*\Omega} = 0 \quad a.e.$$

$[I] = [\complement O] = [\Omega]$ *and* $[O] = [\complement I] = [\complement\Omega]$.

The next theorem links the six invariant sets and describe some of the interesting properties of the measure theoretic representative.

Theorem 2.6. *Let Ω be a Lebesgue measurable set in $\mathbf{R}^N$ and let Ω_0, Ω_1, $\Omega_\bullet$, O, I, and $\partial_*\Omega$ be the sets constructed from Ω in Definition 2.3.*

(i) *By definition, Ω_0, Ω_1, $\Omega_\bullet$, O, I, and $\partial_*\Omega$ are invariant for all members of the equivalence class $[\Omega]$ of Ω and*

$$\Omega_1 \subset I, \quad \Omega_0 \subset O, \quad \Omega_\bullet \supset \partial_*\Omega.$$

(ii) *Ω_0 and Ω_1 are open, and $\Omega_\bullet$ is closed.*

(iii) *The following identities hold:*

$$\partial I = \Omega_\bullet = \partial O, \quad \text{int } I = \Omega_1 = \text{int } \complement O, \quad \text{int } \complement I = \Omega_0 = \text{int } O, \tag{2.20}$$

$$\left(\overline{\Omega}_0 \cap \overline{\Omega}_1\right) \cup \partial_*\Omega \subset \Omega_\bullet, \tag{2.21}$$

$$\text{int } \Omega \subset \text{int } I, \quad \text{int } \complement\Omega \subset \text{int } \complement I, \quad \partial I \subset \partial\Omega. \tag{2.22}$$

(iv) *The condition $\mathrm{m}(\partial\Omega) = 0$ implies $\mathrm{m}(\partial I) = 0$. When $\mathrm{m}(\partial I) = 0$, then int I and $\overline{I}$ can be chosen as the respective open and closed representatives of Ω. In particular, this is true for Lipschitzian sets and hence for convex sets.*

Proof. (i) All the sets involved depend only on χ_Ω almost everywhere.

(ii) We use the proof given by Giusti [1, Prop. 4.1, pp. 42–43]. To show that Ω_0 is open, pick any point x in Ω_0. By construction there exists $\rho > 0$ such that $m\big(\Omega \cap B(x,\rho)\big) = 0$. For any $y \in B(x,\rho)$ (that is, $|y - x| < \rho$), choose $\rho_0 = \rho - |x-y| > 0$. Then $B(y,\rho_0) \subset B(x,\rho)$ and $m\big(B(y,\rho_0)\cap\Omega\big) \le m\big(B(x,\rho)\cap\Omega\big) = 0$. So $B(x,\rho) \subset \Omega_0$ and Ω_0 is open. Similarly, by repeating the argument for $\complement\Omega$, we obtain that Ω_1 is open. By complementarity, $\Omega_\bullet$ is closed.

(iii) *Identity* (2.20). For all $x \in \operatorname{int} I$, there exists $r > 0$ such that $B(x,r) \subset I$ and

$$\mathrm{m}\big(B(x,r)\big) = \mathrm{m}\big(B(x,r) \cap I\big) = \mathrm{m}\big(B(x,r) \cap \Omega\big).$$

Hence $\operatorname{int} I \subset \Omega_1$. But by the definition of I, $I \supset \Omega_1$ and necessarily $\operatorname{int} I = \Omega_1$. Similarly for all $x \in \operatorname{int} O$, there exists $r > 0$ such that $B(x,r) \subset O$ and

$$\mathrm{m}\big(B(x,r)\big) = \mathrm{m}\big(B(x,r) \cap O\big) = \mathrm{m}\big(B(x,r) \cap \complement\Omega\big) \tag{2.23}$$

$$\Rightarrow 0 = \mathrm{m}\big(B(x,r) \cap \Omega\big). \tag{2.24}$$

Hence $\operatorname{int} O \subset \Omega_0$. Again by the definition of O, $O \supset \Omega_0$ and necessarily $\operatorname{int} O = \Omega_0$. The other two identities, $\operatorname{int}\complement I = \Omega_0$ and $\operatorname{int}\complement O = \Omega_1$, are obtained by complementarity using the fact that $\mathrm{m}(\partial_*\Omega) = 0$ and hence $[\complement O] = [I]$ and $[\complement I] = [O]$. Since $(\operatorname{int} I, \operatorname{int}\complement I, \partial I)$, $(\operatorname{int} O, \operatorname{int}\complement O, \partial O)$, and $(\Omega_0, \Omega_1, \Omega_\bullet)$ are three partitions of $\mathbf{R}^N$, then $\partial O = \Omega_\bullet = \partial I$.

Next $I \supset \operatorname{int} I = \Omega_1$, $O \supset \operatorname{int} O = \operatorname{int}\complement I = \Omega_0$, and hence $\partial_*\Omega \subset \Omega_\bullet$ and $\partial I \supset \overline{\Omega}_0 \cap \overline{\Omega}_1$.

Identity (2.22). Finally, by the definition of Ω_0 and Ω_1,

$$\operatorname{int}\Omega \subset \Omega_1, \quad \operatorname{int}\complement\Omega \subset \Omega_0, \quad \partial\Omega \supset \Omega_\bullet$$

and the identities (2.22) follow from the identities (2.20). $\qquad\square$

The inclusions (2.22) indicate that the interior and exterior are enlarged and that the boundary is reduced. In general this operation does not commute with the set theoretic operations. For instance, it does not commute with the closure, as can be seen from the following simple example.

Example 2.2.
Consider the set Ω of all rational numbers in $[0,1]$. Then

$$\left.\begin{aligned} O &= \Omega_0 = \mathbf{R} \\ I &= \Omega_1 = \varnothing \\ \partial_*\Omega &= \Omega_\bullet = \varnothing \end{aligned}\right\} \Rightarrow I(\Omega) = \varnothing,$$

$$\left.\begin{aligned} O(\overline{\Omega}) &= (\overline{\Omega})_0 = \mathbf{R} \setminus [0,1] \\ I(\overline{\Omega}) &= (\overline{\Omega})_1 = \,]0,1[\\ \partial_*(\overline{\Omega}) &= (\overline{\Omega})_\bullet = \{0,1\} \end{aligned}\right\} \Rightarrow I(\overline{\Omega}) = \,]0,1[. \qquad\square$$

However, the complement operation commutes and determines the same sets

$$(\complement\Omega)_0 = \Omega_1, \quad (\complement\Omega)_1 = \Omega_0, \quad (\complement\Omega)_\bullet = \Omega_\bullet, \tag{2.25}$$

$$\mathrm{I}(\complement\Omega) = \mathrm{O}(\Omega), \quad \mathrm{I}(\Omega) = \mathrm{O}(\complement\Omega), \quad \partial_*(\Omega) = \partial_*(\complement\Omega). \tag{2.26}$$

The next theorem is a companion to Lemma 5.2 in Chapter 2.

Theorem 2.7. *The following conditions are equivalent:*

(i) $\overline{\mathrm{int}\,\Omega} = \overline{\Omega}$ *and* $\overline{\mathrm{int}\,\complement\Omega} = \overline{\complement\Omega}$,

(ii) $\mathrm{int}\,\Omega = \Omega_1, \quad \mathrm{int}\,\complement\Omega = \Omega_0, \quad \partial\Omega = \partial\Omega_0 = \partial\Omega_1$.

If any one of the above two conditions are satisfied, then

$$\overline{\mathrm{int}\,\Omega} = \overline{\mathrm{int}\,\mathrm{I}}, \quad \overline{\mathrm{int}\,\complement\Omega} = \overline{\mathrm{int}\,\complement\mathrm{I}}, \quad \partial\Omega = \partial\mathrm{I} = \Omega_\bullet = \partial\Omega_0 = \partial\Omega_1.$$

Proof. (i) $\Rightarrow$ (ii). From the first identity (2.22), $\overline{\Omega} = \overline{\mathrm{int}\,\Omega} \subset \overline{\Omega_1} \subset \Omega_1 \cup \Omega_\bullet$. By complementarity, the second identity (2.22) and the third identity (2.20), $\Omega_0 \subset \complement\overline{\Omega} = \overline{\mathrm{int}\,\complement\Omega} \subset \Omega_0$ and $\mathrm{int}\,\complement\Omega = \Omega_0$. By the same technique $\mathrm{int}\,\Omega = \Omega_1$ and necessarily $\partial\Omega = \Omega_\bullet$. Now from Lemma 5.2 in Chapter 2, $\partial\Omega_1 = \partial\mathrm{int}\,\Omega = \partial\Omega$ and $\partial\Omega = \partial\mathrm{int}\,\complement\Omega = \partial\Omega_0$.

(ii) $\Rightarrow$ (i). By assumption,

$$\overline{\Omega} = \mathrm{int}\,\Omega \cup \partial\Omega \subset \Omega_1 \cup \left(\overline{\Omega_1} \cap \overline{\Omega_0}\right) \subset \overline{\Omega_1} = \overline{\mathrm{int}\,\Omega},$$

$$\overline{\complement\Omega} = \mathrm{int}\,\complement\Omega \cup \partial\Omega \subset \Omega_0 \cup \left(\overline{\Omega_1} \cap \overline{\Omega_0}\right) \subset \overline{\Omega_0} = \overline{\mathrm{int}\,\complement\Omega}$$

$$\Rightarrow \partial\Omega \subset \overline{\Omega_1} \cap \overline{\Omega_0} = \Omega_\bullet \subset \partial\Omega \quad \Rightarrow \partial\Omega = \overline{\Omega_1} \cap \overline{\Omega_0} = \Omega_\bullet = \partial\mathrm{I}$$

from identities (2.20) and (2.22). $\qquad\square$

2.4 The Family of Convex Sets

Convex sets will play a special role here and in the subsequent chapters. We shall say that an equivalence class $[\Omega]$ of Lebesgue measurable subsets of D is *convex* if there exists a convex Lebesgue measurable subset Ω^* of D such that $[\Omega] = [\Omega^*]$. We also introduce the notation

$$\boxed{\mathcal{C}(D) \stackrel{\mathrm{def}}{=} \{\chi_\Omega : \Omega \text{ convex subset of } D\}.}$$

Theorem 2.8.

(i) *If Ω is convex, then $\overline{\Omega}$ and $\mathrm{int}\,\Omega$ are convex.*[3]

(ii) *If $\Omega \neq \mathbf{R}^N$ is convex and $\mathrm{int}\,\Omega \neq \varnothing$, then $\partial\Omega \neq \varnothing$, $\mathrm{int}\,\complement\Omega \neq \varnothing$,*

$$\overline{\mathrm{int}\,\Omega} = \overline{\Omega} \text{ and } \overline{\mathrm{int}\,\complement\Omega} = \overline{\complement\Omega}.$$

---

[3]By convention $\varnothing$ is convex.

(iii) *Given a measurable (resp., bounded measurable) subset D in $\mathbf{R}^N$*

$$\forall 1 \le p < \infty, \quad \mathcal{C}(D) \text{ is closed in } L^p_{\mathrm{loc}}(D) \ (resp., L^p(D)).$$

Proof. (i) For all x and y in $\overline{\Omega}$, there exist $\{x_n\}$ and $\{y_n\}$ in Ω such that $x_n \to x$ and $y_n \to y$. Then for all $\lambda \in [0,1]$

$$\Omega \ni \lambda\, x_n + (1 - \lambda)\, y_n \to \lambda\, x + (1 - \lambda)\, y \in \overline{\Omega}.$$

If $\operatorname{int} \Omega \ne \varnothing$, then for all x and y in $\operatorname{int}\Omega$, there exist $r_x > 0$ and $r_y > 0$ such that $B(x, r_x) \subset \Omega$ and $B(y, r_y) \subset \Omega$. So for all $\lambda \in [0,1]$

$$x_\lambda \stackrel{\mathrm{def}}{=} \lambda x + (1 - \lambda)y \in B(x_\lambda, \lambda r_x + (1 - \lambda)r_y)$$
$$\subset \lambda B(x, r_x) + (1 - \lambda)B(y, r_y) \subset \Omega$$

and $x_\lambda \in \operatorname{int}\Omega$.

(ii) From Theorem 6.2 in Chapter 2, Ω is locally Lipschitzian, and from Theorem 5.4 in Chapter 2, $\partial\Omega \ne \varnothing$, $\operatorname{int}\complement\Omega \ne \varnothing$, and the two identities follow.

(iii) It is sufficient to prove the result for D bounded. For any Cauchy sequence $\{\chi_n\}$ in $\mathcal{C}(D)$, there exists a sequence of convex sets $\{\Omega_n\}$ and χ_Ω in $X(D)$ such that $\chi_{\Omega_n} \to \chi_\Omega$ in $L^p(D)$. In particular there exists a subsequence $\chi_k = \chi_{\Omega_{n_k}}$ such that

$$\chi_k(x) \to \chi_\Omega(x) \quad \text{a.e. in } D.$$

Define

$$\Omega^* \stackrel{\mathrm{def}}{=} \{x \in D : \chi_k(x) \to 1 \text{ as } k \to \infty\}.$$

To show that Ω^* is convex consider x and y in Ω^*. There exists $K \ge 1$ such that

$$\forall k \ge K, \quad |\chi_k(x) - 1| < 1/2, \quad |\chi_k(y) - 1| < 1/2,$$

and since χ_k is either 0 or 1

$$\forall k \ge K, \quad \chi_k(x) = 1 = \chi_k(y).$$

By convexity for all $\lambda \in [0,1]$, $x_\lambda = \lambda x + (1 - \lambda)y \in \Omega_{n_k}$ and

$$\forall k \ge K, \quad \chi_k(\lambda x + (1 - \lambda)y) = 1 \to 1 \quad \Rightarrow \quad x_\lambda \in \Omega^*.$$

But, almost everywhere, $\chi_k(x) \to \chi_\Omega(x)$ and

$$\lim_{k \to \infty} \chi_k(x) \in \{0,1\}.$$

So for $x \in D \backslash \Omega^*$

$$\chi_k(x) \to 0 = \chi_{\Omega^*}(x) \text{ a.e. in } D.$$

By the Lebesgue-dominated convergence theorem $\chi_k \to \chi_{\Omega^*}$ in $L^p(D)$ and necessarily $\chi_\Omega = \chi_{\Omega^*}$. This means that $[\Omega]$ is a convex equivalence class. $\qquad\square$

We shall show in section 5.1 (Corollary 1) that for bounded domains D, $\mathcal{C}(D)$ is also compact for the $L^p(D)$ topology, $1 \le p < \infty$.

2.5 Sobolev Spaces for Measurable Domains

The lack of a priori smoothness on Ω may introduce technical difficulties in the formulation of some boundary value problems. However, it is possible to relax such boundary value problems from smooth bounded open connected domains Ω to measurable domains (cf. Zolésio [7]). For instance, consider the homogeneous Dirichlet boundary value problem

$$-\triangle y = f \text{ in } \Omega, \quad y = 0 \text{ on } \Gamma \tag{2.27}$$

over a bounded open connected domain Ω with a boundary Γ of class C^1 and associate with its solution $y = y(\Omega)$ the objective function and volume constraint

$$J(\Omega) = \frac{1}{2} \int_\Omega |y - g|^2 \, dx, \quad \int_\Omega dx = \pi. \tag{2.28}$$

There is a priori no reason to assume that an optimal (minimizing) domain Ω^* is of class C^1 or is connected. So the problem must be suitably relaxed to a large enough class of domains, which preserves the meaning of the underlying function spaces, the well-posedness of the original problem, and the volume.

To extend problem (2.27)–(2.28) to Lebesgue measurable sets, we first have to make sense of the Sobolev space for measurable subsets Ω of D.

Theorem 2.9. *Let D be an open domain in $\mathbf{R}^N$. For any Lebesgue measurable subset Ω of $\mathbf{R}^N$, the spaces*

$$H_\bullet^1(\Omega; D) \stackrel{\text{def}}{=} \left\{\varphi \in H_0^1(D) : (1 - \chi_\Omega)\nabla\varphi = 0 \text{ a.e. in } D\right\}, \tag{2.29}$$

$$H_\diamond^1(\Omega; D) \stackrel{\text{def}}{=} \left\{\varphi \in H_0^1(D) : (1 - \chi_\Omega)\varphi = 0 \text{ a.e. in } D\right\} \tag{2.30}$$

are closed subspaces of $H_0^1(D)$ and hence Hilbert spaces.[4] Similarly, for any $\chi \in \overline{\text{co}}\,\mathrm{X}(D)$,

$$H_\bullet^1(\chi; D) \stackrel{\text{def}}{=} \left\{\varphi \in H_0^1(D) : (1 - \chi)\nabla\varphi = 0 \text{ a.e. in } D\right\} \tag{2.31}$$

$$H_\diamond^1(\chi; D) \stackrel{\text{def}}{=} \left\{\varphi \in H_0^1(D) : (1 - \chi)\varphi = 0 \text{ a.e. in } D\right\} \tag{2.32}$$

are also closed subspaces of $H_0^1(D)$ and hence Hilbert spaces. Furthermore[5]

$$H_\diamond^1(\Omega; D) \subset H_\bullet^1(\Omega; D), \quad H_\diamond^1(\chi; D) \subset H_\bullet^1(\chi; D).$$

Proof. We only give the proof for $H_\bullet^1(\Omega; D)$. Let $\{\varphi_n\}$ in $H_\bullet^1(\Omega; D)$ be a Cauchy sequence. It converges to an element φ in the $H_0^1(D)$-topology. Hence $\{\nabla\varphi_n\}$ converges to $\nabla\varphi$ in $L^2(D)$. But for all n

$$(1 - \chi_\Omega)\nabla\varphi_n = 0 \quad \text{in} \quad L^2(D).$$

---

[4]Observe that, for two open domains in the same equivalence class, the spaces defined by (2.29), (2.30) coincide. Therefore their functions do not see cracks in the underlying domain. The case of cracks will be handled in Chapter 6 by capacity methods.

[5]From Evans and Gariepy [1, Thm. 4, p. 130] $\nabla\varphi = 0$ almost everywhere on $\{\varphi = 0\}$.

By Schwarz's inequality the map $\varphi \mapsto (1 - \chi_\Omega)\nabla\varphi$ is continuous and

$$(1 - \chi_\Omega)\nabla\varphi = 0 \quad \text{in} \quad L^2(D).$$

So finally $\varphi \in H^1_\bullet(\Omega; D)$ and $H^1_\bullet(\Omega; D)$ is a closed subspace of $H^1_0(D)$. $\square$

Assuming that D is bounded, the variational problems

$$\begin{cases} \text{to find } y = y(\Omega) \in H^1_\bullet(\Omega; D) \text{ such that} \\[2mm] \forall\varphi \in H^1_\bullet(\Omega; D), \quad \int_D \nabla y \cdot \nabla\varphi \, dx = \int_D \chi_\Omega f\varphi \, dx, \end{cases} \qquad (2.33)$$

$$\begin{cases} \text{to find } y = y(\Omega) \in H^1_\diamond(\Omega; D) \text{ such that} \\[2mm] \forall\varphi \in H^1_\diamond(\Omega; D), \quad \int_D \nabla y \cdot \nabla\varphi \, dx = \int_D \chi_\Omega f\varphi \, dx \end{cases} \qquad (2.34)$$

now make sense and have unique solutions for measurable subsets Ω of D (or even $\chi \in \overline{\text{co}}\, \text{X}(D)$), and the associated objective function

$$J(\Omega) = h\big(\chi_\Omega, y(\Omega)\big), \quad h(\chi, \varphi) \stackrel{\text{def}}{=} \frac{1}{2} \int_D \chi |\varphi - g|^2 \, dx \qquad (2.35)$$

can be minimized over all measurable subsets Ω of D (or all $\chi \in \overline{\text{co}}\, \text{X}(D)$) with fixed measure $m(\Omega) = \pi$.

The above problems are now well posed and their restriction to smooth bounded open connected domains coincides with the initial problem (2.27)–(2.28). Indeed, if Ω is a connected open domain with a boundary Γ of class C^1, then Γ has a zero Lebesgue measure and the definition (2.30) of $H^1_\bullet(\Omega; D)$ coincides with $H^1_\diamond(\Omega; D)$. Therefore, the problem specified by (2.33)–(2.35) is a well-defined extension of problem (2.27)–(2.28). In general, when Ω contains holes, the elements of $H^1_\bullet(\Omega; D)$ are not necessarily equal to zero on the boundary of each hole and can be equal to different constants from hole to hole, as in physical problems involving a potential.

Example 2.3.
Let $D = \,]{-2}, 2[$, $\Omega = \,]{-2}, {-1}[\, \cup\,]1, 2[$ and $f = 1$ on D. The solution of (2.33) is given by

$$\begin{cases} -\dfrac{d^2 y}{dx^2} = 1 \quad \text{on }]{-2}, {-1}[, \\[3mm] y(-2) = 0, \quad \dfrac{dy}{dx}(-1) = 0, \end{cases} \qquad \begin{cases} -\dfrac{d^2 y}{dx^2} = 1 \quad \text{on }]1, 2[, \\[3mm] y(2) = 0, \quad \dfrac{dy}{dx}(1) = 0, \end{cases}$$

$$y = \frac{1}{2} \quad \text{on } [-1, 1],$$

$$\Rightarrow y(x) = \begin{cases} -(x + 2)\dfrac{x}{2} & \text{in }]{-2}, {-1}[, \\[3mm] \dfrac{1}{2} & \text{in } [-1, 1], \\[3mm] -(x - 2)\dfrac{x}{2} & \text{in }]1, 2[. \end{cases} \qquad \square$$

Example 2.4.
Let $D = B(0,2)$, the open ball of radius 2 centered in 0 in $\mathbf{R}^2$, $H = \overline{B(0,1)}$, $\Omega = \complement_D H$, and $f = 1$. In polar coordinates the solution of (2.33) is given by

$$
\begin{cases}
-\dfrac{1}{r}\dfrac{d}{dr}\left(r\dfrac{dy}{dr}\right) = 1 & \text{in }]1,2[, \\[2mm]
y(2) = 0, \quad \dfrac{dy}{dr}(1) = 0,
\end{cases}
\qquad \text{and } y(r) = \frac{1}{2}\ln\frac{1}{2} + \frac{3}{4} \ \text{ in } [0,1],
$$

or explicitly

$$
y(r) =
\begin{cases}
\dfrac{1}{2}\ln\dfrac{r}{2} + 1 - \dfrac{r^2}{4} & \text{in }]1,2[, \\[3mm]
\dfrac{1}{2}\ln\dfrac{1}{2} + \dfrac{3}{4} & \text{in } [0,1].
\end{cases}
\qquad\qquad \square
$$

The last example is a special case that retains the characteristics of the one-dimensional example. In higher dimensions the normal derivative is not necessarily zero on the "internal boundary" of Ω.

Example 2.5.
Let $D = B(0,1)$ in $\mathbf{R}^2$, let H be a bounded open connected hole in D such that $\overline{H} \subset D$, and let $\Omega = \complement_D \overline{H}$. Then it can be checked that the solution of (2.33) is of the form

$$
-\Delta y = f \text{ in } \Omega, \quad y = 0 \text{ on } \partial D,
$$

where the constant c on H is determined by the condition

$$
\forall \varphi \in H^1_\bullet(\Omega; D), \quad \int_{\partial\Omega} \frac{\partial y}{\partial n}\varphi\, d\gamma = 0
$$

or equivalently

$$
\int_{\partial H} \frac{\partial y}{\partial n}\, d\gamma = 0. \qquad\qquad \square
$$

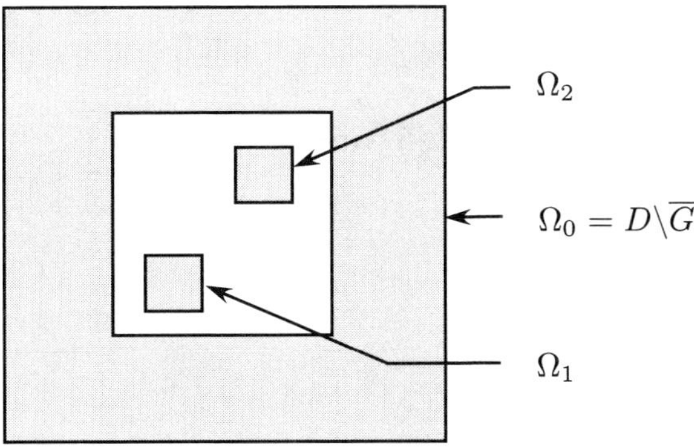

Figure 3.2. *Disconnected domain* $\Omega = \Omega_0 \cup \Omega_1 \cup \Omega_2$.

Example 2.6.
Let $D = \,]{-}2,2[\, \times \,]{-}2,2[$ and let Ω_1 and Ω_2 be two open squares in $G = \,]{-}1,1[\, \times \,]{-}1,1[$ such that $\overline{\Omega_1} \subset G$, $\overline{\Omega_2} \subset G$, and $\overline{\Omega_1} \cap \overline{\Omega_2} = \varnothing$ Define the domain as

$$\Omega = \Omega_0 \cup \Omega_1 \cup \Omega_2, \quad \Omega_0 = D\backslash\overline{G} \quad \text{(cf. Figure 3.2).}$$

Then the solution of (2.33) is characterized by

$$-\Delta y = f \text{ in } \Omega, \quad y = 0 \text{ on } \partial D,$$
$$y = c \text{ on } G\backslash\left[\overline{\Omega_1} \cup \overline{\Omega_2}\right],$$

where the constant c is determined by the condition

$$\int_{\partial\Omega_0^{int}} \frac{\partial y}{\partial n}\, d\gamma + \int_{\partial\Omega_1} \frac{\partial y}{\partial n}\, d\gamma + \int_{\partial\Omega_2} \frac{\partial y}{\partial n}\, d\gamma = 0$$

and $\partial\Omega_0^{int} = \partial G$, the interior boundary of Ω_0. $\qquad\square$

3   Some Compliance Problems with Two Materials

As a first example consider the optimal compliance problem where the optimization variable is the distribution of two materials with different physical characteristics within a fixed domain D. It cannot a priori be assumed that the two regions are separated by a smooth boundary and that each region is connected. The optimal solution may lead to a nonsmooth interface and even to the mixing of the two materials. This type of solution occurs in control or optimization problems over a bounded nonconvex subset of a function space. In general, their relaxed solution lies in the closed convex hull of that subset. In control theory the phenomenon is known as *chattering control*. To illustrate this approach it is best to consider a generic example. In section 3.1 we consider a variation of the optimal design problem studied by Céa and Malanowski [1] in 1970 and then discuss its original version in section 3.2. The variation of the problem is constructed in such a way that the form of the associated variational equation is similar to the one of Example 2.1, which provided a counterexample to the weak continuity of the solution. Yet the maximization of the minimum energy yields a solution which is a characteristic function in both cases. However, this is no longer true for the minimization of the same minimum energy as was shown by Murat and Tartar [1, 3].

3.1   Transmission Problem and Compliance

Let $D \subset \mathbf{R}^N$ be a bounded open domain with Lipschitzian boundary ∂D. Assume that the domain D is partitioned into two subdomains Ω_1 and Ω_2 separated by a smooth boundary $\partial\Omega_1 \cap \partial\Omega_2$, as illustrated in Figure 3.3. Domain Ω_1 (resp., Ω_2) is made up of a material characterized by a constant $k_1 > 0$ (resp., $k_2 > 0$). Let y be the solution of the *transmission problem*

$$\begin{cases} -\,k_1\Delta y = f \text{ in } \Omega_1, \quad -k_2\Delta y = 0 \text{ in } \Omega_2, \\[2mm] y = 0 \text{ on } \partial D, \quad k_1\dfrac{\partial y}{\partial n_1} + k_2\dfrac{\partial y}{\partial n_2} = 0 \text{ on } \partial\Omega_1 \cap \partial\Omega_2, \end{cases} \tag{3.1}$$

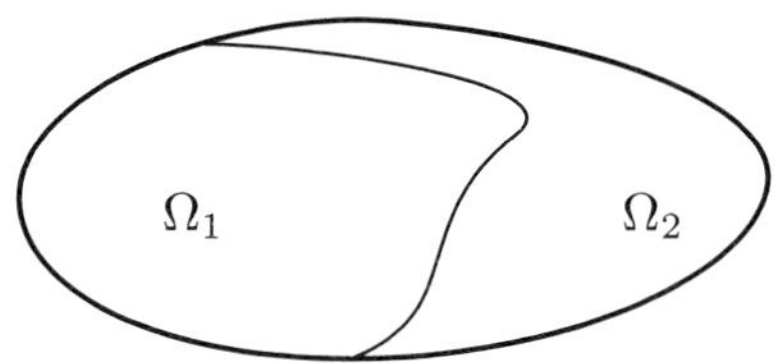

Figure 3.3. *Fixed domain D and its partition into Ω_1 and Ω_2.*

where n_1 (resp., n_2) is the unit outward normal to Ω_1 (resp., Ω_2) and f is a given function in $L^2(D)$. Our objective is to maximize the equivalent of the *compliance*

$$J(\Omega_1) = -\int_{\Omega_1} fy\, dx \tag{3.2}$$

over all domains Ω_1 in D. In mechanics the compliance is associated with the total work of the body forces f.

Denote by Ω the domain Ω_1. By complementarity $\Omega_2 = \complement_D \overline{\Omega}$ and let $\chi = \chi_\Omega$. Problem (3.1) can be rewritten in the following variational form:

$$\boxed{\begin{array}{l} \text{to find } y = y(\chi) \in H_0^1(D) \text{ such that } \forall \varphi \in H_0^1(D), \\[2mm] \displaystyle\int_D [k_1\chi + k_2(1-\chi)]\,\nabla y \cdot \nabla \varphi\, dx = \int_D \chi f \varphi\, dx. \end{array}} \tag{3.3}$$

For $k_1 > 0$ and $k_2 > 0$ and all χ in $\mathrm{X}(D)$

$$\begin{aligned} k(x) &\stackrel{\text{def}}{=} k_1\,\chi(x) + k_2\,(1 - \chi(x)), \\ 0 &< \min\{k_1, k_2\} \le k(x) \le \max\{k_1, k_2\} \text{ a.e. in } D. \end{aligned} \tag{3.4}$$

By the Lax–Milgram theorem, the variational equation (3.3) still makes sense and has a unique solution $y = y(\chi)$ in $H_0^1(D)$ which coincides with the solution of the boundary value problem

$$-\mathrm{div}\,(k\,\nabla y) = \chi f \text{ in } D, \quad y = 0 \text{ on } \partial D. \tag{3.5}$$

Notice that the high-order term has the same form as the term in Example 2.1. As for the objective function, it can be rewritten as

$$\boxed{J(\chi) = -\int_D \chi f y(\chi)\, dx.} \tag{3.6}$$

Thus the initial boundary value problem (3.1) has been transformed into the variational problem (3.3), and the initial objective function (3.2) into (3.6). Both make sense for χ in $\mathrm{X}(D)$ and even in $\overline{\mathrm{co}}\,\mathrm{X}(D)$. The family of characteristic functions $\mathrm{X}(D)$ and its closed convex hull $\overline{\mathrm{co}}\,\mathrm{X}(D)$ have been defined and characterized in (2.2) and (2.3).

The objective function (3.6) can be further rewritten as a minimum

$$J(\chi) = \min_{\varphi \in H_0^1(D)} E(\chi, \varphi) \tag{3.7}$$

for the *energy function*

$$E(\chi, \varphi) \stackrel{\text{def}}{=} \int_D (k_1\,\chi + k_2\,(1 - \chi))\,|\nabla\varphi|^2 - 2\,\chi f\varphi\,dx \tag{3.8}$$

associated with the variational problem (3.3). The initial optimal design problem becomes a max-min problem

$$\boxed{\max_{\chi \in X(D)}\ J(\chi) = \max_{\chi \in X(D)}\ \min_{\varphi \in H_0^1(D)}\ E(\chi, \varphi),} \tag{3.9}$$

where $X(D)$ can be considered as a subset of $L^p(D)$ for some p, $1 \le p < \infty$.
This problem can also be relaxed to functions χ with value in $[0, 1]$:

$$\boxed{\max_{\chi \in \overline{\mathrm{co}}\,X(D)}\ J(\chi) = \max_{\chi \in \overline{\mathrm{co}}\,X(D)}\ \min_{\varphi \in H_0^1(D)}\ E(\chi, \varphi).} \tag{3.10}$$

We shall refer to this problem as the *relaxed problem.*

These formulations have been introduced by Céa and Malanowski [1], who used the variable $k(x)$ and the following *equality constraint* on its integral:

$$\int_D k(x)\,dx = \gamma \quad \text{or} \quad \int_D \chi(x)\,dx = \frac{\gamma - k_2\,m(D)}{k_1 - k_2}$$

for some appropriate $\gamma > 0$. Moreover, the force f in (3.1)–(3.3) was exerted everywhere in D and not only in Ω. So the objective function (3.2)–(3.6) was the integral of fy over all of D. We shall see in section 3.2 that the fact that the support of f is Ω or all of D does not affect the nature of the results.

In both maximization problems (3.1)–(3.2) and (3.3)–(3.6) it is necessary to introduce a *volume constraint* in order to avoid a trivial solution. Notice that by using (3.3) with $\varphi = y(\chi)$ the objective function (3.6) becomes

$$J(\chi) = -\int_D (k_1\chi + k_2(1 - \chi))\,|\nabla y(\chi)|^2\,dx \le 0.$$

So maximizing $J(\chi)$ is equivalent to minimizing the integral of $k(x)\,|\nabla y|^2$. Therefore, $\chi = 0$ (only material k_2) is a maximizer since the corresponding solution of (3.3) is $y = 0$. In order to eliminate this situation we introduce the following constraint on the volume of material k_1:

$$\boxed{\int_D \chi\,dx \ge \alpha > 0} \tag{3.11}$$

for some α, $0 < \alpha \le m(D)$. The case $\alpha = m(D)$ yields the unique solution $\chi = 1$ (only material k_1). So we can further assume that $\alpha < m(D)$.

For $0 < \alpha < m(D)$, the optimal design problem becomes

$$\max_{\substack{\chi \in X(D) \\ \int_D \chi\, dx \geq \alpha}} J(\chi) = \max_{\substack{\chi \in X(D) \\ \int_D \chi\, dx \geq \alpha}} \min_{\varphi \in H_0^1(D)} E(\chi, \varphi) \tag{3.12}$$

and its relaxed version

$$\max_{\substack{\chi \in \overline{\mathrm{co}}\, X(D) \\ \int_D \chi\, dx \geq \alpha}} J(\chi) = \max_{\substack{\chi \in \overline{\mathrm{co}}\, X(D) \\ \int_D \chi\, dx \geq \alpha}} \min_{\varphi \in H_0^1(D)} E(\chi, \varphi). \tag{3.13}$$

We shall now show that problem (3.13) has a unique solution χ^* in $\overline{\mathrm{co}}\, X(D)$ and that χ^* is a characteristic function, $\chi^* \in X(D)$, for which the inequality constraint is saturated

$$\chi^* \in X(D) \quad \text{and} \quad \int_D \chi^*\, dx = \alpha. \tag{3.14}$$

At this juncture it is advantageous to incorporate the volume inequality constraint into the problem formulation by introducing a Lagrange multiplier $\lambda \geq 0$. The formulation of the relaxed problem becomes

$$\max_{\chi \in \overline{\mathrm{co}}\, X(D)} \min_{\substack{\varphi \in H_0^1(D) \\ \lambda \geq 0}} G(\chi, \varphi, \lambda), \tag{3.15}$$

where

$$G(\chi, \varphi, \lambda) \overset{\text{def}}{=} E(\chi, \varphi) + \lambda \left(\int_D \chi\, dx - \alpha \right) \tag{3.16}$$

and $E(\chi, \varphi)$ is the energy function given in (3.7).

We first establish the existence of saddle point solutions to problem (3.15). We use a general result in Ekeland and Temam [1, Prop. 2.4, p. 164]. The set $\overline{\mathrm{co}}\, X(D)$ is a nonempty bounded closed convex subset of $L^2(D)$ and the set $H_0^1(D) \times [\mathbf{R}^+ \cup \{0\}]$ is trivially closed and convex. The function G is concave-convex with the following properties:

$$\begin{cases} \forall \chi \in \overline{\mathrm{co}}\, X(D), (\varphi, \lambda) \mapsto G(\chi, \varphi, \lambda) \text{ is convex, continuous, and} \\ \exists \chi_0 \in \overline{\mathrm{co}}\, X(D) \text{ such that } \lim_{\|\varphi\|_{H_0^1} + |\lambda| \to \infty} G(\chi_0, \varphi, \lambda) = +\infty; \end{cases} \tag{3.17}$$

$$\begin{cases} \forall \varphi \in H_0^1(D), \forall \lambda \geq 0, \quad \text{the map } \chi \mapsto G(\chi, \varphi, \lambda) \text{ is affine and} \\ \text{continuous for } L^2(D)\text{-strong.} \end{cases} \tag{3.18}$$

For the first condition recall that $0 < \alpha < m(D)$ and pick $\chi_0 = 1$ on D. To check the second condition pick any sequence $\{\chi_n\}$ in $\overline{\mathrm{co}}\, X(D)$ which converges to some χ in $L^2(D)$-strong. Then the sequence also converges in $L^2(D)$-weak, and by Lemma 2.1, in $L^\infty(D)$-weak$\star$. Hence $G(\chi_n, \varphi, \lambda)$ converges to $G(\chi, \varphi, \lambda)$.

The set of saddle points $(\hat{\chi}, y, \hat{\lambda})$ is of the form $X \times Y \subset \overline{\mathrm{co}}\, \mathrm{X}(D) \times \{H_0^1(D) \times [\mathbf{R}^+ \cup \{0\}]\}$ and is completely characterized by the following variational equation and inequalities (cf. Ekeland and Temam [1, Prop. 1.6, p. 157]):

$$\forall \varphi \in H_0^1(D), \quad \int_D [k_2 + (k_1 - k_2)\hat{\chi}]\, \nabla y \cdot \nabla \varphi - \hat{\chi} f \varphi \, dx = 0, \tag{3.19}$$

$$\forall \chi \in \overline{\mathrm{co}}\, \mathrm{X}(D), \quad \int_D \left[(k_1 - k_2)|\nabla y|^2 - 2fy + \hat{\lambda}\right] (\chi - \hat{\chi})\, dx \leq 0, \tag{3.20}$$

$$\left(\int_D \hat{\chi}\, dx - \alpha\right)\hat{\lambda} = 0, \quad \int_D \hat{\chi}\, dx - \alpha \geq 0, \quad \hat{\lambda} \geq 0. \tag{3.21}$$

But for each $\chi \in \overline{\mathrm{co}}\, \mathrm{X}(D)$ there exists a unique $y(\chi)$ solution of (3.19) and then

$$\forall \hat{\chi} \in X, \quad \{\hat{\chi}\} \times Y \subset \{\hat{\chi}\} \times \{\{y(\hat{\chi})\} \times [\mathbf{R}^+ \cup \{0\}]\}.$$

Therefore, $y(\hat{\chi})$ is independent of $\hat{\chi} \in X$; that is, $Y = \{y\} \times \Lambda$ and

$$\forall \hat{\chi} \in X, \hat{\lambda} \in \Lambda, \quad y(\hat{\chi}) = y.$$

For each $\hat{\lambda}$, inequality (3.20) is completely equivalent to the following characterization of the maximizer $\hat{\chi} \in X$:

$$\hat{\chi}(x) = \begin{cases} 1, & \text{if } (k_1 - k_2)|\nabla y|^2 - 2fy + \hat{\lambda} > 0, \\ \in [0, 1], & \text{if } (k_1 - k_2)|\nabla y|^2 - 2fy + \hat{\lambda} = 0, \\ 0, & \text{if } (k_1 - k_2)|\nabla y|^2 - 2fy + \hat{\lambda} < 0. \end{cases} \tag{3.22}$$

Associate with an arbitrary $\lambda \geq 0$ the sets

$$\begin{aligned} D_+(\lambda) &= \{x \in D : (k_1 - k_2)|\nabla y|^2 - 2fy + \lambda > 0\}, \\ D_0(\lambda) &= \{x \in D : (k_1 - k_2)|\nabla y|^2 - 2fy + \lambda = 0\}. \end{aligned} \tag{3.23}$$

In order to complete the characterization of the optimal triplets, we need the following general result.

Lemma 3.1. *Consider a function*

$$G : A \times B \to \mathbf{R} \tag{3.24}$$

for some sets A and B. Define

$$g \overset{\text{def}}{=} \inf_{x \in A} \sup_{y \in B} G(x, y), \quad A_0 \overset{\text{def}}{=} \left\{x \in A : \sup_{y \in B} G(x, y) = g\right\}, \tag{3.25}$$

$$h \overset{\text{def}}{=} \sup_{y \in B} \inf_{x \in A} G(x, y), \quad B_0 \overset{\text{def}}{=} \left\{y \in B : \inf_{x \in A} G(x, y) = h\right\}. \tag{3.26}$$

When $g = h$ the set of saddle points (possibly empty) will be denoted by

$$S \overset{\text{def}}{=} \{(x, y) \in A \times B : g = G(x, y) = h\}. \tag{3.27}$$

Then

(i) *In general $h \leq g$ and*

$$\forall (x_0, y_0) \in A_0 \times B_0, \quad h \leq G(x_0, y_0) \leq g. \tag{3.28}$$

(ii) *If $h = g$, then*

$$S = A_0 \times B_0. \tag{3.29}$$

Proof. (i) If $A_0 \times B_0 = \varnothing$ there is nothing to prove. If there exist $x_0 \in A_0$ and $y_0 \in B_0$, then by definition

$$h = \inf_{x \in A} \ G(x, y_0) \leq G(x_0, y_0) \leq \sup_{y \in B} \ G(x_0, y) = g. \tag{3.30}$$

(ii) If $h = g$, then in view of (3.27)–(3.28), $A_0 \times B_0 \subset S$. Conversely, if there exists $(x_0, y_0) \in S$, then $h = G(x_0, y_0) = g$, and by the definitions of A_0 and B_0, $(x_0, y_0) \in A_0 \times B_0$. $\qquad\square$

Associate with an arbitrary solution $(\hat{\chi}, y, \hat{\lambda})$, the characteristic function

$$\chi_{\hat{\lambda}} = \begin{cases} \chi_{D_+(\hat{\lambda})} & \text{if } D_+(\hat{\lambda}) \neq \varnothing, \\ 0 & \text{if } D_+(\hat{\lambda}) = \varnothing. \end{cases} \tag{3.31}$$

Then from (3.22)

$$\hat{\chi}\left\{(k_1 - k_2)|\nabla y|^2 - 2fy + \hat{\lambda}\right\} = \chi_{\hat{\lambda}}\left\{(k_1 - k_2)|\nabla y|^2 - 2fy + \hat{\lambda}\right\} \ \text{ a.e. in } D$$

and

$$\begin{aligned}
G(\hat{\chi}, y, \hat{\lambda}) &= \int_D [\hat{\chi}k_1 + (1 - \hat{\chi})k_2]\,|\nabla y|^2 - 2\hat{\chi}fy\,dx + \hat{\lambda}\left(\int_D \hat{\chi}\,dx - \alpha\right) \\
&= \int_D k_2|\nabla y|^2 + \hat{\chi}\left[(k_1 - k_2)|\nabla y|^2 - 2fy + \hat{\lambda}\right]\,dx - \hat{\lambda}\alpha \\
&= \int_D k_2|\nabla y|^2 + \chi_{\hat{\lambda}}\left[(k_1 - k_2)|\nabla y|^2 - 2fy + \hat{\lambda}\right]\,dx - \hat{\lambda}\alpha \\
&= G(\chi_{\hat{\lambda}}, y, \hat{\lambda}).
\end{aligned}$$

From Lemma 3.1, $(\chi_{\hat{\lambda}}, y, \hat{\lambda})$ is also a saddle point. So there exists a maximizer $\chi_{\hat{\lambda}} \in X(D)$ that is a characteristic function, and necessarily

$$\max_{\substack{\chi \in X(D)}} \min_{\substack{\varphi \in H_0^1(D) \\ \lambda \geq 0}} G(\chi, \varphi, \lambda) = \max_{\substack{\chi \in \overline{co}\,X(D)}} \min_{\substack{\varphi \in H_0^1(D) \\ \lambda \geq 0}} G(\chi, \varphi, \lambda).$$

But we can show more than that. If there exists $\hat{\lambda} \in \Lambda$ such that $\hat{\lambda} > 0$, then by construction $\hat{\chi} \geq \chi_{\hat{\lambda}}$ and by (3.21)

$$\alpha = \int_D \hat{\chi}\,dx \geq \int_D \chi_{\hat{\lambda}}\,dx = \alpha.$$

Since $\hat{\chi} = \chi_{\hat{\lambda}} = 1$ almost everywhere in $D_+(\hat{\lambda})$, then $\hat{\chi} = \chi_{\hat{\lambda}}$. But, always by construction, $\chi_{\hat{\lambda}}$ is independent of $\hat{\chi}$. Thus the maximizer is unique, a characteristic function, and its integral is equal to α.

The case $\Lambda = \{0\}$ is a degenerate one. Set $\varphi = y$ in (3.19) and regroup the terms as follows:

$$0 = \int_D (k_2 + (k_1 - k_2)\hat{\chi}) \, |\nabla y|^2 - fy\hat{\chi} \, dx$$

$$= \int_D \left(k_2 + \frac{k_1 - k_2}{2}\hat{\chi}\right) |\nabla y|^2 \, dx + \int_D \left\{\frac{k_1 - k_2}{2}\hat{\chi} |\nabla y|^2 - fy\right\} \hat{\chi} \, dx.$$

The integrand of the first integral is positive. From the characterization of $\hat{\chi}$, the integrand of the second one is also positive. Hence they are both zero almost everywhere in D. As a result

$$m(D_+(0)) = 0 \quad \text{and} \quad \nabla y = 0.$$

Therefore $y = 0$ in D. Hence the saddle points are of the general form $(\hat{\chi}, y, \hat{\lambda}) = (\hat{\chi}, 0, 0)$, $\hat{\chi} \in X$. In particular, $D_+(0) = \varnothing$ and from our previous considerations $\chi_\varnothing$, that is $\hat{\chi} = 0$ is a solution. But this is impossible since $\int_D \hat{\chi} \, dx \geq \alpha > 0$. Therefore $\hat{\lambda} = 0$ cannot occur.

In conclusion, there exists a (unique for $\alpha > 0$) maximizer χ^* in $\overline{\text{co}}\, X(D)$ that is in fact a characteristic function, and necessarily

$$\max_{\substack{\chi \in X(D) \\ \int_D \chi \, dx \geq \alpha}} \min_{\varphi \in H_0^1(D)} E(\chi, \varphi) = \max_{\substack{\chi \in \overline{\text{co}}\, X(D) \\ \int_D \chi \, dx \geq \alpha}} \min_{\varphi \in H_0^1(D)} E(\chi, \varphi). \tag{3.32}$$

Moreover, for $0 < \alpha \leq m(D)$,

$$\int_D \chi^* \, dx = \alpha \tag{3.33}$$

and χ^* is the unique solution of the problem with an equality constraint

$$\max_{\substack{\chi \in X(D) \\ \int_D \chi \, dx = \alpha}} \min_{\varphi \in H_0^1(D)} E(\chi, \varphi) = \max_{\substack{\chi \in \overline{\text{co}}\, X(D) \\ \int_D \chi \, dx \geq \alpha}} \min_{\varphi \in H_0^1(D)} E(\chi, \varphi). \tag{3.34}$$

As a numerical illustration of the theory consider equation (3.1) over the diamond-shaped domain

$$D = \{(x, y) : |x| + |y| < 1\} \tag{3.35}$$

with the function f of Figure 3.4. This function has a sharp peak in $(0, 0)$, which has been scaled down in the picture. The variational form (3.3) of the boundary value problem was approximated by continuous piecewise linear finite elements on each triangle, and the function χ by a piecewise constant function on each triangle. The constant on each triangle was constrained to lie between 0 and 1 together with the global constraint on its integral over the whole domain D. Figure 3.5 shows

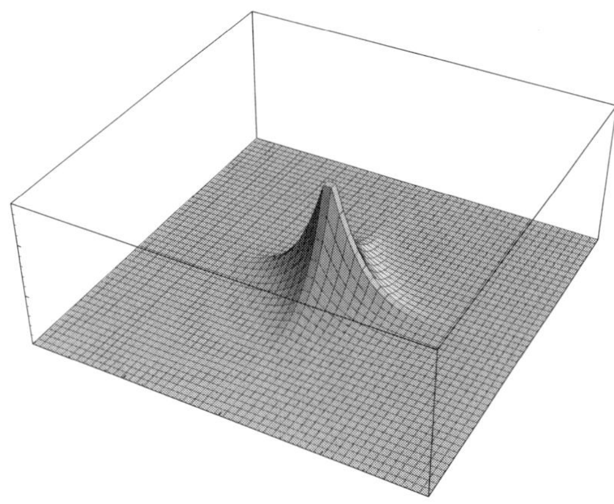

Figure 3.4. *The function* $f(x, y) = 56\,(1 - |x| - |y|)^6$.

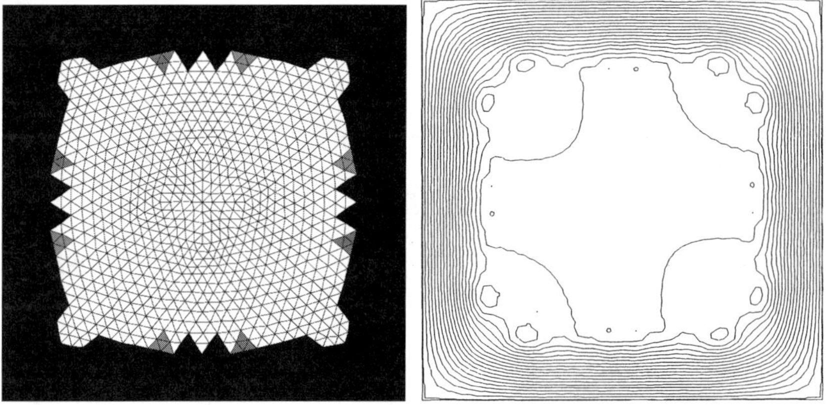

Figure 3.5. *Optimal distribution and isotherms with* $k_1 = 2$ *(black) and* $k_2 = 1$ *(white) for the problem of section* 3.1.

the optimal partition (the domain has been rotated 45 degrees to save space). The grey triangles correspond to the region $D_0(\hat{\lambda}_m)$, where $\hat{\chi} \in [0, 1]$. The presence of this grey zone in the approximated problem is due to the fact that equality for the total area where $\hat{\chi} = 1$ could not be exactly achieved with the chosen triangulation of the domain. Thus the problem had to adjust the value of $\hat{\chi}$ between 0 and 1 in a few triangles in order to achieve equality for the integral of $\hat{\chi}$. For this example the Lagrange multiplier associated with the problem is strictly positive.

3.2 The Original Problem of Céa and Malanowski

We now put the force f everywhere in the fixed domain D. The same technique and the same conclusion can be drawn: there exists at least one maximizer of the compliance, which is a characteristic function. For completeness we give the main elements below.

Fix the bounded open Lipschitzian domain D in $\mathbf{R}^N$ and let Ω be a smooth subset of D. Let y be the solution of the *transmission problem*

$$\begin{cases} -\,k_1\triangle y = f \text{ in } \Omega, \quad -k_2\triangle y = f \text{ in } D\backslash\overline{\Omega}, \\ y = 0 \text{ on } \partial D, \\ k_1\dfrac{\partial y}{\partial n_1} + k_2\dfrac{\partial y}{\partial n_2} = 0 \text{ on } \partial\Omega \cap D, \end{cases} \tag{3.36}$$

where n_1 (resp., n_2) is the unit outward normal to Ω (resp., $\complement_D\overline{\Omega}$) and f is a given function in $L^2(D)$. Again problem (3.36) can be reformulated in terms of the characteristic function $\chi = \chi_\Omega$:

$$\boxed{\begin{aligned} &\text{to find } y = y(\chi) \in H_0^1(D) \text{ such that } \forall\varphi \in H_0^1(D), \\ &\int_D [k_1\chi + k_2(1-\chi)]\,\nabla y \cdot \nabla\varphi\,dx = \int_D f\varphi\,dx, \end{aligned}} \tag{3.37}$$

with the objective function

$$\boxed{J(\chi) = -\int_D fy(\chi)\,dx} \tag{3.38}$$

to be maximized over all $\chi \in X(D)$. As in the previous case, the function $J(\chi)$ can be rewritten as the minimum of the energy function

$$E(\chi,\varphi) \overset{\text{def}}{=} \int_D [k_1\chi + k_2(1-\chi)]\,|\nabla\varphi|^2 - 2f\varphi\,dx \tag{3.39}$$

over $H_0^1(D)$.

$$\boxed{J(\chi) = \min_{\varphi\in H_0^1(D)} E(\chi,\varphi)} \tag{3.40}$$

and we have the relaxed max-min problem

$$\boxed{\max_{\chi\in\overline{\mathrm{co}}\,X(D)} \min_{\varphi\in H_0^1(D)} E(\chi,\varphi).} \tag{3.41}$$

Without constraint on the integral of χ, the problem is trivial and $\chi = 1$ (resp., $\chi = 0$) if $k_1 > k_2$ (resp., $k_2 > k_1$). In other words it is optimal to use only the strong material. In order to make the problem nontrivial, assume that the strong material is k_1, that is, $k_1 > k_2$, and put an upper bound on the volume of material k_1 which occupies the part Ω of D:

$$\int_D \chi \, dx \leq \alpha, \quad 0 < \alpha < m(D). \tag{3.42}$$

The case $\alpha = 0$ trivially yields $\chi = 0$. Under assumption (3.42) the case $\chi = 1$ with only the strong material k_1 is no longer admissible. Thus we consider for $0 < \alpha < m(D)$ the problem

$$\max_{\substack{\chi \in \overline{\mathrm{co}}\, \mathrm{X}(D) \\ \int_D \chi \, dx \leq \alpha}} \min_{\varphi \in H_0^1(D)} E(\chi, \varphi). \tag{3.43}$$

We shall now show that problem (3.43) has a unique solution χ^* in $\overline{\mathrm{co}}\, \mathrm{X}(D)$ and that in fact χ^* is a characteristic function for which the inequality constraint is saturated:

$$\int_D \chi^* \, dx = \alpha. \tag{3.44}$$

This exactly solves the original problem of Céa and Malanowski with equality constraint

$$\max_{\substack{\chi \in \mathrm{X}(D) \\ \int_D \chi \, dx = \alpha}} \min_{\varphi \in H_0^1(D)} E(\chi, \varphi). \tag{3.45}$$

As in the previous section it is convenient to reformulate the problem with a Lagrange multiplier $\lambda \geq 0$ for the constraint inequality (3.42):

$$\max_{\substack{\chi \in \mathrm{X}(D) \\ \lambda \geq 0}} \min_{\varphi \in H_0^1(D)} G(\chi, \varphi, \lambda), \tag{3.46}$$

$$G(\chi, \varphi, \lambda) = E(\chi, \varphi) - \lambda \left[\int_D \chi \, dx - \alpha \right]. \tag{3.47}$$

We then relax the problem to $\overline{\mathrm{co}}\, \mathrm{X}(D)$,

$$\max_{\substack{\chi \in \overline{\mathrm{co}}\, \mathrm{X}(D) \\ \lambda \geq 0}} \min_{\varphi \in H_0^1(D)} G(\chi, \varphi, \lambda). \tag{3.48}$$

By the same arguments as the ones used in section 3.1 we have the existence of saddle points $(\hat{\chi}, y, \hat{\lambda})$, which are completely characterized by

$$\forall \varphi \in H_0^1(D), \quad \int_D [k_2 + (k_1 - k_2)\hat{\chi}] \nabla y \cdot \nabla \varphi - f\varphi \, dx = 0, \tag{3.49}$$

$$\forall \chi \in \overline{\mathrm{co}}\, \mathrm{X}(D), \quad \int_D \left[(k_1 - k_2)|\nabla y|^2 - \hat{\lambda} \right] (\chi - \hat{\chi}) \, dx \leq 0, \tag{3.50}$$

$$\left[\int_D \hat{\chi} \, dx - \alpha \right] \hat{\lambda} = 0, \quad \int_D \hat{\chi} \, dx - \alpha \leq 0, \quad \hat{\lambda} \geq 0. \tag{3.51}$$

As before, y is unique and the set of saddle points of G is a closed convex set of the form

$$X \times \{\{y\} \times \Lambda\} \subset \mathrm{X}(D) \times \{\{y\} \times \{\lambda : \lambda \geq 0\}\}.$$

The closed convex set Λ has a minimal element $\hat{\lambda}_m \geq 0$.

For each $\hat{\lambda}$, each maximizer $\hat{\chi} \in X$ is necessarily of the form

$$\hat{\chi}(x) = \begin{cases} 1 & \text{if } (k_1 - k_2)|\nabla y|^2 - \hat{\lambda} > 0, \\ \in [0,1] & \text{if } (k_1 - k_2)|\nabla y|^2 - \hat{\lambda} = 0, \\ 0 & \text{if } (k_1 - k_2)|\nabla y|^2 - \hat{\lambda} < 0. \end{cases} \tag{3.52}$$

Associate with each $\lambda > 0$ the sets

$$D_+(\lambda) = \{x \in D : (k_1 - k_2)|\nabla y|^2 - \lambda > 0\}, \tag{3.53}$$
$$D_0(\lambda) = \{x \in D : (k_1 - k_2)|\nabla y|^2 - \lambda = 0\}, \tag{3.54}$$
$$D_-(\lambda) = \{x \in D : (k_1 - k_2)|\nabla y|^2 - \lambda < 0\}. \tag{3.55}$$

Define the characteristic function

$$\chi_m = \chi_{D_+(\lambda_m)}. \tag{3.56}$$

By construction all $\hat{\chi} \in X$ are of the form

$$\chi_m \leq \hat{\chi}, \quad \int_D \chi \, dx \leq \alpha. \tag{3.57}$$

Again it is easy to show that $(\chi_m, y, \hat{\lambda}_m)$ is also a saddle point of G. Therefore, we have a maximizer $\chi_m \in \mathrm{X}(D)$ over $\overline{\mathrm{co}}\, \mathrm{X}(D)$ which is a characteristic function and

$$\int_D \chi_m \, dx \leq \int_D \hat{\chi} \, dx \leq \alpha.$$

If there exists $\hat{\lambda} > 0$ in Λ, then for all $\hat{\chi} \in X$

$$\int_D \chi_m \, dx = \alpha = \int_D \hat{\chi} \, dx.$$

As a result the maximizer χ_m is unique, a characteristic function, and its integral is equal to α.

The case $\Lambda = \{0\}$ cannot occur since the triplet $(1, y_1, 0)$ would be a saddle point where y_1 is the solution of the variational equation (3.36) for $\chi = 1$. To see this, first observe that for $k_1 > k_2$

$$\forall \chi, \varphi, \quad G(\chi, \varphi, 0) = E(\chi, \varphi) \leq E(1, \varphi) = G(1, \varphi, 0).$$

As a result

$$\inf_{\varphi,\lambda} G(\chi,\varphi,\lambda) \leq \inf_{\varphi} G(\chi,\varphi,0) \leq \inf_{\varphi} G(1,\varphi,0),$$

$$\inf_{\varphi} G(1,\varphi,0) = G(1,y_1,0) \leq \sup_{\chi}\inf_{\varphi} G(\chi,\varphi,0)$$

$$\Rightarrow \sup_{\chi}\inf_{\varphi,\lambda} G(\chi,\varphi,\lambda) \leq \sup_{\chi}\inf_{\varphi} G(\chi,\varphi,0) = G(1,y_1,0).$$

But we know that there exists a unique $y \in H_0^1(D)$ such that

$$\sup_{\chi}\inf_{\varphi} G(\chi,\varphi,0) \leq \sup_{\chi} G(\chi,y,0) = \inf_{\varphi,\lambda}\sup_{\chi} G(\chi,\varphi,\lambda).$$

So from the above two inequalities,

$$\sup_{\chi}\inf_{\varphi,\lambda} G(\chi,\varphi,\lambda) = G(1,y_1,0) = \inf_{\varphi,\lambda}\sup_{\chi} G(\chi,\varphi,\lambda),$$

$(1,y_1,0)$ is a saddle point of G, and

$$\alpha \geq \int_D \chi\,dx = \mathrm{m}(D).$$

This contradicts the fact that $\alpha < \mathrm{m}(D)$.

In conclusion in all cases there exists a (unique when $0 \leq \alpha < \mathrm{m}(D)$) maximizer χ^* in $\overline{\mathrm{co}}\,X(D)$, which is in fact a characteristic function, and

$$\max_{\substack{\chi\in X(D)\\ \int_D \chi\,dx\geq\alpha}} \min_{\varphi\in H_0^1(D)} E(\chi,\varphi) = \max_{\substack{\chi\in\overline{\mathrm{co}}\,X(D)\\ \int_D \chi\,dx\geq\alpha}} \min_{\varphi\in H_0^1(D)} E(\chi,\varphi).$$

Moreover, for $0 \leq \alpha < \mathrm{m}(D)$

$$\int_D \chi^*\,dx = \alpha.$$

This is precisely *the* solution of the original problem of Céa and Malanowski and

$$\max_{\substack{\chi\in X(D)\\ \int_D \chi\,dx=\alpha}} \min_{\varphi\in H_0^1(D)} E(\chi,\varphi) = \max_{\substack{\chi\in\overline{\mathrm{co}}\,X(D)\\ \int_D \chi\,dx\geq\alpha}} \min_{\varphi\in H_0^1(D)} E(\chi,\varphi).$$

Again, as an illustration of the theoretical results, consider equation (3.36) over the diamond-shaped domain D defined in (3.35) with the function f of Figure 3.4 in section 3.1. The variational form (3.37) of the boundary value problem was approximated in the same way as the variational form (3.3) of equation (3.1) in section 3.1. Figure 3.6 shows the optimal partition of the domain (rotated 45 degrees). The grey triangles correspond to the region $D_0(\hat{\lambda}_m)$, where $\hat{\chi} \in [0,1]$. The black region corresponds to the points where $|\nabla y|^2 > \hat{\lambda}_m/(k_1 - k_2)$, and the white region, to the ones where $|\nabla y|^2 < \hat{\lambda}_m/(k_1 - k_2)$, as can be readily seen on Figure 3.6. For this example the Lagrange multiplier associated with the problem is strictly positive. It is interesting to compare this computation with the one of Figure 3.5 in the previous section, where the support of the force was restricted to Ω.

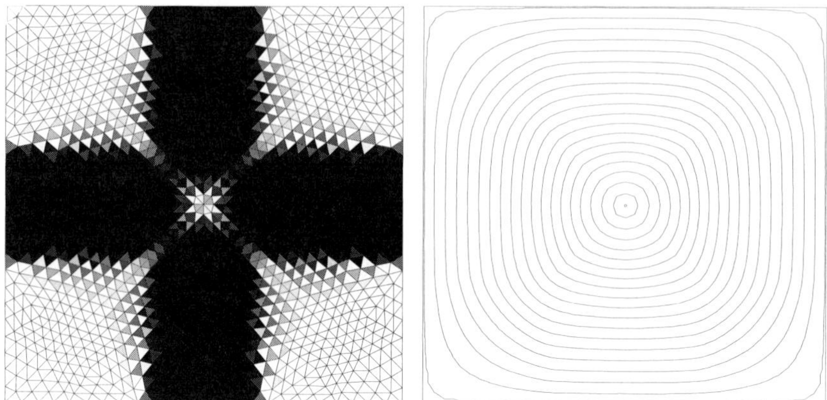

Figure 3.6. *Optimal distribution and isotherms with $k_1 = 2$ (black) and $k_2 = 1$ (white) for the problem of Céa–Malanowski.*

3.3   Relaxation and Homogenization

In sections 3.1 and 3.2 the possible homogenization phenomenon predicted in Example 2.1 of section 2.2 did not take place, and in both examples the solution was a characteristic function. Yet it is sufficient to change the problem of Céa and Malanowski [1] in section 3.2 from a maximization to a minimization of the minimum energy to see it happen. In 1985 Murat and Tartar [1] gave a general framework to study this class of problems by relaxation. It is based on the use of Young's [2] generalized functions (measures). They present a fairly complete analysis of the homogenization theory of second-order elliptic problems of the form

$$-\sum_{ij} \frac{\partial}{\partial x_i}\left(a_{ij}(x)\frac{\partial u}{\partial x_j}\right) = f.$$

They give as examples the maximization and minimization versions of the problem considered in section 3.2. This material would have deserved a whole chapter in this book, but fortunately it is available in English, so the reader is referred to Murat and Tartar [3] to complete this chapter. Other results on composite materials can be found in the book edited by Cherkaev and Kohn [1], which gathers a selection of translations of key papers originally written in French and Russian.

4   Buckling of Columns

One of the very early optimal design problem was formulated by Lagrange in 1770 (cf. Todhunter and Pearson [1]) and later studied by Clausen in 1849. It consists in finding the best profile of a vertical column to prevent buckling. This problem and other problems related to columns have been revisited in a series of papers by

Cox [1], Cox and Overton [1], Cox [2], and Cox and McCarthy [1]. Since Lagrange many authors have proposed solutions, but a complete theoretical and numerical solution for the buckling of a column has only been given in 1992 by Cox and Overton [1].

Consider a normalized column of unit height and unit volume. Denote by ℓ the *magnitude of the normalized axial load* and by u the resulting transverse displacement. Assume that the potential energy is the sum of the bending and elongation energies

$$\int_0^1 EI\,|u''|^2\,dx - \ell \int_0^1 |u'|^2\,dx,$$

where I is the second moment of area of the column's cross section and E is its Young's modulus. For sufficiently small load ℓ the minimum of this potential energy with respect to all admissible u is zero. The *Euler's buckling load* λ of the column is the largest ℓ for which this minimum is zero. This is equivalent to finding the following minimum:

$$\lambda \overset{\text{def}}{=} \inf_{0 \neq u \in V} \frac{\int_0^1 EI\,|u''|^2\,dx}{\int_0^1 |u'|^2\,dx}, \tag{4.1}$$

where $V = H_0^2(0,1)$ corresponds to the clamped case, but other types of boundary conditions can be contemplated. This is an eigenvalue problem with a special Rayleigh quotient.

Assume that E is constant and that the second moment of area $I(x)$ of the column's cross section at the height x, $0 \leq x \leq 1$, is equal to a constant c times its cross-sectional area $A(x)$,

$$I(x) = c\,A(x) \quad \Rightarrow \quad \int_0^1 A(x)\,dx = 1.$$

Normalizing λ by cE and taking into account the engineering constraints

$$\exists\, 0 < A_0 < A_1,\ \forall x \in [0,1], \quad 0 < A_0 \leq A(x) \leq A_1,$$

we finally get

$$\sup_{A \in \mathcal{A}} \lambda(A), \quad \lambda(A) \overset{\text{def}}{=} \inf_{0 \neq u \in V} \frac{\int_0^1 A\,|u''|^2\,dx}{\int_0^1 |u'|^2\,dx}, \tag{4.2}$$

$$\mathcal{A} \overset{\text{def}}{=} \left\{ A \in L^2(0,1) : A_0 \leq A \leq A_1 \text{ and } \int_0^1 A(x)\,dx = 1 \right\}. \tag{4.3}$$

This problem can also be reformulated by rewriting

$$A(x) = A_0 + \chi(x)\,(A_1 - A_0), \quad \int_0^1 \chi(x)\,dx = \alpha \overset{\text{def}}{=} \frac{1 - A_0}{A_1 - A_0}$$

for some $\chi \in \overline{co}\, X([0,1])$. Clearly the problem makes sense only for $0 < A_0 \leq 1$. Then

$$\sup_{\substack{\chi \in \overline{co}\, X([0,1]) \\ \int_0^1 \chi(x)\, dx = \alpha}} \tilde{\lambda}(\chi), \quad \tilde{\lambda}(\chi) \stackrel{\text{def}}{=} \inf_{0 \neq u \in V} \frac{\int_0^1 [A_0 + (A_1 - A_0)\chi]\, |u''|^2\, dx}{\int_0^1 |u'|^2\, dx}. \tag{4.4}$$

Rayleigh's quotient is not a nice convex-concave function with respect to (v, A) and its analysis necessitates different tools than the ones of section 3. One of the original elements of the paper of Cox and Overton [1] was to replace Rayleigh's quotient by Auchmuty's [1] dual variational principle for the eigenvalue problem (4.1). We first recall the existence of solution to the minimization of Rayleigh's quotient. In what follows we shall use the norm $\|u'\|_{L^2}$ for the space $H_0^1(0,1)$ and $\|u''\|_{L^2}$ for the space $H_0^2(0,1)$.

Theorem 4.1. *There exists at least one nonzero solution $u \in V$ to the minimization problem*

$$\lambda(A) \stackrel{\text{def}}{=} \inf_{0 \neq u \in V} \frac{\int_0^1 A\, |u''|^2\, dx}{\int_0^1 |u'|^2\, dx}. \tag{4.5}$$

Then $\lambda(A_0) > 0$, and for all $A \in \mathcal{A}$, $\lambda(A) \geq \lambda(A_0)$, and

$$\forall v \in V, \quad \int_0^1 |v'|^2\, dx \leq \lambda(A_0)^{-1} \int_0^1 A\, |v''|^2\, dx. \tag{4.6}$$

The solutions are completely characterized by the variational equation:

$$\boxed{\exists u \in V, \, \forall v \in V, \quad \int_0^1 A\, u''\, v''\, dx = \lambda(A) \int_0^1 u'\, v'\, dx.} \tag{4.7}$$

Proof. The infimum is bounded below by 0 and is necessarily finite. Let $\{u_n\}$ be a minimizing sequence such that $\|u_n'\|_{L^2} = 1$. Then the sequence $\|u_n''\|_{L^2(D)}$ is bounded. Hence $\{u_n\}$ is a bounded sequence in $H^2(0,1)$ and there exist $u \in H^2(0,1)$ and a subsequence, still indexed by n, such that $u_n \rightharpoonup u$ in $H_0^2(0,1)$-weak. Therefore the subsequence strongly converges in $H_0^1(0,1)$ and

$$1 = \int_0^1 |u_n'|^2\, dx \to \int_0^1 |u'|^2\, dx \quad \text{and} \quad \int_0^1 A\, |u''|^2\, dx \leq \liminf_{n \to \infty} \int_0^1 A\, |u_n''|^2\, dx$$

$$\Rightarrow \frac{\int_0^1 A\, |u''|^2\, dx}{\int_0^1 |u'|^2\, dx} \leq \liminf_{n \to \infty} \frac{\int_0^1 A\, |u_n''|^2\, dx}{\int_0^1 |u_n'|^2\, dx} = \lambda(A),$$

and, by the definition of $\lambda(A)$, $u \in H_0^2(0,1)$ is a minimizing element. Notice that

$$\forall A \in \mathcal{A}, \quad \lambda(A_0) = \inf \left\{ \frac{\int_0^1 A_0\, |u''|^2\, dx}{\int_0^1 |u'|^2\, dx} : \forall u \in H_0^2(0,1),\, u \neq 0 \right\}$$

$$\leq \inf \left\{ \frac{\int_0^1 A\, |u''|^2\, dx}{\int_0^1 |u'|^2\, dx} : \forall u \in H_0^2(0,1),\, u \neq 0 \right\} = \lambda(A)$$

and similarly $\lambda(A) \leq \lambda(A_1)$. If $\lambda(A_0) = 0$, we repeat the above construction and end up with an element $u \in H_0^2(0,1)$ such that

$$\int_0^1 |u'|^2 \, dx = 1 \text{ and } A_0 \int_0^1 |u''|^2 \, dx = \int_0^1 A_0 |u''|^2 \, dx = 0,$$

which is impossible in $H_0^2(0,1)$ since $A_0 \neq 0$ and $u'' = 0$ implies $u = 0$.

If $u \neq 0$, then Rayleigh's quotient is differentiable and its directional semiderivative in the direction v is given by

$$2\frac{\int_0^1 A \, u'' \, v'' \, dx}{\|u'\|_{L^2}^2} - 2\int_0^1 A \, |u''|^2 \, dx \frac{\int_0^1 u' \, v' \, dx}{\|u'\|_{L^2}^4}.$$

A nonzero solution u of the minimization problem is necessarily a stationary point, and for all $v \in H_0^2(0,1)$

$$\int_0^1 A \, u'' \, v'' \, dx = \frac{\int_0^1 A_0 \, |u''|^2 \, dx}{\int_0^1 |u'|^2 \, dx} \int_0^1 u' \, v' \, dx = \lambda(A) \int_D u' \, v' \, dx.$$

Conversely, any nonzero solution of (4.7) is necessarily a minimizer of the Rayleigh quotient. $\qquad\square$

The dual variational principle of Auchmuty [1] for the eigenvalue problem (4.5) can be chosen as

$$\mu(A) \stackrel{\text{def}}{=} \inf\{L(A,v) : v \in H_0^2(0,1)\}, \tag{4.8}$$

$$L(A,v) \stackrel{\text{def}}{=} \frac{1}{2}\int_0^1 A \, |v''|^2 \, dx - \left[\int_0^1 |v'|^2 \, dx\right]^{1/2}. \tag{4.9}$$

Theorem 4.2. *For each $A \in \mathcal{A}$, there exists at least one minimizer of $L(A,v)$,*

$$\mu(A) = -\frac{1}{2\lambda(A)}, \tag{4.10}$$

and the set of minimizers of (4.8) is given by

$$E(A) \stackrel{\text{def}}{=} \left\{ u \in H_0^2(0,1) : \begin{array}{c} u \text{ is solution of (4.7) and} \\ \left\{\int_0^1 |u'|^2 \, dx\right\}^{1/2} = 1/\lambda(A) \end{array} \right\}. \tag{4.11}$$

Proof. The existence of the solution follows from the fact that $v \mapsto L(A,v)$ is weakly lower semicontinuous and coercive. From Theorem 4.1 the set $E(A)$ is not empty, and for any $u \in E(A)$

$$\mu(A) \leq L(A,u) = -\frac{1}{2\lambda(A)} < 0. \tag{4.12}$$

Therefore, the minimizers of (4.7) are different from the zero functions. For $u \neq 0$, the function $L(A, u)$ is differentiable and its directional semiderivative is given by

$$dL(A, u; v) = \int_0^1 A\, u''\, v''\, dx - \frac{1}{\|u'\|_{L^2}} \int_0^1 u'\, v'\, dx, \qquad (4.13)$$

and any minimizer u of $L(A, v)$ is a stationary point of $dL(A, u; v)$, that is,

$$\forall v \in H_0^2(0, 1), \quad \int_0^1 A\, u''\, v''\, dx - \frac{1}{\|u'\|_{L^2}} \int_0^1 u'\, v'\, dx = 0. \qquad (4.14)$$

Therefore u is a solution of the eigenvalue problem with

$$\lambda' = \frac{1}{\|u'\|_{L^2}} \quad \Rightarrow \quad -\frac{1}{2\lambda'} = \mu(A) \leq -\frac{1}{2\lambda(A)}$$

from inequality (4.12). By the minimality of $\lambda(A)$, we necessarily have $\lambda' = \lambda(A)$, and this concludes the proof of the theorem. $\qquad\square$

Theorem 4.3.

   (i) *The set $\mathcal{A}$ is compact in the $L^2(0, 1)$-weak topology.*

  (ii) *The function $A \mapsto \mu(A)$ is concave and upper semicontinuous with respect to the $L^2(0, 1)$-weak topology.*

 (iii) *There exists A in $\mathcal{A}$ which maximizes $\mu(A)$ over $\mathcal{A}$ and, a fortiori, which minimizes $\lambda(A)$ over $\mathcal{A}$.*

Proof. (i) $\mathcal{A}$ is convex, bounded, and closed in $L^2(0, 1)$. Hence it is compact in $L^2(0, 1)$-weak.

(ii) *Concavity.* For all λ in $[0, 1]$ and A and A' in $\mathcal{A}$,

$$L(\lambda A + (1 - \lambda)A', v) = \lambda L(A, v) + (1 - \lambda)L(A', v)$$
$$\geq \lambda \inf_{v \in V} L(A, v) + (1 - \lambda) \inf_{v \in V} L(A', v)$$
$$\Rightarrow \mu(\lambda A + (1 - \lambda)A') = \inf_{v \in V} L(\lambda A + (1 - \lambda)A', v) \geq \lambda\mu(A) + (1 - \lambda)\mu(A').$$

(Upper semicontinuous) Let $u_A \in \mathcal{A}$ be a minimizer of $L(A, v)$, and $\{A_n\}$ a sequence converging to A in the $L^2(0, 1)$-weak topology. By Lemma 2.1, $\{A_n\}$ converges to A in the $L^\infty(0, 1)$-weak$\star$ topology. Then

$$\mu(A_n) - \mu(A) \leq L(A_n, u_A) - L(A, u_A) = \frac{1}{2} \int_0^1 (A_n - A)\, |u_A''|^2\, dx$$

$$\Rightarrow \limsup_{n \to \infty} \mu(A_n) - \mu(A) \leq \lim_{n \to \infty} \frac{1}{2} \int_0^1 (A_n - A)\, |u_A''|^2\, dx = 0.$$

(iii) The existence of solution follows from the fact that μ is concave and upper semicontinuous. Hence it is weakly upper semicontinuous over the weakly compact subset $\mathcal{A}$ of $L^2(0, 1)$. The existence of a minimizer of $\lambda(A)$ follows from identity (4.10). $\qquad\square$

5　Caccioppoli or Finite Perimeter Sets

The notion of *finite perimeter sets* has been introduced and mainly developed by Caccioppoli [1] and De Giorgi [1] in the context of Plateau's problem, named after the Belgian physicist and professor J. A. F. Plateau [1] (1801–1883) who did experimental observations on the geometry of soap films. A modern treatment of this subject can be found in the book of Giusti [1]. One of the difficulties in studying the *minimal surface problem* is the description of such surfaces in the usual language of differential geometry. For instance, the set of possible singularities is not known. Finite perimeter sets provide a geometrically significant solution to Plateau's problem without having to know ahead of time what all the possible singularities of the solution can be. The characterization of all the singularities of the solution is a difficult problem which can be considered separately.

This was a very fundamental contribution to the theory of variational problems, where the optimization variable is the geometry of a domain. This point of view has been expanded, and a variational calculus was developed by Almgren [1]. This is the *theory of varifolds*:

> Perhaps the one most important virtue of varifolds is that it is possible to obtain a geometrically significant solution to a number of variational problems, including Plateau's problem, without having to know ahead of time what all the possible singularities of the solution can be. (cf. Almgren [1, p. viii])

The treatment of this topic is unfortunately much beyond the scope of this book, and the interested reader is referred to the above reference for more details.

5.1　Finite Perimeter Sets

Given an open subset D of $\mathbf{R}^N$, consider L^1-functions f on D with distributional gradient ∇f in the space $M^1(D)^N$ of (vectorial) bounded measures; that is,

$$\vec{\varphi} \mapsto \langle \nabla f, \vec{\varphi} \rangle_D \stackrel{\text{def}}{=} - \int_D f \operatorname{div} \vec{\varphi} \, dx : \mathcal{D}^1(D; \mathbf{R}^N) \to \mathbf{R}$$

is continuous with respect to the topology of uniform convergence in D:

$$\|\nabla f\|_{M^1(D)^N} \stackrel{\text{def}}{=} \sup_{\substack{\vec{\varphi} \in \mathcal{D}^1(D;\mathbf{R}^N) \\ \|\vec{\varphi}\|_C \leq 1}} \langle \nabla f, \vec{\varphi} \rangle_D < \infty, \tag{5.1}$$

where

$$\|\vec{\varphi}\|_C \stackrel{\text{def}}{=} \sup_{x \in D} |\vec{\varphi}(x)|_{\mathbf{R}^N},$$

$M^1(D) \stackrel{\text{def}}{=} \mathcal{D}^0(D)'$ is the topological dual of $\mathcal{D}^0(D)$, and

$$\nabla f \in \mathcal{L}\left(\mathcal{D}^0(D; \mathbf{R}^N), \mathbf{R}\right) \equiv \mathcal{L}\left(\mathcal{D}^0(D), \mathbf{R}\right)^N = M^1(D)^N.$$

Such functions are known as *functions of bounded variation*. The space

$$\boxed{BV(D) \stackrel{\text{def}}{=} \{f \in L^1(D) : \nabla f \in M^1(D)^N\}} \tag{5.2}$$

endowed with the norm

$$\boxed{\|f\|_{BV(D)} = \|f\|_{L^1(D)} + \|\nabla f\|_{M^1(D)^N}} \tag{5.3}$$

is a Banach space.

Definition 5.1.
A measurable function f on $\mathbf{R}^N$ is said to be *locally of bounded variation*, if for all bounded open subset D of $\mathbf{R}^N$, $f \in BV(D)$. The set of all such functions will be denoted $BV_{\text{loc}}(\mathbf{R}^N)$. $\qquad\square$

Theorem 5.1. *f belongs to $BV_{\text{loc}}(\mathbf{R}^N)$ if and only if for each $x \in \mathbf{R}^N$ there exists $\rho > 0$ such that $f \in BV(B(x,\rho))$, $B(x,\rho)$ the open ball of radius ρ in x.*

Proof. If $f \in BV_{\text{loc}}(\mathbf{R}^N)$, then the result is true by specializing to balls in each point of $\mathbf{R}^N$. Conversely, denote by $\rho_x > 0$ the radius associated with the point $x \in \mathbf{R}^N$ for which $f \in BV(B(x,\rho_x))$. Given a bounded open subset D of $\mathbf{R}^N$, the compact set $\overline{D}$ can be covered by a finite number of such balls for points $x_i \in \overline{D}$, $1 \le i \le n$,

$$\overline{D} \subset \bigcup_{i=1}^{n} B_i, \quad B_i \stackrel{\text{def}}{=} B(x_i, \rho_{x_i}).$$

Denote by $\{\psi_i \in \mathcal{D}(B_i)\}_{i=1}^{n}$ a partition of unity for the family $\{B_i\}$ such that

$$\forall i,\ 0 \le \psi_i \le 1 \text{ in } B_i, \text{ and } \sum_{i=1}^{n} \psi_i = 1 \text{ on } \overline{D}.$$

Then for each $\vec{\varphi} \in \mathcal{D}^1(D)^N$

$$\vec{\varphi} = \sum_{i=1}^{n} \vec{\varphi}_i \text{ on } D, \quad \vec{\varphi}_i \stackrel{\text{def}}{=} \psi_i\, \vec{\varphi},$$

and by construction $\vec{\varphi}_i \in \mathcal{D}^1(B_i \cap D)^N$. Therefore, for all $\vec{\varphi} \in \mathcal{D}^1(D)^N$ such that $\|\vec{\varphi}\|_{C(D)} \le 1$

$$-\int_D f \operatorname{div} \vec{\varphi}\, dx = -\sum_{i=1}^{n} \int_D f \operatorname{div} \vec{\varphi}_i\, dx = -\sum_{i=1}^{n} \int_{D \cap B_i} f \operatorname{div} \vec{\varphi}_i\, dx$$

$$\Rightarrow \left| \int_D f \operatorname{div} \vec{\varphi}\, dx \right| \le \sum_{i=1}^{n} \left| \int_{B_i \cap D} f \operatorname{div} \vec{\varphi}_i\, dx \right|$$

$$\Rightarrow \left| \int_D f \operatorname{div} \vec{\varphi}\, dx \right| \le \sum_{i=1}^{n} \|\nabla f\|_{M^1(B_i \cap D)} \le \sum_{i=1}^{n} \|\nabla f\|_{M^1(B_i)} < \infty$$

since $\mathcal{D}^1(B_i \cap D)^N \subset \mathcal{D}^1(B_i)^N$ and $\|\vec{\varphi}_i\|_{C(B_i)} \leq 1$. Therefore, $\|\nabla f\|_{M^1(D)}$ is finite and $f \in BV_{\mathrm{loc}}(\mathbf{R}^N)$. $\qquad\square$

For more details and properties see also Morgan [1, p. 117], Federer [3, section 4.5.9], Evans and Gariepy [1], Ziemer [1], and Temam [1].

As in the previous section, consider measurable subsets Ω of a fixed bounded open subset D of $\mathbf{R}^N$. Their characteristic functions $\chi_\Omega \in X(D)$ are $L^1(D)$-functions

$$\|\chi_\Omega\|_{L^1(D)} = \int_D \chi_\Omega \, dx = \mathrm{m}(\Omega) \leq \mathrm{m}(D) < \infty \tag{5.4}$$

with distributional gradient

$$\forall \vec{\varphi} \in \left(\mathcal{D}(D)\right)^N, \quad \langle \nabla\chi_\Omega, \vec{\varphi}\rangle_{\mathcal{D}} \stackrel{\text{def}}{=} -\int_D \chi_\Omega \mathrm{div}\, \vec{\varphi}\, dx. \tag{5.5}$$

When Ω is an open domain with boundary Γ of class C^1, then by the Stokes divergence theorem

$$-\int_{\Omega \cap D} \mathrm{div}\, \vec{\varphi}\, dx = -\int_{\partial(\Omega \cap D)} \vec{\varphi} \cdot n\, d\Gamma = -\int_{\Gamma \cap D} \vec{\varphi} \cdot n\, d\Gamma,$$

where n is the outward normal field along $\partial(\Omega \cap D)$. Since Γ is of class C^1 the normal field n along Γ belongs to $C^0(\Gamma)$, and from the last identity the maximum is

$$P_D(\Omega) \stackrel{\text{def}}{=} \|\nabla\chi_\Omega\|_{M^1(D)} = \int_{\Gamma \cap D} |n|^2\, d\Gamma = \int_{\Gamma \cap D} d\Gamma = H_{N-1}(\Gamma \cap D),$$

the $(N-1)$-dimensional Hausdorff measure of $\Gamma \cap D$. As a result

$$H_{N-1}(\Gamma) = P_D(\Omega) + H_{N-1}(\Gamma \cap \partial D). \tag{5.6}$$

Thus the norm of the gradient provides a natural relaxation of the notion of perimeter to the following larger class of domains.

Definition 5.2.
Let Ω be a measurable subset of $\mathbf{R}^N$.

(i) Given an open set D in $\mathbf{R}^N$, Ω is said to have *finite perimeter with respect to D* if $\chi_\Omega \in BV(D)$. This *perimeter* denoted by $P_D(\Omega)$ is given by the expression

$$P_D(\Omega) \stackrel{\text{def}}{=} \|\nabla\chi_\Omega\|_{M^1(D)^N}. \tag{5.7}$$

The family of all such characteristic functions is denoted

$$\mathrm{BX}(D) \stackrel{\text{def}}{=} \{\chi_\Omega \in X(D) : \chi_\Omega \in BV(D)\}.$$

(ii) Ω is said to have *locally finite perimeter* if for all bounded open subsets D of $\mathbf{R}^N$, $\chi_\Omega \in BV(D)$, that is, $\chi_\Omega \in BV_{\mathrm{loc}}(\mathbf{R}^N)$.

(iii) Ω is said to have *finite perimeter* if $\chi_\Omega \in BV(\mathbf{R}^N)$. $\square$

Theorem 5.2. *Let Ω be a Lebesgue measurable subset of $\mathbf{R}^N$. Ω has locally finite perimeter, that is, $\chi_\Omega \in BV_{\mathrm{loc}}(\mathbf{R}^N)$, if and only if for each $x \in \partial\Omega$ there exists $\rho > 0$ such that $\chi_\Omega \in BV(B(x,\rho))$, $B(x,\rho)$ the open ball of radius ρ in x.*

Proof. We use Theorem 5.1. In one direction the result is obvious. Conversely, if $x \in \operatorname{int}\Omega$, then there exists $\rho > 0$ such that $B(x,\rho) \subset \operatorname{int}\Omega$, and for all $\vec{\varphi} \in \mathcal{D}^1(B(x,\rho))$

$$-\int_{B(x,\rho)} \chi_\Omega \operatorname{div}\vec{\varphi}\, dx = -\int_{B(x,\rho)} \operatorname{div}\vec{\varphi}\, dx = -\int_{\partial B(x,\rho)} \vec{\varphi}\cdot n\, dH_{N-1} = 0$$

since $\vec{\varphi} = 0$ on the boundary of $B(x,\rho)$. Thus $\chi_\Omega \in BV(B(x,\rho))$. If $x \in \operatorname{int}\complement\Omega$, then there exists $\rho > 0$ such that $B(x,\rho) \subset \operatorname{int}\complement\Omega$, and for all $\vec{\varphi} \in \mathcal{D}^1(B(x,\rho))$

$$-\int_{B(x,\rho)} \chi_\Omega \operatorname{div}\vec{\varphi}\, dx = 0.$$

So again $\chi_\Omega \in BV(B(x,\rho))$. Finally, if $x \in \partial\Omega$, then by assumption there exists $\rho > 0$ such that $\chi_\Omega \in BV(B(x,\rho))$. Therefore, by Theorem 5.1, $\chi_\Omega \in BV_{\mathrm{loc}}(\mathbf{R}^N)$. $\square$

The interest behind this construction is twofold. First, the notion of the perimeter of a set is extended to measurable sets; second, this framework provides a first compactness theorem which will be useful in obtaining existence of optimal domains.

Theorem 5.3. *Assume that D is a bounded open domain in $\mathbf{R}^N$ with a Lipschitzian boundary ∂D. Let $\{\Omega_n\}$ be a sequence of measurable domains in D with finite perimeter. If there exists a constant $c > 0$ such that*

$$\forall n, \quad P_D(\Omega_n) \le c, \tag{5.8}$$

then there exist a measurable set Ω in D and a subsequence $\{\Omega_{n_k}\}$ such that

$$\chi_{\Omega_{n_k}} \to \chi_\Omega \ \text{ in } L^1(D) \text{ as } k \to \infty, \tag{5.9}$$

$$P_D(\Omega) \le \liminf_{k\to\infty} P_D(\Omega_{n_k}) \le c. \tag{5.10}$$

Moreover, $\nabla\chi_{\Omega_{n_k}}$ "converges in measure" to $\nabla\chi_\Omega$ in $M^1(D)^N$; that is, for all $\vec{\varphi}$ in $\mathcal{D}^0(D,\mathbf{R}^N)$,

$$\lim_{k\to\infty} \langle \nabla\chi_{\Omega_{n_k}}, \vec{\varphi}\rangle_{M^1(D)^N} \to \langle \nabla\chi_\Omega, \vec{\varphi}\rangle_{M^1(D)^N}. \tag{5.11}$$

Proof. This follows from the fact that the injection of the space $BV(D)$ endowed with the norm (5.3) into $L^1(D)$ is continuous and compact (cf. Giusti [1, Thm. 1.19, p. 17], Maz'ja [1, Thm. 6.1.4, p. 300; Lem. 1.4.6, p. 62], Morrey Jr. [1, Thm. 4.4.4, p. 75], and Evans and Gariepy [1]). $\square$

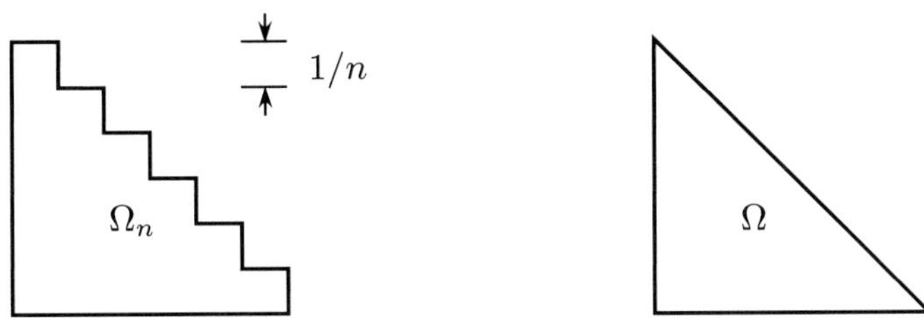

Figure 3.7. *The staircase.*

Example 5.1 (the staircase).
In (5.10) the inequality can be strict, as can be seen from the following example (cf.
Figure 3.7). For each $n \geq 1$, define the set

$$\Omega_n = \bigcup_{j=1}^{n} \left[\frac{j}{n}, \frac{j+1}{n}\right] \times \left[0, 1 - \frac{j}{n}\right].$$

Its limit is the set

$$\Omega = \{(x,y) : 0 \leq x \leq 1, 0 \leq y \leq x\}.$$

It is easy to check that the sets Ω_n are contained in the hold-all $D =]-1, 2[\times]-1, 2[$
and that

$$\forall n \geq 1, \quad P_D(\Omega_n) = 4, \quad P_D(\Omega) = 2 + \sqrt{2},$$
$$\chi_{\Omega_n} \to \chi_\Omega \quad \text{in} \quad L^p(D), \ 1 \leq p < \infty.$$

Each Ω_n is uniformly Lipschitzian but the Lipschitz constant and the two neighborhoods of Definition 2.2 in Chapter 2 cannot be chosen independently of n. $\square$

Corollary 1. *Let D be open bounded and Lipschitzian. There exists a constant
$c > 0$ such that for all convex domains Ω in D*

$$P_D(\Omega) \leq c, \quad m(\Omega) \leq c,$$

and the set $\mathcal{C}(D)$ of convex subsets of D is compact in $L^p(D)$ for all p, $1 \leq p < \infty$.

Proof. If Ω is a convex set with zero volume, $m(\Omega) = 0$, then its perimeter $P_D(\Omega) =
0$ (cf. Giusti [1, Remark 1.7 (iii), p. 6]). If $m(\Omega) > 0$, then int $\Omega \neq \emptyset$. For convex
sets with a nonempty interior the perimeter and the volume enjoy a very nice
monotonicity property. If $C^\bullet$ denotes the set of nonempty closed convex subsets of
$\mathbf{R}^N$, then the map

$$\Omega \mapsto m(\Omega) \colon C^\bullet \to \mathbf{R}$$

is strictly increasing

$$\forall A, B \in C^\bullet, \quad A \subsetneq B \implies m(A) < m(B).$$

(cf. Berger [1, Vol. 3, Prop. 12.9.4.3, p. 141]). Similarly, if $P(\Omega)$ denotes the perimeter of Ω in $\mathbf{R}^N$, the map

$$\Omega \mapsto P(\Omega) \colon \mathcal{C}^\bullet \to \mathbf{R}$$

is strictly increasing (cf. Berger [1, Vol. 3, Prop. 12.10.2, p. 144]). As a result

$$\forall n, \quad m(\Omega_n) \le m(D) < \infty,$$
$$\forall n, \quad P_D(\Omega_n) \le P(\Omega_n) \le P(\operatorname{co} D) < \infty,$$

since the perimeter of a bounded convex set is finite. Then the conditions of the theorem are satisfied and the conclusions of the theorem follow. But we have seen in Theorem 2.8 (iii) that the set $\mathcal{C}(D)$ is closed in $L^1(D)$. Therefore, Ω can be chosen convex in the equivalence class. This completes the proof. $\qquad\square$

This theorem and the lower (resp., upper) semicontinuity of the shape function $\Omega \mapsto J(\Omega)$ will provide existence results for domains in the class of finite perimeter sets in D. For instance, the transmission problem (3.1) to (3.3) of section 3.1 with the objective function

$$J(\Omega) = \frac{1}{2} \int_\Omega |y(\Omega) - g|^2 \, dx + \alpha P_D(\Omega), \quad \alpha > 0, \tag{5.12}$$

for some $g \in L^2(D)$. We shall come back later to the homogeneous Dirichlet boundary value problem (2.27)–(2.28).

It is important to recall that even if a set Ω in D has a finite perimeter $P_D(\Omega)$, its *relative boundary* $\Gamma \cap D$ can have a nonzero N-dimensional Lebesgue measure. To illustrate this point consider the following adaptation of Example 1.10 in Giusti [1, p. 7].

Example 5.2.
Let $D = B(0,1)$ in $\mathbf{R}^2$ be the open ball in 0 of radius 1. For $i \ge 1$, let $\{x_i\}$ be an ordered sequence of all points in D with rational coordinates. Associate with each i the open ball

$$B_i = \{x \in D : |x - x_i| < \rho_i\}, \quad 0 < \rho_i \le \min\{2^{-i}, 1 - |x_i|\}.$$

Define the new sequence of open subsets of D,

$$A_n = \bigcup_{i=1}^{n} B_i,$$

and notice that for all $n \ge 1$,

$$m(\partial A_n) = 0, \quad P_D(A_n) \le 2\pi,$$

where ∂A_n is the boundary of A_n. Moreover, since the sequence of sets $\{A_n\}$ is increasing,

$$\chi_{A_n} \to \chi_A \text{ in } L^1(D), \quad A = \bigcup_{i=1}^{\infty} B_i,$$

$$P_D(A) \le \liminf_{n \to \infty} P_D(A_n) \le 2\pi.$$

Now A is an open subset of D. Observe that $\bar{A} = \bar{D}$ and $\partial A = \bar{A} \cap \complement A \supset \bar{D} \cap \complement A$. Thus

$$m(\partial A) = m(\bar{D} \cap \complement A) \geq m(\bar{D}) - m(A) \geq \frac{2\pi}{3}$$

since

$$m(D) = \pi \quad \text{and} \quad m(A) \leq \sum_{1=1}^{\infty} \pi 2^{-2i} = \frac{\pi}{3}.$$

Recall that

$$m(A_n) \leq \sum_{i=1}^{n} m(B_i) \leq \sum_{i=1}^{n} \pi 2^{-2i} \quad \Rightarrow \quad m(A) \leq \sum_{i=1}^{\infty} m(B_i) \leq \sum_{i=1}^{\infty} \pi 2^{-2i}.$$

For p, $1 \leq p < \infty$, the sequence of characteristic functions $\{\chi_{A_n}\}$ converge to χ_A in $L^p(D)$-strong. However, for all n, $m(\partial A_n) = 0$, but $m(\partial A) > 0$. $\qquad\square$

In fact we can associate with the perimeter $P_D(\Omega)$ the *reduced boundary* $\partial^*\Omega$, which is the set of all $x \in \partial\Omega$ for which the normal $n(x)$ exists. We quote the following interesting theorems from Fleming [1, p. 455] (cf. also De Giorgi [2], Evans and Gariepy [1, Lem. 1, p. 208], and Federer [5]).

Theorem 5.4. *Let Ω have finite perimeter $P(\Omega)$. Let $\partial^*\Omega$ denote the set of $x \in \partial\Omega$ for which $n(x)$ exists. Then*

(i) $\partial^*\Omega \subset \partial_*\Omega \subset \Omega_\bullet \subset \partial\Omega$ *and* $\overline{\partial^*\Omega} = \partial\Omega$,

(ii) $P(\Omega) = H_{N-1}(\partial^*\Omega)$ *(cf. Giusti [1, Chap. 4]),*

(iii) *the Gauss–Green theorem holds with* $\partial^*\Omega$,

(iv) $m(\partial^*\Omega) = m(\partial_*\Omega) = 0, \quad H_{N-1}(\partial_*\Omega - \partial^*\Omega) = 0$.

We quote the following density theorem (cf. Giusti [1, Thm. 1.24 and Lem. 1.25, p. 23]), which complements Theorem 2.3.

Theorem 5.5. *Let Ω be a bounded measurable domain in $\mathbf{R}^N$ with finite perimeter. Then there exists a sequence $\{\Omega_j\}$ of C^∞-domains such that as j goes to ∞*

$$\int_{\mathbf{R}^N} |\chi_{\Omega_j} - \chi_\Omega| \, dx \to 0 \quad \text{and} \quad P(\Omega_j) \to P(\Omega).$$

5.2 Decomposition of the Integral along Level Sets

We complete this section by giving some useful theorems on the decomposition of the integral along the *level sets* of a function. The first one was used by Zolésio [6] (plasma physics) in 1979 and [9, p. 95, section 4.4] in 1981, by Temam [1] (monotone rearrangements) in 1979, by Rakotoson and Temam [1] in 1987, and more recently in 1995, by Delfour and Zolésio [19, 20, 21, 25] (intrinsic formulation of models of shells). We quote the version given in Evans and Gariepy [1, Prop. 3, p. 118]. The original theorem can be found in Federer [2] and Young [1].

Theorem 5.6. *Let $f : \mathbf{R}^N \to \mathbf{R}$ be Lipschitz continuous with*

$$|\nabla f| > 0 \quad a.e.$$

For any Lebesgue summable function $g : \mathbf{R}^N \to \mathbf{R}$, we have the following decomposition of the integral along the level curves of f,

$$\int_{\{f>t\}} g\, dx = \int_t^\infty \left(\int_{\{f=s\}} \frac{g}{|\nabla f|}\, dH^{N-1} \right) ds.$$

The second theorem (Fleming and Rishel [1]) uses a BV-function instead of a Lipschitz function.

Theorem 5.7. *Let D be an open subset of $\mathbf{R}^N$. For any $f \in BV(D)$ and real t, let*

$$E_t \stackrel{\text{def}}{=} \{x : f(x) < t\}.$$

Then

$$\|\nabla f\|_{M^1(D)^N} = \int_{-\infty}^\infty P(E_t)\, dt$$

(cf., for instance, Whitney [1, Chap. 11]).

This is known as the *co-area formula* (see also Giusti [1, Thm. 1.23, p. 20] and De Giorgi [4]).

5.3   Domains of Class $W^{\varepsilon,p}(D)$, $0 \le \varepsilon < 1/p$, $p \ge 1$

There is a general property enjoyed by functions in $BV(D)$.

Theorem 5.8. *Let D be a bounded open Lipschitzian domain in $\mathbf{R}^N$.*

(i) $BV(D) \subset W^{\varepsilon,1}(D), \quad 0 \le \varepsilon < 1.$

(ii) $BV(D) \cap L^\infty(D) \subset W^{\varepsilon,p}(D), \quad 0 \le \varepsilon < \frac{1}{p}, 1 \le p < \infty.$

This theorem says that for a Caccioppoli set Ω in D,

$$\forall p \ge 1, 0 \le \varepsilon < \frac{1}{p}, \quad \chi_\Omega \in W^{\varepsilon,p}(D).$$

The special case $p = 2$ was proved by Baiocchi et al. [1] in the context of the celebrated problem of the dam. They showed that the graph Ω of a continuous monotonically decreasing function on a closed interval is $H^\varepsilon(D) = W^{\varepsilon,2}(D)$, $0 \le \varepsilon < 1/2$, in $\mathbf{R}^2$.

Proof of Theorem 5.8. (i) Given f in $BV(D)$ we want to show that $f \in W^{\varepsilon,1}(D)$, $0 < \varepsilon < 1$. This is equivalent to showing that the double integral

$$I = \int_D dx \int_D dy \, \frac{|f(y) - f(x)|}{|y - x|^{N+\varepsilon}}$$

is finite. Given $\alpha > 0$ small we break the above integral into two parts:

$$I_1 = \int_D dx \int_{D \cap B(x,\alpha)} dy \, \frac{|f(y) - f(x)|}{|y - x|^{N+\varepsilon}},$$

$$I_2 = \int_D dx \int_{\complement_D B(x,\alpha)} dy \, \frac{|f(y) - f(x)|}{|y - x|^{N+\varepsilon}}.$$

The second integral I_2 is bounded since $|y - x| \geq \alpha$. For the first one we change the variable y to $t = y - x$,

$$I_1 = \int_D dx \int_{B(0,\alpha)} dt \, \frac{|(\chi_D f)(x + t) - (\chi_D f)(x)|}{|t|^{N+\varepsilon}},$$

and after a change in the order of integration,

$$I_1 = \int_{B(0,\alpha)} dt \, \frac{1}{|t|^{N+\varepsilon}} \int_D dx \, |(\chi_D f)(x + t) - (\chi_D f)(x)|.$$

But the integral over D is bounded by

$$|t| \, \|\nabla f\|_{M^1(D)}.$$

The result is true for smooth functions (cf. Maz'ja [1, Lem. 1.4.6, p. 62]). Pick a sequence of smooth functions such that

$$f_n \to f \text{ in } L^1(D), \quad \nabla f_n \to \nabla f \text{ in } M^1(D)\text{-weak};$$

then

$$\|\nabla f\|_{M^1(D)} \leq \limsup_{n \to \infty} \|\nabla f_n\|_{M^1(D)} = \limsup_{n \to \infty} \|\nabla f_n\|_{L^1(D)}$$

is bounded, and by going to the limit on both sides of the inequality,

$$\int_D dx \, |(\chi_D f_n)(x + t) - (\chi_D f_n)(x)| \leq |t| \|\nabla f_n\|_{L^1(D)},$$

we obtain the desired result. Now coming back to the estimate of I_1,

$$I_1 \leq \int_{B(x,\alpha)} |t|^{1-N-\varepsilon} \, dt \, \|\nabla f\|_{M^1(D)}$$

$$\leq c(\alpha) \frac{\alpha^{1-\varepsilon}}{1 - \varepsilon} \|\nabla f\|_{M^1(D)}, \quad 0 < \varepsilon < 1.$$

(ii) The function f belongs to $W^{\varepsilon,p}(D)$ if the integral

$$I = \int_D dx \int_D dy \, \frac{|f(y) - f(x)|^p}{|y - x|^{N+\varepsilon p}} < +\infty.$$

For each $f \in BV(D) \cap L^\infty(D)$, there exists $M > 0$ such that

$$|f(x)| \leq M \quad \text{a.e. in } D.$$

Then

$$I = (2\,M)^p \int_D dx \int_D dy \, \frac{\left| \frac{f(y) - f(x)}{2\,M} \right|^p}{|y - x|^{N + \varepsilon p}},$$

and since

$$\left| \frac{f(y) - f(x)}{2\,M} \right| \leq 1,$$

then for each p, $1 \leq p < \infty$,

$$\left| \frac{f(y) - f(x)}{2\,M} \right|^p \leq \left| \frac{f(y) - f(x)}{2\,M} \right|.$$

As a result

$$I \leq (2\,M)^{p-1} \int_D dx \int_D dy \, \frac{|f(y) - f(x)|}{|y - x|^{N + \varepsilon p}}.$$

But we have seen in part (i) that for $f \in BV(D)$, the double integral is finite if

$$0 \leq \varepsilon p < 1 \Rightarrow 0 \leq \varepsilon < \frac{1}{p}.$$

This complete the proof. $\qquad\square$

For a characteristic function χ_Ω of a measurable set Ω and $\varepsilon > 0$, we have $0 \leq \varepsilon < 1/p$ for all $p \geq 1$ and

$$\|\chi_\Omega\|_{W^{\varepsilon,p}(D)}^p = \int_D \int_D \frac{|\chi_\Omega(x) - \chi_\Omega(y)|}{|x - y|^{N + p\varepsilon}} \, dx\, dy + \|\chi_\Omega\|_{L^p(D)}^p. \tag{5.13}$$

The $W^{\varepsilon,p}(D)$-norm is equivalent to the $W^{\varepsilon',p'}(D)$ for all pairs (p', ε'), $p' \geq 1$, $0 \geq \varepsilon' < 1/p'$ such that $p'\,\varepsilon' = p\varepsilon$. For $p = 2$ and a characteristic function, χ_Ω, of a measurable set Ω, we get

$$\|\chi_\Omega\|_{H^\varepsilon(D)}^2 = 2 \int_\Omega \int_{\complement_D \Omega} |x - y|^{-(N + 2\varepsilon)} \, dx\, dy + m(\Omega \cap D). \tag{5.14}$$

This new norm can now be used in various problems or to regularize the objective function to obtain existence of approximate solutions. For instance, to obtain existence results when minimizing a function $J(\Omega)$ (defined for all measurable sets Ω in D), we can consider a regularized problem in the following form:

$$J_\alpha(\Omega) = J(\Omega) + \alpha \|\chi_\Omega\|_{H^\varepsilon(D)}^2, \quad \alpha > 0. \tag{5.15}$$

5.4 Compactness and Uniform Cone Property

In Theorem 5.3 of section 5.1 we have seen a first compactness theorem for a family of sets with a uniformly bounded perimeter. In this section we present a second compactness theorem for measurable domains satisfying a global *uniform cone property* due to Chenais [1, 4, 6]. We provide a proof of this result, which emphasizes the fact that the perimeter of sets in that family is uniformly bounded. It then becomes a special case of Theorem 5.3. Another completely different proof will be given in section 10 of Chapter 5.

Theorem 5.9. *Let D be a bounded open hold-all in $\mathbf{R}^N$ with uniformly Lipschitzian boundary ∂D. For $r > 0$, $\omega > 0$, and $\lambda > 0$ consider the family*

$$L(D, r, \omega, \lambda) \overset{\text{def}}{=} \left\{ \Omega \subset D : \begin{array}{l} \Omega \text{ is Lebesgue measurable and satisfies} \\ \text{the uniform cone property for } (r, \omega, \lambda) \end{array} \right\}.$$

For p, $1 \le p < \infty$, the set

$$X(D, r, \omega, \lambda) \overset{\text{def}}{=} \{ \chi_\Omega : \forall \Omega \in L(D, r, \omega, \lambda) \}$$

is compact in $L^p(D)$ and there exists a constant $p(D, r, \omega, \lambda) > 0$ such that

$$\forall \Omega \in L(D, r, \omega, \lambda), \quad H_{N-1}(\partial \Omega) \le p(D, r, \omega, \lambda).$$

We first show that the perimeter of the sets of the family $L(D, r, \omega, \lambda)$ is uniformly bounded and use Theorem 5.7 to show that any sequence of $X(D, r, \omega, \lambda)$ has a subsequence converging to the characteristic function of some finite perimeter set Ω. We use the full strength of the compactness of the injection of $BV(D)$ into $L^1(D)$ rather than checking directly the conditions under which a subset of $L^p(D)$ is relatively compact. The proof will be completed by showing that the nice representative (Definition 2.3) I of Ω satisfies the same uniform cone property. The proof uses some elements of Chenais's [4, 6] original proof.

Proof of Theorem 5.9. The parameters (r, λ, ω) are the same for all $\Omega \in L(D, r, \omega, \lambda)$ and all points $x \in \partial \Omega$. As a result we can construct the neighborhoods $V(x)$ and $\mathcal{U}(x)$ as in (6.2) for ρ and r_λ in (6.3) and the map f_x satisfying the Lipschitz condition (6.4) from Chapter 2. By construction

$$\begin{aligned}
B(x, \rho) \subset B(x, r_\lambda) \quad &\Rightarrow B(x, \rho) \cap \{ y : P_{H(x)}(y - x) \in V(x) \} \\
&\qquad \subset B(x, r_\lambda) \cap \{ y : P_{H(x)}(y - x) \in V(x) \} \\
&\Rightarrow B(x, \rho) \subset \mathcal{U}(x).
\end{aligned}$$

The family $\{ B(z, \rho/2) : z \in \overline{D} \}$ is an open cover of $\overline{D}$. Since $\overline{D}$ is compact, there exists a finite subcover $\{ B_i \}_{i=1}^m$, $B_i = B(z_i, \rho/2)$, of $\overline{D}$. Now for any $\Omega \in L(D, r, \omega, \lambda)$, $\partial \Omega \subset \overline{D}$ and there is a subcover $\{ B_{i_k} \}_{k=1}^K$ of $\{ B_i \}_{i=1}^m$ such that

$$\partial \Omega \subset \bigcup_{k=1}^K B_{i_k} \quad \text{and} \quad \forall k, \ 1 \le k \le K, \ \partial \Omega \cap B_{i_k} \ne \varnothing.$$

Pick a sequence $\{x_k\}_{k=1}^K$ such that

$$\forall k, \quad 1 \le k \le K, \quad x_k \in \partial\Omega \cap B_{i_k},$$

and notice that

$$B_{i_k} = B\left(z_{i_k}, \frac{\rho}{2}\right) \subset B(x_k, \rho)$$

or

$$\exists K \le m, \quad \exists \{x_k\}_{k=1}^K \subset \partial\Omega, \quad \partial\Omega \subset \bigcup_{k=1}^K B(x_k, \rho).$$

Thus from the estimate (6.6) in Chapter 2,

$$H_{N-1}(\partial\Omega) \le \sum_{k=1}^K H_{N-1}\big(\partial\Omega \cap B(x_k, \rho)\big) \le \sum_{k=1}^K H_{N-1}\big(\partial\Omega \cap \mathcal{U}(x_k)\big)$$
$$\le \sum_{k=1}^K \rho^{N-1}\alpha_{N-1}\sqrt{1+(\tan\omega)^2} \le m\,\rho^{N-1}\alpha_{N-1}\sqrt{1+(\tan\omega)^2},$$

where α_{N-1} is the volume of the unit $(N-1)$-dimensional ball.

So the right-hand side of the above inequality is a constant that is equal to $p = p(D, r, \omega, \lambda) > 0$ and is independent of Ω in $L(D, r, \omega, \lambda)$:

$$\forall \Omega \in L(D, r, \omega, \lambda), \quad P_D(\Omega) = H_{N-1}(\partial\Omega) \le p.$$

Now from Theorem 5.3 for any sequence $\{\Omega_n\} \subset L(D, r, \omega, \lambda)$, there exists Ω such that $\chi_\Omega \in BV(D)$ and a subsequence, still denoted $\{\Omega_n\}$, such that

$$\chi_{\Omega_n} \to \chi_\Omega \text{ in } L^1(D) \text{ and } P_D(\Omega) \le p.$$

(ii) To complete the proof we consider the representative I of Ω (cf. Definition 2.3) and show that $I \in L(D, r, \omega, \lambda)$. We need the following lemma.

Lemma 5.1. *Let $\chi_{\Omega_n} \to \chi_\Omega$ in $L^p(D)$, $1 \le p < \infty$, for $\Omega \subset D$, $\Omega_n \subset D$, and let* I *be the measure theoretic representative of Ω. Then*

$$\forall x \in \bar{I}, \quad \forall R > 0, \quad \exists N(x, R) > 0, \quad \forall n \ge N(x, R), \quad \mathrm{m}\big(B(x, R) \cap \Omega_n\big) > 0.$$

Moreover,

$$\forall x \in \partial I, \forall R > 0, \exists N(x, R) > 0, \forall n \ge N(x, R),$$
$$\mathrm{m}\big(B(x, R) \cap \Omega_n\big) > 0 \quad and \quad \mathrm{m}\big(B(x, R) \cap \complement\Omega_n\big) > 0$$

and

$$B(x, R) \cap \partial\Omega_n \ne \varnothing.$$

Proof. We proceed by contradiction. Assume that

$$\exists x \in \bar{I}, \quad \exists R > 0, \quad \forall N > 0, \quad \exists n \geq N, \quad \mathrm{m}\big(B(x,R) \cap \Omega_n\big) = 0.$$

So there exists a subsequence $\{\Omega_{n_k}\}$, $n_k \to \infty$, such that

$$m\big(B(x,R) \cap \Omega\big) = \lim_{k\to\infty} m\big(B(x,R) \cap \Omega_{n_k}\big) = 0$$

$$\Rightarrow \exists R > 0, \quad m\big(B(x,R) \cap \Omega\big) = 0 \implies x \in \Omega_0 = \mathrm{int}\,\complement I = \overline{\complement I}.$$

But this is in contradiction with the fact that $x \in \bar{I}$. For $x \in \partial I = \bar{I} \cap \overline{\complement I}$, simultaneously apply the first statement to $\bar{I}$ and $\chi_{\Omega_n} \to \chi_\Omega$ and to $\overline{\complement I}$ and $\chi_{\complement \Omega_n} \to \chi_{\complement \Omega}$ since $\chi_{\complement \Omega} = \chi_{\complement I}$ almost everywhere by Theorem 2.5 and choose the largest of the two integers. As for the last property, it follows from the fact that the open ball $B(x,R)$ cannot be partitioned into two nonempty disjoint sets. $\qquad\square$

We wish to prove that

$$\forall x \in \partial I, \quad \exists d, |d| = 1, \quad \forall y \in \bar{I} \cap B(x,r), \quad C_y(\lambda, \omega, d) \subset \mathrm{int}\, I.$$

Since Ω_n is Lipschitzian, $\complement \Omega_n$ is also Lipschitzian and

$$\chi_{\overline{\Omega}_n} = \chi_{\Omega_n} \quad \text{and} \quad \chi_{\overline{\complement \Omega_n}} = \chi_{\complement \Omega_n}$$

almost everywhere. So we apply the second part of the lemma to $x \in \partial I, \chi_{\overline{\Omega}_n} \to \chi_\Omega$ and $\chi_{\overline{\complement \Omega_n}} \to \chi_{\complement \Omega}$. So for each $x \in \partial I, \forall k \geq 1, \exists n_k \geq k$ such that

$$B\left(x, \frac{r}{2^k}\right) \cap \partial \Omega_{n_k} \neq \varnothing.$$

Denote by x_k an element of that intersection:

$$\forall k \geq 1, \quad x_k \in B\left(x, \frac{r}{2^k}\right) \cap \partial \Omega_{n_k}.$$

By construction $x_k \to x$. Next consider $y \in B(x,r) \cap \bar{I}$. By Lemma 5.1, there exists a subsequence of $\{\overline{\Omega}_{n_k}\}$, still denoted $\{\overline{\Omega}_{n_k}\}$, such that

$$\forall k \geq 1, \quad B\left(y, \frac{r}{2^k}\right) \cap \overline{\Omega}_{n_k} \neq \varnothing.$$

For each $k \geq 1$ denote by y_k a point of that intersection. By construction,

$$y_k \in \overline{\Omega}_{n_k} \to y \in \bar{I} \cap B(x,r).$$

There exists $K > 0$ large enough such that

$$\forall k \geq K, \quad y_k \in B(x_k, r).$$

To see this, note that $y \in B(x,r)$ and that

$$\exists \rho > 0, \quad B(y,\rho) \subset B(x,r) \quad \text{and} \quad |y - x| + \frac{\rho}{2} < r.$$

Now

$$|y_k - x_k| \leq |y_k - y| + |y - x| + |x - x_k|$$
$$\leq \frac{r}{2^k} + r - \frac{\rho}{2} + \frac{r}{2^k} \leq r + \left[\frac{r}{2^{k-1}} - \frac{\rho}{2}\right] < r$$

for

$$\frac{r}{2^{k-1}} - \frac{\rho}{2} < 0 \implies k > 2 + \frac{\log(r/\rho)}{\log 2}.$$

So we have constructed a subsequence $\{\Omega_{n_k}\}$ such that for $k \geq K$,

$$x_k \in \partial\Omega_{n_k} \to x \in \partial\mathrm{I},$$
$$y_k \in \overline{\Omega}_{n_k} \cap B(x_k, r) \to y \in \overline{\mathrm{I}} \cap B(x, r).$$

For each k, $\exists d_k \in \mathbf{R}^{\mathrm{N}}$, $|d_k| = 1$, such that

$$C_{y_k}(\lambda, \omega, d_k) \subset \operatorname{int} \Omega_{n_k}.$$

Pick another subsequence of $\{\Omega_{n_k}\}$, still denoted $\{\Omega_{n_k}\}$, such that

$$\exists d \in \mathbf{R}^{\mathrm{N}}, \ |d| = 1, \quad d_k \to d.$$

Now consider $z \in C_y(\lambda, \omega, d)$, and since z is an interior point

$$\exists \rho > 0, \quad B(z, \rho) \subset C_y(\lambda, \omega, d).$$

So there exists $K' \geq K$ such that

$$\forall k \geq K', \quad B\left(z, \frac{\rho}{2}\right) \subset C_{y_k}(\lambda, \omega, d_k) \subset \operatorname{int} \Omega_{n_k}.$$

Therefore, for $k \geq K'$

$$\mathrm{m}\left(B\left(z, \frac{\rho}{2}\right)\right) = \mathrm{m}\left(B\left(z, \frac{\rho}{2}\right) \cap \Omega_{n_k}\right),$$

and as $k \to \infty$

$$\mathrm{m}\left(B\left(z, \frac{\rho}{2}\right)\right) = \mathrm{m}\left(B\left(z, \frac{\rho}{2}\right) \cap \Omega\right),$$

and by Definition 2.3, $z \in \Omega_1$ and $C_y(\lambda, \omega, d) \subset \Omega_1 = \operatorname{int} \mathrm{I}$. This proves that $\mathrm{I} \subset \overline{D}$ satisfies the uniform cone property and $\mathrm{I} \in L(D, \lambda, \omega, r)$. $\square$

6 Existence for the Bernoulli Free Boundary Problem

6.1 An Example: Elementary Modeling of the Water Wave

Consider a fluid in a domain Ω in $\mathbf{R}^3$ and assume that the velocity of the flow u satisfies the *Navier–Stokes equation*

$$u_t + Du\,u - \nu\,\Delta u + \nabla p = -\rho\,g \text{ in } \Omega, \tag{6.1}$$
$$\operatorname{div} u = 0 \text{ in } \Omega, \tag{6.2}$$

where $\nu > 0$ and $\rho > 0$ are the respective *viscosity* and *density* of the fluid, p is the pressure, and g is the gravity constant. The second equation characterizes the incompressibility of the fluid. A standard example considered by physicists is the water wave in a channel. The boundary conditions on $\partial\Omega$ are the *sliding conditions* at the bottom S and on the free boundary Γ, that is,

$$u \cdot n = 0 \text{ on } S \cup \Gamma. \tag{6.3}$$

Assume that a *stationary regime* has been reached so that the velocity of the fluid is no longer a function of the time. Furthermore, assume that the motion of the fluid is *irrotational.* By the classical Hodge's decomposition, the velocity can be written in the form

$$u = \nabla\varphi + \operatorname{curl}\xi. \tag{6.4}$$

As $\operatorname{curl} u = \operatorname{curl}\operatorname{curl}\xi = 0$ we conclude that $\xi = 0$, since $\operatorname{curl}\operatorname{curl}$ is a *good* isomorphism. Then $u = \nabla\varphi$ and the incompressibility condition becomes

$$\Delta\varphi = 0 \text{ in } \Omega. \tag{6.5}$$

Then the Navier–Stokes equation reduces to

$$D(\nabla\varphi)\,\nabla\varphi + \nu\,\nabla(\Delta\varphi) + \nabla p = -\nabla(\rho g z), \tag{6.6}$$

but

$$D(\nabla\varphi)\,\nabla\varphi = \frac{1}{2}\nabla(|\nabla\varphi|^2), \tag{6.7}$$

so that

$$\nabla\left(\frac{1}{2}|\nabla\varphi|^2 + p + \rho g z\right) = 0 \text{ in } \Omega. \tag{6.8}$$

Then if Ω is connected, there exists a constant c such that

$$\frac{1}{2}|\nabla\varphi|^2 + p + \rho g z = c \text{ in } \Omega. \tag{6.9}$$

In the domain Ω, that *Bernoulli condition* explicitly yields the pressure p as a function of the velocity $|\nabla\varphi|$ and the height z in the fluid. The flow is now assumed to be independent of the transverse variable y so that the initial three-dimensional problem in a perfect channel reduces to a two-dimensional one. Since the problem has been reduced to a two-dimensional one, introduce the harmonic conjugate ψ, the so-called *stream function*, so that the boundary condition $\partial\varphi/\partial n = 0$ takes the form $\psi = $ constant on each connected component of the boundary $\partial\Omega$.

On the *free boundary* at the top of the wave, the pressure is related to the existing *atmospheric pressure* p_a through the surface tension $\sigma > 0$ and the mean curvature H of the free boundary. Actually

$$p - p_a = -\sigma\,H \tag{6.10}$$

on the free boundary of the stationary wave, where H is the mean curvature associated with the fluid domain. Also, from the Cauchy conditions, we have $|\nabla\psi| = |\nabla\varphi|$ so that the boundary condition (6.9) on the free boundary of the wave takes the form

$$\frac{1}{2}|\nabla\psi|^2 + \rho g z + \sigma H = p_a. \tag{6.11}$$

In order to simplify the presentation we replace the equation $\Delta\psi = 0$ by $\Delta\psi = f$ in Ω to avoid a *forcing term* on a part of the boundary, and we show that the resulting free boundary problem has the following *shape variational formulation*. Let D be a fixed, sufficiently large, smooth, and bounded open domain in $\mathbf{R}^2$ and a be a real number such that $0 < a < \mathrm{m}(D)$. To find $\Omega \subset D$, $\mathrm{m}(\Omega) = a$ and $\psi \in H^1_0(D)$ such that

$$-\Delta\psi = f \text{ in } \Omega \tag{6.12}$$

and $\psi = constant$ and satisfies the boundary condition (6.11) on the free part $\partial\Omega \cap D$ of the boundary.

For a fixed Ω, the solution of this problem is a minimizing element of the following variational problem:

$$J(\Omega) \stackrel{\text{def}}{=} \inf_{\varphi \in H^1_\diamond(\Omega;D)} \int_\Omega \left(\frac{1}{2}|\nabla\varphi|^2 - f\varphi + \rho g z\right) dx + \sigma\, P_D(\Omega), \tag{6.13}$$

where

$$H^1_\diamond(\Omega;D) \stackrel{\text{def}}{=} \left\{u \in H^1_0(D) : u(x) = 0 \text{ a.e. } x \text{ in } D \setminus \Omega\right\} \tag{6.14}$$

is the relaxation of the definition of the Sobolev space $H^1_0(\Omega)^6$ for any measurable subset Ω of D. Its properties were studied in Theorem 2.9.

With that formulation there exists a measurable $\Omega^* \subset D$, $|\Omega^*| = a$, such that

$$J(\Omega^*) \leq J(\Omega) \quad \forall\Omega \subset D, \; |\Omega| = a. \tag{6.15}$$

By using the methods of Chapter 9 it follows that if $\partial\Omega^*$ is sufficiently smooth the *shape Euler condition* $dJ(\Omega^*; V) = 0$ yields the original free boundary problem and the free boundary condition (6.11). The existence of a solution will now follow from Theorem 5.3 in section 5.1. The case without surface tension is physically important. It occurs in phenomena with "evaporation." For more details see Souli and Zolésio [2]. In that case the previous Bernoulli condition takes the form

$$\left|\frac{\partial\xi}{\partial n}\right|^2 = g^2 \quad (g \geq 0)$$

on the free boundary, so that, if we assume (in the channel setting) that there is no cavitation or recirculation in the fluid, then $\partial\xi/\partial n > 0$ on the free boundary and we get the Neumann-like condition $\partial\xi/\partial n = g$ together with the Dirichlet condition. In section 6.4 we shall consider the case with surface tension.

---

^{6}In Chapter 6 the space $H^1_0(\Omega; D)$ of extensions by zero to D of elements of $H^1_0(\Omega)$ is defined for Ω open. It is then characterized in Lemma 7.2 of Chapter 6 by a capacity condition on the complement $D\backslash\Omega$. This characterization extends to quasi-open sets Ω as defined in section 7 of Chapter 6.

6.2   Existence for a Class of Free Boundary Problems

Consider the following free boundary problem, which has been studied in Zolésio [25, 26, 29]: to find Ω in a fixed hold-all D and a function y on Ω such that

$$-\Delta y = f \text{ in } \Omega, \tag{6.16}$$

$$y = 0 \text{ and } \frac{\partial y}{\partial n} = Q^2 \text{ on } \partial\Omega, \tag{6.17}$$

where f and Q are appropriate functions defined in D. To study this type of problem Alt and Caffarelli [1] introduced the following function:

$$\boxed{J(\varphi) \stackrel{\text{def}}{=} \int_D \frac{1}{2}|\nabla\varphi|^2 - f\varphi\, dx + \int_D Q^2 \chi_{\varphi>0}\, dx} \tag{6.18}$$

to be minimized over

$$\boxed{K \stackrel{\text{def}}{=} \{u \in H_0^1(D) : u(x) \geq 0 \text{ a.e. in } D\},} \tag{6.19}$$

where $\chi_{\varphi>0}$ (resp., $\chi_{\varphi\neq0}$) is the characteristic function of the set $\{x \in D : \varphi(x) > 0\}$[7] (resp., $\{x \in D : \varphi(x) \neq 0\}$). The existence of a solution is based on the following lemma.

Lemma 6.1. *Let $\{u_n\}$ and $\{\chi_n\}$ be two converging sequences such that $u_n \to u$ in $L^2(D)$-strong, and let the χ_n's be characteristic functions, $\chi_n(1 - \chi_n) = 0$, which converge to some function λ in $L^2(D)$-weak. Then*

$$\forall n, \quad (1 - \chi_n)u_n = 0 \quad \Rightarrow \quad \lambda \geq \chi_{u\neq0}. \tag{6.20}$$

Proof. We have $(1 - \chi_n)u_n = 0$, and in the limit $(1 - \lambda)u = 0$. Thus on the set $\{x : u(x) \neq 0\}$ we have $\lambda = 1$; elsewhere λ lies between 0 and 1 as the weak limit of a sequence of characteristic functions. $\square$

Proposition 6.1. *Let f and Q be two elements of $L^2(D)$ such that $f \geq \emptyset$ almost everywhere. There exists u in K which minimizes the function J over the positive cone K of $H_0^1(D)$.*

Proof. Let $\{u_n\} \subset K$ be a minimizing sequence for the function J over the convex set K. Denote by χ_n the characteristic function of the set $\{x \in D : u_n(x) > 0\}$, which is in fact equal to the subset $\{x \in D : u_n(x) \neq 0\}$. It is easy to verify that the sequence $\{u_n\}$ remains bounded in $H_0^1(D)$. Still denote by $\{u_n\}$ a subsequence that weakly converges in $H_0^1(D)$ to an element u of K. That convergence holds in $L^2(D)$-strong so that Lemma 6.1 applies and we get $\lambda \geq \chi_{u\neq0}$ for any weak

---

[7]This set defined up to a set of zero measure is a *quasi-open* set in the sense of section 7 in Chapter 6.

limiting element λ of the sequence $\{\chi_n\}$ (which is bounded in $L^2(D)$). Denote by j the minimum of J over K. Then $J(u_n)$ weakly converges to j. We get

$$\int_D \frac{1}{2}|\nabla u|^2 - fu\, dx \leq \lim_{n\to\infty} \inf \int_D \frac{1}{2}|\nabla u_n|^2 - fu_n\, dx, \tag{6.21}$$

$$\int_D \chi_{u\neq 0} Q^2\, dx \leq \int_D \lambda Q^2 dx = \lim_{n\to\infty} \int_D \chi_n Q^2\, dx. \tag{6.22}$$

Finally, by adding these two estimates we get $J(u) \leq j$. $\square$

Obviously in the upper bound (6.22), Q^2 must be a positive function. Moreover,

(i) the set Ω is given by $\{x \in D : u(x) > 0\}$, and

(ii) the restriction $u|_\Omega$ of u to Ω is a weak solution of the free boundary problem

$$-\Delta u = f \text{ in } \Omega,$$
$$u = 0 \text{ and } \frac{\partial u}{\partial n} = Q^2 \text{ on } \partial\Omega. \tag{6.23}$$

Effectively, the minimization problem (6.18)–(6.19) can be written as a shape optimization problem. First introduce for any measurable subset Ω of D the positive cone

$$H^1_0(\Omega)_+ \overset{\text{def}}{=} \left\{u \in H^1_\diamond(\Omega; D) : u(x) \geq 0 \text{ a.e. } x \in D\right\}$$

in the Hilbert space $H^1_\diamond(\Omega; D)$. Then consider the following shape optimization problem:

$$\boxed{\inf\left\{E(\Omega) : \Omega \text{ is measurable subset in } D\right\},} \tag{6.24}$$

where the energy function E is given by

$$\boxed{E(\Omega) \overset{\text{def}}{=} \min_{\varphi\in H^1_0(\Omega)_+} \left\{\int_\Omega \frac{1}{2}|\nabla\varphi|^2 - f\varphi\, dx\right\} + \int_\Omega Q^2\, dx.} \tag{6.25}$$

The necessary condition associated with the minimum could be obtained by the techniques introduced in section 2.1 of Chapter 9. The important difference with the previous formulation (6.18)–(6.19) is that the independent variable in the objective function is no longer a function but a domain. The shape formulation (6.24)–(6.25) now makes it possible to handle constraints on the volume or the perimeter of the domain, which were difficult to incorporate in the first formulation.

As we have seen in Theorem 2.9, $H^1_\diamond(\Omega; D)$ endowed with the norm of $H^1_0(D)$ is a Hilbert space, and $H^1_0(\Omega)_+$, a closed convex cone in $H^1_\diamond(\Omega; D)$ so that for any measurable subset Ω in D, problem (6.25) has a unique solution y in $H^1_0(\Omega)_+$. Thus we have the following equivalence between problems (6.24)–(6.25) and the minimization (6.18)–(6.19) of J over K.

Proposition 6.2. *Let u be a minimizing element of J over K. Then*

$$\Omega \overset{\text{def}}{=} \{x \in D : u(x) > 0\}$$

is a solution of problem (6.24) *and* $y = u_{|\Omega}$ *is a solution of* (6.25). *Conversely, if Ω is a measurable subset of D and y is a solution of* (6.24)–(6.25) *in* $H_0^1(\Omega)_+$, *then the element u defined by*

$$u(x) \overset{\text{def}}{=} \begin{cases} y(x) & \text{if } x \in \Omega, \\ 0 & \text{if } x \in D \setminus \Omega \end{cases}$$

belongs to K and minimizes J over K.

6.3 Weak Solutions of Some Generic Free Boundary Problems

6.3.1 Problem without Constraint

Problem (6.24)–(6.25) can be relaxed as follows: given any f in $L^2(D)$, G in $L^1(D)$,

$$\boxed{(\mathcal{P}_0) \quad \inf\left\{E(\Omega) : \Omega \subset D \text{ measurable}\right\},} \tag{6.26}$$

where for any measurable subset Ω of D the energy function is now defined by

$$\boxed{E(\Omega) \overset{\text{def}}{=} \min_{\varphi \in H_\diamond^1(\Omega;D)} E_D(\varphi) + \int_\Omega G\,dx,} \tag{6.27}$$

$$E_D(\varphi) \overset{\text{def}}{=} \int_D \frac{1}{2}|\nabla\varphi|^2 - f\varphi\,dx, \tag{6.28}$$

where $H_\diamond^1(\Omega;D)$ is defined in (6.14). By Theorem 2.9 $H_\diamond^1(\Omega;D)$ is contained in $H_\bullet^1(\Omega;D)$, and for any element u in $H_\diamond^1(\Omega;D)$ we have $\nabla u(x) = 0$ for almost all x in $D \setminus \Omega$ so that

$$\forall \varphi \in H_\diamond^1(\Omega;D), \quad E_D(\varphi) = \int_\Omega \frac{1}{2}|\nabla\varphi|^2 - f\varphi\,dx \tag{6.29}$$

$$\Rightarrow E(\Omega) = \min_{\varphi \in H_\diamond^1(\Omega;D)} \left\{\int_\Omega \frac{1}{2}|\nabla\varphi|^2 - f\varphi\,dx + \int_\Omega G\,dx\right\}. \tag{6.30}$$

We have the following existence result for problem $(\mathcal{P}_0)$.

Theorem 6.1. *For any f in $L^2(D)$, $G = Q^2$ in $L^1(D)$, there exists at least one solution to problem $(\mathcal{P}_0)$.*

Proof. Let $\{\Omega_n\}$ be a minimizing sequence for problem $(\mathcal{P}_0)$, and for each n let u_n be the solution to problem (6.27) with $\Omega = \Omega_n$. If χ_n is the characteristic function of the measurable set Ω_n we have u_n in $H_\diamond^1(\Omega_n;D)$, which implies that $(1 - \chi_n)u_n = 0$. On the other hand, the sequence $\{u_n\}$ remains uniformly bounded in $H_0^1(D)$. Taking $\varphi = 0$ in (6.27) we get

$$\int_D \frac{1}{2}|\nabla u_n|^2 - f u_n\,dx \leq 0,$$

and the conclusion follows from the equivalence of norms in $H_0^1(D)$. We can assume that the sequence $\{\chi_n\}$ weakly converges in $L^2(D)$ to an element λ and that the sequence $\{u_n\}$ weakly converges in $H_0^1(D)$ to an element u. From Lemma 6.1 we get $\lambda \geq \chi_{u \neq 0}$ almost everywhere in D. Define

$$\Omega(u) \overset{\text{def}}{=} \{x \in D : u(x) \neq 0\}.$$

Then u belongs to $H_0^1(\Omega(u))$ and we have

$$m(\Omega(u)) = m(\{x \in D : \lambda(x) = 1\}) \leq \alpha,$$

since we have

$$\alpha = m(\{x : \lambda(x) = 1\}) + m(\{x : 0 \leq \lambda(x) < 1\}).$$

In the limit with $\Omega = \Omega(u)$ we get

$$\int_\Omega \frac{1}{2}|\nabla u|^2 - fu\,dx = \int_D \frac{1}{2}|\nabla u|^2 - fu\,dx \leq \liminf \int_D \frac{1}{2}|\nabla u_n|^2 - fu_n\,dx, \quad (6.31)$$

$$\int_\Omega G\,dx \leq \int_D \lambda G\,dx = \lim_{n \to \infty} \int_{\Omega_n} G\,dx. \quad (6.32)$$

By adding (6.31) and (6.32) we get that Ω minimizes E and u minimizes $E(\Omega)$. $\quad\square$

6.3.2   Constraint on the Measure of the Domain Ω

Consider an important variation of the problem (6.24)–(6.25): Given any f in $L^2(D)$, G in $L^1(D)$, and a real number α, $0 < \alpha < m(D)$,

$$\boxed{(\mathcal{P}_0^\alpha) \quad \inf\{E(\Omega) : \Omega \subset D \text{ measurable and } m(\Omega) = \alpha\}.} \quad (6.33)$$

We have the following existence result for problem $(\mathcal{P}_0^\alpha)$.

Theorem 6.2. *For any f in $L^2(D)$, $G = 0$, and any real number α, $0 < \alpha < m(D)$, there exists at least one solution to problem $(\mathcal{P}_0^\alpha)$.*

Proof. Let $\{\Omega_n\}$ be a minimizing sequence for problem $(\mathcal{P}_0)$, and for each n let u_n be the unique solution to problem (6.27). If χ_n is the characteristic function of the measurable set Ω_n, we have u_n in $H_\diamond^1(\Omega_n; D)$, which implies that $(1 - \chi_n)u_n = 0$. On the other hand, by picking $\varphi = 0$ in (6.27), $\{u_n\}$ remains bounded in $H_0^1(D)$:

$$\int_D \frac{1}{2}|\nabla u_n|^2 - fu_n\,dx \leq 0,$$

and the conclusion follows from the equivalence of norms in $H_0^1(D)$. We can assume that $\{\chi_n\}$ weakly converges in $L^2(D)$ to an element λ and that $\{u_n\}$ weakly converges in $H_0^1(D)$ to an element $u \in H_0^1(D)$. In the limit we get

$$\int_D \lambda(x)\,dx = \lim_{n \to \infty} \int_D \chi_n(x)\,dx = \alpha.$$

From Lemma 6.1 we get $\lambda \geq \chi_{u \neq 0}$ almost everywhere in D. Define

$$\Omega(u) \stackrel{\mathrm{def}}{=} \{x \in D : u(x) \neq 0\}.$$

Then u belongs to $H^1_\diamond(\Omega(u); D)$ and we have

$$m(\Omega(u)) = m(\{x \in D : \lambda(x) = 1\}) \leq \alpha$$

(as we have $\alpha = m(\{x : \lambda(x) = 1\}) + m(\{x : 0 \leq \lambda(x) < 1\})$. In the limit we get for $\Omega = \Omega(u)$

$$\int_\Omega \frac{1}{2}|\nabla u|^2 - f u \, dx = \int_D \frac{1}{2}|\nabla u|^2 - f u \, dx \leq \liminf_{n \to \infty} \int_D \frac{1}{2}|\nabla u_n|^2 - f u_n \, dx, \quad (6.34)$$

$$\int_D \lambda G \, dx = \lim_{n \to \infty} \int_{\Omega_n} G \, dx, \quad (6.35)$$

so that

$$\int_\Omega \frac{1}{2}|\nabla u|^2 - f u \, dx + \int_D \lambda G \, dx \leq \inf_{\substack{\Omega' \subset D \\ m(\Omega') = \alpha}} E(\Omega').$$

If $G \geq 0$ almost everywhere in D, we get

$$E(\Omega) \leq \inf_{\substack{\Omega' \subset D \\ m(\Omega') = \alpha}} \{E(\Omega')\},$$

but Ω does not necessarily satisfy the constraint $m(\Omega) = \alpha$. Note that for any measurable set Ω' such that $\Omega \subset \Omega' \subset D$, we have

$$\int_{\Omega'} \frac{1}{2}|\nabla u|^2 - f u \, dx = \int_\Omega \frac{1}{2}|\nabla u|^2 - f u \, dx$$

so that in expression (6.34) Ω can be enlarged to any such Ω'. The inclusion of Ω in Ω' implies the inclusion of $H^1_\diamond(\Omega; D)$ in $H^1_\diamond(\Omega'; D)$. So if $G \leq 0$ almost everywhere in D, then it is readily seen that from (6.27) $E(\Omega') \leq E(\Omega)$. In view of the previous assumption on G, now $G = 0$ almost everywhere in D. To conclude the proof we just have to select Ω' with $m(\Omega') = \alpha$. That measurable set Ω' is admissible and minimizes the objective function in (6.33), and we have

$$E(\Omega') = E(\Omega) = \inf_{\substack{\Omega'' \subset D \\ m(\Omega'') = \alpha}} \{E(\Omega'')\}. \qquad \square$$

Corollary 2. *Assume that $G = 0$ and f in $L^2(D)$ and $f = Q^2$ in $L^1(D)$. Then the following problem has an optimal solution:*

$$(\mathcal{P}_0^{\alpha-}) \qquad \inf_{\substack{\Omega \subset D \\ m(\Omega) \leq \alpha}} E(\Omega). \qquad (6.36)$$

Proof. The proof is similar to the proof of the theorem where the minimizing sequence is chosen such that $m(\Omega_n) \leq \alpha$, so that in the weak limit we get

$$m(\Omega) \leq \int_D \lambda(x) \, dx \leq \alpha. \qquad \square$$

6.4 Weak Existence with Surface Tension

Problems $(\mathcal{P}_0)$, $(\mathcal{P}_0^\alpha)$, and $(\mathcal{P}_0^{\alpha-})$ have optimal solutions, but as they are associated with the homogeneous Dirichlet boundary condition, u in $H^1_\diamond(\Omega; D)$, the optimal domain Ω is in general not allowed to have holes, that is to say, roughly speaking, that the topology of Ω is a priori specified. In many examples it turns out that the solution u physically corresponds to a potential and the classical homogeneous Dirichlet boundary condition is not the appropriate one. The physical condition is that the potential u should be constant on each connected component of the boundary $\partial\Omega$ in D. When Ω is a simply connected domain in $\mathbf{R}^2$, then $\partial\Omega$ has a single connected component so the constant can be taken as zero. In general this constant can be fixed only in one connected component; in the others the constant is an unknown of the problem.

The minimization problems $(\mathcal{P}_0)$, $(\mathcal{P}_0^\alpha)$, and $(\mathcal{P}_0^{\alpha-})$ associated with the Hilbert space $H^1_\bullet(\Omega; D)$ fail (in the sense that the previous techniques for existence of an optimal Ω fail). The main reason is that Lemma 6.1 is no longer true when u_n is replaced by ∇u_n weakly converging in $L^2(D)^N$. The key idea is to recover the equivalent of Lemma 6.1 by imposing the strong $L^2(D)$-convergence of the sequence $\{u_n\}$. In practice $\{u_n\}$ corresponds to the sequence of characteristic functions χ_{Ω_n} of a minimizing sequence $\{\Omega_n\}$. To obtain the strong $L^2(D)$-convergence of a subsequence we add a constraint on the perimeters. Consider the family of finite perimeter sets in D of Definition 5.2(ii) in section 5.1:

$$\mathrm{BPS}(D) \stackrel{\mathrm{def}}{=} \{\Omega \subset D : \chi_\Omega \in BV(D)\},$$

where $BV(D)$ is defined in (5.2). Introduce the following problem indexed by $\sigma > 0$:

$$(\mathcal{P}_\sigma^\alpha) \qquad \inf_{\substack{\Omega \subset D \\ \mathrm{m}(\Omega)=\alpha}} E_\sigma(\Omega), \quad E_\sigma(\Omega) \stackrel{\mathrm{def}}{=} E(\Omega) + \sigma P_D(\Omega), \tag{6.37}$$

$$E(\Omega) \stackrel{\mathrm{def}}{=} \min_{\varphi \in H^1_\bullet(\Omega;D)} \left\{ \int_\Omega \frac{1}{2}|\nabla\varphi|^2 - f\varphi\,dx + \int_\Omega G\,dx \right\}. \tag{6.38}$$

Theorem 6.3. *Let f in $L^2(D)$, G in $L^1(D)$, $\sigma > 0$, $0 \le \alpha < \mathrm{m}(D)$. Then problem $(\mathcal{P}_\sigma^\alpha)$ has at least one optimal solution Ω in $\mathrm{BPS}(D)$.*

Proof. Let $\{\Omega_n\}$ be a minimizing sequence for $(\mathcal{P}_\sigma^\alpha)$ and let $\{\chi_n\}$ be the corresponding sequence of characteristic functions associated with $\{\Omega_n\}$. By picking $\varphi = 0$ in (6.38),

$$P_D(\Omega_n) \le \sigma^{-1}\left[c\|f\|_{L^2(D)}^2 + \int_D |G|\,dx + \sigma P_D(D)\right].$$

By Theorem 5.3 there exists a subsequence of $\{\Omega_n\}$, still indexed by n, and $\Omega \subset D$ with finite perimeter such that $\chi_n \to \chi = \chi_\Omega$ in $L^1(D)$-strong and $\alpha = \mathrm{m}(\Omega_n) = \mathrm{m}(\Omega)$. For each n, let u_n in $H^1_\bullet(\Omega_n; D)$ be the unique minimizer of (6.38). That sequence remains bounded in $H^1_0(D)$. Pick a subsequence, still indexed by n, such that $\{u_n\}$ weakly converges to an element u in $H^1_0(D)$. From the identity

$(1 - \chi_n)\nabla u_n = 0$ almost everywhere in D, we get in the limit $(1 - \chi_\Omega)\nabla u = 0$ almost everywhere in D, so that the limiting element u belongs to $H^1_\bullet(\Omega; D)$. Hence we have

$$\int_\Omega \frac{1}{2}|\nabla u|^2 - fu\,dx = \int_D \frac{1}{2}|\nabla u|^2\,dx - \int_\Omega fu\,dx$$
$$\leq \liminf_{n\to\infty} \int_D \frac{1}{2}|\nabla u_n|^2\,dx - \int_{\Omega_n} fu_n\,dx, \tag{6.39}$$

$$\int_\Omega fu\,dx = \lim_{n\to\infty} \int_{\Omega_n} fu_n\,dx, \quad \int_\Omega G\,dx = \lim_{n\to\infty} \int_{\Omega_n} G\,dx, \tag{6.40}$$

so that

$$\int_\Omega \frac{1}{2}|\nabla u|^2 - fu\,dx + \int_\Omega G\,dx + \sigma P_D(\Omega) \leq \inf_{\substack{\Omega \subset D \\ m(\Omega)=\alpha}} E_\sigma(\Omega). \tag{6.41}$$

$\square$

7 Continuity of the Dirichlet Boundary Value Problem

We have seen in section 2.5 how the homogeneous Dirichlet boundary value problem could be relaxed to measurable domains. In this section we consider the variational problem (2.33) for $\Omega \subset D$ such that $\chi = \chi_\Omega \in X(D)$:

$$y \in H^1_\bullet(\Omega; D), \ \forall \varphi \in H^1_\bullet(\Omega; D), \quad \int_D \nabla y \cdot \nabla \varphi - \chi f \varphi\,dx = 0, \tag{7.1}$$

$$H^1_\bullet(\Omega; D) \overset{\text{def}}{=} \left\{ \varphi \in H^1_0(D) : (1 - \chi_\Omega)\nabla\varphi = 0 \text{ a.e. in } D \right\}. \tag{7.2}$$

Note that this problem can be relaxed further to $\chi \in \overline{\text{co}}\, X(D)$ by replacing χ_Ω by χ in the above definitions. The associated compliance is then given by

$$C(\chi) \overset{\text{def}}{=} \frac{1}{2} \int_D \chi f\, y(\chi)\,dx = - \min_{\varphi \in H^1_\bullet(\chi; D)} E(\chi, \varphi), \tag{7.3}$$

$$E(\chi, \varphi) \overset{\text{def}}{=} \int_D \frac{1}{2}|\nabla\varphi|^2 - \chi f \varphi\,dx. \tag{7.4}$$

One of the difficulties in the study of the continuity of the Dirichlet boundary problem with respect to the underlying domain is the fact that the space of the solution changes with the domain. In this section we consider the continuity for a family of domains living in a fixed bounded open connected subset D of $\mathbf{R}^N$. Each Dirichlet problem corresponding to $\chi = \chi_\Omega \in X(D)$ is first approximated by a transmission problem over D indexed by $\varepsilon > 0$ going to zero:

$$\text{to find } y_\varepsilon = y_\varepsilon(\chi) \in H^1_0(D) \text{ such that } \forall \varphi \in H^1_0(D)$$
$$\int_D \left[\chi + \frac{1}{\varepsilon}(1 - \chi) \right] \nabla y_\varepsilon \cdot \nabla \varphi\,dx = \int_D \chi f \varphi\,dx. \tag{7.5}$$

They are special cases of the general transmission problem (3.3) of section 3.1,

$$\exists y = y(\chi) \in H_0^1(D), \quad \forall \varphi \in H_0^1(D),$$

$$\int_D [k_1\chi + k_2(1 - \chi)]\, \nabla y \cdot \nabla \varphi - \chi f \varphi\, dx = 0 \tag{7.6}$$

for $k_1 = 1$ and $k_2 = 1/\varepsilon$.

Some continuity of the Dirichlet problem with respect to domains is then obtained by combining a continuity result in section 7.1 for the above transmission problem with the approximation (7.5) by transmission problems in section 7.2.

7.1   Continuity of Transmission Problems

Recall that when D is a bounded open connected domain in $\mathbf{R}^N$, the first eigenvalue $\lambda_1 = \lambda_1(D)$ of the Laplacian $-\Delta$ on $H_0^1(D)$ is strictly positive and hence

$$\forall \varphi \in H_0^1(D), \quad \|\varphi\|_{L^2(D)} \le c(D)\,\|\nabla\varphi\|_{L^2(D)}, \quad c(D) \overset{\text{def}}{=} \frac{1}{\sqrt{\lambda_1(D)}}.$$

The following continuity will be sufficient to obtain some existence results.

Proposition 7.1. *Let D be a bounded open connected domain in $\mathbf{R}^N$. Let $k_1 > 0$, $k_2 > 0$, and f in $L^2(D)$ be given such that $\min\{k_1, k_2\} \ge \alpha$ for some constant $\alpha > 0$. For any sequence $\{\chi_n\}$ and χ in $\overline{\mathrm{co}}\, X(D)$ such that $\chi_n \to \chi \in L^2(D)$-strong, there exists a subsequence $\{y_{n_k} = y(\chi_{n_k})\}$ of $\{y_n = y(\chi_n)\}$ such that*

$$y_{n_k} = y(\chi_{n_k}) \to y = y(\chi) \text{ in } H_0^1(D)\text{-strong} \tag{7.7}$$

and any strongly convergent subsequence of $\{y_n\}$ converges to the same limit.

Proof. For each n, let $k_n \overset{\text{def}}{=} k_1\chi_n + k_2(1 - \chi_n)$ and let $y_n \in H_0^1(D)$ be the solution of the equation

$$\forall \varphi \in H_0^1(D), \quad \int_D k_n \nabla y_n \cdot \nabla \varphi - \chi_n f \varphi\, dx = 0. \tag{7.8}$$

The difference $z_n = y_n - y$ is the solution of the variational equation

$$\forall \varphi \in H_0^1(D), \quad \int_D k_n \nabla z_n \cdot \nabla \varphi + (k_n - k)\, \nabla y \cdot \nabla \varphi - (\chi_n - \chi)\, f\varphi\, dx = 0.$$

By assumption, $k_n \ge \alpha$. Setting $\varphi = z_n$, we get

$$\alpha \int_D |\nabla z_n|^2\, dx \le \int_D k_n |\nabla z_n|^2 dx = \int_D -(k_n - k)\, \nabla y \cdot \nabla z_n + (\chi_n - \chi)\, f\, z_n\, dx,$$

$$\|\nabla z_n\|_{L^2(D)}^2 \le \frac{1}{\alpha}\|(k_n - k)\, \nabla y\|_{L^2(D)}\, \|\nabla z_n\|_{L^2(D)} + \|(\chi_n - \chi)\, f\|_{L^2(D)}\, \|z_n\|_{L^2(D)}.$$

If $\lambda_1 = \lambda_1(D)$, the first eigenvalue of the Laplacian $-\Delta$ in $H_0^1(D)$, then

$$\left(\int_D |\nabla z_n|^2 \, dx \right)^{1/2}$$

$$\leq \frac{1}{\alpha} \left[\left(\int_D |k_n - k|^2 \, |\nabla y|^2 \, dx \right)^{1/2} + \frac{1}{\sqrt{\lambda_1}} \left(\int_D |\chi_n - \chi|^2 \, f^2 \, dx \right)^{1/2} \right].$$

On the right-hand side we use the Lebesgue-dominated convergence theorem as follows: there exist some subsequences $\{k_{n_k} - k\}$ (resp., $\{\chi_{n_k} - \chi\}$) which converge to 0 almost everywhere in D while at the same time

$$|k_n - k|^2 \, |\nabla y|^2 \leq |k_2 - k_1|^2 \, |\nabla y|^2 \in L^1(D),$$
$$|\chi_n - \chi|^2 \, f^2 \leq 2f^2 \in L^1(D).$$

The last part of the lemma is straightforward. $\qquad\qquad\square$

7.2 Approximation by Transmission Problems

We now show that, as $\varepsilon \to 0$, the convergence in $H_0^1(D)$ of $y_\varepsilon(\chi)$ to the solution $y(\chi)$ of the relaxed homogeneous Dirichlet problem (7.1)–(7.2) can be controlled through the norm of some *extension operator*.

Theorem 7.1. *Let D be a bounded open connected domain in $\mathbf{R}^N$ and Ω a measurable subset of D such that*

$$\Omega \subset \overline{\Omega} \subset D \text{ and } \Omega^c \overset{\text{def}}{=} D\backslash\overline{\Omega} \text{ is connected,} \tag{7.9}$$

$$H(\Omega^c) \overset{\text{def}}{=} \left\{ \varphi|_{\Omega^c} \, : \, \forall \varphi \in H_0^1(D) \right\}. \tag{7.10}$$

Assume that there exists a continuous linear extension operator

$$P \, : \, H(\Omega^c) \to H_0^1(D), \tag{7.11}$$

$$\forall \varphi \in H(\Omega^c), \quad \|\nabla(P\varphi)\|_{L^2(D)} \leq \|P\| \, \|\nabla\varphi\|_{L^2(\Omega^c)}, \tag{7.12}$$

where $\|P\|$ is the norm of P in $\mathcal{L}(H(\Omega^c), H_0^1(D))$. Given $\chi = \chi_{\overline{\Omega}}$ and ε, $0 < \varepsilon \leq 1$, let y_ε be the solution of the variational equation (7.5) in $H_0^1(D)$ and y the solution of the variational equation (7.1) in the space $H_\bullet^1(\overline{\Omega}; D)$ defined in (7.2) for $\chi = \chi_{\overline{\Omega}}$. Then there exists a constant $c'(D) > 0$ such that

$$\|y_\varepsilon - y\|_{H_0^1(D)} \leq \sqrt{\varepsilon} \, c'(D) \, \|P\|^{3/2} \, \|f\|_{L^2(D)}.$$

Proof. For ε, $0 < \varepsilon \leq 1$, substitute $\varphi = y_\varepsilon$ in (7.5) to get the first bound

$$\|\nabla y_\varepsilon\|_{L^2(D)} \leq c(D) \, \|f\|_{L^2(D)}.$$

In view of the properties of Ω, $\|\nabla\varphi\|_{L^2(\Omega^c)}$ is a norm on $H(\Omega^c)$. Set $\varphi = P(y_\varepsilon|_{\Omega^c})$ in (7.5) to get

$$\frac{1}{\varepsilon}\int_{\Omega^c}|\nabla y_\varepsilon|^2\,dx = \int_\Omega -\nabla y_\varepsilon \cdot \nabla P(y_\varepsilon|_{\Omega^c}) + f\,P(y_\varepsilon|_{\Omega^c})\,dx,$$

$$\int_{\Omega^c}|\nabla y_\varepsilon|^2\,dx$$

$$\leq \varepsilon\left\{ \|\nabla y_\varepsilon\|_{L^2(D)}\,\|\nabla P(y_\varepsilon|_{\Omega^c})\|_{L^2(D)} + \|f\|_{L^2(D)}\,\|P(y_\varepsilon|_{\Omega^c})\|_{L^2(D)} \right\}$$

$$\leq \varepsilon\,2\,c(D)\,\|f\|_{L^2(D)}\,\|\nabla P(y_\varepsilon|_{\Omega^c})\|_{L^2(D)}$$

$$\leq \varepsilon\,2\,c(D)\,\|f\|_{L^2(D)}\,\|P\|\,\|\nabla(y_\varepsilon|_{\Omega^c})\|_{L^2(D\setminus\overline\Omega)} \leq \varepsilon\,2\,c(D)^2\,\|f\|_{L^2(D)}^2\,\|P\|$$

$$\Rightarrow \|\nabla y_\varepsilon\|_{L^2(\Omega^c)} \leq \sqrt{2\varepsilon\,\|P\|}\,c(D)\,\|f\|_{L^2(D)}.$$

Observe that for all $\varphi \in H^1_\bullet(\overline\Omega; D)$, $\nabla\varphi = 0$ in $D\setminus\overline\Omega$. By subtracting (7.5) from (7.1)

$$\forall\varphi \in H^1_\bullet(\overline\Omega; D), \quad \int_D \nabla(y - y_\varepsilon)\cdot\nabla\varphi\,dx = 0.$$

Note that for all $\psi \in H^1_0(D)$

$$\psi|_{D\setminus\overline\Omega} \in H(\Omega^c), \quad \psi - P(\psi|_{D\setminus\overline\Omega}) = 0 \text{ in } D\setminus\overline\Omega,$$

$$\nabla(\psi - P(\psi|_{D\setminus\overline\Omega})) = 0 \text{ a.e. in } D\setminus\overline\Omega$$

$$\Rightarrow \forall\psi \in H^1_0(D), \quad \psi - P(\psi|_{D\setminus\overline\Omega}) \in H^1_\bullet(\overline\Omega; D).$$

In particular, $\varphi = y - y_\varepsilon - P([y - y_\varepsilon]|_{\Omega^c}) \in H^1_\bullet(\overline\Omega; D)$ and

$$\int_D \nabla(y - y_\varepsilon)\cdot\nabla(y - y_\varepsilon - P([y - y_\varepsilon]|_{\Omega^c}))\,dx = 0$$

$$\Rightarrow \|\nabla(y - y_\varepsilon)\|_{L^2(D)}$$

$$\leq \|P\|\,\|\nabla(y - y_\varepsilon)\|_{L^2(\Omega^c)} \leq \|P\|\,\sqrt{2\varepsilon\,\|P\|}\,c(D)\,\|f\|_{L^2(D)}. \qquad \square$$

7.3   Continuity of the Dirichlet Problem

Theorem 7.2. *Let $\{\Omega_n\}$ be a sequence of open sets in D, $\overline\Omega_n \subset D$, satisfying the assumptions of Theorem 7.1. Furthermore, assume that there exists $M > 0$ such that for all n,*

$$\|P_{\Omega_n^c}\|_{\mathcal{L}(H^1(\Omega^c), H^1_0(D))} \leq M \text{ and } \|P_{\Omega^c}\|_{\mathcal{L}(H^1(\Omega^c), H^1_0(D))} \leq M.$$

Then if $\chi_{\overline\Omega_n} \to \chi_{\overline\Omega}$ in $L^2(D)$-strong, there exists a subsequence $\{y_{n_k}\}$ of $\{y_n = y(\chi_{\overline\Omega_n})\}$ such that

$$y_{n_k} \to y = y(\chi_{\overline\Omega}) \text{ in } H^1_0(D)\text{-strong.}$$

If, in addition, all the domains are Lipschitzian, then if $\chi_{\Omega_n} \to \chi_\Omega$ in $L^2(D)$-strong, there exists a subsequence $\{y_{n_k}\}$ of $\{y_n = y(\chi_{\Omega_n})\}$ such that

$$y_{n_k} \to y = y(\chi_\Omega) \quad in \ H_0^1(D)\text{-}strong.$$

Proof. The proof follows from Proposition 7.1 and Theorem 7.1. $\qquad\square$

Remark 7.1.

In Agmon, Douglis, and Nirenberg [1, 2] the extension operator is constructed from the locally uniform cone property of Definition 6.1 in Chapter 2. It can be verified that for a family of domains Ω satisfying the uniform cone property in a bounded hold-all D, the corresponding extension operators P_Ω and P_{Ω^c} are uniformly bounded with respect to all such domains Ω in D. $\qquad\square$

Remark 7.2.

In section 7 of Chapter 6 a sharper continuity of the homogeneous Dirichlet problem for the Laplacian will be given under capacity conditions. Nevertheless, the penalization technique used in this section can be applied to higher-order elliptic, and even parabolic, problems and the Navier–Stokes equation (cf. Dziri and Zolésio [6]). $\qquad\square$

Chapter 4
Topologies Generated by Distance Functions

1 Introduction

In Chapter 3 the characteristic function was used to embed the equivalence classes of measurable subsets of D into $L^p(D)$ or $L^p_{\text{loc}}(D)$, $1 \le p < \infty$, and induce a metric on the equivalence classes of measurable sets. This construction is generic and extends to other set-dependent functions embedded in an appropriate function space. The Hausdorff metric is the result of such a construction, where the distance function plays the role of the characteristic function. The distance function embeds equivalence classes of subsets A of a closed hold-all D with the same closure $\overline{A}$ into the space $C(D)$ of continuous functions. When D is bounded the C^0-norm of the difference of two distance functions is the Hausdorff metric. The Hausdorff topology has many much-desired properties. In particular, for D bounded the set of equivalence classes of nonempty subsets A of D is compact.

Yet the volume and perimeter are not continuous with respect to the Hausdorff topology. This can be fixed by changing the space $C(D)$ for the space $W^{1,p}(D)$ since distance functions also belong to that space. With that metric the volume is again continuous. The price to pay is the loss of compactness even when D is bounded. But new sequentially compact subfamilies can easily be constructed. By analogy with Caccioppoli sets we introduce the sets for which the elements of the Hessian matrix of second-order derivatives of the distance function are bounded measures. They are called sets of *bounded curvature*. Their closure is a Caccioppoli set, and they seem to enjoy other interesting properties. Convex sets belong to that family. General compactness theorems are obtained for such families under global or local conditions. This chapter simultaneously deals with the family of open sets characterized by the distance function to their complement. They are discussed in parallel with the sets described by their distance function.

The properties of distance functions and Hausdorff and Hausdorff complementary metric topologies are studied in section 2. The differentiability of distance functions is discussed in section 3 along with the notions of projections, skeleton, and cracks. $W^{1,p}$-topologies are introduced in section 4. The compact families of sets of bounded and locally bounded curvature are characterized in section 5. The special families of convex sets and Federer's sets of positive reach are studied in sections 6 and 7 and will be further investigated in Chapter 5. Finally, section 8 gives several compactness theorems under global and local conditions on the Hessian matrix of the distance function.

2   Hausdorff Metric Topologies

2.1   The Family of Distance Functions $C_d(D)$

In this section we review some properties of distance functions and present the general approach that will be followed in this chapter.

Definition 2.1.
Given $A \subset \mathbf{R}^N$ the *distance function* from a point x to A is defined as

$$
d_A(x) \overset{\text{def}}{=} \begin{cases} \underset{y \in A}{\inf} \, |y - x|, & A \neq \varnothing, \\ +\infty, & A = \varnothing, \end{cases}
\tag{2.1}
$$

and the family of all distance functions of nonempty subsets of D, as

$$
C_d(D) \overset{\text{def}}{=} \left\{ d_A : \forall A, \, \varnothing \neq A \subset \overline{D} \right\}.
\tag{2.2}
$$

When $D = \mathbf{R}^N$ the family $C_d(\mathbf{R}^N)$ is denoted C_d. $\square$

Observe that d_A is finite in $\mathbf{R}^N$ if and only if $A \neq \varnothing$. We recall the following properties of distance functions.

Theorem 2.1. *Assume that A and B are nonempty subsets of $\mathbf{R}^N$.*

(i) *The map $x \mapsto d_A(x)$ is uniformly Lipschitz continuous in $\mathbf{R}^N$,*

$$
\forall x, y \in \mathbf{R}^N, \quad |d_A(y) - d_A(x)| \leq |y - x|
\tag{2.3}
$$

and $d_A \in C^{0,1}_{\mathrm{loc}}(\overline{\mathbf{R}^N}).$[1]

(ii) *There exists $y \in \overline{A}$ such that $d_A(x) = |y - x|$ and $d_A = d_{\overline{A}}$ in $\mathbf{R}^N$.*

(iii) $\overline{A} = \{x \in \mathbf{R}^N : d_A(x) = 0\}.$

(iv) $d_A = 0$ *in* $\mathbf{R}^N \iff \overline{A} = \mathbf{R}^N.$

(v) $\overline{A} \subset \overline{B} \iff d_A \geq d_B.$

(vi) $d_{A \cup B} = \min\{d_A, d_B\}.$

---

[1] A function f belongs to $C^{0,1}_{\mathrm{loc}}(\mathbf{R}^N)$ if for all bounded open subsets D of $\mathbf{R}^N$ its restriction to D belongs to $C^{0,1}(\overline{D})$.

(vii) d_A *is (Fréchet) differentiable almost everywhere and*

$$|\nabla d_A(x)| \leq 1 \qquad a.e.\ in\ \mathbf{R}^N. \tag{2.4}$$

Proof. (i) For all $z \in A$ and $x, y \in \mathbf{R}^N$

$$|z - y| \leq |z - x| + |y - x|,$$
$$d_A(y) = \inf_{y \in A} |z - y| \leq \inf_{y \in A} |z - x| + |y - x| = d_A(x) + |y - x|,$$

and hence the Lipschitz continuity.

(ii) Let $\{y_n\} \subset A$ be a minimizing sequence

$$d_A(x) = \inf_A |y - x| = \lim_{n \to \infty} |y_n - x|.$$

Then $\{y_n - x\}$ and hence $\{y_n\}$ are bounded sequences. Hence there exist a subsequence converging to some $y \in \overline{A}$ and $d_A(x) = |y - x|$. Clearly by definition $d_{\overline{A}}(x) \leq d_A(x)$ since $A \subset \overline{A}$. So either there exists $y \in A$ such that $d_{\overline{A}}(x) = |y - x|$ and hence $d_{\overline{A}}(x) = d_A(x)$ or there exists $\{y_n\} \subset A$ such that

$$d_{\overline{A}}(x) = |y - x| = \lim_{n \to \infty} |y_n - x| \geq \inf_{z \in A} |z - x| = d_A(x)$$

and we get the identity. By the continuity of d_A, $d_A^{-1}\{0\}$ is closed and

$$A \subset d_A^{-1}\{0\} \quad \Rightarrow \overline{A} \subset d_A^{-1}\{0\}.$$

Conversely, if $d_A(x) = 0$, there exists y in $\overline{A}$ such that $0 = d_A(x) = |y - x|$ and necessarily $x = y \in \overline{A}$. (iii) follows from (ii). (iv) follows from (iii). (v) follows from (ii) for $x \in \overline{A}$, $d_A(x) = 0$, and $d_B(x) \leq d_A(x) = 0$ necessarily implies $d_B(x) = 0$ and $x \in \overline{B}$. Conversely, if $\overline{A} \subset \overline{B}$, then

$$d_B(x) = d_{\overline{B}}(x) = \inf_{y \in \overline{B}} |y - x| \leq \inf_{y \in \overline{A}} |y - x| = d_{\overline{A}}(x) = d_A(x).$$

(vi) is obvious. (vii) follows from Rademacher's theorem. $\qquad\qquad\square$

2.2 Hausdorff Metric Topology

Let D be a nonempty subset of $\mathbf{R}^N$ and associate with each nonempty subset A of $\overline{D}$ the equivalence class

$$[A] \stackrel{\text{def}}{=} \left\{ B : \forall B, B \subset \overline{D} \text{ and } \overline{B} = \overline{A} \right\},$$

since from Theorem 2.1(v) $d_A = d_B$ if and only if $\overline{A} = \overline{B}$. So $\overline{A}$ is the *closed representative* of the class $[A]$. Consider the set

$$\boxed{\mathcal{F}(D) \stackrel{\text{def}}{=} \left\{ [A] : \forall A, \varnothing \neq A \subset \overline{D} \right\}.}$$

By the definition of $[A]$ the map

$$[A] \mapsto d_A \colon \mathcal{F}(D) \to C_d(D) \subset C(\overline{D})$$

is injective. So $\mathcal{F}(D)$ can be identified with the subset of distance functions $C_d(D)$ in $C(\overline{D})$. The distance function plays the same role as the set $X(D)$ in $L^p(D)$ of equivalence classes of characteristic functions of measurable sets.

When D is bounded, $C(\overline{D})$ is a Banach space when endowed with the norm

$$\|f\|_{C(D)} = \sup_{x \in D} |f(x)|.$$

As for the characteristic functions of Chapter 3, this will induce a complete metric

$$\boxed{\rho([A],[B]) \stackrel{\text{def}}{=} \|d_A - d_B\|_{C(D)} = \sup_{x \in D} |d_A(x) - d_B(x)|} \qquad (2.5)$$

on $\mathcal{F}(D)$, which turns out to be equal to the classical Hausdorff metric

$$\boxed{\rho_H([A],[B]) \stackrel{\text{def}}{=} \max \left\{ \sup_{x \in B} d_A(x), \sup_{y \in A} d_B(y) \right\}}$$

(cf. Dugundji [1, p. 205, Chap. IX, Prob. 4.8] for the definition of ρ_H).

When D is open but not necessarily bounded, the space $C(D)$ of continuous functions on D is endowed with the Fréchet topology of uniform convergence on compact subsets K of D, which is defined by the family of seminorms

$$\forall K \text{ compact} \subset D, \quad q_K(f) \stackrel{\text{def}}{=} \max_{x \in K} |f(x)|. \qquad (2.6)$$

It is metrizable since the topology induced by the family of seminorms $\{q_K\}$ is equivalent to the one generated by the subfamily $\{q_{K_k}\}_{k \geq 1}$, where the compact sets $\{K_k\}_{k \geq 1}$ are chosen as follows:

$$K_k \stackrel{\text{def}}{=} \left\{ x \in D : d_{\complement D}(x) \geq \frac{1}{k} \text{ and } |x| \leq k \right\}, \quad k \geq 1 \qquad (2.7)$$

(cf., for instance, Horváth [1, Example 3, p. 116]). Thus the Fréchet topology on $C(D)$ is equivalent to the topology defined by the metric

$$\delta(f, g) \stackrel{\text{def}}{=} \sum_{k=1}^{\infty} \frac{1}{2^k} \frac{q_{K_k}(f - g)}{1 + q_{K_k}(f - g)}. \qquad (2.8)$$

In that case we use the notation $C_{\text{loc}}(D)$. It will be shown below that $C_d(D)$ is a closed subset of $C_{\text{loc}}(D)$ and that this will induce the following complete metric on $\mathcal{F}(D)$:

$$\boxed{\rho_\delta([A],[B]) \stackrel{\text{def}}{=} \delta(d_A, d_B) = \sum_{k=1}^{\infty} \frac{1}{2^k} \frac{q_{K_k}(d_A - d_B)}{1 + q_{K_k}(d_A - d_B)},} \qquad (2.9)$$

which is a natural extension of the Hausdorff metric to an unbounded domain D.

Theorem 2.2. *Let D be a nonempty open (resp., bounded open) subset of $\mathbf{R}^N$.*

(i) *The set $C_d(D)$ is closed in $C_{\mathrm{loc}}(D)$ (resp., $C(\overline{D})$) and ρ_δ (resp., ρ) defines a complete metric topology on $\mathcal{F}(D)$.*

(ii) *When D is a bounded open subset of $\mathbf{R}^N$, the set $C_d(D)$ is compact in $C(\overline{D})$ and the metrics ρ and ρ_H are equal.*

Proof. (i) We use the proof given by Dellacherie [1, p. 42, Thm. 2 and p. 43, Remark 1]. The basic constructions will again be used in Chapter 5. Consider a sequence $\{A_n\}$ of nonempty subsets of $\overline{D}$ such that d_{A_n} converges to some element f of $C_{\mathrm{loc}}(D)$. We wish to prove that $f = d_A$ for

$$A \stackrel{\text{def}}{=} \{x \in \overline{D} : f(x) = 0\}$$

and that the closed subset A of $\overline{D}$ is nonempty. Fix x in $\overline{D}$; then

$$\forall n, \exists y_n \in \overline{A}_n, \quad |y_n - x| = \inf_{z \in A_n} |z - x| = d_{A_n}(x)$$

$$\Rightarrow \lim_{n \to \infty} |y_n - x| = \lim_{n \to \infty} d_{A_n}(x) = f(x).$$

Hence $\{y_n - x\}$ and $\{y_n\} \subset \overline{D}$ are bounded, and there exists a subsequence, still indexed by n, which converges to some $y \in \overline{D}$,

$$y_n \to y, \text{ and } |y - x| = f(x).$$

In particular, $f(y) = 0$, since in the inequality

$$f(y) \le f(y) - d_{A_n}(y) + d_{A_n}(y) - d_{A_n}(y_n) + d_{A_n}(y_n),$$

the last term is zero and d_{A_n} is Lipschitz continuous with constant equal to 1:

$$|f(y)| \le |f(y) - d_{A_n}(y)| + |y - y_n|,$$

and both terms go to zero. By the definition of A, $y \in A$ and A is not empty. Therefore for each $x \in \overline{D}$, there exists $y \in A$ such that

$$f(x) = |y - x| \ge \inf_{z \in A} |z - x| = d_A(x).$$

Next we prove the inequality in the other direction. By construction for any $A_n \subset \overline{D}$

$$\forall x, y \in \overline{D}, \quad |d_{A_n}(x) - d_{A_n}(y)| \le |x - y|$$

and

$$|f(x) - f(y)| \le |f(x) - d_{A_n}(x)| + |d_{A_n}(x) - d_{A_n}(y)| + |d_{A_n}(y) - f(y)|.$$

By uniform convergence the first and last terms converge to zero, and by Lipschitz continuity of d_{A_n},

$$\forall x, y \in \overline{D}, \quad |f(x) - f(y)| \le |x - y|.$$

Hence for all $x \in \overline{D}$ and $y \in A$, $f(y) = 0$ and

$$f(x) \le f(y) + |x - y| = |x - y| \quad \Rightarrow \quad \forall x \in \overline{D} \quad f(x) \le \inf_{y \in A} |x - y| = d_A(x).$$

This proves the reverse equality.

(ii) Observe that on a compact set $\overline{D}$ and for any $A \subset \overline{D}$

$$d_A(x) = \inf_{y \in A} |y - x| \le \sup_{y \in A} |y - x| \le c \overset{\text{def}}{=} \sup_{y, x \in \overline{D}} |y - x| < \infty$$

$$\forall x, y \in \overline{D}, \quad |d_A(y) - d_A(x)| \le |y - x|.$$

The compactness of $C_d(D)$ now follows by the Ascoli–Arzelà Theorem 2.1 of Chapter 2 and part (i). By construction ρ and ρ_δ are metrics on $\mathcal{F}(D)$. By definition, for A and B in the compact $\overline{D}$,

$$\rho(A, B) = \max_{x \in D} |d_A(x) - d_B(x)|$$

$$\ge \max \left\{ \max_{x \in B} |d_A(x) - d_B(x)|, \max_{x \in A} |d_A(x) - d_B(x)| \right\}$$

$$\ge \max \left\{ \max_{x \in B} d_A(x), \max_{x \in A} d_B(x) \right\} = \rho_H(A, B).$$

Conversely, for any $x \in \overline{D}$ and y in A,

$$d_A(x) - d_B(x) \le |y - x| - d_B(x),$$

and there exists $x_B \in \overline{B}$ such that $d_B(x) = |x - x_B|$. Therefore,

$$\forall y \in A, d_A(x) - d_B(x) \le |y - x| - |x - x_B| \le |y - x_B|$$
$$\Rightarrow d_A(x) - d_B(x) \le \inf_{y \in A} |y - x_B| = d_A(x_B) \le \max_{x \in B} d_A(x).$$

Similarly

$$d_B(x) - d_A(x) \le \max_{x \in A} d_B(x),$$

and for all x in $\overline{D}$

$$|d_B(x) - d_A(x)| \le \max \left\{ \max_{x \in B} d_A(x), \max_{x \in A} d_B(x) \right\} \quad \Rightarrow \quad \rho(A, B) \le \rho_H(A, B). \quad \square$$

When D is bounded the family $\mathcal{F}(D)$ enjoys many more interesting properties.

Theorem 2.3. *Let D be a nonempty open (resp., bounded open) subset of $\mathbf{R}^N$. Define for a subset S of $\mathbf{R}^N$ the sets*

$$H(S) \overset{\text{def}}{=} \left\{ d_A \in C_d(D) : S \subset \overline{A} \right\},$$

$$I(S) \overset{\text{def}}{=} \left\{ d_A \in C_d(D) : \overline{A} \subset S \right\},$$

$$J(S) \overset{\text{def}}{=} \left\{ d_A \in C_d(D) : \overline{A} \cap S \ne \varnothing \right\}.$$

(i) *Let S be a subset of $\mathbf{R}^N$. Then $H(S)$ is closed in $C_{\mathrm{loc}}(D)$ (resp., $C(\overline{D})$).*

(ii) *Let S be a closed subset of $\mathbf{R}^N$. Then $I(S)$ is closed in $C_{\mathrm{loc}}(D)$ (resp., $C(\overline{D})$). If, in addition, $S \cap D$ is compact, then $J(S)$ is closed in $C_{\mathrm{loc}}(D)$ (resp., $C(\overline{D})$).*

(iii) *Let S be an open subset of $\mathbf{R}^N$. Then $J(S)$ is open in $C_{\mathrm{loc}}(D)$ (resp., $C(\overline{D})$). If, in addition, $\complement S \cap D$ is compact, then $I(S)$ is open in $C_{\mathrm{loc}}(D)$ (resp., $C(\overline{D})$).*

(iv) *For D bounded, associate with an equivalence class $[A]$*

$$\#([A]) \overset{\mathrm{def}}{=} \text{number of connected components of } \overline{A}.$$

Then the map

$$[A] \mapsto \#([A]) \colon \mathcal{F}(D) \to \mathbf{R}$$

is lower semicontinuous. In particular, for a fixed number $c \geq 0$ the subset

$$\{d_A \in C_d(D) : \#([A]) \leq c\}$$

is compact in $C(\overline{D})$.

Proof. (i) If $S \not\subset \overline{D}$, then $H(S) = \varnothing$ and there is nothing to prove. Assume that $S \subset \overline{D}$. From Theorem 2.1(v), $S \subset \overline{A} \Rightarrow d_S \geq d_A$. So for any sequence $\{d_{A_n}\}$ in $H(S)$ converging to d_A in $C_{\mathrm{loc}}(D)$,

$$\forall n, \quad d_S \geq d_{A_n} \quad \Rightarrow d_S \geq d_A \quad \Rightarrow S \subset \overline{S} \subset \overline{A} \quad \Rightarrow d_A \in H(S).$$

(ii) We use the same technique for $I(S)$ as for $H(S)$, but here we need $S = \overline{S}$ to conclude. For $\overline{D} \cap S = \varnothing$, $J(S) = \varnothing$ and there is nothing to prove. Assume that $\overline{D} \cap S \neq \varnothing$ and consider a sequence $\{d_{A_n}\}$ in $J(S)$ converging to d_A in $C_{\mathrm{loc}}(D)$. Assume that $\overline{A} \cap S = \varnothing$. Then $\overline{A} \subset \overline{D}$ implies $\overline{A} \cap [S \cap \overline{D}] = \overline{A} \cap S = \varnothing$. By assumption, $S \cap \overline{D}$ is compact and

$$\exists \delta > 0, \, \forall x \in S \cap \overline{D}, \quad d_A(x) \geq \delta,$$
$$\exists N \geq 1, \, \forall n \geq N, \quad \|d_{A_n} - d_A\|_{C(D)} \leq \delta/2.$$

So for all $x \in \overline{D} \cap S$

$$d_{A_n}(x) \geq d_A(x) - |d_{A_n}(x) - d_A(x)| \geq \delta - \delta/2 > 0$$
$$\Rightarrow \overline{A}_n \cap S = \overline{A}_n \cap [\overline{D} \cap S] = \varnothing,$$

and this contradicts the fact that $A_n \in J(S)$.

(iii) By definition, for any S, $\complement J(S) = I(\complement S)$:

$$\complement J(S) = \{d_A \in C_d(D) : \overline{A} \cap S = \varnothing\} = \{d_A \in C_d(D) : \overline{A} \subset \complement S\} = I(\complement S).$$

Since $\complement S$ is closed, $\complement J(S)$ is closed from part (ii) and $J(S)$ is open. Similarly, replacing S by $\complement S$ in the previous identity, $I(S) = \complement J(\complement S)$. So from part (ii) if $\overline{D} \cap \complement S$ is compact, then $\complement I(S)$ is closed and $I(S)$ is open. From part (ii) if $\overline{D} \cap \complement S$ is compact, then $\complement I(S)$ is closed and $I(S)$ is open.

(iv) Let $\{A_n\}$ and A be nonempty subsets of $\overline{D}$ such that d_{A_n} converges to d_A in $C(\overline{D})$. Assume that $\#([A]) = k$ is finite. Then there exists a family of disjoint open sets $G_1, \ldots, G_k$ such that

$$\overline{A} \subset G = \cup_{i=1}^{k} G_i \quad \text{and} \quad \forall i, \ \overline{A} \cap G_i \neq \varnothing.$$

In view of the definitions of $I(S)$ and $J(S)$ in part (i),

$$\overline{A} \in \mathcal{U} = \cap_{i=1}^{k} J(G_i) \cap I(G).$$

But $\mathcal{U}$ is not empty and open as the finite intersection of $k + 1$ open sets. As a result there exists $\varepsilon > 0$ and an open neighborhood of $[A]$,

$$N_\varepsilon([A]) = \{[B] : \|d_B - d_A\| < \varepsilon\} \subset \mathcal{U}.$$

Hence since d_{A_n} converges to d_A, there exists $\overline{n} > 0$ such that

$$\forall n \geq \overline{n}, \quad [A_n] \in \mathcal{U},$$

and necessarily

$$\forall n \geq \overline{n}, \quad \overline{A}_n \subset G, \quad \overline{A}_n \cap G_i \neq \varnothing, \forall i$$
$$\Rightarrow \#([A_n]) \geq \#([A]).$$

Therefore, $[A] \mapsto \#([A])$ is lower semicontinuous. Now for $\#([A]) = +\infty$, we repeat the above procedure and refine the open covering. $\qquad\square$

This theorem has many interesting corollaries. For instance, part (iii) says something about the function that gives the number of connected components of $\overline{A}$ (cf. Richardson [1] for an application to image segmentation).

2.3   Hausdorff Complementary Metric Topology and $C_d^c(D)$

In the previous section we dealt with a theory of closed sets, since the equivalence class of the subsets of $\overline{D}$ was completely determined by their unique closure. In partial differential equations the underlying domain is usually open. To accommodate this point of view, consider the set of open subsets Ω of a fixed nonempty open hold-all D in $\mathbf{R}^N$, endowed with the Hausdorff topology generated by the distance functions $d_{\complement\Omega}$ to the complement of Ω. This approach has been used in several contexts, for instance, in Zolésio [6, section 1.3, p. 405] in the context of free boundary problems. Also, Šverák [2] uses the family $C_d^c(D)$ for open sets Ω such that $\#([\complement\Omega]) \leq c$ for some fixed $c > 0$. His main result is that in dimension 2 the convergence of a sequence $\{\Omega_n\}$ to Ω of such sets implies the convergence of the corresponding projection operators $\{P_{\Omega_n} : H_0^1(D) \to H_0^1(\Omega_n)\}$ to $P_\Omega : H_0^1(D) \to H_0^1(\Omega)$, where the projection operators are directly related to homogeneous Dirichlet linear boundary value problems on the corresponding domains $\{\Omega_n\}$ and Ω. In dimension 1 the constraint on the number of components can be dropped. This result will be discussed in Theorem 8.2 of section 8 in Chapter 6.

By analogy with the constructions of the previous section, consider for a nonempty subset D of $\mathbf{R}^N$ the family

$$F(D) \stackrel{\text{def}}{=} \left\{ d_{\complement A} \, : \, \varnothing \neq \complement A \text{ and } A \subset \overline{D} \right\}.$$

By definition,

$$\complement A \supset \complement \overline{D} \quad \Rightarrow \quad d_{\complement A} = 0 \text{ in } \complement \overline{D},$$

and, by Lipschitz continuity,

$$d_{\complement A} = 0 \text{ in } \overline{\complement \overline{D}} \quad \Rightarrow \quad d_{\complement A} \in C(\text{int } \overline{D})$$

and in $C_0(\text{int } \overline{D})$ if D is bounded. If $\text{int } \overline{D} = \varnothing$, then $d_{\complement A} = 0$ in $\mathbf{R}^N$. If $\text{int } \overline{D} \neq \varnothing$, associate with each A the open set

$$\Omega \stackrel{\text{def}}{=} \complement \overline{\complement A} \quad \Rightarrow \quad \complement \Omega = \overline{\complement A}.$$

By definition and the previous considerations

$$\complement A \supset \complement \overline{D} \quad \Rightarrow \quad \Omega = \complement \overline{\complement A} \subset \complement \overline{\complement \overline{D}} = \text{int } \overline{D}.$$

So finally, for $\text{int } \overline{D} \neq \varnothing$,

$$\left\{ d_{\complement A} \, : \, \varnothing \neq \complement A \text{ and } A \subset \overline{D} \right\} = \left\{ d_{\complement \Omega} \, : \, \Omega \neq \mathbf{R}^N \text{ and } \Omega \text{ open } \subset \text{int } \overline{D} \right\}.$$

From this analysis it will be sufficient to consider the family of open subsets of an open hold-all.

Definition 2.2.
Let D be a nonempty open subset of $\mathbf{R}^N$. Define the family of functions

$$\boxed{C_d^c(D) \stackrel{\text{def}}{=} \left\{ d_{\complement \Omega} \, : \, \forall \Omega \text{ open in } D \text{ and } \Omega \neq \mathbf{R}^N \right\}} \tag{2.10}$$

corresponding to the family of open sets

$$\mathcal{G}(D) \stackrel{\text{def}}{=} \left\{ \Omega \subset D : \forall \Omega \text{ open and } \Omega \neq \mathbf{R}^N \right\}. \tag{2.11}$$

For D bounded define the *Hausdorff complementary metric* ρ_H^c on $\mathcal{G}(D)$:

$$\boxed{\rho_H^c(\Omega_2, \Omega_1) \stackrel{\text{def}}{=} \| d_{\complement \Omega_2} - d_{\complement \Omega_1} \|_{C(D)}.} \tag{2.12}$$

$\square$

Note that for $\Omega_1 \subset D$ and $\Omega_2 \subset D$, $d_{\complement \Omega_1} = d_{\complement \Omega_2} = 0$ in $\complement D$. Therefore,

$$d_{\complement \Omega_1} = d_{\complement \Omega_2} \text{ in } D \iff \complement \Omega_1 = \complement \Omega_2 \iff \Omega_1 = \Omega_2.$$

Theorem 2.4. *Let D be a nonempty open subset of $\mathbf{R}^N$.*

(i) *The set $C_d^c(D)$ is closed in $C_{\mathrm{loc}}(D)$.*

(ii) *If, in addition, D is bounded, then $C_d^c(D)$ is compact in $C_0(D)$ and $(\mathcal{G}(D), \rho_H^c)$ is a compact metric space.*

(iii) *(Compactivorous property) Let $\{\Omega_n\}$ and Ω be sets in $\mathcal{G}(D)$ such as*

$$d_{\complement\Omega_n} \to d_{\complement\Omega} \quad in \quad C_{\mathrm{loc}}(D).$$

Then for any compact subset $K \subset \Omega$, there exists an integer $N(K) > 0$ such that

$$\forall n \ge N(K), \quad K \subset \Omega_n.$$

Proof. (i) Let $\{\Omega_n\}$ be a sequence of open subsets in D such that $\{d_{\complement\Omega_n}\}$ is a Cauchy sequence in $C_{\mathrm{loc}}(D)$. For all n, $d_{\complement\Omega_n} = 0$ in $\complement D$ and $\{d_{\complement\Omega_n}\}$ is also a Cauchy sequence in $C_{\mathrm{loc}}(\mathbf{R}^N)$. By Theorem 2.2 (i) the sequence $\{d_{\complement\Omega_n}\}$ converges in $C_{\mathrm{loc}}(\mathbf{R}^N)$ to some distance function d_A. By construction

$$\Omega_n \subset D \implies \complement\Omega_n \supset \complement D \implies d_{\complement\Omega_n} \le d_{\complement D}$$
$$\implies d_A \le d_{\complement D} \implies \overline{A} \supset \complement D \implies \complement\overline{A} \subset D.$$

By choosing the open set $\Omega = \complement\overline{A}$ in D we get

$$d_{\complement\Omega_n} \to d_{\complement\Omega} \quad \text{in } C_{\mathrm{loc}}(\mathbf{R}^N) \text{ and } C_{0,\mathrm{loc}}(D).$$

(ii) To prove the compactness we use the compactness of $C_d(D)$ from Theorem 2.2 (ii) and the fact that $C_d^c(D)$ is closed in $C_0(D)$. Observe that since $\complement\Omega_n \supset \complement D$, $d_{\complement\Omega_n} = 0$ in $\complement D$, $d_{\complement\Omega_n} \in C_0(D)$, and

$$\complement\Omega_n = [\complement\Omega_n \cap \overline{D}] \cup [\complement\Omega_n \cap \complement D] \subset [\complement\Omega_n \cap \overline{D}] \cup \complement D \subset \complement\Omega_n$$
$$\Rightarrow \quad d_{\complement\Omega_n} = d_{[\complement\Omega_n \cap \overline{D}] \cup \complement D} = \min\{d_{\complement\Omega_n \cap \overline{D}}, d_{\complement D}\}.$$

There exists a subsequence, still indexed by n, and a set A, $\varnothing \ne \overline{A} \subset \overline{D}$ such that

$$d_{\complement\Omega_n \cap \overline{D}} \to d_{\overline{A}} \quad \text{in } C_0(D)$$
$$\Rightarrow \quad d_{\complement\Omega_n} = \min\{d_{\complement\Omega_n \cap \overline{D}}, d_{\complement D}\} \to \min\{d_{\overline{A}}, d_{\complement D}\} = d_{\overline{A} \cup \complement D} \quad \text{in } C(\overline{D})$$
$$\Rightarrow \quad d_{\complement\Omega_n} \to d_{\overline{A} \cup \complement D} \quad \text{in } C(\overline{\mathbf{R}^N})$$

since $d_{\complement\Omega_n} = 0 = d_{\overline{A} \cup \complement D}$ in $\complement D$. The result now follows by choosing the open set

$$\Omega \stackrel{\text{def}}{=} \complement[\overline{A} \cup \complement D] = \complement\overline{A} \cap D \subset D.$$

(iii) Define

$$m \stackrel{\text{def}}{=} \inf_{x \in K} \inf_{y \in \complement\Omega} |x - y| = \inf_{x \in K} d_{\complement\Omega}(x).$$

Since $K \subset \Omega \subset D$, there exists $\hat{x} \in K$, $\hat{y} \in \partial\Omega$ such that

$$m = \inf_{x \in K} \inf_{y \in \partial\Omega} |x - y| = |\hat{x} - \hat{y}|.$$

Necessarily, $m > 0$ since $\hat{x} \in K \subset \Omega$ and $\hat{y} \in \partial\Omega$. Now

$$\exists N > 0, \forall n \geq N, \quad \|d_{\complement\Omega_n} - d_{\complement\Omega}\|_{C(K)} < m/2$$
$$\Rightarrow \forall x \in K, \quad d_{\complement\Omega_n}(x) \geq d_{\complement\Omega}(x) - |d_{\complement\Omega_n}(x) - d_{\complement\Omega}(x)| \geq m - m/2 > 0$$
$$\Rightarrow x \notin \overline{\complement\Omega_n} \quad \Rightarrow x \in \complement\overline{\complement\Omega_n} = \operatorname{int}\Omega_n = \Omega_n \quad \Rightarrow K \subset \Omega_n. \qquad \square$$

Notation 2.1.

It will be convenient to write

$$\Omega_n \xrightarrow{H^c} \Omega$$

for the Hausdorff complementary convergence of open sets of $\mathcal{G}(D)$,

$$d_{\complement\Omega_n} \to d_{\complement\Omega} \quad \text{in } C_{\mathrm{loc}}(D). \qquad \square$$

3   Projection, Skeleton, Crack, and Differentiability

In this section we study the connection between the gradient of d_A and $d_{\complement A}$ and the projections and the characteristic functions associated with $\overline{A}$ and $\overline{\complement A}$. We further relate the set of singularities of the gradients and the notions of skeleton and set of cracks.

Definition 3.1.

(i) Given $A \subset \mathbf{R}^N$, $\varnothing \neq A$ (resp., $\varnothing \neq \complement A$), and $x \in \mathbf{R}^N$, the *set of projections of x on A* (resp., $\complement A$) is given by

$$\Pi_A(x) \stackrel{\text{def}}{=} \left\{ p \in \overline{A} : |p - x| = d_A(x) \right\} \tag{3.1}$$
$$\left(\text{resp., } \Pi_{\complement A}(x) \stackrel{\text{def}}{=} \left\{ p \in \overline{\complement A} : |p - x| = d_{\complement A}(x) \right\} \right). \tag{3.2}$$

The elements of $\Pi_A(x)$ (resp., $\Pi_{\complement A}(x)$) are called *projections* onto $\overline{A}$ (resp., $\overline{\complement A}$) and denoted by $p_A(x)$ (resp., $p_{\complement A}(x)$).

(ii) Given $A \subset \mathbf{R}^N$, $\varnothing \neq A$ (resp., $\varnothing \neq \complement A$), the set of points where the projection on A (resp., $\complement A$) is not unique,

$$\mathrm{Sk}_{\mathrm{ext}}(A) \stackrel{\text{def}}{=} \left\{ x \in \mathbf{R}^N : \Pi_A(x) \text{ is not a singleton} \right\} \tag{3.3}$$
$$\left(\text{resp., } \mathrm{Sk}_{\mathrm{int}}(A) \stackrel{\text{def}}{=} \left\{ x \in \mathbf{R}^N : \Pi_{\complement A}(x) \text{ is not a singleton} \right\} \right) \tag{3.4}$$

is called the *exterior (resp., interior) skeleton*. Since for $x \in \partial A$ the sets $\Pi_A(x)$ and $\Pi_{\complement A}(x)$ are singletons,

$$\mathrm{Sk}_{\mathrm{ext}}(A) \subset \operatorname{int}\complement A = \complement\overline{A} \quad \text{and} \quad \mathrm{Sk}_{\mathrm{int}}(A) \subset \operatorname{int} A. \qquad \square$$

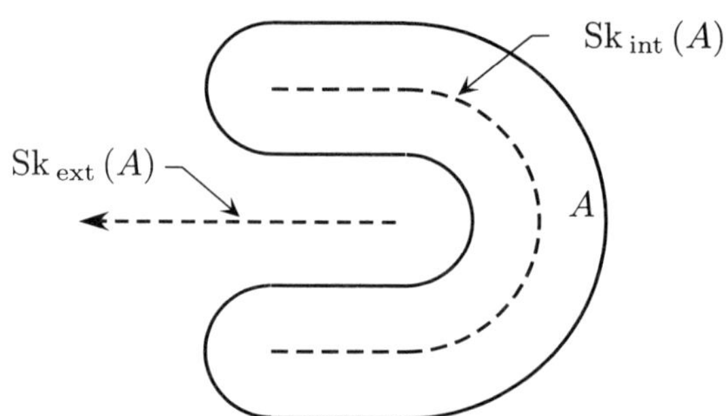

Figure 4.1. *Nonuniqueness of the projection and skeleton.*

Intuitively, the smoothness of the boundary ∂A of A is related to the smoothness of $\nabla d_A(x)$ in a small exterior neighborhood of the boundary ∂A of A. However, even for very smooth sets, $\nabla d_A(x)$ may not exist far from the boundary, as shown in Figure 4.1, where $\nabla d_A(x)$ exists outside $\overline{A}$ except on a semi-infinite line. The directional derivative of the square of the distance function can be computed by using a theorem on the differentiability of the min with respect to a parameter (cf. Chapter 9, section 2.3, Thm. 2.1). It is related to the support function of the set $\Pi_A(x)$.

Theorem 3.1. *Let A, $\varnothing \neq A \subset \mathbf{R}^{\mathrm{N}}$, and $x \in \mathbf{R}^{\mathrm{N}}$. Define*

$$\boxed{f_A(x) \stackrel{\text{def}}{=} \frac{1}{2}\left(|x|^2 - d_A^2(x)\right).}$$

(i) *The set $\Pi_A(x)$ is nonempty, compact, and*

$$\forall x \notin \overline{A}, \quad \Pi_A(x) \subset \partial A \text{ and } \forall x \in \overline{A} \quad \Pi_A(x) = \{x\}.$$

(ii) *For all x and v in $\mathbf{R}^{\mathrm{N}}$*

$$dd_A^2(x;v) \stackrel{\text{def}}{=} \lim_{t \searrow 0} \frac{d_A^2(x+tv) - d_A^2(x)}{t} = \min_{z \in \Pi_A(x)} 2(x-z)\cdot v$$

$$= 2(x\cdot v - \sigma_{\Pi_A(x)}(v)),$$

$$df_A(x;v) \stackrel{\text{def}}{=} \lim_{t \searrow 0} \frac{f_A(x+tv) - f_A(x)}{t} = \sigma_{\Pi_A(x)}(v) = \sigma_{\mathrm{co}\,\Pi_A(x)}(v),$$

where σ_B is the support function of the set B,

$$\sigma_B(v) = \sup_{z \in B} z\cdot v,$$

and $\mathrm{co}\,B$ is the convex hull of B.

(iii) *The following statements are equivalent:*

(a) $d_A^2(x)$ *is (Fréchet) differentiable at* x,

(b) $d_A^2(x)$ *is Gâteaux differentiable at* x,

(c) $\Pi_A(x)$ *is a singleton.*

Henceforth

$$\mathrm{Sk}_{\mathrm{ext}}(A) = \left\{x \in \mathbf{R}^N : \nabla d_A^2(x) \not\exists\right\} \subset \complement\overline{A} = \mathrm{int}\,\complement A,$$
$$\mathrm{Sing}\,(\nabla d_A) = \mathrm{Sk}_{\mathrm{ext}}(A) \cup \mathrm{C}_{\mathrm{ext}}(A),$$

where

$$\boxed{\mathrm{C}_{\mathrm{ext}}(A) \stackrel{\mathrm{def}}{=} \left\{x \in \mathbf{R}^N : \nabla d_A^2(x) \exists \text{ but } \nabla d_A(x) \not\exists\right\}.}$$

When d_A^2 *is differentiable at* x, $\Pi_A(x) = \{p_A(x)\}$ *is a singleton and*

$$\boxed{p_A(x) = x - \frac{1}{2}\nabla d_A^2(x) = \nabla f_A(x).} \tag{3.5}$$

For all $x \in \overline{A}$, $\Pi_A(x) = \{x\}$, d_A^2 *is differentiable at* x *and* $\nabla d_A^2(x) = 0$.

(iv) *For* $\complement A \neq \varnothing$, *the conclusions of parts* (i)–(iii) *are true with* $\complement A$ *in place of* A *and*

$$\boxed{\mathrm{C}_{\mathrm{int}}(A) \stackrel{\mathrm{def}}{=} \left\{x \in \mathbf{R}^N : \nabla d_{\complement A}^2(x) \exists \text{ but } \nabla d_{\complement A}(x) \not\exists\right\}}$$

in place of $\mathrm{C}_{\mathrm{ext}}(A)$.

Proof. (i) Existence. The function

$$x \mapsto |z - x|^2 \colon \mathbf{R}^N \to \mathbf{R}$$

is continuous and $|z - x|^2 \to \infty$ as $|z| \to \infty$. Hence for any

$$k > m = \inf_{z \in A} |z - x|^2 \geq 0$$

the set

$$F_k = \{z \in A : |z - x|^2 \leq k\}$$

is nonempty and compact and

$$\inf_{z \in A} |z - x|^2 = \inf_{z \in F_k} |z - x|^2.$$

Thus $\Pi_A(x)$ is nonempty and

$$\Pi_A(x) = \bigcap_{k > m} F_k$$

is compact. By definition, $\Pi_A(x) \subset \overline{A}$. If $x \in \partial A$, $p_A(x) = x \in \partial A$. If $x \in \text{int}\,\complement A$, then $d_A(x) > 0$. Now if $p_A(x) \in \text{int}\,A$, then there exists an open ball B of radius r, $0 < r < d_A(x)/2$, at $p_A(x)$, which is contained in A. Choose

$$y = p_A(x) + r\,\frac{x - p_A(x)}{|x - p_A(x)|}.$$

Then y belongs to $\overline{A}$ and

$$\begin{aligned}
y - x &= p_A(x) - x + r\,\frac{x - p_A(x)}{|x - p_A(x)|} \\
&= \left[1 - \frac{r}{d_A(x)}\right](p_A(x) - x).
\end{aligned}$$

So there exists $y \in \overline{A}$ such that

$$|y - x| = \left[1 - \frac{r}{d_A(x)}\right]|p_A(x) - x| < d_A(x).$$

This contradicts the minimality of $d_A(x)$. Hence $x \in \partial A$.

(ii) Differentiability of f_A. It is sufficient to prove the semidifferentiability of d_A^2; that is, for each x and v the existence of the limit

$$dd_A^2(x; v) \overset{\text{def}}{=} \lim_{t \searrow 0} \frac{d_A^2(x + tv) - d_A^2(x)}{t}.$$

Recall that for $t \geq 0$,

$$\Pi_A(x + t\,v) = \left\{p_t \in \overline{A} : |x + tv - p_t|^2 = d_A^2(x + tv)\right\}$$

and consider for $t > 0$ the quotient

$$q_t = \frac{d_A^2(x + tv) - d_A^2(x)}{t}.$$

For all $p \in \Pi_A(x)$ and $p_t \in \Pi_A(x + t\,v)$

$$\begin{aligned}
q_t &= \frac{|x + tv - p_t|^2 - |x - p|^2}{t} \\
&\leq \frac{|x + tv - p|^2 - |x - p|^2}{t} = t\,|v|^2 + 2\,v \cdot (x - p),
\end{aligned}$$

and for all $p \in \Pi_A(x)$

$$\limsup_{t \searrow 0} q_t \leq 2\,v \cdot (x - p) \quad \Rightarrow \quad \overline{q} = \limsup_{t \searrow 0} q_t \leq 2 \inf_{p \in \Pi_A(x)} v \cdot (x - p).$$

In the other direction choose a sequence $t_k > 0$ such that $t_k \to 0$ and $q_{t_k} \to \underline{q} = \liminf_{t \searrow 0}$. The corresponding sequence p_{t_k} in $\overline{A}$ is uniformly bounded since

$$\begin{aligned}
|p_{t_k}| &\leq |x + t_k v - p_{t_k}| + |x + t_k v| \\
&\leq d_A(x + t_k v) - d_A(x) + d_A(x) + |x + t_k v| \\
&\leq t_k|v| + d_A(x) + |x + t_k v|,
\end{aligned}$$

and we can find $p_0 \in \bar{A}$ and a subsequence of $\{p_{t_k}\}$, still denoted $\{p_{t_k}\}$, such that $p_{t_k} \to p_0$. But by continuity

$$d_A(x + t_k v) \to d_A(x) \text{ and } |x + t_k v - p_{t_k}| \to |x - p_0|$$
$$\Rightarrow \exists p_0 \in \bar{A} \text{ such that } d_A(x) = |x - p_0| \quad \Rightarrow p_0 \in \Pi_A(x).$$

Therefore,

$$
\begin{aligned}
q_{t_k} &= \frac{|x + t_k v - p_{t_k}|^2 - |x - p_0|^2}{t_k} \\
&\geq \frac{|x + t_k v - p_{t_k}|^2 - |x - p_{t_k}|^2}{t_k} = t_k |v|^2 + 2\, v \cdot (x - p_{t_k}) \\
&\Rightarrow \underline{q} \geq 2\, v \cdot (x - p_0) \geq 2 \inf_{p \in \Pi_A(x)} v \cdot (x - p).
\end{aligned}
$$

Hence

$$d_A^2(x; v) = 2 \inf_{p \in \Pi_A(x)} v \cdot (x - p).$$

(iii) From part (ii) $d_A^2(x)$ is Gâteaux differentiable at x if and only if $\Pi_A(x)$ is a singleton. The (Fréchet) differentiability is a standard part of Rademacher's theorem, but we reproduce it here for completeness as a lemma.

Lemma 3.1. *Let* $f : \mathbf{R}^N \to \mathbf{R}$ *be a locally Lipschitzian function. Then* f *is (Fréchet) differentiable at* x*; that is,*

$$\lim_{y \to x} \frac{f(y) - f(x) - \nabla f(x) \cdot (y - x)}{|y - x|} = 0$$

if and only if it is Gâteaux differentiable at x*; that is,*

$$\forall v \in \mathbf{R}^N, \quad df(x; v) = \lim_{t \searrow 0} \frac{f(x + tv) - f(x)}{t} \quad \text{exists}$$

and the map $v \mapsto df(x; v)$ *is linear and continuous.*

Proof. It is sufficient to prove that the Gâteaux differentiability implies that

$$\lim_{y \to x} \frac{f(y) - f(x) - \nabla f(x) \cdot (y - x)}{|y - x|} \to 0.$$

The function f is locally Lipschitzian and there exist $\delta_1 > 0$ and $c_1 > 0$ such that f is uniformly Lipschitzian in the open ball $B(x, \delta_1)$ with Lipschitz constant c_1. Therefore, $|\nabla f(x)| \leq c_1$. The unit sphere $S(0, 1) = \{v \in \mathbf{R}^N : |v| = 1\}$ is compact and can be covered by the family of open balls $\{B(v, \varepsilon/(4c_1)) : v \in S(0, 1)\}$ for an arbitrary $\varepsilon > 0$. Hence there exists a finite subcover $\{B(v_n, \varepsilon/(4c_1)) : 1 \leq n \leq N\}$ of $S(0, 1)$. As a result, given any $v \in S(0, 1)$, there exists n, $1 \leq n \leq N$, such that $|v - v_n| \leq \varepsilon/(4c_1)$. Define the function

$$g(x, v, t) = \frac{f(x + tv) - f(x)}{t} - \nabla f(x) \cdot v$$

for $x \in \mathbf{R}^N$, $v \in S(0,1)$, and $t > 0$. Since $f(x)$ is Gâteaux differentiable

$$\exists \delta, \ 0 < \delta < \delta_1, \ \forall n, \ 1 \le n \le N, \ \forall t < \delta, \quad |g(x, v_n, t)| < \varepsilon/2.$$

Hence for $|y - x| < \delta$

$$\left| g\left(x, \frac{y-x}{|y-x|}, |y-x|\right) \right|$$

$$\le |g(x, v_n, |y-x|)| + \left| g\left(x, \frac{y-x}{|y-x|}, |y-x|\right) - g(x, v_n, |y-x|) \right|$$

$$\le |g(x, v_n, |y-x|)| + 2\, c_1 \left| \frac{y-x}{|y-x|} - v_n \right| \le \frac{\varepsilon}{2} + 2\, c_1\, \frac{\varepsilon}{(4c_1)} = \varepsilon,$$

and we conclude that $f(x)$ is differentiable at x. $\qquad\qquad\qquad\qquad\square$

From the above equivalences, when d_A^2 is differentiable at x, then $\Pi_A(x) = \{p_A(x)\}$ is a singleton, and from part (ii)

$$\nabla d_A^2(x) = 2(x - p_A(x)),$$

which yields the expression for $p_A(x)$. When $x \in \overline{A}$, $\Pi_A(x) = \{x\}$, and by substitution, $\nabla d_A^2(x) = 0$. This completes the proof of the theorem. $\qquad\square$

Remark 3.1.
In general, for each $v \in \mathbf{R}^N$,

$$\exists p \in \Pi_A(x), \quad df(x; v) = 2(x - p) \cdot v,$$

but the choice of p depends on the direction v and is not necessarily unique. The set-valued map

$$x \mapsto \Pi_A(x) \colon \mathbf{R}^N \to \mathcal{K}(\mathbf{R}^N),$$

($\mathcal{K}(\mathbf{R}^N)$, the set of nonempty compact subsets of $\mathbf{R}^N$) contains a lot of information about the set A. $\qquad\square$

We now turn to the differentiability of d_A. The distance function is uniformly Lipschitzian of constant 1 and differentiable almost everywhere. When it is differentiable, d_A^2 is also differentiable, and from Theorem 3.1, $\Pi_A(x)$ contains a unique element $p_A(x)$. However, the smoothness of the boundary ∂A does not imply that $\Pi_A(x)$ is a singleton everywhere in $\mathbf{R}^N$. For instance, if

$$A = \{x \in \mathbf{R}^N : |x| = 1\},$$

then ∂A is C^∞ but $\Pi_A(0) = A$. However, we shall see in section 6 that for convex sets A, $\Pi_A(x)$ is always a singleton.

We now explicitly compute the gradients of d_A and $d_{\complement A}$ and relate them to the characteristic functions of the closures of A and $\complement A$.

Theorem 3.2.

(i) *Let A, $\varnothing \neq A \subset \mathbf{R}^{\mathrm{N}}$. If $\nabla d_A(x)$ exists at a point x in $\mathbf{R}^{\mathrm{N}}$, then $\Pi_A(x) = \{p_A(x)\}$ is a singleton,*

$$d_A(x) = |p_A(x) - x| \quad and \quad \nabla d_A(x) = \begin{cases} 0 & if\ x \in \overline{A}, \\ \dfrac{x - p_A(x)}{|x - p_A(x)|} & if\ x \notin \overline{A}. \end{cases}$$

(ii) *For A, $\varnothing \neq A \subset \mathbf{R}^{\mathrm{N}}$, and $x \in \mathbf{R}^{\mathrm{N}} \setminus \partial A$, d_A is differentiable at x if and only if d_A^2 is differentiable at x.*

(iii) *Given $A \subset \mathbf{R}^{\mathrm{N}}$, $\varnothing \neq A$ (resp., $\varnothing \neq \complement A$),*

$$\mathrm{m}(\mathrm{Sing}\,(\nabla d_A)) = 0 = \mathrm{m}(\mathrm{Sing}\,(\nabla d_{\complement A})),$$

and for almost all x

$$\chi_{\overline{A}}(x) = 1 - |\nabla d_A(x)|$$
$$and\ \chi_{\mathrm{int}\,\complement A}(x) = |\nabla d_A(x)|\ in\ \mathbf{R}^{\mathrm{N}} \setminus \mathrm{Sing}\,(\nabla d_A)$$
$$(resp.,\ \chi_{\overline{\complement A}}(x) = 1 - |\nabla d_{\complement A}(x)|$$
$$and\ \chi_{\mathrm{int}\,A}(x) = |\nabla d_{\complement A}(x)|\ in\ \mathbf{R}^{\mathrm{N}} \setminus \mathrm{Sing}\,(\nabla d_{\complement A})).$$

(iv) *Given $x \in \mathbf{R}^{\mathrm{N}}$, $\alpha \in [0,1]$, $p \in \Pi_A(x)$, and $x_\alpha \stackrel{\mathrm{def}}{=} p + \alpha\,(x - p)$,*

$$d_A(x_\alpha) = |x_\alpha - p| = \alpha\,|x - p| = \alpha\,d_A(x)$$
$$\forall \alpha \in [0,1], \quad \Pi_A(x_\alpha) \subset \Pi_A(x).$$

In particular, if $\Pi_A(x)$ is a singleton, then $\Pi_A(x_\alpha)$ is a singleton and $\nabla d_A^2(x_\alpha)$ exists for all α, $0 \leq \alpha \leq 1$. If, in addition, $x \neq \overline{A}$, $\nabla d_A(x_\alpha)$ exists for all $0 < \alpha \leq 1$.

Proof. (i) If $\nabla d_A(x)$ exists for $x \in \overline{A}$, then for all $v \in \mathbf{R}^{\mathrm{N}}$

$$\nabla d_A(x) \cdot v = \lim_{t \searrow 0} \frac{d_A(x + tv) - d_A(x)}{t} = \lim_{t \searrow 0} \frac{d_A(x + tv)}{t} \geq 0,$$

which implies that $\nabla d_A(x) = 0$. If $\nabla d_A(x)$ exists for $x \notin \overline{A}$, then

$$\nabla d_A^2(x) = 2\,d_A(x)\,\nabla d_A(x)$$

exists, and by Theorem 3.1, $\Pi_A(x) = \{p_A(x)\}$ is a singleton and $\nabla d_A^2(x) = 2(x - p_A(x))$. But for $x \notin \overline{A}$,

$$d_A(x) = |p_A(x) - x| > 0,$$

and it is sufficient to divide both sides by $2\,|p_A(x) - x|$.

(ii) If $\nabla d_A(x)$ exists, then $\nabla d_A^2(x) = 2\, d_A(x)\, \nabla d_A(x)$ exists. Conversely, assume that $\nabla d_A^2(x)$ exists. If $x \in \operatorname{int} A$, then $d_A = 0$ in some neighborhood of x and hence $\nabla d_A(x)$ exists and is equal to zero. If $x \notin \overline{A}$, then for $t \neq 0$

$$\frac{d_A(x + tv) - d_A(x)}{t} = \frac{d_A^2(x + tv) - d_A^2(x)}{t} \, \frac{1}{d_A(x + tv) + d_A(x)},$$

and since $d_A(x) > 0$ the limit exists as $t \to 0$:

$$dd_A(x; v) = dd_A^2(x; v)\, \frac{1}{2\, d_A(x)} = \nabla d_A^2(x) \cdot v\, \frac{1}{2\, d_A(x)}.$$

So $\nabla d_A(x)$ exists and is equal to $\nabla d_A^2(x)/(2\, d_A(x))$.

(iii) From part (i) if $\nabla d_A(x)$ exists for $x \in \overline{A}$, it is equal to 0. Hence $\chi_{\overline{A}}(x) = 1 = 1 - |\nabla d_A(x)|$ on $\overline{A}\backslash\operatorname{Sing}(\nabla d_A)$. From the definition of $\operatorname{Sk}_{\mathrm{ext}}(A)$, $\nabla d_A(x)$ exists and $|\nabla d_A(x)| = 1$ for all points in $\operatorname{int}\mathbb{C}A\backslash\operatorname{Sk}_{\mathrm{ext}}(A)$. Hence from part (i) $\chi_{\overline{A}}(x) = 0 = 1 - |\nabla d_A(x)| = 0$ on $\operatorname{int}\mathbb{C}A\backslash\operatorname{Sk}_{\mathrm{ext}}(A)$. Since $\mathrm{m}(\operatorname{Sing}(\nabla d_A)) = 0$, the above identities are satisfied almost everywhere in $\mathbf{R}^N$. We get the same results with $\mathbb{C}A$ in place of A.

(iv) If $x \in \overline{A}$, $\Pi_A(x) = \{x\}$ and there is nothing to prove. If $x \notin \overline{A}$, then $d_A(x) = |x - p| > 0$ and $p \neq x$ for all $p \in \Pi_A(x)$. First

$$d_A(x_\alpha) \leq |x_\alpha - p| = \alpha\, |x - p|.$$

If the inequality is strict, then there exists $p_\alpha \in \overline{A}$ such that $|x_\alpha - p_\alpha| = d_A(x_\alpha)$ and

$$|x - p_\alpha| \leq |x - x_\alpha| + |x_\alpha - p_\alpha| = (1 - \alpha)|x - p| + d_A(x_\alpha)$$
$$< |x - p| = d_A(x).$$

This contradicts the minimality of $d_A(x)$ with respect to A. Therefore,

$$d_A(x_\alpha) = \alpha\, |x - p| = |x_\alpha - p| \quad \Rightarrow p \in \Pi_A(x_\alpha).$$

Now for any $p_\alpha \in \Pi_A(x_\alpha)$, $p_\alpha \in \overline{A}$ and

$$|p_\alpha - x| \leq |p_\alpha - x_\alpha| + |x_\alpha - x| = \alpha\, |x - p| + (1 - \alpha)\, |p - x| = |p - x| = d_A(x)$$

and $p_\alpha \in \Pi_A(x)$. Hence $\Pi_A(x_\alpha) \subset \Pi_A(x)$. This completes the proof. $\qquad\square$

Observe that for points outside of $\overline{A}$ the norm of $\nabla d_A(x)$ is equal to 1. It coincides with the outward unit normal to $\overline{A}$ at the point $p(x)$ when A is sufficiently smooth. When A is not smooth this normal is not always unique, as can be seen in Figure 4.2.

We now complete the description of the singularities of the gradients of d_A and $d_{\mathbb{C}A}$,

$$\operatorname{Sing}(\nabla d_A) \stackrel{\mathrm{def}}{=} \left\{ x \in \mathbf{R}^N \,:\, \nabla d_A(x)\,\nexists \right\}, \tag{3.6}$$

$$\operatorname{Sing}(\nabla d_{\mathbb{C}A}) \stackrel{\mathrm{def}}{=} \left\{ x \in \mathbf{R}^N \,:\, \nabla d_{\mathbb{C}A}(x)\,\nexists \right\} \tag{3.7}$$

by giving a name to the sets $\mathrm{C}_{\mathrm{ext}}(A)$ and $\mathrm{C}_{\mathrm{int}}(A)$ introduced in Theorem 3.2.

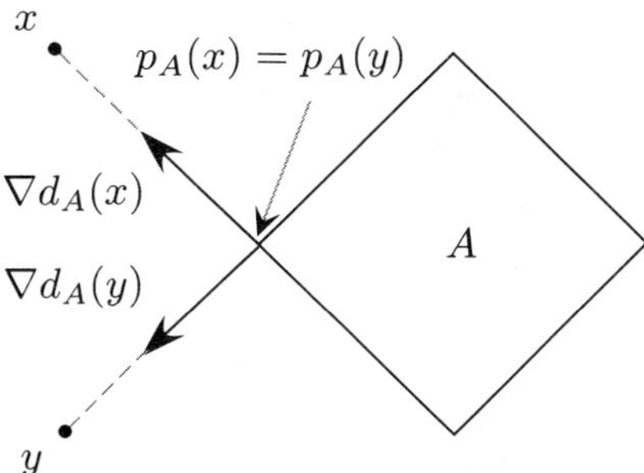

Figure 4.2. *Nonuniqueness of the exterior normal.*

Definition 3.2.

(i) For $A \subset \mathbf{R}^N$, $A \neq \varnothing$, the *set of exterior cracks* is defined as

$$\mathrm{C}_{\text{ext}}(A) = \left\{ x \in \mathbf{R}^N \ : \ \nabla d_A^2(x) \exists \text{ and } \nabla d_A(x) \not\exists \right\}. \tag{3.8}$$

(ii) For $A \subset \mathbf{R}^N$, $\complement A \neq \varnothing$, the *set of interior cracks* is defined as

$$\mathrm{C}_{\text{int}}(A) = \left\{ x \in \mathbf{R}^N \ : \ \nabla d_{\complement A}^2(x) \exists \text{ and } \nabla d_{\complement A}(x) \not\exists \right\}. \tag{3.9}$$

$\square$

Points of $\mathrm{C}_{\text{ext}}(A)$ cannot belong to $\mathbf{R}^N \setminus \partial A$ since, from Theorem 3.2 (ii), for such points the existence of $\nabla d_A^2(x)$ implies the existence of $\nabla d_A(x)$. The same is true for $\mathrm{C}_{\text{int}}(A)$. Therefore,

$$\mathrm{C}_{\text{ext}}(A) \cup \mathrm{C}_{\text{int}}(A) \subset \partial A \text{ and } \mathrm{m}(\mathrm{C}_{\text{ext}}(A)) = 0 = \mathrm{m}(\mathrm{C}_{\text{int}}(A)). \tag{3.10}$$

The sets $\mathrm{C}_{\text{ext}}(A)$ and $\mathrm{C}_{\text{int}}(A)$ are generally not equal to all of ∂A, as can be seen from the next example.

Example 3.1.
Let $B(0,1)$ be the open ball of radius 1 at the origin and let

$$A = \{x \in B(0,1) \ : \ x \text{ has rational coordinates}\}.$$

Then

$$\partial A = \overline{B(0,1)}, \quad \nabla d_A(x) = \nabla d_A^2(x) = 0 \text{ in } B(0,1),$$
$$\mathrm{Sk}_{\text{ext}(A)} = \varnothing \quad \mathrm{Sk}_{\text{int}}(A) = \{0\},$$
$$C_{\text{ext}}(A) = C_{\text{int}}(A) = \partial B(0,1).$$

Note that in the equivalence class $[A]$, the boundary can be different for two members of the class. In fact, $\partial \overline{A} \subset \partial A$ for the closed representative of $\overline{A}$ and this inclusion can be strict, as shown from this example. $\square$

4　$W^{1,p}$-Topology and Characteristic Functions

Distance functions are locally Lipschitzian. They belong to $C_{\text{loc}}^{0,1}(\mathbf{R}^N)$ and hence to $W_{\text{loc}}^{1,p}(\mathbf{R}^N)$ for all $p \geq 1$. Thus the previous constructions can be repeated with $W_{\text{loc}}^{1,p}(D)$ in place of $C_{\text{loc}}(D)$ to generate new $W^{1,p}$-metric topologies on the family $C_d(D)$. One big advantage is that the $W^{1,p}$-convergence of sequences will imply the L^p-convergence of the corresponding characteristic functions of the closure of the sets and hence the convergence of volumes (cf. Theorem 3.2 (iii)). The convergence of volumes and perimeters is lost in the Hausdorff topology. The next example shows that the Hausdorff metric convergence is not sufficient to get the L^p-convergence of the characteristic functions of the closure of the corresponding sets in the sequence. The volume function is only upper semicontinuous with respect to the Hausdorff topology. The perimeters do not converge either. This is also illustrated by the example of the staircase of Example 5.1 and Figure 3.7 in Chapter 3, where the volumes converge but not the perimeters.

Example 4.1.
Denote by $D = [-1, 2] \times [-1, 2]$ the unit square in $\mathbf{R}^2$, and for each $n \geq 1$ define the sequence of closed sets

$$A_n = \left\{ (x_1, x_2) \in D : \frac{2k}{2n} \leq x_1 \leq \frac{2k+1}{2n}, 0 \leq k < n \right\}.$$

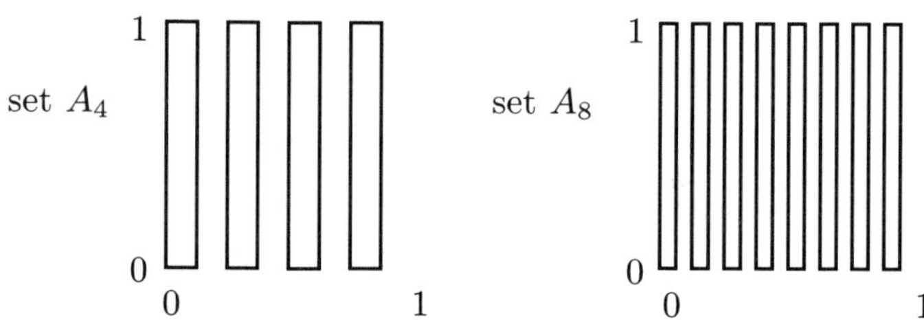

Figure 4.3.　*Vertical stripes of Example* 4.1.

This defines n vertical stripes of equal width $1/2n$ each distant of $1/2n$ (cf. Figure 4.3). Clearly, for all $n \geq 1$,

$$m(A_n) = \frac{1}{2}, \quad P_D(A_n) = 2n + 1$$

$$\forall x \in S, \quad d_{A_n}(x) \leq \frac{1}{4n}, \quad \|\nabla d_{A_n}\|_{L^p(D)} \geq 2^{-1/p},$$

where $S = [0, 1] \times [0, 1]$ and $P_D(A_n)$ is the perimeter of A_n. Hence

$$d_{A_n} \to d_S \quad \text{in } C(\overline{D}), \quad m(S) = 1, \quad P_D(S) = 4.$$

But

$$m(\overline{A_n}) = m(A_n) = \frac{1}{2} \nrightarrow 1 = m(S) \implies \chi_{A_n} \nrightarrow \chi_S \text{ in } L^2(D),$$
$$P_D(A_n) \nrightarrow P_D(S),$$
$$\chi_{A_n} \rightharpoonup \frac{1}{2}\chi_S \text{ in } L^2(D)\text{-weak.}$$

Since the characteristic functions do not converge, the sequence $\{\nabla d_{A_n}\}$ does not converge in $L^p(D)^N$. $\qquad\square$

Theorem 4.1. *Let D be an open (resp., bounded open) subset of $\mathbf{R}^N$.*

(i) *The topologies induced by $W^{1,p}_{\mathrm{loc}}(D)$ (resp., $W^{1,p}(D)$) on $C_d(D)$ and $C_d^c(D)$ are all equivalent for p, $1 \le p < \infty$.*

(ii) *$C_d(D)$ is closed in $W^{1,p}_{\mathrm{loc}}(D)$ (resp., $W^{1,p}(D)$) for p, $1 \le p < \infty$, and*

$$\rho_D([A_2],[A_1]) \stackrel{\text{def}}{=} \sum_{n=1}^{\infty} \frac{1}{2^n} \frac{\|d_{A_2} - d_{A_1}\|_{W^{1,p}(B(0,n))}}{1 + \|d_{A_2} - d_{A_1}\|_{W^{1,p}(B(0,n))}}$$

$$\left(\text{resp., } \rho_D([A_2],[A_1]) \stackrel{\text{def}}{=} \|d_{A_2} - d_{A_1}\|_{W^{1,p}(D)}\right)$$

defines a complete metric structure on $\mathcal{F}(D)$. For p, $1 \le p < \infty$, the map

$$d_A \mapsto \chi_{\overline{A}} = 1 - |\nabla d_A| \,:\, C_d(D) \subset W^{1,p}_{\mathrm{loc}}(D) \to L^p_{\mathrm{loc}}(D)$$

is "Lipschitz continuous": for all bounded open subsets K of D and nonempty subsets A_1 and A_2 of D,

$$\|\chi_{\overline{A_2}} - \chi_{\overline{A_1}}\|_{L^p(K)} \le \|\nabla d_{A_2} - \nabla d_{A_1}\|_{L^p(K)} \le \|d_{A_2} - d_{A_1}\|_{W^{1,p}(K)}.$$

(iii) *$C_d^c(D)$ is closed in $W^{1,p}_{\mathrm{loc}}(D)$ (resp., $W^{1,p}_0(D)$) for p, $1 \le p < \infty$, and*

$$\rho_D(\Omega_2,\Omega_1) \stackrel{\text{def}}{=} \sum_{n=1}^{\infty} \frac{1}{2^n} \frac{\|d_{\complement\Omega_2} - d_{\complement\Omega_1}\|_{W^{1,p}(B(0,n))}}{1 + \|d_{\complement\Omega_2} - d_{\complement\Omega_1}\|_{W^{1,p}(B(0,n))}}$$

$$\left(\text{resp., } \rho_D(\Omega_2,\Omega_1) \stackrel{\text{def}}{=} \|d_{\complement\Omega_2} - d_{\complement\Omega_1}\|_{W^{1,p}(D)}\right)$$

defines a complete metric structure on the family $\mathcal{G}(D)$ of open subsets of D. For p, $1 \le p < \infty$, the map

$$d_{\complement\Omega} \mapsto \chi_\Omega = |\nabla d_{\complement\Omega}| \,:\, C_d(D) \subset W^{1,p}_{\mathrm{loc}}(D) \to L^p_{\mathrm{loc}}(D)$$

is "Lipschitz continuous": for all bounded open subsets K of D and open subsets $\Omega_1 \ne \mathbf{R}^N$ and $\Omega_2 \ne \mathbf{R}^N$ of D

$$\|\chi_{\Omega_2} - \chi_{\Omega_1}\|_{L^p(K)} \le \|\nabla d_{\complement\Omega_2} - \nabla d_{\complement\Omega_1}\|_{L^p(K)} \le \|d_{\complement\Omega_2} - d_{\complement\Omega_1}\|_{W^{1,p}(K)}.$$

Proof. (i) The proof is similar to the proof of Theorem 2.2 in Chapter 3. It is sufficient to prove it for D bounded open. Since D is bounded it is contained in a sufficiently large ball of radius c. Therefore,

$$d_A(x) = \inf_{y \in A} |x - y| \le |x| + |y| \le 2c$$

since $y \in \overline{A} \subset \overline{D}$, and by Theorem 2.1 (vii)

$$|\nabla d_A(x)| \le 1 \text{ a.e. in } D.$$

For all $p > 1$ the injection of $W^{1,p}(D)$ into $W^{1,1}(D)$ is continuous since

$$\|d_A\|_{W^{1,1}(D)} \le \|d_A\|_{W^{1,p}(D)}\, \mathrm{m}(D)^{1/q}$$

with $1/p + 1/q = 1$. Conversely, we have the continuity in the other direction. For $p > 1$ and any d_A and d_B in $C_d(D)$,

$$\int_D |d_A - d_B|^p + |\nabla d_A - \nabla d_B|^p \, dx$$

$$= \int_D |d_A - d_B|\, |d_A - d_B|^{p-1} + |\nabla d_A - \nabla d_B|\, |\nabla d_A - \nabla d_B|^{p-1} \, dx$$

$$\le \max\{(2c)^{p-1}, 2^{p-1}\} \int_D |d_A - d_B| + |\nabla d_A - \nabla d_B| \, dx$$

$$\Rightarrow \|d_A - d_B\|^p_{W^{1,p}(D)} \le (2 \max\{c, 1\})^{p-1} \|d_A - d_B\|_{W^{1,1}(D)}.$$

Therefore, for any $\varepsilon > 0$, pick $\delta = \varepsilon^p / (2 \max\{c, 1\})^{p-1}$ and

$$\|d_A - d_B\|_{W^{1,1}(D)} \le \delta \quad \Rightarrow \quad \|d_A - d_B\|_{W^{1,p}(D)} \le \varepsilon.$$

(ii) It is sufficient to prove it for D bounded open. Let $\{d_{A_n}\}$ be a Cauchy sequence in $C_d(D)$ which converges to some f in $W^{1,p}(D)$-strong. By Theorem 2.2 (ii), $C_d(D)$ is compact in $C(\overline{D})$ and there exist a subsequence, still denoted $\{d_{A_n}\}$, and $\varnothing \ne A \subset \overline{D}$ such that

$$d_{A_n} \to d_A \text{ in } C(\overline{D})$$

and, a fortiori, in $L^p(D)$-strong since D is bounded. By uniqueness of the limit in $L^p(D)$-strong, $f = d_A$ and d_{A_n} converge to d_A in $W^{1,p}(D)$-strong. Therefore, $C_d(D)$ is closed in $W^{1,p}(D)$. For the Lipschitz continuity, recall that the distance function d_A is differentiable almost everywhere in $\mathbf{R}^N$ for $A \ne \varnothing$. In view of Theorem 3.2 (iv)

$$\chi_{\overline{A}} = 1 - |\nabla d_A(x)| \text{ a.e. in } \mathbf{R}^N.$$

Given two nonempty subsets A_1 and A_2 of D

$$|\nabla d_{A_2}| \le |\nabla d_{A_1}| + |\nabla d_{A_2} - \nabla d_{A_1}|$$

$$\Rightarrow \chi_{A_1} \le \chi_{A_2} + |\nabla d_{A_2} - \nabla d_{A_1}|$$

$$\Rightarrow \int_D |\chi_{A_1} - \chi_{A_2}|^p \, dx \le \|d_{A_2} - d_{A_1}\|^p_{W^{1,p}(D)}$$

for $1 \leq p < \infty$ and with the ess-sup norm for $p = \infty$.

(iii) Again it is sufficient to prove the result for D bounded. In that case $\complement\Omega \neq \varnothing$ for all open subsets Ω of D. Let $\{\Omega_n\}$ be a sequence of open subsets of D such that $\{d_{\complement\Omega_n}\}$ is Cauchy in $W^{1,p}(D)$. By assumption $\Omega_n \subset D$, $\complement\Omega_n \supset \complement D$,

$$\forall n \geq 1, \quad d_{\complement\Omega_n} = 0 \text{ in } \complement D \quad \Rightarrow \quad d_{\complement\Omega_n} \in W_0^{1,p}(D),$$

and the Cauchy sequence converges to some $f \in W_0^{1,p}(D)$. By Theorem 2.4 (ii), $C_d^c(D)$ is compact in $C(\overline{D})$ and there exist a subsequence, still denoted $\{d_{\complement\Omega_n}\}$, and an open set $\Omega \subset D$ such that

$$d_{\complement\Omega_n} \to d_{\complement\Omega} \text{ in } C(\overline{D})$$

and hence in $L^p(D)$-strong, since D is bounded. By uniqueness of the limit in $L^p(D)$, $f = d_{\complement\Omega}$ and the Cauchy sequence $d_{\complement\Omega_n}$ converges to $d_{\complement\Omega}$ in $W_0^{1,p}(D)$. The other part of the proof is similar to that of part (ii). $\qquad\square$

We have the following general result.

Theorem 4.2. *Let D be a bounded open domain in $\mathbf{R}^N$.*

(i) *If $\{d_{A_n}\}$ weakly converges in $W^{1,p}(D)$ for some p, $1 \leq p < \infty$, then it weakly converges in $W^{1,p}(D)$ for all p, $1 \leq p < \infty$.*

(ii) *If $\{d_{A_n}\}$ converges in $C(\overline{D})$, then it weakly converges in $W^{1,p}(D)$ for all p, $1 \leq p < \infty$. Conversely if $\{d_{A_n}\}$ weakly converges in $W^{1,p}(D)$ for some p, $1 \leq p < \infty$, it converges in $C(\overline{D})$.*

(iii) *$C_d(D)$ is compact in $W^{1,p}(D)$-weak for all p, $1 \leq p < \infty$.*[2]

(iv) *Parts (i)–(iii) also apply to $C_d^c(D)$.*

Proof. (i) Recall that for D bounded there exists a constant $c > 0$ such that for all $d_A \in C_d(D)$

$$d_A(x) \leq c \text{ and } |\nabla d_A(x)| \leq 1 \text{ a.e. in } D.$$

If $\{d_{A_n}\}$ weakly converges in $W^{1,p}(D)$, then

$$\{d_{A_n}\} \text{ weakly converges in } L^p(D),$$
$$\{\nabla d_{A_n}\} \text{ weakly converges in } L^p(D)^N.$$

By Lemma 2.1 (iii) in Chapter 3 both sequences weakly converge for all $p \geq 1$, and hence $\{d_{A_n}\}$ weakly converges in $W^{1,p}(D)$ for all $p \geq 1$.

---

[2] In a metric space the compactness is equivalent to the sequential compactness. For the weak topology we use the fact that if E is a separable normed space, then, in its topological dual E', any closed ball is a compact metrizable space for the weak topology. Since $C_d(D)$ is a bounded subset of the normed reflexive separable Banach space $W^{1,p}(D)$, $1 \leq p < \infty$, the weak compactness of $C_d(D)$ coincides with the weak sequential compactness (cf. Dieudonne [1, II, Chap. XII, section 12.15.9, p. 75]).

(ii) If $\{d_{A_n}\}$ converges in $C(\overline{D})$, then by Theorem 2.2 (i) there exists $d_A \in C_b(D)$ such that $d_{A_n} \to d_A$ in $C(\overline{D})$ and hence in $L^p(D)$. So for all $\varphi \in \mathcal{D}(D)^N$,

$$\int_D \nabla d_{A_n} \cdot \varphi\, dx = -\int_D d_{A_n} \operatorname{div} \varphi\, dx \to -\int_D d_A \operatorname{div} \varphi\, dx = \int_D \nabla d_A \cdot \varphi\, dx.$$

By density of $\mathcal{D}(D)$ in $L^2(D)$, $\nabla d_{A_n} \to \nabla d_A$ in $L^2(D)^N$-weak and hence $d_{A_n} \to d_A$ in $W^{1,2}(D)$-weak. From part (i) it converges in $W^{1,p}(D)$-weak for all p, $1 \le p < \infty$. Conversely, the weakly convergent sequence converges to some f in $W^{1,p}(D)$. By compactness of $C_b(D)$ there exist a subsequence, still indexed by n, and d_A such that $d_{A_n} \to d_A$ in $C(\overline{D})$ and hence in $W^{1,p}(D)$-weak. By uniqueness of the limit, $d_A = f$. Therefore, all convergent subsequences in $C(\overline{D})$ converge to the same limit, so the whole sequence converges in $C(\overline{D})$. This concludes the proof.

(iii) Consider an arbitrary sequence $\{d_{A_n}\}$ in $C_d(D)$. From Theorem 2.2 (ii) $C_d(D)$ is compact and there exists a subsequence $\{d_{A_{n_k}}\}$ and $d_A \in C_d(D)$ such that $d_{A_{n_k}} \to d_A$ in $C(\overline{D})$. From part (ii) the subsequence weakly converges in $W^{1,p}(D)$ and hence $C_d(D)$ is compact in $W^{1,p}(D)$-weak. $\qquad\square$

Theorem 4.3. *Let D be a closed domain in $\mathbf{R}^N$ and $\{A_n\}$ a sequence of nonempty sets converging to a nonempty set A of D in the Hausdorff topology:*

$$d_{A_n} \to d_A \text{ in } C_{\mathrm{loc}}(D).$$

Then

$$\forall x \in \mathbf{R}^N \setminus \overline{A}, \quad \lim_{n \to \infty} \chi_{\overline{A_n}}(x) = \chi_{\overline{A}}(x) = 0,$$

$$\forall x \in \mathbf{R}^N, \quad \limsup_{n \to \infty} \chi_{\overline{A_n}}(x) \le \chi_{\overline{A}}(x),$$

and for all compact subsets K of D

$$\limsup_{n \to \infty} \int_K \chi_{\overline{A_n}}\, dx \le \int_K \chi_{\overline{A}}\, dx.$$

Corollary 1. *Let $D \ne \varnothing$ be a bounded open subset of $\mathbf{R}^N$ and $\{\Omega_n\}$, $\Omega_n \ne \varnothing$, a sequence of open subsets of D converging to an open subset Ω, $\Omega \ne \varnothing$, of D in the Hausdorff complementary topology:*

$$d_{\complement\Omega_n} \to d_{\complement\Omega} \text{ in } C(\bar{D}).$$

Then

$$\forall x \in \Omega, \quad \lim_{n \to \infty} \chi_{\Omega_n}(x) = \chi_\Omega(x) = 1,$$

$$\forall x \in \mathbf{R}^N, \quad \liminf_{n \to \infty} \chi_{\Omega_n}(x) \ge \chi_\Omega(x),$$

and

$$\liminf_{n \to \infty} \int_D \chi_{\Omega_n}\, dx \ge \int_D \chi_\Omega\, dx.$$

Proof of Theorem 4.3. For all $x \notin \overline{A}$, $d_A(x) > 0$ and

$$\exists N_x > 0, \forall n \geq N_x, \quad \|d_A - d_{A_n}\| \leq d_A(x)/2$$
$$\Rightarrow \exists N_x > 0, \forall n \geq N_x, \quad d_{A_n}(x) \geq d_A(x)/2 > 0$$
$$\Rightarrow \exists N_x > 0, \forall n \geq N_x, \quad x \notin \overline{A_n} \text{ and } \chi_{\overline{A_n}}(x) = 0.$$

So for all $x \notin \overline{A}$

$$\lim_{n \to \infty} \chi_{\overline{A_n}}(x) = \chi_{\overline{A}}(x) = 0.$$

Finally, for all $x \in \overline{A}$ and all $n \geq 1$

$$\chi_{\overline{A_n}}(x) \leq 1 = \chi_{\overline{A}}(x)$$

and the result follows trivially by taking the limsup of each term.
We conclude that

$$\limsup_{n \to \infty} \chi_{\overline{A_n}} \leq \chi_{\overline{A}},$$

and by using the analogue of Fatou's lemma for the limsup we get for all compact subsets K of D

$$\limsup_{n \to \infty} \int_K \chi_{\overline{A_n}} \, dx \leq \int_K \limsup_{n \to \infty} \chi_{\overline{A_n}} \, dx \leq \int_K \chi_{\overline{A}} \, dx. \qquad \square$$

In order to get the L^p-convergence of the characteristic functions of the closure of the sets in the sequence, we need the L^p-convergence of the gradients of the distance functions which are related to the characteristic functions of the closure of the sets (cf. Theorem 3.2 (iv)).

We have seen in Example 4.1 that the weak convergence of the characteristic functions is not sufficient to obtain the strong convergence of the sequence $\{d_{A_n}\}$ to d_A in $W^{1,2}(D)$. However, if we assume that $\{\chi_{\overline{A_n}}\}$ is strongly convergent, it converges to the characteristic function χ_B of some measurable subset B of D. Is this sufficient to conclude that $\chi_B = \chi_{\overline{A}}$? The answer is negative. The counterexample is provided by Example 5.2 in Chapter 3, where

$$d_{A_n} \rightharpoonup d_D \text{ in } W^{1,2}(D)\text{-weak} \quad \text{and} \quad \chi_{A_n} \to \chi_B \text{ in } L^2(D)\text{-strong}$$

for some $B \subset D$ such that

$$m(D) = \pi > \frac{\pi}{3} \geq m(B) \quad \Rightarrow \chi_D \neq \chi_B.$$

Remark 4.1.

In view of part (ii) of Theorem 4.1 an optimization problem with respect to the characteristic functions χ_Ω of open sets Ω in D for which we have the continuity with respect to χ_Ω can be transformed into an optimization problem with respect to $d_{\complement\Omega}$ in $W^{1,1}(D)$ since

$$\chi_\Omega = |\nabla d_{\complement\Omega}|.$$

For instance, this would apply to the transmission problem (3.3)–(3.6) in section 3.1 of Chapter 3. $\qquad \square$

5 Sets of Bounded and Locally Bounded Curvature

Going from the uniform Hausdorff metric topology to the $W^{1,p}$-topology readily extends the applicability of the distance function to problems involving the characteristic function. However, even for a bounded open hold-all D, the families $C_d(D)$ and $C_d^c(D)$ are closed but not compact in $W^{1,p}(D)$-strong (cf. Example 4.1). The situation is similar to that encountered for the set of characteristic functions $X(D)$ in the $L^p(D)$-topology, where we have introduced the Caccioppoli sets. Their natural analogue in $C_d(D)$ and $C_d^c(D)$ are the sets introduced by Delfour and Zolésio [17, 32], which include C^k domains, convex sets, and Federer's [2] sets of positive reach. They lead to compactness theorems for $C_d(D)$ and $C_d^c(D)$ in $W^{1,p}(D)$.

5.1 Definitions and Properties

Definition 5.1.

(i) Given a bounded open set D in $\mathbf{R}^N$, a subset Ω of $\overline{D}$, $\Omega \neq \varnothing$ (resp., $\complement\Omega \neq \varnothing$), is said to be of *bounded exterior* (resp., *interior*) *curvature* with respect to D if

$$\nabla d_\Omega \in \mathrm{BV}(D)^N \quad (\text{resp.,} \ \nabla d_{\complement\Omega} \in \mathrm{BV}(D)^N) . \tag{5.1}$$

Denote those families as follows:

$$\mathrm{BC}_d(D) \overset{\text{def}}{=} \left\{ d_\Omega \in C_d(D) \ : \ \nabla d_\Omega \in \mathrm{BV}(D)^N \right\},$$

$$\mathrm{BC}_d^c(D) \overset{\text{def}}{=} \left\{ d_{\complement\Omega} \in C_d^c(D) \ : \ \nabla d_{\complement\Omega} \in \mathrm{BV}(D)^N \right\}.$$

(ii) A subset Ω of $\mathbf{R}^N$, $\Omega \neq \varnothing$ (resp., $\complement\Omega \neq \varnothing$), is said to be of *locally bounded exterior* (*resp., interior*) *curvature* if

$$\forall x \in \partial\Omega, \ \exists \rho > 0 \ \text{such that} \ \nabla d_\Omega \ (\text{resp.,} \ \nabla d_{\complement\Omega}) \in \mathrm{BV}(B(x, \rho))^N, \tag{5.2}$$

where $B(x, \rho)$ is the open ball of radius $\rho > 0$ in x. $\qquad\square$

From Theorem 5.1 and Definition 5.1 in Chapter 3, a function belongs to $\mathrm{BV}_{\mathrm{loc}}(\mathbf{R}^N)$ if and only if for each $x \in \mathbf{R}^N$ it belongs to $\mathrm{BV}(B(x, \rho))$ for some $\rho > 0$. As in Theorem 5.2 of Chapter 3 for Caccioppoli sets, it is sufficient to satisfy this condition for points of the boundary $\partial\Omega$. Thus property (5.2) only depends on the *properties* of the boundary.

Theorem 5.1. *Let Ω, $\Omega \neq \varnothing$ (resp., $\complement\Omega \neq \varnothing$)), be a subset of $\mathbf{R}^N$. Then Ω is of locally bounded exterior (resp., interior) curvature if and only if ∇d_Ω (resp., $\nabla d_{\complement\Omega}$) belongs to $\mathrm{BV}_{\mathrm{loc}}(\mathbf{R}^N)^N$.*

This theorem will be proved later as part (iii) of Theorem 5.3.

Since we simultaneously deal with sets Ω and their complement $\complement\Omega$, we shall often use the notation A to cover both cases. Recall that the relaxation of the perimeter of a set was obtained from the norm of the gradient of its characteristic function for Caccioppoli sets. For distance functions $\chi_{\overline{A}} = 1 - |\nabla d_A|$ and the gradient of d_A also has a jump discontinuity along the boundary ∂A of magnitude at most 1. So it is not too surprising to discover that the closure of a set A with bounded interior or exterior curvature is a Caccioppoli set.

Theorem 5.2.

(i) *Let D be a bounded open Lipschitzian subset of $\mathbf{R}^N$. For any $\Omega \subset \overline{D}$, $\varnothing \neq \Omega$ (resp., $\varnothing \neq \complement\Omega$) such that*

$$\nabla d_\Omega (resp., \nabla d_{\complement\Omega}) \in BV(D)^N,$$

$\overline{\Omega}$ (resp., $\overline{\complement\Omega}$) has finite perimeter; that is,

$$\chi_{\overline{\Omega}} \in BV(D) \ and \ \|\nabla\chi_{\overline{\Omega}}\|_{M^1(D)} \leq 2\,\|D^2 d_\Omega\|_{M^1(D)}$$
$$(resp., \ \chi_{\overline{\complement\Omega}} \in BV(D) \ and \ \|\nabla\chi_{\overline{\complement\Omega}}\|_{M^1(D)} \leq 2\,\|D^2 d_{\complement\Omega}\|_{M^1(D)}).$$

(ii) *For any subset Ω of $\mathbf{R}^N$, $\varnothing \neq \Omega$ (resp., $\varnothing \neq \complement\Omega$) such that*

$$\nabla d_\Omega \ (resp., \nabla d_{\complement\Omega}) \in BV_{\mathrm{loc}}(\mathbf{R}^N)^N,$$

$\overline{\Omega}$ (resp., $\overline{\complement\Omega}$) has locally finite perimeter; that is,

$$\chi_{\overline{\Omega}} \ (resp., \chi_{\overline{\complement\Omega}}) \in BV_{\mathrm{loc}}(\mathbf{R}^N).$$

Proof. Given ∇d_A in $BV(D)^N$, there exists a sequence $\{u_k\}$ in $C^\infty(D)^N$ such that

$$u_k \to \nabla d_A \ \text{in} \ L^1(D)^N,$$
$$\|Du_k\|_{M^1(D)} \to \|D^2 d_A\|_{M^1(D)},$$

as k goes to infinity, and since $|\nabla d_A(x)| \leq 1$, this sequence can be chosen in such a way that

$$\forall k \geq 1, \quad |u_k(x)| \leq 1.$$

This follows from the use of mollifiers (cf. Giusti [1, Thm. 1.17, p. 15]). For all V in $\mathcal{D}(D)^N$

$$-\int_D \chi_{\overline{A}}\mathrm{div}\,V\,dx = \int_D (|\nabla d_A|^2 - 1)\mathrm{div}\,V\,dx = \int_D |\nabla d_A|^2 \mathrm{div}\,V\,dx.$$

For each u_k

$$\int_D |u_k|^2 \mathrm{div}\,V\,dx = -2\int_D {}^\top[Du_k]u_k \cdot V\,dx = -2\int_D u_k \cdot [Du_k]V\,dx,$$

where ${}^\top Du_k$ is the transpose of the Jacobian matrix Du_k and

$$\left|\int_D |u_k|^2 \mathrm{div}\,V\,dx\right| \leq 2\int_D |u_k|\,|Du_k|\,|V|\,dx$$
$$\leq 2\,\|Du_k\|_{L^1}\|V\|_{C(\overline{D})} \leq 2\,\|Du_k\|_{M^1}\|V\|_{C(\overline{D})}$$

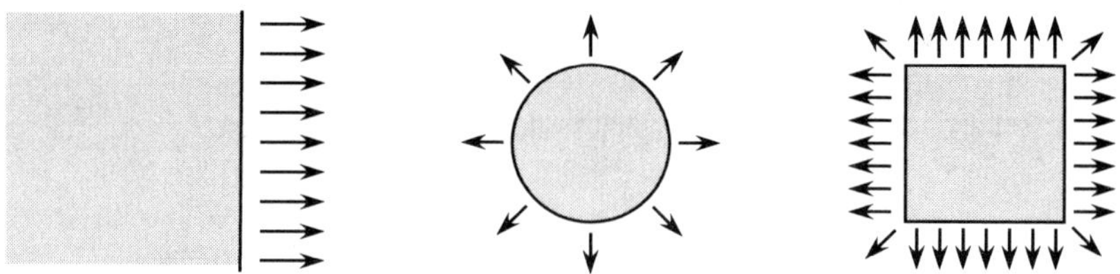

Figure 4.4. ∇d_A *for Examples* 5.1, 5.2, *and* 5.3.

since for $W^{1,1}(D)$-functions $\|\nabla f\|_{L^1(D)^N} = \|\nabla f\|_{M^1(D)^N}$. Therefore, as k goes to infinity

$$\left| \int_D \chi_{\overline{A}} \operatorname{div} V \, dx \right| = \left| \int_D |\nabla d_A|^2 \operatorname{div} V \, dx \right| \leq 2\|D^2 d_A\|_{M^1} \|V\|_{C(\overline{D})},$$

where $D^2 d_A$ is the Hessian matrix of second-order partial derivatives of d_A. Therefore, $\nabla \chi_{\overline{A}} \in M^1(D)^N$. $\qquad\square$

5.2 Examples

It is useful to consider the following three simple illustrative examples (cf. Figure 4.4).

Example 5.1 (half-plane in $\mathbf{R}^2$).
Consider the domain

$$A = \{(x_1, x_2) : x_1 \leq 0\}, \quad \partial A = \{(x_1, x_2) : x_1 = 0\}.$$

It is readily seen that

$$d_A(x_1, x_2) = \max\{x_1, 0\}, \quad \nabla d_A(x_1, x_2) = \begin{cases} (0,0), & x_1 < 0, \\ (1,0), & x_1 > 0, \end{cases}$$

$$\langle \partial_{11} d_A, \varphi \rangle = \int_{\partial A} \varphi \, dH_1, \quad \partial_{12} d_A = \partial_{21} d_A = \partial_{22} d_A = 0,$$

$$\langle \Delta d_A, \varphi \rangle = \int_{\partial A} \varphi \, dH_1 \quad \Rightarrow \quad \Delta d_A = H_1.$$

Thus Δd_A is the one-dimensional Hausdorff measure of ∂A. $\qquad\square$

Example 5.2 (ball of radius $R > 0$ in $\mathbf{R}^2$).
Consider the domain

$$A = \{x \in \mathbf{R}^2 : |x| \leq R\}, \quad \partial A = \{x \in \mathbf{R}^2 : |x| = R\}.$$

Clearly,

$$d_A(x) = \max\{0, |x| - R\}, \quad \nabla d_A(x) = \begin{cases} x/|x|, & |x| > R, \\ (0,0), & |x| < R, \end{cases}$$

$$\langle \partial_{11} d_A, \varphi \rangle = \int_0^{2\pi} R\cos^2(\theta)\varphi\, d\theta + \int_{\complement A} \frac{x_2^2}{(x_1^2 + x_2^2)^{3/2}}\, \varphi\, dx,$$

$$\langle \partial_{22} d_A, \varphi \rangle = \int_0^{2\pi} R\sin^2(\theta)\varphi\, d\theta + \int_{\complement A} \frac{x_1^2}{(x_1^2 + x_2^2)^{3/2}}\, \varphi\, dx,$$

$$\langle \partial_{12} d_A, \varphi \rangle = \langle \partial_{21} d_A, \varphi \rangle = \int_0^{2\pi} R\cos(\theta)\sin(\theta)\varphi\, d\theta + \int_{\complement A} \frac{x_2 x_1}{(x_1^2 + x_2^2)^{3/2}}\, \varphi\, dx,$$

$$\langle \Delta d_A, \varphi \rangle = \int_{\partial A} \varphi\, dH_1 + \int_{\complement A} \frac{1}{(x_1^2 + x_2^2)^{1/2}}\, \varphi\, dx,$$

where H_1 is the one-dimensional Hausdorff measure. We see that Δd_A contains the one-dimensional Hausdorff measure of the boundary ∂A plus a term that corresponds to the volume integral of the mean curvature over the level sets of d_A in $\complement A$. $\qquad\square$

Example 5.3 (unit square in $\mathbf{R}^2$).
Consider the domain

$$A = \{x = (x_1, x_2) : |x_1| \le 1, |x_2| \le 1\}.$$

Since A is symmetrical with respect to both axes, it is sufficient to specify d_A in the first quadrant. We use the notation Q_1, Q_2, Q_3, and Q_4 for the four quadrants in the counterclockwise order and c_1, c_2, c_3, and c_4 for the four corners of the square in the same order. We also divide the plane into three regions:

$$D_1 = \{(x_1, x_2) : |x_2| \le \min\{1, |x_1|\}\},$$
$$D_2 = \{(x_1, x_2) : |x_1| \le \min\{1, |x_2|\}\},$$
$$D_3 = \{(x_1, x_2) : |x_1| \ge 1 \text{ and } |x_2| \ge 1\}.$$

Hence

$$d_A(x) = \begin{cases} \min\{x_2 - 1, 0\}, & x \in D_2 \cap Q_1, \\ |x - c_1|, & x \in D_3 \cap Q_1, \\ \min\{x_1 - 1, 0\}, & x \in D_1 \cap Q_1, \end{cases}$$

$$\nabla d_A(x) = \begin{cases} (0,1) & x \in D_2 \cap Q_1 \text{ and } x_2 > 1, \\ \frac{x - c_1}{|x - c_1|}, & x \in D_3 \cap Q_1, \\ (1,0), & x \in D_1 \cap Q_1 \text{ and } x_1 > 1, \\ (0,0), & x \in Q_1, x_1 < 1 \text{ and } x_2 < 1. \end{cases}$$

$$\langle \partial_{11} d_A, \varphi \rangle = \sum_{i=1}^4 \int_{D_3 \cap Q_i} \frac{(x_2 - c_{i,2})^2}{|x - c_i|^2}\, \varphi\, dx + \int_{\partial A \cap Q_i \cap D_1} \varphi\, dH_1,$$

$$\langle \partial_{22} d_A, \varphi \rangle \sum_{i=1}^{4} \int_{D_3 \cap Q_i} \frac{(x_1 - c_{i,1})^2}{|x - c_i|^2} \, \varphi \, dx + \int_{\partial A \cap Q_i \cap D_2} \varphi \, dH_1,$$

$$\langle \partial_{12} d_A, \varphi \rangle = \langle \partial_{21} d_A, \varphi \rangle = \sum_{i=1}^{4} \int_{D_3 \cap Q_i} \frac{(x_2 - c_{i,2})(x_1 - c_{i,1})}{|x - c_i|^2} \, \varphi \, dx,$$

$$\langle \Delta d_A, \varphi \rangle = \sum_{i=1}^{4} \int_{D_3 \cap Q_i} \frac{1}{|x - c_i|} \, \varphi \, dx + \int_{\partial A} \varphi \, dH_1.$$

Notice that the structure of the Laplacian is similar to that observed in the previous examples. $\qquad \square$

The norm $\|D^2 d_A\|_{M^1(U_h(A))}$ is decreasing as h goes to zero. The limit is particularly interesting since it singles out the behavior of the singular part of the Hessian matrix in a shrinking neighborhood of the boundary ∂A.

Example 5.4.
Let $N = 2$. For the finite square and the ball of finite radius,

$$\lim_{h \searrow 0} \|\Delta d_A\|_{M^1(U_h(\partial A))} = H_1(\partial A),$$

where H_1 is the one-dimensional Hausdorff measure. $\qquad \square$

We now turn to the proof of Theorem 5.1, which is part of a more general theorem that contains several important technical results that will be used later for sets of positive reach and in Chapter 5. It is based on a slight extension of a result in Fu [1, Prop. 1.2]. We first give a few additional definitions.

Definition 5.2.

(i) Given any set A, $\varnothing \neq A \subset \mathbf{R}^N$, and a real number $h > 0$, the *h-tubular neighborhood* of A is defined as

$$U_h(A) \overset{\text{def}}{=} \{y \in \mathbf{R}^N : d_A(y) < h\}$$

and the *closed h-tubular neighborhood* of A as

$$A_h \overset{\text{def}}{=} \{y \in \mathbf{R}^N : d_A(y) \leq h\}.$$

(ii) A function $f : U \to \mathbf{R}$ defined in a convex subset U of $\mathbf{R}^N$ is *convex* if for all x and y in U and all $\lambda \in [0, 1]$,

$$f(\lambda x + (1 - \lambda)y) \leq \lambda f(x) + (1 - \lambda)f(y).$$

It is *concave* if the function $-f$ is convex. A function $f : U \to \mathbf{R}$ defined in a convex subset U of $\mathbf{R}^N$ is *semiconvex* (resp., *semiconcave*) if

$$\exists c \geq 0, \quad f_c(x) = c\,|x|^2 + f(x) \quad (\text{resp.,} \ f_c(x) = c\,|x|^2 - f(x))$$

is convex in U.

(iii) A function $f : U \to \mathbf{R}$ defined in a subset U of $\mathbf{R}^N$ is *locally convex* (resp., *locally concave*) if it is convex (resp., concave) in every convex subset of U. A function $f : U \to \mathbf{R}$ defined in a subset U of $\mathbf{R}^N$ is *locally semiconvex* (resp., *locally semiconcave*) if it is semiconvex (resp., semiconcave) in every convex subset of U.

When U is convex the local definitions coincide with the global ones. $\qquad\square$

In general, $\overline{U_h(A)} \subset A_h$, but the equality does not necessarily hold.

Lemma 5.1. *Given a subset A, $A \neq \varnothing$, of $\mathbf{R}^N$ and $h > 0$, the function*

$$k(x) \overset{\text{def}}{=} \begin{cases} |x|^2 - 2\,h\,d_A(x), & d_A(x) \geq h, \\ |x|^2 - d_A^2(x) - h^2, & d_A(x) < h, \end{cases}$$

is convex in $\mathbf{R}^N$. In particular,

$$k_{A,h}(x) \overset{\text{def}}{=} \frac{|x|^2}{2\,h} - d_A(x) \ \text{ and } \ |x|^2 - d_A^2(x)$$

are, respectively, locally convex in $\mathbf{R}^N \setminus A_h$ and $U_h(A)$.

Proof. For all $p \in A$ define the convex function

$$\ell_p(x) = \begin{cases} |x - p| - h, & |x - p| \geq h, \\ 0, & |x - p| < h. \end{cases}$$

Since ℓ_p is nonnegative, ℓ_p^2 is convex and

$$\ell_p^2(x) = \begin{cases} |x|^2 - 2\,h\,|x - p| + |p|^2 + h^2 - 2\,x \cdot p, & |x - p| \geq h, \\ 0, & |x - p| < h. \end{cases}$$

By subtracting the constant term $|p|^2 + h^2$ and the linear term $-2\,x \cdot p$ from ℓ_p^2, we get the new convex function

$$m_p(x) = \begin{cases} |x|^2 - 2\,h\,|x - p|, & |x - p| \geq h, \\ |x|^2 - |p - x|^2 - h^2, & |x - p| < h. \end{cases}$$

Then the function

$$k(x) \overset{\text{def}}{=} \sup_{p \in A} m_p(x)$$

is finite for each $x \in \mathbf{R}^N$, convex in x, and

$$k(x) = \begin{cases} |x|^2 - 2\,h\,d_A(x), & d_A(x) \geq h, \\ |x|^2 - d_A^2(x) - h^2, & d_A(x) < h. \end{cases}$$

If x is such that $d_A(x) \geq h$, then for all $p \in A$, $|x - p| \geq d_A(x) \geq h$ and $k(x) = |x|^2 - 2 h\, d_A(x)$. If $d_A(x) < h$, then there exists $p \in \overline{A}$ such that $|x - p| < h$ and

$$\inf_{\substack{p \in A \\ |x-p| < h}} |x - p| = d_A(x).$$

Then, either for all $p \in A$, $|x - p| < h$ and $k(x) = |x|^2 - h^2 - d_A^2(x)$ or, else, there exists $p \in \overline{A}$ such that $|x - p| \geq h$,

$$\inf_{\substack{p \in A \\ |x-p| \geq h}} |p - x| \geq h$$

$$\Rightarrow \quad |x|^2 - 2 h \inf_{\substack{p \in A \\ |x-p| \geq h}} |x - p| \leq |x|^2 - 2 h^2 \leq |x|^2 - h^2 - d_A^2(x)$$

and $k(x) = |x|^2 - d_A^2(x) - h^2$. We recover the result of Fu [1, Prop. 1.2] by observing that the restriction of k to $\mathbf{R}^N \backslash A_h$ is locally convex. In addition, the function $|x|^2 - d_A^2(x)$ is locally convex in $U_h(A)$. $\qquad\square$

This lemma has far-reaching consequences.

Theorem 5.3. *Let A be a nonempty subset of $\mathbf{R}^N$.*

(i) *The function $f_A(x) = \frac{1}{2}\left(|x|^2 - d_A^2(x)\right)$ is convex in $\mathbf{R}^N$ and $\nabla d_A^2 \in \mathrm{BV}_{\mathrm{loc}}(\mathbf{R}^N)^N$. For all x and y in $\mathbf{R}^N$,*

$$\forall p \in \Pi_A(x), \quad \frac{1}{2}\left(|y|^2 - d_A^2(y)\right) \geq \frac{1}{2}\left(|x|^2 - d_A^2(x)\right) + p \cdot (y - x), \tag{5.3}$$

or equivalently,

$$\forall p \in \Pi_A(x), \quad d_A^2(y) - d_A^2(x) - 2\,(x - p) \cdot (y - x) \leq |x - y|^2. \tag{5.4}$$

The set of projections $\Pi_A(x)$ onto $\overline{A}$ is a singleton $p_A(x)$ if and only if the gradient of f_A exists. In that case

$$p_A(x) = \frac{1}{2}\nabla\left(|x|^2 - d_A^2(x)\right). \tag{5.5}$$

For all x and y in $\mathbf{R}^N$

$$\forall p(x) \in \Pi_A(x),\ \forall p(y) \in \Pi_A(y), \quad (p(y) - p(x)) \cdot (y - x) \geq 0. \tag{5.6}$$

The function $d_A^2(x)$ is the difference of two convex functions

$$d_A^2(x) = |x|^2 - \left(|x|^2 - d_A^2(x)\right). \tag{5.7}$$

(ii) $\nabla d_A \in \mathrm{BV}_{\mathrm{loc}}(\mathbf{R}^N \backslash \overline{A})^N$. *More precisely, for all $x \in \mathbf{R}^N \backslash \overline{A}$, there exists $\rho > 0$, $0 < 3\rho < d_A(x)$, such that $k_{A,\rho}$ is convex in $B(x, 2\rho)$ and hence $\nabla d_A \in \mathrm{BV}(B(x, \rho))^N$.*

(iii) *A subset $A \neq \varnothing$ of $\mathbf{R}^N$ is of locally bounded exterior curvature if and only if $\nabla d_A \in \mathrm{BV}_{\mathrm{loc}}(\mathbf{R}^N)^N$. An open subset $\Omega \neq \mathbf{R}^N$ of $\mathbf{R}^N$ is of locally bounded interior curvature if and only if $\nabla d_{\complement\Omega} \in \mathrm{BV}_{\mathrm{loc}}(\mathbf{R}^N)^N$.*

Proof. (i) From Lemma 5.1 for all $h > 0$, $|x|^2 - d_A^2$ is locally convex in $U_h(A)$ and hence in $\mathbf{R}^N = \cup_{h>0} U_h(A)$. Therefore, $|x|^2 - d_A^2$ is convex in $\mathbf{R}^N$, and hence from Evans and Gariepy [1, Thm. 3, p. 240; Thm. 2, p. 239; and Aleksandrov's Theorem, p. 242], $\nabla d_A^2 \in \mathrm{BV}_{\mathrm{loc}}(\mathbf{R}^N)^N$. Inequality (5.3) follows directly from the inequality

$$\forall x \text{ and } y \in \mathbf{R}^N, \ \forall p(x) \in \Pi_A(x), \quad d_A^2(y) \leq |p(x) - y|^2$$

since

$$\begin{aligned}
d_A^2(y) - |y|^2 &\leq |p(x) - x + x - y|^2 - |y|^2 \\
&\leq |p(x) - x|^2 + |x - y|^2 + 2\,(p(x) - x) \cdot (x - y) - |y|^2 \\
&\leq |p(x) - x|^2 - |x|^2 + 2\,(p(x) - x) \cdot (x - y) + |x - y|^2 + |x|^2 - |y|^2
\end{aligned}$$

and

$$\begin{aligned}
&- 2f_A(y) \\
&\leq -2f_A(x) + 2\,p(x) \cdot (x - y) - 2\,x \cdot (x - y) + |x - y|^2 + |x|^2 - |y|^2 \\
&\leq -2f_A(x) - 2\,p(x) \cdot (y - x) \ \Rightarrow\ f_A(y) \geq f_A(x) + 2\,p(x) \cdot (y - x).
\end{aligned}$$

Inequality (5.4) is (5.3) rewritten. Inequality (5.6) follows by adding inequality (5.3) to the same inequality with x and y permuted.

(ii) For any point x in $\mathbf{R}^N \backslash \overline{A}$, there exists ρ, $0 < 3\rho < d_A(x)$ such that the open ball $B(x, 2\rho)$ is contained in $\mathbf{R}^N \backslash A_\rho$, where $k_{A,\rho}$ is locally convex by Lemma 5.1. Hence $k_{A,\rho}$ is convex in $B(x, 2\rho)$ and $\nabla k_{A,\rho}$, and a fortiori ∇d_A belong to $\mathrm{BV}(B(x, \rho))^N$.

(iii) Clearly, if $d_A \in \mathrm{BV}_{\mathrm{loc}}(\mathbf{R}^N)^N$, then property (5.2) is satisfied and A is of locally bounded exterior curvature. Conversely, by Theorem 5.1 of Chapter 3, it is sufficient to establish that for each x there exists $\rho > 0$ such that $\nabla d_A \in \mathrm{BV}(B(x, \rho))^N$. This is true in $\mathrm{int}\, A$, where $\nabla d_A = 0$, and in ∂A by assumption. It is also true for any point x in $\mathbf{R}^N \backslash \overline{A}$ by part (ii). Therefore, ∇d_A belongs to $\mathrm{BV}_{\mathrm{loc}}(\mathbf{R}^N)^N$ and A is of locally bounded exterior curvature. Similarly, for an open subset Ω of $\mathbf{R}^N$ we only need to prove that for each x there exists $\rho > 0$ such that $\nabla d_{\complement\Omega} \in \mathrm{BV}(B(x, \rho))^N$. This is true in $\mathrm{int}\,\complement\Omega$, where $\nabla d_{\complement\Omega} = 0$, and in $\partial\Omega$ by assumption. It is also true for any point x in $\mathbf{R}^N \backslash \complement\Omega$ by part (ii). Therefore, $\nabla d_{\complement\Omega}$ belongs to $\mathrm{BV}_{\mathrm{loc}}(\mathbf{R}^N)^N$ and Ω is of locally bounded interior curvature. $\qquad\square$

6   Characterization of Convex Sets

In the convex case the squared distance function is differentiable everywhere and this property can be used to characterize the convexity of a set.

Theorem 6.1. *Let A be a nonempty subset of $\mathbf{R}^N$ with convex closure. Then*

(i) *for each $x \in \mathbf{R}^N$, $\Pi_A(x) = \{p_A(x)\}$ is a singleton and*

$$\forall x, \forall y \in \mathbf{R}^N, \quad |p_A(y) - p_A(x)| \leq |y - x|;$$

(ii) *d_A^2 belongs to $C_{\mathrm{loc}}^{1,1}(\mathbf{R}^N)$ (and a fortiori to $W_{\mathrm{loc}}^{2,\infty}(\mathbf{R}^N)$).*

Proof. (i) By definition

$$d_A^2(x) = \inf_{z \in A} |z - x|^2 = \inf_{z \in \overline{A}} |z - x|^2,$$

and since $\overline{A}$ is convex and $z \mapsto |z - x|^2$ is strictly convex, there exists a unique $p_A(x)$ in $\overline{A}$ such that

$$\forall z \in \overline{A}, \quad 2(p_A(x) - x) \cdot (z - p_A(x)) \geq 0$$

(cf., for instance, Zarantonello [1, pp. 237–246] or Aubin [2, p. 24, Example 4.3]). So, for any two points x and y,

$$(p_A(x) - x) \cdot (p_A(y) - p_A(x)) \geq 0, \quad (p_A(y) - y) \cdot (p_A(x) - p_A(y)) \geq 0.$$

By adding up the above two inequalities,

$$|p_A(y) - p_A(x)|^2 \leq (y - x) \cdot (p_A(y) - p_A(x)) \leq |y - x| |p_A(y) - p_A(x)|.$$

(ii) From part (i) the map $x \mapsto p_A(x)$ is Lipschitz continuous and hence the map $x \mapsto \nabla d_A^2(x) = 2[x - p_A(x)]$ is also Lipschitz continuous. Therefore, d_A^2 belongs to $C_{\mathrm{loc}}^{1,1}(\mathbf{R}^N)$. $\qquad\square$

Theorem 6.2.

(i) *Let A be a nonempty subset of $\mathbf{R}^N$. Then*

     (a) *A convex $\Rightarrow d_A$ convex;*

     (b) *d_A convex $\Rightarrow \overline{A}$ convex;*

     (c) *$\forall x \in \mathbf{R}^N$, $\Pi_A(x)$ is a singleton $\Leftrightarrow \overline{A}$ is convex.*[3]

(ii) *Let D be a nonempty open subset of $\mathbf{R}^N$. The subfamily*

$$\boxed{\,\mathcal{C}_d(D) \overset{\mathrm{def}}{=} \{d_A \in C_d(D) \; : \; A \neq \varnothing \text{ convex}\}\,}$$

of $C_d(D)$ is closed in $C_{\mathrm{loc}}(D)$. It is compact in $C(\overline{D})$ when D is bounded. The subfamily

$$\boxed{\,\mathcal{C}_d^c(D) \overset{\mathrm{def}}{=} \{d_{\complement\Omega} \in C_d^c(D) \; : \; \Omega \text{ open, convex}\}\,}$$

of $C_d^c(D)$ is closed in $C_{\mathrm{loc}}(D)$. It is compact in $C(\overline{D})$ when D is bounded.

---

[3]This part of the theorem is related to deeper results on the convexity of *Chebyshev sets* in metric spaces. A subset A of a metric space X is called a Chebyshev set provided that every point x of X has a unique projection $p_A(x)$ in A. The reader is referred to Klee [1] for details and background material.

(iii) *For all subsets A of $\mathbf{R}^N$ such that $\partial A \neq \varnothing$ and $\overline{A}$ is convex, ∇d_A belongs to $\mathrm{BV}_{\mathrm{loc}}(\mathbf{R}^N)^N$, and the Hessian matrix $D^2 d_A$ of second-order derivatives is a matrix of signed Radon measures that are nonnegative on the diagonal. Moreover, d_A has a second-order derivative almost everywhere, and for almost all x and y in $\mathbf{R}^N$,*

$$|d_A(y) - d_A(x) - \nabla d_A(x) \cdot (y - x) - \tfrac{1}{2}(y - x) \cdot D^2 d_A(x)\,(y - x)|$$
$$= o(|y - x|^2)$$

as $y \to x$.

Proof. (i) (a) Given x and y in $\mathbf{R}^N$, there exist $\overline{x}$ and $\overline{y}$ in $\overline{A}$ such that $d_A(x) = |x - \overline{x}|$ and $d_A(y) = |y - \overline{y}|$. By convexity of A, $\overline{A}$ is convex, and for all λ, $0 \leq \lambda \leq 1$, $\lambda \overline{x} + (1 - \lambda)\overline{y} \in \overline{A}$ and

$$d_A(\lambda x + (1 - \lambda)y) \leq |\lambda x + (1 - \lambda)y - (\lambda \overline{x} + (1 - \lambda)\overline{y})|$$
$$\leq \lambda |x - \overline{x}| + (1 - \lambda)|y - \overline{y}|$$
$$= \lambda d_A(x) + (1 - \lambda) d_A(y)$$

and d_A is convex in $\mathbf{R}^N$.

(b) If d_A is convex, then

$$\forall \lambda \in [0, 1], \forall x, y \in \overline{A}, \quad d_A(\lambda x + (1 - \lambda)y) \leq \lambda d_A(x) + (1 - \lambda)d_A(y).$$

But x and y in $\overline{A}$ imply that $d_A(x) = d_A(y) = 0$ and hence

$$\forall \lambda \in [0, 1], d_A(\lambda x + (1 - \lambda)y) = 0.$$

Thus $\lambda x + (1 - \lambda)y \in \overline{A}$ and $\overline{A}$ is convex.

(c) By Theorem 6.1 (i) and for the converse, see in Valentine [1, p. 179, Prop. 7.1] the general result of Klee [1].

(ii) The set of all convex functions in $C_{\mathrm{loc}}(D)$ is a closed convex cone with vertex at 0. Hence its intersection with $C_d(D)$ is closed. Any Cauchy sequence in $\mathcal{C}_d(D)$ converges to some convex d_A. From part (i) $\overline{A}$ is convex. But $d_A = d_{\overline{A}}$ and the nonempty convex set $\overline{A}$ can be chosen as the limit set. For the complementary distance function, it is sufficient to prove the compactness for D bounded. Given any sequence $\{d_{\complement\Omega_n}\}$, there exist an open subset Ω of D and a subsequence, still indexed by n, such that $d_{\complement\Omega_n} \to d_{\complement\Omega}$ in $C(\overline{D})$. If Ω is empty, there is nothing to prove. If Ω is not empty, consider two points x and y in Ω and $\lambda \in [0, 1]$. There exists $r > 0$ such that

$$B(x, r) \subset \Omega \text{ and } B(y, r) \subset \Omega \quad \Rightarrow d_{\complement\Omega}(x) \geq r \text{ and } d_{\complement\Omega}(y) \geq r.$$

There exists $N > 0$ such that for all $n > N$, $\|d_{\complement\Omega_n} - d_{\complement\Omega}\|_{C(D)} < r/2$ and

$$d_{\complement\Omega_n}(x) \geq d_{\complement\Omega}(x) - r/2 = r/2 \Rightarrow B(x, r/2) \subset \Omega_n,$$
$$d_{\complement\Omega_n}(y) \geq d_{\complement\Omega}(y) - r/2 = r/2 \Rightarrow B(y, r/2) \subset \Omega_n.$$

By convexity of Ω_n, $x_\lambda = \lambda x + (1 - \lambda)y \in \Omega_n$,

$$B(x_\lambda, r/2) \subset \lambda B(x, r/2) + (1 - \lambda)B(y, r/2) \subset \Omega_n$$
$$\Rightarrow \forall n > N, \quad d_{\complement\Omega_n}(x_\lambda) \geq r/2$$
$$\Rightarrow d_{\complement\Omega}(x_\lambda) \geq r/2 \quad \Rightarrow B(x_\lambda, r/2) \subset \Omega.$$

Therefore, $x_\lambda \in \Omega$, Ω is convex, and $\mathcal{C}_d^c(D)$ is compact.

(iii) From part (i), d_A is convex and continuous, and the result follows from Evans and Gariepy [1, Thm. 3, p. 240; Thm. 2, p. 239; and Aleksandrov's Theorem, p. 242]. $\qquad\square$

Remark 6.1.

It is important to observe that the convexity of d_A is equivalent to the convexity of the closure of A. Notice that the set A of all points of the unit open ball with rational coordinates has a convex closure but is not convex. $\qquad\square$

Theorem 6.3. *Let A be a nonempty subset of $\mathbf{R}^N$:*

(i) *The Fenchel transform*

$$d_A^*(x^*) = \sup_{x \in \mathbf{R}^N} x^* \cdot x - d_A(x)$$

of d_A is given by

$$d_A^*(x^*) = \sigma_A(x^*) + I_{\overline{B(0,1)}}(x^*), \tag{6.1}$$

where $\sigma_A(x^*)$ is the support function of A and

$$I_{\overline{B(0,1)}}(x^*) = \begin{cases} +\infty, & x \notin \overline{B(0,1)}, \\ 0, & x \in \overline{B(0,1)}, \end{cases}$$

is the indicator function of the closed unit ball $\overline{B(0,1)}$ at the origin.

(ii) *The Fenchel transform of d_A^* is given by*

$$d_A^{**} = d_{\mathrm{co}\,A} = d_{\overline{\mathrm{co}}\,A}. \tag{6.2}$$

In particular, $d_{\mathrm{co}\,A}$ is the convex envelope of d_A.

The function d_A and the indicator function $I_{\overline{A}}$ are both zero on $\overline{A}$ and the Fenchel transform of $I_{\overline{A}}$,

$$I_{\overline{A}}^* = \sigma_{\overline{A}} = \sigma_A,$$

coincides with d_A^* on $\overline{B(0,1)}$.

Proof. (i) From the definition,

$$d_A^*(x^*) = \sup_{x \in \mathbf{R}^N} x^* \cdot x - d_A(x)$$

$$= \sup_{x \in \mathbf{R}^N} \left[x^* \cdot x - \inf_{p \in A} |x - p| \right]$$

$$= \sup_{x \in \mathbf{R}^N} \sup_{p \in A} x^* \cdot x - |x - p|$$

$$= \sup_{p \in A} \sup_{x \in \mathbf{R}^N} x^* \cdot x - |x - p|$$

$$= \sup_{p \in A} \sup_{x \in \mathbf{R}^N} x^* \cdot p + x^* \cdot (x - p) - |x - p|$$

$$= \sup_{p \in A} x^* \cdot p + \sup_{x \in \mathbf{R}^N} \left[x^* \cdot (x - p) - |x - p| \right]$$

$$= \sup_{p \in A} x^* \cdot p + \sup_{x \in \mathbf{R}^N} \left[x^* \cdot x - |x| \right] = \sigma_A(x^*) + I_{\overline{B(0,1)}}(x^*).$$

(ii) We now use the property that $\sigma_A = \sigma_{\mathrm{co}\,A}$:

$$d_A^{**}(x^{**}) = \sup_{x^* \in \mathbf{R}^N} x^{**} \cdot x^* - d_A^*(x^*)$$

$$= \sup_{x^* \in \mathbf{R}^N} x^{**} \cdot x^* - \sigma_A(x^*) - I_{\overline{B(0,1)}}(x^*)$$

$$= \sup_{|x^*| \leq 1} x^{**} \cdot x^* - \sigma_A(x^*)$$

$$= \sup_{|x^*| \leq 1} \left[x^{**} \cdot x^* - \sup_{x \in \mathrm{co}\,A} x^* \cdot x \right]$$

$$= \sup_{|x^*| \leq 1} \inf_{x \in \mathrm{co}\,A} (x^{**} - x) \cdot x^*$$

$$\leq \sup_{|x^*| \leq 1} (x^{**} - p(x^{**})) \cdot x^* \leq |x^{**} - p(x^{**})| = d_{\mathrm{co}\,A}(x^{**}).$$

But in fact we have a saddle point. If $x^{**} \in \overline{\mathrm{co}}\,A$, pick $x^* = 0$ and

$$d_A^{**}(x^{**}) = \sup_{|x^*| \leq 1} \inf_{x \in \mathrm{co}\,A} (x^{**} - x) \cdot x^* \geq 0 = d_{\mathrm{co}\,A}(x^{**}).$$

For $x^{**} \notin \overline{\mathrm{co}}\,A$ rewrite the function as

$$(x^{**} - x) \cdot x^* = (x^{**} - p(x^{**})) \cdot x^* + (p(x^{**}) - x) \cdot x^*.$$

Since $p(x^{**})$ is the minimizing point of the function $|x - x^{**}|^2$ over the closed convex set $\overline{\mathrm{co}}\,A$ it is completely characterized by the variational inequality

$$\forall x \in \overline{\mathrm{co}}\,A, \quad (p(x^{**}) - x^{**}) \cdot (x - p(x^{**})) \geq 0. \tag{6.3}$$

By choosing the following special x^*,

$$x^* = \frac{x^{**} - p(x^{**})}{|x^{**} - p(x^{**})|},$$

we get

$$d_A^{**}(x^{**}) = \sup_{|x^*| \leq 1} \inf_{x \in \mathrm{co}\, A} (x^{**} - x) \cdot x^*$$

$$\geq |x^{**} - p(x^{**})| + (p(x^{**}) - x) \cdot \frac{x^{**} - p(x^{**})}{|x^{**} - p(x^{**})|}$$

$$\geq |x^{**} - p(x^{**})| = d_{\mathrm{co}\, A}(x^{**})$$

in view of (6.3). Therefore

$$\sup_{|x^*| \leq 1} \inf_{x \in \mathrm{co}\, A} (x^{**} - x) \cdot x^* = d_{\mathrm{co}\, A}(x^{**}) = \inf_{x \in \mathrm{co}\, A} \sup_{|x^*| \leq 1} (x^{**} - x) \cdot x^*.$$

Finally, from the previous theorem, $d_{\mathrm{co}\, A}$ is a convex function that coincides with the convex envelope since it is the bidual of d_A. $\qquad \square$

Since the properties in Theorem 6.2 (iii) arise solely from the convexity of the function d_A, the distance function d_A can be convex or semiconvex. However, it is never semiconcave in $\mathbf{R}^N$ for nontrivial sets.

Theorem 6.4.

(i) *Given a subset A, $A \neq \varnothing$ of $\mathbf{R}^N$,*

$$\exists c \geq 0, \quad f_c(x) = c\,|x|^2 - d_A(x) \text{ is convex in } \mathbf{R}^N$$

if and only if

$$\exists h > 0, \exists c \geq 0, \quad f_c(x) = c\,|x|^2 - d_A(x)$$

is locally convex in $U_h(A)$.

(ii) *Given a subset A of $\mathbf{R}^N$ such that $\varnothing \neq \overline{A} \neq \mathbf{R}^N$,*

$$\nexists c \geq 0, \quad f_c(x) = c\,|x|^2 - d_A(x) \text{ be convex in } \mathbf{R}^N.$$

Proof. (i) If f_c is convex in $\mathbf{R}^N$, it is locally convex in $U_h(A)$. Conversely, assume that there exists $h > 0$ and $c \geq 0$ such that f_c is locally convex in $U_h(A)$. From Lemma 5.1,

$$\frac{1}{h}\,|x|^2 - d_A(x)$$

is locally convex in $\mathbf{R}^N \setminus A_{h/2}$. Therefore, for $\overline{c} = \max\{c, 1/h\}$, the function $f_{\overline{c}}$ is locally convex in $\mathbf{R}^N$ and hence convex on $\mathbf{R}^N$.

(ii) Assume the existence of a $c \geq 0$ for which f_c is convex in $\mathbf{R}^N$. For each $y \in \overline{A}$ the function

$$x \mapsto F_c(x, y) = c\,|x - y|^2 - d_A(x)$$

is also convex since it differs from $f_c(x)$ by the linear term

$$x \mapsto c\left(|y|^2 - 2\,x \cdot y\right).$$

Since $\varnothing \neq \overline{A} \neq \mathbf{R}^N$, there exist $x \in \complement \overline{A}$ and $p \in \overline{A}$ such that

$$0 < d_A(x) = |x - p| \leq \frac{1}{2c}.$$

For any $t > 0$, define $x_t = p - t\,(x - p)$ and $\lambda = t/(1+t) \in \,]0, 1[$ and observe that

$$x_\lambda \overset{\text{def}}{=} \lambda\,x + (1 - \lambda)\,x_t = p \text{ and } F_c(x_\lambda, p) = 0.$$

But

$$
\begin{aligned}
&\lambda\, F_c(x, p) + (1 - \lambda)\, F_c(x_t, p) \\
&= \frac{t}{1+t}\,[c\,|x - p|^2 - d_A(x)] + \frac{1}{1+t}\,[c\,|x_t - p|^2 - d_A(x_t)] \\
&\leq \frac{t}{1+t}\,[c\,d_A(x)^2 - d_A(x) + c\,t d_A(x)^2] \\
&\leq \frac{t}{1+t}\,d_A(x)\,[(1+t)\,c\,d_A(x) - 1] \\
&\leq \frac{t}{1+t}\,d_A(x)\left[(1+t)\,\frac{1}{2} - 1\right] = \frac{d_A(x)\,t}{1+t}\,\frac{t-1}{2},
\end{aligned}
$$

since by construction $d_A(x) > 0$ and $c\,d_A(x) < 1/2$. Therefore for t, $0 < t < 1$, the above quantity is strictly negative, and we have constructed two points x and x_t and a λ, $0 < \lambda < 1$, such that

$$F_c(\lambda\,x + (1 - \lambda)\,x_t, p) = 0 > \lambda\, F_c(x, p) + (1 - \lambda)\, F_c(x_t, p).$$

This contradicts the convexity of the function $x \mapsto F_c(x, p)$ and a fortiori of f_c. $\quad\square$

Theorem 6.5. *Let A be a nonempty subset of $\mathbf{R}^N$ for which the following condition is satisfied:*

$$\forall x \in \partial A, \ \exists \rho > 0 \text{ such that } d_A \text{ is semiconvex in } B(x, \rho).$$

Then the gradient of d_A belongs to $\mathrm{BV}_{\mathrm{loc}}(\mathbf{R}^N)^N$.

Proof. For all $x \in \operatorname{int} A$, $d_A = 0$ and there exists $\rho > 0$ such that $B(x, \rho) \subset \operatorname{int} A$ and the result is trivial. For each $x \in \mathbf{R}^N \setminus \overline{A}$, $d_A(x) > 0$. Pick $h = d_A(x)/2$. Then for all $y \in B(x, h)$ and all $a \in A$,

$$|y - a| \geq |x - a| - |y - x| \geq d_A(x) - |y - x| > 2h - h = h$$

and $B(x, h) \subset \mathbf{R}^N \setminus A_h$. By Lemma 5.1 the function $|x|^2/(2h) - d_A(x)$ is locally convex in $\mathbf{R}^N \setminus A_h$ and a fortiori convex in $B(x, h)$. Finally, by assumption for all $x \in \partial A$, there exists $\rho > 0$ such that d_A is semiconvex. Hence for each $x \in \mathbf{R}^N$, ∇d_A belongs to $\mathrm{BV}(B(x, \rho/2))$ for some $\rho > 0$. Therefore, ∇d_A belongs to $\mathrm{BV}_{\mathrm{loc}}(\mathbf{R}^N)^N$. $\quad\square$

7　Federer's Sets of Positive Reach

For closed submanifolds of codimension larger than 1, $\Omega = \partial\Omega$, $d_\Omega = d_{\partial\Omega}$, $d_{\complement\Omega} = 0$, and the gradient of d_Ω has a discontinuity along Ω. In that case it is natural to go to the square $d_{\partial\Omega}^2 = d_\Omega^2$ of the function and relate the properties of d_Ω^2 to the smoothness of Ω. This is directly connected with the sets of positive reach as introduced by Federer [1].

Definition 7.1.
A nonempty set $A \subset \mathbf{R}^N$ is said to have *positive reach* if there exists $h > 0$ such that $\Pi_A(x) = \{p_A(x)\}$ is a singleton for every $x \in U_h(A)$. The maximum h for which the property holds is called the *reach* of A and denoted reach (A).　　　□

For a convex set A, reach $(A) = +\infty$. All nonempty convex sets have positive reach. Bounded domains and submanifolds of class C^2 also have positive reach. A quite impressive result of Federer was to make sense of the classical Steiner–Minkowski formula for that class of sets. He also showed that for all r, $0 < r < h$, the boundaries of the closed tubular neighborhoods A_r are $C^{1,1}$-submanifolds of codimension 1 in $\mathbf{R}^N$. The next theorem summarizes several characterizations of sets with positive reach. Note that condition (vii) is a global condition on the smoothness of d_A^2 in the tubular neighborhood $U_h(A)$ that is similar to the one of Definition 5.1(ii) on ∇d_Ω in a neighborhood of $\partial\Omega$.

Theorem 7.1. *Given a nonempty subset A of $\mathbf{R}^N$, the following conditions are equivalent.*

   (i) *$\exists h > 0$ such that d_A belongs to $C^{1,1}_{\mathrm{loc}}(U_h(A)\backslash\overline{A})$.*

   (ii) *$\exists h > 0$ such that d_A belongs to $C^1(U_h(A)\backslash\overline{A})$.*

   (iii) *$\exists h > 0$ such that $\forall x \in U_h(A)\backslash\overline{A}$, $\Pi_A(x)$ is a singleton.*

   (iv) *$\exists h > 0$ such that $\forall x \in U_h(A)$, $\Pi_A(x)$ is a singleton.*

   (v) *A has positive reach, that is, reach $(A) > 0$.*

   (vi) *$\exists h > 0$ such that p_A belongs to $C^{0,1}_{\mathrm{loc}}(U_h(A))$.*

   (vii) *$\exists h > 0$ such that d_A^2 belongs to $C^{1,1}_{\mathrm{loc}}(U_h(A))$.*

Proof. The elements of the proof can be found in Federer [1].

   (i) $\Rightarrow$ (ii) $\Rightarrow$ (iii) $\Rightarrow$ (iv) $\Rightarrow$ (v) are obvious.

   (v) $\Rightarrow$ (vi) For each $x \in U_h(A)$, $\Pi_A(x) = \{p(x)\}$ is a singleton, and for all $t \geq 0$ such that $t\,d_A(x) < h$,

$$d_A(p(x) + t(x - p(x))) \leq t\,|x - p(x)| = t\,d_A(x) < h.$$

By assumption for all $t \geq 0$, $t\,d_A(x) < h$, and the projection of $p(x) + t(x - p(x))$ onto $\overline{A}$ is unique and equal to $p(x)$. Otherwise we could reduce $d_A(x)$, leading to a contradiction. Hence for all $t \geq 0$, $t\,d_A(x) < h$,

$$d_A(p(x) + t(x - p(x))) = t\,d_A(x).$$

For all $a \in \overline{A}$, $y \in U_h(A)$ and $t \geq 0$ such that $t\, d_A(y) < h$

$$|a - (p(x) + t(x - p(x)))|^2 \geq d_A(p(x) + t(x - p(x)))^2 = t^2 d_A(x)^2,$$
$$|a - p(x)|^2 + t^2 |x - p(x)|^2 + 2t\, (a - p(x)) \cdot (p(x) - x) \geq t^2 d_A(x)^2,$$
$$(a - p(x)) \cdot (p(x) - x) \geq -\frac{1}{2t} |a - p(x)|^2.$$

So for any y_1 and y_2 in $\mathbf{R}^N$ and $t \geq 0$ such that $t\, d_A(y_1) < t$ and $t\, d_A(y_2) < t$

$$(p(y_2) - p(y_1)) \cdot (p(y_1) - y_1) \geq -\frac{1}{2t} |p(y_2) - p(y_1)|^2,$$
$$(p(y_1) - p(y_2)) \cdot (p(y_2) - y_2) \geq -\frac{1}{2t} |p(y_1) - p(y_2)|^2,$$
$$(p(y_2) - p(y_1)) \cdot (y_2 - y_1) \geq \frac{t - 1}{t} |p(y_2) - p(y_1)|^2.$$

For any $x \in U_h(A)$, there exists $\rho > 0$ such that $d_A(x) + \rho < h$. Let $t = h/(d_A(x) + \rho)$, which is strictly greater than 1. For all $y \in B(x, \rho)$, $d_A(y) < d_A(x) + \rho$ and $t\, d_A(y) < h$. Therefore, for all y_1 and y_2 in $B(x, \rho)$

$$|p(y_2) - p(y_1)| \leq \frac{h}{h - (d_A(x) + \rho)} |y_2 - y_1|.$$

The result follows from the fact that any compact subset K of $U_h(A)$ can be covered by a finite number of neighborhoods $B(x_i, \rho_i)$, $x_i \in K$, $\rho_i > 0$.

(vi) $\Rightarrow$ (vii) The proof follows from the identity $p_A(x) = x - \frac{1}{2}\nabla d_A^2(x)$.

(vii) $\Rightarrow$ (i) For each $x \in U_h(A) \backslash \overline{A}$, there exists ρ, $0 < \rho$, such that $d_A(x) + \rho < h$. Therefore, for all $y \in B(x, \rho)$, $h > d_A(x) + \rho > d_A(y) \geq d_A(x) - \rho > 0$, and $d_A^2 \in C^{1,1}(B(x, \rho))$. For any y_1 and y_2 in $B(x, \rho)$,

$$\nabla d_A(y_2) - \nabla d_A(y_1) = \frac{1}{d_A(y_2)}[d_A(y_2)\nabla d_A(y_2) - d_A(y_2)\nabla d_A(y_1)]$$
$$= \frac{1}{2\, d_A(y_2)}[\nabla d_A^2(y_2) - \nabla d_A^2(y_1)]$$
$$+ \frac{1}{d_A(y_2)}\, [d_A(y_2) - d_A(y_1)]\nabla d_A(y_1).$$

So, from the proof of (v) $\Rightarrow$ (vi) and $d_A(y_2) \geq d_A(x) - \rho > 0$,

$$|\nabla d_A(y_2) - \nabla d_A(y_1)| \leq \frac{1}{2\,(d_A(x) - \rho)} \frac{2h - (d_A(x) + \rho)}{h - (d_A(x) + \rho)} |y_2 - y_1|$$
$$+ \frac{1}{d_A(x) - \rho}\, |y_2 - y_1|$$

and $\nabla d_A \in C^{0,1}(B(x, \rho))$, $d_A \in C^{1,1}(B(x, \rho))$. Therefore, the property holds in any compact subset of $U_h(A) \backslash \overline{A}$ and $d_A \in C^{1,1}_{\text{loc}}(U_h(A) \backslash \overline{A})$. $\qquad\square$

8   Compactness Theorems

Subsets of a bounded hold-all $\overline{D}$ with a positive reach greater than or equal to some $h > 0$ form a compact family of sets (cf. Federer [1, Thm. 4.13]).

Theorem 8.1. *Let D be a fixed bounded open subset of $\mathbf{R}^N$. Let $\{A_n\}$, $A_n \neq \varnothing$, be a sequence of subsets of $\overline{D}$. Assume that there exists $h > 0$ such that*

$$\forall n, \quad d_{A_n}^2 \in C_{\mathrm{loc}}^{1,1}(U_h(A_n)). \tag{8.1}$$

Then there exist a subsequence $\{A_{n_k}\}$ and $A \subset \overline{D}$, $A \neq \varnothing$, such that $d_A^2 \in C_{\mathrm{loc}}^{1,1}(U_h(A))$ and

$$d_{A_{n_k}}^2 \to d_A^2 \ \ in \ C^1(\overline{U_h(A)}).$$

This is a compactness theorem similar to the compactness of $C_d(D)$ in $C(D)$ for a bounded open subset of $\mathbf{R}^N$.

For the family of sets with bounded interior or exterior curvature, the key result is the compactness of the embeddings

$$\mathrm{BC}_d(D) = \left\{ d_\Omega \in C_d(D) \ : \ \nabla d_\Omega \in \mathrm{BV}(D)^N \right\} \to W^{1,p}(D), \tag{8.2}$$

$$\mathrm{BC}_d^c(D) = \left\{ d_\Omega \in C_d^c(D) \ : \ \nabla d_{\complement\Omega} \in \mathrm{BV}(D)^N \right\} \to W^{1,p}(D) \tag{8.3}$$

for bounded open Lipschitzian subsets D of $\mathbf{R}^N$ and p, $1 \leq p < \infty$. It is the analogue of the compactness theorem (Theorem 5.3 of Chapter 3) for Caccioppoli sets

$$\mathrm{BX}(D) = \{ \chi \in \mathrm{X}(D) \ : \ \chi \in \mathrm{BV}(D) \} \to L^p(D), \tag{8.4}$$

which is a consequence of the compactness of the embedding

$$\mathrm{BV}(D) \to L^1(D) \tag{8.5}$$

for bounded open Lipschitzian subsets D of $\mathbf{R}^N$ (cf. Morrey [1, Def. 3.4.1, p. 72; Thm. 3.4.4, p. 75] and Evans and Gariepy [1, Thm. 4, p. 176]).

As for characteristic functions in Chapter 3, we give a first version involving global conditions on a fixed bounded open Lipschitzian hold-all D. In the second version the sets are contained in a bounded open hold-all D with local conditions in their tubular neighborhood or the tubular neighborhood of their boundary.

8.1   Global Conditions on D

Theorem 8.2. *Let D be a nonempty bounded open Lipschitzian subset of $\mathbf{R}^N$. The embedding (8.2) is compact. Thus for any sequence $\{\Omega_n\}$, $\varnothing \neq \Omega_n$, of subsets of $\overline{D}$ such that*

$$\exists c > 0, \forall n \geq 1, \quad \|D^2 d_{\Omega_n}\|_{M^1(D)} \leq c, \tag{8.6}$$

there exist a subsequence $\{\Omega_{n_k}\}$ and a set Ω, $\Omega \neq \varnothing$, such that $\nabla d_\Omega \in \mathrm{BV}(D)^N$ and

$$d_{\Omega_{n_k}} \to d_\Omega \ \ in \ W^{1,p}(D)\text{-strong}$$

for all p, $1 \leq p < \infty$. Moreover, for all $\varphi \in \mathcal{D}^0(D)$,

$$\lim_{n \to \infty} \langle \partial_{ij} d_{\Omega_{n_k}}, \varphi \rangle = \langle \partial_{ij} d_\Omega, \varphi \rangle, \quad 1 \leq i,j \leq N, \tag{8.7}$$

$$\|D^2 d_\Omega\|_{M^1(D)} \leq c,$$

and $\chi_{\overline{\Omega}} \in \mathrm{BV}(D)$.

Proof. Given $c > 0$ consider the set

$$S_c \overset{\mathrm{def}}{=} \left\{ d_\Omega \in C_d(D) \ : \ \|D^2 d_\Omega\|_{M^1(D)} \leq c \right\}.$$

By compactness of the embedding (8.5), given any sequence $\{d_{\Omega_n}\}$, there exist a subsequence, still denoted $\{d_{\Omega_n}\}$, and $f \in \mathrm{BV}(D)^N$ such that $\nabla d_{\Omega_n} \to f$ in $L^1(D)^N$. But by Theorem 2.2 (ii), $C_d(D)$ is compact in $C(\overline{D})$ for bounded D and there exist another subsequence $\{d_{\Omega_{n_k}}\}$ and $d_\Omega \in C_d(D)$ such that $d_{\Omega_{n_k}} \to d_\Omega$ in $C(\overline{D})$ and, a fortiori, in $L^1(D)$. Therefore, $d_{\Omega_{n_k}}$ converges in $W^{1,1}(D)$ and also in $L^1(D)$. By uniqueness of the limit $f = \nabla d_\Omega$ and $d_{\Omega_{n_k}}$ converges in $W^{1,1}(D)$ to d_Ω. For $\Phi \in \mathcal{D}^1(D)^{N \times N}$ as k goes to infinity

$$\int_D \nabla d_{\Omega_{n_k}} \cdot \overrightarrow{\mathrm{div}}\,\Phi \, dx \to \int_D \nabla d_\Omega \cdot \overrightarrow{\mathrm{div}}\,\Phi \, dx,$$

$$\Rightarrow \left| \int_D \nabla d_\Omega \cdot \overrightarrow{\mathrm{div}}\,\Phi \, dx \right| = \lim_{k \to \infty} \left| \int_D \nabla d_{\Omega_{n_k}} \cdot \overrightarrow{\mathrm{div}}\,\Phi \, dx \right| \leq c \|\Phi\|_{C(D)},$$

$\|D^2 d_\Omega\|_{M^1(D)} \leq c$, and $\nabla d_\Omega \in \mathrm{BV}(D)^N$. This proves the compactness of the embedding for $p = 1$ and properties (8.7). The conclusions remain true for $p \geq 1$ by the equivalence of the $W^{1,p}$-topologies on $C_d(D)$ in Theorem 4.1(i). $\qquad \square$

When D is bounded open, $C_d^c(D)$ is compact in $C(\overline{D})$ and closed in $W^{1,p}(D)$, $1 \leq p < \infty$, and we have the analogue of the previous two compactness theorems.

Theorem 8.3. *Let D be a nonempty bounded open Lipschitzian subset of $\mathbf{R}^N$. The embedding (8.3) is compact. Thus for any sequence $\{\Omega_n\}$ of open subsets of D such that*

$$\exists h > 0, \ \exists c > 0, \ \forall n, \quad \|D^2 d_{\complement\Omega_n}\|_{M^1(D)} \leq c, \tag{8.8}$$

there exist a subsequence $\{\Omega_{n_k}\}$ and an open subset Ω of D such that $\nabla d_{\complement\Omega} \in \mathrm{BV}(D)^N$ and

$$d_{\complement\Omega_{n_k}} \to d_{\complement\Omega} \ \ in \ W_0^{1,p}(D) \tag{8.9}$$

for all p, $1 \leq p < \infty$. Moreover, for all $\varphi \in \mathcal{D}^0(D)^{N \times N}$,

$$\langle D^2 d_{\complement\Omega_n}, \varphi \rangle \to \langle D^2 d_{\complement\Omega}, \varphi \rangle \tag{8.10}$$

$$\|D^2 d_{\complement\Omega}\|_{M^1(D)} \leq c, \tag{8.11}$$

and $\chi_\Omega \in \mathrm{BV}(D)$.

Proof. Given $c > 0$ consider the set

$$S_c^c \overset{\text{def}}{=} \left\{ d_{\complement\Omega} \in C_d^c(D) \: : \: \|D^2 d_{\complement\Omega}\|_{M^1(D)} \le c \right\}.$$

By compactness of the embedding (8.5), given any sequence $\{d_{\complement\Omega_n}\}$ there exist a subsequence, still denoted $\{d_{\complement\Omega_n}\}$, and $f \in \mathrm{BV}(D)^N$ such that $\nabla d_{\complement\Omega_n} \to f$ in $L^1(D)^N$. But by Theorem 2.4 (ii), $C_d^c(D)$ is compact in $C_0(D)$ for bounded D and there exist another subsequence $\{d_{\complement\Omega_{n_k}}\}$ and $d_{\complement\Omega} \in C_d^c(D)$ such that $d_{\complement\Omega_{n_k}} \to d_{\complement\Omega}$ in $C_0(D)$ and, a fortiori, in $L^1(D)$. Therefore, $d_{\complement\Omega_{n_k}}$ converges in $W_0^{1,1}(D)$ and also in $L^1(D)$. By uniqueness of the limit, $f = \nabla d_{\complement\Omega}$ and $d_{\complement\Omega_{n_k}}$ converges in $W_0^{1,1}(D)$ to $d_{\complement\Omega}$. For all $\Phi \in \mathcal{D}^1(D)^{N \times N}$ as k goes to infinity

$$\int_D \nabla d_{\complement\Omega_{n_k}} \cdot \overrightarrow{\operatorname{div}}\Phi \, dx \to \int_D \nabla d_{\complement\Omega} \cdot \overrightarrow{\operatorname{div}}\Phi \, dx$$

$$\Rightarrow \left| \int_D \nabla d_{\complement\Omega} \cdot \overrightarrow{\operatorname{div}}\Phi \, dx \right| = \lim_{k \to \infty} \left| \int_D \nabla d_{\complement\Omega_{n_k}} \cdot \overrightarrow{\operatorname{div}}\Phi \, dx \right| \le c\|\Phi\|_{C(D)},$$

$\|D^2 d_{\complement\Omega}\|_{M^1(D)} \le c$, $\nabla d_{\complement\Omega} \in \mathrm{BV}(D)^N$, and $\chi_{\complement\Omega} \in \mathrm{BV}(D)$. This proves the compactness of the embedding for $p = 1$ and properties (8.10)–(8.11). The conclusions remain true for $p \ge 1$ by the equivalence of the $W^{1,p}$-topologies on $C_d^c(D)$ in Theorem 4.1 (i). $\qquad\square$

8.2   Local Conditions in Tubular Neighborhoods

The global conditions (8.6) and (8.8) can be weakened to a local one in a neighborhood of each set of the sequence. Simultaneously the Lipschitzian condition on D can be removed since only the uniform boundedness of the sets of the sequence is required.

Theorem 8.4. *Let D be a nonempty bounded open subset of $\mathbf{R}^N$ and $\{\Omega_n\}$, $\varnothing \ne \Omega_n$, be a sequence of subsets of $\overline{D}$. Assume that there exist $h > 0$ and $c > 0$ such that*

$$\forall n, \quad \|D^2 d_{\Omega_n}\|_{M^1(U_h(\partial\Omega_n))} \le c. \tag{8.12}$$

Then there exist a subsequence $\{\Omega_{n_k}\}$ and a subset Ω, $\varnothing \ne \Omega$, of $\overline{D}$ such that $\nabla d_\Omega \in \mathrm{BV}_{\mathrm{loc}}(\mathbf{R}^N)^N$, and for all p, $1 \le p < \infty$,

$$d_{\Omega_{n_k}} \to d_\Omega \quad in \quad W^{1,p}(U_h(D))\text{-strong.} \tag{8.13}$$

Moreover, for all $\varphi \in \mathcal{D}^0(U_h(\Omega))$,

$$\lim_{k \to \infty} \langle \partial_{ij} d_{\Omega_{n_k}}, \varphi \rangle = \langle \partial_{ij} d_\Omega, \varphi \rangle, \quad 1 \le i, j \le N,$$

$$\|D^2 d_\Omega\|_{M^1(U_h(\Omega))} \le c, \tag{8.14}$$

and $\chi_{\overline{\Omega}}$ belongs to $\mathrm{BV}_{\mathrm{loc}}(\mathbf{R}^N)$.

Proof. First notice that since Ω_n is bounded and nonempty, $\partial\Omega_n \neq \varnothing$. Further, $U_h(\partial\Omega_n)$ can be replaced by $U_h(\Omega_n)$ in condition (8.12) and it is sufficient to prove the theorem for that case.

Lemma 8.1. *For any* Ω, $\partial\Omega \neq \varnothing$ *and* $h > 0$,

$$\|D^2 d_\Omega\|_{M^1(U_h(\Omega))} = \|D^2 d_\Omega\|_{M^1(U_h(\partial\Omega))}.$$

The proof of the lemma will be given after the proof of the theorem.

(i) The assumption $\Omega_n \subset \overline{D}$ implies that $U_h(\Omega_n) \subset U_h(D)$. Since $U_h(D)$ is bounded, there exist a subsequence, still indexed by n, and a set Ω, $\varnothing \neq \Omega \subset \overline{D}$, such that

$$d_{\Omega_n} \to d_\Omega \quad \text{in } C(\overline{U_h(D)})\text{-strong}$$

and another subsequence, still denoted $\{d_{\Omega_n}\}$, such that

$$d_{\Omega_n} \to d_\Omega \quad \text{in } H^1(U_h(D))\text{-weak}.$$

For all $\varepsilon > 0$, $0 < 3\varepsilon < h$, there exists $N > 0$ such that for all $n \geq N$ and x in $U_h(D)$

$$d_{\Omega_n}(x) \leq d_\Omega(x) + \varepsilon, \quad d_\Omega(x) \leq d_{\Omega_n}(x) + \varepsilon.$$

Therefore,

$$\overline{\Omega}_n \subset U_{h-2\varepsilon}(\Omega_n) \subset U_{h-\varepsilon}(\Omega) \subset U_h(\Omega_n), \tag{8.15}$$

$$\complement U_{h-\varepsilon}(\Omega) \subset \complement U_{h-2\varepsilon}(\Omega_n) \subset \complement\overline{\Omega}_n. \tag{8.16}$$

From (8.12) and (8.15)

$$\forall n \geq N, \quad \|D^2 d_{\Omega_n}\|_{M^1(U_{h-\varepsilon}(\Omega))} \leq c.$$

In order to use the compactness of the embedding (8.5) as in the proof of Theorem 8.2, we would need $U_{h-\varepsilon}(\Omega)$ to be Lipschitzian. To get around this, we construct a bounded Lipschitzian set between $U_{h-2\varepsilon}(\Omega)$ and $U_{h-\varepsilon}(\Omega)$. Indeed, by definition,

$$U_{h-\varepsilon}(\Omega) = \cup_{x\in\overline{\Omega}} B(x, h - \varepsilon) \text{ and } \overline{U_{h-2\varepsilon}(\Omega)} \subset U_{h-\varepsilon}(\Omega),$$

and by compactness, there exists a finite sequence of points $\{x_i\}_{i=1}^n$ in $\overline{\Omega}$ such that

$$\overline{U_{h-2\varepsilon}(\Omega)} \subset U_B \stackrel{\text{def}}{=} \cup_{i=1}^n B(x_i, h - \varepsilon) \subset U_h(D).$$

Since U_B is Lipschitzian as the union of a finite number of balls, it now follows by compactness of the embedding (8.5) for U_B that there exists a subsequence, still denoted $\{d_{\Omega_n}\}$, and $f \in \text{BV}(U_B)^N$ such that $\nabla d_{\Omega_n} \to f$ in $L^1(U_B)^N$. Since $U_h(D)$ is bounded, $C_d(U_h(D))$ is compact in $C(\overline{U_h(D)})$ and there exists another subsequence, still denoted $\{d_{\Omega_n}\}$, and $\varnothing \neq \Omega \subset \overline{D}$ such that $d_{\Omega_n} \to d_\Omega$ in $C(\overline{U_h(D)})$ and, a fortiori, in $L^1(U_h(D))$. Therefore, d_{Ω_n} converges in $W^{1,1}(U_B)$ and also in $L^1(U_B)$. By uniqueness of the limit, $f = \nabla d_\Omega$ on U_B and d_{Ω_n} converges to d_Ω in

$W^{1,1}(U_B)$. By Definition 5.2 and Theorem 5.3 (iii), ∇d_Ω and ∇d_{Ω_n} all belong to $\mathrm{BV}_{\mathrm{loc}}(\mathbf{R}^N)^N$ since they are BV in tubular neighborhoods of their respective boundaries. Moreover, by Theorem 5.2 (ii), $\chi_{\overline{\Omega}} \in \mathrm{BV}_{\mathrm{loc}}(\mathbf{R}^N)$. The above conclusions also hold for the subset $U_{h-2\varepsilon}(\Omega)$ of U_B.

(ii) *Convergence in* $W^{1,p}(U_h(D))$. Consider the integral

$$\int_{U_h(D)} |\nabla d_{\Omega_n} - \nabla d_\Omega|^2 \, dx$$

$$= \int_{U_{h-2\varepsilon}(\Omega)} |\nabla d_{\Omega_n} - \nabla d_\Omega|^2 \, dx + \int_{U_h(D)\backslash U_{h-2\varepsilon}(\Omega)} |\nabla d_{\Omega_n} - \nabla d_\Omega|^2 \, dx.$$

From part (i) the first integral on the right-hand side converges to zero as n goes to infinity. The second integral is on a subset of $\complement U_{h-2\varepsilon}(\Omega)$. From (8.16) for all $n \geq N$,

$$|\nabla d_{\Omega_n}(x)| = 1 \text{ a.e. in } \complement\overline{\Omega}_n \supset \complement U_{h-3\varepsilon}(\Omega_n) \supset \complement U_{h-2\varepsilon}(\Omega),$$

$$|\nabla d_\Omega(x)| = 1 \text{ a.e. in } \complement\overline{\Omega} \supset \complement U_{h-2\varepsilon}(\Omega).$$

The second integral reduces to

$$\int_{U_h(D)\backslash U_{h-2\varepsilon}(\Omega)} |\nabla d_{\Omega_n} - \nabla d_\Omega|^2 \, dx = \int_{U_h(D)\backslash U_{h-2\varepsilon}(\Omega)} 2\,(1 - \nabla d_{\Omega_n} \cdot \nabla d_\Omega)\, dx,$$

which converges to zero since $\nabla d_{\Omega_n} \rightharpoonup \nabla d_\Omega$ in $L^2(U_h(D))^N$-weak in part (i) and the fact that $|\nabla d_\Omega| = 1$ almost everywhere in $U_h(D)\backslash U_{h-2\varepsilon}(\Omega)$. Therefore, since $d_{\Omega_n} \to d_\Omega$ in $C(\overline{U_h(D)})$,

$$d_{\Omega_n} \to d_\Omega \text{ in } H^1(U_h(D))\text{-strong},$$

and by Theorem 4.1(i) the convergence is true in $W^{1,p}(U_h(D))$ for all $p \geq 1$.

(iii) *Properties* (8.14). Consider the initial subsequence $\{d_{\Omega_n}\}$ which converges to d_Ω in $H^1(U_h(D))$-weak constructed at the beginning of part (i). This sequence is independent of ε and the subsequent constructions of other subsequences. By convergence of d_{Ω_n} to d_Ω in $H^1(U_h(D))$-weak for each $\Phi \in \mathcal{D}^1(U_h(\Omega))^{N\times N}$,

$$\lim_{n\to\infty} \int_{U_h(\Omega)} \nabla d_{\Omega_n} \cdot \overrightarrow{\mathrm{div}}\, \Phi \, dx = \int_{U_h(\Omega)} \nabla d_\Omega \cdot \overrightarrow{\mathrm{div}}\, \Phi \, dx.$$

Each such Φ has compact support in $U_h(\Omega)$, and there exists $\varepsilon = \varepsilon(\Phi) > 0$, $0 < 3\varepsilon < h$, such that

$$\overline{\mathrm{supp}\,\Phi} \subset U_{h-2\varepsilon}(\Omega).$$

From part (ii) there exists $N(\varepsilon) > 0$ such that

$$\forall n \geq N(\varepsilon), \quad U_{h-2\varepsilon}(\Omega_n) \subset U_{h-\varepsilon}(\Omega) \subset U_h(\Omega_n).$$

For $n \geq N(\varepsilon)$ consider the integral

$$\int_{U_h(\Omega)} \nabla d_{\Omega_n} \cdot \overrightarrow{\operatorname{div}} \, \Phi \, dx = \int_{U_{h-2\varepsilon}(\Omega)} \nabla d_{\Omega_n} \cdot \overrightarrow{\operatorname{div}} \, \Phi \, dx = \int_{U_h(\Omega_n)} \nabla d_{\Omega_n} \cdot \overrightarrow{\operatorname{div}} \, \Phi \, dx$$

$$\Rightarrow \left| \int_{U_h(\Omega)} \nabla d_{\Omega_n} \cdot \overrightarrow{\operatorname{div}} \, \Phi \, dx \right| \leq \| D^2 d_{\Omega_n} \|_{M^1(U_h(\Omega_n))} \, \| \Phi \|_{C(U_h(\Omega_n))}$$

$$\leq c \, \| \Phi \|_{C(U_{h-2\varepsilon}(\Omega))} = c \, \| \Phi \|_{C(U_h(\Omega))}.$$

By convergence of ∇d_{Ω_n} to ∇d_Ω in $L^2(D \cup U_h(\Omega))$-weak, for all $\Phi \in \mathcal{D}^1(U_h(\Omega))^{N \times N}$

$$\left| \int_{U_h(\Omega)} \nabla d_\Omega \cdot \overrightarrow{\operatorname{div}} \, \Phi \, dx \right| \leq c \, \| \Phi \|_{C(U_h(\Omega))} \quad \Rightarrow \quad \| D^2 d_\Omega \|_{M^1(U_h(\Omega))} \leq c.$$

Finally, the convergence remains true for all subsequences constructed in parts (i) and (ii). This completes the proof. $\qquad\square$

Proof of Lemma 8.1. First check that

$$U_h(\Omega) = U_h(\partial\Omega) \cup_{x \in \Omega_{-h}} B(x, h), \quad \Omega_{-h} \overset{\text{def}}{=} \left\{ x \in \mathbf{R}^N \, : \, B(x, h) \subset \Omega \right\}.$$

It is sufficient to show that all points x of $\operatorname{int} \Omega$ are contained in the right-hand side of the above expression. If $d_{\partial\Omega}(x) < h$, then $x \in U_h(\partial\Omega)$; if $d_{\partial\Omega}(x) \geq h$, then $B(x, h) \subset \Omega$ and $x \in \Omega_{-h}$. For all $\Phi \in \mathcal{D}^1(U_h(\Omega))^{N \times N}$, Φ has compact support and there exists $\varepsilon > 0$, $0 < 2\varepsilon < h$, such that

$$\overline{\operatorname{supp} \Phi} \subset U_{h-2\varepsilon}(\Omega).$$

But $\overline{U_{h-\varepsilon}(\Omega)} \subset U_h(\Omega)$ and there exists $\{x_j \in \Omega_{-h} \, : \, 1 \leq j \leq m\}$ such that

$$\overline{U_{h-\varepsilon}(\Omega)} \subset U_h(\partial\Omega) \cup_{j=1}^m B(x_j, h).$$

Let $\psi_0 \in \mathcal{D}(U_h(\partial\Omega))$ and $\psi_j \in \mathcal{D}(B(x_j, h))$ be a partition of unity such that

$$0 \leq \psi_j \leq 1, \quad \sum_{j=0}^m \psi_j = 1 \text{ on } \overline{U_{h-2\varepsilon}(\Omega)}.$$

Consider the integral

$$\int_{U_h(\Omega)} \nabla d_\Omega \cdot \overrightarrow{\operatorname{div}} \, \Phi \, dx = \int_{U_{h-2\varepsilon}(\Omega)} \nabla d_\Omega \cdot \overrightarrow{\operatorname{div}} \, \Phi \, dx = \int_{U_{h-2\varepsilon}(\Omega)} \nabla d_\Omega \cdot \overrightarrow{\operatorname{div}} \left(\sum_{j=0}^m \psi_j \Phi \right) dx$$

$$= \int_{U_{h-2\varepsilon}(\Omega)} \nabla d_\Omega \cdot \overrightarrow{\operatorname{div}} \left(\psi_0 \Phi \right) dx + \sum_{j=1}^m \int_{U_{h-2\varepsilon}(\Omega)} \nabla d_\Omega \cdot \overrightarrow{\operatorname{div}} \left(\psi_j \Phi \right) dx.$$

By construction for $\|\Phi\|_{C(U_h(\Omega))} \leq 1$

$$\psi_0\Phi \in \mathcal{D}^1(U_{h-2\varepsilon}(\Omega) \cap U_h(\partial\Omega))^{N\times N}, \quad \|\psi_0\Phi\|_{C(U_h(\partial\Omega))} \leq 1,$$
$$\psi_j\Phi \in \mathcal{D}^1(U_{h-2\varepsilon}(\Omega) \cap B(x_j,h))^{N\times N}, \quad \|\psi_j\Phi\|_{C(B(x_j,h))} \leq 1, \quad 1 \leq j \leq m.$$

But for j, $1 \leq j \leq m$, $B(x_j,h) \subset \operatorname{int}\Omega$, where both d_Ω and ∇d_Ω are identically zero. As a result the above integral reduces to

$$\int_{U_h(\Omega)} \nabla d_\Omega \cdot \overrightarrow{\operatorname{div}} \Phi \, dx = \int_{U_{h-2\varepsilon}(\Omega)} \nabla d_\Omega \cdot \overrightarrow{\operatorname{div}} (\psi_0\Phi) \, dx = \int_{U_h(\partial\Omega)} \nabla d_\Omega \cdot \overrightarrow{\operatorname{div}} (\psi_0\Phi) \, dx$$

$$\Rightarrow \left| \int_{U_h(\Omega)} \nabla d_\Omega \cdot \overrightarrow{\operatorname{div}} \Phi \, dx \right| \leq \|D^2 d_\Omega\|_{M^1(U_h(\partial\Omega))} \|\psi_0\Phi\|_{C(U_h(\partial\Omega))}$$

$$\Rightarrow \|D^2 d_\Omega\|_{M^1(U_h(\Omega))} \leq \|D^2 d_\Omega\|_{M^1(U_h(\partial\Omega))},$$

and this completes the proof. $\qquad\square$

Theorem 8.5. *Let D be a nonempty bounded open subset of $\mathbf{R}^N$. Let $\{\Omega_n\}$ be a sequence of nonempty open subsets of D and assume that there exist $h > 0$ and $c > 0$ such that*

$$\forall n, \quad \|D^2 d_{\complement\Omega_n}\|_{M^1(U_h(\partial\Omega_n))} \leq c. \tag{8.17}$$

Then there exist a subsequence $\{\Omega_{n_k}\}$ and an open subset Ω of D such that $\nabla d_{\complement\Omega} \in \mathrm{BV}_{\mathrm{loc}}(\mathbf{R}^N)^N$, and for all p, $1 \leq p < \infty$,

$$d_{\complement\Omega_{n_k}} \to d_{\complement\Omega} \text{ in } W_0^{1,p}(D)\text{-strong.} \tag{8.18}$$

Moreover, for all $\varphi \in \mathcal{D}^0(U_h(\complement\Omega))$,

$$\lim_{k\to\infty} \langle \partial_{ij} d_{\complement\Omega_{n_k}}, \varphi \rangle = \langle \partial_{ij} d_{\complement\Omega}, \varphi \rangle, \quad 1 \leq i,j \leq N,$$
$$\|D^2 d_{\complement\Omega}\|_{M^1(U_h(\complement\Omega))} \leq c, \tag{8.19}$$

and χ_Ω belongs to $\mathrm{BV}_{\mathrm{loc}}(\mathbf{R}^N)$.

Proof. First note that $\partial\Omega_n \neq \varnothing$ since Ω_n is bounded and nonempty. By Lemma 8.1, $U_h(\partial\Omega_n)$ can be replaced by $U_h(\complement\Omega_n)$ in condition (8.17). Moreover, since $\complement\Omega_n$ can be unbounded, we shall work with the bounded neighborhoods $U_h(\complement_{\bar{D}}\Omega_n)$ of $\complement_{\bar{D}}\Omega_n$ rather than with $U_h(\complement\Omega_n)$.

Lemma 8.2. *Let $h > 0$ and let Ω be an open subset of a nonempty bounded open subset D of $\mathbf{R}^N$. Then*

$$\|D^2 d_{\complement\Omega}\|_{M^1(U_h(\complement_{\bar{D}}\Omega))} = \|D^2 d_{\complement\Omega}\|_{M^1(U_h(\complement\Omega))}.$$

If $\partial\Omega \neq \varnothing$,

$$\|D^2 d_{\complement\Omega}\|_{M^1(U_h(\partial\Omega))} = \|D^2 d_{\complement\Omega}\|_{M^1(U_h(\complement\Omega))}.$$

The proof of this lemma will be given after the proof of the theorem.

(i) By Theorem 2.4 (ii) since D is bounded, there exists a subsequence, still denoted $\{d_{\complement\Omega_n}\}$, and an open subset Ω of D such that

$$d_{\complement\Omega_n} \to d_{\complement\Omega} \quad \text{in } C_0(D) \text{ and } C_c(D \cup U_h(\complement\Omega)),$$

and another subsequence, still denoted $\{d_{\complement\Omega_n}\}$, such that

$$d_{\complement\Omega_n} \to d_{\complement\Omega} \quad \text{in } H_0^1(D)\text{-weak and } H_0^1(D \cup U_h(\complement\Omega))\text{-weak}.$$

For all $\varepsilon > 0$, $0 < 3\varepsilon < h$, there exists $N > 0$ such that for all $n \geq N$

$$d_{\complement\Omega_n}(x) \leq d_{\complement\Omega}(x) + \varepsilon, \quad d_{\complement\Omega}(x) \leq d_{\complement\Omega_n}(x) + \varepsilon.$$

Clearly, since $\Omega \subset D$ and $\Omega_n \subset D$ and D is open, $\complement_{\overline{D}}\Omega_n \neq \varnothing$ and $\complement_{\overline{D}}\Omega \neq \varnothing$. Furthermore,

$$\complement_{\overline{D}}\Omega_n \subset U_{h-2\varepsilon}(\complement_{\overline{D}}\Omega_n) \subset U_{h-\varepsilon}(\complement_{\overline{D}}\Omega) \subset U_h(\complement_{\overline{D}}\Omega_n), \tag{8.20}$$
$$\complement_{\overline{D}}U_{h-\varepsilon}(\complement_{\overline{D}}\Omega) \subset \complement_{\overline{D}}U_{h-2\varepsilon}(\complement_{\overline{D}}\Omega_n) \subset \Omega_n. \tag{8.21}$$

From (8.17) and (8.20),

$$\forall n \geq N, \quad \|D^2 d_{\complement\Omega_n}\|_{M^1(U_{h-\varepsilon}(\complement_{\overline{D}}\Omega))} \leq c. \tag{8.22}$$

As in part (i) of the proof of Theorem 8.4, we can construct a bounded open Lipschitzian set U_B between $U_{h-2\varepsilon}(\complement_{\overline{D}}\Omega)$ and $U_{h-\varepsilon}(\complement_{\overline{D}}\Omega)$ such that

$$\overline{U_{h-2\varepsilon}(\complement_{\overline{D}}\Omega)} \subset U_B \subset U_{h-\varepsilon}(\complement_{\overline{D}}\Omega).$$

Since U_B is bounded and Lipschitzian, it now follows by compactness of the embedding (8.5) for U_B that there exists a subsequence, still denoted $\{d_{\complement\Omega_n}\}$, and $f \in \mathrm{BV}(U_B)^N$ such that $\nabla d_{\complement\Omega_n} \to f$ in $L^1(U_B)^N$. Since $D \cup U_B$ is bounded, $C_d^c(D \cup U_B)$ is compact in $C_0(D \cup U_B)$, and there exists another subsequence, still denoted $\{d_{\complement\Omega_n}\}$, and an open subset Ω of D such that $d_{\complement\Omega_n} \to d_{\complement\Omega}$ in $C_0(D \cup U_B)$ and, a fortiori, in $L^1(D \cup U_B)$. Therefore $d_{\complement\Omega_n}$ converges in $W^{1,1}(U_B)$ and also in $L^1(U_B)$. By uniqueness of the limit, $f = \nabla d_{\complement\Omega}$ on B_D and $d_{\complement\Omega_n}$ converges to $d_{\complement\Omega}$ in $W^{1,1}(U_B)$. By Definition 5.2 and Theorem 5.3 (iii), $\nabla d_{\complement\Omega}$ and $\nabla d_{\complement\Omega_n}$ all belong to $\mathrm{BV}_{\mathrm{loc}}(\mathbf{R}^N)^N$ since they are BV in tubular neighborhoods $U_h(\complement_{\overline{D}}\Omega)$ and $U_h(\complement_{\overline{D}}\Omega_n)$ of their respective boundaries $\partial\Omega$ and $\partial\Omega_n$. Moreover, by Theorem 5.2 (ii), $\chi_{\complement\Omega} \in \mathrm{BV}_{\mathrm{loc}}(\mathbf{R}^N)$ and, a fortiori, χ_Ω since Ω is open. The above conclusions also hold for the subset $U_{h-2\varepsilon}(\complement_{\overline{D}}\Omega)$ of U_B.

(ii) *Convergence in $W_0^{1,p}(D)$.* Consider the integral

$$\int_D |\nabla d_{\complement\Omega_n} - \nabla d_{\complement\Omega}|^2 \, dx$$

$$= \int_{D \cap U_{h-2\varepsilon}(\complement_{\overline{D}}\Omega)} |\nabla d_{\complement\Omega_n} - \nabla d_{\complement\Omega}|^2 \, dx + \int_{D \setminus U_{h-2\varepsilon}(\complement_{\overline{D}}\Omega)} |\nabla d_{\complement\Omega_n} - \nabla d_{\complement\Omega}|^2 \, dx.$$

From part (i) the first integral on the right-hand side converges to zero as n goes to infinity. The second integral is on a subset of $\complement_{\overline{D}} U_{h-2\varepsilon}(\complement_{\overline{D}}\Omega)$. From the relations (8.21) for all $n \geq N$,

$$|\nabla d_{\complement\Omega_n}(x)| = 1 \text{ a.e. in } \Omega_n \supset \complement_{\overline{D}} U_{h-3\varepsilon}(\complement_{\overline{D}}\Omega_n) \supset \complement_{\overline{D}} U_{h-2\varepsilon}(\complement_{\overline{D}}\Omega),$$
$$|\nabla d_\Omega(x)| = 1 \text{ a.e. in } \Omega \supset \complement_{\overline{D}} U_{h-2\varepsilon}(\complement_{\overline{D}}\Omega).$$

The second integral reduces to

$$\int_{D\setminus U_{h-2\varepsilon}(\complement_{\overline{D}}\Omega)} |\nabla d_{\complement\Omega_n} - \nabla d_{\complement\Omega}|^2\, dx = \int_{D\setminus U_{h-2\varepsilon}(\complement_{\overline{D}}\Omega)} 2\left(1 - \nabla d_{\complement\Omega_n} \cdot \nabla d_{\complement\Omega}\right) dx,$$

which converges to zero by weak convergence of $\nabla d_{\complement\Omega_n}$ to $\nabla d_{\complement\Omega}$ in $L^2(D)^N$ and the fact that $|\nabla d_{\complement\Omega}| = 1$ almost everywhere in $D\setminus U_{h-2\varepsilon}(\complement_{\overline{D}}\Omega)$. Therefore, since $d_{\complement\Omega_n} \to d_{\complement\Omega}$ in $C_0(D)$

$$d_{\complement\Omega_n} \to d_{\complement\Omega} \text{ in } H_0^1(D)\text{-strong},$$

and by Theorem 4.1 (i) the convergence is true in $W_0^{1,p}(D)$ for all $p \geq 1$.

(iii) *Properties* (8.19). Consider the initial subsequence $\{d_{\complement\Omega_n}\}$, which converges to $d_{\complement\Omega}$ in $H_0^1(D \cup U_h(\complement\Omega))$-weak constructed at the beginning of part (i). This sequence is independent of ε and the subsequent constructions of other subsequences. By convergence of $d_{\complement\Omega_n}$ to $d_{\complement\Omega}$ in $H_0^1(D \cup U_h(\complement\Omega))$-weak for each $\Phi \in \mathcal{D}^1(U_h(\complement\Omega))^{N\times N}$,

$$\lim_{n\to\infty} \int_{U_h(\complement\Omega)} \nabla d_{\complement\Omega_n} \cdot \overrightarrow{\operatorname{div}} \Phi\, dx = \int_{U_h(\complement\Omega)} \nabla d_{\complement\Omega} \cdot \overrightarrow{\operatorname{div}} \Phi\, dx.$$

Now each $\Phi \in \mathcal{D}^1(U_h(\complement\Omega))^{N\times N}$ has compact support in $U_h(\complement\Omega)$ and there exists $\varepsilon = \varepsilon(\Phi) > 0$, $0 < 3\varepsilon < h$, such that

$$\overline{\operatorname{supp} \Phi} \subset U_{h-2\varepsilon}(\complement\Omega).$$

From part (ii) there exists $N(\varepsilon) > 0$ such that

$$\forall n \geq N(\varepsilon), \quad U_{h-2\varepsilon}(\complement\Omega_n) \subset U_{h-\varepsilon}(\complement\Omega) \subset U_h(\complement\Omega_n).$$

For $n \geq N(\varepsilon)$ consider the integral

$$\int_{U_h(\complement\Omega)} \nabla d_{\complement\Omega_n} \cdot \overrightarrow{\operatorname{div}} \Phi\, dx = \int_{U_{h-2\varepsilon}(\complement\Omega)} \nabla d_{\complement\Omega_n} \cdot \overrightarrow{\operatorname{div}} \Phi\, dx = \int_{U_h(\complement\Omega_n)} \nabla d_{\Omega_n} \cdot \overrightarrow{\operatorname{div}} \Phi\, dx$$

$$\Rightarrow \left| \int_{U_h(\complement\Omega)} \nabla d_{\complement\Omega_n} \cdot \overrightarrow{\operatorname{div}} \Phi\, dx \right| \leq \|D^2 d_{\complement\Omega_n}\|_{M^1(U_h(\complement_D\Omega_n))} \|\Phi\|_{C(U_h(\complement_D\Omega_n))}$$

$$\leq c\,\|\Phi\|_{C(U_{h-2\varepsilon}(\complement_D\Omega))} = c\,\|\Phi\|_{C(U_h(\complement\Omega))}.$$

By convergence of $\nabla d_{\complement\Omega_n}$ to $\nabla d_{\complement\Omega}$ in $L^2(D\cup U_h(\complement\Omega))$-weak, for all $\Phi \in \mathcal{D}^1(U_h(\complement\Omega))^{N\times N}$

$$\left| \int_{U_h(\complement\Omega)} \nabla d_{\complement\Omega} \cdot \overrightarrow{\operatorname{div}} \Phi\, dx \right| \leq c\,\|\Phi\|_{C(U_h(\complement\Omega))} \quad \Rightarrow \quad \|D^2 d_{\complement\Omega}\|_{M^1(U_h(\complement\Omega))} \leq c.$$

Finally, the convergence remains true for all subsequences constructed in parts (i) and (ii). This completes the proof. $\qquad\square$

Proof of Lemma 8.2. The proof is similar to that of Lemma 8.1 with

$$U_h(\complement\Omega) = U_h(\complement_{\overline{D}}\Omega) \cup \operatorname{int}\complement D = U_h(\complement_{\overline{D}}\Omega)\cup_{x\in(\complement\overline{D})_{-h}} B(x,h),$$

$$(\complement\overline{D})_{-h} \stackrel{\mathrm{def}}{=} \left\{x \in \mathbf{R}^N \ : \ B(x,h) \subset \complement\overline{D}\right\},$$

since both $d_{\complement\Omega}$ and $\nabla d_{\complement\Omega}$ are identically zero on $\overline{\complement D}$. $\qquad\square$

Chapter 5
Oriented Distance Function and Smoothness of Sets

1 Introduction

In this chapter we study *oriented distance functions* and their role in the description of the geometric properties of domains and their boundaries. They are also known as *algebraic* or *signed distance functions*. Our choice of terminology emphasizes the fact that, for a smooth domain, the associated oriented distance function defines an *orientation* of the normal to the boundary. They enjoy many interesting properties. For instance, they retain the nice properties of the distance functions but also generate the classical geometric properties associated with sets and their boundaries. The smoothness of the oriented boundary functions in a neighborhood of the boundary of the set is equivalent to the smoothness of its boundary. Similarly, the convexity of the function is equivalent to the convexity of the closure of the set. In addition, their respective gradient and Hessian matrix, respectively, coincide with the *unit outward normal* and the *second fundamental form* on the boundary of the set. Finally, they provide a framework for the classification of domains and sets according to their degree of smoothness, much like Sobolev spaces and spaces of continuous and Hölderian functions do for functions.

The first part of the chapter deals with basic definitions, constructions, and results. The second part specializes to specific subfamilies of oriented distance functions. The last part concentrates on compact families of subsets of oriented distance functions. In the first part, section 2 presents the basic properties, introduces the uniform metric topology, and shows its connection with the Hausdorff and complementary Hausdorff topologies of Chapter 4. Section 3 is devoted to the differentiability properties, the associated set of projections onto the boundary, and completes the treatment of skeletons and cracks. Section 4 gives the equivalence of the smoothness of a set and the smoothness of its oriented distance function in a neighborhood of its boundary for sets of class $C^{1,1}$ or better. When the domain is sufficiently smooth the trace of the Hessian matrix of second-order partial derivatives on the boundary is the classical second fundamental form of geometry. Section 5 deals with the $W^{1,p}$-topology on the set of oriented distance functions and the closed subfamily of sets for which the volume of the boundary is zero.

In the second part of the chapter, in section 6, we study the subfamily of sets for which the gradient of the oriented distance function is a vector of functions of bounded variation: the sets with global or local bounded curvature. They are rather large classes for which quite general compactness theorems will be proved in section 9. Some examples are given to illustrate the behavior of the norms in tubular neighborhoods as the thickness of the neighborhood goes to zero in section 6.2. Section 6.3 introduces *Sobolev or $W^{s,p}$-domains*, which provide a framework for the classification of sets according to their degree of smoothness. For smooth sets this classification intertwines with the classical classification of C^k- and Hölderian domains. Section 7 extends the characterization of closed convex sets by the distance function to the oriented distance function and introduces the notion of semiconvex sets. The set of equivalence classes of convex subsets of a compact hold-all D is again compact, as it was for all the topologies considered in Chapters 3 and 4. Section 8 shows that sets of positive reach introduced in Chapter 4 are sets of locally bounded curvature and that their boundary has zero volume.

In the last part of the chapter, section 9 gives compactness theorems for sets of global and local bounded curvatures from a uniform bound in tubular neighborhoods of their boundary. They are the analogues of the theorems of Chapter 4. Finally, section 10 gives the compactness of the sets of Lipschitzian domains in a bounded hold-all under the uniform cone property for the $W^{1,p}$-topology associated with the oriented distance functions. We recover as a corollary the compactness of Theorem 5.9 of section 5.4 in Chapter 3 for the associated family of characteristic functions in $L^p(D)$. Furthermore, we get the compactness of the set of characteristic functions of the complements in $L^p(D)$ and the compactness of the associated families of distance functions and distance functions of the complement in $W^{1,p}(D)$ and $C(\overline{D})$.

2 Uniform Metric Topology

2.1 The Family of Oriented Distance Functions $C_b(D)$

The distance function d_A provides a good description of a domain A from the "outside." However, its gradient undergoes a jump discontinuity at the boundary of A which prevents the extraction of information about the smoothness of A from the smoothness of d_A in a neighborhood of ∂A. To get around this, it is natural to take into account the negative of the distance function of its complement $\complement A$, which somehow cancels the jump discontinuity at the boundary.

Definition 2.1.
Given a subset A of $\mathbf{R}^N$, the *oriented distance function* from x to A is defined as

$$\boxed{b_A(x) \overset{\text{def}}{=} d_A(x) - d_{\complement A}(x) \quad \forall x \in \mathbf{R}^N.}$$

$$(2.1)$$

Associate with a nonempty subset D of $\mathbf{R}^N$ the family

$$\boxed{C_b(D) \overset{\text{def}}{=} \left\{ b_A : A \subset \overline{D} \text{ and } \partial A \neq \varnothing \right\}.} \tag{2.2}$$

$\square$

This new function gives a level set description of a set whose boundary coincides with the zero level set. From Chapter 4 the function b_A is finite in $\mathbf{R}^N$ if and only if $\varnothing \neq A \neq \mathbf{R}^N$, since $b_A = d_A = +\infty$ when $A = \varnothing$ and $b_A = -d_{\complement A} = -\infty$ when $\complement A = \varnothing$. This condition is completely equivalent to $\partial A \neq \varnothing$, in which case b_A coincides with the *algebraic distance function* to the boundary of A:

$$\boxed{b_A(x) = \begin{cases} d_A(x) = d_{\partial A}(x), & x \in \operatorname{int} \complement A, \\ 0, & x \in \partial A, \\ -d_{\complement A}(x) = -d_{\partial A}(x), & x \in \operatorname{int} A. \end{cases}} \tag{2.3}$$

Noting that $b_{\complement A} = -b_A$, it means that we have implicitly chosen the negative sign for the interior of A and the positive sign for the interior of its complement. We shall see later that for sets with a smooth boundary, the restriction of the gradient of b_A to ∂A coincides with the outward unit normal to the boundary of A. Changing the sign of b_A gives the inward orientation to the gradient and the normal.

Theorem 2.1. *Let A be a subset of $\mathbf{R}^N$. Then*

(i) *$A \neq \varnothing$ and $\complement A \neq \varnothing \iff \partial A \neq \varnothing$.*

(ii) *Given b_A and b_B in $C_b(D)$,*

$$A \supset B \Rightarrow b_A \leq b_B \text{ and } A = B \Rightarrow b_A = b_B,$$

$$b_A \leq b_B \text{ in } D \iff \overline{B} \subset \overline{A} \text{ and } \overline{\complement A} \subset \overline{\complement B},$$

$$b_A = b_B \text{ in } D \iff \overline{B} = \overline{A} \text{ and } \overline{\complement A} = \overline{\complement B} \iff \overline{B} = \overline{A} \text{ and } \partial A = \partial B.$$

In particular, $b_{\overline{A}} \leq b_A$ and $b_{\overline{A}} = b_A \iff \partial \overline{A} = \partial A$.

(iii) *$|b_A| = d_A + d_{\complement A} = \max\{d_A, d_{\complement A}\} = d_{\partial A}$ and $\partial A = \{x \in \mathbf{R}^N : b_A(x) = 0\}$.*

(iv) *$b_A \geq 0 \iff \overline{\complement A} \supset \partial A \supset \overline{A} \iff \partial A = \overline{A}$.*

(v) *$b_A = 0 \iff \overline{\complement A} = \partial A = \overline{A} \iff \partial A = \mathbf{R}^N$.*

(vi) *If $\partial A \neq \varnothing$, the function b_A is uniformly Lipschitz continuous in $\mathbf{R}^N$ and*

$$\forall x, y \in \mathbf{R}^N, \quad |b_A(y) - b_A(x)| \leq |y - x|. \tag{2.4}$$

Moreover, b_A is (Fréchet) differentiable almost everywhere and

$$|\nabla b_A(x)| \leq 1 \quad a.e. \text{ in } \mathbf{R}^N. \tag{2.5}$$

Proof. (i) The proof is obvious. (ii) By assumption,

$$d_A = b_A^+ \le b_B^+ = d_B \text{ and } d_{\complement A} = b_A^- \ge b_B^- = d_{\complement B} \text{ in } \overline{D}.$$

Since $A \subset \overline{D}$ and $B \subset \overline{D}$, then $\complement A \supset \complement \overline{D}$ and $\complement B \supset \complement \overline{D}$ and

$$d_{\complement A} = d_{\complement B} = 0 \text{ in } \complement \overline{D} \quad \Rightarrow \quad d_{\complement A} \ge d_{\complement B} \text{ in } \mathbf{R}^N \quad \Rightarrow \quad \overline{\complement A} \subset \overline{\complement B}.$$

Also, since $\overline{B} \subset \overline{D}$ for all $x \in \overline{B}$, $d_A(x) \le d_B(x) = 0$ and $x \in \overline{A}$. Therefore, $\overline{\complement A} \subset \overline{\complement B}$ and $\overline{B} \subset \overline{A}$. Conversely,

$$\overline{B} \subset \overline{A} \quad \Rightarrow \quad d_A \le d_B \text{ in } \mathbf{R}^N \quad \text{and} \quad \overline{\complement A} \subset \overline{\complement B} \quad \Rightarrow \quad d_{\complement B} \le d_{\complement A} \text{ in } \mathbf{R}^N$$

and a fortiori in D. The equality case follows from the fact that $b_A = b_B$ if and only if $b_A \ge b_B$ and $b_A \le b_B$.

(iii) For x in $\overline{A}$

$$b_A(x) = -d_{\complement A}(x) \implies |b_A(x)| = d_{\complement A}(x) \le d_{\partial A}(x)$$

since $\overline{\complement A} \supset \partial A$ and the inf over $\overline{\complement A}$ is smaller than the inf over its subset ∂A. Similarly, for x in $\overline{\complement A}$

$$b_A(x) = d_A(x) \implies |b_A(x)| = d_A(x) \le d_{\partial A}(x),$$

and finally

$$|b_A(x)| \le \max\{d_{\complement A}(x), d_A(x)\} \le d_{\partial A}(x).$$

Conversely, for each x in $\overline{\complement A}$, the set of projections $\Pi_A(x) \subset \overline{A} \cap \overline{\complement A} = \partial A$ is not empty. Hence

$$|b_A(x)| = d_A(x) = \min_{y \in \Pi_A(x)} |x - y| \ge \inf_{y \in \partial A} |x - y| = d_{\partial A}(x),$$

and similarly for all x in $\overline{A}$,

$$|b_A(x)| = d_{\complement A}(x) \ge d_{\partial A}(x).$$

Therefore $|b_A(x)| \ge d_{\partial A}(x)$.

(iv) If $b_A = d_A - d_{\complement A} \ge 0$, then $d_A \ge d_{\complement A}$ and $\overline{A} \subset \overline{\complement A}$, and necessarily $\overline{A} \subset \partial A$ and $\overline{A} = \partial A$. Conversely, if $x \in \partial A$, then by definition $b_A(x) = 0$. If $x \notin \partial A$, then $x \in \complement \partial A = \complement \overline{A} = \operatorname{int} \complement A$ and $b_A(x) = d_A(x) \ge 0$.

(v) $b_A = 0$ is equivalent to $b_A \ge 0$ and $b_{\complement A} = -b_A \ge 0$. Then we apply (v) twice. But $\overline{\complement A} = \partial A = \overline{A} \iff \operatorname{int} \complement A = \varnothing = \operatorname{int} A \iff \partial A = \mathbf{R}^N$.

(vi) Clearly,

$$\forall x, y \in \overline{A}, \quad |b_A(y) - b_A(x)| = |d_{\complement A}(y) - d_{\complement A}(x)| \le |y - x|,$$
$$\forall x, y \in \overline{\complement A}, \quad |b_A(y) - b_A(x)| = |d_A(y) - d_A(x)| \le |y - x|.$$

For $x \in \overline{A}$ and $y \in \text{int}\,\complement A = \complement\overline{A}$, $d_A(y) > 0$ and

$$b_A(y) - b_A(x) = d_A(y) + d_{\complement A}(x) > 0 \quad \Rightarrow \quad |b_A(y) - b_A(x)| = d_A(y) + d_{\complement A}(x).$$

By assumption $B(y, d_A(y)) \subset \text{int}\,\complement A$. Define the point

$$\overline{x} = y + \frac{d_A(y)}{|x - y|}(x - y) \in \overline{\complement A}$$

$$\Rightarrow d_{\complement A}(x) \leq |x - \overline{x}| = \left|\left(1 - \frac{d_A(y)}{|x - y|}\right)(y - x)\right| = |x - y| - d_A(y)$$

$$\Rightarrow |b_A(y) - b_A(x)| = d_A(y) + d_{\complement A}(x) \leq |x - y|.$$

The argument is similar for $x \in \text{int}\,A$ and $y \in \overline{\complement A}$. The differentiability follows from Theorem 2.1 (vii) in Chapter 4. $\qquad\qquad\square$

2.2   Uniform Metric Topology

From Theorem 2.1 (i) the function b_A is finite at each point when $\partial A \neq \varnothing$. This excludes $A = \varnothing$ and $A = \mathbf{R}^N$. The zero function $b_A(x) = 0\ \forall x \in \mathbf{R}^N$ corresponds to the equivalence class of sets A such that

$$\overline{A} = \partial A = \overline{\complement A} \text{ or } \partial A = \mathbf{R}^N.$$

This class of sets is not empty. For instance, choose the subset of points of $\mathbf{R}^N$ with rational coordinates or the set of all lines parallel to one of the coordinate axes with rational coordinates.

Let D be a nonempty subset of $\mathbf{R}^N$ and associate with each subset A of $\overline{D}$, $\partial A \neq \varnothing$, the equivalence class

$$[A]_b \overset{\text{def}}{=} \left\{B \,:\, \forall B,\ B \subset \overline{D},\ \overline{B} = \overline{A} \text{ and } \partial A = \partial B\right\}$$

and the family of equivalence classes

$$\boxed{\mathcal{F}_b(D) \overset{\text{def}}{=} \left\{[A]_b \,:\, \forall A,\ A \subset \overline{D} \text{ and } \partial A \neq \varnothing\right\}.}$$

The equivalence classes induced by b_A are finer than those induced by d_A since both the closures and the boundaries of the respective sets must coincide. As in the case of d_A we identify $\mathcal{F}_b(D)$ with the family $C_b(D)$ through the embedding

$$[A]_b \mapsto b_A \colon \mathcal{F}_b(D) \to C_b(D) \subset C(\overline{D}).$$

When D is bounded, the space $C(\overline{D})$ endowed with the norm $\|f\|_{C(D)}$ is a Banach space. Moreover, for each $A \subset \overline{D}$, b_A is bounded, uniformly continuous on D, and $b_A \in C(\overline{D})$. This will induce the following complete metric:

$$\boxed{\rho([A]_b, [B]_b) \overset{\text{def}}{=} \|b_A - b_B\|_{C(D)}} \qquad\qquad (2.6)$$

on $\mathcal{F}_b(D)$.

When D is open but not necessarily bounded, we use the space $C_{\mathrm{loc}}(D)$ defined in section 2.2 of Chapter 4 endowed with the complete metric ρ_δ defined in (2.8) for the family of seminorms $\{q_K\}$ defined in (2.7). It will be shown below that $C_b(D)$ is a closed subset of $C_{\mathrm{loc}}(D)$ and that this induces the following complete metric on $\mathcal{F}_b(D)$:

$$\rho_\delta([A]_b, [B]_b) \stackrel{\mathrm{def}}{=} \sum_{k=1}^{\infty} \frac{1}{2^k} \frac{q_{K_k}(b_A - b_B)}{1 + q_{K_k}(b_A - b_B)}. \tag{2.7}$$

The following subfamilies:

$$\mathcal{F}_b^0(D) \stackrel{\mathrm{def}}{=} \{[A]_b \ : \ \forall A, \ A \subset D, \ \partial A \neq \varnothing, \ \mathrm{m}(\partial A) = 0\},$$

$$C_b^0(D) \stackrel{\mathrm{def}}{=} \{b_A \in C_b(D) \ : \ \mathrm{m}(\partial A) = 0\}$$

of $\mathcal{F}_b(D)$ and $C_b(D)$ will be important. We now have the equivalent of Theorem 2.2 of Chapter 4.

Theorem 2.2. *Let D, $\varnothing \neq D$, be an open (resp., bounded open) subset of $\mathbf{R}^N$.*

(i) *The set $C_b(D)$ is closed in $C_{\mathrm{loc}}(D)$ (resp., $C(\overline{D})$), and ρ_δ (resp., ρ) defines a complete metric topology on $\mathcal{F}_b(D)$.*

(ii) *For a bounded open subset D of $\mathbf{R}^N$, the set $C_b(D)$ is compact in $C(\overline{D})$.*

(iii) *For a bounded open subset D of $\mathbf{R}^N$, the map*

$$b_A \mapsto (b_A^+, b_A^-, |b_A|) = (d_A, d_{\complement A}, d_{\partial A}) : C_b(D) \subset C(\overline{D}) \to C(\overline{D})^3$$

is continuous: for all b_A and b_B in $C_b(D)$,

$$\max \left\{ \|d_B - d_A\|_{C(D)}, \ \|d_{\complement B} - d_{\complement A}\|_{C(D)}, \ \|d_{\partial B} - d_{\partial A}\|_{C(D)} \right\} \tag{2.8}$$
$$\leq \|b_B - b_A\|_{C(D)}.$$

Proof. (i) It is sufficient to consider the case for which D is bounded. Consider a sequence $\{A_n\} \subset \overline{D}$, $\partial A_n \neq \varnothing$ such that $b_{A_n} \to f$ in $C(\overline{D})$ for some f in $C(\overline{D})$. Associate with each g in $C(D)$ its positive and negative parts

$$g^+(x) = \max\{g(x), 0\}, \quad g^-(x) = \max\{-g(x), 0\}.$$

Then by continuity of this operation

$$d_{A_n} = b_{A_n}^+ \to f^+ \text{ and } d_{\complement A_n} = b_{A_n}^- \to f^- \text{ in } C(\overline{D}),$$
$$d_{\partial A_n} = |b_{A_n}| \to |f| \text{ in } C(\overline{D}).$$

By Theorem 2.2 (i) of Chapter 4, there exists a closed subset F, $\varnothing \neq F \subset \overline{D}$, such that

$$f^+ = d_F \text{ in } \mathbf{R}^N \text{ and } f^+ > 0 \text{ in } \complement\overline{D}.$$

Moreover, $F \neq \mathbf{R}^N$ since D is bounded. By Theorem 2.4 and the remark at the beginning of section 2.3 of Chapter 4, there exists an open subset $G \subset D$, $G \neq \mathbf{R}^N$, such that

$$f^- = d_{\complement G} \text{ in } \mathbf{R}^N \text{ and } d_{\complement G} \in C_0(D).$$

Therefore,

$$f = f^+ - f^- = d_F - d_{\complement G},$$

$$f = f^+ - f^- = d_F \text{ in } \overline{\complement D} \text{ and } f > 0 \text{ in } \complement \overline{D}.$$

Define the sets

$$A^+ \overset{\text{def}}{=} \left\{ x \in \mathbf{R}^N : f(x) > 0 \right\} = \left\{ x \in \overline{D} : f(x) > 0 \right\} \cup \complement \overline{D},$$

$$A^- \overset{\text{def}}{=} \left\{ x \in \mathbf{R}^N : f(x) < 0 \right\} = \left\{ x \in \overline{D} : f(x) < 0 \right\},$$

$$A^0 \overset{\text{def}}{=} \left\{ x \in \mathbf{R}^N : f(x) = 0 \right\} = \left\{ x \in \overline{D} : f(x) = 0 \right\}.$$

They form a partition of $\mathbf{R}^N$, $\mathbf{R}^N = A^- \cup A^0 \cup A^+$, and

$$\mathbf{R}^N \neq A^0 \cup A^- = F \neq \varnothing, \quad A^0 \cup A^+ = \complement G \neq \varnothing, \quad \Rightarrow A^0 = F \cap \complement G \neq \varnothing$$

$$\complement \overline{D} \subset A^+ \text{ and } F = A^0 \cup A^- \subset \overline{D},$$

$$f = d_{A^0 \cup A^-} - d_{A^0 \cup A^+}.$$

If A^0 was empty, $\mathbf{R}^N$ could be partitioned into two nonempty disjoint closed subsets. Since A^0 is closed

$$\partial A^0 = A^0 \cap \overline{\complement A^0} = A^0 \cap \overline{A^+ \cup A^-} = A^0 \cap \left[\overline{A^+} \cup \overline{A^-} \right]$$

$$= A^0 \cap \left[A^+ \cup A^- \cup \partial A^- \cup \partial A^+ \right] = A^0 \cap \left[\partial A^- \cup \partial A^+ \right].$$

Moreover, since A^- and A^+ are open, $\overline{A^-} \subset A^- \cup A^0$, $\overline{A^+} \subset A^+ \cup A^0$,

$$\partial A^- = \complement A^- \cap \overline{A^-} \subset [A^0 \cup A^+] \cap [A^0 \cup A^-] = A^0,$$

$$\partial A^+ = \complement A^+ \cap \overline{A^+} \subset [A^0 \cup A^-] \cap [A^0 \cup A^+] = A^0$$

$$\Rightarrow \partial A^- \cup \partial A^+ = [\partial A^- \cup \partial A^+] \cap A^0 = \partial A^0$$

$$\Rightarrow \boxed{\partial A^- \cup \partial A^+ = \partial A^0} \quad \Rightarrow \boxed{\partial A^0 = \partial A^- \cup (\partial A^+ \backslash \partial A^-).}$$

Let $\mathbb{Q}$ be the subset of points in $\mathbf{R}^N$ with rational coordinates. Define

$$B^0 \overset{\text{def}}{=} \text{int } A^0, \quad B^0_+ \overset{\text{def}}{=} B^0 \cap \mathbb{Q}, \quad B^0_- \overset{\text{def}}{=} B^0 \cap \complement \mathbb{Q},$$

and notice that by density of $\mathbb{Q}$ and $\complement \mathbb{Q}$ in $\mathbf{R}^N$,

$$\overline{B^0_+} = \overline{B^0} = \overline{B^0_-}.$$

Consider the following new partition of $\mathbf{R}^N$:

$$\mathbf{R}^N = A^+ \cup (\partial A^+ \backslash \partial A^-) \cup A^- \cup \partial A^- \cup B^0_+ \cup B^0_-.$$

Define

$$\boxed{A \stackrel{\text{def}}{=} A^- \cup (\partial A^+ \backslash \partial A^-) \cup B^0_+} \quad \Rightarrow \complement A = A^+ \cup \partial A^- \cup B^0_-$$

and

$$\overline{A} = \overline{A^-} \cup (\overline{\partial A^+ \backslash \partial A^-}) \cup \overline{B^0_+} = A^- \cup \partial A^- \cup (\overline{\partial A^+ \backslash \partial A^-}) \cup \overline{B^0},$$
$$\overline{\complement A} = \overline{A^+} \cup \overline{\partial A^-} \cup \overline{B^0_-} = A^+ \cup \partial A^+ \cup \partial A^- \cup \overline{B^0}.$$

But

$$\partial A^- \cup \partial A^+ = \partial A^- \cup (\partial A^+ \backslash \partial A^-) \subset \partial A^- \cup (\overline{\partial A^+ \backslash \partial A^-}) \subset \partial A^- \cup \partial A^+,$$

and since $\partial A^- \cup \partial A^+ = \partial A^0$

$$\left. \begin{aligned} \overline{A} &= A^- \cup \partial A^0 \cup \overline{\operatorname{int} A^0} = A^- \cup A^0 \\ \overline{\complement A} &= A^+ \cup \partial A^0 \cup \overline{\operatorname{int} A^0} = A^+ \cup A^0 \end{aligned} \right| \Rightarrow \partial A = A^0 \neq \varnothing$$

$$\Rightarrow b_A = d_A - d_{\complement A} = b_{\overline{A}} - d_{\overline{\complement A}} = d_F - d_{\complement G} = f.$$

(ii) For the compactness, consider any sequence $\{b_{A_n}\} \subset C_b(D) \subset C^{0,1}(\overline{D})$. Since D is bounded, b_{A_n} and ∇b_{A_n} are both pointwise uniformly bounded in $\overline{D}$. From Theorem 2.2 in Chapter 2 the injection of $C^{0,1}(\overline{D})$ into $C(\overline{D})$ is compact and there exist $f \in C(\overline{D})$ and a subsequence $\{b_{A_{n_k}}\}$ such that $b_{A_{n_k}} \to f$ in $C(\overline{D})$. From the proof of the closure in part (i), there exists $A \subset \overline{D}$, $\partial A \neq \varnothing$, such that $f = b_A$.

(iii) For all b_A and b_B in $C_b(D)$ and x in $\overline{D}$,

$$|b_B(x)| \le |b_A(x)| + |b_B(x) - b_A(x)|$$
$$\Rightarrow \big| |b_B(x)| - |b_A(x)| \big| \le |b_B(x) - b_A(x)|$$
$$\Rightarrow \|d_{\partial B} - d_{\partial A}\|_{C(D)} = \big\| |b_B| - |b_A| \big\|_{C(D)} \le \|b_B - b_A\|_{C(D)}.$$

Moreover, $d_A = b_A^+ = (|b_A| + b_A)/2$ and $d_{\complement A} = b_A^- = (|b_A| - b_A)/2$, and necessarily

$$\|b_B^\pm - b_A^\pm\|_{C(D)} \le \|b_B - b_A\|_{C(D)}.$$

By combining the above three inequalities we get (2.8). $\qquad\qquad \square$

We have the analogue of Theorem 2.3 of Chapter 4 and its corollary for A, $\complement A$ and ∂A. In the last case it takes the following form (to be compared with Richardson [1, Lem. 3.2, p. 44] and Kulkarni, Mitter, and Richardson [1] for an application to image segmentation):

Corollary 1. *Let D be a nonempty open (resp., bounded open) subset of $\mathbf{R}^N$. Define for a subset S of $\mathbf{R}^N$ the sets*

$$H_b(S) \stackrel{\text{def}}{=} \{b_A \in C_b(D) \ : \ S \subset \partial A\},$$

$$I_b(S) \stackrel{\text{def}}{=} \{b_A \in C_b(D) \ : \ \partial A \subset S\},$$

$$J_b(S) \stackrel{\text{def}}{=} \{b_A \in C_b(D) \ : \ \partial A \cap S \neq \varnothing\}.$$

(i) *Let S be a subset of $\mathbf{R}^N$. Then $H_b(S)$ is closed in $C_{\text{loc}}(D)$ (resp., $C(\overline{D})$).*

(ii) *Let S be a closed subset of $\mathbf{R}^N$. Then $I_b(S)$ is closed in $C_{\text{loc}}(D)$ (resp., $C(\overline{D})$). If, in addition, $S \cap \overline{D}$ is compact, then $J_b(S)$ is closed in $C_{\text{loc}}(D)$ (resp., $C(\overline{D})$).*

(iii) *Let S be an open subset of $\mathbf{R}^N$. Then $J_b(S)$ is open in $C_{\text{loc}}(D)$ (resp., $C(\overline{D})$). If, in addition, $\complement S \cap \overline{D}$ is compact, then $I_b(S)$ is open in $C_{\text{loc}}(D)$ (resp., $C(\overline{D})$).*

(iv) *For D bounded, associate with an equivalent class $[A]_b$ the number*

$$\#_b([A]_b) = \text{number of connected components of } \partial A.$$

    Then the map

$$[A]_b \mapsto \#_b([A]_b) \colon \mathcal{F}_b(D) \to \mathbf{R}$$

is lower semicontinuous.

Proof. This proof is the same proof as for Theorem 2.3 of Chapter 4 using the Lipschitz continuity of the map $b_A \mapsto d_{\partial A} = |b_A|$ from $C(\overline{D})$ to $C(\overline{D})$. $\qquad\square$

3   Projection, Skeleton, and Differentiability of b_A

In this section we study the connection between the gradient of b_A and the projection onto ∂A and the characteristic functions associated with ∂A. We further relate the set of singularities of the gradients and the notions of skeleton[1] and set of cracks.

Definition 3.1.
Let A be a subset of $\mathbf{R}^N$ such that $\varnothing \neq \partial A$.

(i) The *set of projections of x onto ∂A* is given by

$$\Pi_{\partial A}(x) \stackrel{\text{def}}{=} \{z \in \partial A : |z - x| = d_{\partial A}(x)\}. \tag{3.1}$$

    The elements of $\Pi_{\partial A}(x)$ are called *projections* onto ∂A and denoted by $p_{\partial A}(x)$.

(ii) The set of points where the projection onto ∂A is not unique,

$$\text{Sk}\,(A) \stackrel{\text{def}}{=} \{x \in \mathbf{R}^N \ : \ \Pi_{\partial A}(x) \text{ is not a singleton}\}, \tag{3.2}$$

    is called the *skeleton* of A. Since $\Pi_{\partial A}(x)$ is a singleton for $x \in \partial A$, then

$$\text{Sk}\,(A) \subset \mathbf{R}^N \backslash \partial A.$$

---

[1]Our definition of a skeleton does not exactly coincide with the one used in *morphological mathematics* (cf., for instance, Matheron [1] or Rivière [1]).

(iii) The *set of cracks* is defined as

$$C(A) \stackrel{\text{def}}{=} \left\{ x \in \mathbf{R}^N \ : \ \nabla b_A^2(x) \exists \text{ and } \nabla b_A(x) \not\exists \right\}. \tag{3.3}$$

The *set of singularities*, $\text{Sing}(\nabla b_A)$, of ∇b_A has zero N-dimensional Lebesgue measures since b_A is Lipschitz continuous and hence differentiable almost everywhere. $\qquad\square$

The function b_A enjoys properties similar to the ones of d_A and $d_{\complement A}$ since $|b_A| = d_{\partial A}$ and we have the analogue of Theorem 3.1 in Chapter 4.

Theorem 3.1. *Let A be a subset of $\mathbf{R}^N$ such that $\varnothing \neq \partial A$, and let $x \in \mathbf{R}^N$. Define*

$$\boxed{f_{\partial A}(x) \stackrel{\text{def}}{=} \frac{1}{2}\left(|x|^2 - b_A^2(x)\right).}$$

(i) *The set $\Pi_{\partial A}(x)$ is nonempty, compact, and*

$$\forall x \notin \partial A \quad \Pi_{\partial A}(x) \subset \partial A, \quad \text{and} \quad \forall x \in \partial A, \quad \Pi_{\partial A}(x) = \{x\}.$$

(ii) *For all x and v in $\mathbf{R}^N$*

$$db_A^2(x; v) = \lim_{t \searrow 0} \frac{b_A^2(x + tv) - b_A^2(x)}{t} = \min_{z \in \Pi_{\partial A}(x)} 2(x - z) \cdot v,$$

$$df_{\partial A}(x; v) = \lim_{t \searrow 0} \frac{f_{\partial A}(x + tv) - f_{\partial A}(x)}{t} = \sigma_{\Pi_{\partial A}(x)}(v) = \sigma_{\text{co}\,\Pi_{\partial A}(x)}(v),$$

where σ_B is the support function of the set B,

$$\sigma_B(v) = \sup_{z \in B} z \cdot v,$$

and $\text{co}\,B$ is the convex hull of B.

(iii) *The following statements are equivalent:*

(a) *$b_A^2(x)$ is (Fréchet) differentiable at x,*
(b) *$b_A^2(x)$ is Gâteaux differentiable at x,*
(c) *$\Pi_{\partial A}(x)$ is a singleton.*

Henceforth

$$\text{Sk}(A) = \left\{ x \in \mathbf{R}^N \ : \ \nabla b_A^2(x) \not\exists \right\},$$
$$\text{Sing}(\nabla b_A) = \text{Sk}(A) \cup C(A).$$

(iv) *When b_A^2 is differentiable at x, $\Pi_{\partial A}(x) = \{p_{\partial A}(x)\}$ is a singleton, and*

$$\boxed{p_{\partial A}(x) = x - \frac{1}{2}\nabla b_A^2(x).} \tag{3.4}$$

For all $x \in \partial A$, $\Pi_{\partial A}(x) = \{x\}$, b_A^2 is differentiable at x and $\nabla b_A^2(x) = 0$. Hence

$$\Pi_{\partial A}(x) = \begin{cases} \Pi_A(x), & \text{if } x \in \text{int } \complement A, \\ \{x\}, & \text{if } x \in \partial A, \\ \Pi_{\complement A}(x), & \text{if } x \in \text{int } A, \end{cases} \tag{3.5}$$

$$\text{Sk}\,(A) = \text{Sk}_{\text{int}}\,(A) \cup \text{Sk}_{\text{ext}}\,(A). \tag{3.6}$$

Proof. The result follows from Theorem 3.1 of Chapter 4 using the identity $b_A^2 = d_{\partial A}^2$. $\qquad\square$

Remark 3.1.

The uniqueness of the projection $p_{\partial A}(x)$ at x is equivalent to the existence of $\nabla b_A^2(x)$. In both cases the identity

$$p_{\partial A}(x) = x - \frac{1}{2}\nabla b_A^2(x) = \nabla f_{\partial A}(x) \tag{3.7}$$

is satisfied, and all the properties of $p_{\partial A}$ can be obtained from those of b_A^2 or $f_{\partial A}$. $\quad\square$

We now compute the gradient of b_A and relate it to the characteristic functions of ∂A.

Theorem 3.2. *Let A be a subset of $\mathbf{R}^N$ such that $\varnothing \neq \partial A$.*

(i) *If $\nabla b_A(x)$ exists at a point x in $\mathbf{R}^N$, then $\Pi_{\partial A}(x) = \{p_{\partial A}(x)\}$ is a singleton,*

$$b_A(x) = \begin{cases} |p_{\partial A}(x) - x|, & \text{if } x \in \text{int } \complement A, \\ 0, & \text{if } x \in \partial A, \\ -|p_{\partial A}(x) - x|, & \text{if } x \in \text{int } A, \end{cases} \tag{3.8}$$

$$\nabla b_A(x) = \begin{cases} \dfrac{x - p_{\partial A}(x)}{b_A(x)}, & \text{if } x \notin \partial A, \\ 0, & \text{a.e. in } \partial A. \end{cases} \tag{3.9}$$

In particular, for almost all $x \in \partial A$, $\nabla b_A(x) = \nabla d_{\partial A}(x) = 0$,

$$\boxed{\chi_{\partial A}(x) = 1 - |\nabla b_A(x)| \ \text{a.e. in } \mathbf{R}^N} \tag{3.10}$$

and the set

$$\boxed{\partial_b A \overset{\text{def}}{=} \{x \in \partial A \ : \ \nabla b_A(x) \text{ exists and } |\nabla b_A(x)| \neq 0\}}$$

has zero N-dimensional Lebesgue measure.

(ii) *For all $x \in \partial A$, $\nabla b_A^2(x)$ exists and is equal to 0. For all $x \notin \partial A$, b_A is differentiable at x if and only if b_A^2 is differentiable at x. In particular,*

$$\text{Sk}\,(A) = \{x \in \mathbf{R}^N \ : \ \nabla b_A^2(x) \not\exists\} \subset \mathbf{R}^N \setminus \partial A, \tag{3.11}$$

$$\frac{1}{2}\nabla b_A^2(x) = x - p_{\partial A}(x) \ \text{in } \mathbf{R}^N \setminus \text{Sk}\,(A), \tag{3.12}$$

and the last identity is satisfied almost everywhere in $\mathbf{R}^N$.

(iii) *Given $x \in \mathbf{R}^N$, $\alpha \in [0,1]$, $p \in \Pi_{\partial A}(x)$, and $x_\alpha \stackrel{\text{def}}{=} p + \alpha(x - p)$, then*

$$b_A(x_\alpha) = \alpha\, b_A(x),$$
$$\forall \alpha \in [0,1], \quad \Pi_{\partial A}(x_\alpha) \subset \Pi_{\partial A}(x).$$

In particular if $\Pi_{\partial A}(x)$ is a singleton, then $\Pi_{\partial A}(x_\alpha)$ is a singleton and $\nabla b_A^2(x_\alpha)$ exists for all α, $0 \le \alpha \le 1$. If, in addition, $x \ne \partial A$, then $\nabla b_A(x_\alpha)$ exists for all $0 < \alpha \le 1$.

Proof. (i) If $x \in \operatorname{int} A$, use the fact that $b_A = -d_{\complement A}$ in $\operatorname{int} A$ and repeat the proof of Theorem 3.2 (i) in Chapter 4 to obtain

$$-\nabla b_A(x) = \nabla d_{\complement A}(x) = \frac{x - p_{\partial A}(x)}{|x - p_{\partial A}(x)|}$$

and $p_{\partial A}(x) \in \partial A$ is unique. Similarly, for $x \in \operatorname{int} \complement A$, $b_A = d_A$,

$$\nabla b_A(x) = \nabla d_A(x) = \frac{x - p_{\partial A}(x)}{|x - p_{\partial A}(x)|}$$

and $p_{\partial A}(x) \in \partial A$ is unique. Finally, when $x \in \partial A$, $p_{\partial A}(x) = x$. The distance function $d_{\partial A}$ is differentiable almost everywhere. From Theorem 3.2 (i) in Chapter 4, whenever it is differentiable at a point $x \in \partial A$, $\nabla d_{\partial A}(x) = 0$. But for all v in $\mathbf{R}^N$ and $t > 0$

$$\left| \frac{b_A(x + tv) - b_A(x)}{t} \right| = \frac{|b_A(x + tv)|}{t} = \frac{d_{\partial A}(x + tv) - d_{\partial A}(x)}{t}$$

since $|b_A| = d_{\partial A}$, and for almost all $x \in \partial A$, $\nabla b_A(x) = \nabla d_{\partial A}(x) = 0$.

Since b_A is a Lipschitzian function, it is differentiable almost everywhere in $\mathbf{R}^N$, and in view of the previous considerations, when it is differentiable

$$|\nabla b_A(x)| = \begin{cases} 1, & x \notin \partial A, \\ 0, & \text{a.e. in } \partial A \end{cases}$$

$$\Rightarrow \chi_{\partial A}(x) = 1 - |\nabla b_A(x)|, \quad \text{a.e. in } \mathbf{R}^N.$$

Therefore, the set $\partial_b A$ has at most a zero measure.

(ii) For $x \in \partial A$ and any $y \in \mathbf{R}^N$, consider the differential quotient

$$\Delta(y) = \frac{b_A^2(y) - b_A^2(x)}{|y - x|} = [b_A(y) + b_A(x)] \frac{[b_A(y) - b_A(x)]}{|y - x|}.$$

Then

$$|\Delta(y)| \le |b_A(y)| \frac{|y - x|}{|y - x|} \to 0 \text{ as } y \to x.$$

Hence

$$\forall x \in \partial A, \quad \nabla b_A^2(x) = 0 = \frac{1}{2}(x - x) = \frac{1}{2}(x - p_{\partial A}(x)).$$

If $x \notin \partial A$ and $\nabla b_A^2(x)$ exists, then $b_A(x) \neq 0$ and

$$\nabla b_A(x) = \frac{1}{2} \frac{\nabla b_A^2(x)}{b_A(x)}.$$

To see this consider the new differential quotient

$$q(y) = \left[b_A(y) - b_A(x) - \frac{1}{2} \frac{\nabla b_A^2(x)}{b_A(x)} \cdot (y - x) \right] \frac{1}{|y - x|}.$$

Then

$$\left(b_A(y) + b_A(x) \right) q(y)$$
$$= \left[b_A^2(y) - b_A^2(x) - \frac{b_A(x) + b_A(y)}{2 b_A(x)} \nabla b_A^2(x) \cdot (y - x) \right] \frac{1}{|y - x|}$$
$$= \left[b_A^2(y) - b_A^2(x) - \nabla b_A^2(x) \cdot (y - x) \right] \frac{1}{|y - x|}$$
$$+ \frac{b_A(x) - b_A(y)}{2 b_A(x)} \nabla b_A^2(x) \cdot \frac{(y - x)}{|y - x|}.$$

Since b_A^2 is differentiable at x, the first term goes to zero as y goes to x. As for the second term it is bounded by

$$\frac{|\nabla b_A^2(x)|}{2 |b_A(x)|} |x - y|$$

and hence goes to zero as y goes to x. So the term

$$[b_A(y) + b_A(x)] q(y) \to 0 \text{ as } y \to x,$$

and since $b_A(x) \neq 0$, $q(y) \to 0$ as $y \to x$ and b_A is differentiable at x. In the other direction the result is trivial.

    (iii) The proof of (iii) is straightforward. $\square$

Remark 3.2.
For sufficiently smooth domains, the subset $\partial_b A$ of the boundary ∂A coincides with the reduced boundary of finite perimeter sets. $\square$

Remark 3.3.
In general, $\nabla d_A(x)$ and $\nabla d_{\complement A}(x)$ do not exist for $x \in \partial A$. This is readily seen by constructing the directional derivatives for the half-space

$$A = \{ (x_1, x_2) \in \mathbf{R}^2 : x_1 \leq 0 \} \tag{3.13}$$

at the point $(0, 0)$. Nevertheless, $\nabla b_A(0, 0)$ exists and is equal to $(1, 0)$, which is the outward unit normal at $(0, 0) \in \partial A$ to A. Note also that for all $x \in \partial A$, $|\nabla b_A(x)| = 1$. This is possible since $m_N(\partial A) = 0$. If $\nabla b_A(x)$ exists, then it is easy to check that $|\nabla b_A(x)| \leq 1$. $\square$

4   Boundary Smoothness and Smoothness of b_A

The smoothness of the boundary ∂A is directly related to the smoothness of the function b_A in a neighborhood of ∂A. The next two theorems establish the equivalence for domains with a $C^{1,1}$- or smoother boundary. Their proofs will use part (iv) and (vii) of the following specialization of Theorem 7.1 of Chapter 4 to $d_{\partial A} = |b_A|$. Tubular neighborhoods are defined in Definition 5.2 of Chapter 4.

Theorem 4.1. *Given a subset A of $\mathbf{R}^N$ such that $\partial A \neq \varnothing$, the following conditions are equivalent.*

  (i) *$\exists h > 0$ such that b_A belongs to $C^{1,1}_{\mathrm{loc}}(U_h(\partial A) \backslash \partial A)$.*

  (ii) *$\exists h > 0$ such that b_A belongs to $C^1(U_h(\partial A) \backslash \partial A)$.*

  (iii) *$\exists h > 0$ such that $\forall x \in U_h(\partial A) \backslash \partial A$, $\Pi_{\partial A}(x)$ is a singleton.*

  (iv) *$\exists h > 0$ such that $\forall x \in U_h(\partial A)$, $\Pi_{\partial A}(x)$ is a singleton.*

  (v) *∂A has positive reach, that is, $\mathrm{reach}\,(\partial A) > 0$.*

  (vi) *$\exists h > 0$ such that $p_{\partial A}$ belongs to $C^{0,1}_{\mathrm{loc}}(U_h(\partial A))$.*

  (vii) *$\exists h > 0$ such that b_A^2 belongs to $C^{1,1}_{\mathrm{loc}}(U_h(\partial A))$.*

In the first direction the smoothness of b_A in a neighborhood of ∂A implies the smoothness of the domain.

Theorem 4.2. *Let $k \geq 1$ be an integer, ℓ, $0 \leq \ell \leq 1$, be a real number, and A be a subset of $\mathbf{R}^N$ such that $\varnothing \neq \partial A \neq \mathbf{R}^N$. If for each x in ∂A there exists an open neighborhood $V(x)$ of x where $b_A \in C^{k,\ell}(V(x))$, then A is a set of class $C^{k,\ell}$ and $\mathrm{m}(\partial A) = 0$. Moreover, $\Pi_{\partial A}(y) = \{p_{\partial A}(y)\}$ is a singleton in the open neighborhood*

$$V \overset{\mathrm{def}}{=} \cup_{x \in \partial A} V(x)$$

of ∂A, and

$$b_A^2 \in C^{1,1}_{\mathrm{loc}}(V) \cap C^{k,\ell}_{\mathrm{loc}}(V) \ \text{and} \ p_{\partial A} \in C^{0,1}_{\mathrm{loc}}(V)^N \cap C^{k-1,\ell}_{\mathrm{loc}}(V)^N.$$

Proof. By construction V is an open neighborhood of ∂A and $\varnothing \neq \partial A \neq V$ since $\varnothing \neq \partial A \neq \mathbf{R}^N$. By assumption, $b_A \in C^{k,\ell}_{\mathrm{loc}}(V)$ and hence b_A^2 belongs to $C^{k,\ell}_{\mathrm{loc}}(V)$. By Theorem 3.1 (iii), $\Pi_{\partial A}(y) = \{p_{\partial A}(y)\}$ is a singleton in V. By Theorem 4.1, $b_A^2 \in C^{1,1}_{\mathrm{loc}}(V)$. From Theorem 3.2 (i), $|\nabla b_A| = 1$ in $V \backslash \partial A$ and $\nabla b_A = 0$ almost everywhere in ∂A. But $|\nabla b_A| \in C(V)$ and hence either $|\nabla b_A| = 0$ or $|\nabla b_A| = 1$ in V. In the first case this would imply that $b_A = 0$ in V since $b_A = 0$ on ∂A. As a result, $\partial A \subset V \subset \partial A$ and ∂A would be open and closed, that is, equal to $\varnothing$ or $\mathbf{R}^N$, which is not possible by assumption. Hence $|\nabla b_A| = 1$ in V. So the set A is characterized by the level sets of the function b_A and $|\nabla b_A| = 1$ on $\partial A = b_A^{-1}\{0\}$. By Theorem 4.2 in Chapter 2 A is of class $C^{k,\ell}$, and by Theorem 3.2 (i) $\mathrm{m}(\partial A) = 0$ since $|\nabla b_A| = 1$ on ∂A. $\qquad\square$

In the other direction the smoothness of the domain implies the smoothness of b_A in a neighborhood of ∂A when the domain is at least of class $C^{1,1}$. A counterexample will be given in dimension 2.

Theorem 4.3. *Let A be a subset of $\mathbf{R}^N$ such that $\partial A \neq \varnothing$.*

(i) *If A is of class $C^{1,1}$, then for each $x \in \partial A$, there exists a neighborhood $W(x)$ of x such that $b_A \in C^{1,1}(W(x))$. Moreover, $\nabla b_A = n \circ p_{\partial A}$ in $W(x)$.*

(ii) *If A is of class $C^{k,\ell}$ for an integer k, $k \geq 2$, and a real number ℓ, $0 \leq \ell \leq 1$, then for each $x \in \partial A$, there exists a neighborhood $W(x)$ of x such that $b_A \in C^{k,\ell}(W(x))$ and $b_A^2 \in C^{1,1}(W(x))$.*

Proof. (i) From the definition of a set of class $C^{1,1}$ for each $x \in \partial A$, there exists a bounded open neighborhood $U(x)$ of x where the set A can be locally described by the level sets of the $C^{1,1}$-function

$$f(y) \stackrel{\text{def}}{=} g_x(y) \cdot e_N, \tag{4.1}$$

since by definition

$$\text{int } A \cap U(x) = \{y \in U(x) : f(y) > 0\},$$
$$\partial A \cap U(x) = \{y \in U(x) : f(y) = 0\}.$$

Denote by c_x the Lipschitz constant of ∇f in $U(x)$. The boundary ∂A is the zero level set of f and the gradient

$$\nabla f(y) = Dg_x(y)^* e_N \neq 0 \text{ in } U(x)$$

is normal to that level set. Thus the *outward normal* to A on ∂A is given by

$$\boxed{n(y) = -\frac{\nabla f(y)}{|\nabla f(y)|}} = -\frac{Dg_x(y)^* e_N}{|Dg_x(y)^* e_N|} = -\frac{Dh_x(g_x(y))^{-*} e_N}{|Dh_x(g_x(y))^{-*} e_N|}. \tag{4.2}$$

There exists a bounded open neighborhood $V(x)$ of x, and $\alpha > 0$ such that $\overline{V(x)} \subset U(x)$,

$$\forall y \in \overline{V(x)}, \quad |\nabla f(y)| \geq \alpha, \quad \text{and}$$
$$\forall z_1, z_2 \in \overline{V(x)}, \ |\nabla f(z_2) - \nabla f(z_1)| \leq c_x |z_2 - z_1|.$$

By the characterization of ∂A for all $y \in V(x)$,

$$d_{\partial A}^2(y) = \inf_{\substack{z \in \mathbf{R}^N \ f(z)=0}} |z - y|^2.$$

We know that the set of minimizers

$$\Pi_{\partial A}(y) = \{p \in \mathbf{R}^N : f(p) = 0 \text{ and } |p - y| = d_{\partial A}(y)\}$$

is nonempty and compact. By using the Lagrange multiplier theorem for the Lagrangian

$$|z - y|^2 + \lambda f(z),$$

the elements p of $\Pi_{\partial A}(y)$ are characterized by the existence of a $\lambda(p) \in \mathbf{R}$ such that

$$2(p - y) + \lambda(p)\nabla f(p) = 0 \text{ and } f(p) = 0$$

$$\Rightarrow 2d_{\partial A}(y) = |\lambda(p)| \, |\nabla f(p)| \quad \Rightarrow p = y + b_A(y)\frac{\nabla f(p)}{|\nabla f(p)|} = y - b_A(y)n(p)$$

since $\nabla f(p) \neq 0$ on ∂A. The next question is the uniqueness of p. Construct the function

$$F(y, z) \stackrel{\text{def}}{=} y + b_A(y)\frac{\nabla f(z)}{|\nabla f(z)|} \tag{4.3}$$

and show that, for y sufficiently close to ∂A, $F(y, p)$ has a unique fixed point p. Consider for z_2 and z_1 in $V(x)$ the difference

$$\left| \frac{\nabla f(z_2)}{|\nabla f(z_2)|} - \frac{\nabla f(z_1)}{|\nabla f(z_1)|} \right|$$

$$= \left| \frac{1}{|\nabla f(z_2)|}(\nabla f(z_2) - \nabla f(z_1)) + \left(\frac{1}{|\nabla f(z_2)|} - \frac{1}{|\nabla f(z_1)|} \right)\nabla f(z_1) \right|$$

$$= \frac{1}{|\nabla f(z_2)|}\left| \nabla f(z_2) - \nabla f(z_1) + (|\nabla f(z_1)| - |\nabla f(z_2)|)\frac{\nabla f(z_1)}{|\nabla f(z_1)|} \right|$$

$$\leq 2\frac{1}{|\nabla f(z_2)|}|\nabla f(z_2) - \nabla f(z_1)| \leq \frac{2}{\alpha}c_x|z_2 - z_1|.$$

In particular, the outward unit normal is Lipschitzian

$$\forall z_2,\, z_1 \in V(x) \cap \partial A, \quad |n(z_2) - n(z_1)| \leq \frac{2}{\alpha}c_x|z_2 - z_1|.$$

Coming back to (4.3),

$$|F(y, z_2) - F(y, z_1)| = |b_A(y)|\left| \frac{\nabla f(z_2)}{|\nabla f(z_2)|} - \frac{\nabla f(z_1)}{|\nabla f(z_1)|} \right|$$

$$\leq \frac{2}{\alpha}c_x\, |b_A(y)|\, |z_2 - z_1|.$$

Choose the new neighborhood of x:

$$W(x) \stackrel{\text{def}}{=} \left\{ y \in V(x) : d_{\partial A}(y) < d_{\complement V(x)}(y) \text{ and } |b_A(y)| < \frac{\alpha}{4\, c_x} \right\}.$$

So for all $y \in W(x)$

$$|F(y, z_2) - F(y, z_1)| \leq \frac{1}{2}|z_2 - z_1|$$

is a contraction and there is a unique $p(y) \in V(x)$ such that

$$p(y) = y + b_A(y) \frac{\nabla f(p(y))}{|\nabla f(p(y))|}.$$

By construction of $W(x)$

$$\forall y \in W(x), \quad d_{\partial A}(y) < d_{\complement V(x)}(y) \quad \Rightarrow \quad \Pi_{\partial A}(y) \subset V(x) \cap \partial A.$$

Since $\Pi_{\partial A}(y) \neq \varnothing$ and $p(y)$ is unique, $\Pi_{\partial A}(y) = \{p(y)\}$ is necessarily a singleton. By Theorem 4.1, $b_A^2 \in C^{1,1}(W(x))$ and $p_{\partial A} \in C^{0,1}(W(x))$. By Theorem 3.1 (iii) for $y \in W(x)\backslash \partial A$, $\nabla b_A(y)$ exists and

$$\nabla b_A(y) = \frac{\nabla b_A^2(y)}{2 b_A(y)} = \frac{y - p(y)}{b_A(y)} = -\frac{\nabla f(p(y))}{|\nabla f(p(y))|} = N(p(y)),$$

where

$$N(z) \stackrel{\text{def}}{=} -\frac{\nabla f(z)}{|\nabla f(z)|}.$$

Both N and p are Lipschitzian and hence the composition $N \circ p$ is Lipschitzian. Moreover, since b_A is differentiable almost everywhere and $\mathrm{m}(\partial A) = 0$

$$\nabla b_A(y) = N(p(y)) \quad \text{for almost all } y \text{ in } W(x), \tag{4.4}$$

and necessarily $\nabla b_A \in C^{0,1}(W(x))^N$, $b_A \in C^{1,1}(W(x))$. In view of (4.2), $\nabla b_A = n \circ p$. This proves the theorem when A is of class $C^{1,1}$.

(ii) When A is of class $C^{k,\ell}$ for an integer k, $k \geq 2$, and a real number ℓ, $0 \leq \ell \leq 1$, it is $C^{1,1}$, and the previous constructions and conclusions remain true. Consider the map

$$(y, z) \mapsto G(y, z) \stackrel{\text{def}}{=} z - y - b_A(y) \frac{\nabla f(z)}{|\nabla f(z)|} \; : \; W(x) \times V(x) \to \mathbf{R}^N.$$

From the first part, there exists a unique $p(y) \in V(x)$ such that

$$G(y, p(y)) = 0$$
$$\Rightarrow \; DG_y(y, p(y)) + DG_z(y, p(y)) \, Dp(y) = 0,$$

where

$$DG_z(y, z) = I - \frac{b_A(y)}{|\nabla f(z)|} \left[D^2 f(z) - \frac{\nabla f(z)}{|\nabla f(z)|} \,^* \left(D^2 f(z) \frac{\nabla f(z)}{|\nabla f(z)|} \right) \right].$$

By assumption, f is at least C^2 in $V(x)$ and hence bounded in some smaller neighborhood $V'(x)$, $\overline{V'(x)} \subset V(x)$. By putting a smaller bound on $|b_A(y)|$ there exists a smaller neighborhood $W'(x)$ of x in $W(x)$ such that $D_z G(y, z)$ is invertible for all $(y, z) \in W'(x) \times V'(x)$. So the conditions of the implicit function theorem are met. There exists a neighborhood $Y(x) \subset V'(x)$ of x and a unique C^1- (and hence $C^{k-1,\ell}$-) mapping

$$p : Y(x) \to \mathbf{R}^N \text{ such that } \forall y \in Y(x), \quad G(y, p(y)) = 0$$

(cf., for instance, Lang [1, Prop. 5.3, p. 15] and Schwartz [3, Thm. 31, p. 299]). In view of the fact that (4.4) is satisfied everywhere in $Y(x)$,

$$\forall y \in Y(x), \quad \nabla b_A(y) = N(p(y)),$$

then ∇b_A belongs to $C^{k-1,\ell}(Y(x))^N$ as the composition of two $C^{k-1,\ell}$ maps. Therefore, b_A is $C^{k,\ell}$ in $Y(x)$. $\qquad\square$

Example 4.1.
Consider the two-dimensional domain

$$\Omega \overset{\text{def}}{=} \{(x, z) : f(x) > z, \forall x \in \mathbf{R}\}$$

defined as the epigraph of the function

$$f(x) \overset{\text{def}}{=} |x|^{2-\frac{1}{n}}$$

for some arbitrary integer $n \geq 1$. We claim that Ω is a set of class $C^{1,1-1/n}$ and

$$\text{Sk}(\Omega) \overset{\text{def}}{=} \{(0, y) : y > 0\}.$$

In view of the presence of the absolute value the point $(0, 0)$ is the point where the smoothness of $\partial\Omega$ will be minimum. Therefore, there exists no neighborhood of $(0, 0)$ where the derivative ∇b_Ω exists and b_Ω is not C^1 and a fortiori not $C^{1,1-1/n}$. Indeed, for each $(0, y)$

$$b_\Omega^2(0, y) = \inf_{x \in \mathbf{R}} \left\{ x^2 + |f(x) - y|^2 \right\}.$$

The point $(0, y)$, $y > 0$, belongs to $\text{Sk}(\Omega)$ if there exist two different points $\hat{x}$ that minimize the function

$$F(x, y) \overset{\text{def}}{=} x^2 + |f(x) - y|^2. \tag{4.5}$$

Since f is symmetric with respect to the y-axis it is sufficient to show that there exists a strictly positive minimizer $\hat{x} > 0$. A locally minimizing point $\hat{x} \geq 0$ must satisfy the conditions

$$F'_x(\hat{x}, y) = 2\hat{x} + 2\left(|\hat{x}|^{2-\frac{1}{n}} - y\right)(2 - 1/n)\,\hat{x}^{1-\frac{1}{n}} = 0, \tag{4.6}$$

$$\frac{1}{2}F''_x(\hat{x}, y) = 1 + \left(2 - \frac{1}{n}\right)\left(3 - \frac{2}{n}\right)\hat{x}^{2-\frac{2}{n}} - \left(2 - \frac{1}{n}\right)\left(1 - \frac{1}{n}\right)y\,\hat{x}^{-\frac{1}{n}} \geq 0. \tag{4.7}$$

Equation (4.6) can be rewritten

$$2\hat{x}^{1-\frac{1}{n}}\left[\hat{x}^{\frac{1}{n}} + \left(\hat{x}^{2-\frac{1}{n}} - y\right)(2 - 1/n)\right] = 0,$$

and $\hat{x} = 0$ is a solution. The second factor can be written as an equation in the new variable $X \overset{\text{def}}{=} \hat{x}^{1/n}$ and

$$g(X) \overset{\text{def}}{=} \frac{n}{2n-1} X + X^{2n-1} - y = 0.$$

It has exactly one solution $\hat{X}$ since

$$\forall X, \quad \frac{dg}{dX}(X) = \frac{n}{2n-1} + (2n-1)X^{2n-2} > 0,$$

$g(0) = -y$ and $g(x)$ goes to infinity as X goes to infinity. In particular, $\hat{X} > 0$ for $y > 0$ and $\hat{X} = 0$ for $y = 0$. For $y > 0$ and $\hat{x} = \hat{X}^n$

$$\frac{1}{2}F'_x(\hat{x}, y) = 0 \quad \Rightarrow \quad \boxed{y = \hat{x}^{2-\frac{1}{n}} + \frac{n}{2n-1}\hat{x}^{\frac{1}{n}}} \tag{4.8}$$

and

$$\begin{aligned}
\frac{1}{2}F''_x(\hat{x}, y) &= 1 + \left(2 - \frac{1}{n}\right)\left(3 - \frac{2}{n}\right)\hat{x}^{2-\frac{2}{n}} \\
&\quad - \left(2 - \frac{1}{n}\right)\left(1 - \frac{1}{n}\right)\left[\hat{x}^{2-\frac{2}{n}} + \frac{n}{2n-1}\right] \\
&= 1 - \left(\frac{n-1}{n}\right) + \left(\frac{2n-1}{n}\right)^2 \hat{x}^{2-\frac{2}{n}} = \frac{1}{n} + \left(\frac{2n-1}{n}\hat{x}^{1-\frac{1}{n}}\right)^2 > 0.
\end{aligned}$$

Therefore, $\hat{x}$ is a *local minimum*. To complete the proof for $y > 0$ we must compare $F(0, y) = y^2$ for the solution corresponding to $\hat{x} = 0$ with

$$F(\hat{x}, y) = \hat{x}^2 + |\hat{x}^{2-\frac{1}{n}} - y|^2$$

for the solution $\hat{x} > 0$. Again using identity (4.8),

$$F(\hat{x}, y) = \hat{x}^2 + \left(\frac{n}{2n-1}\right)^2 \hat{x}^{\frac{2}{n}},$$

$$F(0, y) = |\hat{x}^{2-\frac{1}{n}} + \frac{n}{2n-1}\hat{x}^{\frac{1}{n}}|^2$$

$$= \left(\hat{x}^{2-\frac{1}{n}}\right)^2 + \left(\frac{n}{2n-1}\right)^2 \hat{x}^{\frac{2}{n}} + 2\frac{n}{2n-1}\hat{x}^2$$

$$\Rightarrow F(0, y) - F(\hat{x}, y) = \left(\frac{2n}{2n-1} - 1\right)\hat{x}^2 + \left(\hat{x}^{2-\frac{1}{n}}\right)^2$$

$$= \frac{1}{2n-1}\hat{x}^2 + \left(\hat{x}^{2-\frac{1}{n}}\right)^2 > 0.$$

This proves that for $y > 0$, $\hat{x} > 0$ is the minimizing positive solution. Therefore, for all $n \geq 1$, $f(x) = |x|^{2-1/n}$ yields a domain Ω for which the strictly positive part of the y-axis is the skeleton. It is interesting to note that, for each point x, the points $(x, f(x)) \in \partial\Omega$ are the projections of the point

$$(0, y) = \left(0, |x|^{2 - \frac{1}{n}} + \frac{n}{2n - 1} |x|^{\frac{1}{n}} \right)$$

and that the square of the distance function is equal to

$$b_\Omega^2(0, y) = x^2 + \left(\frac{n}{2n - 1} \right)^2 x^{\frac{2}{n}},$$

$$\nabla b_\Omega(x, f(x)) = \frac{1}{\sqrt{1 + f'(x)^2}} (f'(x), -1)$$

$$= \frac{1}{\sqrt{1 + \left(\frac{2n-1}{n} \right)^2 |x|^{2 - \frac{2}{n}}}} ((2 - 1/n) x |x|^{-\frac{1}{n}}, -1).$$

Finally, for all $n \geq 1$, $f \notin C^{1,1}$. For $n = 1$, $f \in C^{0,1}$, and for $n > 1$, $f \in C^{1,1-1/n}$. This gives an example in dimension 2 of a domain Ω of class $C^{1,\lambda}$, $0 \leq \lambda < 1$, with skeleton converging to the boundary. $\square$

From Theorem 4.3 (i) additional information can be obtained on the flow of ∇b_Ω.

Theorem 4.4. *Let Ω be a subset of $\mathbf{R}^N$ such that $\Gamma \stackrel{\text{def}}{=} \partial\Omega \neq \varnothing$ and assume that Ω is of class $C^{1,1}$.*

(i) *For each $x \in \Gamma$ there exists a neighborhood $W(x)$ of x such that $b_\Omega \in C^{1,1}(W(x))$,*

$$\boxed{\nabla b_\Omega = n \circ p_\Gamma = \nabla b_\Omega \circ p_\Gamma,} \quad |\nabla b_\Omega(x)| = 1 \ \text{in} \ W(x). \tag{4.9}$$

(ii) *Let $T_t(x)$ be the flow of ∇b_Ω in $W(x)$, that is,*

$$T_t(x) \stackrel{\text{def}}{=} x(t), \quad \frac{dx}{dt}(t) = \nabla b_\Omega(x(t)), \quad x(0) = x \in W(x). \tag{4.10}$$

Then for t in a neighborhood of 0 we have the following properties:

$$\boxed{p_\Gamma \circ T_t = p_\Gamma,} \quad \boxed{\nabla b_\Omega \circ T_t = \nabla b_\Omega,} \quad \boxed{T_t(x) = x + t\,\nabla b_\Omega(x).} \tag{4.11}$$

(iii) *Almost everywhere in $W(x)$*

$$DT_t = I + t\,D^2 b_\Omega, \quad \frac{d}{dt} DT_t = D^2 b_\Omega \tag{4.12}$$

$$\Rightarrow D^2 b_\Omega \circ T_t \ DT_t = D^2 b_\Omega, \ D^2 b_\Omega \circ T_t = D^2 b_\Omega \left[I + t D^2 b_\Omega \right]^{-1}, \tag{4.13}$$

$$\boxed{(DT_t)^{-1} \nabla b_\Omega = \nabla b_\Omega = {}^*(DT_t)^{-1} \nabla b_\Omega.} \tag{4.14}$$

Proof. For simplicity, we use the notation $b = b_\Omega$ and $p = p_\Gamma$.

(i) The result follows from Theorem 4.3 (i). (ii) By assumption from part (i) the function b belongs to $C^{1,1}(W(x))$ and hence $p \in C^{0,1}(W(x); \mathbf{R}^N)$. Thus, almost everywhere in $W(x)$,

$$Dp = I - \nabla b \,^*\nabla b - bD^2 b \quad \Rightarrow \quad Dp\nabla b = 0.$$

Now

$$\frac{d}{dt}(p \circ T_t) = Dp \circ T_t \frac{dT_t}{dt} = Dp \circ T_t \nabla b \circ T_t = (Dp\,\nabla b) \circ T_t = 0$$

$$\Rightarrow \; p \circ T_t = p \text{ in } W(x).$$

But since p and T_t belong to $C^{0,1}(W(x); \mathbf{R}^N)$, the identity necessarily holds everywhere in $W(x)$. Moreover,

$$\nabla b \circ T_t = n \circ p \circ T_t = n \circ p = \nabla b \text{ in } W(x).$$

Finally

$$\frac{dT_t}{dt} = \nabla b \circ T_t = \nabla b \quad \Rightarrow \quad T_t(y) = y + t\nabla b(y) \text{ in } W(x).$$

In particular, almost everywhere in $W(x)$,

$$DT_t = I + t\,D^2 b \quad \Rightarrow \quad \frac{d}{dt}DT_t = D^2 b,$$

$$DT_t \nabla b = \nabla b \quad \Rightarrow \quad (DT_t)^{-1}\nabla b = \nabla b = \,^*(DT_t)^{-1}\nabla b. \qquad \square$$

5 $W^{1,p}(D)$-Topology and the Family $C_b^0(D)$

From Theorem 2.1 (vi) b_A is locally Lipschitzian and belongs to $W^{1,p}_{\text{loc}}(\mathbf{R}^N)$ for all $p \geq 1$. So the previous constructions for d_A and $d_{\complement A}$ can be repeated with the space $W^{1,p}_{\text{loc}}(\mathbf{R}^N)$ in place of $C_{\text{loc}}(D)$ to generate new $W^{1,p}$ metric topologies on the family $C_b(D)$. Moreover the other distance functions can be recovered from the map

$$b_A \mapsto (b_A^+, b_A^-, |b_A|) = (d_A, d_{\complement A}, d_{\partial A})$$

and the characteristic functions from the maps

$$b_A \mapsto b_A^- = d_{\complement A} \mapsto \chi_{\text{int } A} = |\nabla d_{\complement A}|, \tag{5.1}$$

$$b_A \mapsto b_A^+ = d_A \mapsto \chi_{\text{int } \complement A} = |\nabla d_A|, \tag{5.2}$$

$$b_A \mapsto \chi_{\partial A} = 1 - |\nabla b_A|. \tag{5.3}$$

One of the advantages of the function b_A is that the $W^{1,p}$-convergence of sequences will imply the L^p-convergence of the corresponding characteristic functions of int A, int $\complement A$, and ∂A, that is, continuity of the volume of these sets. In this section we give the analogues of Theorems 4.1 and 4.2 in Chapter 4.

Theorem 5.1. *Let D be an open (resp., bounded open) subset of $\mathbf{R}^N$.*

(i) *The topologies induced by $W^{1,p}_{\mathrm{loc}}(D)$ (resp., $W^{1,p}(D)$) on $C_b(D)$ are all equivalent for p, $1 \le p < \infty$.*

(ii) *$C_b(D)$ is closed in $W^{1,p}_{\mathrm{loc}}(D)$ (resp., $W^{1,p}(D)$) for p, $1 \le p < \infty$, and*

$$\rho_D([A_2]_b, [A_1]_b) \overset{\text{def}}{=} \sum_{n=1}^{\infty} \frac{1}{2^n} \frac{\|b_{A_2} - b_{A_1}\|_{W^{1,p}(B(0,n))}}{1 + \|b_{A_2} - b_{A_1}\|_{W^{1,p}(B(0,n))}}$$

$$\left(\textit{resp., } \rho_D([A_2]_b, [A_1]_b) \overset{\text{def}}{=} \|b_{A_2} - b_{A_1}\|_{W^{1,p}(D)}\right)$$

defines a complete metric structure on $\mathcal{F}_b(D)$.

(iii) *For p, $1 \le p < \infty$, the map*

$$b_A \mapsto \chi_{\partial A} = 1 - |\nabla b_A| \;:\; C_b(D) \subset W^{1,p}_{\mathrm{loc}}(D) \to L^p_{\mathrm{loc}}(D)$$

is "Lipschitz continuous": for all bounded open subsets K of D and b_{A_1} and b_{A_2} in $C_b(D)$,

$$\|\chi_{\partial A_2} - \chi_{\partial A_1}\|_{L^p(K)} \le \|\nabla b_{A_2} - \nabla b_{A_1}\|_{L^p(K)} \le \|b_{A_2} - b_{A_1}\|_{W^{1,p}(K)}.$$

In particular, the family of sets whose boundary has a zero N-dimensional measure,

$$\boxed{C_b^0(D) = \{b_A \;:\; \forall A,\, A \subset \overline{D},\, \partial A \neq \varnothing \text{ and } \mathrm{m}(\partial A) = 0\},}$$

is a closed subset of $C_b(\overline{D})$ for the $W^{1,p}$-topology.

(iv) *Let D be a bounded open subset of $\mathbf{R}^N$. For each $b_A \in C_b(D)$*

$$\|b_A\|_{W^{1,p}(D)} = \|\,|b_A|\,\|_{W^{1,p}(D)} = \|d_{\partial A}\|_{W^{1,p}(D)},$$
$$\|d_A\|_{W^{1,p}(D)} \le \|b_A\|_{W^{1,p}(D)}, \quad \|d_{\complement A}\|_{W^{1,p}(D)} \le \|b_A\|_{W^{1,p}(D)}. \tag{5.4}$$

The map

$$b_A \mapsto (b_A^+, b_A^-, |b_A|) = (d_A, d_{\complement A}, d_{\partial A}) \;:\; C_b(D) \subset W^{1,p}(D) \to W^{1,p}(D)^3 \tag{5.5}$$

is continuous.

(v) *Let D be an open (resp., bounded open) subset of $\mathbf{R}^N$. For all p, $1 \le p < \infty$, the map*

$$b_A \mapsto (\chi_{\partial A}, \chi_{\mathrm{int}\, A}, \chi_{\mathrm{int}\, \complement A})$$
$$: W^{1,p}_{\mathrm{loc}}(D) \to L^p_{\mathrm{loc}}(D) \;(\textit{resp., } W^{1,p}(D) \to L^p(D))$$

is continuous.

Proof. The proof of (i) and (ii) is essentially the same as the proof of Theorem 4.1 of Chapter 4 using the properties established for $C_b(D)$ in $C(\overline{D})$ of Theorem 2.2.

(iii) From Theorem 3.2 (i) for any two subsets A_1 and A_2 of D with nonempty boundaries and for any open U in D

$$|\nabla b_{A_2}| \leq |\nabla b_{A_1}| + |\nabla b_{A_2} - \nabla b_{A_1}|$$

$$\Rightarrow \chi_{\partial A_1} \leq \chi_{\partial A_2} + |\nabla b_{A_2} - \nabla b_{A_1}| \quad \Rightarrow \quad |\chi_{\partial A_1} - \chi_{\partial A_2}| \leq |\nabla b_{A_2} - \nabla b_{A_1}|$$

$$\Rightarrow \int_U |\chi_{\partial A_2} - \chi_{\partial A_1}|^p \, dx \leq \|\nabla b_{A_2} - \nabla b_{A_1}\|_{L^p(U)}^p \leq \|b_{A_2} - b_{A_1}\|_{W^{1,p}(U)}^p$$

for p, $1 \leq p < \infty$, and with the ess sup norm for $p = \infty$. The closure of $C_b^0(D)$ follows from the continuity of the map

$$b_A \mapsto m(\partial A \cap U) = \int_{\partial A \cap U} \chi_{\partial A} \, dx \; : \; W^{1,p}(U) \to \mathbf{R}$$

for all bounded open subsets U of D.

(iv) First observe that

$$|b_A(x)| = d_A(x) + d_{\complement A}(x) \text{ and } \nabla |b_A(x)| = \nabla d_A(x) + \nabla d_{\complement A}(x),$$
$$b_A(x) = d_A(x) - d_{\complement A}(x) \text{ and } \nabla b_A(x) = \nabla d_A(x) - \nabla d_{\complement A}(x),$$

and since $\nabla b_A = \nabla d_A = \nabla d_{\complement A} = 0$ almost everywhere on $\partial A = \overline{A} \cap \overline{\complement A}$, then

$$|\nabla b_A(x)| = |\nabla d_A(x)| + |\nabla d_{\complement A}(x)| = |\nabla |b_A(x)|| \text{ for almost all } x \text{ in } \mathbf{R}^N.$$

From this we readily get (5.4). However, this is not sufficient to prove the continuity since the map (5.5) is not linear. To get around this consider a sequence $\{b_{A_n}\} \subset C_b(D)$ converging to b_A in $W^{1,p}(D)$. From Theorem 2.2 (iii), $|b_{A_n}| = d_{\partial A_n}$, $b_{A_n}^+ = d_{A_n}$, and $b_{A_n}^- = d_{\complement A_n}$ converge to $|b_A| = d_{\partial A}$, $b_A^+ = d_A$, and $b_A^- = d_{\complement A}$ in $C(D)$ and hence in $W^{1,p}(D)$-weak. To prove that the convergence is strong, consider the L^2-norm

$$\int_D |\nabla d_{A_n} - \nabla d_A|^2 + |\nabla d_{\complement A_n} - \nabla d_{\complement A}|^2 \, dx$$

$$= \int_D |\nabla d_{A_n}|^2 + |\nabla d_A|^2 + |\nabla d_{\complement A_n}|^2 + |\nabla d_{\complement A}|^2$$

$$- 2 \nabla d_{A_n} \cdot \nabla d_A - 2 \nabla d_{\complement A_n} \cdot \nabla d_{\complement A} \, dx$$

$$= \int_D |\nabla b_{A_n}|^2 + |\nabla b_A|^2 - 2 \nabla d_{A_n} \cdot \nabla d_A - 2 \nabla d_{\complement A_n} \cdot \nabla d_{\complement A} \, dx$$

$$\to \int_D 2|\nabla b_A|^2 - 2|\nabla d_A|^2 - 2|\nabla d_{\complement A}|^2 \, dx = \int_D 2|\nabla b_A|^2 - 2|\nabla b_A|^2 \, dx = 0$$

by weak L^2-convergence of ∇d_{A_n} and $\nabla d_{\complement A_n}$ to ∇d_A and $\nabla d_{\complement A}$. So both sequences d_{A_n} and $d_{\complement A_n}$ converge in $W^{1,2}(D)$-strong, and hence from part (i) in $W^{1,p}(D)$-strong for all p, $1 \leq p < \infty$.

(v) It is sufficient to prove the result for D bounded open. From part (iii) it is true for $\chi_{\partial A}$, and from part (iv) and Theorem 4.1 (ii) and (iii) of Chapter 4 for the other two. $\qquad \square$

Theorem 5.2. *Let D be a bounded open domain in $\mathbf{R}^N$.*

(i) *If $\{b_{A_n}\}$ weakly converges in $W^{1,p}(D)$ for some p, $1 \le p < \infty$, then it weakly converges in $W^{1,p}(D)$ for all p, $1 \le p < \infty$.*

(ii) *If $\{b_{A_n}\}$ converges in $C(\overline{D})$, then it weakly converges in $W^{1,p}(D)$ for all p, $1 \le p < \infty$. Conversely, if $\{b_{A_n}\}$ weakly converges in $W^{1,p}(D)$ for some p, $1 \le p < \infty$, it converges in $C(\overline{D})$.*

(iii) *$C_b(D)$ is compact in $W^{1,p}(D)$-weak for all p, $1 \le p < \infty$.*

Proof. (i) Recall that for D bounded there exists a constant $c > 0$ such that for all $b_A \in C_b(D)$

$$|b_A(x)| \le c \text{ and } |\nabla b_A(x)| \le 1 \text{ a.e. in } D.$$

If $\{b_{A_n}\}$ weakly converges in $W^{1,p}(D)$ for some $p \ge 1$, then

$$\{b_{A_n}\} \text{ weakly converges in } L^p(D),$$
$$\{\nabla b_{A_n}\} \text{ weakly converges in } L^p(D)^N.$$

By Lemma 2.1 (iii) in Chapter 3 both sequences weakly converge for all $p \ge 1$ and hence $\{b_{A_n}\}$ weakly converges in $W^{1,p}(D)$ for all $p \ge 1$.

(ii) If $\{b_{A_n}\}$ converges in $C(\overline{D})$, then by Theorem 2.2 (ii) there exists $b_A \in C_b(D)$ such that $b_{A_n} \to b_A$ in $C(\overline{D})$ and hence in $L^p(D)$. So for all $\varphi \in \mathcal{D}(D)^N$

$$\int_D \nabla b_{A_n} \cdot \varphi \, dx = -\int_D b_{A_n} \operatorname{div} \varphi \, dx \to -\int_D b_A \operatorname{div} \varphi \, dx = \int_D \nabla b_A \cdot \varphi \, dx.$$

By density of $\mathcal{D}(D)$ in $L^2(D)$, $\nabla b_{A_n} \to \nabla b_A$ in $L^2(D)^N$-weak and hence $b_{A_n} \to b_A$ in $W^{1,2}(D)$-weak. From part (i) it converges in $W^{1,p}(D)$-weak for all p, $1 \le p < \infty$. Conversely, the weakly convergent sequence converges to some f in $W^{1,p}(D)$. By compactness of $C_b(D)$, there exist a subsequence, still indexed by n, and b_A such that $b_{A_n} \to b_A$ in $C(\overline{D})$ and hence in $W^{1,p}(D)$-weak. By uniqueness of the limit, $b_A = f$. Therefore, all convergent subsequences in $C(\overline{D})$ converge to the same limit. So the whole sequence converges in $C(\overline{D})$. This concludes the proof.

(iii) Consider an arbitrary sequence $\{b_{A_n}\}$ in $C_d(D)$. From Theorem 2.2 (ii) $C_b(D)$ is compact and there exists a subsequence $\{b_{A_{n_k}}\}$ and $b_A \in C_b(D)$ such that $b_{A_{n_k}} \to b_A$ in $C(\overline{D})$. From part (ii) the subsequence weakly converges in $W^{1,p}(D)$ and hence $C_b(D)$ is compact in $W^{1,p}(D)$-weak. $\square$

6   Sets of Bounded and Locally Bounded Curvature

We introduce the family of sets with bounded or locally bounded curvature which are the analogues for b_A of the sets of exterior and interior bounded curvature associated with d_A and $d_{\complement A}$ in section 5 of Chapter 4. They include $C^{1,1}$-domains, convex sets, and the sets of positive reach of Federer [2]. They lead to compactness theorems for $C_b(D)$ in $W^{1,p}(D)$.

6.1 Definitions and Main Properties

Definition 6.1.

(i) Given a bounded open nonempty subset D of $\mathbf{R}^N$, a subset A of $\overline{D}$, $\partial A \neq \varnothing$, is said to be of *bounded curvature* with respect to D if

$$\nabla b_A \in \mathrm{BV}(D)^N. \tag{6.1}$$

This family of sets will be denoted as follows:

$$\boxed{\mathrm{BC}_b(D) \stackrel{\mathrm{def}}{=} \left\{ b_A \in C_b(D) \; : \; \nabla b_A \in \mathrm{BV}(D)^N \right\}.}$$

(ii) A subset A of $\mathbf{R}^N$, $\partial A \neq \varnothing$, is said to be of *locally bounded curvature* if

$$\forall x \in \partial A, \; \exists \rho > 0 \text{ such that } \nabla b_A \in \mathrm{BV}(B(x,\rho))^N, \tag{6.2}$$

where $B(x,\rho)$ is the open ball of radius $\rho > 0$ in x. $\square$

From Theorem 5.1 and Definition 5.1 in Chapter 3, a function belongs to $\mathrm{BV}_{\mathrm{loc}}(\mathbf{R}^N)$ if and only if for each $x \in \mathbf{R}^N$ it belongs to $\mathrm{BV}(B(x,\rho))$ for some $\rho > 0$. As in Theorem 5.2 of Chapter 3 for Caccioppoli sets, it is sufficient to satisfy this condition for points of the boundary.

Theorem 6.1. *Let A, $\partial A \neq \varnothing$, be a subset of $\mathbf{R}^N$. Then A is of locally bounded curvature if and only if ∇b_A belongs to $\mathrm{BV}_{\mathrm{loc}}(\mathbf{R}^N)^N$.*

Corollary 2. *All sets A, $\partial A \neq \varnothing$, in $\mathbf{R}^N$ of class $C^{1,1}$ are of locally bounded curvature.*

The theorem will be proved as part (iii) of Theorem 6.3, and the corollary follows from Theorem 4.3 (i).

Theorem 6.2.

(i) *Let D be a nonempty bounded open Lipschitzian subset of $\mathbf{R}^N$. For any subset A of $\overline{D}$, $\partial A \neq \varnothing$, such that*

$$\nabla b_A \in \mathrm{BV}(D)^N,$$

∂A has finite perimeter, that is,

$$\chi_{\partial A} \in \mathrm{BV}(D).$$

(ii) *For any subset A of $\mathbf{R}^N$, $\partial A \neq \varnothing$, such that*

$$\nabla b_A \in \mathrm{BV}_{\mathrm{loc}}(\mathbf{R}^N)^N,$$

∂A has locally finite perimeter, that is,

$$\chi_{\partial A} \in \mathrm{BV}_{\mathrm{loc}}(\mathbf{R}^N).$$

Proof. Given ∇b_A in $\mathrm{BV}(D)^N$, there exists a sequence $\{u_k\}$ in $C^\infty(D)^N$ such that

$$u_k \to \nabla b_A \text{ in } L^1(D)^N,$$

$$\|Du_k\|_{M^1(D)} \to \|D^2 b_A\|_{M^1(D)}$$

as k goes to infinity, and since $|\nabla b_A(x)| \le 1$, this sequence can be chosen in such a way that

$$\forall k \ge 1, \quad |u_k(x)| \le 1.$$

This follows from the use of mollifiers (cf. Giusti [1, Thm. 1.17, p. 15]). For all V in $\mathcal{D}(D)^N$

$$-\int_D \chi_{\partial A} \operatorname{div} V \, dx = \int_D (|\nabla b_A|^2 - 1) \operatorname{div} V \, dx = \int_D |\nabla b_A|^2 \operatorname{div} V \, dx.$$

For each u_k

$$\int_D |u_k|^2 \operatorname{div} V \, dx = -2 \int_D [\,^*Du_k] u_k \cdot V \, dx = -2 \int_D u_k \cdot [Du_k] V \, dx,$$

where $^*Du_k$ is the transpose of the Jacobian matrix Du_k and

$$\left| \int_D |u_k|^2 \operatorname{div} V \, dx \right| \le 2 \int_D |u_k| \, |Du_k| \, |V| \, dx$$

$$\le 2 \|Du_k\|_{L^1} \|V\|_{C(D)} \le 2 \|Du_k\|_{M^1} \|V\|_{C(D)}$$

since for $W^{1,1}(D)$-functions, $\|\nabla f\|_{L^1(D)^N} = \|\nabla f\|_{M^1(D)^N}$. Therefore, as k goes to infinity,

$$\left| \int_D \chi_{\partial A} \operatorname{div} V \, dx \right| = \left| \int_D |\nabla b_A|^2 \operatorname{div} V \, dx \right| \le 2 \|D^2 b_A\|_{M^1} \|V\|_{C(D)}.$$

Therefore, $\nabla \chi_{\partial A} \in M^1(D)^N$. $\qquad\square$

Theorem 6.3. *Let A be a subset of $\mathbf{R}^N$ such that $\partial A \ne \varnothing$.*

(i) *The function $f_{\partial A}(x) = \frac{1}{2}\left(|x|^2 - b_A^2(x)\right)$ is convex in $\mathbf{R}^N$ and $\nabla b_A^2 \in \mathrm{BV}_{\mathrm{loc}}(\mathbf{R}^N)^N$. For all x and y in $\mathbf{R}^N$*

$$\forall p \in \Pi_{\partial A}(x), \quad \frac{1}{2}\left(|y|^2 - b_A^2(y)\right) \ge \frac{1}{2}\left(|x|^2 - b_A^2(x)\right) + p \cdot (y - x) \tag{6.3}$$

or equivalently

$$\forall p \in \Pi_{\partial A}(x), \quad b_A^2(y) - b_A^2(x) - 2(x - p) \cdot (y - x) \le |x - y|^2. \tag{6.4}$$

The set of projections $\Pi_{\partial A}(x)$ onto ∂A is a singleton $p_{\partial A}(x)$ if and only if the gradient of $f_{\partial A}$ exists. In both cases

$$p_{\partial A}(x) = \frac{1}{2} \nabla \left(|x|^2 - b_A^2(x)\right). \tag{6.5}$$

For all x and y in $\mathbf{R}^N$

$$\forall p(x) \in \Pi_{\partial A}(x), \ \forall p(y) \in \Pi_{\partial A}(y), \quad (p(y) - p(x)) \cdot (y - x) \geq 0. \tag{6.6}$$

The function $b_A^2(x)$ is the difference of two convex functions

$$b_A^2(x) = |x|^2 - \left(|x|^2 - b_A^2(x)\right). \tag{6.7}$$

(ii) $\nabla b_A \in \mathrm{BV}_{\mathrm{loc}}(\mathbf{R}^N \setminus \partial A)^N$. *More precisely, for all $x \in \mathbf{R}^N \setminus \partial A$ there exists $\rho > 0$, $0 < 3\rho < d_{\partial A}(x)$, such that $k_{\partial A, \rho}$ is convex in $B(x, 2\rho)$ and hence $\nabla b_A \in \mathrm{BV}(B(x, \rho))^N$.*

(iii) *A subset A of $\mathbf{R}^N$, $\partial A \neq \varnothing$, is of locally bounded curvature if and only if $\nabla b_A \in \mathrm{BV}_{\mathrm{loc}}(\mathbf{R}^N)^N$.*

Proof. The proof follows from Lemma 5.1 and Theorem 5.3 in Chapter 4 by applying it to $d_{\partial A} = |b_A|$. Use the fact that $b_A = d_A$ and $b_A = -d_{\complement A}$ outside of ∂A. $\qquad\square$

6.2   Examples and Limits of the Tubular Norms as h Goes to Zero

It is informative to compute the Laplacian of b_A for a few examples. The first three are illustrated in Figure 5.1.

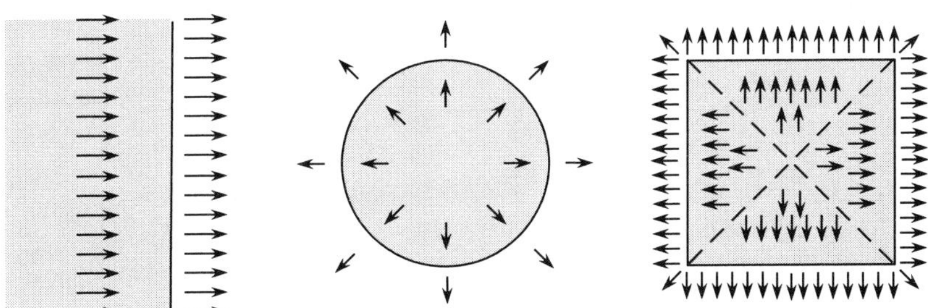

Figure 5.1. ∇b_A *for Examples* 6.1, 6.2, *and* 6.3.

Example 6.1 (half-plane in $\mathbf{R}^2$; cf. Example 5.1 in Chapter 4).
Consider the domain

$$A = \{(x_1, x_2) : x_1 \leq 0\}, \quad \partial A = \{(x_1, x_2) : x_1 = 0\}.$$

It is readily seen that

$$b_A(x_1, x_2) = x_1, \quad \nabla b_A(x_1, x_2) = (1, 0), \quad \Delta b_A(x) = 0,$$

and

$$b_{\partial A}(x_1, x_2) = |x_1|, \quad \nabla b_{\partial A}(x_1, x_2) = \left(\frac{x_1}{|x_1|}, 0\right),$$

$$\langle \Delta b_{\partial A}, \varphi \rangle = 2 \int_{\partial A} \varphi \, dx. \qquad\square$$

Example 6.2 (ball of radius $R > 0$ in $\mathbf{R}^2$; cf. Example 5.2 in Chapter 4).
Consider the domain

$$A = \{x \in \mathbf{R}^2 : |x| \le R\}, \quad \partial A = \{x \in \mathbf{R}^2 : |x| = R\}.$$

Clearly,

$$b_A(x) = |x| - R, \quad \nabla b_A(x) = \frac{x}{|x|},$$

$$\langle \Delta b_A, \varphi \rangle = \int_{\mathbf{R}^2} \frac{1}{|x|} \varphi \, dx.$$

Also

$$b_{\partial A}(x) = \big||x_1| - R\big|, \quad \nabla b_{\partial A}(x) = \begin{cases} \frac{x}{|x|}, & |x| > R, \\ -\frac{x}{|x|}, & 0 < |x| < R, \end{cases}$$

$$\langle \Delta b_{\partial A}, \varphi \rangle = 2 \int_{\partial A} \varphi \, ds - \int_A \frac{1}{|x|} \varphi \, dx + \int_{\complement A} \frac{1}{|x|} \varphi \, dx.$$

Again $\Delta b_{\partial A}$ contains twice the boundary measure on ∂A. $\square$

Example 6.3 (unit square in $\mathbf{R}^2$; cf. Example 5.3 in Chapter 4).
Consider the domain

$$A = \{x = (x_1, x_2) : |x_1| \le 1, |x_2| \le 1\}.$$

Since A is symmetrical with respect to both axes, it is sufficient to specify b_A in the first quadrant. We use the notation Q_1, Q_2, Q_3, and Q_4 for the four quadrants in the counterclockwise order and c_1, c_2, c_3, and c_4 for the four corners of the square in the same order. We also divide the plane into three regions:

$$D_1 = \{(x_1, x_2) : |x_2| \le \min\{1, |x_1|\}\},$$
$$D_2 = \{(x_1, x_2) : |x_1| \le \min\{1, |x_2|\}\},$$
$$D_3 = \{(x_1, x_2) : |x_1| \ge 1 \text{ and } |x_2| \ge 1\}.$$

Hence for $1 \le i \le 4$

$$b_A(x) = \begin{cases} |x_2| - 1, & x \in D_2 \cap Q_i, \\ |x - c_i|, & x \in D_3 \cap Q_i, \\ |x_1| - 1, & x \in D_1 \cap Q_i, \end{cases} \quad \nabla b_A(x) = \begin{cases} (0, 1), & x \in D_2 \cap Q_i, \\ \frac{x - c_i}{|x - c_i|}, & x \in D_3 \cap Q_i, \\ (1, 0), & x \in D_1 \cap Q_i, \end{cases}$$

and for the whole plane

$$\langle \Delta b_A, \varphi \rangle = \sum_{i=1}^{4} \int_{D_3 \cap Q_i} \frac{1}{|x - c_i|} \varphi \, dx + \sqrt{2} \int_{D_1 \cap D_2} \varphi \, dx,$$

where $D_1 \cap D_2$ is made up of the two diagonals of the square where ∇b_A has a singularity, that is, the skeleton. Moreover, for $1 \le i \le 4$

$$b_{\partial A}(x) = \begin{cases} \big|\,|x_2| - 1\,\big|, & x \in D_2 \cap Q_i, \\ |x - c_i|, & x \in D_3 \cap Q_i, \\ \big|\,|x_1| - 1\,\big|, & x \in D_1 \cap Q_i, \end{cases}$$

and

$$\langle \Delta b_{\partial A}, \varphi \rangle = \sum_{i=1}^{4} \int_{D_3 \cap Q_i} \frac{1}{|x - c_i|}\, \varphi\, dx - \sqrt{2} \int_{D_1 \cap D_2} \varphi\, dx + 2 \int_{\partial A} \varphi\, dx.$$

Notice that the structures of the Laplacian are similar to the ones observed in the previous examples except for the presence of a singular term along the two diagonals of the square. $\qquad\square$

All C^2-domains with a compact boundary belong to all the categories of Definition 6.1 and Definition 5.1 of Chapter 4. The h-dependent norms $\|D^2 b_A\|_{M^1(U_h(\partial A))}$, $\|D^2 d_{\complement A}\|_{M^1(U_h(\complement A))}$, and $\|D^2 d_A\|_{M^1(U_h(A))}$ are all decreasing as h goes to zero. The limit is particularly interesting since it singles out the behavior of the singular part of the Hessian matrix in a shrinking neighborhood of the boundary ∂A.

Example 6.4.
If $A \subset \mathbf{R}^N$ is of class C^2 with compact boundary, then

$$\lim_{h \searrow 0} \|D^2 b_A\|_{M^1(U_h(\partial A))} = 0. \qquad\qquad\qquad\square$$

Example 6.5.
Let $A = \{x_i\}_{i=1}^{I}$ be I distinct points in $\mathbf{R}^N$. Then $\partial A = A$ and

$$\lim_{h \searrow 0} \|D^2 b_A\|_{M^1(U_h(\partial A))} = \begin{cases} 2I - 1, & N = 1, \\ 0, & N \ge 2. \end{cases} \qquad\square$$

Example 6.6.
Let A be a closed line in $\mathbf{R}^N$ of length $L > 0$. Then $\partial A = A$, and

$$\lim_{h \searrow 0} \|D^2 b_A\|_{M^1(U_h(\partial A))} = \begin{cases} 2L, & N = 2, \\ 0, & N \ge 3. \end{cases} \qquad\square$$

Example 6.7.
Let $N = 2$. For the finite square and the ball of finite radius,

$$\lim_{h \searrow 0} \|\Delta d_A\|_{M^1(U_h(\partial A))} = H_1(\partial A),$$

where H_1 is the one-dimensional Hausdorff measure (cf. Examples 5.2 and 5.3 in Chapter 4). $\qquad\square$

Also, looking at Δb_A in $U_h(A)$ as h goes to zero provides information about $\mathrm{Sk}\,(A)$.

Example 6.8.

Let A be the unit square in R^2. Then

$$\lim_{h\searrow 0} \|\Delta b_A\|_{M^1(U_h(A))} = \frac{\sqrt{2}}{2}\, H_1(\mathrm{Sk}\,(A)),$$

where H_1 is the one-dimensional Hausdorff measure and $\mathrm{Sk}(A) = \mathrm{Sk}_{\mathrm{int}}(A)$ is the skeleton of A made up of the two interior diagonals (cf. Example 6.3). It would seem that in general

$$\lim_{h\searrow 0} \|\Delta b_A\|_{M^1(U_h(A))} = \int_{\mathrm{Sk}(A)} |\,[\nabla b_A]\cdot n|\,dH_1,$$

where $[\nabla b_A]$ is the jump in ∇b_A and n is the unit normal to $\mathrm{Sk}(A)$ (if it exists!). $\qquad\square$

6.3 (m, p)-Sobolev Domains (or $W^{m,p}$-Domains)

The N-dimensional Lebesgue measure of the boundary of a Lipschitzian subset of $\mathbf{R}^N$ is zero. However, this is generally not true for sets of locally bounded curvature, as can be seen from the following example.

Example 6.9.

Let B be the open unit ball centered in 0 of $\mathbf{R}^2$ and define

$$A = \{x \in B \,:\, x \text{ with rational coordinates}\}.$$

Then $\partial A = \overline{B}$, $b_A = d_B$, and for all $h > 0$

$$\nabla b_A \in BV(U_h(\partial A))^2,$$

$$\langle \Delta b_A, \varphi \rangle = \int_{\partial B} \varphi\, dx + \int_{\complement B} \frac{1}{|x|}\varphi\, dx. \qquad\square$$

The next natural question that comes to mind is whether the boundary or the skeleton have locally finite $(N-1)$-Hausdorff measure. Then come questions about the mean curvature of the boundary. For the function b_A, $\Delta b_A = \mathrm{tr}\, D^2 b_A$ is proportional to the mean curvature of the boundary, which is only a measure in $\mathbf{R}^N$ for sets of bounded local curvature. This calls for the introduction of some classification of sets that would complement the classical Hölderian terminology introduced in Chapter 2, would fill the gap in between, and would possibly say something about sets whose boundary is not even continuous. This seems to have been first introduced in Delfour and Zolésio [32].

Definition 6.2 (Sobolev domains).

Given $m > 1$ and $p \geq 1$, a subset A of $\mathbf{R}^N$ is said to be an (m, p)-*Sobolev domain* or simply a $W^{m,p}$-*domain* if $\partial A \neq \varnothing$ and there exists $h > 0$ such that

$$b_A \in W^{m,p}_{\mathrm{loc}}(U_h(\partial A)). \qquad\square$$

This classification gives an analytical description of the smoothness of boundaries in terms of the derivatives of b_A. The definition is obviously vacuous for $m = 1$ since $b_A \in W_{\text{loc}}^{1,\infty}(\mathbf{R}^N)$ for any set A such that $\partial A \neq \varnothing$. For $1 < m < 2$ we shall see that it encompasses sets that are less smooth than sets of locally bounded curvature. For $m = 2$ the $W^{2,p}$-domains are of class $C^{1,1-N/p}$ for $p > N$. For $m > 2$ they are intertwined with sets of class $C^{k,\ell}$.

Theorem 6.4. *Given any subset A of $\mathbf{R}^N$, $\partial A \neq \varnothing$,*

$$\nabla b_A \in \mathrm{BV}_{\text{loc}}(\mathbf{R}^N)^N$$

$$\Rightarrow \forall p, \ 1 \leq p < \infty, \quad \forall \eta, \ 0 \leq \eta < \frac{1}{p}, \quad b_A \in W_{\text{loc}}^{1+\eta,p}(\mathbf{R}^N).$$

Proof. Since $|\nabla b_A(x)| \leq 1$ almost everywhere, $\nabla b_A \in \mathrm{BV}_{\text{loc}}(\mathbf{R}^N)^N \cap L_{\text{loc}}^\infty(\mathbf{R}^N)^N$ and the theorem follows directly from Theorem 5.8 in Chapter 3. $\qquad\square$

It is quite interesting that a domain of locally bounded curvature is a $W^{2-\varepsilon,1}$-domain for any arbitrary small $\varepsilon > 0$. So it is *almost* a $W^{2,1}$-domain, and domains of class $W^{2-\varepsilon,1}$ seem to be a larger class than domains of locally bounded curvature. In addition, their boundary does not generally have zero measure.

Now consider the boundary case $m = 2$.

Theorem 6.5. *Given an integer $N \geq 1$, let A be a subset of $\mathbf{R}^N$ such that $\partial A \neq \varnothing$.*

(i) *If there exist $p > N$ and $h > 0$ such that*

$$b_A \in W_{\text{loc}}^{2,p}(U_h(\partial A)), \tag{6.8}$$

then

$$b_A \in C_{\text{loc}}^{1,1-N/p}(U_h(\partial A)),$$

that is, A is a Hölderian set of class $C^{1,1-N/p}$. Moreover, the function b_A^2 belongs to $C_{\text{loc}}^{1,1}(U_h(\partial A))$, ∂A has positive reach, and the N-dimensional Lebesgue measure of ∂A is zero.

(ii) *In dimension $N = 2$ the condition $b_A \in W_{\text{loc}}^{2,p}(U_h(\partial A))$ is equivalent to $\Delta b_A \in L_{\text{loc}}^p(U_h(\partial A))$.*

Proof. (i) Consider the function $|\nabla b_A|^2$. Since $|\nabla b_A| \leq 1$, then

$$b_A \in W_{\text{loc}}^{2,p}(U_h(\partial A)) \cap W_{\text{loc}}^{1,\infty}(U_h(\partial A))$$

$$\Rightarrow b_A^2 \in W_{\text{loc}}^{2,p}(U_h(\partial A)) \cap W_{\text{loc}}^{1,\infty}(U_h(\partial A)).$$

In particular, for all $x \in \partial A$,

$$b_A, \ b_A^2 \in W^{2,p}(B(x,h)) \cap W^{1,\infty}(B(x,h)).$$

For $p > N$, from Adams [1, Thm. 5.4, Part III, and Remark 5.5 (3), p. 98],

$$b_A, \ b_A^2 \in C^{1,\lambda}(B(x,h)), \quad 0 < \lambda \leq 1 - N/p$$
$$\Rightarrow b_A, \ b_A^2 \in C^{1,\lambda}_{\text{loc}}(U_h(A)), \quad 0 < \lambda < 1 - N/p.$$

From Theorem 4.1, $b_A^2 \in C^{1,1}_{\text{loc}}(U_h(\partial A))$ and ∂A has positive reach. From Theorem 4.2 A is of class $C^{1,1-N/p}$ and $\mathrm{m}(\partial A) = 0$.

(ii) From part (i), $|\nabla b_A(x)|^2 = 1$ in $U_h(\partial A)$, and $D^2 b_A(x)\, \nabla b_A(x) = 0$ almost everywhere. Hence in dimension $N = 2$,

$$\partial^2_{12} b_A(x) = \partial^2_{21} b_A(x) = -\partial_1 b_A(x)\, \partial_2 b_A(x)\, \Delta b_A(x),$$

$$\partial^2_{11} b_A(x) = (\partial_2 b_A(x))^2\, \Delta b_A(x),$$

$$\partial^2_{22} b_A(x) = (\partial_1 b_A(x))^2\, \Delta b_A(x). \qquad \square$$

As can be seen, a certain amount of work would be necessary to characterize all the Sobolev domains and answer the many associated open questions.

7　Characterization of Convex and Semiconvex Sets

We have seen in Chapter 4 that the convexity of d_A is equivalent to the convexity of $\overline{A}$. We shall see that this characterization remains true with $b_{\overline{A}}$ in place of $d_A = d_{\overline{A}}$. However, the next example shows that the convexity of b_A is not sufficient to get the convexity of A.

Example 7.1.
Let B be the open unit ball in $\mathbf{R}^N$ and A the set B minus all the points in B with rational coordinates. By definition,

$$\overline{A} = \overline{B}, \quad \overline{\complement A} = \mathbf{R}^N, \quad \partial A = \overline{B}, \quad \partial\overline{A} = \partial\overline{B} \neq \partial A,$$
$$d_A = d_B = d_{\overline{A}}, \quad d_{\complement A} = d_{\overline{\complement A}} = d_{\mathbf{R}^N} = 0$$
$$\Rightarrow b_A = d_B \neq b_{\overline{B}} = b_{\overline{A}}.$$

By Theorem 6.2 (i) of Chapter 4 the function d_B and, a fortiori b_A, are convex, but A is not convex and $\partial\overline{A} \neq \partial A$. $\qquad \square$

Theorem 7.1.

(i) *Let A be a subset of $\mathbf{R}^N$ such that $\partial A \neq \varnothing$. Then*

$$b_A \text{ convex in } \mathbf{R}^N \quad \Rightarrow \quad \overline{A} \text{ convex.} \tag{7.1}$$

(ii) *Let A be a subset of $\mathbf{R}^N$ such that $\partial\overline{A} \neq \varnothing$. Then*

$$b_{\overline{A}} \text{ convex in } \mathbf{R}^N \quad \Longleftrightarrow \quad \overline{A} \text{ convex.} \tag{7.2}$$

(iii) *Let A be a subset of $\mathbf{R}^N$ such that $\partial A \neq \varnothing$. Then*

$$A \text{ convex} \quad \Rightarrow \quad b_A = b_{\overline{A}} \text{ convex in } \mathbf{R}^N \text{ and } \partial A = \partial \overline{A}. \qquad (7.3)$$

(iv) *For all convex sets A such that $\partial A \neq \varnothing$, ∇b_A belongs to $\mathrm{BV}_{\mathrm{loc}}(\mathbf{R}^N)^N$ and the Hessian matrix $D^2 b_A$ of second-order derivatives is a matrix of signed Radon measures which are nonnegative on the diagonal. Moreover, b_A has a second-order derivative almost everywhere, and for almost all x and y in $\mathbf{R}^N$,*

$$\left| b_A(y) - b_A(x) - \nabla b_A(x) \cdot (y - x) - \frac{1}{2}(y - x) \cdot D^2 b_A(x)\,(y - x) \right| = o(|y - x|^2)$$

as $y \to x$.

Proof. (i) Denote by x_λ the convex combination $\lambda x + (1 - \lambda)y$ of two points x and y in $\overline{A}$ for some $\lambda \in [0, 1]$. By convexity of b_A,

$$b_A(x_\lambda) \leq \lambda b_A(x) + (1 - \lambda)b_A(y) = -[\lambda d_{\complement A}(x) + (1 - \lambda)d_{\complement A}(y)] \leq 0$$
$$\Rightarrow \quad d_A(x_\lambda) = b_A^+(x_\lambda) = \max\{b_A(x_\lambda), 0\} = 0 \quad \Rightarrow \quad x_\lambda \in \overline{A}$$

and $\overline{A}$ is convex.

(ii) It is sufficient to prove the result for A closed. In the first direction the result follows from part (i). In the other direction we consider three cases. The first one deserves a lemma.

Lemma 7.1. *Let A be a subset of $\mathbf{R}^N$ such that $\partial \overline{A} \neq \varnothing$ and $\overline{A}$ be convex. Then*

$$b_{\overline{A}} = -d_{\complement \overline{A}} \text{ is convex in } \overline{A}.$$

Proof. Again we can assume that A is closed. Associate with x and y in A, the radii $r_x = d_{\complement A}(x)$, $r_y = d_{\complement A}(y)$, $r_\lambda = \lambda r_x + (1 - \lambda)r_y$, and the closed balls $\overline{B}_x$ of center x and radius r_x, $\overline{B}_y$ of center y and radius r_y, and $\overline{B}_\lambda$ of center x_λ and radius r_λ. By the definition of $d_{\complement A}$, $\overline{B}_x \subset A$ and $\overline{B}_y \subset A$ since A is closed. Associate with each $z \in \overline{B}_\lambda$ the points $z_x = x$ and $z_y = y$ if $r_\lambda = 0$, and if $r_\lambda > 0$ the points

$$z_x \stackrel{\text{def}}{=} x + \frac{r_x}{r_\lambda}(z - x_\lambda) \quad \Rightarrow \quad z_x \in \overline{B}_x,$$
$$z_y \stackrel{\text{def}}{=} y + \frac{r_y}{r_\lambda}(z - x_\lambda) \quad \Rightarrow \quad z_y \in \overline{B}_y$$
$$\Rightarrow \quad \lambda z_x + (1 - \lambda)z_y = x_\lambda + \frac{\lambda r_x + (1 - \lambda)r_y}{r_\lambda}(z - x_\lambda) = z.$$

Obviously $z = x_\lambda$ if $r_\lambda = 0$. Therefore,

$$\overline{B}_\lambda \subset \lambda \overline{B}_x + (1 - \lambda)\overline{B}_y \subset A,$$

since A is closed and convex. In particular

$$d_{\complement A}(x_\lambda) \geq d_{\complement \overline{B}_\lambda}(x_\lambda) \geq r_\lambda = \lambda d_{\complement A}(x) + (1 - \lambda)d_{\complement A}(y),$$

and this proves the lemma. $\qquad \square$

For the second case consider x and y in $\overline{\complement A}$. By definition

$$b_A(x) = d_A(x) \text{ and } b_A(y) = d_A(y).$$

By Theorem 6.2 (i) of Chapter 4, d_A is convex when A is convex and

$$b_A(x_\lambda) \le d_A(x_\lambda) \le \lambda d_A(x) + (1-\lambda)d_A(y) = \lambda b_A(x) + (1-\lambda)b_A(y).$$

The third and last case is the mixed one for $x \in \overline{\complement A}$ and $y \in A$. Define

$$x_\lambda \stackrel{\text{def}}{=} \lambda x + (1-\lambda)y, \quad p_\lambda \stackrel{\text{def}}{=} p_{\partial A}(x_\lambda).$$

Since A is convex, denote by H the tangent hyperplane to A through p_λ, by H^+ the closed half-space containing A, and by H^- the other closed subspace associated with H. By the definition of H,

$$d_{\partial A}(x_\lambda) = d_H(x_\lambda), \quad H \subset H^\pm,$$

and by convexity of A, $A \subset H^+$ and $H^- \subset \overline{\complement A}$. The projection onto H is a linear operator and

$$p_\lambda = p_H(x_\lambda) = \lambda p_H(x) + (1-\lambda)p_H(y),$$
$$p_\lambda - x_\lambda = \lambda(p_H(x) - x) + (1-\lambda)(p_H(y) - y).$$

If $y \in H^+$ and $x \in H^+$

$$\boxed{d_H(x_\lambda) = \lambda d_H(x) + (1-\lambda)d_H(y),} \tag{7.4}$$

and if $y \in H^+$ and $x \in H^-$

$$\boxed{d_H(x_\lambda) = \begin{cases} \lambda d_H(x) - (1-\lambda)d_H(y) & \text{if } x_\lambda \in H^-, \\ (1-\lambda)d_H(y) - \lambda d_H(x) & \text{if } x_\lambda \in H^+. \end{cases}} \tag{7.5}$$

By convexity any y belongs to $\overline{A} \subset H^+$ and since $H^- \subset \overline{\complement A}$, we readily have

$$d_H(y) = d_{H^-}(y) \ge d_{\complement A}(y) = -b_A(y).$$

First consider the case $x_\lambda \in \complement A$. Then $d_A(x_\lambda) > 0$, $x_\lambda \in H^-$, and necessarily $x \in H^-$. From (7.5)

$$b_A(x_\lambda) = d_A(x_\lambda) = d_H(x_\lambda) = \lambda d_H(x) - (1-\lambda)d_H(y).$$

But since $x \in H^-$ and $A \subset H^+$

$$d_H(x) = d_{H^+}(x) \le d_A(x)$$

and

$$b_A(x_\lambda) = d_H(x_\lambda) \le \lambda d_A(x) - (1-\lambda)d_{\complement A}(y) = \lambda b_A(x) + (1-\lambda)b_A(y).$$

Next consider the case $x_\lambda \in A$ for which $x_\lambda \in H^+$ and

$$-b_A(x_\lambda) = d_{\complement A}(x_\lambda) = d_H(x_\lambda).$$

If $x \in H^-$, then since $A \subset H^+$,

$$d_H(x) = d_{H^+}(x) \le d_A(x),$$

and from (7.5)

$$\begin{aligned}
-b_A(x_\lambda) &= d_H(x_\lambda) \\
&= (1-\lambda)d_H(y) - \lambda d_H(x) \\
&\ge (1-\lambda)d_{\complement A}(y) - \lambda d_A(x) = -[(1-\lambda)b_A(y) + \lambda b_A(x)].
\end{aligned}$$

If, on the other hand, $x \in H^+$, then $x, y \in H^+$ and $x_\lambda \in A \subset H^+$. So from (7.4)

$$\begin{aligned}
-b_A(x_\lambda) &= d_H(x_\lambda) \\
&= (1-\lambda)d_H(y) + \lambda d_H(x) \\
&\ge (1-\lambda)d_{\complement A}(y) - \lambda d_A(x) = -[(1-\lambda)b_A(y) + \lambda b_A(x)].
\end{aligned}$$

This covers all cases and concludes the proof.

(iii) Consider two cases. For any convex set with a nonempty interior, $\operatorname{int} \overline{A} = \operatorname{int} A$ and hence $\complement\overline{\complement\overline{A}} = \complement\overline{\complement A}$. Therefore, $\overline{\complement A} = \overline{\complement A}$ and $d_{\complement \overline{A}} = d_{\complement A}$. For a convex set with an empty interior, $\overline{A} = \partial A$, $\overline{\complement A} = \mathbf{R}^N$, and $\overline{A}$ is contained in an affine subspace of $\mathbf{R}^N$ of dimension strictly less than N. Therefore, $\overline{\complement A} = \mathbf{R}^N = \overline{\complement \overline{A}}$ and $d_{\complement \overline{A}} = d_{\complement A}$. Thus in both cases $b_A = b_{\overline{A}}$ since $d_A = d_{\overline{A}}$ and $\partial A = \overline{A} \cap \overline{\complement A} = \overline{A} \cap \overline{\complement \overline{A}} = \partial \overline{A}$. Finally, since A is convex by Theorem 2.8 in Chapter 3 so is $\overline{A}$, and from part (ii) so is $b_A = b_{\overline{A}}$.

(iv) Same argument as in the proof of Theorem 6.2 (iii) of Chapter 4. $\qquad \square$

We have seen in Theorem 7.1 (iii) that for all convex subsets A of $\overline{D}$ such that $\partial A \ne \varnothing$, $b_A = b_{\overline{A}}$ and hence $\overline{A}$ is the unique closed representative in the equivalence class. We now prove that this subfamily of $C_b(D)$ is closed in $C(\overline{D})$. If we want to work with open convex subsets rather than closed subsets, it is necessary to add a constraint.

Theorem 7.2. *Let D be a nonempty open (resp., bounded open) subset of $\mathbf{R}^N$.*

(i) *The family*

$$\boxed{\mathcal{C}_b(D) \overset{\mathrm{def}}{=} \{b_A \,:\, A \subset \overline{D}, \ \partial A \ne \varnothing, \ A \ \text{convex}\}} \tag{7.6}$$

is closed in $C_{\mathrm{loc}}(D)$ (resp., compact in $C(\overline{D})$).

(ii) *Let E be a nonempty open subset of $\overline{D}$. Then the family*

$$\boxed{\mathcal{C}_b(D; E) \overset{\mathrm{def}}{=} \{b_\Omega \,:\, E \subset \Omega \subset \overline{D}, \ \partial \Omega \ne \varnothing, \ \Omega \ \text{convex and open}\}} \tag{7.7}$$

is closed in $C_{\mathrm{loc}}(D)$ (resp., compact in $C(\overline{D})$).

(iii) *The conclusions of parts* (i) *and* (ii) *remain true with* $W^{1,p}_{\text{loc}}(D)$- *(resp.,* $W^{1,p}(D)$-*) strong in place of* $C_{\text{loc}}(D)$ *(resp.,* $C(\overline{D})$*).*

(iv) *For D bounded, the families $C_d(D)$ and $C^c_d(D)$ as defined in Theorem 6.2* (ii) *of Chapter 4 and the corresponding subfamilies $C_d(D; E)$ and $C^c_d(D; E)$ are also compact in $W^{1,p}(D)$, $1 \le p < \infty$. Moreover,*

$$\boxed{\mathcal{C}_d(D) = \left\{d_A \ : \ A \subset \overline{D}, \ \partial A \neq \varnothing, \ A \text{ convex}\right\},}$$

$$\boxed{\mathcal{C}^c_d(D) = \left\{d_{\complement A} \ : \ A \subset \overline{D}, \ \partial A \neq \varnothing, \ A \text{ convex}\right\}.}$$

Proof. It is sufficient to prove the result for D bounded. Furthermore, by compactness of $C_b(D)$ in $C(\overline{D})$, it is sufficient to prove that $\mathcal{C}_b(D)$ and $\mathcal{C}_b(D; E)$ are closed in $C(\overline{D})$.

(i) Let $\{b_{A_n}\}$ be a Cauchy sequence in $\mathcal{C}_b(D)$. It converges to some $b_A \in \mathcal{C}_b(D)$. By Theorem 7.1 (iii), the b_{A_n}'s are convex and so is the limit b_A, and by Theorem 7.1 (i), $\overline{A}$ is convex, and by Theorem 7.1 (ii), $b_{\overline{A}}$ is convex. Consider the following two cases: $\text{int } \overline{A} = \varnothing$ and $\text{int } \overline{A} \neq \varnothing$.

(1) In the first case

$$\varnothing = \text{int } \overline{A} = \complement \overline{\complement \overline{A}} \quad \Rightarrow \quad \overline{\complement \overline{A}} = \mathbf{R}^N \quad \Rightarrow \quad d_{\complement \overline{A}} = 0,$$

$$A \subset \overline{A} \quad \Rightarrow \quad \complement A \supset \complement \overline{A} \quad \Rightarrow \quad d_{\complement A} \le d_{\complement \overline{A}} = 0 \quad \Rightarrow \quad d_{\complement A} = 0.$$

Therefore

$$b_A = d_A - d_{\complement A} = d_{\overline{A}} - d_{\complement \overline{A}} = b_{\overline{A}},$$

$$\overline{\complement A} \supset \overline{\complement \overline{A}} = \mathbf{R}^N \quad \Rightarrow \quad \partial A = \partial \overline{A},$$

and $b_{A_n} \to b_A = b_{\overline{A}}$, where $\overline{A}$ is convex: $b_A = b_{\overline{A}} \in \mathcal{C}_b(D)$.

(2) For the second case, $\text{int } \overline{A} \neq \varnothing$ and $\overline{A}$ convex imply that $\overline{\text{int } \overline{A}} = \overline{A}$ by Theorems 5.4 and 5.3 in Chapter 2. Consider the following two subcases: $\overline{\text{int } A} = \overline{A}$ and $\overline{\text{int } A} \neq \overline{A}$.

(2a) If $\overline{\text{int } A} = \overline{\text{int } \overline{A}} = \overline{A}$, then, using the fact that $\complement \text{int } A = \overline{\complement A}$,

$$b_{\text{int } A} = d_{\text{int } A} - d_{\complement \text{int } A} = d_A - d_{\complement A} = b_A,$$

and $\partial \, \text{int } A = \partial A \neq \varnothing$. Thus $b_{A_n} \to b_A = b_{\text{int } A}$, where $\text{int } A$ is convex by Theorem 6.2 (ii).

(2b) We show that $\overline{\text{int } A} = \overline{A}$ leads to a contradiction and cannot occur. If $\overline{\text{int } A} \neq \overline{A}$, then either $\text{int } A = \varnothing$ or $\text{int } A \neq \varnothing$. If $\text{int } A = \varnothing$, then $\partial A = \overline{A}$, and since $\text{int } \overline{A} \neq \varnothing$, there exist x and $r > 0$ such that $B(x, r) \subset \partial A$. If $\text{int } A \neq \varnothing$

$$\exists a \in \overline{A} \text{ such that } a \notin \overline{\text{int } A} \quad \Rightarrow \quad d = d_{\text{int } A}(a) > 0.$$

Since $\overline{A}$ is convex and $a \in \overline{A} \backslash \overline{\text{int } A}$, the convex hull

$$C \stackrel{\text{def}}{=} \text{co} \, \{a, \text{int } A\} - \{a\} \subset \text{int } \overline{A}$$

is open (convex) and not empty. Moreover, since $B(a,d) \cap \overline{\operatorname{int} A} = \varnothing$,

$$\varnothing \neq B(a,d) \cap C \subset \operatorname{int} \overline{A} \backslash \overline{\operatorname{int} A} \subset \partial A$$

and there exists a ball $B(x,r)$, $r > 0$, such that $B(x,r) \subset \partial A$. Thus in both cases there exist x and $r > 0$ such that $B(x,r) \subset \partial A$. We prove that this leads to a contradiction. Choose N such that for all $n \geq N$

$$d_A \geq d_{A_n} - r/8 \text{ and } d_{\complement A} \geq d_{\complement A_n} - r/8 \text{ in } \mathbf{R}^N. \tag{7.8}$$

We shall prove that for all $n \geq N$ there exists $x_n \in B(x,r)$ such that $d_{A_n}(x_n) \geq 3r/8$. Hence $x_n \in B(x,r) \subset \partial A \subset \overline{A}$ and from (7.8)

$$0 = d_A(x_n) \geq d_{A_n}(x_n) - r/8 \quad \Rightarrow \quad d_A(x_n) \leq r/8,$$

which yields a contradiction and proves that necessarily $\overline{\operatorname{int} A} = \overline{A}$, and we are back to case (2a).

Let $p_n \in \partial A_n$ be a projection of x onto ∂A_n and choose

$$\nu_n \overset{\text{def}}{=} \begin{cases} (x - p_n)/b_{A_n}(x), & x \notin \partial A_n, \\ \text{a unit exterior normal to } A_n \text{ in } x, & x \in \partial A_n, \end{cases}$$

since A_n is convex and bounded. Choose

$$x_n \overset{\text{def}}{=} x + \nu_n r/2 \quad \Rightarrow \quad |x_n - x| = r/2 \quad \Rightarrow \quad x_n \in B(x,r).$$

Let

$$H_n^+ \overset{\text{def}}{=} \left\{ y \in \mathbf{R}^N : (y - p_n) \cdot \nu_n > 0 \right\}, \; H_n^- \overset{\text{def}}{=} \left\{ y \in \mathbf{R}^N : (y - p_n) \cdot \nu_n < 0 \right\}.$$

We claim that $\overline{A_n} \subset \overline{H_n^-}$. This is obvious if $x \in \complement \overline{A_n}$. For $x \in \operatorname{int} A_n$ we can show by contradiction that if there exists $z \in H_n^+$ such that $z \in \overline{A_n}$, then $(z - p_n) \cdot \nu_n > 0$, and since $B(x, d_{\complement A_n(x)}(x)) \subset A_n$

$$\operatorname{co} \left\{ z, \overline{B(x, d_{\complement A_n(x)}(x))} \right\} \subset \overline{A_n}.$$

From this it is easy to show that $p_n \in \operatorname{int} A_n$. But this would contradict the fact that by definition $p_n \in \partial A_n$. As a result

$$x_n - p_n = x_n - x + x - p_n = \left[\frac{r}{2} + b_{A_n}(x) \right] \nu_n,$$

$$(x_n - p_n) \cdot \nu_n = \frac{r}{2} + b_{A_n}(x) \geq \frac{r}{2} - |b_{A_n}(x)| = \frac{r}{2} - \frac{r}{8} = 3\frac{r}{8}.$$

By definition of H_n^+, $x_n \in H_n^+$ and

$$d_{H_n^-}(x_n) = |x_n - p_n| = \frac{r}{2} + b_{A_n}(x) \geq \frac{r}{2} - \frac{r}{8} = 3\frac{r}{8}.$$

However, $\overline{A_n} \subset \overline{H_n^-}$ and necessarily

$$d_{A_n}(x_n) \geq d_{H_n^-}(x_n) \geq 3\frac{r}{8},$$

and this concludes the proof of part (i).

(ii) From part (i), given a Cauchy sequence $\{b_{\Omega_n}\}$ in $\mathcal{C}_b(D; E)$, there exists a convex subset A of $\overline{D}$ such that $b_{\Omega_n} \to b_A$. Moreover, since $d_{\complement\Omega_n} \to d_{\complement A}$,

$$E \subset \Omega_n \Rightarrow \complement E \supset \complement\Omega_n \Rightarrow \complement E \supset \overline{\complement A} = \complement \text{int } A \Rightarrow \varnothing \neq E \subset \text{int } A.$$

Since A is convex and int $A \neq \varnothing$, $\overline{\text{int } A} = \overline{A}$ and $b_A = b_{\text{int } A}$. Thus $\Omega = \text{int } A$ can be chosen as the limit set.

(iii) Let $\{b_{A_n}\} \subset \mathcal{C}_b(D)$ be a Cauchy sequence in $W^{1,p}(D)$. From Theorem 5.2 (ii) it is also Cauchy in $C(\overline{D})$. Hence from part (i) there exist a convex set A, $\partial A \neq \varnothing$, such that $b_{A_n} \to b_A$ in $C(\overline{D})$ and, a fortiori, in $W^{1,p}(D)$. This shows that $\mathcal{C}_b(D)$ is closed in $W^{1,p}(D)$. To show that it is compact consider a sequence $\{b_{A_n}\} \subset \mathcal{C}_b(D)$. From part (i) there exists a subsequence, still indexed by n, and a convex subset A, $\partial A \neq \varnothing$, of $\overline{D}$ such that $b_{A_n} \to b_A$ in $C(\overline{D})$. Since D is bounded it also converges in $L^p(D)$. Since $|\nabla b_{A_n}|$ is pointwise bounded by one, there exists another subsequence of $\{b_{A_n}\}$, still indexed by n, such that ∇b_{A_n} converges to ∇b_A in $W^{1,p}(D)^N$-weak. But A_n and A are convex. Thus $\text{m}(\partial A) = 0 = \text{m}(\partial A_n)$ and $|\nabla b_{A_n}| = 1 = |\nabla b_A|$ almost everywhere in $\mathbf{R}^N$. As a result, for $p = 2$

$$\int_D |\nabla b_{A_n} - \nabla b_A|^2\, dx = \int_D |\nabla b_{A_n}|^2 + |\nabla b_A|^2\, dx - 2\nabla b_{A_n} \cdot \nabla b_A\, dx$$

$$= \int_D 2 - 2\nabla b_{A_n} \cdot \nabla b_A\, dx \to 0.$$

Hence we get the compactness in $W^{1,2}(D)$-strong and by Theorem 5.1 (i) in $W^{1,p}(D)$-strong p, $1 \leq p < \infty$.

(iv) The result follows from the continuity of the map (5.5) in Theorem 5.1 (iv). $\qquad\square$

We have seen in Theorem 7.1 (iv) that in view of the convexity of the distance (resp., oriented distance) functions of a closed convex set A, closed convex sets are of locally bounded exterior curvature and of locally finite perimeter (resp., locally of bounded curvature). When A is convex and of class C^2, then for any $X \in \partial A$, there exists a strictly convex neighborhood $N(X)$ of X such that

$$b_A \in C^2(N(X)) \text{ and } \forall x \in N(X),\, \forall \xi \in \mathbf{R}^N, \quad D^2 b_A(x)\xi \cdot \xi \geq 0, \tag{7.9}$$

or, since $D^2 b_A(x)\,\nabla b_A(x) = 0$ in $N(X)$, then for all $x \in N(X)$,

$$\forall \xi \in \mathbf{R}^N \text{ such that } \xi \cdot \nabla b_A(x) = 0, \quad D^2 b_A(x)\xi \cdot \xi \geq 0. \tag{7.10}$$

This is related to the notion of *strong elliptic midsurface* in the theory of shells: there exists $c > 0$

$$\forall x \in \partial A, \forall \xi \in \mathbf{R}^N \text{ such that } \xi \cdot \nabla b_A(x) = 0, \quad D^2 b_A(x)\,\xi \cdot \xi \geq c|\xi|^2.$$

All this motivates the introduction of the following notions.

Definition 7.1.

Let A be a closed subset of $\mathbf{R}^N$ such that $\partial A \neq \varnothing$.

  (i) The set A is *locally convex* (resp., *locally strictly convex*) if for each $X \in \partial A$ there exists a strictly convex neighborhood $N(X)$ of X such that

$$b_A \text{ is convex (resp., strictly convex) in } N(X).$$

  (ii) The set A is *semiconvex* if

$$\exists \alpha \geq 0, \quad b_A(x) + \alpha |x|^2 \text{ is convex in } \mathbf{R}^N.$$

  (iii) The set A is *locally semiconvex* if for each $X \in \partial A$ there exists a strictly convex neighborhood $N(X)$ of X and

$$\exists \alpha \geq 0, \quad b_A(x) + \alpha |x|^2 \text{ is convex in } N(X). \qquad \square$$

Remark 7.1.

When a closed set A has a compact C^2 boundary, $D^2 b_A$ is bounded in a bounded neighborhood of ∂A and A is necessarily locally semiconvex. When A is a locally semiconvex set, for each $X \in \partial A$ there exists a strictly convex neighborhood $N(X)$ of X such that $\nabla b_A \in BV(N(X))^N$. If, in addition, ∂A is compact, then there exists $h > 0$ such that $\nabla b_A \in BV(U_h(\partial A))^N$. $\qquad \square$

Remark 7.2.

Given a fixed constant $\beta > 0$, consider all the subsets of D that are semiconvex with constant $0 \leq \alpha \leq \beta$. Then this set is closed for the uniform and the $W^{1,p}$-topologies, $1 \leq p < \infty$. $\qquad \square$

8 Federer's Sets of Positive Reach

The distance function d_{A_r} of the dilated set $A_r = \{x \in \mathbf{R}^N : d_A(x) \leq r\}$, $r > 0$, of A provides a uniform approximation of d_A, as can be seen from the following theorem.

Theorem 8.1. *Assume that A is a nonempty subset of $\mathbf{R}^N$. For $r > 0$*

$$\forall x \in \mathbf{R}^N, \quad 0 \leq d_A(x) - d_{A_r}(x) \leq r, \tag{8.1}$$

and, as r goes to zero,

$$\boxed{d_{A_r} \to d_A \text{ uniformly in } \mathbf{R}^N.}$$

Proof. For any $q \in A_r$, $d_A(q) \leq r$, and for $p \in \Pi_A(q)$,

$$d_A(x) \leq |p - x| \leq |p - q| + |q - x| \leq r + |q - x|$$

$$\Rightarrow d_A(x) \leq r + \inf_{q \in A_r} |q - x| = r + d_{A_r}(x). \qquad \square$$

When, in addition, A has positive reach, $\text{reach}(A) \geq r$, and $A_r \neq \mathbf{R}^N$ for some $r > 0$, the dilated sets A_r are of class $C^{1,1}$ and $b_{\overline{A}}$ can be approximated by b_{A_r} in the $W^{1,p}$-topology.

Theorem 8.2. *Let A be a nonempty subset of $\mathbf{R}^N$. Assume that there exists $h > 0$ such that*

$$d_A^2 \in C_{\mathrm{loc}}^{1,1}(U_h(A)) \quad \text{and} \quad A_h \neq \mathbf{R}^N.$$

(i) *For all r, $0 < r < h$, $\partial A_r \neq \varnothing$ and A_r is a set of class $C^{1,1}$ such that*

$$\overline{U_r(A)} = A_r, \quad \varnothing \neq \partial A_r = \{q : d_A(q) = r\},$$

and for each $x \in \partial A_r$ there exists a neighborhood $V(x)$ of x in $U_h(A) \backslash \overline{A}$ such that

$$b_{A_r} \in C^{1,1}(V(x)).$$

(ii) *Moreover, $\partial \overline{A} \neq \varnothing$, and for all r, $0 < r < h$,*

$$\boxed{b_{A_r} = b_{\overline{A}} - r \ \text{in } U_h(A) \ \text{and} \ 0 \leq b_{\overline{A}} - b_{A_r} \leq r \ \text{in } \mathbf{R}^N.}$$

In particular,

$$\nabla b_{A_r} = \nabla b_{\overline{A}} \ \text{in } U_h(A).$$

(iii) *As r goes to zero, $b_{A_r} \to b_{\overline{A}}$ uniformly in $\mathbf{R}^N$, and for $p \geq 1$,*

$$\boxed{b_{A_r} \to b_{\overline{A}} \ \text{in } W_{\mathrm{loc}}^{1,p}(U_h(A)),}$$

$$d_{A_r} \to d_{\overline{A}}, \quad d_{\complement A_r} \to d_{\complement \overline{A}}, \quad d_{\partial A_r} \to d_{\partial \overline{A}} \ \text{in } W_{\mathrm{loc}}^{1,p}(U_h(A)).$$

(iv) *Furthermore, $\overline{A}$ is of local bounded curvature and $\partial \overline{A}$ has zero volume*

$$\nabla b_{\overline{A}} \in \mathrm{BV}_{\mathrm{loc}}(\mathbf{R}^N)^N, \quad \mathrm{m}(\partial \overline{A}) = 0. \tag{8.2}$$

The proof of the theorem requires the following two lemmas.

Lemma 8.1. *Assume that A is a nonempty subset of $\mathbf{R}^N$. For $r > 0$ such that $A_r \neq \mathbf{R}^N$*

$$r \leq d_{\complement A_r}(x) + d_A(x) \quad \text{in } \mathbf{R}^N,$$
$$d_{\complement A_r}(x) \geq d_{\complement \overline{A}}(x) + r \quad \text{in } \overline{A}.$$

Proof of Lemma 8.1. (i) *First inequality.* For all y such that $d_A(y) > r$ and all $p \in \overline{A}$,

$$r \leq |y - p| \leq |y - x| + |x - p|,$$
$$r \leq \inf_{d_A(y)>r} |y - x| + \inf_{p \in A} |x - p| = d_{\complement A_r}(x) + d_A(x).$$

Second inequality. Given $x \in \bar{A}$, by definition of $d_{\complement\bar{A}}(x)$, $B_x \overset{\text{def}}{=} B(x, d_{\complement\bar{A}}(x))$ is a subset of $\bar{A}$. For $r > 0$ there exists $p_r \in \complement\bar{A}_r$ such that

$$d_{\complement A_r}(x) = |p_r - x|.$$

Define

$$p \overset{\text{def}}{=} x + d_{\complement\bar{A}}(x)\frac{p_r - x}{|p_r - x|} \in \bar{B_x} \quad \Rightarrow \quad |p - x| = d_{\complement\bar{A}}(x).$$

By construction,

$$d_{\complement A_r}(x) = |x - p_r| = |x - p| + |p - p_r| = d_{\complement\bar{A}}(x) + |p - p_r|,$$
$$|p - p_r| \geq \inf_{p \in B_x} |p' - p_r| \geq \inf_{p' \in \bar{A}} |p' - p_r| = d_A(p_r) \geq r$$

and $d_{\complement A_r}(x) \geq d_{\complement\bar{A}}(x) + r$. $\qquad\qquad\qquad\qquad\qquad\qquad\qquad\quad\square$

Lemma 8.2. *Let A be a nonempty subset of $\mathbf{R}^N$ and assume that there exists $h > 0$ such that $d_A^2 \in C_{\text{loc}}^{1,1}(U_h(A))$. For all $x \in U_h(A)\backslash\bar{A}$ and $0 < r < h$,*

$$y = p_A(x) + r\nabla d_A(x) = p_A(x) + r\frac{x - p_A(x)}{d_A(x)},$$
$$d_A(y) = r \text{ and } p_A(y) = p_A(x).$$

Proof of Lemma 8.2. In the region $U_h(A)\backslash\bar{A}$ the gradient ∇d_A is locally Lipschitz. For any point x, $0 < d_A(x) < h$, consider the flow

$$\frac{dy}{dt}(t) = \nabla d_A(y(t)), \quad y(0) = x.$$

There exists a unique local solution through x. Moreover,

$$\frac{d}{dt}d_A(y(t)) = \nabla d_A(y(t)) \cdot \frac{dy}{dt}(t) = 1 \quad \Rightarrow \quad d_A(y(t)) = d_A(x) + t.$$

Therefore, the solution exists and is unique for s, $0 \leq s < h - d_A(x)$. For any r, $d_A(x) < r < h$, let $y = y(r - d_A(x))$ and define

$$x(s) = p_A(y) + (s + d_A(x))\,\nabla d_A(y), \quad 0 \leq s \leq r - d_A(x).$$

It is readily seen that $d_A(x(s)) \leq s + d_A(x)$ and that

$$d_A(y) - d_A(x(s)) \leq |y - x(s)| = r - (s + d_A(x))$$
$$\Rightarrow \quad d_A(x(s)) \geq s + d_A(x) \quad \Rightarrow \quad d_A(x(s)) = s + d_A(x)$$
$$\Rightarrow \quad p_A(x(s)) = p_A(y) \text{ and } \nabla d_A(x(s)) = \nabla d_A(y).$$

As a result

$$\frac{dx}{ds}(s) = \nabla d_A(y) = \nabla d_A(x(s)), \quad x(r - d_A(x)) = y = y(r - d_A(x)),$$

and by uniqueness of solution $y(s) = x(s)$, $0 \le s \le r - d_A(x)$. In particular, $x(0) = y(0) = x$

$$p_A(y) = p_A(y(0)) = p_A(x) \text{ and } \nabla p_A(y) = \nabla p_A(y(0)) = \nabla p_A(x),$$

and finally

$$y(s) = p_A(x) + (s + d_A(x)) \nabla d_A(x) = x + s \nabla d_A(x)$$
$$= p_A(x) + (s + d_A(x)) \nabla d_A(x).$$

Hence for all r and x such that $0 < d_A(x) < r < h$,

$$y = p_A(x) + r \nabla d_A(x), \quad d_A(y) = r, \text{ and } p_A(y) = p_A(x).$$

From Theorem 3.2 in Chapter 4, $\nabla d_A(x) = (x - p_A(x))/d_A(x)$, and we get the equivalent form of the above identity. $\qquad\qquad \square$

Proof of Theorem 8.2. (i) First observe that for $0 < r < h$, $\varnothing \ne \overline{A} \subset A_r \subset A_h \ne \mathbf{R}^N$ implies that $\partial A_r \ne \varnothing$ and $\partial A_r = A_r \cap \overline{\complement A_r} \subset d_A^{-1}\{r\}$. By assumption $d_A^2 \in C_{\text{loc}}^{1,1}(U_h(A))$. So by Theorem 7.1 (iv) of Chapter 4 for all $x \in U_h(A)$, $\Pi_A(x)$ is a singleton. By part (i) of the same theorem $d_A \in C_{\text{loc}}^{1,1}(U_h(A)\backslash\overline{A})$, and by Theorem 3.2 (i), $|\nabla d_A(x)| = 1$ in $U_h(A)\backslash\overline{A}$. Define

$$f(x) \stackrel{\text{def}}{=} d_A(x) - r.$$

Therefore, f is Lipschitz continuous in $\mathbf{R}^N$,

$$f \in C_{\text{loc}}^{1,1}(U_h(A)\backslash\overline{A}), \quad |\nabla f(x)| = |\nabla d_A(x)| = 1 \text{ in } U_h(A)\backslash\overline{A},$$

and $U_h(A)\backslash\overline{A}$ is a neighborhood of $f^{-1}\{0\} = d_A^{-1}\{r\}$ since $d_A^{-1}\{r\} \subset U_h(A)\backslash A_{r/4} \subset U_h(A)\backslash\overline{A}$. By Theorem 4.2 of Chapter 2,

$$U_r(A) = \left\{x \in \mathbf{R}^N : d_A(x) < r\right\} = \text{int } U_r(A)$$

is a set of class $C^{1,1}$ with boundary

$$\partial U_r(A) = \left\{x \in \mathbf{R}^N : d_A(x) = r\right\}.$$

Hence by Theorem 5.3 of Chapter 2,

$$\overline{U_r(A)} = A_r \text{ and } \partial A_r = \partial U_r(A) = \left\{x \in \mathbf{R}^N : d_A(x) = r\right\}.$$

By Theorem 4.3 (i), for each $x \in \partial A_r$, there exists a neighborhood $V(x)$ of x such that $b_{A_r} \in C^{1,1}(V(x))$ and this neighborhood can be chosen in $U_h(A)\backslash\overline{A}$.

    (ii) By assumption, $\varnothing \ne \overline{A} \subset A_h \ne \mathbf{R}^N$ implies that $\partial\overline{A} \ne \varnothing$ and $b_{\overline{A}}$ is finite in $\mathbf{R}^N$. For all $x \in \mathbf{R}^N$ such that $d_A(x) \ge h$ and $0 < r < h$,

$$d_{\complement\overline{A}}(x) = d_{\complement A_r}(x) = 0,$$

and from inequality (8.1) in Theorem 8.1,

$$0 \le d_A(x) - d_{A_r}(x) \le r \quad \Rightarrow \quad \boxed{0 \le b_{\overline{A}}(x) - b_{A_r}(x) \le r \text{ in } \mathbf{R}^N \setminus U_h(A).}$$

For all $x \in \mathbf{R}^N$ such that $r < d_A(x) < h$, $d_{\complement\overline{A}}(x) = d_{\complement A_r}(x) = 0$. Moreover, from Lemma 8.2 and part (i)

$$y = p_A(x) + r\frac{x - p_A(x)}{d_A(x)} \in \partial A_r,$$

$$d_{A_r}(x) \le |x - y| = \left|x - p_A(x) - r\frac{x - p_A(x)}{d_A(x)}\right| = d_A(x) - r.$$

From inequality (8.1) in Theorem 8.1,

$$0 \le d_A(x) - d_{A_r}(x) \le r \Rightarrow d_{A_r}(x) = d_A(x) - r \Rightarrow \boxed{b_{A_r}(x) = b_A(x) - r.}$$

For all $x \in \mathbf{R}^N$ such that $0 < d_A(x) \le r$, $d_{\complement\overline{A}}(x) = 0 = d_{A_r}(x)$. From Lemma 8.2 and part (i)

$$y = p_A(x) + r\frac{x - p_A(x)}{d_A(x)} \in \partial A_r,$$

$$d_{\complement A_r}(x) \le |y - x| = r - d_A(x).$$

From the first inequality in Lemma 8.1

$$d_{\complement A_r}(x) \ge r - d_A(x) \Rightarrow d_{\complement A_r}(x) = r - d_A(x) \Rightarrow \boxed{b_{A_r}(x) = b_{\overline{A}}(x) - r.}$$

Finally, for all $x \in \mathbf{R}^N$ such that $0 = d_A(x)$, that is, $x \in \overline{A}$, there exists $p = p_{\partial \overline{A}}(x) \in \partial \overline{A}$ such that $d_{\complement\overline{A}}(x) = |p - x|$. But, by assumption, $\overline{A} \ne \varnothing$, $\complement\overline{A} \supset \complement A_h \ne \varnothing$ and, by Theorem 2.1 (i), $\partial\overline{A} \ne \varnothing$. There exists a sequence $\{y_n\} \subset \complement\overline{A}$, $0 < d_A(y_n) < h$, such that $y_n \to p$. By Lemma 8.2 and part (i), associate with each y_n

$$q_n = p_A(y_n) + r\frac{x - p_A(y_n)}{d_A(y_n)} \in \partial A_r,$$

for which $p_A(q_n) = p_A(y_n)$. By Theorem 7.1 in Chapter 4, p_A belongs to $C^{0,1}_{\text{loc}}(U_h(A))$. In particular, $p_A \in C^{0,1}(\overline{B(p, (r + h)/2)})$ and there exists $c > 0$ such that

$$|p_A(y_n) - p_A(p)| \le c\,|y_n - p|,$$
$$|q_n - p| \le |q_n - p_A(y_n)| + |p_A(y_n) - p_A(p)| \le r + c\,|y_n - p|.$$

Since $\{q_n\}$ is bounded there exists q and a subsequence, still indexed by n, such that

$$q_n \to q \quad \Rightarrow r = d_A(q_n) \to d_A(q) \quad \Rightarrow p_A(y_n) \to p_A(p) = p.$$

Therefore,

$$\boxed{\forall p \in \partial \overline{A}, \ \exists q \in \partial A_r \ \text{such that} \ p_A(q) = p.}$$

Henceforth, for each $x \in \overline{A}$ we can associate with $p \in \Pi_{\partial A}(x)$ a $q \in \partial A_r$ such that $p_A(q) = p$ and

$$d_{\complement A_r}(x) = d_{\partial A_r}(x) \le |q - x| \le |q - p| + |p - x|$$
$$\le r + d_{\partial \overline{A}}(x) = r + d_{\complement \overline{A}}(x).$$

From the second inequality in Lemma 8.1,

$$d_{\complement A_r}(x) \ge d_{\complement \overline{A}}(x) + r \quad \Rightarrow \quad d_{\complement A_r}(x) = d_{\complement \overline{A}}(x) + r,$$

and since $d_A(x) = 0 = d_{A_r}(x)$,

$$\boxed{b_{A_r}(x) = b_{\overline{A}}(x) - r.}$$

(iii) The uniform convergence of b_{A_r} to $b_{\overline{A}}$ in $\mathbf{R}^N$ is a direct consequence of part (ii). For all bounded open subsets D of $U_h(A)$ as $r \to 0$,

$$\|b_{A_r} - b_{\overline{A}}\|_{W^{1,p}(D)} = \|b_{A_r} - b_{\overline{A}}\|_{L^p(D)} \le r \, \mathrm{m}(D)^{1/p} \to 0,$$

since from part (ii), $\nabla b_{A_r} = \nabla b_{\overline{A}}$ in $U_h(A)$. The convergence of $d_{A_r} = b_{A_r}^+$, $d_{\complement A_r} = b_{A_r}^-$, and $d_{\partial A_r} = |b_{A_r}|$ to $b_{\overline{A}}^+ = d_{\overline{A}}$, $b_{\overline{A}}^- = d_{\complement \overline{A}}$, and $|b_{\overline{A}}| = d_{\partial \overline{A}}$ in $W^{1,p}_{\mathrm{loc}}(U_h(A))$ is now a consequence of Theorem 5.1 (iv).

(iv) From part (i) for all $0 < r < h$, A_r is a set of class $C^{1,1}$, and for each $x \in \partial A_r$ there exists a neighborhood $V(x)$ of x in $U_h(A) \backslash \overline{A}$ such that $b_{A_r} \in C^{1,1}(V(x))$. By Theorem 6.1, $\nabla b_{A_r} \in \mathrm{BV}_{\mathrm{loc}}(\mathbf{R}^N)^N$. Therefore, by Definition 5.1 in Chapter 3, for all $x \in \partial \overline{A}$,

$$\nabla b_{\overline{A}} = \nabla b_{A_r} \in \mathrm{BV}(B(x, h))^N,$$

since from part (ii) $\nabla b_{A_r} = \nabla b_{\overline{A}}$ in $U_h(A)$. By Theorem 6.1 (ii), $\overline{A}$ is of locally bounded curvature and, by Theorem 6.1, in $\mathrm{BV}_{\mathrm{loc}}(\mathbf{R}^N)^N$. Finally, by continuity of the map

$$b_{A_r} \mapsto \chi_{\partial A_r} = 1 - |\nabla b_{A_r}| = 0 \ : \ W^{1,p}_{\mathrm{loc}}(U_h(A)) \to L^p_{\mathrm{loc}}(U_h(A)),$$

$\chi_{\partial \overline{A}} = 0$ and $\mathrm{m}(\partial \overline{A}) = 0$. $\qquad\qquad\square$

9   Compactness Theorems for Sets of Bounded Curvature

For the family of sets with bounded curvature, the key result is the compactness of the embedding

$$\mathrm{BC}_b(D) = \{ b_A \in C_b(D) \ : \ \nabla b_A \in \mathrm{BV}(D)^N \} \to W^{1,p}(D) \tag{9.1}$$

for bounded open Lipschitzian subsets of $\mathbf{R}^N$ and p, $1 \leq p < \infty$. It is the analogue of the compactness Theorem 5.3 of Chapter 3 for Caccioppoli sets

$$\mathrm{BX}(D) = \{\chi \in \mathrm{X}(D) : \chi \in \mathrm{BV}(D)\} \rightarrow L^p(D), \tag{9.2}$$

which is a consequence of the compactness of the embedding

$$\mathrm{BV}(D) \rightarrow L^1(D) \tag{9.3}$$

for bounded open Lipschitzian subsets of $\mathbf{R}^N$ (cf. Morrey [1, Def. 3.4.1, p. 72, Thm. 3.4.4, p. 75] and Evans and Gariepy [1, Thm. 4, p. 176]).

As for characteristic functions in Chapter 3, we give a first version involving global conditions on a fixed bounded open Lipschitzian hold-all D. In the second version the sets are contained in a bounded open hold-all D with local conditions in the tubular neighborhood of their boundary.

9.1 Global Conditions on D

Theorem 9.1. *Let D be a nonempty bounded open Lipschitzian subset of $\mathbf{R}^N$. The embedding (9.1) is compact. Thus for any sequence $\{A_n\}$, $\partial A_n \neq \varnothing$, of subsets of $\overline{D}$ such that*

$$\exists c > 0, \forall n \geq 1, \quad \|D^2 b_{A_n}\|_{M^1(D)} \leq c, \tag{9.4}$$

there exist a subsequence $\{A_{n_k}\}$ and a set A, $\partial A \neq \varnothing$, such that $\nabla b_A \in \mathrm{BV}(D)^N$ and

$$b_{A_{n_k}} \rightarrow b_A \text{ in } W^{1,p}(D)\text{-strong}$$

for all p, $1 \leq p < \infty$. Moreover, for all $\varphi \in \mathcal{D}^0(D)$,

$$\lim_{n \to \infty} \langle \partial_{ij} b_{A_{n_k}}, \varphi \rangle = \langle \partial_{ij} b_A, \varphi \rangle, \quad 1 \leq i, j \leq N,$$
$$\|D^2 b_A\|_{M^1(D)} \leq c. \tag{9.5}$$

Proof. Given $c > 0$ consider the set

$$S_c \stackrel{\text{def}}{=} \left\{ b_A \in C_b(D) : \|D^2 b_A\|_{M^1(D)} \leq c \right\}.$$

By compactness of the embedding (9.3), given any sequence $\{b_{A_n}\}$ there exist a subsequence, still denoted $\{b_{A_n}\}$, and $f \in \mathrm{BV}(D)^N$ such that $\nabla b_{A_n} \rightarrow f$ in $L^1(D)^N$. But by Theorem 2.2 (ii), $C_b(D)$ is compact in $C(\overline{D})$ for bounded D and there exist another subsequence $\{b_{A_{n_k}}\}$ and $b_A \in C_b(D)$ such that $b_{A_{n_k}} \rightarrow b_A$ in $C(\overline{D})$ and, a fortiori, in $L^1(D)$. Therefore, $b_{A_{n_k}}$ converges in $W^{1,1}(D)$ and also in $L^1(D)$. By uniqueness of the limit, $f = \nabla b_A$ and $b_{A_{n_k}}$ converges in $W^{1,1}(D)$ to b_A. For $\Phi \in \mathcal{D}^1(D)^{N \times N}$ as k goes to infinity

$$\int_D \nabla b_{A_{n_k}} \cdot \overrightarrow{\mathrm{div}}\, \Phi \, dx \rightarrow \int_D \nabla b_A \cdot \overrightarrow{\mathrm{div}}\, \Phi \, dx$$

$$\Rightarrow \left| \int_D \nabla b_A \cdot \overrightarrow{\mathrm{div}}\, \Phi \, dx \right| = \lim_{k \to \infty} \left| \int_D \nabla b_{A_{n_k}} \cdot \overrightarrow{\mathrm{div}}\, \Phi \, dx \right| \leq c \|\Phi\|_{C(D)},$$

$\|D^2 b_A\|_{M^1(D)} \leq c$, and $\nabla b_A \in \mathrm{BV}(D)^N$. This proves the sequential compactness of the embedding (9.1) for $p = 1$ and properties (9.5). The conclusions remain true for $p \geq 1$ by the equivalence of the $W^{1,p}$-topologies on $C_b(D)$ in Theorem 5.1 (i). $\qquad \square$

9.2 Local Conditions in Tubular Neighborhoods

The global condition (9.4) is now weakened to a local one in a neighborhood of each set of the sequence. Simultaneously, the Lipschitzian condition on D is removed since only the uniform boundedness of the sets of the sequence is required.

Theorem 9.2. *Let D be a nonempty bounded open subset of $\mathbf{R}^N$ and $\{A_n\}$, $\varnothing \neq \partial A_n$, be a sequence of subsets of $\overline{D}$. Assume that there exist $h > 0$ and $c > 0$ such that*

$$\forall n, \quad \|D^2 b_{A_n}\|_{M^1(U_h(\partial A_n))} \leq c. \tag{9.6}$$

Then there exist a subsequence $\{A_{n_k}\}$ and a subset A, $\varnothing \neq \partial A$, of $\overline{D}$ such that $\nabla b_A \in \mathrm{BV}_{\mathrm{loc}}(\mathbf{R}^N)^N$, and for all p, $1 \leq p < \infty$,

$$b_{A_{n_k}} \to b_A \text{ in } W^{1,p}(U_h(D))\text{-strong}. \tag{9.7}$$

Moreover, for all $\varphi \in \mathcal{D}^0(U_h(\partial A))$,

$$\lim_{k \to \infty} \langle \partial_{ij} b_{A_{n_k}}, \varphi \rangle = \langle \partial_{ij} b_A, \varphi \rangle, \quad 1 \leq i, j \leq N,$$
$$\|D^2 b_A\|_{M^1(U_h(\partial A))} \leq c, \tag{9.8}$$

and $\chi_{\partial A}$ belongs to $\mathrm{BV}_{\mathrm{loc}}(\mathbf{R}^N)$.

Proof. (i) By assumption, $A_n \subset \overline{D}$ implies that $U_h(A_n) \subset U_h(D)$. Since $U_h(D)$ is bounded, there exist a subsequence, still indexed by n, and a subset A of $\overline{D}$, $\partial A \neq \varnothing$, such that

$$b_{A_n} \to b_A \quad \text{in } C(\overline{U_h(D)})\text{-strong}$$

and another subsequence, still indexed by n, such that

$$b_{A_n} \to b_A \quad \text{in } H^1(U_h(D))\text{-weak}.$$

For all $\varepsilon > 0$, $0 < 3\varepsilon < h$, there exists $N > 0$ such that for all $n \geq N$ and all $x \in \overline{U_h(D)}$,

$$d_{\partial A_n}(x) \leq d_{\partial A}(x) + \varepsilon, \quad d_{\partial A}(x) \leq d_{\partial A_n}(x) + \varepsilon. \tag{9.9}$$

Therefore,

$$\partial A_n \subset U_{h-2\varepsilon}(\partial A_n) \subset U_{h-\varepsilon}(\partial A) \subset U_h(\partial A_n), \tag{9.10}$$
$$\complement U_{h-\varepsilon}(\partial A) \subset \complement U_{h-2\varepsilon}(\partial A_n) \subset \complement \partial A_n. \tag{9.11}$$

From (9.6) and (9.10)

$$\forall n \geq N, \quad \|D^2 b_{A_n}\|_{M^1(U_{h-\varepsilon}(\partial A))} \leq c.$$

In order to use the compactness of the embedding (9.3) as in the proof of Theorem 9.1, we would need $U_{h-\varepsilon}(\partial A)$ to be Lipschitzian. To get around this we construct a bounded Lipschitzian set between $U_{h-2\varepsilon}(\partial A)$ and $U_{h-\varepsilon}(\partial A)$. Indeed by definition,

$$U_{h-\varepsilon}(\partial A) = \cup_{x \in \partial A} B(x, h - \varepsilon) \text{ and } \overline{U_{h-2\varepsilon}(\partial A)} \subset U_{h-\varepsilon}(\partial A),$$

and by compactness there exists a finite sequence of points $\{x_i\}_{i=1}^n$ in ∂A such that

$$\overline{U_{h-2\varepsilon}(\partial A)} \subset U_B \stackrel{\text{def}}{=} \cup_{i=1}^n B(x_i, h - \varepsilon) \subset U_h(D).$$

Since U_B is Lipschitzian as the union of a finite number of balls, it now follows by compactness of the embedding (9.3) for U_B that there exists a subsequence, still denoted $\{b_{A_n}\}$, and $f \in \mathrm{BV}(U_B)^N$ such that $\nabla b_{A_n} \to f$ in $L^1(U_B)^N$. Since $U_h(D)$ is bounded, $C_b(U_h(D))$ is compact in $C(\overline{U_h(D)})$ and there exists another subsequence, still denoted $\{b_{A_n}\}$, and $A \subset \overline{D}$, $\partial A \neq \varnothing$, such that $b_{A_n} \to b_A$ in $C(\overline{U_h(D)})$ and, a fortiori, in $L^1(U_h(D))$. Therefore, b_{A_n} converges in $W^{1,1}(U_B)$ and also in $L^1(U_B)$. By uniqueness of the limit, $f = \nabla b_A$ on U_B and b_{A_n} converges to b_A in $W^{1,1}(U_B)$. By Definition 6.1 and Theorem 6.3 (iii), ∇b_A and ∇b_{A_n} all belong to $\mathrm{BV}_{\mathrm{loc}}(\mathbf{R}^N)^N$ since they are of bounded variation in tubular neighborhoods of their respective boundaries. Moreover, by Theorem 6.2 (ii), $\chi_{\partial A} \in \mathrm{BV}_{\mathrm{loc}}(\mathbf{R}^N)$. The above conclusions also hold for the subset $U_{h-2\varepsilon}(\partial A)$ of U_B.

(ii) *Convergence in $W^{1,p}(U_h(D))$.* Consider the integral

$$\int_{U_h(D)} |\nabla b_{A_n} - \nabla b_A|^2 \, dx$$

$$= \int_{U_{h-2\varepsilon}(\partial A)} |\nabla b_{A_n} - \nabla b_A|^2 \, dx + \int_{U_h(D) \setminus U_{h-2\varepsilon}(\partial A)} |\nabla b_{A_n} - \nabla b_A|^2 \, dx.$$

From part (i) the first integral on the right-hand side converges to zero as n goes to infinity. The second integral is on a subset of $\complement U_{h-2\varepsilon}(\partial A)$. From (9.11) for all $n \geq N$,

$$|\nabla b_{A_n}(x)| = 1 \text{ a.e. in } \complement \partial A_n \supset \complement U_{h-3\varepsilon}(\partial A_n) \supset \complement U_{h-2\varepsilon}(\partial A),$$
$$|\nabla b_A(x)| = 1 \text{ a.e. in } \complement \partial A \supset \complement U_{h-2\varepsilon}(\partial A).$$

The second integral reduces to

$$\int_{U_h(D) \setminus U_{h-2\varepsilon}(\partial A)} |\nabla b_{A_n} - \nabla b_A|^2 \, dx = \int_{U_h(D) \setminus U_{h-2\varepsilon}(\partial A)} 2 \left(1 - \nabla b_{A_n} \cdot \nabla b_A\right) dx,$$

which converges to zero by weak convergence of ∇b_{A_n} to ∇b_A in the space $L^2(U_h(D))^N$ in part (i) and the fact that $|\nabla b_A| = 1$ almost everywhere in $U_h(D) \setminus U_{h-2\varepsilon}(\partial A)$. Therefore, since $b_{A_n} \to b_A$ in $C(\overline{U_h(D)})$

$$b_{A_n} \to b_A \text{ in } H^1(U_h(D))\text{-strong,}$$

and by Theorem 5.1 (i) the convergence is true in $W^{1,p}(U_h(D))$ for all $p \geq 1$.

(iii) *Properties* (9.8). Consider the initial subsequence $\{b_{A_n}\}$ which converges to b_A in $H^1(U_h(D))$-weak constructed at the beginning of part (i). This sequence is independent of ε and the subsequent constructions of other subsequences. By convergence of b_{A_n} to b_A in $H^1(U_h(D))$-weak for each $\Phi \in \mathcal{D}^1(U_h(\partial A))^{N \times N}$,

$$\lim_{n \to \infty} \int_{U_h(\partial A)} \nabla b_{A_n} \cdot \overrightarrow{\operatorname{div}} \Phi \, dx = \int_{U_h(\partial A)} \nabla b_A \cdot \overrightarrow{\operatorname{div}} \Phi \, dx.$$

Each such Φ has compact support in $U_h(\partial A)$, and there exists $\varepsilon = \varepsilon(\Phi) > 0$, $0 < 3\varepsilon < h$, such that

$$\overline{\operatorname{supp} \Phi} \subset U_{h-2\varepsilon}(\partial A).$$

From part (ii) there exists $N(\varepsilon) > 0$ such that

$$\forall n \geq N(\varepsilon), \quad U_{h-2\varepsilon}(\partial A_n) \subset U_{h-\varepsilon}(\partial A) \subset U_h(\partial A_n).$$

For $n \geq N(\varepsilon)$ consider the integral

$$\int_{U_h(\partial A)} \nabla b_{A_n} \cdot \overrightarrow{\operatorname{div}} \Phi \, dx = \int_{U_{h-2\varepsilon}(\partial A)} \nabla b_{A_n} \cdot \overrightarrow{\operatorname{div}} \Phi \, dx = \int_{U_h(\partial A_n)} \nabla b_{A_n} \cdot \overrightarrow{\operatorname{div}} \Phi \, dx$$

$$\Rightarrow \left| \int_{U_h(\partial A)} \nabla b_{A_n} \cdot \overrightarrow{\operatorname{div}} \Phi \, dx \right| \leq \|D^2 b_{A_n}\|_{M^1(U_h(\partial A_n))} \|\Phi\|_{C(U_h(\partial A_n))}$$

$$\leq c \|\Phi\|_{C(U_{h-2\varepsilon}(\partial A))} = c \|\Phi\|_{C(U_h(\partial A))}.$$

By convergence of ∇b_{A_n} to ∇b_A in the space $L^2(U_h(D))$-weak, then for all $\Phi \in \mathcal{D}^1(U_h(\partial A))^{N \times N}$

$$\left| \int_{U_h(\partial A)} \nabla b_A \cdot \overrightarrow{\operatorname{div}} \Phi \, dx \right| \leq c \|\Phi\|_{C(U_h(\partial A))} \quad \Rightarrow \quad \|D^2 b_A\|_{M^1(U_h(\partial A))} \leq c.$$

Finally the convergence remains true for all subsequences constructed in parts (i) and (ii). This completes the proof. $\qquad\square$

10   Compactness and Uniform Cone Property

In Theorem 5.9 of section 5.4 in Chapter 3, we have seen a compactness theorem for the family of subsets of a bounded hold-all D satisfying the uniform cone property. In this section we give a direct proof of the compactness for the $C(\overline{D})$- and $W^{1,p}(D)$-topologies associated with the oriented distance function b_Ω. As a consequence we get the compactness for the $C(\overline{D})$- and $W^{1,p}(D)$-topologies associated with d_Ω and $d_{\complement\Omega}$. Furthermore, we recover the compactness of Theorem 5.9 of Chapter 3 for χ_Ω in $L^p(D)$. Finally we also get the compactness for $\chi_{\complement\Omega}$ in $L^p(D)$.

Recall the following notation and definition

$$L(D, r, \omega, \lambda) \overset{\text{def}}{=} \left\{ \Omega \subset \overline{D} : \begin{array}{l} \Omega \text{ satisfies the uniform} \\ \text{cone property for } (r, \omega, \lambda) \end{array} \right\}.$$

We have seen in Theorem 6.1 of Chapter 2 that for a Lipschitzian set Ω, $\partial\Omega \neq \varnothing$, $\mathrm{m}(\partial\Omega) = 0$, $\mathrm{int}\,\Omega \neq \varnothing$, $\mathrm{int}\,\complement\Omega \neq \varnothing$. Moreover, if the properties of Definition 6.1 of Chapter 2 are satisfied for a set Ω, they are satisfied for all sets in the equivalence class

$$[\Omega]_b = \left\{\Omega' : \overline{\Omega'} = \overline{\Omega} \text{ and } \partial\Omega' = \partial\Omega\right\}.$$

Theorem 10.1. *Let D be a nonempty bounded open subset of $\mathbf{R}^N$ and $1 \leq p < \infty$. For $r > 0$, $\omega > 0$, and $\lambda > 0$, the family*

$$B(D, r, \omega, \lambda) \stackrel{\text{def}}{=} \{b_\Omega : \forall\Omega \in L(D, r, \omega, \lambda)\}$$

is compact in $C(\overline{D})$ and $W^{1,p}(D)$. As a consequence the families

$$B_d(D, r, \omega, \lambda) \stackrel{\text{def}}{=} \{d_\Omega : \forall\Omega \in L(D, r, \omega, \lambda)\},$$

$$B_d^c(D, r, \omega, \lambda) \stackrel{\text{def}}{=} \{d_{\complement\Omega} : \forall\Omega \in L(D, r, \omega, \lambda)\},$$

$$B_d^\partial(D, r, \omega, \lambda) \stackrel{\text{def}}{=} \{d_{\partial\Omega} : \forall\Omega \in L(D, r, \omega, \lambda)\}$$

are compact in $C(\overline{D})$ and $W^{1,p}(D)$, and the families

$$X(D, r, \omega, \lambda) \stackrel{\text{def}}{=} \{\chi_\Omega : \forall\Omega \in L(D, r, \omega, \lambda)\},$$

$$X^c(D, r, \omega, \lambda) \stackrel{\text{def}}{=} \{\chi_{\complement\Omega} : \forall\Omega \in L(D, r, \omega, \lambda)\}$$

are compact in $L^p(D)$.

The proof of this theorem is similar to the proof of Theorem 5.9 in Chapter 3 and uses a key lemma.

Proof of Theorem 10.1. (i) *Compactness in $C(\overline{D})$.* Consider an arbitrary sequence $\{\Omega_n\}$ in $L(D, r, \omega, \lambda)$. For $\overline{D}$ compact $C_b(D)$ is compact in $C(\overline{D})$ and there exists $\Omega \subset \overline{D}$ and a subsequence $\{\Omega_{n_k}\}$ such that

$$b_{\Omega_{n_k}} \to b_\Omega \text{ in } C(\overline{D}).$$

It remains to prove that $\Omega \in L(D, r, \omega, \lambda)$. This requires the equivalent of Lemma 5.1 in Chapter 3.

Lemma 10.1. *Given a sequence $\{b_{\Omega_n}\} \subset C_b(D)$ such that $b_{\Omega_n} \to b_\Omega$ in $C(\overline{D})$ for some $b_\Omega \in C_b(D)$, we have the following properties:*

$$\forall x \in \overline{\Omega}, \quad \forall R > 0, \quad \exists N(x, R) > 0, \quad \forall n \geq N(x, R), \quad B(x, R) \cap \Omega_n \neq \varnothing,$$

and for all $x \in \overline{\complement\Omega}$,

$$\forall R > 0, \quad \exists N(x, R) > 0, \quad \forall n \geq N(x, R), \quad B(x, R) \cap \complement\Omega_n \neq \varnothing. \tag{10.1}$$

Moreover,

$$\forall x \in \partial\Omega, \quad \forall R > 0, \quad \exists N(x, R) > 0, \quad \forall n \geq N(x, R),$$
$$B(x, R) \cap \Omega_n \neq \emptyset \quad and \quad B(x, R) \cap \complement\Omega_n) \neq \emptyset,$$

and

$$B(x, R) \cap \partial\Omega_n \neq \emptyset.$$

Proof. We proceed by contradiction. Assume that

$$\exists x \in \overline{\Omega}, \quad \exists R > 0, \quad \forall N > 0, \quad \exists n \geq N, \quad B(x, R) \cap \Omega_n = \emptyset.$$

So there exists a subsequence $\{\Omega_{n_k}\}$, $n_k \to \infty$, such that

$$b_{\Omega_{n_k}}(x) = d_{\Omega_{n_k}}(x) \geq R \neq 0 = d_\Omega(x) \geq b_\Omega(x),$$

which contradicts the fact that $b_{\Omega_{n_k}} \to b_\Omega$. Of course, the same assertion is true for the complements and for all $x \in \overline{\complement\Omega}$,

$$\forall R > 0, \quad \exists N(x, R) > 0, \quad \forall n \geq N(x, R), \quad B(x, R) \cap \complement\Omega_n \neq \emptyset,$$

when $x \in \partial\Omega$, $x \in \overline{\Omega} \cap \overline{\complement\Omega}$, and we combine the two results. For the last result use the fact that the open ball cannot be partitioned into two nonempty disjoint open subsets. $\square$

Coming back to the proof of the theorem, we wish to show that Ω has the uniform cone property

$$\forall x \in \partial\Omega, \quad \exists d, |d| = 1, \quad \forall y \in \overline{\Omega} \cap B(x, r), \quad C_y(\lambda, \omega, d) \subset \operatorname{int}\Omega.$$

Since Ω_n is Lipschitzian, so is $\complement\Omega_n$, and from the second part of the lemma for each $x \in \partial\Omega$, $\forall k \geq 1$, $\exists n_k \geq k$ such that

$$B\left(x, \frac{r}{2^k}\right) \cap \partial\Omega_{n_k} \neq \emptyset.$$

Denote by x_k an element of that intersection:

$$\forall k \geq 1, \quad x_k \in B\left(x, \frac{r}{2^k}\right) \cap \partial\Omega_{n_k}.$$

By construction $x_k \to x$. Next consider $y \in B(x, r) \cap \overline{\Omega}$. From the first part of the lemma, there exists a subsequence of $\{\Omega_{n_k}\}$, still denoted $\{\Omega_{n_k}\}$, such that

$$\forall k \geq 1, \quad B\left(y, \frac{r}{2^k}\right) \cap \Omega_{n_k} \neq \emptyset.$$

For each $k \geq 1$ denote by y_k a point of that intersection. By construction

$$y_k \in \overline{\Omega}_{n_k} \to y \in \overline{\Omega} \cap B(x, r).$$

There exists $K > 0$ large enough such that

$$\forall k \geq K, \quad y_k \in B(x_k, r).$$

To see this, note that $y \in B(x, r)$ and that

$$\exists \rho > 0, \quad B(y, \rho) \subset B(x, r) \quad \text{and} \quad |y - x| + \frac{\rho}{2} < r.$$

Now

$$|y_k - x_k| \leq |y_k - y| + |y - x| + |x - x_k|$$
$$\leq \frac{r}{2^k} + r - \frac{\rho}{2} + \frac{r}{2^k} \leq r + \left[\frac{r}{2^{k-1}} - \frac{\rho}{2}\right] < r.$$

Since $r/\rho > 1$ the result is true for

$$\frac{r}{2^{k-1}} - \frac{\rho}{2} < 0 \quad \Rightarrow \quad k > 2 + \log(r/\rho).$$

So we have constructed a subsequence $\{\Omega_{n_k}\}$ such that for $k \geq K$

$$x_k \in \partial\Omega_{n_k} \to x \in \partial\Omega,$$
$$y_k \in \overline{\Omega}_{n_k} \cap B(x_k, r) \to y \in \overline{\Omega} \cap B(x, r).$$

For each k, $\exists d_k \in \mathbf{R}^N$, $|d_k| = 1$, such that

$$C_{y_k}(\lambda, \omega, d_k) \subset \text{int } \Omega_{n_k}.$$

Pick another subsequence of $\{\Omega_{n_k}\}$, still denoted $\{\Omega_{n_k}\}$, such that

$$\exists d \in \mathbf{R}^N, \quad |d| = 1, \quad d_k \to d.$$

Now consider $z \in C_y(\lambda, \omega, d)$. Since z is an interior point

$$\exists \rho > 0, \quad B(z, \rho) \subset C_y(\lambda, \omega, d),$$

and there exists $K' \geq K$ such that

$$\forall k \geq K', \quad B(z, \rho/2) \subset C_{y_k}(\lambda, \omega, d_k) \subset \text{int } \Omega_{n_k} = \overline{\complement\complement\Omega_{n_k}},$$
$$\complement B(z, \rho/2) \supset \overline{\complement\Omega_{n_k}} \quad \Rightarrow \quad 0 < \rho/2 = d_{\complement B(z, \rho/2)}(z) \leq d_{\complement\Omega_{n_k}}(z) \to d_{\complement\Omega}(z)$$
$$\Rightarrow \quad 0 < d_{\complement\Omega}(z) \quad \Rightarrow \quad z \in \overline{\complement\complement\Omega} = \text{int } \Omega \quad \Rightarrow \quad C_y(\lambda, \omega, d) \subset \text{int } \Omega.$$

This proves that $\Omega \subset \overline{D}$ satisfies the uniform cone property and $\Omega \in L(D, \lambda, \omega, r)$.

(ii) *Compactness in $W^{1,p}(D)$.* From Theorem 5.1 (i) it is sufficient to prove the result for $p = 2$. Consider the subsequence $\{\Omega_{n_k}\} \subset L(D, \lambda, \omega, r)$ and let $\Omega \in L(D, \lambda, \omega, r)$ be the set previously constructed such as $b_{\Omega_{n_k}} \to b_\Omega$ in $C(\overline{D})$. Hence $b_{\Omega_{n_k}} \to b_\Omega$ in $L^2(D)$. Since $\overline{D}$ is compact for all $\Omega \subset \overline{D}$

$$\int_D |b_{\Omega_{n_k}}|^2 \, dx \leq \int_D \text{diam}\,(D)^2 \, dx \leq \text{diam}\,(D)^2 \text{m}(D),$$

$$\int_D |\nabla b_{\Omega_{n_k}}|^2 \, dx \leq \int_D dx = \text{m}(D),$$

and there exists a subsequence, still denoted $\{b_{\Omega_{n_k}}\}$, which converges weakly to b_Ω. Since all the sets are Lipschitzian, $\mathrm{m}(\partial\Omega_{n_k}) = 0 = \mathrm{m}(\partial\Omega)$ and $|\nabla b_\Omega| = 1 = |\nabla b_{\Omega_{n_k}}|$ almost everywhere in D, and

$$\int_D |\nabla b_{\Omega_{n_k}} - \nabla b_\Omega|^2 \, dx = \int_D |\nabla b_{\Omega_{n_k}}|^2 + |\nabla b_\Omega|^2 - 2\nabla b_{\Omega_{n_k}} \cdot \nabla b_\Omega \, dx$$

$$= 2 \int_D (1 - \nabla b_{\Omega_{n_k}} \cdot \nabla b_\Omega) \, dx \to 2 \int_D (1 - |\nabla b_\Omega|^2) \, dx = 0.$$

Therefore $b_{\Omega_{n_k}} \to b_\Omega$ in $W^{1,2}(D)$-strong, since the convergence was already established in $L^2(D)$-strong.

(iii) The compactness of the other families follows from the continuity of the maps

$$b_\Omega \mapsto (b_\Omega^+, b_\Omega^-, |b_\Omega|) = (d_\Omega, d_{\complement\Omega}, d_{\partial\Omega}) \; : \; C_b(D) \subset C(\overline{D}) \to C(\overline{D})^3$$

in Theorem 2.2 (iii) and

$$b_\Omega \mapsto (b_\Omega^+, b_\Omega^-, |b_\Omega|) = (d_\Omega, d_{\complement\Omega}, d_{\partial\Omega}) \; : \; C_b(D) \subset W^{1,p}(D) \to W^{1,p}(D)^3,$$

$$b_\Omega \mapsto (\chi_{\partial\Omega}, \chi_{\mathrm{int}\,\Omega}, \chi_{\mathrm{int}\,\complement\Omega}) : W^{1,p}(D) \to L^p(D)^3$$

in Theorem 5.1 (iv) and (v) and the fact that $\mathrm{m}(\partial\Omega) = 0$ implies $\chi_{\mathrm{int}\,\Omega} = \chi_\Omega$ and $\chi_{\mathrm{int}\,\complement\Omega} = \chi_{\complement\Omega}$ almost everywhere for $\Omega \in L(D, \lambda, \omega, r)$. $\square$

11   Compactness and Uniform Cusp Property

The compactness theorem, Theorem 10.1, is no longer true when the uniform cone property is replaced by a uniform segment property, that is, when the cone $C(\lambda, \omega, d)$ is replaced by the segment $(0, \lambda d)$. This is readily seen by considering the following example.

Example 11.1.
Given an integer $n \geq 1$, consider the following sequence of open domains in $\mathbf{R}^2$:

$$\Omega_n \stackrel{\text{def}}{=} \left\{ (x, y) \in \mathbf{R}^2 \; : \; |x| < 1 \text{ and } |x|^{1/n} < y < 2 \right\}.$$

They satisfy the uniform segment property of Definition 7.1 (ii) of Chapter 2 by choosing $\lambda = r = 1/4$. The sequence $\{\overline{\Omega_n}\}$ converges to the closed set

$$A \stackrel{\text{def}}{=} \left\{ (x, y) \in \mathbf{R}^2 \; : \; |x| \leq 1 \text{ and } 1 \leq y \leq 2 \right\} \cup L$$

$$L \stackrel{\text{def}}{=} \left\{ (0, y) \in \mathbf{R}^2 \; : \; 0 \leq y \leq 1 \right\}$$

in the uniform topologies associated with d_{Ω_n} and b_{Ω_n} or in the L^p-topologies associated with χ_{Ω_n} and $\chi_{\complement\Omega_n}$. However, the segment property is not satisfied along the line L and the corresponding family of subsets of the hold-all $D = B(0, 4)$ satisfying the uniform segment property with $r = \lambda = 1/4$ is not closed and, a fortiori, not compact. $\square$

This example shows that a uniform segment property is too *meager* to make the corresponding family compact. Looking back at the proof of Theorem 10.1, everything goes through with the uniform segment property except in the last seven lines of the proof of part (i), where the fact that the cone is an open set is critically used. This suggests that the cone could be replaced by an open *cusp* or *horn* that would yield larger families than those of Lipschitzian domains considered in the previous section. For instance, the cone $C(\lambda, \omega, d)$ can be replaced by the cuspidal region

$$H(\lambda, h, d) \stackrel{\text{def}}{=} \left\{ y \in \mathbf{R}^N : P_H(y) \in B_H(0, \rho) \text{ and } h(|P_H(y)|) < y \cdot d < \lambda \right\}, \quad (11.1)$$

where $B_H(0, \rho)$ is the open ball of radius $\rho > 0$ and center 0 in the hyperplane H through 0 orthogonal to the direction d and $h : [0, \rho] \to \mathbf{R}$ is a continuous function such that

$$h(0) = 0, \quad h(\rho) = \lambda, \quad \forall \theta, \, 0 < \theta < \rho, \quad 0 < h(\theta) < \lambda. \quad (11.2)$$

This will be referred to as a *uniform cusp property*. Hence, we can now choose functions $h : [0, \rho] \to \mathbf{R}$ of the form

$$h(\theta) = \lambda \, (\theta/\rho)^\alpha, \quad 0 < \alpha \le 1.$$

The special case of the cone corresponds to

$$\alpha = 1, \, \rho = \lambda \tan \omega \text{ and } h(\theta) = \theta/\tan \omega.$$

The following theorem is now a corollary to Theorem 10.1.

Theorem 11.1. *Let the assumptions of Theorem* 10.1 *be verified with the cone* $C(\lambda, \omega, d)$ *replaced by the cuspidal region* $H(\lambda, h, d)$ *defined in* (11.1) *under the conditions* (11.2) *on the continuous function h. Then the compactness properties of Theorem* 10.1 *remain true.*

In some applications, it might be interesting to relax the uniform cusp property by permitting the axis of the cuspidal region to bend: this makes the region look like a horn and the corresponding property becomes a *horn condition* or *property*. Horn-shaped domains have been studied in several contexts in the literature. In particular, conditions on domains have been introduced in the context of extension operators and embedding theorems: the domains of F. John;[2] the (ε, δ)-domains of P. W. John;[3] and domains satisfying a *flexible horn condition* (which is a broader notion than the previous two) by O. V. Besov.[4]

---

[2] F. John, *Rotation and strain*, Comm. Pure Appl. Math. **14** (1961), 391–413.

[3] P. W. Jones, *Quasiconformal mappings and extendability of functions in Sobolev spaces*, Acta Math. **147** (1981), 71–88.

[4] O. V. Besov, *Integral representations of functions in a domain with the flexible horn condition, and embedding theorems* (Russian), Dokl. Akad. Nauk SSSR **273** (1983), 1294–1297; English translation: Soviet Math. Dokl. **28** (1983), 769–772. O. V. Besov, *Embeddings of an anisotropic Sobolev space for a domain with a flexible horn condition* (Russian), in "Studies in the Theory of Differentiable Functions of Several Variables and Its Applications, XII," Trudy Mat. Inst. Steklov **181** (1988), 3–14; 269; English translation: Proc. Steklov Inst. Math. (1989), 1–13.

Chapter 6
Optimization of Shape Functions

1 Introduction and Generic Examples

In Chapter 3 we have seen several examples of optimization problems involving the L^p-topology on measurable sets via the characteristic function. In this chapter we consider optimization problems where the underlying topology is specified by the distance functions of Chapters 4 and 5. We give a brief account of two generic optimization problems: the minimization of a quadratic objective function which depends on the solution of the homogeneous Dirichlet boundary value problem associated with an elliptic operator, and the optimization of the first eigenvalue of the same operator.

In both problems the strong continuity of solutions of the elliptic equation or the eigenvector equation with respect to the underlying domain is the key element in the proof of the existence of optimal domains. To get that continuity, some extra conditions have to be imposed on the family of open domains.

Those questions have received a lot of attention for the Laplace equation with homogeneous Dirichlet boundary conditions. To a sequence of domains in a fixed hold-all D, associate a sequence of extensions by zero of the solutions in the fixed space $H_0^1(D)$. The classical Poincaré inequality is uniform for that sequence as the first eigenvalue of the Laplace equation in each domain is dominated by the one associated with the larger hold-all D. By a classical compactness argument, the sequence of extensions converges to some limiting element y in $H_0^1(D)$.

To complete the proof of the continuity, two more fundamental questions remain. Is y the solution of the Laplace equation for the limit domain? Does y satisfy the Dirichlet boundary condition on the boundary of the limit domain? The first question can be resolved by assuming that the Sobolev spaces associated with the moving domains converges in the Kuratowski sense, i.e., with the property that any element in the Sobolev space associated with the limit domain can be approached by a sequence of elements in the moving Sobolev spaces. That property is obtained in examples where the *compactivorous property*[1] follows from the choice of

---

[1] A sequence of open sets converging to a limit open set in some topology has the compactivorous property if any compact subset of the limit set is contained in all sets of the sequence after a certain rank: the sequence *eats up* the compact set after a certain rank.

259

the definition of convergence of the domains. If the domains are open subsets of D, then the complementary Hausdorff topology has that property. For the second question it is necessary to impose constraints at least on the limit domain. The stability condition introduced by Rauch and Taylor [1] and used by Dancer [1] and Daners [1] precisely assumes that the limit domain is sufficiently smooth, just enough to have the solution in $H_0^1(\Omega)$. Of course assuming such a regularity on the limit domain makes things easier. The more fundamental issue is to identify the families of domains for which this stability property is preserved in the limit.

The uniform cone property has been used in the context of shape optimization by Chenais [1, 2] in 1973. Capacity conditions have been introduced in 1994 by Bucur and Zolésio [3, 5, 6, 8, 9] in order to construct compact subfamilies of domains with respect to the Hausdorff complementary topology (see section 2.3 of Chapter 4). Furthermore, Bucur [1] proved that the condition given in 1993 by Šverák [1, 2] in dimension 2 involving a bound on the number of connected components of the complement of the domain can be recovered from the more general capacity conditions that are sufficient in the case of the Laplacian with homogeneous Dirichlet boundary conditions. In a more recent paper, Bucur [5] proved that they are *almost necessary*. Intuitively those capacity conditions are such that, locally, the complement of the domains in the family under consideration has *enough capacity* to preserve and retain the homogeneous Dirichlet boundary condition in the limit.

As a first example consider the minimization of the objective function

$$J(\Omega) \stackrel{\text{def}}{=} \frac{1}{2} \int_\Omega |u_\Omega - g|^2 \, dx \tag{1.1}$$

over the family of open subsets Ω of a bounded open hold-all D of $\mathbf{R}^N$, where g in $L^2(D)$, $u_\Omega \in H_0^1(\Omega)$ is the solution of the variational problem

$$\forall \varphi \in H_0^1(\Omega), \quad \int_\Omega A\nabla u_\Omega \cdot \nabla \varphi \, dx = \langle f|_\Omega, \varphi \rangle_{H^{-1}(\Omega) \times H_0^1(\Omega)}, \tag{1.2}$$

$A \in L^\infty(D; \mathcal{L}(\mathbf{R}^N, \mathbf{R}^N))$ is a matrix function on D such that

$$^*A = A \text{ and } \alpha I \leq A \leq \beta I \tag{1.3}$$

for some coercivity and continuity constants $0 < \alpha \leq \beta$, and $f|_\Omega$ denotes the restriction of f in $H^{-1}(D)$ to $H^{-1}(\Omega)$. The second example is the optimization of the first eigenvalue of the associated differential operator

$$\left.\begin{array}{c} \sup\limits_{\Omega \text{ open } \subset D} \lambda^A(\Omega) \\[2ex] \inf\limits_{\Omega \text{ open } \subset D} \lambda^A(\Omega) \end{array}\right| \quad \lambda^A(\Omega) \stackrel{\text{def}}{=} \inf_{0 \neq \varphi \in H_0^1(\Omega)} \frac{\int_\Omega A\nabla\varphi \cdot \nabla\varphi \, dx}{\int_\Omega |\varphi|^2 \, dx}. \tag{1.4}$$

We shall not provide a parallel treatment of the above problems for the homogeneous Neumann boundary conditions. However, the techniques are similar and most results remain true. In section 5 we introduce the fundamental H^1-density property to establish the continuity of the solution of the Neumann boundary value problem with respect to the domain. This property is satisfied when the segment property is verified. Such domains are C^0-epigraphs. Then it is a matter of constructing compact families of subsets of a bounded hold-all with that property (for instance, the uniform cone property of section 10 and the uniform cusp property of section 11 in Chapter 5).

2 Embedding of $H_0^1(\Omega)$ into $H_0^1(D)$ and First Eigenvalue

Let Ω and D be two open subsets of $\mathbf{R}^N$ such that $\Omega \subset D$. Denote by $e_0(\varphi)$ the extension by zero of an element φ of $\mathcal{D}(\Omega)$ to D and consider the linear injection

$$\varphi \mapsto e_0(\varphi) : \mathcal{D}(\Omega) \to \mathcal{D}(D).$$

By definition, for $k \geq 1$,

$$\|\varphi\|_{H^k(\Omega)} = \|e_0(\varphi)\|_{H^k(D)}$$

and e_0 extends by continuity and density to a linear isometric map

$$\boxed{e_0 \: : \: H_0^k(\Omega) \overset{\text{def}}{=} \overline{\mathcal{D}(\Omega)}^{H^k} \to H_0^k(D) \overset{\text{def}}{=} \overline{\mathcal{D}(D)}^{H^k}.}$$

Denote by $H_0^k(\Omega; D)$ the image of $H_0^k(\Omega)$ by e_0.

Theorem 2.1. *Let Ω and D be two bounded open subsets of $\mathbf{R}^N$. The linear subspace $H_0^k(\Omega; D)$ of $H_0^k(D)$ is closed and isometrically isomorphic to $H_0^k(\Omega)$ with the following properties: for all $\psi \in H_0^k(\Omega; D)$*

$$\psi|_\Omega \in H_0^k(\Omega) \ \text{and} \ \forall\alpha, \ |\alpha| \leq k, \quad \partial^\alpha \psi = 0 \ \text{a.e. in} \ D\backslash\Omega. \tag{2.1}$$

Any convergent sequence in $H_0^k(\Omega)$-weak converges in $H_0^{k-1}(\Omega)$-strong.

Proof. (i) It is sufficient to prove that $H_0^k(\Omega; D)$ is closed. The other properties are easy to check. Pick a sequence $\{\varphi_n\}$ in $H_0^k(\Omega)$ such that $\{e_0(\varphi_n)\}$ is Cauchy in $H_0^k(D)$ and denote by Φ its limit in $H_0^k(D)$. Since e_0 is an isometry, then $\{\varphi_n\}$ is also Cauchy in $H_0^k(\Omega)$. Denote by φ its limit in $H_0^k(\Omega)$. Then

$$\|\Phi - e_0(\varphi)\|_{H^k(D)} \leq \|\Phi - e_0(\varphi_n)\|_{H^k(D)} + \|e_0(\varphi_n - \varphi)\|_{H^k(D)}$$
$$= \|\Phi - e_0(\varphi_n)\|_{H^k(D)} + \|\varphi_n - \varphi\|_{H^k(\Omega)} \to 0$$

and there exists $\varphi \in H_0^k(\Omega)$ such that $\Phi = e_0(\varphi)$ in $H_0^k(D)$. Thence $\Phi \in H_0^k(\Omega; D)$.

(ii) Since Ω is bounded, there exists a sufficiently large open ball B such that $\Omega \subset B$. By the embedding of $H_0^k(\Omega)$ into $H_0^k(B)$, if a sequence φ_n converges to φ in $H_0^k(\Omega)$-weak, then $e_\Omega(\varphi_n)$ converges to $e_\Omega(\varphi)$ in $H_0^k(B)$-weak. Since the ball is sufficiently smooth, by Rellich's theorem, the sequence converges in $H_0^{k-1}(B)$-strong, and in view of the linear isometric isomorphism, φ_n converges to φ in $H_0^{k-1}(\Omega)$-strong. $\qquad\square$

Going back to the restriction of an element $f \in H^{-1}(D)$ to $\Omega \subset D$, define

$$\forall \varphi \in \mathcal{D}(\Omega), \quad \langle f|_\Omega, \varphi \rangle_{H^{-1}(\Omega) \times H_0^1(\Omega)} \stackrel{\text{def}}{=} \langle f, \varphi \rangle_{H^{-1}(D) \times H_0^1(D)},$$

which makes sense since

$$\forall \varphi \in \mathcal{D}(\Omega), \quad \|\varphi\|_{H_0^1(\Omega)} = \|\varphi\|_{H_0^1(D)},$$

and by density and continuity $f|_\Omega$ readily extends to a continuous linear form on the subspace $H_0^1(\Omega; D)$ of $H_0^1(D)$ and a fortiori on $H_0^1(\Omega)$. Furthermore, the homogeneous Dirichlet boundary value problem (1.2) in Ω is completely equivalent to the variational problem: to find $u \in H_0^1(\Omega; D)$ such that

$$\forall \varphi \in H_0^1(\Omega; D), \quad \int_D A\nabla u \cdot \nabla \varphi \, dx = \langle f, \varphi \rangle$$

and $u = e_\Omega(u_\Omega)$.

Define the following continuous symmetrical linear operator on D and its restriction to Ω:

$$\varphi \mapsto \mathcal{A}(\varphi) \stackrel{\text{def}}{=} \operatorname{div}(A\,\nabla\varphi) : H_0^1(D) \to H^{-1}(D),$$

$$\varphi \mapsto \mathcal{A}_\Omega(\varphi) \stackrel{\text{def}}{=} \operatorname{div}(A\,\nabla\varphi) : H_0^1(\Omega) \to H^{-1}(\Omega). \tag{2.2}$$

With the above notation, equation (1.2) reduces to

$$-\mathcal{A}_\Omega(u_\Omega) = f|_\Omega \text{ in } H^{-1}(\Omega).$$

The following technical lemma will be useful.

Theorem 2.2. *Given a bounded open domain Ω in $\mathbf{R}^N$, the minimization problem* (1.4) *has nonzero solutions in $H_0^1(\Omega)$ which are solution of the eigenvector equation*

$$\exists u_\Omega \in H_0^1(\Omega), \, \forall \varphi \in H_0^1(\Omega), \quad \int_\Omega A\nabla u_\Omega \cdot \nabla \varphi - \lambda^A(\Omega)\, u_\Omega \, \varphi \, dx = 0. \tag{2.3}$$

Given a bounded open nonempty domain D, there exists $\lambda_D^A > 0$ (which only depends on the diameter of D and A) such that for all open subsets Ω of D,

$$\lambda^A(\Omega) = \min_{0 \neq \varphi \in H_0^1(\Omega; D)} \frac{\int_D A\nabla\varphi \cdot \nabla\varphi \, dx}{\int_D |\varphi|^2 \, dx} \geq \lambda^A(D) \geq \lambda_D^A > 0.$$

Proof. (i) For any bounded open Ω, the infimum is bounded below by 0 and hence finite. Let $\{\varphi_n\}$ be a minimizing sequence such that $\|\varphi_n\|_{L^2(\Omega)} = 1$. Then the sequence $\|\nabla\varphi_n\|_{L^2(\Omega)}$ is bounded. Hence $\{\varphi_n\}$ is a bounded sequence in $H_0^1(\Omega)$ and there exist $\varphi \in H_0^1(\Omega)$ and a subsequence, still indexed by n, such that $\varphi_n \rightharpoonup \varphi$ in $H^1(\Omega)$-weak. By Theorem 2.1 the subsequence strongly converges in $L^2(\Omega)$. Then

$$1 = \int_\Omega \varphi_n^2 \, dx \to \int_\Omega \varphi^2 \, dx \text{ and } \int_\Omega A\nabla\varphi \cdot \nabla\varphi \, dx \leq \liminf_{n \to \infty} \int_\Omega A\nabla\varphi_n \cdot \nabla\varphi_n \, dx$$

$$\Rightarrow \frac{\int_\Omega A\nabla\varphi \cdot \nabla\varphi \, dx}{\int_\Omega \varphi^2 \, dx} \leq \liminf_{n \to \infty} \frac{\int_\Omega A\nabla\varphi_n \cdot \nabla\varphi_n \, dx}{\int_\Omega \varphi_n^2 \, dx} = \lambda^A(\Omega),$$

and, by definition of $\lambda^A(\Omega)$, $0 \neq \varphi \in H_0^1(\Omega)$ is a minimizing element. Since $\varphi \neq 0$, the Rayleigh quotient is differentiable and its directional derivative in the direction ψ is given by

$$2\frac{\int_\Omega A\nabla\varphi \cdot \nabla\psi \, dx}{\|\varphi\|_{L^2(\Omega)}^2} - 2\int_\Omega A\nabla\varphi \cdot \nabla\varphi \, dx \frac{\int_\Omega \varphi\psi \, dx}{\|\varphi\|_{L^2(\Omega)}^4}.$$

A nonzero solution φ of the minimization problem on Ω is necessarily a stationary point, and for all $\psi \in H_0^1(\Omega)$,

$$\int_\Omega A\nabla\varphi \cdot \nabla\psi \, dx = \frac{\int_\Omega A\nabla\varphi \cdot \nabla\varphi \, dx}{\|\varphi\|_{L^2(\Omega)}^2} \int_\Omega \varphi\psi \, dx = \lambda^A(\Omega) \int_\Omega \varphi\psi \, dx.$$

Conversely any nonzero solution of (2.3) is necessarily a minimizer of the Rayleigh quotient.

(ii) Let $\mathrm{diam}\,(D)$ be the diameter of D and $x \in \mathbf{R}^N$ be a point such that $D \subset B_x = B(x, \mathrm{diam}\,(D))$. Therefore from Theorem 2.1,

$$H_0^1(\Omega; B_x) \subset H_0^1(D; B_x) \subset H_0^1(B_x), \quad H_0^1(\Omega; D) \subset H_0^1(D)$$

with the associated isometries. Hence, by definition,

$$\lambda^A(\Omega) = \min_{0\neq\varphi\in H_0^1(\Omega;B_x)} \frac{\int_{B_x} A\nabla\varphi \cdot \nabla\varphi \, dx}{\int_{B_x} |\varphi|^2 \, dx} \geq \lambda^A(D) \geq \lambda^A(B_x).$$

If $\lambda^A(B_x) = 0$, we repeat the above construction with $H_0^1(B_x)$ in place of $H_0^1(\Omega; B_x)$ and end up with an element $\varphi \in H_0^1(B_x)$ such that

$$\int_{B_x} \varphi^2 \, dx = 1 \text{ and } \alpha \int_{B_x} |\nabla\varphi|^2 \, dx \leq \int_{B_x} A\nabla\varphi \cdot \varphi \, dx = 0$$

which is impossible in $H_0^1(B_x)$. Finally $\lambda^A(B_x)$ is independent of the choice of x since the eigenvalue is invariant under a translation of the domain: it only depends on the diameter of D. $\qquad\square$

Remark 2.1.
For $A = I$ denote by $\lambda(\Omega)$, $\lambda(D)$, and λ_D, the quantities $\lambda^A(\Omega)$, $\lambda^A(D)$, and λ_D^A of Theorem 2.2. For any open subset Ω of D and any $\varphi \in H_0^1(\Omega; D)$

$$\|\varphi\|_{L^2(D)} \leq \frac{1}{\sqrt{\lambda_D}}\|\nabla\varphi\|_{L^2(D)}$$

$$\Rightarrow \forall\Omega \text{ open } \subset D, \quad \|\varphi\|_{H^1(\Omega)}^2 \leq \frac{1}{\lambda_D}\|\nabla\varphi\|_{L^2(\Omega)}^2 + \|\nabla\varphi\|_{L^2(\Omega)}^2. \qquad\square$$

As a consequence, in what follows, the spaces $H_0^1(\Omega)$ and $H_0^1(D)$ will be endowed with the respective equivalent norms

$$\|\nabla\varphi\|_{L^2(\Omega)} \text{ and } \|\nabla\varphi\|_{L^2(D)}.$$

3   Hausdorff Complementary Topology

One way to prove the existence of extremal domains is to construct compact families in some topology of the space of domains and to prove that the map $\Omega \mapsto J(\Omega)$ is continuous. If we use a very strong topology on the family of domains, we easily get the continuity of the function, but the compact sets will very likely be trivial. In this chapter we use the Hausdorff complementary topology defined in section 2.3 of Chapter 4, but this will not be sufficient and some extra conditions will have to be imposed on the family of domains, for instance, the *uniform cone property* of section 6 in Chapter 2 for Lipschitzian domains or the more general *capacity conditions* which will be introduced in the next sections. The latter include the *flat cone condition* which generalizes the uniform cone property for Lipschitzian domains to a much larger class of bounded open domains.

Consider for a bounded open domain D in $\mathbf{R}^N$ the *Hausdorff complementary metric* (2.12) in section 2.3 of Chapter 4:

$$\boxed{\rho_H^c(\Omega_2, \Omega_1) \stackrel{\text{def}}{=} \|d_{\complement\Omega_2} - d_{\complement\Omega_1}\|_{C(D)}} \tag{3.1}$$

on the family $\mathcal{G}(D) \stackrel{\text{def}}{=} \{\Omega \subset D : \forall\Omega \text{ open}\}$ of all open subsets of D. It was proved in Theorem 2.4 of section 2.3 of Chapter 4 that ρ_H^c induces a complete metric topology called the *Hausdorff complementary topology*, denoted H^c. The Hausdorff complementary convergence is denoted as $\Omega_n \xrightarrow{H^c} \Omega$. For convenience we recall that theorem below.

Theorem 3.1. *Let D be a nonempty open subset of $\mathbf{R}^N$.*

  (i) *The set $C_d^c(D)$ is closed in $C_{\text{loc}}(D)$.*

  (ii) *If, in addition, D is bounded, $C_d^c(D)$ is compact in $C_0(D)$ and the space $(\mathcal{G}(D), \rho_H^c)$ is a compact metric space.*

  (iii) *(Compactivorous property) Given a sequence $\{\Omega_n\}$ and a set Ω in $\mathcal{G}(D)$ such that*

$$d_{\complement\Omega_n} \to d_{\complement\Omega} \quad \text{in } C_{\text{loc}}(D),$$

for any compact subset $K \subset \Omega$, there exists an integer $N(K) > 0$ such that

$$\forall n \geq N(K), \quad K \subset \Omega_n.$$

4   Continuity of u_Ω with Respect to Ω for the Dirichlet Problem

Consider a sequence of open subsets of a bounded open nonempty hold-all D of $\mathbf{R}^N$. By Theorem 3.1 (ii) there exists a subsequence $\{\Omega_n\}$ and Ω in $\mathcal{G}(D)$ such that

$$d_{\complement\Omega_n} \to d_{\complement\Omega} \text{ in } C(\overline{D}) \quad (\text{that is, } \Omega_n \xrightarrow{H^c} \Omega).$$

Denote by u_{Ω_n} the solution in $H_0^1(\Omega_n)$ of the variational equation

$$\forall \varphi \in H_0^1(\Omega_n), \quad \int_{\Omega_n} A\nabla u_{\Omega_n} \cdot \nabla \varphi \, dx = \langle f|_{\Omega_n}, \varphi \rangle_{H^{-1}(\Omega_n) \times H_0^1(\Omega_n)}, \tag{4.1}$$

and by u_Ω the solution in $H_0^1(\Omega)$ of the variational equation (1.2). Denote by $u_n = e_{\Omega_n}(u_{\Omega_n})$ and $u_0 = e_\Omega(u_\Omega)$ the extensions by zero in $H_0^1(D)$ of u_{Ω_n} and u_Ω. They are the respective solutions of the variational equations

$$\exists u_n \in H_0^1(\Omega_n; D), \ \forall \varphi \in H_0^1(\Omega_n; D), \quad \int_D A\nabla u_n \cdot \nabla \varphi \, dx = \langle f, \varphi \rangle, \tag{4.2}$$

$$\exists u_0 \in H_0^1(\Omega; D), \ \forall \varphi \in H_0^1(\Omega; D), \quad \int_D A\nabla u_0 \cdot \nabla \varphi \, dx = \langle f, \varphi \rangle, \tag{4.3}$$

where $\langle \cdot, \cdot \rangle$ denotes the duality pairing between $H^{-1}(D)$ and $H_0^1(D)$.

Lemma 4.1. *Assume that D is a bounded open nonempty domain in $\mathbf{R}^N$. For any open subset Ω of D,*

$$\|\nabla e_\Omega(u_\Omega)\|_{L^2(D)} \le \alpha^{-1} \|f\|_{H^{-1}(D)},$$

where u_Ω is the solution of (1.2) in Ω.

Proof. Let u be the solution in $H_0^1(\Omega; D)$ of the variational equation (4.3). By assumption on A

$$\alpha \|\nabla u\|_{L^2(D)}^2 \le \int_D A\nabla u \cdot \nabla u \, dx = \langle f, u \rangle \le \|f\|_{H^{-1}(D)} \|\nabla u\|_{L^2(D)}$$

$$\Rightarrow \quad \|\nabla u\|_{L^2(D)} \le \alpha^{-1} \|f\|_{H^{-1}(D)}$$

and this completes the proof. $\qquad\square$

In view of Lemma 4.1 for all n

$$\|\nabla u_n\|_{L^2(D)} \le \alpha^{-1} \|f\|_{H^{-1}(D)},$$

where u_n is the solution of (4.2) in $H_0^1(\Omega_n; D)$. Therefore there exists a subsequence, still indexed by n, and $u \in H_0^1(D)$ such that

$$u_n \rightharpoonup u \text{ in } H_0^1(D)\text{-weak}.$$

Given $\varphi \in \mathcal{D}(\Omega)$, its support $K \overset{\text{def}}{=} \operatorname{supp} \varphi$ is a compact subset of Ω. By Theorem 3.1 (iii) there exists $N > 0$ such that, for all $n \ge N$, $K \subset \Omega_n$ and

$$\int_D A\nabla u_n \cdot \nabla \varphi \, dx = \langle f, \varphi \rangle.$$

By going to the limit in the above equation,

$$u \in H_0^1(D), \ \forall \varphi \in \mathcal{D}(\Omega), \quad \int_D A\nabla u \cdot \nabla \varphi \, dx = \langle f, \varphi \rangle.$$

By density this extends to all φ in $H_0^1(\Omega; D)$. We have proved the following result.

Theorem 4.1. *Let D be a bounded open nonempty domain in $\mathbf{R}^N$. Assume that $\{\Omega_n\}$ is a sequence of open subsets of D which converges to Ω in the Hausdorff complementary topology and denote by u_n the solution of (4.2) in $H_0^1(\Omega_n; D)$. Then there exists a subsequence of $\{\Omega_n\}$, still indexed by n, and $u \in H_0^1(D)$ such that*

$$u_n \rightharpoonup u \text{ in } H_0^1(D)\text{-weak}, \quad u = 0 \text{ a.e. in } D\backslash\bar{\Omega} \tag{4.4}$$

$$\forall \varphi \in H_0^1(\Omega; D), \quad \int_D A\nabla u \cdot \nabla\varphi \, dx = \langle f, \varphi \rangle \tag{4.5}$$

or equivalently

$$\forall \varphi \in H_0^1(\Omega), \quad \int_\Omega A\nabla u \cdot \nabla\varphi \, dx = \langle f|_\Omega, \varphi \rangle. \tag{4.6}$$

If, in addition, $u \in H_0^1(\Omega; D)$, then the convergence is strong in $H_0^1(D)$. This condition is satisfied for locally Lipschitzian domains.

Remark 4.1.

In dimension $N = 1$, $u \in H_0^1(\Omega)$ and the convergence is strong. This follows from the fact that $H_0^1(D) \subset C(\bar{D})$ and Ω is at most the union of a countable number of open intervals. Hence $\partial\Omega \cap D$ is made up of at most a countable number of isolated points which are the limit of points of int $\complement\Omega$ where u is zero. $\square$

Proof. (i) The first part of the theorem follows from the preceding discussion. If $u \in H_0^1(\Omega; D)$, then by letting $\varphi = u$ in (4.5),

$$\int_D A\nabla u \cdot \nabla u \, dx = \langle f, u \rangle$$

and the strong convergence follows:

$$\alpha\|\nabla(u_n - u)\|_{L^2(D)}^2 \le \int_D A\nabla(u_n - u) \cdot \nabla(u_n - u)\, dx$$

$$= \int_D A\nabla u_n \cdot \nabla u_n + A\nabla u \cdot \nabla u - 2\, A\nabla u_n \cdot \nabla u \, dx$$

$$= \langle f, u_n \rangle + \langle f, u \rangle - 2\int_D A\nabla u_n \cdot \nabla u \, dx$$

$$\rightarrow \quad \langle f, u \rangle + \langle f, u \rangle - 2\int_D A\nabla u \cdot \nabla u \, dx = 0.$$

(ii) Recall that

$$\chi_{\Omega_n} = |\nabla d_{\complement\Omega_n}| \text{ and } \chi_{\complement\bar{\Omega}_n} = |\nabla d_{\Omega_n}| \quad \text{a.e. in } D.$$

By Theorem 2.1, the subsequence $\{u_n\}$ converges in $L^2(D)$-strong and $u_n = 0$ almost everywhere in $D\backslash\Omega_n$:

$$\int_D \chi_{\complement\bar{\Omega}_n}|u|^2 \, dx = \int_{D\backslash\bar{\Omega}_n} |u_n - u|^2 \, dx \le \int_D |u_n - u|^2 \, dx \rightarrow 0$$

$$\Rightarrow 0 = \liminf_{n\to\infty} \int_D \chi_{\complement\bar{\Omega}_n}|u_n - u|^2 \, dx \ge \int_D \chi_{\complement\bar{\Omega}}|u|^2 \, dx$$

from Corollary 1 to Theorem 4.3 in Chapter 4. Therefore $u \in H_0^1(D)$, $u = 0$ almost everywhere in $D \backslash \overline{\Omega}$. For Lipschitzian domains the trace of u is well defined on $\partial \Omega$. It is zero since u and ∇u are zero almost everywhere in the locally Lipschitzian domain $\complement \overline{\Omega}$ by using the Gauss–Green formula. $\square$

To get the continuity of u_Ω with respect to Ω, it would remain to show that u belongs to $H_0^1(\Omega; D)$. However, this is generally not true and u is not necessarily the solution of (4.3) in Ω. To get around this difficulty the attention shifts to finding a family $\mathcal{O}$ of open subsets of D for which the following continuity property holds:

$$\Omega_n \in \mathcal{O}, \; \Omega_n \xrightarrow{H^c} \Omega \quad \Rightarrow \quad \Omega \in \mathcal{O}, \; e_{\Omega_n}(u_{\Omega_n}) \xrightarrow{H_0^1(D)} e_\Omega(u_\Omega).$$

This problem is directly related to the sequential continuity of the *projection* $P_\Omega v$ of an element $v \in H_0^1(D)$ onto $H_0^1(\Omega; D)$ defined as

$$\forall \varphi \in H_0^1(\Omega; D), \quad \int_D A\nabla(P_\Omega v) \cdot \nabla \varphi \, dx = \int_D A\nabla v \cdot \nabla \varphi \, dx.$$

Therefore Theorem 4.1 applies with

$$\langle f, \varphi \rangle \overset{\text{def}}{=} \int_D A\nabla v \cdot \nabla \varphi \, dx$$

and there exist v_0, an open subset Ω of D, and a subsequence $\{\Omega_n\}$ of open subsets of D such that

$$d_{\complement \Omega_n} \to d_{\complement \Omega} \text{ in } C(\overline{D}), \quad P_{\Omega_n} v \rightharpoonup v_0 \text{ in } H_0^1(D)\text{-weak}.$$

If $v_0 \in H_0^1(\Omega; D)$, then $v_0 = P_\Omega v$ and the convergence is strong.

5 Continuity of the Neumann Problem

In the case of the Neumann problem associated with the Laplacian, the most important condition for obtaining the continuity with respect to the domain is the following density property.

Definition 5.1.
An open set Ω is said to have the H^1-*density property* if the set

$$\{\varphi|_\Omega \; : \; \varphi \in C^1(\mathbf{R}^N)\}$$

is dense in $H^1(\Omega)$. $\square$

Notice that this property is not verified for a domain Ω with a crack in $\mathbf{R}^2$, since elements of $H^1(\Omega)$ have different traces on each side of the crack.

Consider a sequence $\{\Omega_n\}$ of domains and a domain Ω satisfying the H^1-density property. Assume that the sequence converges to Ω in both the complementary Hausdorff metric and the L^p-topology of their characteristic functions.

Given $f \in L^2(\Omega)$, let y_n and y be the respective solutions of the Neumann problems on Ω_n and Ω:

$$\begin{cases} -\Delta y_n + y_n = f \text{ in } \Omega_n, \\ \dfrac{\partial y_n}{\partial n} = 0 \text{ on } \partial\Omega_n, \end{cases} \qquad \begin{cases} -\Delta y + y = f \text{ in } \Omega, \\ \dfrac{\partial y}{\partial n} = 0 \text{ on } \partial\Omega. \end{cases}$$

Let y_n^0 and $(\nabla y_n)^0$ be the respective extensions by zero of y_n and ∇y_n to $\mathbf{R}^N$ as elements of $L^2(\mathbf{R}^N)$ and $L^2(\mathbf{R}^N)^N$. Then we get the following continuity:

$$y_n^0 \rightharpoonup y^0 \text{ weakly in } L^2(\mathbf{R}^N) \text{ and } (\nabla y_n)^0 \rightharpoonup (\nabla y)^0 \text{ weakly in } L^2(\mathbf{R}^N)^N,$$

where y^0 and $(\nabla y)^0$ are the respective extensions by zero of y and ∇y to $\mathbf{R}^N$. The proof is based on the assumption that Ω_n and Ω satisfy the H^1-density property. This implies that

$$\forall \varphi \in C^1(\mathbf{R}^N), \quad \int_{\mathbf{R}^N} \chi_{\Omega_n} \left(\nabla y_n \cdot \nabla \varphi + y_n\, \varphi - f\, \varphi \right) dx = 0.$$

Obviously

$$\|y_n^0\|_{L^2(\mathbf{R}^N)} \leq \|f\|_{L^2(\mathbf{R}^N)} \text{ and } \|(\nabla y_n)^0\|_{L^2(\mathbf{R}^N)} \leq \|f\|_{L^2(\mathbf{R}^N)}.$$

Then there exist $z \in L^2(\mathbf{R}^N)$ and $Z \in L^2(\mathbf{R}^N)^N$, and subsequences, still indexed by n, such that

$$y_n^0 \rightharpoonup z \text{ in } L^2(\mathbf{R}^N)\text{-weak and } (\nabla y_n)^0 \rightharpoonup Z \text{ in } L^2(\mathbf{R}^N)^N\text{-weak.}$$

From the compactivorous property (cf. Theorem 2.4 (iii) in Chapter 4), it follows that $Z = (\nabla y)^0$. Going to the limit as n goes to infinity in the previous integral, we get

$$\forall \varphi \in C^1(\mathbf{R}^N), \quad \int_{\mathbf{R}^N} \chi_\Omega \left(\nabla z \cdot \nabla \varphi + z\, \varphi - f\, \varphi \right) dx = 0.$$

As we have also assumed that Ω had the H^1-density property, the variational equation yields $z = y$. This is the approach followed by Liu and Rubio [1] in 1992.

We know that the H^1-density condition holds for the family of domains satisfying the *segment condition* of Definition 7.1 in Chapter 2 and, not only for H^1-spaces, but also $W^{m,p}$-spaces (cf. Theorem 7.5 in Chapter 2), thus extending the property and the results to a broader class of partial differential equations. We have shown in Theorem 7.3 of Chapter 2 that a domain satisfying the segment property is locally a C^0-epigraph in the sense of Definition 5.2 of Chapter 2.

We have seen in section 11 of Chapter 5 that the uniform segment property of Definition 7.1 (ii) in Chapter 2 is not sufficient to get the compactness of the corresponding family of subsets of a bounded hold-all D. However, from Theorem 11.1 in section 11 of Chapter 5, we have the compactness for the family of subsets of D verifying a *uniform cusp property* for which the segment property is also satisfied.

In an unpublished report Tiba [2], and in a note Liu, Neittaanmäki, and Tiba [1], introduce a subfamily of domains which are locally a C^0-epigraph in the sense of Definition 5.2 of Chapter 2. A compactness result is obtained by introducing the following conditions on the local C^0-epigraphs: the boundaries of the domains belong (after a rotation and a translation) to a family $\mathcal{F}$ of uniformly equicontinuous functions defined in a fixed neighborhood V of 0 in $\mathbf{R}^{N-1}$. This means that there exists a modulus of continuity $\mu(\varepsilon) > 0$ such that

$$\forall \varepsilon > 0, \ \forall f \in \mathcal{F}, \ \forall x, y \in V, \quad |x - y| < \mu(\varepsilon) \quad \Rightarrow \quad |f(y) - f(x)| < \varepsilon.$$

The modulus μ effectively *controls* the singularities of the boundary. Then the compactness follows by the Ascoli–Arzelà theorem, Theorem 2.1 of Chapter 2. By Theorem 7.3 (ii) of Chapter 2 such domains satisfy the segment property and even the uniform segment property. Hence they satisfy the H^1-density property and we have the continuity of the solutions of the Neumann problems with respect to domains in that family.

Nevertheless, it is important to recall that the subsets of a bounded hold-all verifying a uniform segment property do not include the very important domains with cracks for which some specific results have been obtained by Bucur and Zolésio [2, 4] (and by Bucur and Varchon [1] in dimension 2).

6 Optimization over Families of Lipschitzian or Convex Domains

Consider a family of bounded open Lipschitzian domains satisfying the *uniform cone property* of Definition 6.1 in section 6 of Chapter 2, for which we recall several compactness properties in Theorem 10.1 of section 10 in Chapter 5.

Theorem 6.1. *Let D be a bounded open subset of $\mathbf{R}^{\mathrm{N}}$ and $1 \le p < \infty$. For $r > 0$, $\omega > 0$, and $\lambda > 0$, the family*

$$B(D, r, \omega, \lambda) \overset{\text{def}}{=} \{b_\Omega \ : \ \forall \Omega \in L(D, r, \omega, \lambda)\}$$

is compact in $C(\overline{D})$ and $W^{1,p}(D)$. As a consequence the families

$$B_d(D, r, \omega, \lambda) \overset{\text{def}}{=} \{d_\Omega \ : \ \forall \Omega \in L(D, r, \omega, \lambda)\},$$

$$B_d^c(D, r, \omega, \lambda) \overset{\text{def}}{=} \{d_{\complement\Omega} \ : \ \forall \Omega \in L(D, r, \omega, \lambda)\},$$

$$B_d^\partial(D, r, \omega, \lambda) \overset{\text{def}}{=} \{d_{\partial\Omega} \ : \ \forall \Omega \in L(D, r, \omega, \lambda)\}$$

are compact in $C(\overline{D})$ and $W^{1,p}(D)$ and the families

$$X(D, r, \omega, \lambda) \overset{\text{def}}{=} \{\chi_\Omega \ : \ \forall \Omega \in L(D, r, \omega, \lambda)\},$$

$$X^c(D, r, \omega, \lambda) \overset{\text{def}}{=} \{\chi_{\complement\Omega} \ : \ \forall \Omega \in L(D, r, \omega, \lambda)\}$$

are compact in $L^p(D)$.

Therefore, we have the strong convergence in all the above topologies. In particular the characteristic functions converge in $L^1(D)$ and the volume function is continuous. Also recall from Theorem 7.2 in Chapter 5 that the above properties associated with $L(D, r, \omega, \lambda)$ are also true for

$$\boxed{\mathcal{C}_E(D) \overset{\text{def}}{=} \{\Omega \ : \ E \subset \Omega \text{ convex open } \subset D\},} \tag{6.1}$$

where E is some nonempty open subset of D.

6.1   Minimization of an Objective Function under State Equation Constraint

Let $\{\Omega_n\}$ be a sequence of open subsets of $L(D, r, \omega, \lambda)$ such that the objective function $J(\Omega)$ defined by (1.1) converges to its infimum with respect to $L(D, r, \omega, \lambda)$. In view of Theorem 6.1 the elements of the sequence $\{\Omega_n\}$ are bounded Lipschitzian domains in D satisfying the corresponding uniform cone property. Moreover, there exists a subsequence, still indexed by n, and Ω in $L(D, r, \omega, \lambda)$ such that $b_{\Omega_n} \to b_\Omega$ in $C(\overline{D})$ and $\chi_{\Omega_n} \to \chi_\Omega$ in $L^p(D)$, $1 \le p < \infty$. Then by Lemma 2.1 in Chapter 3

$$d_{\complement\Omega_n} \to d_{\complement\Omega}, \quad d_{\Omega_n} \to d_\Omega \text{ in } C(\overline{D}), \text{ and } \chi_{\Omega_n} \to \chi_\Omega \text{ in } L^\infty(D)\text{-weak}\star.$$

Since Ω is Lipschitzian, $\partial\Omega = \partial \operatorname{int}\complement\Omega$ by Theorem 6.1 in Chapter 2. Therefore from Theorem 4.1, $u \in H_0^1(\Omega; D)$ and the convergence of the corresponding u_n to u is strong in $H_0^1(D)$ and $u = e_\Omega(u_\Omega)$.

Theorem 6.2. *Let D be a bounded open domain in $\mathbf{R}^N$. The infimum of the function (1.1) over $L(D, r, \omega, \lambda)$ subject to the state equation constraint (1.2) has a minimizer Ω in $L(D, r, \omega, \lambda)$ and u_Ω satisfies equation (1.2). The conclusions remain true when $L(D, r, \omega, \lambda)$ is replaced by $\mathcal{C}_E(D)$ for some nonempty open subset E of D.*

Proof. (i) It is sufficient to show that the limiting element u is a minimizer of (1.1). For n

$$J(\Omega_n) = \int_{\Omega_n} |u_n - g|^2 \, dx = \int_D |u_n|^2 - 2g\, u_n + \chi_{\Omega_n} |g|^2 \, dx.$$

Take the liminf of both sides using Corollary 1 to Theorem 4.3 in Chapter 4 for χ_{Ω_n} and the strong convergence for u_n:

$$\liminf_{n\to\infty} J(\Omega_n) \ge \int_D |u|^2 - 2g\, u + \chi_\Omega |g|^2 \, dx = \int_\Omega |u - g|^2 \, dx = J(\Omega).$$

By choice of the sequence, Ω is a minimizing element in $L(D, r, \omega, \lambda)$ and u_Ω satisfies the state equation by Theorem 4.1 for Lipschitzian domains such that $\operatorname{int}\Omega \ne \varnothing$ and $\Omega \ne \mathbf{R}^N$.

(ii) Convex case. This proof has the same argument using the fact that $\mathcal{C}_E(D) = I(\complement E) \cap \mathcal{C}_d^c(D)$, where

$$I(\complement E) \overset{\text{def}}{=} \{d_{\complement\Omega} \in C_d^c(D) : \complement\Omega \subset \complement E\}.$$

The result follows from the compactness of $\mathcal{C}_d^c(D)$ (Theorem 6.2 (ii) in section 6 of Chapter 4) and the fact that $I(\complement E)$ is closed in $C(\overline{D})$ (Theorem 2.3 (ii) of Chapter 4). Since $E \subset \Omega \subset D$, Ω is not empty and not equal to $\mathbf{R}^N$. Therefore Ω is locally Lipschitzian and the last condition of Theorem 4.1 is satisfied. $\qquad\square$

6.2 Optimization of the First Eigenvalue

Consider the first eigenvalue $\lambda^A(\Omega)$ associated with the operator $\mathcal{A}$ in (2.2). By Theorem 2.2 for any open subset Ω of D,

$$\lambda^A(\Omega) = \inf_{0 \neq \varphi \in H_0^1(\Omega;D)} \frac{\int_D A\nabla\varphi \cdot \nabla\varphi \, dx}{\int_D |\varphi|^2 \, dx}$$

and any minimizer is solution of equation (2.3).

Theorem 6.3. *Let D be a bounded open domain in $\mathbf{R}^N$ and A a matrix function satisfying assumption* (1.3). *The optimization problems*

$$\sup_{\Omega \in L(D,r,\omega,\lambda)} \lambda(\Omega) \quad and \quad \inf_{\Omega \in L(D,r,\omega,\lambda)} \lambda(\Omega) \tag{6.2}$$

have solutions in $L(D,r,\omega,\lambda)$. The conclusions of the theorem remain true when $L(D,r,\omega,\lambda)$ is replaced by $\mathcal{C}_E(D)$ defined in (6.1).

Proof. (i) *Maximization.* By the definition of $L(D,r,\omega,\lambda)$, for all $\Omega \subset D$, there exists $x \in \partial\Omega$ and d_x, $|d_x| = 1$, such that $C_x = x + C(\lambda,\omega,d_x) \subset \Omega$ and

$$\lambda^A(\Omega) \leq \lambda^A(C_x) \overset{\text{def}}{=} \inf_{0 \neq \varphi \in H_0^1(C_x;D)} \frac{\int_D A\nabla\varphi \cdot \nabla\varphi \, dx}{\int_D |\varphi|^2 \, dx}$$

$$\leq \beta \inf_{0 \neq \varphi \in H_0^1(C_x;D)} \frac{\int_D \nabla\varphi \cdot \nabla\varphi \, dx}{\int_D |\varphi|^2 \, dx}$$

$$= \beta \inf_{0 \neq \varphi \in H_0^1(C(\lambda,\omega,d;D)} \frac{\int_D \nabla\varphi \cdot \nabla\varphi \, dx}{\int_D |\varphi|^2 \, dx}$$

$$\leq \beta \lambda(C_x) = \beta \lambda(C(\lambda,\omega,d))$$

for any d, $|d| = 1$ since the inf over $H_0^1(C_x; D)$ only depends on the small cone and not its origin or direction. This provides an upper bound λ_C and

$$\forall \Omega \text{ open}, \quad \lambda^A(D) \leq \lambda^A(\Omega) \leq \lambda_C = \beta \lambda(C(\lambda,\omega,d)).$$

From that point on the proof is similar to what we have done before for the previous minimization problem. Let $\{\Omega_n\}$ be a maximizing sequence in $L(D,r,\omega,\lambda)$. By Theorem 6.1 there exists a subsequence, still indexed by n, $\Omega \in L(D,r,\omega,\lambda)$, λ^*, $\lambda^A(D) \leq \lambda^* \leq \lambda_C$, $u \in H_0^1(D)$, such that $b_{\Omega_n} \to b_\Omega$, and

$$\lambda_n \overset{\text{def}}{=} \lambda(\Omega_n) \to \lambda^*,$$

$$u_n \overset{\text{def}}{=} u_{\Omega_n}, \quad \|u_n\|_{L^2(\Omega_n)} = 1, \quad u_n \rightharpoonup u \text{ in } H_0^1(D)\text{-weak}$$

$$\Rightarrow d_{\complement\Omega_n} \to d_{\complement\Omega} \text{ and } d_{\Omega_n} \to d_\Omega \text{ in } C(\overline{D}).$$

By the same argument as for equation (1.2) applied to the eigenvector equation (2.3),

$$\forall \varphi \in H_0^1(\Omega_n; D), \quad \int_D A\nabla u_n \cdot \nabla \varphi \, dx = \lambda_n \int_D u_n \, \varphi \, dx,$$

and it is readily seen that $u \in H_0^1(\Omega; D)$ satisfies the equation

$$\forall \varphi \in H_0^1(\Omega; D), \quad \int_D A\nabla u \cdot \nabla \varphi \, dx = \lambda^* \int_D u \, \varphi \, dx,$$

$$\|u\|_{L^2(\Omega)} = 1 \text{ and } u_n \to u \text{ in } H_0^1(D)\text{-strong.}$$

Therefore, $\lambda^* = \lambda^A(\Omega)$ and Ω is a maximizer.

    (ii) *Minimization.* Same proof using the fact that, for all $\Omega \in L(D, r, \omega, \lambda)$, $\lambda^A(\Omega)$ is bounded below by the strictly positive constant $\lambda^A(D) > 0$.

    (iii) *Convex case.* Same argument as in the proof of Theorem 6.2. $\square$

Corollary 1. *Under the assumptions of Theorem 6.3, the maximization and minimization problems (6.2) also have solutions in $L(D, r, \omega, \lambda)$ under an equality or an inequality constraint of the form*

$$\mathrm{m}(\Omega) = \alpha, \, \mathrm{m}(\Omega) \leq \alpha, \; or \; \mathrm{m}(\Omega) \geq \alpha$$

provided that there exists $\Omega' \in L(D, r, \omega, \lambda)$ such that

$$\mathrm{m}(\Omega') = \alpha, \, \mathrm{m}(\Omega') \leq \alpha \; or \; \mathrm{m}(\Omega') \geq \alpha.$$

The conclusions remain true when $L(D, r, \omega, \lambda)$ is replaced by $\mathcal{C}_E(D)$ defined in (6.1).

Proof. This follows from Theorem 6.1, where it is shown that the convergence is strong not only in the Hausdorff topologies, but also for the associated characteristic functions in $L^1(D)$, thus making the volume function continuous with respect to open domains in $L(D, r, \omega, \lambda)$. For the case of $\mathcal{C}_E(D)$, recall from Theorem 7.2 (iii) of Chapter 5 that $\mathcal{C}_b(D; E)$ is compact in $W^{1,1}(D)$, and hence the volume functional is continuous. $\square$

7   Elements of Capacity Theory

We have seen in the previous section that the key element in the proof of the continuity of the solution of (1.2) with respect to its underlying domain is that the limiting element satisfies the homogeneous Dirichlet boundary condition for the limiting domain. This section introduces capacity constraints which generalize the property to a broad class of open domains.

7.1   Definition and Basic Properties

The following definition of *capacity* and a number of basic technical results can be found in Hedberg [1] and are summarized below.

Definition 7.1.

Let D be a fixed bounded open subset of $\mathbf{R}^N$.

(i) The *capacity*[2] (*with respect to D*) is defined as follows:

— for a compact subset $K \subset D$

$$\operatorname{cap}_D(K) \overset{\text{def}}{=} \inf \left\{ \int_D |\nabla \varphi|^2 \, dx \; : \; \varphi \in C_0^\infty(D), \; \varphi > 1 \text{ on } K \right\};$$

— for an open subset $G \subset D$

$$\operatorname{cap}_D(G) \overset{\text{def}}{=} \sup \left\{ \operatorname{cap}_D(K) \; : \; \forall K \subset G, \; K \text{ compact} \right\};$$

— for an arbitrary subset $E \subset D$

$$\operatorname{cap}_D(E) \overset{\text{def}}{=} \inf \left\{ \operatorname{cap}_D(G) \; : \; \forall G \supset E, \; G \text{ open} \right\}.$$

(ii) A function f on D is said to be *quasi-continuous* if

$$\forall \varepsilon > 0, \quad \exists G_\varepsilon \text{ open such that } \operatorname{cap}_D(G_\varepsilon) < \varepsilon$$

$$\text{and } f \text{ is continuous on } D \backslash G_\varepsilon.$$

(iii) A set E in $\mathbf{R}^N$ is said to be *quasi-open* if

$$\forall \varepsilon > 0, \quad \exists G_\varepsilon \text{ open such that } \operatorname{cap}_D(G_\varepsilon) < \varepsilon$$

$$\text{and } E \cup G_\varepsilon \text{ is open.} \qquad \square$$

It can easily be shown that for a quasi-open set there exists a decreasing sequence $\{\Omega_n\}$ of open sets, such that $\Omega_n \supset E$ and $\operatorname{cap}_D(\Omega_n \backslash E)$ goes to zero as n goes to infinity.

We say that a property holds *quasi-everywhere* (q.e.) in D if it holds in the complement $D \backslash E$ of a set E of zero capacity. A set of zero capacity has zero measure, but the converse is not true. The capacity is a countably subadditive set function, but it is not additive even for disjoint sets. Hence the union of a countable number of sets of zero capacity has zero capacity.

7.2 Quasi-Continuous Representative and H^1-Functions

Lemma 7.1 (Hedberg [1]). *Any f in $H^1(D)$ has a quasi-continuous representative: there exists a quasi-continuous function f_1 defined on D such that $f_1 = f$ almost everywhere in D (hence f_1 is a representative of f in $H^1(D)$). Any two quasi-continuous representatives of the same element of $H^1(D)$ are equal quasi-everywhere in D.*

The following key lemma completes the picture.

---

[2]This definition is also referred to as the *exterior capacity*.

Lemma 7.2 (Hedberg [1]). *Let Ω and D be two bounded open subsets of $\mathbf{R}^N$ such that $\Omega \subset D$ and consider an element u of $H_0^1(D)$. Then $u|_\Omega \in H_0^1(\Omega)$ if and only if $u_1 = 0$ quasi-everywhere on $D\backslash\Omega$, where u_1 is a quasi-continuous representative of u.*

A function φ in $H^1(D)$ is said to be *zero quasi-everywhere* in a subset E of D if there exists a quasi-continuous representative of φ which is zero quasi-everywhere in E. This makes sense since any two quasi-continuous representatives of an element φ of $H^1(D)$ are equal quasi-everywhere. For any $\varphi \in H_0^1(D)$ and $t \in \mathbf{R}$, the set $\{x \in D : \varphi(x) > t\}$ is quasi-open. Moreover, the subspace $H_0^1(\Omega; D)$ introduced in section 2 can now be characterized by the capacity.

Corollary 2. *Under the assumptions of Lemma 7.2,*

$$\boxed{H_0^1(\Omega; D) = \left\{\varphi \in H_0^1(D) : \varphi = 0 \ q.e. \ in \ D\backslash\Omega\right\}.} \tag{7.1}$$

The characterization of $H_0^1(\Omega; D)$ requires the notion of *capacity* in a very essential way. It cannot be obtained by saying that the function and its derivatives are zero almost everywhere in $D\backslash\Omega$. Recall the definition of the other extension $H_\diamond^1(\Omega; D)$ of $H^1(\Omega)$ to a measurable set D containing Ω (cf. Chapter 3, section 2.5, Theorem 2.9, identity (2.29)):

$$H_\diamond^1(\Omega; D) \stackrel{\text{def}}{=} \left\{\psi \in H_0^1(D) : \psi = 0 \ \text{a.e. in } D\backslash\Omega\right\}.$$

By definition $H_0^1(\Omega; D) \subset H_\diamond^1(\Omega; D)$, but the two spaces are generally not equal, as can be seen from the following example. Denote by B_r, $r > 0$, the open ball of radius $r > 0$ in $\mathbf{R}^2$. Define $\Omega = B_2\backslash\partial B_1$ and $D = B_3$. The circular crack ∂B_1 in Ω has zero measure but nonzero capacity. Since ∂B_1 has zero measure, $H_\diamond^1(\Omega; D)$ contains functions $\psi \in H_0^1(B_2)$ whose restriction to B_2 are not zero on the circle ∂B_1 and hence do not belong to $H_0^1(\Omega)$.

Yet for Lipschitzian domains Ω the two spaces are indeed equal. The following terminology is due to Rauch and Taylor [1].

Definition 7.2.
Ω is said to be *stable* with respect to D if $H_0^1(\Omega; D) = H_\diamond^1(\Omega; D)$. $\qquad\square$

7.3　Transport of Sets of Zero Capacity

Any Lipschitz continuous transformation of $\mathbf{R}^N$ which has a Lipschitz continuous inverse transports sets of zero capacity onto sets of zero capacity.

Lemma 7.3. *Let D be an open subset of $\mathbf{R}^N$ and T an invertible transformation of $\overline{D}$ such that both T and T^{-1} are Lipschitz continuous. For any $E \subset D$*

$$\mathrm{cap}_D(E) = 0 \iff \mathrm{cap}_D(T(E)) = 0.$$

Proof. (a) $E = K$, K *compact in* D. Observe that, in the definition of the capacity, it is not necessary to choose the functions φ in $C_0^\infty(D)$. They can be chosen in a larger space as long as

$$\int_D |\nabla\varphi|^2 \, dx < \infty \text{ and } \varphi > 1 \text{ on } K.$$

So the capacity is also given by

$$\operatorname{cap}_D(K) \stackrel{\text{def}}{=} \inf \left\{ \int_D |\nabla\varphi|^2 \, dx \; : \; \varphi \in H_0^1(D) \cap C(\overline{D}), \; \varphi > 1 \text{ on } K \right\}.$$

But the space $\mathcal{E} \stackrel{\text{def}}{=} H_0^1(D) \cap C(\overline{D})$ is stable under the action of T:

$$\varphi \in \mathcal{E} \iff T \circ \varphi \in \mathcal{E}.$$

Consider the matrix

$$A(x) \stackrel{\text{def}}{=} |\det(DT(x))| \; {}^*DT(x)({}^*DT(x))^{-1}.$$

By assumption on T, the elements of the matrix A belong to $L^\infty(D)$ and

$$\exists \alpha > 0, \quad \alpha I \le A(x) \le \alpha^{-1} I \text{ a.e. in } D.$$

Define $E(K) = \{\varphi \in \mathcal{E} : \varphi > 1 \text{ on } K\}$. For any $\varphi \in E(T(K))$, $\varphi \circ T \in E(K)$ and

$$\int_D |\nabla\varphi|^2 \, dx = \int_D A \, \nabla(\varphi \circ T) \cdot \nabla(\varphi \circ T) \, dx \ge \alpha \int_D |\nabla(\varphi \circ T)|^2 \, dx.$$

Let K be a compact subset such that $\operatorname{cap}_D(K) = 0$. For each $\varepsilon > 0$ there exists $\varphi \in E(K)$ such that

$$\int_D |\nabla\varphi|^2 \, dx \le \varepsilon \quad \Rightarrow \forall \varepsilon, \quad \int_D |\nabla(\varphi \circ T)|^2 \, dx \le \varepsilon/\alpha \text{ or } \varphi \circ T \in E(T^{-1}(K))$$

$$\Rightarrow \forall \varepsilon, \quad \operatorname{cap}_D(T^{-1}(K)) \le \varepsilon/\alpha \quad \Rightarrow \operatorname{cap}_D(T^{-1}(K)) = 0,$$

and we can repeat the proof with T^{-1} in place of T.

(b) $E = G$, G *open*. Let

$$\operatorname{cap}_D(G) = \sup\{\operatorname{cap}_D(K) : \forall K \subset G, \; K \text{ compact}\} = 0.$$

Therefore $\operatorname{cap}_D(K) = 0$, which implies that $\operatorname{cap}_D(T(K)) = 0$ and hence

$$\operatorname{cap}_D(T(G)) = \sup\{\operatorname{cap}_D(K') : \forall K' \subset T(G), \; K' \text{ compact}\}$$
$$= \sup\{\operatorname{cap}_D(T(K)) : \forall K \subset G, \; K \text{ compact}\} = 0.$$

(c) *General case.* Let

$$\operatorname{cap}_D(E) \stackrel{\text{def}}{=} \inf\{\operatorname{cap}_D(G) : \forall G \supset E, \; G \text{ open}\}.$$

Therefore for all $\varepsilon > 0$, there exists $G \supset E$ open such that

$$\mathrm{cap}_D(E) \leq \mathrm{cap}_D(G) \leq \varepsilon \quad \Rightarrow \quad \mathrm{cap}_D(T(E)) \leq \mathrm{cap}_D(T(G))$$

since $T(E) \subset T(G)$ is open. By definition of $\mathrm{cap}_D(G)$ we have

$$\forall K \subset G, \quad \mathrm{cap}_D(K) \leq \varepsilon \quad \Rightarrow \quad \mathrm{cap}_D(T(K)) \leq \alpha \varepsilon$$
$$\Rightarrow \mathrm{cap}_D(T(G)) = \sup\{\mathrm{cap}_D(T(K)) \, : \, K \subset G\} \leq \alpha \varepsilon.$$

Finally, for all $\varepsilon > 0$, $\mathrm{cap}_D(T(E)) \leq \alpha \varepsilon$ and hence $\mathrm{cap}_D(T(E)) = 0$. $\qquad\square$

8 Continuity under Capacity Constraints

In order to preserve the homogeneous Dirichlet boundary condition for u on the boundary of the limit domain Ω in section 4, it is necessary to restrict our attention to smaller families of open domains. We have to be careful since the relative capacity of the complement of the domains near the boundary must not vanish. In order to handle this point, we introduce the following concepts and terminology related to the local capacity of the complement near the boundary points. The following definition is due to Heinonen, Kilpelainen, and Martio [1].

Definition 8.1.
For $r > 0$ and a compact $K \subset \mathbf{R}^N$, the *capacity condenser* of K in the ball $B(x, r)$ is defined as

$$\mathrm{cap}_{x,r}(K) \overset{\mathrm{def}}{=} \mathrm{cap}_{B(x,2r)}(K \cap B(x,r)). \qquad\square$$

The reader is referred to Hedberg [1] and Bucur and Zolésio [5] for detailed properties.

Definition 8.2.

(i) Given $r > 0$, $c > 0$, and an open set, Ω is said to satisfy the *(r, c)-capacity density condition* if

$$\forall x \in \partial\Omega, \quad \frac{\mathrm{cap}_{B(x,2r)}(\complement\Omega \cap B(x,r))}{\mathrm{cap}_{B(x,2r)}(B(x,r))} \geq c.$$

(ii) For $0 < r < 1$ define the following family of open subsets of D:

$$\mathcal{O}_{c,r}(D) \overset{\mathrm{def}}{=} \left\{ \Omega \subset D : \begin{array}{l} \forall r_0, \quad 0 < r_0 < r, \\ \Omega \text{ has the } (r_0, c) \text{ capacity density condition.} \end{array} \right\} \qquad\square$$

Remark 8.1.
There are different definitions of the capacity condenser. The definition used here is given in Hedberg [1]. The following slightly different definition is given by Heinonen, Kilpelainen, and Martio [1]:

$$\overline{\mathrm{cap}}_{x,r}(K) \overset{\mathrm{def}}{=} \mathrm{cap}_{B(x,2r)}(K \cap \overline{B}(x,r)). \qquad (8.1)$$

For a given compact set K we generally have

$$\mathrm{cap}_{B(x,2r)}(K \cap \overline{B}(x,r)) \neq \mathrm{cap}_{B(x,2r)}(K \cap B(x,r)),$$

but the following two conditions are equivalent:

$$\forall\, 0 < \delta < r, \qquad \frac{\mathrm{cap}_{B(x,2\delta)}(K \cap B(x,\delta))}{\mathrm{cap}_{B(x,2\delta)}(B(x,\delta))} \geq c \qquad (8.2)$$

and

$$\forall\, 0 < \delta < r, \qquad \frac{\mathrm{cap}_{B(x,2\delta)}(K \cap \overline{B}(x,\delta))}{\mathrm{cap}_{B(x,2\delta)}(\overline{B}(x,\delta))} \geq c \qquad (8.3)$$

and so from the $\mathcal{O}_{c,r}$ families viewpoint, the two definitions are equivalent. $\qquad\square$

Definition 8.3.
Let x be a point on $\partial\Omega$. The set $\complement\Omega$ is said to be *thick* at x if the complement of Ω locally has "enough" capacity, that is, if

$$\int_0^1 \frac{\mathrm{cap}_{B(x,2r)}(\complement\Omega \cap B(x,r))}{\mathrm{cap}_{B(x,2r)}(B(x,r))} \, \frac{dr}{r} = \infty. \qquad\square$$

Remark 8.2.
If Ω satisfies the (r,c)-capacity density condition, the complement is thick in any point of the boundary. $\qquad\square$

Recall Theorem 4.1 and the notation u_n for the solution of equation (4.2) in $H_0^1(\Omega; D)$ and u for the weak limit in $H_0^1(D)$ which satisfies equation (4.6). The following theorem gives the main continuity result.

Theorem 8.1. *Assume that D is a bounded open nonempty subset of $\mathbf{R}^N$ and that D is of class C^2 for $N \geq 3$. Assume that A is a matrix function that satisfies the conditions (1.3) and that the elements of A belong to $C^1(\overline{D})$. Let $\{\Omega_n\}$ be a sequence in $\mathcal{O}_{c,r}(D)$ which converges in the H^c-topology to an open set Ω. Then $u_n \to u$ in $H_0^1(D)$-strong and $u_\Omega = u|_\Omega \in H_0^1(\Omega)$ is the solution of equation (1.2).*

Proof. When the dimension N of the space is equal to 1 or 2, $H^1(D) \subset C(\overline{D})$ and no approximation of f is necessary. When $N \geq 3$ we first prove the result for $f \in H^s(D)$, $s > N/2 - 2$ (since for D of class C^2 the corresponding solution will belong to $H^{s+2}(D) \subset C(\overline{D})$) and then, by approximation of f, we prove it for $f \in H^{-1}(D)$.

(i) First consider $f \in H^s(D)$, $s > N/2 - 2$ for $N \geq 3$ ($s = -1$ for $N = 1$ or 2). The proof will make use of the following results from Heinonen, Kilpelainen, and Martio [1].

Lemma 8.1. *Let the assumptions of Theorem 8.1 on the open domain D and the matrix function A be satisfied. Let v be an $\mathcal{A}$-harmonic function, that is, the weak solution of equation (1.2) in the open set Ω for $f = 0$. Then v can be redefined on a set of zero measure, so that it becomes continuous in Ω.*

Note that the continuous representative from Lemma 8.1 is in fact a quasi-continuous $H_0^1(\Omega)$ representative. Indeed, let $v_1 = v$ almost everywhere and v_1 be continuous on Ω. We want to show that v_1 is a quasi-continuous representative of v. There exists a quasi-continuous representative v_2 of v, which is equal to v almost everywhere. So v_1 is continuous, v_2 is quasi-continuous, and $v_1 = v_2$ almost everywhere. Using Heinonen, Kilpelainen, and Martio [1, Thm. 4.12], we get $v_1 = v_2$ quasi-everywhere.

Lemma 8.2. *Let the assumptions of Theorem 8.1 on the open domain D and the matrix function A be satisfied. Let Ω belong to $\mathcal{O}_{c,r}(D)$. If $\theta \in H^1(\Omega) \cap C(\Omega)$ and if h is an $\mathcal{A}$-harmonic function in Ω such that $h - \theta \in H_0^1(\Omega)$, then*

$$\forall x_0 \in \partial\Omega, \quad \lim_{x \to x_0} h(x) = \theta(x_0).$$

Note that the fact that Ω belongs to $\mathcal{O}_{c,r}(D)$ involves the notion of *thickness in any point* of its boundary, which necessarily occurs in the proof of the lemma (cf. Definition 8.3 and Hedberg [1]). Returning to Theorem 8.1, it will be sufficient to prove the continuity for a subsequence of $\{\Omega_n\}$. By Theorem 4.1 there exists a subsequence of $\{\Omega_n\}$, still indexed by n, such that u_n weakly converges to u in $H_0^1(D)$ and u satisfies equation (4.6) in Ω. We now prove that under our assumptions, $u_\Omega = u|_\Omega \in H_0^1(\Omega)$, which implies that $u = e_\Omega(u_\Omega)$. For that purpose we use Lemma 7.2 which says that it is sufficient to prove that $u = 0$ quasi-everywhere in $D\backslash\Omega$ for some quasi-continuous representative u. From Theorem 4.1 we already know that $u = 0$ quasi-everywhere in $D\backslash\overline{\Omega}$. So it remains to show that $u = 0$ quasi-everywhere in $\partial\Omega \cap D$. From the Banach–Saks theorem (cf. Ekeland and Temam [1]), there exists a sequence of averages

$$\psi_n \stackrel{\text{def}}{=} \sum_{k=n}^{N_n} \alpha_k^n u_n, \quad 0 \leq \alpha_k^n \leq 1, \quad \sum_{k=n}^{N_n} \alpha_k^n = 1$$

such that $\psi_n \to u$ in $H_0^1(D)$. Because of the strong convergence of $\{\psi_n\}$ to u in $H_0^1(D)$, we have

$$\psi_n(x) \to u(x) \quad \text{q.e. in } D$$

for a subsequence of $\{\psi_n\}$, still indexed by n. Let G_0 be the set of zero capacity on which $\{\psi_n(x)\}$ does not converge to $u(x)$. Given $x \in D\backslash(\Omega \cup G_0)$, we prove that, for all $\varepsilon > 0$, $|u(x)| < \varepsilon$. We have

$$|u(x)| \leq |u(x) - \psi_n(x)| + |\psi_n(x)|.$$

There exists $N_{\varepsilon,x} > 0$ such that for all $n > N_{\varepsilon,x}$,

$$|u(x) - \psi_n(x)| < \varepsilon/2.$$

It remains to show that there exists N' such that for all $n \geq N'$, $|u_n(x)| < \varepsilon/2$ implies $\psi_n(x)| < \varepsilon/2$. Denote by u_D the solution of (1.2) in D. By assumption on D and f, the solution u_D is continuous in $\overline{D}$. Subtracting the corresponding equations, we obtain

$$\forall \varphi \in H_0^1(\Omega_n; D), \quad \int_D A\nabla(u_D - u_n) \cdot \nabla\varphi \, dx = 0,$$

$$\forall \varphi \in H_0^1(\Omega_n), \quad \int_{\Omega_n} A\nabla(u_D - u_{\Omega_n}) \cdot \nabla\varphi \, dx = 0.$$

Define

$$\tilde{h}_n \stackrel{\text{def}}{=} u_D - u_n, \quad h_n \stackrel{\text{def}}{=} u_D|_{\Omega_n} - u_{\Omega_n}, \quad \theta_n \stackrel{\text{def}}{=} u_D|_{\Omega_n}.$$

Therefore, the restriction h_n of $\tilde{h}_n$ to Ω_n is $\mathcal{A}$-harmonic in Ω_n and, from Lemma 8.1, continuous in Ω_n. Moreover, we have the continuity of u_{Ω_n} in the closure $\overline{\Omega}_n$, and u_{Ω_n} is zero on the boundary. To show that, we use Lemma 8.2 with h_n and θ_n. By definition, $h_n - \theta_n = -u_{\Omega_n}$ belongs to $H_0^1(\Omega_n)$. From the continuity of u_D we obtain that the continuous extension $\tilde{h}_n$ of h_n is equal to u_D on $\partial\Omega_n$. Hence the extension u_n of u_{Ω_n} to the boundary is zero. Using Heinonen, Kilpelainen, and Martio [1, Thm. 6.44], we obtain that, if h_n is δ-Hölderian on $\partial\Omega_n$, then there exists δ_1, $\delta \leq \delta_1 < 1$, such that it is δ_1-Hölderian on all Ω_n. We have that u_D is $(s + 2 - N/2)$-Hölderian on D (with $s + 2 - N/2 > 0$), with a constant M, because of the assumption on f and D. Finally, we get

$$\forall x, y \in \partial\Omega_n, \quad |h_n(x) - h_n(y)| = |u_D(x) - u_D(y)| \leq M|x - y|^{s - N/2 + 2}.$$

So there exists $\delta_1 = \delta_1(N, \beta/\alpha, c)$ and

$$M_{1,n} = 80Mr^{-2} \max(1, (\operatorname{diam}(\Omega_n))^2 \leq 80Mr^{-2} \max(1, (\operatorname{diam}(B))^2 = M_1$$

such that

$$\forall x, y \in \Omega_n, \quad |h_n(x) - h_n(y)| \leq M_{1,n}|x - y|^{\delta_1} \leq M_1|x - y|^{\delta_1}.$$

By a simple argument we obtain that this inequality holds in D, and hence there exists δ_2, $\delta_1 \leq \delta_2 < 1$, such that for all $x, y \in D$ we have

$$|u_n(x) - u_n(y)| \leq |\tilde{h}_n(x) - \tilde{h}_n(y)| + |u_D(x) - u_D(y)|$$
$$\leq M_1|x - y|^{\delta_1} + M|x - y| \leq M_2|x - y|^{\delta_2}.$$

Choose $R > 0$ such that $M_2 R^{\delta_2} < \varepsilon/2$. Because of the H^c-convergence of Ω_n to Ω there exists an integer $n_R > 0$ such that for all $n \geq n_R$ we have $(D\backslash\Omega_n) \cap B(x, R) \neq \varnothing$. For $x_n \in (D\backslash\Omega_n) \cap B(x, R)$,

$$|u_n(x)| = |u_n(x) - u_n(x_n)| \leq M_2|x - x_n|^{\delta_2} \leq M_2 R^{\delta_2} \leq \varepsilon/2$$

because $u_n(x_n) = 0$. So

$$\forall n > n_R, \quad |\psi_n(x)| = \left| \sum_{k=n}^{N_n} \alpha_k^n u_n(x) \right| \leq \sum_{k=n}^{N_n} \alpha_k^n \frac{\varepsilon}{2} = \frac{\varepsilon}{2}.$$

Finally we obtain $|u(x)| \leq \varepsilon$. Because ε was arbitrary, we have $u(x) = 0$ quasi-everywhere on $D\backslash\Omega$, which implies that $u \in H_0^1(\Omega; D)$ and $u_\Omega = u|_\Omega \in H_0^1(\Omega)$. The strong convergence of $\{u_n\}$ to u now follows from Theorem 4.1 and $u = e_\Omega(u_\Omega)$.

(ii) In the next step we prove that the continuity result is preserved for $f \in H^{-1}(D)$. The main idea is to use the continuous dependence of the solution $u_{\Omega,f} \in H_0^1(\Omega; D)$ with respect to f, which is uniform in Ω. Indeed, let $\Omega \subset D$ and $f, g \in H^{-1}(D)$. Then, by a simple subtraction of the equations, we get

$$\int_D |\nabla(u_{\Omega,f} - u_{\Omega,g})|^2 \, dx = \langle f - g, u_{\Omega,f} - u_{\Omega,g} \rangle$$
$$\Rightarrow \|u_{\Omega,f} - u_{\Omega,g}\|_{H_0^1(D)} \leq \|f - g\|_{H^{-1}(D)}.$$

So, let $f \in H^{-1}(D)$ and $f_\varepsilon = f * \rho_\varepsilon$ for some mollifier ρ_ε. Letting $\varepsilon \to 0$ we have $\|f_\varepsilon - f\|_{H^{-1}(D)} \to 0$. Let $\{\Omega_n\} \subset \mathcal{O}_{c,r}(D)$ be a sequence that converges in the H^c-topology to an open set Ω. Then we have

$$u_{\Omega_n, f_\varepsilon} \xrightarrow{H_0^1(D)} u_{\Omega, f_\varepsilon}$$

because $f_\varepsilon \in H^s(D)$ and, from the previous considerations,

$$u_{\Omega_n, f_\varepsilon} \xrightarrow{H_0^1(D)} u_{\Omega_n, f}$$

uniformly in Ω_n. Given $\delta > 0$, we get

$$\|u_{\Omega_n, f} - u_{\Omega, f}\|_{H_0^1(D)} \leq \|u_{\Omega_n, f} - u_{\Omega_n, f_\varepsilon}\|_{H_0^1(D)}$$
$$+ \|u_{\Omega_n, f_\varepsilon} - u_{\Omega, f_\varepsilon}\|_{H_0^1(D)} + \|u_{\Omega, f_\varepsilon} - u_{\Omega, f}\|_{H_0^1(D)}.$$

Choose ε sufficiently small such that $\|f_\varepsilon - f\|_{H^{-1}(D)} < \delta/4$, and for each ε choose $n_{\varepsilon,\delta} > 0$ such that

$$\forall n > n_{\varepsilon,\delta}, \quad \|u_{\Omega_n, f_\varepsilon} - u_{\Omega, f_\varepsilon}\|_{H_0^1(D)} < \delta/2$$
$$\Rightarrow \forall n > n_{\varepsilon,\delta}, \quad \|u_{\Omega_n, f} - u_{\Omega, f}\|_{H_0^1(D)} \leq \delta.$$

As δ was arbitrary, the proof is complete. $\qquad\square$

We can now recover the result of Šverák [2] in dimension $N = 2$.

Theorem 8.2. *Let the assumptions of Theorem 8.1 on the matrix function A be satisfied. Let $N = 2$ and $l > 0$ be a positive integer. Define the set*

$$\mathcal{O}_l \stackrel{\mathrm{def}}{=} \{\Omega \subset D : \#(D\backslash\Omega) \leq l\},$$

where $\#$ denotes the number of connected components. Then the set $\mathcal{O}_l$ is compact in the H^c-topology and the map

$$\Omega \mapsto e_\Omega(u_\Omega) : \mathcal{O}_l \to H_0^1(D)$$

is continuous.

A proof of the result of Šverák [2] is given in Bucur [1] as a consequence of Theorem 8.1. The main idea of the proof is to consider $f \geq 0$ (because of the decomposition of $f = f^+ - f^-$) and a sequence $\{\Omega_n\} \subset \mathcal{O}_l$ such that Ω_n converges in the H^c-topology to an open set Ω, and $u_{\Omega_n} \rightharpoonup u$ in $H_0^1(D)$. He constructs extensions Ω_n^+ of the domains Ω_n with the following properties:

$$\Omega_n^+ \xrightarrow{H^c} \Omega^+, \quad \mathrm{cap}(\Omega^+ \backslash \Omega) = 0, \quad \{\Omega_n^+\} \subset \mathcal{O}_{c,r},$$

where c and r are suitable constants. In fact it can be proved that Ω_n^+ satisfy this capacity density condition, because in an open connected set any two points are linked by a continuous curve that lies in the set, and in the bidimensional case a curve has a positive capacity. From the previous theorem we get $u_{\Omega_n^+} \rightharpoonup u_{\Omega^+}$. We easily obtain that $u_{\Omega^+} = u_\Omega \geq u \geq 0$, which will imply that $u = 0$ quasi-everywhere on $\complement\Omega$, and this concludes the proof. For details, see Bucur [1].

9   Flat Cone Condition and the Compact Family $\mathcal{O}_{c,r}(D)$

In this section we introduce families of domains which have a simpler geometric description and satisfy the capacity density condition for some constants c and r. For a weaker result in that direction see also Bucur and Zolésio [1]. We next prove that the set $\mathcal{O}_{c,r}(D)$ is compact in the H^c-topology. From this we prove existence of optimal solutions for several shape functions which continuously depend on the solution of the state equation.

Definition 9.1.
Let $x \in \mathbf{R}^N$, $0 < \omega < \pi/2$, $\lambda > 0$, and $\nu, d \in \mathbf{R}^N$ such that $|\nu| = |d| = 1$. The *flat cone* is defined as

$$\overline{C}_{x\nu}(\lambda, \omega, d) \stackrel{\text{def}}{=} \overline{C_x(\lambda, \omega, d)} \cap H_x(\nu)$$

where $C_x(\lambda, \omega, d)$ is the cone defined in Notation 6.1 of Chapter 2 and

$$H_x(\nu) \stackrel{\text{def}}{=} \left\{ y \in \mathbf{R}^N \ : \ (y - x) \cdot \nu = 0 \right\}. \qquad \square$$

The cone $\overline{C}_{x\nu}(\lambda, \omega, d)$ is said to be *flat* since it is contained in the hyperplane $H_x(\nu)$, and ω, λ, and d are, respectively, the aperture, the height, and the orientation of the axis of symmetry of the cone. So its N-dimensional Lebesgue measure is zero.

Definition 9.2 (flat cone condition).
Let $0 < \omega < \pi/2$ and $\lambda > 0$ be real numbers.

  (i) An open set Ω in D, is said to satisfy the (ω, λ)-*flat cone condition* (or simply (ω, λ)-f.c.c.) if, for each $x \in \partial\Omega$, there exist unit vectors $d_x, \nu_x \in \mathbf{R}^N$ such that

$$\overline{C}_{x\nu_x}(\lambda, \omega, d_x) \subset \complement\Omega$$

  (ii) The family of open subsets of D which satisfy the (ω, λ)-f.c.c. will be denoted $\mathcal{O}(\omega, \lambda, D)$. $\qquad \square$

Theorem 9.1. *There exist constants $c > 0$ and $r > 0$ such that*

$$\mathcal{O}(\omega, \lambda, D) \subset \mathcal{O}_{c,r}(D).$$

Proof. This readily follows with $r = \lambda$ and for each $x \in \partial\Omega$

$$c = \frac{\operatorname{cap}_{B(x,2\lambda)}(\overline{C}_{x\nu_x}(\lambda, \omega, d_x) \cap \overline{B}(x, \lambda))}{\operatorname{cap}_{B(x,2\lambda)}(\overline{B}(x, \lambda))}$$

and the properties of the capacity with respect to translation and similarity. $\qquad\square$

Theorem 9.2. *The family $\mathcal{O}(\omega, \lambda, D)$ is compact in the H^c-topology.*

Proof. It is sufficient to prove that $\mathcal{O}(\omega, \lambda, D)$ is closed. For that, consider a sequence $\{\Omega_n\} \subset \mathcal{O}(\omega, \lambda, D)$ and $\Omega_n \xrightarrow{H^c} \Omega$. We shall prove that $\Omega \in \mathcal{O}(\omega, \lambda, D)$. Given $x \in \partial\Omega$, there exists a sequence of points, $x_n \in \Omega_n$ such that $x_n \to x$. As $\Omega_n \in \mathcal{O}(\omega, \lambda, D)$, there exists a flat cone $\overline{C}_{x_n\nu_n}(\lambda, \omega, d_n) \subset \complement\Omega_n$. It is easy to see that there exists subsequences $\{d_{n_k}\}$ and $\{\nu_{n_k}\}$ of $\{d_n\}$ and $\{\nu_n\}$, respectively, such that $d_{n_k} \to d$ and $\nu_{n_k} \to \nu$. Because of the properties of the H^c-topology we obtain that $\overline{C}_{x\nu}(\lambda, \omega, d) \subset \complement\Omega$. Finally $\mathcal{O}(\omega, \lambda, D)$ is closed and hence compact. $\qquad\square$

Corollary 3. *Given a sequence $\{\Omega_n\}$ of open sets in $\mathcal{O}(\omega, \lambda, D)$, then*

$$\Omega_n \xrightarrow{H^c} \Omega \quad \Rightarrow \quad u_{\Omega_n} \xrightarrow{H_0^1(D)} u_\Omega.$$

Theorem 9.3. *The family $\mathcal{O}_{c,r}(D)$ of Definition 8.2 is compact.*

Proof. It is sufficient to prove that

$$\forall c > 0, \ \forall r, \ 0 < r < 1, \quad \overline{\mathcal{O}_{c,r}(D)}^{H^c} = \mathcal{O}_{c,r}(D).$$

Pick a sequence $\{\Omega_n\} \subset \mathcal{O}_{c,r}(D)$ such that $\Omega_n \xrightarrow{H^c} \Omega$. Define

$$K_n \stackrel{\text{def}}{=} \overline{D}\backslash\Omega_n, \quad K \stackrel{\text{def}}{=} \overline{D}\backslash\Omega \quad \text{and} \quad \forall\varepsilon > 0, \ K_\varepsilon \stackrel{\text{def}}{=} \overline{\bigcup_{x\in K} B(x, \varepsilon)}^{\mathbf{R}^N}.$$

Let $x \in \partial\Omega$. We now prove that the capacity density condition for Ω is satisfied in the point x. Because of the H^c-convergence,

$$\forall\varepsilon > 0, \ \exists n_\varepsilon, \ \forall n > n_\varepsilon, \quad K_n \subset K_\varepsilon.$$

Given $\varepsilon > 0$ pick $n_{\varepsilon/2}$ and $x_n \in \partial\Omega_n$ such that $|x - x_n| < \varepsilon/2$. Denote by τ the translation by the vector $x_n - x$. As $\tau B(x, r) = B(x_n, r)$, and the capacity condenser is invariant under the translation of the two arguments,

$$\operatorname{cap}_{B(x,2r_0)}(K_\varepsilon \cap B(x, r_0)) = \operatorname{cap}_{B(x_n,2r_0)}(\tau K_\varepsilon \cap B(x_n, r_0)).$$

Because $K_n \subset K_{\varepsilon/2}$ and $|x - x_n| < \varepsilon/2$ we have $\tau K_\varepsilon \supset K_n$. Then, from the monotonicity in the first argument, we get

$$\operatorname{cap}_{B(x,2r_0)}(K_\varepsilon \cap B(x, r_0)) \geq \operatorname{cap}_{B(x_n,2r_0)}(K_n \cap B(x_n, r_0)).$$

Using the capacity density condition for Ω_n we have

$$\mathrm{cap}_{B(x,2r_0)}\big(K_\varepsilon \cap B(x,r_0)\big) \geq c\ \mathrm{cap}_{B(x,2r_0)}\big(B(x,r_0)\big).$$

Letting $\varepsilon \to 0$, and using the continuity of the capacity for decreasing sequences of compact sets and the fact that

$$\bigcap_{\varepsilon>0} K_\varepsilon = K,$$

we get the capacity density condition for Ω in x. Hence $\Omega \in \mathcal{O}_{c,r}(D)$. $\qquad\square$

A first example of extremal domain follows directly from Theorem 9.3.

Theorem 9.4. *Let the assumptions of Theorem 8.1 on the open domain D and the matrix function A be satisfied. Further assume that E is a nonempty open subset of D. Further assume that A satisfies conditions (1.3). Then the maximization problem*

$$\sup\big\{\lambda^A(\Omega) : \forall \Omega \in \mathcal{O}_{c,r}(D) \text{ such that } E \subset \Omega \subset D\big\} \tag{9.1}$$

has solutions.

In some sense the maximizing solution is an $\mathcal{O}_{c,r}(D)$-approximation of the set E. The result follows from the upper semicontinuity of the map $\Omega \mapsto \lambda^A(\Omega)$ in the H^c-topology and the fact that the family of admissible domains is compact (Theorem 2.3 in Chapter 4 and Theorem 9.3) for the Hausdorff complementary topology.

We now give the following general existence theorem.

Theorem 9.5. *Let the assumptions of Theorem 8.1 on the open domain D and the matrix function A be satisfied. Let u_Ω be the solution of (1.2) for Ω in $\mathcal{O}_{c,r}(D)$. If h is continuously defined from $H_0^1(D)$ into $\mathbf{R}$, then $J(\Omega) = h(e_\Omega(u_\Omega))$ is continuously defined from $\mathcal{O}_{c,r}(D)$ into $\mathbf{R}$ and reaches its extremal values on that set.*

Example 9.1.
Consider the following shape function for which we can get the existence of minimizing domains even if the assumptions of Theorem 9.5 on h are not satisfied. Given $\alpha > 0$,

$$J(\Omega) \overset{\mathrm{def}}{=} \frac{1}{2} \int_\Omega |\nabla u_\Omega - \bar{z}|^2 dx + \alpha\, \frac{1}{\int_\Omega |u_\Omega|^2 dx}, \tag{9.2}$$

where u_Ω is the solution of equation (1.2) and $\bar{z} \in L^2(D; \mathbf{R}^N)$. From Theorem 2.1 $u \overset{\mathrm{def}}{=} e_\Omega(u_\Omega)$ and ∇u_Ω are zero almost everywhere in $D\backslash\Omega$, and the function can be rewritten as

$$J(\Omega) = \frac{1}{2} \int_D |\nabla u - \bar{z}|^2\, dx + \alpha\, \frac{1}{\int_D |u|^2\, dx} - \frac{1}{2} \int_{D\backslash\Omega} |\bar{z}|^2\, dx. \tag{9.3}$$

The last term is not of the form $h(u_\Omega)$ for some h, but simply of the form $h(\Omega)$. It turns out that J is not continuous but only lower semicontinuous for the H^c-topology. Hence it can be minimized. $\qquad\square$

More precisely, we need the following result.

Lemma 9.1. *Let μ be a positive finite measure on D. The map $\Omega \mapsto \mu(\Omega)$ is lower semicontinuous with respect to the H^c-topology.*

Proof. Let $\{\Omega_n\}$ be a sequence of open sets in B, $K_n = \bar{D}\backslash\Omega_n$, $K = \bar{D}\backslash\Omega$, and assume that $\Omega_n \xrightarrow{H^c} \Omega$. Given $\varepsilon > 0$, we have $K_\varepsilon \supset K_n$ for all $n > n_\varepsilon$. Then for all $n > n_\varepsilon$, $\mu(K_\varepsilon) \geq \mu(K_n)$ and

$$\forall \varepsilon > 0, \quad \mu(K_\varepsilon) \geq \limsup_{n \to \infty} \mu(K_n).$$

But K_ε is a monotonically decreasing sequence and then $\mu(K_\varepsilon) \to \mu(K)$ as $\varepsilon \to 0$, and the result follows. $\qquad\square$

The minimization problem of Example 9.1 can now be formulated as follows on D. Let D be a bounded open nonempty domain in $\mathbf{R}^N$, f an element of $H^{-1}(D)$, and B a ball containing D. Let μ be a positive measure on D and a a positive constant such that $0 < a < \mu(D) < +\infty$. Let u_Ω be the solution of equation (1.2), and $u = e_\Omega(u_\Omega)$, its extension by zero in $H_0^1(D)$. For $J(\Omega)$ defined in (9.2) consider the following problem:

$$\min\{J(\Omega) : \Omega \in \mathcal{O}_{c,r}(B), \ \Omega \subset D, \ \mu(\Omega) \leq a\}. \tag{9.4}$$

Theorem 9.6. *Let the assumptions of Theorem 8.1 on the open domain D and the matrix function A be satisfied. For any constants $\alpha > 0$, $c > 0$ and r, $0 < r < 1$, (9.4) has at least one solution.*

Proof. It is sufficient to notice that the first term in (9.3) is continuous under the assumptions of Theorem 9.5. The second term is lower semicontinuous from Lemma 9.1 (with the measure of density $|\bar{z}|^2$). To complete the proof, recall that, if $\Omega_n \subset D$ and $\Omega_n \xrightarrow{H^c} \Omega$, then $\Omega \subset D$. As $\alpha > 0$, a minimizing sequence cannot converge to $\varnothing$. In fact, for any admissible domain Ω_0 and any optimal solution u_{Ω_0},

$$\int_\Omega |u_{\Omega_0}|^2 dx \geq \frac{\alpha}{J(\Omega_0)}. \qquad\qquad\square$$

10 Examples with a Constraint on the Gradient

Let D be a fixed bounded smooth open domain in $\mathbf{R}^N$, f in $H^{-1}(D)$, and g in $L^2(D)$. For any open subset Ω of D, let u_Ω be the solution of the Dirichlet problem (1.2) in $H_0^1(\Omega)$. Given $\alpha > 0$, $M > 0$, and an open subset E of D, consider the following minimization problem:

$$\inf\{J(\Omega) : \Omega \text{ open }, E \subset \Omega \subset D, \ \operatorname{ess\,sup}|\nabla u_\Omega| \leq M\}, \tag{10.1}$$

$$J(\Omega) \stackrel{\mathrm{def}}{=} \mathrm{m}(\Omega) + \alpha \int_\Omega |u_\Omega - g|^2 dx.$$

It is understood that if $|\nabla u|$ is not in $L^\infty(D)$, then the $\operatorname{ess\,sup}|\nabla u|$ is $+\infty$. In a first step it is easy to check that the problem

$$\inf\left\{ m(\Omega) + \alpha\int_\Omega |u_\Omega - g|^2\,dx \ :\ \operatorname{ess\,sup}|\nabla u_\Omega| \le M,\ E \subset \Omega,\ \Omega \in \mathcal{O}_{c,r}(D)\right\} \quad (10.2)$$

has minimizing solutions. Let $\{\Omega_n\}$ be a minimizing sequence. By assumption there exists Ω and a subsequence, still indexed by n, such that $\Omega_n \to \Omega$ in the H^c-topology and

$$\operatorname{ess\,sup}|\nabla u_{\Omega_n}| \le M.$$

Then $|\nabla u_n|$ converges in $L^2(D)$ to $|\nabla u|$ from the previous sections. For any $\varphi \in L^2_+(D)$ we have

$$\int_D (|\nabla u_n| - M)\,\varphi\,dx \le 0$$

and then, in the limit, we get the same inequality with u. Then, the function being lower semicontinuous with respect to that topology, we get the existence.

Lemma 10.1. *For any $f \in H^{-1}(D)$ and any $r > 0$ and $c > 0$, problem (10.2) has solutions in $\mathcal{O}_{c,r}(D)$*

In fact, the presence of the constraint on the gradient of u is helpful to get the continuity of the application $\Omega \mapsto u = e_\Omega(u_\Omega)$, and we directly get the existence of solutions to the original problem (10.1). In the proof of Theorem 8.1 we only needed the equicontinuity of the family of solutions $\{u_n = e_{\Omega_n}(u_{\Omega_n})\}$ and the fact that if $x \in \complement\Omega$ then $u(x) = 0$ for a quasi-continuous representative. If $\Omega \in \mathcal{O}_{c,r}(D)$, those two assumptions are readily satisfied. Notice that the boundedness of the gradients in (10.1) implies the equicontinuity of any minimizing sequence $\{u_n\}$. So, in order to obtain a continuity result with constraints on the boundedness of the gradient, we only have to notice that if $u \in H^1_0(\Omega)$, $\operatorname{ess\,sup}|\nabla u| \le M$. Then $u \in W^{1,\infty}(\Omega)$ and u is Lipschitzian with the constant M, or more exactly, there exists a Lipschitzian function almost everywhere equal to u, which is also a quasi-continuous representative of u. To obtain that $u(x) = 0$ in any point of the complement of Ω, we have to introduce the following capacity constraints on Ω: we require that Ω be capacity extended (see Bucur [1]), i.e., that $\Omega = \Omega^*$, where

$$\Omega^* \stackrel{\mathrm{def}}{=} \left\{x \in \mathbf{R}^N \ :\ \exists \varepsilon_x > 0,\ \text{such that } \operatorname{cap}_D(B(x,\varepsilon_x) \cap \complement\Omega) = 0\right\}. \quad (10.3)$$

Indeed, let u be continuous on D, $u = 0$ quasi-everywhere on $\complement\Omega$, and $\Omega = \Omega^*$. Then $\forall x \in \complement\Omega$, $\forall \varepsilon > 0$ we have $\operatorname{cap}_D(B(x,\varepsilon_x) \cap \complement\Omega) > 0$ and there exists a point x_ε in $B(x,\varepsilon_x) \cap \complement\Omega$ where $u(x_\varepsilon) = 0$. By continuity we get $u(x) = 0$. We recall from Bucur [1] the main properties of the capacity extension.

Proposition 10.1. *For any open $\Omega \subset D$, the set Ω^* is open,*

$$\Omega \subset \Omega^*, \quad \operatorname{cap}_D(\Omega^* \setminus \Omega) = 0 \quad \text{and} \quad (\Omega^*)^* = \Omega^*.$$

As $\mathrm{cap}_D(\Omega^* \setminus \Omega) = 0$ we get $e_\Omega(u_\Omega) = e^*_\Omega(u_{\Omega^*})$ and any minimizing sequence can only be made up of capacity-extended domains.

Theorem 10.1. *Given* $f, g \in L^2(D)$, $\alpha > 0$, $M > 0$, *problem* (10.1) *has minimizing solutions.*

Proof. From the boundedness of the gradient we obtain that $\{u_{\Omega_n}\}$ are uniformly Lipschitz continuous, and we can apply the same arguments as in Theorem 8.1, avoiding the capacity conditions. $\qquad\square$

Consider the penalized version of problem (10.1). Given $M > 0$ and $\beta > 0$, define

$$J_\beta(\Omega) \stackrel{\text{def}}{=} J(\Omega) + \beta \sup_D \mathrm{ess}\, (|\nabla u_\Omega| - M)^+ \tag{10.4}$$

and consider the existence of solutions to the following problem:

$$\inf \{ J_\beta(\Omega) \,:\, E \subset \Omega \text{ open } \subset D \}. \tag{10.5}$$

Theorem 10.2. *Given* f *and* g *in* $L^2(D)$, $\alpha > 0$, $\beta > 0$, $M \geq 0$, *problem* (10.5) *has minimizing solutions.*

Proof. Let $\{\Omega_n\}$ be a minimizing sequence for the function J_β, $\Omega_n = \Omega^*_n$. We have

$$\mathrm{ess\,sup}\, |\nabla u_{\Omega_n}| \leq \max \left\{ M, \beta^{-1} J(\Omega_1) \right\} = M'$$

for all $n \geq 1$. Then, from the previous considerations we have the strong convergence in $H^1_0(D)$ of $u_n = e_{\Omega_n}(u_{\Omega_n})$ to $u = e_\Omega(u_\Omega)$, for some subsequence $\{\Omega_n\}$ that converges to Ω in the H^c-topology. Hence Ω is a minimizing domain. To complete the proof note that the map

$$\Omega \mapsto \sup_D \mathrm{ess}\, (|\nabla u| - M)^+$$

is not lower semicontinuous from the H^c-topology to $\mathbf{R}$. Nevertheless, if $\{\Omega_n\}$ is a minimizing sequence for problem (10.5), the required property is satisfied. We can assume that the sequence is chosen such that

$$\sup_D \mathrm{ess}\, |\nabla u_n| \to c = \liminf_{n \to \infty} \left(\sup_D \mathrm{ess}\, |\nabla u_n| \right).$$

For any $\varepsilon > 0$ there exists $n_\varepsilon > 0$ such that for $n \geq n_\varepsilon$, $\sup_D \mathrm{ess}\, |\nabla u_{\Omega_n}| \leq c + \varepsilon$. Because of the $H^1_0(D)$-strong convergence of u_n to u, we get

$$\forall \varepsilon > 0, \quad \sup_D \mathrm{ess}\, |\nabla u| \leq c + \varepsilon$$

$$\Rightarrow \sup_D \mathrm{ess}\, |\nabla u| \leq \liminf \left(\sup_D \mathrm{ess}\, |\nabla u| \right). \qquad\square$$

Remark 10.1.
We can change the L^∞-norm for a differentiable one; that is, we need

$$\|\nabla u\|_{1,p} \leq M \quad \forall p > N.$$

By the continuous inclusion $W^{1,p}(D) \subset W^{\varepsilon,\infty}(D)$, ε small, the result still holds. $\qquad\square$

Chapter 7
Transformations versus Flows of Velocities

1 Introduction

In the previous chapters we have constructed examples of metric spaces of sets, which, of course, are not topological vector spaces. Studying continuity and semi-differentiability of a function defined on such spaces is analogous to studying continuity and semidifferentiability on a manifold. The two notions can be associated with the behavior of the function along continuous one-dimensional curves or along flows generated by the solutions of a differential system. Such flows are easy to generate not only in the Euclidean space but also in smooth submanifolds where line paths between two points can no longer be used. The *generic result* of this chapter is that continuity of a shape function with respect to the Courant metric is equivalent to its continuity along the flows of velocity vector fields in the class of transformations associated with the Courant metric. In practice this condition is much easier to satisfy on specific examples. In the next chapter flows of velocities will be adopted as the natural framework for defining shape semiderivatives.

In section 2 we motivate and adapt constructions and definitions of *Gâteaux* and *Hadamard* semiderivatives in topological vector spaces to shape functions defined on shape spaces. The analogue of the Gâteaux semiderivative for sets is obtained by the *method of perturbation of the identity operator*, while the analogue of the Hadamard semiderivative comes from the *velocity (speed) method*. In section 3 the velocity and transformation viewpoints will be emphasized through a series of examples of commonly used families of transformations of sets. They include C^k-domains, Cartesian graphs, polar coordinates, and level sets.

In section 4 we establish the equivalence between deformations obtained by a family of transformations and deformations obtained by the flow of a velocity field. Section 4.1 gives the equivalence under relatively general conditions. Section 4.2 shows that Lipschitzian perturbations of the identity operator can be generated by the flow of a nonautonomous velocity field. In section 4.3 the conditions of section 4.2 are sharpened for the special families of velocity fields in $C_0^k(\mathbf{R}^N, \mathbf{R}^N)$, $C^k(\overline{\mathbf{R}^N}, \mathbf{R}^N)$, and $C^{k,1}(\overline{\mathbf{R}^N}, \mathbf{R}^N)$. The *constrained case* where the family of domains are subsets of a fixed *hold-all* or *universe* is studied in section 5. In both sections 4 and 5 we show that under appropriate conditions, starting from a family of transformations is equivalent to starting from a family of velocity fields. This is a key result, which bridges the gap between the two points of view.

Section 6 contains the central generic result of this chapter: the equivalence of the continuity of a shape function with respect to the Courant metric and its continuity along the flows of all velocity fields for the families $C_0^k(\mathbf{R}^N, \mathbf{R}^N)$, $C^k(\overline{\mathbf{R}^N}, \mathbf{R}^N)$, and $C^{k,1}(\overline{\mathbf{R}^N}, \mathbf{R}^N)$ associated with the metric. This result is of both intrinsic and practical interest since it is generally easier to check the continuity along flows than with respect to the metric. In view of this equivalence the subsequent developments of this book will be based on the velocity (speed) method.

2   Shape Functions and Choice of Shape Derivatives

Recall that for a real function $f : E \to \mathbf{R}$ defined on a topological vector space E, the *Gâteaux semiderivative* at $x \in E$ in the direction $v \in E$ is given as the limit (when it exists)

$$df(x; v) \stackrel{\text{def}}{=} \lim_{t \searrow 0} \frac{f(x + tv) - f(x)}{t},$$

and the *Hadamard semiderivative* as

$$d_H f(x; v) \stackrel{\text{def}}{=} \lim_{\substack{w \to v \\ t \searrow 0}} \frac{f(x + tw) - f(x)}{t}.$$

The analogue of the Gâteaux semiderivative for shape functions will be obtained by the *method of perturbation of the identity operator*, while the analogue of the Hadamard semiderivative will come from the *velocity (speed) method*. The Hadamard notion of semiderivative is important in situations where the shape is parametrized and the chain rule is necessary. These basic notions and constructions will now be formally extended to deal with shape functions.

We first give the definition of a shape function.

Definition 2.1.
Given a nonempty subset D of $\mathbf{R}^N$, consider the set $\mathcal{P}(D) = \{\Omega : \Omega \subset D\}$ of subsets of D. The set D will be referred to as the underlying *hold-all* or *universe*. A *shape function* is a map

$$J : \mathcal{A} \to E \tag{2.1}$$

from some *admissible family* $\mathcal{A}$ of domains in $\mathcal{P}(D)$ into a topological space E such that for any homeomorphism T of $\overline{D}$ such that $T(\Omega) = \Omega$, $J(\Omega) = J(T(\Omega))$ for all elements $\Omega \in \mathcal{A}$. $\qquad \square$

This universe D can represent some physical or mechanical constraint, a submanifold of $\mathbf{R}^N$, or some mathematical constraint. In many instances it can be chosen large and as smooth as necessary for the analysis. In the unconstrained case D is equal to $\mathbf{R}^N$.

Consider a real-valued *shape function*

$$\Omega \mapsto J(\Omega) : \mathcal{A} \subset \mathcal{P}(\mathbf{R}^N) \to \mathbf{R} \tag{2.2}$$

defined over a family $\mathcal{A}$ in $\mathcal{P}(\mathbf{R}^N)$. All the spaces of domains considered in the previous chapters are nonlinear and nonconvex and their elements are equivalence classes of domains or transformations. So it is important to a priori specify the equivalence class under consideration and make sure that the value of the shape function $J(\Omega)$ is well defined: it has the same value for all elements in the equivalence class. Even if $\mathcal{A}$ does not generally have a vector space structure, it is possible to consider differential quotients and their limit along one-dimensional paths around Ω. Consider perturbations of the identity operator

$$\forall X \in \mathbf{R}^N,\ t \geq 0, \quad T(t, X) \stackrel{\text{def}}{=} X + t\,\theta(X) \tag{2.3}$$

for some family of vector fields $\theta : \mathbf{R}^N \to \mathbf{R}^N$. The semiderivative $dJ(\Omega; \theta)$ at Ω in the direction θ is the limit (when it exists) of the differential quotient

$$\frac{J(T_t(\Omega)) - J(\Omega)}{t}$$

as $t > 0$ goes to zero, where

$$\Omega_t \stackrel{\text{def}}{=} T_t(\Omega) = \{T_t(X) : \forall X \in \Omega\} \text{ and } T_t(X) \stackrel{\text{def}}{=} T(t, X). \tag{2.4}$$

When θ is restricted to some topological vector space, the continuity and the linearity of the map

$$\theta \mapsto dJ(\Omega; \theta) \stackrel{\text{def}}{=} \lim_{t \searrow 0} \frac{J(T_t(\Omega)) - J(\Omega)}{t} \tag{2.5}$$

can be studied, and the results are analogous to the ones in Banach spaces. Of course this approach is limited by the fact that perturbations occur along lines between $I + \theta$ and the identity operator I. It is generally not applicable when D is a smooth submanifold of $\mathbf{R}^N$ with nonzero curvature.

This first approach to the definition of a shape semiderivative is not completely satisfactory since the perturbations of the identity are *nonlocal* transformations: the velocity field $dx(t)/dt = \theta(X)$ at the point $x(t) = T_t(X)$ depends on the point X instead of $x(t)$. In general, this will not affect first-order semiderivatives of J, but will introduce an "acceleration term" in the definition of second-order semiderivatives in Chapter 8. This can be easily fixed, and a more natural approach followed. Go back to our initial choice of deformation in (2.3) and consider $t \geq 0$ as an *artificial time*. Rewrite expression (2.3) as a difference

$$x(t) - x(0) = (t - 0)\theta\big(x(0)\big), \quad x(0) = X, \tag{2.6}$$

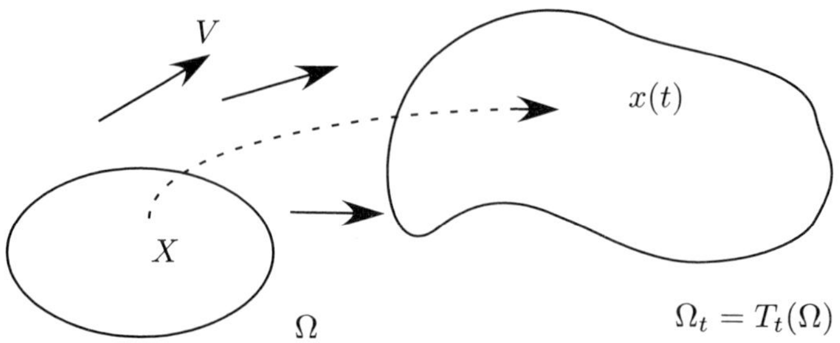

Figure 7.1. *Transport of Ω by the velocity field V.*

and in differential form

$$dx(t) = dt\,\theta\big(x(0)\big) \cong dt\,\theta\big(x(t)\big). \tag{2.7}$$

In the limit this yields a new *local deformation* $T_t : \mathbf{R}^N \to \mathbf{R}^N$,

$$T_t(X) = x(t), \quad t \ge 0, \tag{2.8}$$

defined by the solution $x(t)$ of the differential equation

$$\frac{dx}{dt}(t) = \theta\big(x(t)\big), \quad t > 0, \ x(0) = X. \tag{2.9}$$

This is the basis of the *velocity (speed) method*. Here the velocity of the point $x(t)$ at time t is equal to the velocity field θ evaluated at $x(t)$: it is now a local deformation. The construction readily extends to transformations

$$T(t, X) \stackrel{\text{def}}{=} x(t, X), \quad t \ge 0, \ X \in \mathbf{R}^N, \tag{2.10}$$

defined by the flow of the differential equation

$$\frac{dx}{dt}(t, X) = V(t, x(t)), \ t \ge 0, \quad x(t, X) = X, \tag{2.11}$$

for nonautonomous velocity fields $V(t)(x) \stackrel{\text{def}}{=} V(t, x)$ (cf. Figure 7.1).

The choice of the terminology "velocity" to describe this method is accurate but may become ambiguous in problems where the variables involved are themselves "physical velocities": this situation is commonly encountered in continuum mechanics and material sciences. In such cases it may be useful to distinguish between the "artificial velocity" and the "physical velocity." This is at the origin of the terminology *speed method* which has often been used in the literature. The latter terminology is convenient, but not as accurate and descriptive as *velocity method*. We shall keep both terminologies and use the one that is most suitable in the context of the problem at hand.

The shape semiderivative of J at Ω in the direction V will be defined as

$$dJ(\Omega; V) \stackrel{\text{def}}{=} \lim_{t \searrow 0} \frac{J(\Omega_t) - J(\Omega)}{t} \tag{2.12}$$

(when the limit exists), where

$$\Omega_t \stackrel{\text{def}}{=} T_t(\Omega) = \{T_t(x) : x \in \Omega\} \tag{2.13}$$

and T_t is the transformation of $\mathbf{R}^N$ defined by (2.10)–(2.11). *In this book we have chosen to give all the basic definitions, constructions, and theorems within the context of the velocity method.* This choice is motivated by the fact that most results based on perturbations of the identity or asymptotic developments can be readily recovered by direct application of the velocity method with *nonautonomous velocity fields* $V(t, x)$. To support this assertion we now construct the time-dependent velocity field $V(t, x)$ associated with the perturbation of the identity

$$T_t(X) = x(t) = X + t\,\theta(X). \tag{2.14}$$

The field V is to be chosen such that the function x in (2.14) is the solution of the differential equation

$$\frac{dx}{dt}(t) = V\big(t, x(t)\big), \quad x(0) = X. \tag{2.15}$$

It is readily seen that the appropriate choice is $V(t) = \theta \circ T_t^{-1}$ or

$$\boxed{V(t, x) \stackrel{\text{def}}{=} \theta(T_t^{-1}(x)) \quad \forall x \in \mathbf{R}^N, \ \forall t \geq 0.} \tag{2.16}$$

Under appropriate continuity and differentiability assumptions:

$$\forall x \in \mathbf{R}^N, \quad V(0)(x) \stackrel{\text{def}}{=} V(0, x) = \theta(x) \tag{2.17}$$

$$\forall x \in \mathbf{R}^N, \quad \dot{V}(0)(x) \stackrel{\text{def}}{=} \frac{\partial V}{\partial t}(t, x)\Big|_{t=0} = -\,[D\theta(x)]\,\theta(x) \tag{2.18}$$

where $D\theta(x)$ is the Jacobian matrix of θ at the point x. In compact notation,

$$V(0) = \theta \quad \text{and} \quad \dot{V}(0) = -\,[D\theta]\,\theta. \tag{2.19}$$

This last computation shows that at time 0, the points of the domain Ω are simultaneously affected by the velocity field $V(0) = \theta$ and the acceleration field $\dot{V}(0) = -[D\theta]\theta$. Under suitable assumptions the two methods will produce the same first-order semiderivative. However, second-order semiderivatives will differ by an acceleration term which will appear in the expression obtained by the method of perturbation of the identity.

3   Families of Transformations of Domains

In section 2 we have given a formal presentation of the main steps and options leading to the definition of semiderivatives of a shape function. We have emphasized perturbations of the identity operator and flows of velocity fields. Before proceeding with a more abstract treatment, we present several examples of definitions of shape semiderivatives that can be found in the literature. We consider special classes of domains (C^∞, C^k, Lipschitzian), Cartesian graphs, polar coordinates, and level sets that provide classical examples of parametrized and/or constrained deformations. In each case we construct the associated underlying family (not necessarily unique) of transformations $\{T_t : 0 \le t \le \tau\}$.

3.1 C^∞-Domains

Let Ω be an open domain of class C^∞ in $\mathbf{R}^N$. Recall from section 5 of Chapter 2 that in any point $x \in \Gamma$ the unit outward normal is given by

$$\forall y \in \Gamma_x \stackrel{\text{def}}{=} \mathcal{U}(x) \cap \Gamma, \quad n(y) = \frac{m_x(y)}{|m_x(y)|}, \tag{3.1}$$

where

$$m_x(y) = - {}^*(Dh_x)^{-1}(h_x^{-1}(y))\, e_N \text{ in } \mathcal{U}(x) \tag{3.2}$$

$$\Rightarrow n = -\frac{{}^*(Dh_x)^{-1}e_N}{|\,{}^*(Dh_x)^{-1}e_N|} \circ h_x^{-1}. \tag{3.3}$$

When Γ is compact it is possible to find a finite sequence of points $\{x_j : 1 \le j \le J\}$ in Γ such that

$$\Gamma \subset \mathcal{U} \stackrel{\text{def}}{=} \bigcup_{j=1}^{J} \mathcal{U}_j, \quad \mathcal{U}_j \stackrel{\text{def}}{=} \mathcal{U}(x_j).$$

As in the definition of the boundary integral, associate with $\{\mathcal{U}_j\}$ a partition of unity $\{r_j\}$:

$$r_j \in \mathcal{D}(\mathcal{U}_j), \quad 0 \le r_j \le 1, \quad \sum_{j=1}^{J} r_j = 1 \text{ in } \mathcal{U}_0$$

for some neighborhood $\mathcal{U}_0$ of Γ such that

$$\Gamma \subset \mathcal{U}_0 \subset \overline{\mathcal{U}_0} \subset \mathcal{U}.$$

For the C^∞-domain Ω the normal satisfies

$$n = \sum_{j=1}^{J} r_j n = \sum_{j=1}^{J} r_j \frac{m_j}{|m_j|} \in C^\infty(\Gamma; \mathbf{R}^N)$$

since

$$\forall j, \quad \frac{m_j}{|m_j|} \circ h_j \in C^\infty(B; \mathbf{R}^N).$$

Given any $\rho \in C^\infty(\Gamma)$ and $t \geq 0$, consider the following perturbation Γ_t of Γ along the normal field n:

$$\Gamma_t \stackrel{\text{def}}{=} \{x \in \mathbf{R}^N : x = X + t\rho(X)n(X), \forall X \in \Gamma\}. \tag{3.4}$$

We claim that for τ sufficiently small and all t, $0 \leq t \leq \tau$, the set Γ_t is the boundary of a C^∞-domain Ω_t by constructing a transformation T_t of $\mathbf{R}^N$ which maps Ω onto Ω_t and Γ onto Γ_t. First construct an extension $N \in \mathcal{D}(\mathbf{R}^N, \mathbf{R}^N)$ of the normal field n on Γ. Define

$$m \stackrel{\text{def}}{=} \sum_{j=1}^{J} r_j m_j \in \mathcal{D}(\mathcal{U}, \mathbf{R}^N). \tag{3.5}$$

By construction

$$m \in C^\infty(\Gamma; \mathbf{R}^N), \tag{3.6}$$

$m \neq 0$ on Γ, and there exists a neighborhood $\mathcal{U}_1$ of Γ contained in $\mathcal{U}_0$ where $m \neq 0$ since m is at least C^1.

Now construct a function r_0 in $\mathcal{D}(\mathcal{U}_1)$, $0 \leq r_0(x) \leq 1$, and a neighborhood $\mathcal{V}$ of Γ such that

$$r_0 = 1 \text{ in } \mathcal{V} \quad \text{and} \quad \Gamma \subset \mathcal{V} \subset \overline{\mathcal{V}} \subset \mathcal{U}_1.$$

Define the vector field

$$\forall x \in \mathbf{R}^N, \quad N(x) \stackrel{\text{def}}{=} r_0(x) \frac{m(x)}{|m(x)|}. \tag{3.7}$$

Hence N belongs to $\mathcal{D}(\mathbf{R}^N, \mathbf{R}^N)$ since $\operatorname{supp} N \subset \overline{\mathcal{V}}$ is compact. Moreover,

$$N = \frac{m}{|m|} \text{ in } \mathcal{V} \quad \Rightarrow \quad N(x) = n(x) \text{ on } \Gamma = \Gamma \cap \mathcal{V}.$$

For each j, $\rho \circ h_j \in C^\infty(\overline{B_0})$ and the extensions

$$\rho_j^h(\zeta) = \rho_j^h(\zeta', \zeta_N) \stackrel{\text{def}}{=} \rho(h_j(\zeta', 0)) \quad \forall \zeta \in B, \quad \tilde{\rho}_j \stackrel{\text{def}}{=} \rho_j^h \circ g_j,$$

belong, respectively, to $C^\infty(\overline{B})$ and $C^\infty(\overline{\mathcal{U}_j})$. Then

$$\tilde{\rho} \stackrel{\text{def}}{=} \sum_{j=1}^{J} r_j \tilde{\rho}_j \in \mathcal{D}(\mathbf{R}^N)$$

is an extension of ρ from Γ to $\mathbf{R}^N$ with compact support since $\operatorname{supp} r_j \subset \mathcal{U}_j \subset \mathcal{U}$.

Define the following transformation of $\mathbf{R}^N$:

$$T_t(X) \stackrel{\text{def}}{=} X + t\,\tilde{\rho}(X)N(X), \quad t \geq 0. \tag{3.8}$$

By construction, $\tilde{\rho}\,N$ is uniformly Lipschitzian in $\mathbf{R}^N$, and by Theorem 4.2 there exists $0 < \tau$ such that T_t is bijective and bicontinuous from $\mathbf{R}^N$ onto itself. As a result from Dugundji [1] for $0 \leq t \leq \tau$,

$$\Omega_t = T_t(\Omega) = [I + t\tilde{\rho}\,N](\Omega),$$
$$\partial\Omega_t = T_t(\partial\Omega) = [I + t\tilde{\rho}\,N](\partial\Omega) = [I + t\rho\,n](\partial\Omega) = \Gamma_t.$$

Since the domain Ω_t is specified by its boundary Γ_t, it only depends on ρ and not on its extension $\tilde{\rho}$. The special transformation T_t introduced here is of class C^∞, that is, $T_t \in C^\infty(\mathbf{R}^N, \mathbf{R}^N)$, and $\frac{1}{t}\,[T_t - I]$ is proportional to the normal field n on Γ, but it is not proportional to the normal n_t on Γ_t for $t > 0$. In other words, at $t = 0$ the deformation is along n, but at $t > 0$ the deformation is generally not along n_t.

If $J(\Omega)$ is a real-valued shape function defined on C^∞-domains in D, the semiderivative (if it exists) is defined as follows: for all $\rho \in C^\infty(\Gamma)$

$$d_n J(\Omega; \tilde{\rho}) \stackrel{\text{def}}{=} \lim_{t \searrow 0} \frac{J\big((I + t\tilde{\rho}N)\Omega\big) - J(\Omega)}{t}. \tag{3.9}$$

It turns out that this limit only depends on ρ and not on its extension $\tilde{\rho}$.

3.2　C^k-Domains

When Ω is a domain of class C^k with boundary Γ, the normal field n belongs to $C^{k-1}(\Gamma, \mathbf{R}^N)$. Therefore, choosing deformations along the normal would yield transformations $\{T_t\}$ mapping C^k-domains Ω onto C^{k-1}-domains $\Omega_t = T_t(\Omega)$. The obvious way to deal with C^k-domains is to relax the constraint that the perturbation $\tilde{\rho}N$ be carried by the normal. Choose vector fields Θ in $\mathcal{D}^k(\mathbf{R}^N, \mathbf{R}^N)$ and consider the family of transformations

$$T_t = I + t\Theta, \quad \Omega_t = T_t(\Omega), \quad t \geq 0. \tag{3.10}$$

This is a generalization of the family of transformations (3.8) in section 3.1 from $\tilde{\rho}N$ to Θ. For $k \geq 1$, Θ is again uniformly Lipschitzian in $\mathbf{R}^N$, and by Theorem 4.2 there exists $\tau > 0$ such that T_t is bijective and bicontinuous from $\mathbf{R}^N$ onto itself. Thus for $0 \leq t \leq \tau$,

$$\Omega_t = T_t(\Omega) = [I + t\Theta](\Omega) \text{ and } \partial\Omega_t = T_t(\partial\Omega) = [I + t\Theta](\partial\Omega).$$

A more restrictive approach to get around the lack of sufficient smoothness of the normal n to Γ would be to introduce a transverse field p on Γ such that

$$p \in C^k(\Gamma; \mathbf{R}^N), \quad \forall x \in \Gamma, \quad p(x) \cdot n(x) > 0. \tag{3.11}$$

Given p and $\rho \in C^k(\Gamma)$ define for $t \geq 0$

$$\Gamma_t \stackrel{\text{def}}{=} \big\{x \in \mathbf{R}^N : x = X + t\rho(X)p(X), \forall X \in \Gamma\big\}. \tag{3.12}$$

Choosing C^k-extensions $\tilde{\rho}$ and $\tilde{p}$ of ρ and p we can go back to the case where

$$\Theta = \tilde{\rho}\tilde{p} \in \mathcal{D}^k(\mathbf{R}^N, \mathbf{R}^N). \tag{3.13}$$

For any $\Theta \in \mathcal{D}^k(\mathbf{R}^N, \mathbf{R}^N)$ the semiderivative is defined as

$$d_k J(\Omega; \Theta) \overset{\text{def}}{=} \lim_{t \searrow 0} \frac{J\big((I + t\Theta)(\Omega)\big) - J(\Omega)}{t}. \tag{3.14}$$

3.3 Cartesian Graphs

In many applications it is convenient to work with domains Ω which are the hypograph of some positive function γ in Cartesian coordinates. Such domains are typically of the form

$$\Omega \overset{\text{def}}{=} \left\{ (x', x_N) \in \mathbf{R}^N : x' \in U \subset \mathbf{R}^{N-1} \text{ and } 0 < x_N < \gamma(x') \right\}, \tag{3.15}$$

where U is a connected open set in $\mathbf{R}^{N-1}$ and $\gamma \in C(\overline{U}; \mathbf{R}_+)$ is a positive function. Many free boundary and contact problems are formulated over such domains. Usually the domains Ω (and hence the functions γ) will be constrained. The function γ can be specified on $\overline{U}\backslash U$ or not. In some examples the derivative of γ could also be specified, that is, $\partial\gamma/\partial\nu = g$ on ∂U, where U is smooth and ν is the outward unit normal field along ∂U.

When γ (resp., $\partial\gamma/\partial\nu$) is specified along ∂U, the directions of deformation μ are chosen in $C(U; \mathbf{R}_+)$ such that

$$\mu = 0 \ (\text{resp.}, \ \partial\mu/\partial\nu = 0) \ \text{ on } \partial U. \tag{3.16}$$

For small $t \geq 0$ and each such μ define the perturbed domain

$$\Omega_t = \left\{ (x', x_N) \in \mathbf{R}^N : 0 < x_N < \gamma(x') + t\mu(x') \right\} \tag{3.17}$$

and the obvious family of transformations

$$(x', x_N) \mapsto T_t(x', x_N) = \left(x', x_N + \frac{x_N}{\gamma(x')} t\mu(x') \right) : \Omega \to \Omega_t, \tag{3.18}$$

which can be extended to a neighborhood D of Ω containing the perturbed domains Ω_t, $0 \leq t \leq t_1$, for some small $t_1 > 0$. In general, D will be such that $\overline{D} = \overline{U} \times [0, L]$ for some $L > 0$.

This construction is also appropriate for domains that are Lipschitzian or of class C^k ($1 \leq k \leq \infty$), depending on whether γ is a Lipschitzian or a C^k-function. Again the transformation T_t is equal to $I + t\Theta$, where

$$\Theta(x', x_N) \overset{\text{def}}{=} \left(0, x_N \frac{\mu(x')}{\gamma(x')} \right). \tag{3.19}$$

But of course this Θ is not the only choice for which $\Omega_t = T_t(\Omega)$. For instance, let $\lambda : \mathbf{R}_+ \to \mathbf{R}_+$ be any smooth increasing function such that $\lambda(0) = 0$ and $\lambda(1) = 1$. Then we could consider the transformations

$$T_t(x', x_N) \stackrel{\text{def}}{=} \left(x', x_N + t\lambda \left(\frac{x_N}{\gamma(x')} \right) \mu(x') \right). \tag{3.20}$$

This example illustrates the following general principle: *the transformation T_t of $\overline{D}$ such that $\Omega_t = T_t(\Omega)$ is not unique.*

This means that, at least for smooth domains, only the trace $T_t|_{\Gamma_t}$ on Γ_t is important, while the displacement of the inner points does not contribute to the definition of Ω_t. Nevertheless this statement is to be interpreted with caution. Here we implicitly assume that the objective function $J(\Omega)$ and the associated constraints are only a function of the shape of Ω. However, in some problems involving singularities at inner points of Ω (e.g., when the solution $y(\Omega)$ of the state equation has a singularity or when some constraints on the domain are active), the situation might require a finer analysis. One such example is the internal displacement of the interior nodes of a triangularization τ_h when the solution $y(\Omega)$ of a partial differential equation is approximated by a piecewise polynomial solution over the triangularized domain Ω_h in the finite element method. Such a displacement does not change the shape of Ω_h, but it does change the solution y_h of the problem. When the displacement of the interior nodes is a priori parametrized by the boundary nodes the solution y_h will only depend on the position of the boundary nodes, but the interior nodes will contribute through the choice of the specified parametrization.

3.4   Polar Coordinates

In some examples domains are star-shaped with respect to a point. Since a domain can always be translated, there is no loss of generality in assuming that this point is the origin. Then such domains Ω can be parametrized as follows:

$$\Omega \stackrel{\text{def}}{=} \left\{ x \in \mathbf{R}^N : x = \rho\zeta, \, \zeta \in S_{N-1}, \, 0 \le \rho < f(\zeta) \right\}, \tag{3.21}$$

where S_{N-1} is the unit sphere in $\mathbf{R}^N$,

$$S_{N-1} \stackrel{\text{def}}{=} \left\{ x \in \mathbf{R}^N : |x| = 1 \right\}, \tag{3.22}$$

and $f : S_{N-1} \to \mathbf{R}_+$ is a positive continuous mapping from S_{N-1} such that

$$m \stackrel{\text{def}}{=} \min \left\{ f(\zeta) : \zeta \in S_{N-1} \right\} > 0. \tag{3.23}$$

Given any $g \in C(S_{N-1})$ and a sufficiently small $t \ge 0$ the perturbed domains are defined as

$$\Omega_t \stackrel{\text{def}}{=} \left\{ x \in \mathbf{R}^N : x = \rho\zeta, \, \zeta \in S_{N-1}, \, 0 \le \rho < f(\zeta) + tg(\zeta) \right\}. \tag{3.24}$$

For example, choose t, $0 \le t \le t_1$, for some

$$t_1 = \frac{m}{\|g\|_{C(S_{N-1})}} > 0 \tag{3.25}$$

and define the transformation T_t as

$$\begin{cases} T_t(X) = 0, & \text{if } X = 0, \\ T_t(X) = \left[\rho + t\frac{\rho}{f(\zeta)}g(\zeta)\right]\zeta, & \text{if } X = \rho\zeta \neq 0. \end{cases} \tag{3.26}$$

As in the previous example T_t is not unique, and for any continuous increasing function $\lambda : \mathbf{R}_+ \to \mathbf{R}_+$ such that $\lambda(0) = 0$ and $\lambda(t) = 1$ the transformation

$$\begin{cases} T_t(X) = 0, & \text{if } X = 0, \\ T_t(X) = \left[\rho + t\lambda\left(\frac{\rho}{f(\zeta)}\right)g(\zeta)\right]\zeta, & \text{if } X = \rho\zeta \neq 0, \end{cases} \tag{3.27}$$

yields the same domain Ω_t.

3.5   Level Sets

In sections 3.1–3.4 the perturbed domain Ω_t always appears in the form $\Omega_t = T_t(\Omega)$, where T_t is a bijective transformation of $\mathbf{R}^N$ and T_t is of the form $I + t\Theta$. In some free boundary problems (e.g., plasma physics, propagation of fronts) the free boundary Γ is a level curve of a smooth function u defined over an open domain D. Assume that D is bounded open with smooth boundary ∂D. Let $u \in C^2(\overline{D})$ be a positive function on $\overline{D}$ such that

$$\begin{cases} u \geq 0 \text{ in } \overline{D}, \quad u = 0 \text{ on } \partial D, \\ \exists \text{ a unique } x_u \in D \text{ such that } \forall x \in \overline{D} - \{x_u\} \ \ |\nabla u(x)| > 0. \end{cases} \tag{3.28}$$

If $m = \max\left\{u(x) : x \in \overline{D}\right\}$, then for each t in $[0, m[$ the level set

$$\Gamma_t = u^{-1}(t) \tag{3.29}$$

is a C^2-submanifold of $\mathbf{R}^N$ in D, which is the boundary of the open set

$$\Omega_t = \{x \in D : u(x) > t\}. \tag{3.30}$$

By definition, $\Omega_0 = D$, for all $t_1 > t_2$, $\Omega_{t_1} \subset \Omega_{t_2}$, and the domains Ω_t converge in the Hausdorff topology to the point x_u. The outward unit normal field on Γ_t is given by

$$x \in \Gamma_t, \quad n_t(x) = -|\nabla u(x)|^{-1}\nabla u(x). \tag{3.31}$$

This suggests introducing the velocity field

$$\forall x \in D - \{x_u\}, \quad V(x) \stackrel{\text{def}}{=} |\nabla u(x)|^{-2}\nabla u(x), \tag{3.32}$$

which is continuous everywhere but at $x = x_u$. If V was continuous everywhere, then for each X, the trajectory $x(t; X)$ of the differential equation

$$\begin{cases} \dfrac{dx}{dt}(t) = V\big(x(t)\big), \\ x(0) = X \end{cases} \tag{3.33}$$

would have the property that

$$u\big(x(t)\big) = u(X) + t \tag{3.34}$$

since formally

$$\frac{d}{dt}u\big(x(t)\big) = \nabla u\big(x(t)\big) \cdot \frac{dx}{dt}(t) = 1. \tag{3.35}$$

This means that the map $X \mapsto T_t(X) = x(t; X)$ constructed from (3.33) would map the level sets

$$\Gamma_0 = \big\{X \in \overline{D} : u(X) = 0\big\} \tag{3.36}$$

onto the level set

$$\Gamma_t = \big\{x \in \overline{D} : u(x) = t\big\} \tag{3.37}$$

and eventually Ω_0 onto Ω_t. Unfortunately, it is easy to see that this last property fails on the function $u(x) = 1 - x^2$ defined on the unit disk.

To get around this difficulty, introduce for some arbitrarily small $\varepsilon, 0 < \varepsilon < m/2$, an infinitely differentiable function $\rho_\varepsilon : \mathbf{R}^N \to [0, 1]$ such that

$$\rho_\varepsilon(x) = \begin{cases} 0, & \text{if } |\nabla u(x)| < \varepsilon, \\ 1, & \text{if } |\nabla u(x)| > 2\varepsilon, \end{cases} \tag{3.38}$$

and the velocity

$$V_\varepsilon(x) = \rho_\varepsilon(x)V(x), \quad x \in \overline{D}. \tag{3.39}$$

As above, define the transformation

$$X \mapsto T_t^\varepsilon(X) = x(t; X), \tag{3.40}$$

where $x(t; X)$ is the solution of the differential equation

$$\begin{cases} \dfrac{dx}{dt}(t) = V_\varepsilon\big(x(t)\big), & t \geq 0, \\ x(0) = X \in \overline{D}. \end{cases} \tag{3.41}$$

For $0 \leq t < m - 2\varepsilon$, T_t maps Γ_0 onto Γ_t; for $0 \leq s < m - \varepsilon$ such that $s + t < m - \varepsilon$, T_t maps Γ_s onto Γ_{t+s}. However, for $s > m - \varepsilon$, T_t is the identity operator. As a result, for $0 \leq t < m - 2\varepsilon$

$$T_t(\Omega_0) = \Omega_t \text{ and } T_t(\Gamma_0) = \Gamma_t. \tag{3.42}$$

Of course $\varepsilon > 0$ is arbitrary and we can make the construction for t's arbitrary close to m. This is an example that can be handled by the velocity (speed) method and not by a perturbation of the identity. Here the domains Ω_t are implicitly constrained to stay within the larger domain D. We shall see in section 5 how to introduce and characterize such a constraint.

Another example of description by level sets is provided by the oriented distance function for some open domain Ω of class C^2 with compact boundary Γ. We have seen in Chapter 5 that there exists $h > 0$ and a neighborhood

$$U_h(\Gamma) = \left\{ x \in \mathbf{R}^N : |b_\Omega(x)| < h \right\}$$

such that $b_\Omega \in C^2(U_h(\Gamma))$. Then for $0 \leq t < h$ the flow corresponding to the velocity field $V = \nabla b_\Omega$ maps Ω and its boundary Γ onto

$$T_t(\Omega) = \Omega_t = \left\{ x \in \mathbf{R}^N : b_\Omega(x)| < t \right\},$$
$$T_t(\Gamma) = \Gamma_t = \left\{ x \in \mathbf{R}^N : b_\Omega(x)| = t \right\}.$$

4 Unconstrained Families of Domains

In this section we study equivalences between the velocity method (cf. Zolésio [12, 8]) and methods using a family of transformations. In section 4.1 we give some general conditions to construct a family of transformations of $\mathbf{R}^N$ from a nonautonomous velocity field. Conversely, we show how to construct a nonautonomous velocity field from a family of transformations of $\mathbf{R}^N$. This construction is applied to Lipschitzian perturbations of the identity in section 4.2. In section 4.3 the equivalences of section 4.1 are specialized to velocities in $C_0^{k+1}(\mathbf{R}^N, \mathbf{R}^N)$, $C^{k+1}(\overline{\mathbf{R}^N}, \mathbf{R}^N)$, and $C^{k,1}(\overline{\mathbf{R}^N}, \mathbf{R}^N)$, $k \geq 0$.

4.1 Equivalence between Velocities and Transformations

Let the real number $\tau > 0$ and the map $V : [0, \tau] \times \mathbf{R}^N \to \mathbf{R}^N$ be given. The map V can be viewed as a (time-dependent) nonautonomous velocity field $\{V(t) : 0 \leq t \leq \tau\}$ defined on $\mathbf{R}^N$:

$$x \mapsto V(t)(x) \stackrel{\text{def}}{=} V(t, x) : \mathbf{R}^N \mapsto \mathbf{R}^N. \tag{4.1}$$

Assume that

$$\text{(V)} \quad \begin{aligned} \forall x \in \mathbf{R}^N, \quad &V(\cdot, x) \in C\left([0, \tau]; \mathbf{R}^N\right), \\ \exists c > 0, \, \forall x, y \in \mathbf{R}^N, \quad &\|V(\cdot, y) - V(\cdot, x)\|_{C([0,\tau];\mathbf{R}^N)} \leq c|y - x|, \end{aligned} \tag{4.2}$$

where $V(\cdot, x)$ is the function $t \mapsto V(t, x)$. Note that V is continuous on $[0, \tau] \times \mathbf{R}^N$. Hence it is uniformly continuous on $[0, \tau] \times D$ for any bounded open subset D of $\mathbf{R}^N$ and

$$V(\cdot) \in C([0, \tau]; C(\overline{D}; \mathbf{R}^N)). \tag{4.3}$$

Associate with V the solution $x(t; V)$ of the vector ordinary differential equation

$$\frac{dx}{dt}(t) = V\left(t, x(t)\right), \quad t \in [0, \tau], \quad x(0) = X \in \mathbf{R}^N, \tag{4.4}$$

and define the transformations

$$X \mapsto T_t(V)(X) \overset{\text{def}}{=} x_V(t; X) : \mathbf{R}^N \to \mathbf{R}^N \tag{4.5}$$

and the maps (whenever the inverse of T_t exists)

$$(t, X) \mapsto T_V(t, X) \overset{\text{def}}{=} T_t(V)(X) : [0, \tau] \times \mathbf{R}^N \to \mathbf{R}^N, \tag{4.6}$$

$$(t, x) \mapsto T_V^{-1}(t, x) \overset{\text{def}}{=} T_t^{-1}(V)(x) : [0, \tau] \times \mathbf{R}^N \to \mathbf{R}^N. \tag{4.7}$$

Notation 4.1.
In what follows we shall drop the V in $T_V(t, X)$, $T_V^{-1}(t, x)$, and $T_t(V)$ whenever no confusion arises. □

Theorem 4.1.

(i) *Under assumption* (V) *the map* T *specified by* (4.4)–(4.6) *has the following properties:*

(T1)
$$\forall X \in \mathbf{R}^N, \quad T(\cdot, X) \in C^1\big([0, \tau]; \mathbf{R}^N\big) \text{ and } \exists c > 0,$$
$$\forall X, Y \in \mathbf{R}^N, \quad \|T(\cdot, Y) - T(\cdot, X)\|_{C^1([0,\tau];\mathbf{R}^N)} \le c|Y - X|,$$

(T2) $\quad \forall t \in [0, \tau], \ X \mapsto T_t(X) = T(t, X) : \mathbf{R}^N \to \mathbf{R}^N$ *is bijective,* $\tag{4.8}$

(T3)
$$\forall x \in \mathbf{R}^N, \quad T^{-1}(\cdot, x) \in C\big([0, \tau]; \mathbf{R}^N\big) \text{ and } \exists c > 0,$$
$$\forall x, y \in \mathbf{R}^N, \quad \|T^{-1}(\cdot, y) - T^{-1}(\cdot, x)\|_{C([0,\tau];\mathbf{R}^N)} \le c|y - x|.$$

(ii) *Given a real number* $\tau > 0$ *and a map* $T : [0, \tau] \times \mathbf{R}^N \to \mathbf{R}^N$ *satisfying assumptions* (T1) *to* (T3), *the map*

$$(t, x) \mapsto V(t, x) \overset{\text{def}}{=} \frac{\partial T}{\partial t}\big(t, T_t^{-1}(x)\big) : [0, \tau] \times \mathbf{R}^N \to \mathbf{R}^N \tag{4.9}$$

satisfies conditions (V), *where* T_t^{-1} *is the inverse of* $X \mapsto T_t(X) = T(t, X)$. *If, in addition,* $T(0, \cdot) = I$, *then* $T(\cdot, X)$ *is the solution of* (4.4) *for that* V.

(iii) *Given a real number* $\tau > 0$ *and a map* $T : [0, \tau] \times \mathbf{R}^N \to \mathbf{R}^N$ *satisfying assumptions* (T1) *and* (T2) *and* $T(0, \cdot) = I$, *then there exists* $\tau' > 0$ *such that the conclusions of part* (ii) *hold on* $[0, \tau']$.

A more general version of this theorem for constrained domains (Theorem 5.1) will be given and proved in section 5.1.

Proof. (i) *Conditions* (T1) follow by standard arguments.
Conditions (T2): Associate with X in $\mathbf{R}^N$ the function

$$y(s) = T_{t-s}(X), \quad 0 \le s \le t.$$

Then

$$\frac{dy}{ds}(s) = -V(t - s, y(s)), \quad 0 \le s \le t, \quad y(0) = T_t(X). \tag{4.10}$$

For each $x \in \mathbf{R}^N$, the differential equation

$$\frac{dy}{ds}(s) = -V(t - s, y(s)), \quad 0 \le s \le t, \quad y(0) = x \in \mathbf{R}^N \qquad (4.11)$$

has a unique solution in $C^1([0, t]; \mathbf{R}^N)$. The solutions of (4.11) define the map

$$x \mapsto S_t(x) \stackrel{\text{def}}{=} y(t) : \mathbf{R}^N \to \mathbf{R}^N$$

such that

$$\exists c > 0, \ \forall t \in [0, \tau], \ \forall x, y \in \mathbf{R}^N, \quad |S_t(y) - S_t(x)| \le c|y - x|. \qquad (4.12)$$

In view of (4.10) and (4.11)

$$S_t(T_t(X)) = y(t) = T_{t-t}(X) = X \quad \Rightarrow \quad S_t \circ T_t = I \text{ on } \mathbf{R}^N.$$

To obtain the other identity, consider the function

$$z(r) = y(t - r; x),$$

where $y(\cdot, x)$ is the solution of (4.11) for some arbitrary x in $\mathbf{R}^N$. By definition

$$\frac{dz}{dr}(r) = V(r, z(r)), \quad z(0) = y(t, x),$$

and necessarily

$$x = y(0; x) = z(t) = T_t(y(t; x)) = T_t(S_t(x))$$
$$\Rightarrow T_t \circ S_t = I \text{ on } \mathbf{R}^N \quad \Rightarrow \quad S_t = T_t^{-1} : \mathbf{R}^N \to \mathbf{R}^N.$$

Conditions (T3). The uniform Lipschitz continuity in (T3) follows from (4.12), and we only need to show that

$$\forall x \in \mathbf{R}^N, \quad T^{-1}(\cdot, x) \in C([0, \tau]; \mathbf{R}^N).$$

Given t in $[0, \tau]$ pick an arbitrary sequence $\{t_n\}$, $t_n \to t$. Then for each $x \in \mathbf{R}^N$ there exists $X \in \mathbf{R}^N$ such that

$$T_t(X) = x \quad \text{and} \quad T_{t_n}(X) \to T_t(X) = x$$

from the first condition (T1). But

$$T_{t_n}^{-1}(x) - T_t^{-1}(x) = T_{t_n}^{-1}(T_t(X)) - T_t^{-1}(T_t(X))$$
$$= T_{t_n}^{-1}(T_t(X)) - T_{t_n}^{-1}(T_{t_n}(X)).$$

By the uniform Lipschitz continuity of T_t^{-1}

$$|T_{t_n}^{-1}(x) - T_t^{-1}(x)| = |T_{t_n}^{-1}(T_t(X)) - T_{t_n}^{-1}(T_{t_n}(X))| \le c|T_t(X) - T_{t_n}(X)|,$$

and the last term converges to zero as t_n goes to t.

(ii) The first part of condition (V) is satisfied since, for each $x \in \mathbf{R}^N$ and t, s in $[0, \tau]$,

$$|V(t,x) - V(s,x)|$$

$$\leq \left| \frac{\partial T}{\partial t}(t, T_t^{-1}(x)) - \frac{\partial T}{\partial t}(t, T_s^{-1}(x)) \right| + \left| \frac{\partial T}{\partial t}(t, T_s^{-1}(x)) - \frac{\partial T}{\partial t}(s, T_s^{-1}(x)) \right|$$

$$\leq c|T_t^{-1}(x) - T_s^{-1}(x)| + \left| \frac{\partial T}{\partial t}(t, T_s^{-1}(x)) - \frac{\partial T}{\partial t}(s, T_s^{-1}(x)) \right|.$$

Thus from (T3) and (T1) $t \mapsto V(t,x)$ is continuous at $s = t$, and hence for all x in $\mathbf{R}^N$, $V(.,x) \in C([0,\tau]; \mathbf{R}^N)$. The Lipschitzian property follows directly from the Lipschitzian properties (T1) and (T3): for all x and y in $\mathbf{R}^N$,

$$|V(t,y) - V(t,x)| = \left| \frac{\partial T}{\partial t}(t, T_t^{-1}(y)) - \frac{\partial T}{\partial t}(t, T_t^{-1}(x)) \right|$$

$$\leq c \left| T_t^{-1}(y) - T_t^{-1}(x) \right| \leq cc'|y - x|.$$

This proves that V satisfies condition (V).

(iii) From (T1) and (T2) for $f(t) = T_t - I$ and $t \geq s$,

$$f(t)(y) - f(t)(x) = \int_0^t \frac{\partial T}{\partial t}(r, T_r(y)) - \frac{\partial T}{\partial t}(r, T_r(x))\, dr,$$

$$|f(t)(y) - f(t)(x)| \leq \int_0^t c\,|T_r(y) - T_r(x)|\, dr \leq c^2\, t\, |y - x|.$$

For $\tau' = \min\{\tau, 1/(2c^2)\}$ and $0 \leq t \leq \tau'$, $c(f(t)) \leq 1/2$,

$$g(t) = T_t^{-1} - I = (I - T_t) \circ T_t^{-1} = -f(t) \circ [I + g(t)],$$

$$(1 - c(f(t)))\, c(g(t)) \leq c(f(t)) \quad \Rightarrow \quad c(g(t)) \leq 1 \text{ and } c(T_t^{-1}) \leq 2,$$

and the second condition (T3) is satisfied on $[0, \tau']$. The first one follows by the same argument as in part (i). Therefore the conclusions of part (ii) are true on $[0, \tau']$. $\qquad \square$

This equivalence theorem says that we can start from either a family of velocity fields $\{V(t)\}$ on $\mathbf{R}^N$ or a family of transformations $\{T_t\}$ of $\mathbf{R}^N$ provided that the map V, $V(t,x) = V(t)(x)$, satisfies (V) or the map T, $T(t,X) = T_t(X)$, satisfies (T1) to (T3).

Starting from V, the family of homeomorphisms $\{T_t(V)\}$ generates the family

$$\Omega_t \stackrel{\text{def}}{=} T_t(V)(\Omega) = \{T_t(V)(X) : X \in \Omega\} \tag{4.13}$$

of perturbations of the initial domain Ω. Interior (resp., boundary) points of Ω are mapped onto interior (resp., boundary) points of Ω_t. This is the basis of the *velocity method* which will be used to define shape derivatives.

4.2 Perturbations of the Identity

In examples it is usually possible to show that the transformation T satisfies assumptions (T1) to (T3) and construct the corresponding velocity field V defined in (4.9). For instance, consider perturbations of the identity to the first ($A = 0$) or second order: for $t \geq 0$ and $X \in \mathbf{R}^N$,

$$T_t(X) \stackrel{\text{def}}{=} X + tU(X) + \frac{t^2}{2}A(X), \qquad (4.14)$$

where U and A are transformations of $\mathbf{R}^N$. It turns out that for Lipschitzian transformations U and A, assumptions (T1) to (T3) are satisfied in some interval $[0, \tau]$.

Theorem 4.2. *Let U and A be two uniform Lipschitzian transformations of $\mathbf{R}^N$: $\exists c > 0$ such that for all $X, Y \in \mathbf{R}^N$,*

$$|U(Y) - U(X)| \leq c|Y - X| \ and \ |A(Y) - A(X)| \leq c|Y - X|.$$

There exists $\tau > 0$ such that the map T given by (4.14) satisfies conditions (T1) to (T3) on $[0, \tau]$. The associated velocity V given by

$$(t, x) \mapsto V(t, x) = U\big(T_t^{-1}(x)\big) + tA\big(T_t^{-1}(x)\big) : [0, \tau] \times \mathbf{R}^N \to \mathbf{R}^N \qquad (4.15)$$

satisfies condition (V) on $[0, \tau]$.

Remark 4.1.
Observe that from (4.14) and (4.15)

$$V(0) = U, \quad \dot{V}(0)(x) = \frac{\partial V}{\partial t}(t, x)\Big|_{t=0} = A - [DU]U, \qquad (4.16)$$

where DU is the Jacobian matrix of U. The term $\dot{V}(0)$ is an *acceleration* at $t = 0$ which will always be present even when $A = 0$, but can be eliminated by choosing $A = [DU]U$. $\qquad \square$

Proof. (i) By the definition of T in (4.14), $t \mapsto T(t, X)$ and $t \mapsto \frac{\partial T}{\partial t}(t, X) = U(X) + tA(X)$ are continuous on $[0, \infty[$. Moreover, for all X and Y,

$$|T(t, Y) - T(t, X)| \leq \left[1 + tc + \frac{t^2}{2}c\right]|Y - X|$$

and

$$\left|\frac{\partial T}{\partial t}(t, Y) - \frac{\partial T}{\partial t}(t, X)\right| \leq [c + tc]\,|Y - X|.$$

Thus condition (T1) is satisfied for any finite $\tau > 0$. To check condition (T2), consider for any $Y \in \mathbf{R}^N$ the mapping $h(X) \stackrel{\text{def}}{=} Y - [T_t(X) - X]$. For any X_1 and X_2

$$|h(X_2) - h(X_1)| \le t|U(X_2) - U(X_1)| + \frac{t^2}{2}|A(X_2) - A(X_1)|$$

$$\le tc|X_2 - X_1| + \frac{t^2}{2}c|X_2 - X_1| = tc\,[1 + t/2]\,|X_2 - X_1|. \tag{4.17}$$

For $\tau = \min\{1, 1/(4c)\}$, and any t, $0 \le t \le \tau$, $tc\,[1 + t/2] < 1/2$ and h is a contraction. So for all $0 \le t \le \tau$ and $Y \in \mathbf{R}^N$, there exists a unique $X \in \mathbf{R}^N$ such that

$$Y - [T_t(X) - X] = h(X) = X \iff T_t(X) = Y$$

and T_t is bijective. Therefore, (T2) is satisfied in $[0, \tau']$. The last part of the proof is the uniform Lipschitzian property of T_t^{-1}. In view of (4.17), for all t, $0 \le t \le \tau$, $tc\,[1 + t/2] < 1/2$ and

$$|T_t(X_2) - X_2 - (T_t(X_1) - X_1)| = |h(X_2) - h(X_1)| \le \frac{1}{2}|X_2 - X_1|$$

$$\Rightarrow |X_2 - X_1| - |T_t(X_2) - T_t(X_1)| \le \frac{1}{2}|X_2 - X_1|$$

$$\Rightarrow |X_2 - X_1| \le 2\,|T_t(X_2) - T_t(X_1)|.$$

In view of condition (T2) for all x and y,

$$|T_t^{-1}(y) - T_t^{-1}(x)| \le 2\,|T_t(T_t^{-1}(y)) - T_t(T_t^{-1}(x))| = 2\,|y - x|. \tag{4.18}$$

To complete our argument we prove the continuity with respect to t for each x. Let $X = T_t^{-1}(x)$. For any s in $[0, \tau]$

$$T_s^{-1}(x) - T_t^{-1}(x) = T_s^{-1}(T_t(X)) - T_t^{-1}(T_t(X)) = T_s^{-1}(T_t(X)) - T_s^{-1}(T_s(X)),$$

and in view of (4.18)

$$|T_s^{-1}(x) - T_t^{-1}(x)| \le 2|T_t(X) - T_s(X)|.$$

The continuity of $T_s^{-1}(x)$ at $s = t$ now follows from the continuity of $T_s(X)$ at $s = t$. Thus condition (T3) is satisfied. $\qquad\square$

4.3　Equivalence for Special Families of Velocities

In this section we specialize Theorem 4.1 to velocities in $C^{k,1}(\overline{\mathbf{R}^N}, \mathbf{R}^N)$, $C_0^{k+1}(\mathbf{R}^N, \mathbf{R}^N)$, and $C^{k+1}(\overline{\mathbf{R}^N}, \mathbf{R}^N)$, $k \ge 0$. The following notation will be helpful:

$$f(t) \stackrel{\text{def}}{=} T_t - I, \quad f'(t) = \frac{dT_t}{dt}, \quad g(t) \stackrel{\text{def}}{=} T_t^{-1} - I,$$

whenever T_t^{-1} exists and the identities

$$g(t) = -f(t) \circ T_t^{-1} = -f(t) \circ [I + g(t)],$$

$$V(t) = \frac{dT_t}{dt} \circ T_t^{-1} = f'(t) \circ T_t^{-1} = f'(t) \circ [I + g(t)].$$

Recall also for a function $F : \mathbf{R}^N \to \mathbf{R}^N$ the notation

$$c(F) \stackrel{\text{def}}{=} \sup_{y \neq x} \frac{|F(y) - F(x)|}{|y - x|} \quad \text{and} \quad \forall k \geq 1, \quad c_k(F) \stackrel{\text{def}}{=} \sum_{|\alpha|=k} c(\partial^\alpha F).$$

Theorem 4.3. *Let $k \geq 0$ be an integer.*

(i) *Given $\tau > 0$ and a velocity field V such that*

$$V \in C([0,\tau]; C^k(\overline{\mathbf{R}^N}, \mathbf{R}^N)) \text{ and } c_k(V(t)) \leq c \tag{4.19}$$

for some constant $c > 0$ independent of t, the map T given by (4.4)–(4.6) satisfies conditions (T1), (T2), and

$$f \in C^1([0,\tau]; C^k(\overline{\mathbf{R}^N}, \mathbf{R}^N)) \cap C([0,\tau]; C^{k,1}(\overline{\mathbf{R}^N}, \mathbf{R}^N)), \quad c_k(f'(t)) \leq c \tag{4.20}$$

for some constant $c > 0$ independent of t. Moreover, condition (T3) is satisfied and there exists $\tau' > 0$ such that

$$g \in C([0,\tau']; C^k(\overline{\mathbf{R}^N}, \mathbf{R}^N)), \quad c_k(g(t)) \leq ct \tag{4.21}$$

for some constant c independent of t.

(ii) *Given $\tau > 0$ and $T : [0,\tau] \times \mathbf{R}^N \to \mathbf{R}^N$ satisfying conditions (4.20) and $T(0, \cdot) = I$, there exists $\tau' > 0$ such that the velocity field $V(t) = f'(t) \circ T_t^{-1}$ satisfies conditions (V) and (4.19) in $[0,\tau']$.*

Proof. We prove the theorem for $k = 0$. The general case is obtained by induction over k. (i) By assumption on V, the conditions (V) given by (4.2) are satisfied and by Theorem 4.1 the corresponding family T satisfies conditions (T1) to (T3).

Conditions (4.20) on f. For any x and $s \leq t$

$$T_t(x) - T_s(x) = \int_s^t V(r) \circ T_r(x) \, dr,$$

$$|T_t(x) - T_s(x)| \leq \int_s^t c|T_r(x) - T_s(x)| + |V(r) \circ T_s(x)| \, dr,$$

$$|f(t)(x) - f(s)(x)| \leq \int_s^t c|f(r)(x) - f(s)(x)| + \|V(r)\|_C \, dr.$$

By assumption on V and Gronwall's inequality

$$\forall t, s \in [0,\tau], \quad \|f(t) - f(s)\|_C \leq c\,|t - s|$$

for another constant c independent of t. Moreover,

$$|(f(t) - f(s))(y) - (f(t) - f(s))(x)| = |(T_t - T_s)(y) - (T_t - T_s)(x)|$$

$$\leq \int_s^t |V(r) \circ T_r(y) - V(r) \circ T_r(x)| \, dr$$

$$\leq \int_s^t c|(T_r - T_s)(y) - (T_r - T_s)(x)| + c|T_s(y) - T_s(x)| \, dr$$

$$\leq \int_s^t c|(f(r) - f(s))(y) - (f(r) - f(s))(x)| + cc'|y - x| \, dr$$

for some other constant c' by the second condition (T1). Again by Gronwall's inequality, there exists another constant c such that

$$|(f(t) - f(s))(y) - (f(t) - f(s))(x)| \leq c|t - s| \, |y - x|$$
$$\Rightarrow c(f(t) - f(s)) \leq c\,|t - s|$$

$$\Rightarrow \boxed{f \in C([0, \tau]; C^{0,1}(\overline{\mathbf{R}^N}, \mathbf{R}^N)) \text{ and } \|f(t) - f(s)\|_{C^{0,1}} \leq c\,|t - s|.} \qquad (4.22)$$

Moreover, $f'(t) = V(t) \circ T_t$ and

$$|f'(t)(x) - f'(s)(x)| \leq |V(t)(T_t(x)) - V(s)(T_t(x))|$$
$$+ |V(s)(T_t(x)) - V(s)(T_s(x))|$$
$$\leq \|V(t) - V(s)\|_C + c(V(s)) \|T_t - T_s\|_C$$
$$\leq \|V(t) - V(s)\|_C + c \|f(t) - f(s)\|_C.$$

Finally,

$$|f'(t)(y) - f'(t)(x)| \leq |V(t)(T_t(y)) - V(t)(T_t(x))|$$
$$\leq c(V(t)) |T_t(y) - T_s(x)| \leq c\,c(T_t) |y - x|$$

and $c(f'(t)) \leq c$ for some new constant c independent of t. Therefore,

$$\boxed{f \in C^1([0, \tau]; C(\overline{\mathbf{R}^N}, \mathbf{R}^N)) \text{ and } c(f'(t)) \leq c.}$$

Conditions (4.21) *on g.* Since conditions (T1) and (T2) are satisfied there exists $\tau' > 0$ such that conditions (T3) are satisfied by Theorem 4.1 (iii). Moreover, from conditions (4.20)

$$|g(t)(y) - g(t)(x)| \leq |f(t)(T_t^{-1}(y)) - f(t)(T_t^{-1}(x))|$$
$$\leq c(f(t))|T_t^{-1}(y) - T_t^{-1}(x)|$$
$$\leq c(f(t)) \left(|g(t)(y) - g(t)(x)| + |y - x|\right)$$
$$\Rightarrow (1 - c(f(t))) |g(t)(y) - g(t)(x)| \leq c(f(t)) |y - x| \leq ct \, |y - x|.$$

Choose a new $\tau'' = \min\{\tau', 1/(2c)\}$. Then for $0 \le t \le \tau''$, $c(g(t)) \le 2ct$. Now

$$g(t) - g(s) = -f(t) \circ [I + g(t)] + f(s) \circ [I + g(s)]$$
$$\|g(t) - g(s)\|_C \le \|f(t) \circ [I + g(t)] - f(t) \circ [I + g(s)]\|_C$$
$$+ \|f(t) \circ [I + g(s)] - f(s) \circ [I + g(s)]\|_C$$
$$\le c(f(t))\|g(t) - g(s)\|_C + \|f(t) - f(s)\|_C$$
$$\le ct\,\|g(t) - g(s)\|_C + \|f(t) - f(s)\|_C.$$

For t in $[0, \tau'']$, $ct \le 1/2$, and

$$\|g(t) - g(s)\|_C \le 2\|f(t) - f(s)\|_C$$

$$\Rightarrow \boxed{g \in C([0, \tau'']; C(\overline{\mathbf{R}^N}, \mathbf{R}^N)) \text{ and } c(g(t)) \le 2ct.}$$

The conditions (4.20) on f are satisfied for $k = 0$. For $k = 1$ we start from the equation

$$DT_t - DT_s = \int_s^t DV(r) \circ T_r\, DT_r\, dr$$

and use the fact that $DT_t^{-1} = [DT_t]^{-1} \circ T_t^{-1}$ in connection with the identity

$$Dg(t) = -Df(t) \circ T_t^{-1}\, DT_t^{-1} = -(Df(t)[DT_t]^{-1}) \circ T_t^{-1}.$$

(ii) From conditions (4.20) on f, the transformation T satisfies conditions (T1). To check condition (T2) we consider two cases: $k \ge 1$ and $k = 0$. For $k \ge 1$ the function $t \mapsto Df(t) = DT_t - I : [0, \tau] \to C^{k-1}(\overline{\mathbf{R}^N}, \mathbf{R}^N)^N$ is continuous. Hence $t \mapsto \det DT_t : [0, \tau] \to \mathbf{R}$ is continuous and $\det DT_0 = 1$. So there exists $\tau' > 0$ such that T_t is invertible for all t in $[0, \tau']$ and (T2) is satisfied in $[0, \tau']$. In the case $k = 0$ consider for any Y the map $h(X) = Y - f(t)(X)$. For any X_1 and X_2, $|h(X_2) - h(X_1)| \le c(f(t))\,|X_2 - X_1|$. But by assumption $f \in C([0, \tau]; C^{0,1}(\overline{\mathbf{R}^N}))$ and $c(f(0)) = 0$ since $f(0) = 0$. Hence there exists $\tau' > 0$ such that $c(f(t)) \le 1/2$ for all t in $[0, \tau']$ and h is a contraction. So for all Y in $\mathbf{R}^N$ there exists a unique X such that

$$Y - [T_t(X) - X] = h(X) = X \iff T_t(X) = Y,$$

T_t is bijective, and condition (T2) is satisfied in $[0, \tau']$. By Theorem 4.1 (iii) from (T1) and (T2), there exists another $\tau' > 0$ for which conditions (T3) on g and (V) on $V(t) = f'(t) \circ T_t^{-1}$ are also satisfied. Moreover, we have seen in the proof of part (i) that conditions (4.21) on g follow from (T2) and (4.20). Using conditions (4.20) and (4.21),

$$|V(t)(y) - V(t)(x)| \le |f'(t)(T_t^{-1}(y)) - f'(t)(T_t^{-1}(x))|$$
$$\le c(f'(t))\,|T_t^{-1}(y) - T_t^{-1}(x)|$$
$$\le c(f'(t))\,[1 + c(g(t))]\,|y - x| \le c'|y - x|$$

and $c(V(t)) \le c'$. Also

$$
\begin{aligned}
|V(t)(x) - V(s)(x)| &= |f'(t)(T_t^{-1}(x)) - f'(s)(T_s^{-1}(x))| \\
&\le |f'(t)(T_t^{-1}(x)) - f'(t)(T_s^{-1}(x))| \\
&\quad + |f'(t)(T_s^{-1}(x)) - f'(s)(T_s^{-1}(x))| \\
&\le c(f'(t)) \, |T_t^{-1}(x) - T_s^{-1}(x)| + \|f'(t) - f'(s)\|_C \\
&\le c \, \|g(t) - g(s)\|_C + \|f'(t) - f'(s)\|_C.
\end{aligned}
$$

Therefore, since both g and f' are continuous,

$$
\boxed{V \in C([0, \tau']; C(\overline{\mathbf{R}^N}, \mathbf{R}^N)) \text{ and } c(V(t)) \le c}
$$

for some constant c independent of t. This proves the result for $k = 0$. As in part (i), for $k = 1$ we use the identity

$$
DV(t) = Df'(t) \circ T_t^{-1} \, DT_t^{-1} = (Df'(t)[DT_t]^{-1}) \circ T_t^{-1}
$$

and proceed in the sane way. The general case is obtained by induction over k. $\quad\square$

We now turn to the case of velocities in $C_0^k(\mathbf{R}^N, \mathbf{R}^N)$. As in Chapter 2, it will be convenient to use the notation $\mathcal{C}_0^k$ for the space $C_0^k(\mathbf{R}^N, \mathbf{R}^N)$, $\mathcal{C}^k(\overline{\mathbf{R}^N})$ for the space $C^k(\overline{\mathbf{R}^N}, \mathbf{R}^N)$ and $\mathcal{C}^{k,1}(\overline{\mathbf{R}^N})$ for the space $C^{k,1}(\overline{\mathbf{R}^N}, \mathbf{R}^N)$.

Theorem 4.4. *Let $k \ge 1$ be an integer.*

(i) *Given $\tau > 0$ and a velocity field V such that*

$$
V \in C([0, \tau]; C_0^k(\mathbf{R}^N, \mathbf{R}^N)), \tag{4.23}
$$

the map T given by (4.4)–(4.6) satisfies conditions (T1), (T2), and

$$
f \in C^1([0, \tau]; C_0^k(\mathbf{R}^N, \mathbf{R}^N)). \tag{4.24}
$$

Moreover, conditions (T3) is satisfied and there exists $\tau' > 0$ such that

$$
g \in C([0, \tau']; C_0^k(\mathbf{R}^N, \mathbf{R}^N)). \tag{4.25}
$$

(ii) *Given $\tau > 0$ and $T : [0, \tau] \times \mathbf{R}^N \to \mathbf{R}^N$ satisfying conditions (4.24) and $T(0, \cdot) = I$, there exists $\tau' > 0$ such that the velocity field $V(t) = f'(t) \circ T_t^{-1}$ satisfies conditions (V) and (4.23) on $[0, \tau']$.*

Proof. As in the proof of Theorem 4.3 we only prove the theorem for $k = 1$. The general case is obtained by induction on k, the various identities on f, g, f', and V, and the techniques of Theorem 8.1 and Lemmas 8.2 and 8.3 in Chapter 2.

(i) By the embedding $C_0^1(\mathbf{R}^N, \mathbf{R}^N) \subset C^1(\overline{\mathbf{R}^N}, \mathbf{R}^N) \subset C^{0,1}(\overline{\mathbf{R}^N}, \mathbf{R}^N)$, it follows from (4.23) that $V \in C([0, \tau]; C^{0,1}(\overline{\mathbf{R}^N}, \mathbf{R}^N))$ and condition (4.19) of Theorem 4.3 are satisfied. Therefore, conditions (4.20) and (4.21) of Theorem 4.3 are also satisfied in some interval $[0, \tau']$, $\tau' > 0$.

Conditions (4.24) *on* f. It remains to show that $f(t)$ and $f'(t)$ belong to the subspace $C_0(\mathbf{R}^N, \mathbf{R}^N)$ of $C(\overline{\mathbf{R}^N}, \mathbf{R}^N)$ and to prove the appropriate properties for $Df(t)$ and $Df'(t)$. Recall from the proof of the previous theorems that there exists $c > 0$ such that

$$|f(t)(x)| \le c \int_0^t |V(r)(x)|\, dr \le c \int_0^t |(V(r) - V(0))(x)|\, dr + ct\, |V(0)(x)|.$$

By assumption on $V(0)$, for $\varepsilon > 0$ there exists a compact set K such that

$$\forall x \in \complement K, \quad |V(0)(x)| \le \varepsilon/(2c)$$

and there exists δ, $0 < \delta < 1$, such that

$$\forall t,\ 0 \le t \le \delta, \quad \|V(r) - V(0)\|_C \le \varepsilon/(2c)$$
$$\Rightarrow\ \forall t,\ 0 \le t \le \delta,\ \forall x \in \complement K, \quad |f(t)(x)| \le \varepsilon \quad \Rightarrow\ f(t) \in \mathcal{C}_0.$$

Proceeding in this fashion from the interval $[0, \delta]$ to the next interval $[\delta, 2\delta]$ using the inequality

$$|(f(t) - f(s))(x)| \le c \int_s^t |(V(r) - V(\delta))(x)|\, dr + c\,|t - \delta|\,|V(\delta)(x)|,$$

the uniform continuity of V

$$\forall t, s, \quad |t - s| < \delta, \quad \|V(t) - V(s)\|_C \le \varepsilon/(2c)$$

and the fact that $V(\delta) \in \mathcal{C}_0$, that is, that there exists a compact set $K(\delta)$ such that

$$\forall x \in \complement K(\delta), \quad |V(\delta)(x)| \le \varepsilon/(2c),$$

we get $f(t) \in \mathcal{C}_0$, $\delta \le t \le 2\delta$, and hence $f \in C([0, \tau]; \mathcal{C}_0)$. For $f'(t)$ we make use of the identity $f'(t) = V(t) \circ T_t$. Again by assumption for any $\varepsilon > 0$ there exists a compact set $K(t)$ such that $|V(t)(x)| \le \varepsilon$ on $\complement K(t)$. Thus by choosing the compact $K'_t = T_t^{-1}(K(t))$, $|f'(t)(x)| \le \varepsilon$ on $\complement K'_t$, and $f' \in C([0, \tau]; \mathcal{C}_0)$. In order to complete the proof, it remains to establish the same properties for $Df(t)$ and $Df'(t)$. The matrix $Df(t)$ is solution of the equations

$$\frac{d}{dt} Df(t) = DV(t) \circ T_t\, DT_t, \quad Df(0) = 0$$
$$\Rightarrow\ \boxed{Df'(t) = DV(t) \circ T_t\, Df(t) + DV(t) \circ T_t.} \tag{4.26}$$

From the proof of Theorem 8.1 in Chapter 2 for each t the elements of the matrix

$$A(t) \overset{\text{def}}{=} DV(t) \circ T_t = DV(t) \circ [I + f(t)]$$

belong to $\mathcal{C}_0$ since $DV(t)$ and $f(t)$ do. By assumption, $V \in C([0, \tau]; \mathcal{C}_0^k)$ and V and all its derivatives $\partial^\alpha V$ are uniformly continuous in $[0, \tau] \times \mathbf{R}^N$. Therefore, for each $\varepsilon > 0$ there exists $\delta > 0$ such that

$$\forall\, |t - s| < \delta,\ \forall\, |y' - x'| < \delta, \quad |DV(t)(y') - DV(s)(x')| < \varepsilon.$$

Pick $0 < \delta' < \delta$ such that

$$\forall\, |t - s| < \delta', \quad \|T_t - T_s\|_C = \|f(t) - f(s)\|_C < \delta$$
$$\Rightarrow \forall x, \ \forall\, |t - s| < \delta', \quad |T_t(x) - T_s(x)| < \delta$$
$$\Rightarrow |DV(t)(T_t(x)) - DV(s)(T_s(x))| < \varepsilon$$
$$\Rightarrow \|A(t) - A(s)\|_C < \varepsilon \quad \Rightarrow A \in C([0, \tau]; (\mathcal{C}_0)^N).$$

For each x, $Df(t)(x)$ is the unique solution of the linear matrix equation (4.26). To show that $Df(t) \in (\mathcal{C}_0)^N$ we first show that $Df(t)(x)$ is uniformly continuous for x in $\mathbf{R}^N$. For any x and y

$$|Df(t)(y) - Df(t)(x)| \leq \int_0^t |V(r, T_r(y)) - V(r, T_r(x))|\, dr$$
$$\leq \int_0^t c|T_r(y) - T_r(x)|\, dr$$
$$\leq c \int_0^t |f(r)(y) - f(r)(x)| + |y - x|\, dr.$$

But $f \in C([0, \tau]; \mathcal{C}_0)$ is uniformly continuous in (t, x): for each $\varepsilon > 0$ there exists δ, $0 < \delta < \varepsilon/(2c\tau)$, such that

$$\forall\, |t - s| < \delta, \ \forall\, |y - x| < \delta, \quad |f(t)(y) - f(s)(x)| < \varepsilon/(2c\tau).$$

Substituting in the previous inequality for each $\varepsilon > 0$, there exists $\delta > 0$ such that

$$\forall\, t, \ \forall\, |y - x| < \delta, \quad |Df(t)(y) - Df(t)(x)| < \varepsilon.$$

Hence $Df(t)$ is uniformly continuous in $\mathbf{R}^N$. Furthermore, from the equation (4.26) we have the following inequality:

$$|Df(t)(x)| \leq \int_0^t |DV(r)(T_r(x))|\, |Df(r)(x)| + |DV(r)(T_r(x))|\, dr$$
$$\leq c \int_0^t (|Df(r)(x)| + 1)\, dr \tag{4.27}$$

since $V \in C([0, \tau]; \mathcal{C}_0^1)$. By Gronwall's inequality,

$$|Df(t)(x)| \leq ct$$

for some other constant c independent of t. Thence $Df(t) \in C(\overline{\mathbf{R}^N}, \mathbf{R}^N)^N$. Finally to show that $Df(t)$ vanishes at infinity we start from the integral form of (4.26):

$$Df(t)(x) = \int_0^t DV(r)(T_r(x))\, DT_r(x)\, dr,$$
$$|Df(t)(x)| \leq c \int_0^t |DV(r)(T_r(x)) - DV(r)(x)| + |DV(r)(x)|\, dr$$
$$\leq c' \int_0^t |f(r)(x)| + |DV(r)(x)|\, dr.$$

By the same technique as before for $f(t)$, it follows that the elements of $Df(t)$ belong to $\mathcal{C}_0$ since both $f(s)$ and $DV(r)$ do. Finally for the continuity with respect to t

$$Df(t) - Df(s) = \int_s^t A(r)Df(r) + A(r)\,dr,$$

$$\|Df(t) - Df(s)\|_C \leq \int_s^t \|A(r)\|_C \|Df(r) - Df(s)\|_C$$
$$+ \|A(r)\|_C \left(1 + \|Df(r)\|_C\right) dr.$$

Again, by Gronwall's inequality, there exists another constant c such that

$$\|Df(t) - Df(s)\|_C \leq c\,|t - s|.$$

Therefore, $Df \in C([0,\tau];(\mathcal{C}_0)^N)$ and $f \in C([0,\tau];\mathcal{C}_0^1)$. For Df' we repeat the proof for f' using the identity

$$Df'(t) = DV(t) \circ T_t\, Df(t) + DV(t) \circ T_t$$

to get

$$Df' \in C([0,\tau];(\mathcal{C}_0)^N) \quad \Rightarrow \quad f' \in C([0,\tau];\mathcal{C}_0^1).$$

Conditions (4.25) *on* g. From the remark at the beginning of part (i) of the proof, the conclusions of Theorem 4.3 are true for g, and it remains to check the remaining properties for g and Dg using the identities

$$g(t) = -f(t) \circ [I + g(t)], \quad Dg(t) = -Df(t) \circ [I + g(t)]\,(I + Dg(t)).$$

By the proof of Theorem 8.1 in Chapter 2, $g(t) \in \mathcal{C}_0$ since $Df(t)$ and $g(t)$ do. Therefore, $g(t) \in \mathcal{C}_0^1$. The continuity follows by the same argument as for f' and $g \in C([0,\tau];\mathcal{C}_0^1)$.

(ii) By assumption from conditions (4.24), conditions (T1) are satisfied. For (T2) observe that for $k \geq 1$ the function $t \mapsto Df(t) = DT_t - I : [0,\tau] \to C^{k-1}(\overline{\mathbf{R}^N},\mathbf{R}^N)^N$ is continuous. Hence $t \mapsto \det DT_t : [0,\tau] \to \mathbf{R}$ is continuous and $\det DT_0 = 1$. So there exists $\tau' > 0$ such that T_t is invertible for all t in $[0,\tau']$ and (T2) is satisfied in $[0,\tau']$. Furthermore, from the proof of part (i) conditions (T3) and (4.25) on g are also satisfied in some interval $[0,\tau']$, $\tau' > 0$. Therefore, the velocity field

$$V(t) = f'(t) \circ T_t^{-1} = f'(t) \circ [I + g(t)]$$

satisfies the conditions (V) specified by (4.2) in $[0,\tau']$. By the proof of Theorem 8.1 in Chapter 2, $V(t) \in \mathcal{C}_0^k$ since $f'(t)$ and $g(t)$ belong to $\mathcal{C}_0^k$. By assumption, $f \in C^1([0,\tau];\mathcal{C}_0^k)$. Hence f' and all its derivatives $\partial^\alpha f'$, $|\alpha| \leq k$, are uniformly continuous on $[0,\tau] \times \mathbf{R}^N$; that is, given $\varepsilon > 0$, there exists $\delta > 0$ such that

$$\forall t,s,\ |t-s| < \delta,\ \forall y',x',\ |y'-x'| < \delta,\quad |\partial^\alpha f'(t)(y') - \partial^\alpha f'(s)(x')| < \varepsilon.$$

Similarly $g \in C([0,\tau']; \mathcal{C}_0^k)$ and there exists $0 < \delta' \le \delta$ such that

$$\forall t, s, \ |t - s| < \delta', \ \forall y, x, \ |y - x| < \delta', \quad |\partial^\alpha g(t)(y) - \partial^\alpha g(s)(x)| < \delta.$$

Therefore, for $|t - s| < \delta'$

$$\|T_t^{-1} - T_s^{-1}\|_C = \|g(t) - g(s)\|_C < \delta,$$

and since $\delta' < \delta$

$$\forall x, \quad |f'(t)(T_t^{-1}(x)) - f'(t)(T_s^{-1}(x))| < \varepsilon$$
$$\Rightarrow \|V(t) - V(s)\|_C < \varepsilon \quad \Rightarrow V \in C([0,\tau']; \mathcal{C}_0).$$

We then proceed to the first derivative of V,

$$DV(t) = Df'(t) \circ T_t^{-1} DT_t^{-1} = Df'(t) \circ [I + g(t)] [I + Dg(t)],$$

and by uniform continuity of the right-hand side $V \in C([0,\tau']; \mathcal{C}_0^1)$. By induction on k, we finally get $V \in C([0,\tau']; \mathcal{C}_0^k)$. $\qquad\qquad\square$

The proof of the last theorem is based on the fact that the vector functions involved are uniformly continuous. The fact that they vanish at infinity is not an essential element of the proof. Therefore, the theorem is valid with $C^k(\overline{\mathbf{R}^N}, \mathbf{R}^N)$ in place of $C_0^k(\mathbf{R}^N, \mathbf{R}^N)$.

Theorem 4.5. *Let $k \ge 1$ be an integer.*

(i) *Given $\tau > 0$ and a velocity field V such that*

$$V \in C([0,\tau]; C^k(\overline{\mathbf{R}^N}, \mathbf{R}^N)), \qquad\qquad (4.28)$$

the map T given by (4.4)–(4.6) satisfies conditions (T1), (T2), and

$$f \in C^1([0,\tau]; C^k(\overline{\mathbf{R}^N}, \mathbf{R}^N)). \qquad\qquad (4.29)$$

Moreover, conditions (T3) are satisfied and there exists $\tau' > 0$ such that

$$g \in C([0,\tau']; C^k(\overline{\mathbf{R}^N}, \mathbf{R}^N)). \qquad\qquad (4.30)$$

(ii) *Given $\tau > 0$ and $T : [0,\tau] \times \mathbf{R}^N \to \mathbf{R}^N$ satisfying conditions (4.29) and $T(0,\cdot) = I$, there exists $\tau' > 0$ such that the velocity field $V(t) = f'(t) \circ T_t^{-1}$ satisfies conditions (V) and (4.28) on $[0,\tau']$.*

5   Constrained Families of Domains

We now turn to the case where the family of admissible domains Ω is constrained to lie in a fixed larger subset D of $\mathbf{R}^N$ or its closure. For instance, D can be an open set or a closed submanifold of $\mathbf{R}^N$.

5.1   Equivalence between Velocities and Transformations

Given a nonempty subset D of $\mathbf{R}^N$, consider a family of transformations

$$T : [0, \tau] \times \overline{D} \to \mathbf{R}^N \tag{5.1}$$

with the following properties:

$$(\mathrm{T1}_D) \quad \begin{aligned} &\forall X \in \overline{D}, \quad T(\cdot, X) \in C^1\big([0, \tau]; \mathbf{R}^N\big) \text{ and } \exists c > 0, \\ &\forall X, Y \in \overline{D}, \quad \|T(\cdot, Y) - T(\cdot, X)\|_{C^1([0,\tau];\mathbf{R}^N)} \le c|Y - X|, \end{aligned}$$

$$(\mathrm{T2}_D) \quad \forall t \in [0, \tau], \quad X \mapsto T_t(X) = T(t, X) : \overline{D} \to \overline{D} \text{ is bijective,} \tag{5.2}$$

$$(\mathrm{T3}_D) \quad \begin{aligned} &\forall x \in \overline{D}, \quad T^{-1}(\cdot, x) \in C\big([0, \tau]; \mathbf{R}^N\big) \text{ and } \exists c > 0, \\ &\forall x, y \in \overline{D}, \quad \|T^{-1}(\cdot, y) - T^{-1}(\cdot, x)\|_{C([0,\tau];\mathbf{R}^N)} \le c|y - x|, \end{aligned}$$

where under assumption $(\mathrm{T2}_D)$, T^{-1} is defined from the inverse of T_t as

$$(t, x) \mapsto T^{-1}(t, x) \overset{\text{def}}{=} T_t^{-1}(x) : [0, \tau] \times \overline{D} \to \mathbf{R}^N . \tag{5.3}$$

Those three properties are the analogue for $\overline{D}$ of the same three properties obtained for $\mathbf{R}^N$. In fact, Theorem 4.1 extends from $\mathbf{R}^N$ to $\overline{D}$ by adding one assumption to (V). Specifically we shall consider for $\tau > 0$ velocities

$$V : [0, \tau] \times \overline{D} \to \mathbf{R}^N \tag{5.4}$$

such that

$$(\mathrm{V1}_D) \quad \begin{aligned} &\forall x \in \overline{D}, \quad V(\cdot, x) \in C\big([0, \tau]; \mathbf{R}^N\big), \quad \exists c > 0, \\ &\forall x, y \in \overline{D}, \quad \|V(\cdot, y) - V(\cdot, x)\|_{C([0,\tau];\mathbf{R}^N)} \le c|y - x|, \end{aligned} \tag{5.5}$$

$$(\mathrm{V2}_D) \quad \forall x \in \overline{D}, \forall t \in [0, \tau], \quad \pm V(t, x) \in T_{\overline{D}}(x),$$

where $T_{\overline{D}}(x)$ is the Bouligand contingent cone to $\overline{D}$ at the point x in $\overline{D}$

$$\boxed{\, T_{\overline{D}}(X) \overset{\text{def}}{=} \bigcap_{\varepsilon > 0} \bigcap_{\alpha > 0} \bigcup_{0 < h < \alpha} \left[\frac{1}{h}(\overline{D} - X) + \varepsilon B \right] \,} \tag{5.6}$$

and B is the unit disk in $\mathbf{R}^N$ (cf. Aubin and Cellina [1, p. 176]). This definition is equivalent to

$$T_{\overline{D}}(X) = \limsup_{t \searrow 0} \left\{ \frac{\overline{D} - X}{t} \right\} = \left\{ v \,\middle|\, \liminf_{t \searrow 0} \frac{1}{t} d_D(X + tv) = 0 \right\} \tag{5.7}$$

(cf. Aubin and Frankowska [1, pp. 121–122; 17; 21]). Note that when D is bounded in $\mathbf{R}^N$,

$$\boxed{\, V(\cdot) \in C\big([0, \tau]; C(\overline{D}; \mathbf{R}^N) \cap \mathrm{Lip}\,(\overline{D}; \mathbf{R}^N)\big) = C\big([0, \tau]; C^{0,1}(\overline{D}; \mathbf{R}^N)\big). \,}$$

When D is equal to $\mathbf{R}^{\mathrm{N}}$, $T_{\overline{D}}(x) = \mathbf{R}^{\mathrm{N}}$ for all x and condition $(V2_D)$ can be dropped. When D is equal to the boundary ∂A of a set A of class $C^{1,1}$ in $\mathbf{R}^{\mathrm{N}}$, ∂A is a $C^{1,1}$-submanifold of $\mathbf{R}^{\mathrm{N}}$ and

$$\forall x \in \partial A, \quad \pm V(t,x) \in T_{\partial A}(x) \iff \forall x \in \partial A, \quad V(t,x) \cdot \nabla b_A(x) = 0;$$

that is, at each point of ∂A, the velocity field is tangent to ∂A: it belongs to the tangent linear space of ∂A.

The next theorem is a generalization of Theorem 4.1 from $\mathbf{R}^{\mathrm{N}}$ to an arbitrary set D which shows the equivalence between velocity and transformation viewpoints.

Theorem 5.1.

(i) *Let $\tau > 0$ and V be a family of velocity fields satisfying conditions $(V1_D)$ and $(V2_D)$ and consider the family of transformations*

$$(t,X) \mapsto T(t,X) = x(t;X) : [0,\tau] \times \overline{D} \to \mathbf{R}^{\mathrm{N}}, \tag{5.8}$$

where $x(\cdot, X)$ is the solution of

$$\frac{dx}{dt}(t) = V(t, x(t)), \quad 0 \le t \le \tau, \; x(0) = X. \tag{5.9}$$

Then the family of transformations T satisfies conditions $(T1_D)$ to $(T3_D)$.

(ii) *Conversely, given a family of transformations T satisfying conditions $(T1_D)$ to $(T3_D)$, the family of velocity fields*

$$(t,x) \mapsto V(t,x) = \frac{\partial T}{\partial t}\left(t, T_t^{-1}(x)\right) : [0,\tau] \times \overline{D} \to \mathbf{R}^{\mathrm{N}} \tag{5.10}$$

satisfies conditions $(V1_D)$ and $(V2_D)$. If, in addition, $T(0,\cdot) = I$, then $T(\cdot, X)$ is the solution of (5.9) for that V.

(iii) *Given a real number $\tau > 0$ and a map $T : [0,\tau] \times \overline{D} \to \overline{D}$ satisfying assumptions $(T1_D)$ and $(T2_D)$ and $T(0,\cdot) = I$, then there exists $\tau' > 0$ such that the conclusions of part (ii) hold on $[0,\tau']$.*

Remark 5.1.
Assumption $(V2_D)$ is a *double viability condition*. Nagumo's [1] usual viability condition

$$V(t,x) \in T_{\overline{D}}(x) \quad \forall t \in [0,\tau], \forall x \in \overline{D} \tag{5.11}$$

is a necessary and sufficient condition for a *viable solution* to (5.9):

$$\forall t \in [0,\tau], \forall X \in \overline{D}, \quad x(t;X) \in \overline{D} \; \left(\text{or } T_t(\overline{D}) \subset \overline{D}\right) \tag{5.12}$$

(cf. Aubin and Cellina [1, pp. 174 and 180]). Condition $(V2_D)$:

$$\forall t \in [0,\tau], \forall x \in \overline{D}, \quad \pm V(t,x) \in T_{\overline{D}}(x) \tag{5.13}$$

is a *strict viability condition* which says that T_t maps $\overline{D}$ into $\overline{D}$ and

$$\forall t \in [0, \tau], \quad T_t : \overline{D} \to \overline{D} \quad \text{is a homeomorphism.} \tag{5.14}$$

In particular, it maps interior points onto interior points and boundary points onto boundary points (cf. Dugundji [1, pp. 87–88]). $\square$

Remark 5.2.
Condition $(V2_D)$ is a generalization to an arbitrary set D of the following condition used by Zolésio [12] in 1979: For all x in ∂D,

$$\begin{cases} V(t, x) \cdot n(x) = 0 & \text{if the outward normal } n(x) \text{ exists,} \\ 0 & \text{otherwise.} \end{cases} \quad \square$$

Proof of Theorem 5.1. (i) *Existence and uniqueness of viable solutions to* (5.9). Apply Nagumo's [1] theorem to the augmented system on $[0, \tau]$:

$$\begin{cases} \dfrac{dx}{dt}(t) = V\big(t, x(t)\big), & x(0) = X \in \overline{D}, \\[2mm] \dfrac{dx_0}{dt}(t) = 1, & x_0(0) = 0, \end{cases} \tag{5.15}$$

that is,

$$\begin{cases} \dfrac{d\hat{x}}{dt}(t) = \hat{V}\big(\hat{x}(t)\big), \\[2mm] \hat{x}(0) = (0, X) \in \hat{D} \overset{\text{def}}{=} \mathbf{R}^+ \times \overline{D}, \end{cases} \tag{5.16}$$

where $\hat{x}(t) = \big(x_0(t), x(t)\big) \in \mathbf{R}^{N+1}$, $\hat{V}(\hat{x}) = \big(1, \tilde{V}(\hat{x})\big)$, and

$$\tilde{V}(x_0, x) = \begin{cases} V(x_0, x), & 0 \le x_0 \le \tau \\ V(\tau, x), & \tau < x_0 \end{cases}, \quad x \in \overline{D}. \tag{5.17}$$

It is easy to check that systems (5.15) and (5.16) are equivalent on $[0, \tau]$ and that $\hat{x}(t) = \big(t, x(t)\big)$. The new velocity field on $\hat{V} \subset \mathbf{R}^{N+1}$ is continuous at each point $\hat{x} \in \hat{D}$ by the first assumption $(V1_D)$ since

$$\hat{V}(\hat{y}) - \hat{V}(\hat{x}) = \Big(0, \tilde{V}(y_0, y) - \tilde{V}(x_0, x)\Big),$$

and for $0 \le x_0, y_0 \le \tau$

$$|V(y_0, y) - V(x_0, x)| \le |V(y_0, y) - V(y_0, x)| + |V(y_0, x) - V(x_0, x)| \\ \le n|y - x| + |V(y_0, x) - V(x_0, x)|.$$

In addition,

$$T_{\hat{D}}(\hat{x}) = T_{\mathbf{R}^+}(x_0) \times T_{\overline{D}}(x)$$
$$\Rightarrow \hat{V}(\hat{x}) = \big(1, V(\hat{x})\big) \in T_{\mathbf{R}^+}(x_0) \times T_{\overline{D}}(x).$$

Moreover, $\hat{V}(\hat{D})$ is bounded and $\mathbf{R}^{N+1}$ is finite-dimensional. By using the version of Nagumo's theorem given in Aubin and Cellina [1, Thm. 3, part (b), pp. 182–183], there exists a viable solution $\hat{x}$ to (5.16) for all $t \geq 0$. In particular,

$$\forall t \in [0, \tau], \quad \hat{x}(t) \in \hat{D} = \mathbf{R}^+ \times \overline{D},$$

which is necessarily of the form $\hat{x}(t) = (t, x(t))$. Hence there exists a viable solution $x, x(t) \in \overline{D}$ on $[0, \tau]$, to (5.9). The uniqueness now follows from the Lipschitz condition $(V1_D)$. The Lipschitzian continuity $(T1_D)$ can be established by a standard argument.

Condition $(T2_D)$. Associate with X in $\overline{D}$ the function

$$y(s) = T_{t-s}(X), \quad 0 \leq s \leq t.$$

Then

$$\frac{dy}{ds}(s) = -V\big(t - s, y(s)\big), \quad 0 \leq s \leq t, \quad y(0) = T_t(X). \tag{5.18}$$

For each $x \in \overline{D}$, the differential equation

$$\frac{dy}{ds}(s) = -V\big(t - s, y(s)\big), \quad 0 \leq s \leq t, \quad y(0) = x \in \overline{D} \tag{5.19}$$

has a unique viable solution in $C^1([0, t]; \mathbf{R}^N)$:

$$\forall s \in [0, t], \quad y(s) \in \overline{D}, \tag{5.20}$$

since by assumption $(V2_D)$

$$\forall t \in [0, \tau], \forall x \in \overline{D}, \quad -V(t, x) \in T_{\overline{D}}(x).$$

The proof is the same as above. The solutions of (5.19) define a Lipschitzian mapping

$$x \mapsto S_t(x) = y(t) : \overline{D} \to \overline{D}$$

such that

$$\exists c > 0, \forall t \in [0, \tau], \forall x, y \in \overline{D}, \quad |S_t(y) - S_t(x)| \leq c|y - x|. \tag{5.21}$$

Now in view of (5.18) and (5.19)

$$S_t\big(T_t(X)\big) = y(t) = T_{t-t}(X) = X \quad \Rightarrow \quad S_t \circ T_t = I \text{ on } \overline{D}.$$

To obtain the other identity, consider the function

$$z(r) = y(t - r; x),$$

where $y(\cdot, x)$ is the solution of equation (5.19). By definition,

$$\frac{dz}{dr}(r) = V\big(r, z(r)\big), \quad z(0) = y(t, x)$$

and necessarily

$$x = y(0; x) = z(t) = T_t\big(y(t; x)\big) = T_t\big(S_t(x)\big)$$
$$\Rightarrow T_t \circ S_t = I \text{ on } \overline{D} \Rightarrow S_t = T_t^{-1} : \overline{D} \to \overline{D}.$$

Condition $(T3_D)$. The uniform Lipschitz continuity in $(T3_D)$ follows from (5.21) and $(T2_D)$, and we need only to show that

$$\forall x \in \overline{D}, \quad T^{-1}(\cdot, x) \in C\big([0, \tau]; \mathbf{R}^N\big).$$

Given t in $[0, \tau]$, choose an arbitrary sequence $\{t_n\}$, $t_n \to t$. Then for each $x \in \overline{D}$ there exists $X \in \overline{D}$ such that

$$T_t(X) = x \quad \text{and} \quad T_{t_n}(X) \to T_t(X) = x$$

from $(T1_D)$. But

$$T_{t_n}^{-1}(x) - T_t^{-1}(x) = T_{t_n}^{-1}\big(T_t(X)\big) - T_t^{-1}\big(T_t(X)\big)$$
$$= T_{t_n}^{-1}\big(T_t(X)\big) - T_{t_n}^{-1}\big(T_{t_n}(X)\big).$$

By the uniform Lipschitz continuity of T_t^{-1}

$$|T_{t_n}^{-1}(x) - T_t^{-1}(x)| = |T_{t_n}^{-1}\big(T_t(X)\big) - T_{t_n}^{-1}\big(T_{t_n}(X)\big)|$$
$$\leq c|T_t(X) - T_{t_n}(X)|,$$

and the last term converges to zero as t_n goes to t.

(ii) The first condition $(V1_D)$ is satisfied since for each $x \in \overline{D}$ and t, s in $[0, \tau]$

$$|V(t, x) - V(s, x)| \leq \left| \frac{\partial T}{\partial t}\big(t, T_t^{-1}(x)\big) - \frac{\partial T}{\partial t}\big(t, T_s^{-1}(x)\big) \right|$$
$$+ \left| \frac{\partial T}{\partial t}\big(t, T_s^{-1}(x)\big) - \frac{\partial T}{\partial t}\big(s, T_s^{-1}(x)\big) \right|$$
$$\leq c|T_t^{-1}(x) - T_s^{-1}(x)|$$
$$+ \left| \frac{\partial T}{\partial t}\big(t, T_s^{-1}(x)\big) - \frac{\partial T}{\partial t}\big(s, T_s^{-1}(x)\big) \right|.$$

The second condition $(V1_D)$ follows from $(T1_D)$ and $(T3_D)$ and the following inequality: for all x and y in $\overline{D}$

$$|V(t, y) - V(t, x)| = \left| \frac{\partial T}{\partial t}\big(t, T_t^{-1}(y)\big) - \frac{\partial T}{\partial t}\big(t, T_t^{-1}(x)\big) \right|$$
$$\leq c \left| T_t^{-1}(y) - T_t^{-1}(x) \right| \leq cc'|y - x|.$$

To check condition $(V2_D)$, recall the definition (5.6) of Bouligand contingent cone:

$$T_{\overline{D}}(X) = \bigcap_{\varepsilon > 0} \bigcap_{\alpha > 0} \bigcup_{0 < h < \alpha} \left[\frac{1}{h}(\overline{D} - X) + \varepsilon B \right],$$

where B is the unit disk in $\mathbf{R}^N$. We first show that

$$\forall x \in \overline{D}, \quad V(t,x) = \frac{\partial T}{\partial t}\big(t, T_t^{-1}(x)\big) \in T_{\overline{D}}(x).$$

By (T2$_D$), T_t is bijective. So it is equivalent to show that

$$\forall X \in \overline{D}, \quad \frac{\partial T}{\partial t}(t, X) \in T_{\overline{D}}\big(T_t(X)\big).$$

For simplicity we use the notation

$$x(t) = T_t(X) = T(t, X) \quad \text{and} \quad x'(t) = \frac{\partial T}{\partial t}(t, X). \tag{5.22}$$

By the definition of $T_{\overline{D}}(x(t))$, we must prove that

$$\forall \varepsilon > 0, \forall \alpha > 0, \exists h \in\ ]0, \alpha[\,, \exists u$$

$$\text{such that } x'(t) \in u + \varepsilon B \text{ and } x(t) + hu \in \overline{D}.$$

Choose $\delta, 0 < \delta < \alpha$, such that

$$\forall s, \quad |s - t| < \delta \quad \Rightarrow \quad |x'(s) - s'(t)| < \varepsilon.$$

Then fix $t', 0 < t' - t < \delta$,

$$x(t') - x(t) = \int_t^{t'} \big[x'(s) - x'(t)\big]\, ds + (t' - t)x'(t)$$

$$\Rightarrow \left| \frac{x(t') - x(t)}{t' - t} - x'(t) \right| < \varepsilon.$$

Therefore choose $u = [x(t') - x(t)]/(t' - t)$. Now

$$x(t) + hu = x(t) + \frac{h}{t' - t}\big[x(t') - x(t)\big]$$

and choose $h = t' - t$ since $0 < t' - t < \delta < \alpha$:

$$T(t, X) + (t' - t)u = T(t, X) + \big[T(t', X) - T(t, X)\big] = T(t', X) \in \overline{D}$$

by assumption on T. This proves (5.22). The second part of (V2$_D$) is

$$\forall x \in \overline{D}, \quad -V(t,x) = -\frac{\partial T}{\partial t}\big(t, T_t^{-1}(x)\big) \in T_{\overline{D}}(x),$$

which is equivalent to proving that

$$\forall X \in \overline{D}, \quad -\frac{\partial T}{\partial t}(t, X) \in T_{\overline{D}}\big(T_t(X)\big),$$

or with the simplified notation

$$-x'(t) \in T_{\overline{D}}\big(x(t)\big). \tag{5.23}$$

We proceed exactly as in the proof of (5.22) except that we choose t' such that $0 < t - t' < \delta, h = t - t'$, and $u = -[x(t) - x(t')]/(t - t')$. Then

$$|u + x'(t)| < \varepsilon$$

$$\text{and} \quad x(t) + hu = x(t) + (t - t')\left[-\left(\frac{x(t) - x(t')}{t - t'}\right)\right] = x(t') \in \overline{D},$$

and we get (5.23).

(iii) From $(T1_D)$ and $(T2_D)$

$$f(t)(y) - f(t)(x) = \int_0^t \frac{\partial T}{\partial t}(r, T_r(y)) - \frac{\partial T}{\partial t}(r, T_r(x))\, dr$$

$$|f(t)(y) - f(t)(x)| \leq \int_0^t c\,|T_r(y) - T_r(x)|\, dr \leq c^2\, t\, |y - x|.$$

For $\tau' = \min\{\tau, 1/(2c^2)\}$ and $0 \leq t \leq \tau'$, $c(f(t)) \leq 1/2$,

$$g(t) = T_t^{-1} - I = (I - T_t) \circ T_t^{-1} = -f(t) \circ [I + g(t)],$$
$$(1 - c(f(t)))\, c(g(t)) \leq c(f(t)) \quad \Rightarrow \quad c(g(t)) \leq 1 \text{ and } c(T_t^{-1}) \leq 2,$$

and the second condition (T3) is satisfied on $[0, \tau']$. The first one follows by the same argument as in part (i). Therefore, the conclusions of part (ii) are true on $[0, \tau']$. This completes the proof of the theorem. $\square$

5.2 Transformation of Condition $(V2_D)$ into a Linear Constraint

Condition $(V2_D)$ is equivalent to

$$\boxed{\forall t \in [0, \tau], \forall x \in \overline{D}, \quad V(t, x) \in \{-T_D(x)\} \cap T_D(x),} \tag{5.24}$$

since $T_{\overline{D}}(x) = T_D(x)$. If $T_D(x)$ were convex, then the above intersection would be a closed linear subspace of $\mathbf{R}^N$. This is true when D is convex. In that case $T_D(x) = C_D(x)$, where $C_D(x)$ is Clarke tangent cone and

$$\boxed{L_D(x) = \{-C_D(x)\} \cap C_D(x)} \tag{5.25}$$

is a closed linear subspace of $\mathbf{R}^N$. This means that $(V2_D)$ reduces to

$$\boxed{\forall t \in [0, \tau], \forall x \in \overline{D}, \quad V(t, x) \in L_D(x).} \tag{5.26}$$

It turns out that for continuous vector fields $V(t, \cdot)$, the equivalence of $(V2_D)$ and (5.26) extends to arbitrary domains D. This equivalence generally fails for discontinuous vector fields. Other equivalences might be possible between T_D and some intermediary convex cone between C_D and T_D, but there is no evidence so far of that fact. For smooth bounded open domains Ω, the two cones coincide and the condition reduces to $V \cdot n = 0$, n the normal to $\partial\Omega$, and $V(t, x)$ belongs to the tangent space to $\partial\Omega$ in each point of $\partial\Omega$.

Theorem 5.2.

   (i) *Given a velocity field V satisfying* (V1$_D$), *condition* (V2$_D$) *is equivalent to*

     (V2$_C$) $\forall t \in [0, \tau], \forall x \in \overline{D}, \quad V(t, x) \in L_D(x) = \{-C_D(x)\} \cap C_D(x),$

    where $C_D(x)$ is the (*closed convex*) *Clarke tangent cone to $\overline{D}$ at x,*

$$C_D(x) = \left\{ v \in \mathbf{R}^N : \lim_{\substack{t \searrow 0 \\ y \xrightarrow{\overline{D}} x}} \frac{d_D(y + tv)}{t} = 0 \right\}, \tag{5.27}$$

    and $\xrightarrow{\overline{D}}$ denotes the convergence in $\overline{D}$.

   (ii) *$L_D(x)$ is a closed linear subspace of $\mathbf{R}^N$.*

Proof. (i) The equivalence of (V2$_D$) and (V2$_C$) is a direct consequence of the following lemma.

Lemma 5.1. *Given a vector field $W \in C(\overline{D}; \mathbf{R}^N)$, the following two conditions are equivalent:*

$$\forall x \in \overline{D}, \quad W(x) \in T_D(x), \tag{5.28}$$

$$\forall x \in \overline{D}, \quad W(x) \in C_D(x). \tag{5.29}$$

   (ii) The set $L_D(x)$ is closed as the intersection of two closed sets. To show that it is linear, we show that for all $\alpha \in \mathbf{R}$ and $V \in L_D(x)$, $\alpha V \in L_D(x)$, and for all V and W in $L_D(x)$, $V + W \in L_D(x)$. Since $\pm C_D(x)$ are cones,

$$\forall \alpha \in \mathbf{R}, \forall V \in L_D(x), \quad \pm |\alpha| V \in C_D(x)$$
$$\Rightarrow \pm \alpha V \in C_D(x) \Rightarrow \alpha V \in L_D(x).$$

By convexity of $\pm C_D(x)$

$$\forall V, W \in L_D(x) \quad \pm (V + W) \in C_D(x) \Rightarrow V + W \in L_D(x).$$

This completes the proof of the theorem. $\square$

Proof of Lemma 5.1. Assume that (5.29) is verified. By definition $C_D(x) \subset T_D(x)$ and (5.29) $\Rightarrow$ (5.28). Conversely, either x is an isolated point and $T_D(x) = \{0\} = C_D(x)$, or there are points $x \neq y \in \overline{D}$ such that $y \xrightarrow{\overline{D}} x$. In the latter case we know that

$$\liminf_{y \xrightarrow{\overline{D}} x} T_D(y) = C_D(x)$$

(cf., for instance, Aubin and Frankowska [1, Thm. 4.1.10, section 4.1.5, p. 130]). Since W is continuous in $\overline{D}$ and (5.28) is satisfied, then for each $x \in \overline{D}$

$$W(x) = \lim_{y \xrightarrow{\overline{D}} x} W(y) \in \liminf_{y \xrightarrow{\overline{D}} x} T_D(y) = C_D(x)$$

and (5.28) implies (5.29). $\square$

Remark 5.3.
Lemma 5.1 essentially says that for continuous vector fields we can relax the condition of Nagumo's [1] theorem from $(V2_D)$ involving the Bouligand contingent cone to $(V2_C)$ involving the smaller Clarke convex tangent cone. In dimension $N = 3, L_D(x)$ is $\{0\}$, a line, a plane, or the whole space. $\qquad\square$

Notation 5.1.
In what follows, it will be convenient to introduce the following spaces and subspaces:

$$\boxed{\mathcal{L} = \left\{V : [0,\tau] \times \mathbf{R}^{\mathrm{N}} \to \mathbf{R}^{\mathrm{N}} : V \text{ satisfies (V) on } \mathbf{R}^{\mathrm{N}}\right\},} \qquad (5.30)$$

and for an arbitrary domain D in $\mathbf{R}^{\mathrm{N}}$

$$\boxed{\begin{aligned} \mathcal{L}_D = \{V : [0,\tau] \times \overline{D} \to \mathbf{R}^{\mathrm{N}} : \\ V \text{ satisfies } (V1_D) \text{ and } (V2_C) \text{ on } \overline{D}\}. \end{aligned}} \qquad (5.31)$$

For any integers $k \geq 0$ and $m \geq 0$ and any compact subset K of $\mathbf{R}^{\mathrm{N}}$ define the following subspaces of $\mathcal{L}$:

$$\boxed{\mathcal{V}_K^{m,k} = C^m\left([0,\tau], \mathcal{D}^k(K, \mathbf{R}^{\mathrm{N}})\right) \cap \mathcal{L},} \qquad (5.32)$$

where $\mathcal{D}^k(K, \mathbf{R}^{\mathrm{N}})$ is the space of k-times continuously differentiable transformations of $\mathbf{R}^{\mathrm{N}}$ with compact support in K. In all cases $\mathcal{V}_K^{m,k} \subset \mathcal{L}_K$. As usual $\mathcal{D}^\infty(K, \mathbf{R}^{\mathrm{N}})$ will de written $\mathcal{D}(K, \mathbf{R}^{\mathrm{N}})$. $\qquad\square$

6 Continuity of Shape Functions

In this section we give a characterization of the continuity of a shape function

$$\Omega \mapsto J(\Omega) : \mathcal{A} \subset \mathcal{P}(\mathbf{R}^{\mathrm{N}}) \to B \qquad (6.1)$$

defined on a family $\mathcal{A}$ in $\mathcal{P}(\mathbf{R}^{\mathrm{N}})$ (cf. Definition 2.1) with values in a Banach space B with respect to the Courant metric in terms of its continuity along the flows generated by a family of velocity fields using the equivalence Theorems 4.3, 4.4, and 4.5. Checking the continuity along flows is usually easier and more natural. We specifically consider the continuity of shape functions with respect to the Courant metric associated with the quotient spaces of transformations $\mathcal{F}_0^k/\mathcal{G}(\Omega)$ of section 8 and $\mathcal{F}^k(\overline{\mathbf{R}^{\mathrm{N}}})/\mathcal{G}(\Omega)$ and $\mathcal{F}^{k,1}(\overline{\mathbf{R}^{\mathrm{N}}})/\mathcal{G}(\Omega)$ of section 9.2 in Chapter 2 corresponding to the families of velocity fields $C_0^k(\mathbf{R}^{\mathrm{N}}, \mathbf{R}^{\mathrm{N}})$, $C^k(\overline{\mathbf{R}^{\mathrm{N}}}, \mathbf{R}^{\mathrm{N}})$, and $C^{k,1}(\overline{\mathbf{R}^{\mathrm{N}}}, \mathbf{R}^{\mathrm{N}})$.

6.1 Courant Metrics and Flows of Velocities

We start with the space $C_0^k(\mathbf{R}^{\mathrm{N}}) = C_0^k(\mathbf{R}^{\mathrm{N}}, \mathbf{R}^{\mathrm{N}})$ used in Micheletti [1].

Theorem 6.1. *Let $k \geq 1$ be an integer, B a Banach space, and Ω a nonempty open subset of $\mathbf{R}^N$. Consider a shape function $J : N_\Omega([I]) \to B$ defined in a neighborhood $N_\Omega([I])$ of $[I]$ in $\mathcal{F}_0^k/\mathcal{G}(\Omega)$. Then J is continuous at Ω for the Courant metric if and only if*

$$\lim_{t \searrow 0} J(T_t(\Omega)) = J(\Omega) \tag{6.2}$$

for all families of velocity fields $\{V(t) : 0 \leq t \leq \tau\}$ satisfying the condition

$$V \in C([0,\tau]; C_0^k(\mathbf{R}^N, \mathbf{R}^N)). \tag{6.3}$$

Proof. It is sufficient to prove the theorem for a real-valued function J. The Banach space case is readily obtained by considering the new real-valued function $j(T) = |J(T(\Omega)) - J(\Omega)|$.

(i) If J is δ-continuous at Ω, then for all $\varepsilon > 0$ there exists $\delta > 0$ such that

$$\forall T, \ [T] \in N_\Omega([I]), \quad \delta([T],[I]) < \delta, \quad |J(T(\Omega)) - J(\Omega)| < \varepsilon.$$

Condition (6.3) on V coincides with condition (4.23) of Theorem 4.4, which implies conditions (4.24) and (4.25):

$$f \in C^1([0,\tau]; C_0^k(\mathbf{R}^N, \mathbf{R}^N)) \text{ and } g \in C([0,\tau]; C_0^k(\mathbf{R}^N, \mathbf{R}^N))$$
$$\Rightarrow \|T_t - I\|_{C^k(\mathbf{R}^N)} \to 0 \text{ and } \|T_t^{-1} - I\|_{C^k(\mathbf{R}^N)} \to 0 \text{ as } t \to 0.$$

But by definition of the metric δ,

$$\delta([T_t],[I]) \leq \|T_t^{-1} - I\|_{C^k} + \|T_t - I\|_{C^k} \to 0 \text{ as } t \to 0,$$

and we get the convergence (6.2) of the function $J(T_t(\Omega))$ to $J(\Omega)$ as t goes to zero for all V satisfying (6.3).

(ii) Conversely, it is sufficient to prove that for any sequence $\{[T_n]\}$ such that $\delta([T_n],[I])$ goes to zero, there exists a subsequence such that

$$J(T_{n_k}(\Omega)) \to J(I(\Omega)) = J(\Omega) \text{ as } k \to \infty.$$

Indeed let

$$\ell = \liminf_{n \to \infty} J(T_n(\Omega)) \text{ and } L = \limsup_{n \to \infty} J(T_n(\Omega)).$$

By definition of the liminf, there is a subsequence, still indexed by n, such that $\ell = \liminf_{n \to \infty} J(T_n(\Omega))$. But since there exists a subsequence $\{T_{n_k}\}$ of $\{T_n\}$ such that $J(T_{n_k}(\Omega)) \to J(\Omega)$, then necessarily $\ell = J(\Omega)$. The same reasoning applies to the limsup and hence the whole sequence $J(T_n(\Omega))$ converges to $J(\Omega)$ and we have the continuity of J at Ω.

We prove that we can construct a velocity V associated with a subsequence of $\{T_n\}$ verifying conditions (4.23) of Theorem 4.4 and hence conditions (6.3). By Corollary 1 of Theorem 8.2 in Chapter 2 and the same technique as in the proofs of Lemma 8.7 and Theorem 8.3 in Chapter 2, associate with a sequence $\{T_n\}$ such that $\delta([T_n], [I]) \to 0$ a subsequence, still denoted $\{T_n\}$, such that

$$\|f_n\|_{C^k} + \|g_n\|_{C^k} = \|T_n^{-1} - I\|_{C^k} + \|T_n - I\|_{C^k} \le 2^{-2(n+2)}.$$

For $n \ge 1$ set $t_n = 2^{-n}$ and observe that $t_n - t_{n+1} = -2^{-(n+1)}$. Define the following C^1-interpolation in $(0, 1/2]$: for t in $[t_{n+1}, t_n]$

$$\boxed{T_t(X) \stackrel{\text{def}}{=} T_n(X) + p\left(\frac{t_{n+1} - t}{t_{n+1} - t_n}\right)(T_{n+1}(X) - T_n(X)), \quad T_0(X) \stackrel{\text{def}}{=} X,}$$

where $p \in P^3[0, 1]$ is the polynomial of order 3 on $[0, 1]$ such that $p(0) = 1$ and $p(1) = 0$ and $p^{(1)}(0) = 0 = p^{(1)}(1)$.

Conditions on f. By definition for all t, $0 \le t \le 1/2$, $f(t) = T_t - I \in C_0^k(\mathbf{R}^N)$. Moreover, for $0 < t \le 1/2$

$$T_{t_n}(X) = T_n(X), \ T_{t_{n+1}}(X) = T_{n+1}(X), \ \frac{\partial T}{\partial t}(t_n, X) = 0 = \frac{\partial T}{\partial t}(t_{n+1}, X),$$

$$\frac{\partial T}{\partial t}(t, X) = \frac{T_{n+1}(X) - T_n(X)}{|t_n - t_{n+1}|}p^{(1)}\left(\frac{t_{n+1} - t}{t_{n+1} - t_n}\right),$$

$f'(t) = \partial T/\partial t(t, \cdot) \in C_0^k(\mathbf{R}^N)$ and $f(\cdot)(X) = T(\cdot, X) - I \in C^1((0, 1/2]; \mathbf{R}^N)$. By definition, $f(0) = 0$. For each $0 < t \le 1/2$ there exists $n \ge N$ such that $t_{n+1} \le t \le t_n$ and

$$\|f(t) - f(0)\|_{C^k} = \|f(t)\|_{C^k} = \left\|f_n + p\left(\frac{t_{n+1} - t}{t_{n+1} - t_n}\right)(f_{n+1} - f_n)\right\|_{C^k}$$

$$\le 2\|f_n\|_{C^k} + \|f_{n+1}\|_{C^k} \le 2\,2^{-2(n+2)} + 2^{-2(n+3)} \le 2^{-(n+1)} \le t.$$

Define at $t = 0$, $f'(t) = 0$. By the same technique there exists a constant $c > 0$, and for each $0 < t \le 1/2$ there exists $n \ge N$ such that $t_{n+1} \le t \le t_n$ and

$$\|f'(t) - f'(0)\|_{C^k} = \|f'(t)\|_{C^k}$$

$$= \left\|\frac{\partial T}{\partial t}(t, \cdot)\right\|_{C^k} \le c\frac{\|T_{n+1} - T_n\|_{C^k}}{|t_{n+1} - t_n|} = c\frac{\|f_{n+1} - f_n\|_{C^k}}{2^{-(n+1)}}$$

$$\le c\,2\,2^{-2(n+2)}/2^{-(n+1)} \le c\,2^{-1}2^{-(n+1)} \le c\,2^{-(n+1)} \le ct$$

$$\Rightarrow \boxed{\|f'(t)\|_{C^k} \le ct.}$$

So for each X the functions $t \mapsto f(t)(X)$ and $t \mapsto T_t(X)$ belong to $C^1([0, 1/2]; \mathbf{R}^N)$. By uniform C^k-continuity of the T_n's and the continuity with respect to t for each X, it follows that $f \in C^1([0, 1/2]; C_0^k(\mathbf{R}^N))$ and the condition (4.24) of Theorem 4.4 is satisfied. Hence the corresponding velocity V satisfies conditions (4.23). Finally V satisfies conditions (6.3) and by (6.2) for all $\varepsilon > 0$ there exists $\delta > 0$ such that

$$\forall t, \ 0 \le t \le \delta, \quad |J(T_t(\Omega)) - J(\Omega)| < \varepsilon.$$

In particular there exists $N > 0$ such that for all $n \geq N$, $t_n \leq \delta$, and

$$\forall n \geq N, \quad |J(T_n(\Omega)) - J(\Omega)| = |J(T_{t_n}(\Omega)) - J(\Omega)| < \varepsilon$$

and this proves the δ-continuity for the subsequence $\{T_n\}$. $\qquad\qquad\square$

The case of the Courant metric associated with the space $\mathcal{C}^k(\overline{\mathbf{R}^N}) = C^k(\overline{\mathbf{R}^N}, \mathbf{R}^N)$ is a corollary to Theorem 6.1.

Theorem 6.2. *Let $k \geq 1$ be an integer, B a Banach space, and Ω a nonempty open subset of $\mathbf{R}^N$. Consider a shape function $J : N_\Omega([I]) \to B$ defined in a neighborhood $N_\Omega([I])$ of $[I]$ in $\mathcal{F}^k(\overline{\mathbf{R}^N})/\mathcal{G}(\Omega)$. Then J is continuous at Ω for the Courant metric if and only if*

$$\lim_{t \searrow 0} J(T_t(\Omega)) = J(\Omega) \tag{6.4}$$

for all families of velocity fields $\{V(t) : 0 \leq t \leq \tau\}$ satisfying the condition

$$V \in C([0, \tau]; C^k(\overline{\mathbf{R}^N}, \mathbf{R}^N)). \tag{6.5}$$

The proof of the theorem for the Courant metric topology associated with the space $\mathcal{C}^{k,1}(\overline{\mathbf{R}^N}) = C^{k,1}(\overline{\mathbf{R}^N}, \mathbf{R}^N)$ is similar to the proof of Theorem 6.1 with obvious changes.

Theorem 6.3. *Let $k \geq 0$ be an integer, Ω a nonempty open subset of $\mathbf{R}^N$, and B a Banach space. Consider a shape function $J : N_\Omega([I]) \to B$ defined in a neighborhood $N_\Omega([I])$ of $[I]$ in $\mathcal{F}^{k,1}(\overline{\mathbf{R}^N})/\mathcal{G}(\Omega)$. Then J is continuous at Ω for the Courant metric if and only if*

$$\lim_{t \searrow 0} J(T_t(\Omega)) = J(\Omega) \tag{6.6}$$

for all families $\{V(t) : 0 \leq t \leq \tau\}$ of velocity fields in $C^{k,1}(\overline{\mathbf{R}^N}, \mathbf{R}^N)$ satisfying the conditions

$$V \in C([0, \tau]; C^k(\overline{\mathbf{R}^N}, \mathbf{R}^N)) \ \ and \ \ c_k(V(t)) \leq c \tag{6.7}$$

for some constant c independent of t.

Proof. As in the proof of Theorem 6.1, it is sufficient to prove the theorem for a real-valued function J.

(i) If J is δ-continuous at Ω, then for all $\varepsilon > 0$ there exists $\delta > 0$ such that

$$\forall T, \ [T] \in N_\Omega([I]), \quad \delta([T], [I]) < \delta, \quad |J(T(\Omega)) - J(\Omega)| < \varepsilon.$$

Under condition (6.7), from Theorem 4.3

$$f = T - I \in C([0, \tau]; \mathcal{C}^{k,1}(\overline{\mathbf{R}^N})) \ \ and \ \ \|T_t - I\|_{C^{k,1}} \to 0 \ as \ t \to 0,$$

$$g(t) = T_t^{-1} - I \in \mathcal{C}^{k,1}(\overline{\mathbf{R}^N}) \ \ and \ \ \|T_t^{-1} - I\|_{C^{k,1}} \leq ct \to 0 \ as \ t \to 0.$$

But by definition of the metric δ

$$\delta([T_t], [I]) \leq \|T_t^{-1} - I\|_{C^{k,1}} + \|T_t - I\|_{C^{k,1}} \to 0 \text{ as } t \to 0,$$

and we get the convergence (6.6) of the function $J(T_t(\Omega))$ to $J(\Omega)$ as t goes to zero for all V satisfying (6.7).

(ii) Conversely, as in the proof of Theorem 6.1, it is sufficient to prove that given any sequence $\{[T_n]\}$ such that $\delta([T_n], [I]) \to 0$ there exists a subsequence such that

$$J(T_{n_k}(\Omega)) \to J(I(\Omega)) = J(\Omega) \text{ as } k \to \infty.$$

By Theorem 9.1 (i) and the same technique as in the proofs of Lemma 8.7 and Theorem 8.3 of Chapter 2, associate with a sequence $\{T_n\}$ such that $\delta([T_n], [I]) \to 0$ a subsequence, still denoted $\{T_n\}$, such that

$$\|f_n\|_{C^{k,1}} + \|g_n\|_{C^{k,1}} = \|T_n^{-1} - I\|_{C^{k,1}} + \|T_n - I\|_{C^{k,1}} \leq 2^{-2(n+2)}.$$

For $n \geq 1$ set $t_n = 2^{-n}$ and observe that $t_n - t_{n+1} = -2^{-(n+1)}$. Define the following C^1-interpolation in $(0, 1/2]$: for t in $[t_{n+1}, t_n]$,

$$\boxed{T_t(X) \overset{\text{def}}{=} T_n(X) + p\left(\frac{t_{n+1} - t}{t_{n+1} - t_n}\right)(T_{n+1}(X) - T_n(X)), \quad T_0(X) \overset{\text{def}}{=} X,}$$

where $p \in P^3[0, 1]$ is the polynomial of order 3 on $[0, 1]$ such that $p(0) = 1$ and $p(1) = 0$ and $p^{(1)}(0) = 0 = p^{(1)}(1)$.

Conditions on f. By definition for all t, $0 \leq t \leq 1/2$, $f(t) = T_t - I \in C^{k,1}(\overline{\mathbf{R}^N})$. Moreover, for $0 < t \leq 1/2$

$$T_{t_n}(X) = T_n(X), \ T_{t_{n+1}}(X) = T_{n+1}(X), \ \frac{\partial T}{\partial t}(t_n, X) = 0 = \frac{\partial T}{\partial t}(t_{n+1}, X),$$

$$\frac{\partial T}{\partial t}(t, X) = \frac{T_{n+1}(X) - T_n(X)}{|t_n - t_{n+1}|} p^{(1)}\left(\frac{t_{n+1} - t}{t_{n+1} - t_n}\right),$$

$f'(t) = \partial T/\partial t \in C^{k,1}(\overline{\mathbf{R}^N})$ and $f(\cdot)(X) = T(\cdot, X) - I \in C^1((0, 1/2]; \mathbf{R}^N)$. By definition, $f(0) = 0$. For each $0 < t \leq 1/2$ there exists $n \geq N$ such that $t_{n+1} \leq t \leq t_n$ and

$$\|f(t) - f(0)\|_{C^{k,1}} = \|f(t)\|_{C^{k,1}} = \left\| f_n + p\left(\frac{t_{n+1} - t}{t_{n+1} - t_n}\right)(f_{n+1} - f_n) \right\|_{C^{k,1}}$$

$$\leq 2\|f_n\|_{C^{k,1}} + \|f_{n+1}\|_{C^{k,1}} \leq 2\, 2^{-2(n+2)} + 2^{-2(n+3)} \leq 2^{-(n+1)} \leq t.$$

Define at $t = 0$ $f'(t) = 0$. By the same technique there exists a constant $c > 0$, and for each $0 < t \leq 1/2$ there exists $n \geq N$ such that $t_{n+1} \leq t \leq t_n$ and

$$\|f'(t) - f'(0)\|_{C^{k,1}} = \|f'(t)\|_{C^{k,1}}$$

$$= \left\| \frac{\partial T}{\partial t}(t, \cdot) \right\|_{C^{k,1}} \leq c\, \frac{\|T_{n+1} - T_n\|_{C^{k,1}}}{|t_{n+1} - t_n|} = c\, \frac{\|f_{n+1} - f_n\|_{C^{k,1}}}{2^{-(n+1)}}$$

$$\leq c\, 2\, 2^{-2(n+2)}/2^{-(n+1)} \leq c\, 2^{-1} 2^{-(n+1)} \leq c\, 2^{-(n+1)} \leq c\, t$$

$$\Rightarrow \boxed{\|f'(t)\|_{C^{k,1}} \leq c\, t \text{ and } c_k(f'(t)) \leq c\, t}$$

and for each X the functions $t \mapsto f(t)(X)$ and $t \mapsto T_t(X)$ belong to $C^1([0, 1/2]; \mathbf{R}^N)$. By uniform C^k-continuity of the T_n's and the continuity with respect to t for each X, it follows that $f \in C^1([0, 1/2]; C^k(\overline{\mathbf{R}^N}))$. Moreover, it can be shown that

$$c_k(f'(t)) \leq ct \quad \Rightarrow \quad \forall t, s \in [0, \tau], \quad c_k(f(t) - f(s)) \leq c'|t - s|$$

for some $c' > 0$. The result is straightforward for $k = 0$ and then the general case follows by induction on k. As a result $f \in C([0, 1/2]; C^{k,1}(\overline{\mathbf{R}^N}))$ and the condition (4.20) of Theorem 4.3 is satisfied. Hence the corresponding velocity V satisfies conditions (4.19). Finally the velocity field V satisfies conditions (6.7), and by (6.6) for all $\varepsilon > 0$ there exists $\delta > 0$ such that

$$\forall t \leq \delta, \quad |J(T_t(\Omega)) - J(\Omega)| < \varepsilon.$$

In particular there exists $N > 0$ such that for all $n \geq N$, $t_n \leq \delta$ and

$$\forall n \geq N, \quad |J(T_n(\Omega)) - J(\Omega)| = |J(T_{t_n}(\Omega)) - J(\Omega)| < \varepsilon,$$

and we have the δ-continuity for the subsequence $\{T_n\}$. $\qquad\square$

Remark 6.1.
The conclusions of Theorems 6.1, 6.2, and 6.3 are generic. They also have their counterpart in the constrained case. For instance, a generalization of Theorem 6.1 in the constrained case has been announced by Boisgerault and Gomez [1] for an open subset D of $\mathbf{R}^N$ of class C^2. The difficulty lies in the second part of the theorem, which requires a special construction to make sure that the family of transformations $\{T_t : 0 \leq t \leq \tau\}$ constructed from the sequence $\{T_n\}$ are homeomorphisms of $\overline{D}$. $\quad\square$

6.2　Shape Continuity and Velocity Method

These results point to a natural notion of *directional shape continuity* associated with the velocity method which is very much in the spirit of geometry. This intimate relationship between the Courant metric on quotient spaces of transformations and the velocity method is a nice mathematical result and a powerful tool for the analysis of shape problems. It will be particularly useful for the differentiability of shape functions in Chapter 8.

Definition 6.1.
Let f be a shape function with values in a Banach space B.

(i) f is *shape continuous at Ω in the direction V* (resp., θ) if

$$\lim_{t \searrow 0} f(T_t(\Omega)) = f(\Omega),$$

where T_t is specified by the solutions of

$$T_t(X) \stackrel{\text{def}}{=} x(t), \quad \frac{dx}{dt}(t) = V(t, x(t)), \quad x(0) = X,$$

where $\{V(t) : 0 \leq t \leq \tau\}$ is a family of velocity fields which satisfies condition (V) given by (4.2) for some $\tau > 0$ (resp., T_t is specified by

$$T_t(X) \stackrel{\text{def}}{=} X + t\theta(X), \quad t \geq 0,$$

where $\theta \in \text{Lip}\,(\mathbf{R}^{\text{N}}, \mathbf{R}^{\text{N}})$ is a uniform Lipschitzian transformation of $\mathbf{R}^{\text{N}}$:

$$\exists c > 0 \text{ such that } \forall X, Y \in \mathbf{R}^{\text{N}}, \quad |\theta(Y) - \theta(X)| \leq c|Y - X|).$$

(ii) f is said to be *shape continuous at* Ω with respect to a class $\mathcal{V}$ of families of velocity fields if it is shape continuous at Ω in the direction V for all V in the class $\mathcal{V}$ satisfying conditions (V) given by (4.2).

(iii) f is said to be *shape continuous at* Ω with respect to a family Θ of transformations in $\text{Lip}\,(\mathbf{R}^{\text{N}}, \mathbf{R}^{\text{N}})$ if it is shape continuous at Ω in the direction θ for all $\theta \in \Theta$. $\qquad\square$

Chapter 8
Shape Derivatives and Calculus, and Tangential Differential Calculus

1   Introduction

Chapter 7 has studied the equivalence between the two points of view of transformations and velocities. Equivalent characterizations of shape continuity by the Courant metric and continuity along flows of velocities have been established. Since the velocity approach readily extends to the constrained case, and in particular to submanifolds of $\mathbf{R}^N$, this chapter adopts the velocity framework to study shape semiderivatives.

In section 2 we give a self-contained review of semiderivatives and derivatives in vector spaces in order to prepare the ground for shape derivatives. Section 3 gives the definitions and the main properties of first-order semiderivatives and derivatives of shape functions. Section 3.2 specializes the definitions and results of section 3 to the generic shape spaces endowed with the Courant metric. Finally, a canonical definition of the *shape gradient* is given along with the main structure theorem in section 3.3. Before going to second-order derivatives, the main elements of the *shape calculus* are introduced in section 4 and the basic formulae for domain and boundary integrals are given in sections 4.1 and 4.2. Their application is illustrated in a series of examples in section 4.3.

The final expressions of the gradients in the examples of section 4 always lead to a domain and a boundary expression. The boundary expression contains fundamental properties of the gradients and the natural way to untangle some of the resulting terms is to use the *tangential calculus*. Section 5 gives the main elements of that calculus for a C^2-submanifold of $\mathbf{R}^N$ of codimension 1 including Stokes's and Green's formulae in section 5.5 and the relationship between tangential and covariant derivatives in section 5.6. This is applied to the derivative of the integral of the square of the normal derivative of section 4.3.3.

Section 6 extends definitions and structure theorems to second-order derivatives. In order to develop a better feeling for the abstract definitions the second-order derivative of the domain integral is computed in section 6.1 using the combined strengths of the shape and tangential calculi. A basic formula for the second-order

329

semiderivative of the domain integral is given in section 6.2. This is completed with structure theorems in the nonautonomous case in section 6.3 and in the autonomous case in section 6.4. The shape Hessian is decomposed into a symmetrical term plus the gradient acting on the first half of the Lie bracket in section 6.5. This symmetrical part is itself decomposed into a symmetrical part that only depends on the normal component of the velocity field and a symmetrical term made up of the gradient acting on a generic group of terms that occurs in all examples considered in section 6.

2　Review of Differentiation in Banach Spaces

In this section we review some elements of derivatives and semiderivatives in vector spaces. In that context we shall start with the weaker notion of Gâteaux semiderivative and emphasize the key role played by the Hadamard semiderivative in the semiderivative of the composition of two functions, since in most applications the extension of the *chain rule* is a central ingredient of a good differential calculus.

2.1　Definitions of Semiderivatives and Derivatives

Definition 2.1.

Let f be a real-valued function defined in a neighborhood of a point x of a topological vector space E.

(i) We say that f has a *Gâteaux semiderivative* at a point $x \in U$ in the direction $v \in E$ if the following limit exists:

$$\lim_{\varepsilon \searrow 0} \frac{f(x + \varepsilon v) - f(x)}{\varepsilon}; \tag{2.1}$$

when it exists it will be denoted by $df(x; v)$.

(ii) We say that f has a *Hadamard semiderivative* at $x \in U$ in the direction $v \in E$ if the following limit exists:

$$\lim_{\substack{\varepsilon \searrow 0 \\ w \to v}} \frac{f(x + \varepsilon w) - f(x)}{\varepsilon}; \tag{2.2}$$

when it exists it will be denoted by $d_H f(x; v)$.

The above definitions extend from a real-valued function to a function f into another topological vector space F. □

It is clear that $df(x; v)$ exists and is equal to $d_H f(x; v)$ whenever the semiderivative $d_H f(x; v)$ exists, but the converse is not true without additional assumptions, as can be seen from the following example.

Example 2.1.
Consider the function $f : \mathbf{R}^2 \to \mathbf{R}$:

$$f(x, y) = \begin{cases} \frac{x^6}{(y - x^2)^2 + x^8} & \text{if } (x, y) \neq (0, 0), \\ 0 & \text{if } (x, y) = (0, 0). \end{cases} \tag{2.3}$$

It is readily seen that f has a Gâteaux semiderivative at $(0,0)$ in all directions v in $\mathbf{R}^2$ and that

$$\forall v \in \mathbf{R}^2, \quad df\big((0,0); v\big) = 0, \tag{2.4}$$

which is trivially linear and continuous with respect to v. However, if for $\varepsilon > 0$ we choose the directions

$$w(\varepsilon) = (1, \varepsilon) \to v = (1, 0) \text{ as } \varepsilon \to 0,$$

then

$$\frac{f\big(\varepsilon w(\varepsilon)\big) - f\big((0,0)\big)}{\varepsilon} = \frac{1}{\varepsilon^3} \to +\infty$$

and $d_H f(0, 0; 1, 0)$ does not exist. $\qquad\square$

We shall also need the following notions of full derivatives.

Definition 2.2.
Let f be a real-valued function defined in a neighborhood of a point x of a normed vector space E.

(i) f has a *Gâteaux derivative* at x if

$$\forall v \in E, \quad df(x; v) \text{ exists and}$$
$$v \mapsto df(x; v) : E \to \mathbf{R} \text{ is linear and continuous.} \tag{2.5}$$

Whenever it exists the linear map (2.5) will be denoted $\nabla f : E \to E'$.

(ii) f has a *Fréchet derivative* at x if it has a *Gâteaux derivative* at x and

$$\lim_{h \to 0} \frac{|f(x + h) - f(x) - \langle \nabla f(x), h \rangle_E|}{|h|_E} \to 0,$$

where $\langle \cdot, \cdot \rangle_E$ denotes the duality pairing between E' and E.

The above definitions extend from a real-valued function to a function f into another normed vector space F. $\qquad\square$

The Gâteaux semiderivative is generally neither linear nor continuous with respect to the direction, even in finite dimension.

Example 2.2.
Consider the following function $f : \mathbf{R}^2 \to \mathbf{R}$:

$$f(x,y) = \begin{cases} \frac{x^5}{(y-x^2)^2+x^8} & \text{if } (x,y) \neq (0,0), \\ 0 & \text{if } (x,y) = (0,0). \end{cases} \tag{2.6}$$

It is Gâteaux semidifferentiable at $(0,0)$ in all directions $v = (v_1, v_2)$ in $\mathbf{R}^2$, and

$$df(0,0); (v_1, v_2) = \begin{cases} v_1 & \text{if } v_2 = 0, \\ 0 & \text{if } v_2 \neq 0 \end{cases} \tag{2.7}$$

is neither linear nor continuous with respect to v. $\qquad\square$

As for the Gâteaux semiderivative the Hadamard semiderivative is generally neither linear nor continuous with respect to the direction.

Example 2.3.
The Hadamard semiderivative of the norm $f(x) = |x|_E$ at $x = 0$ is given by

$$\forall v \in E, \quad d_H f(0; v) = |v|_E. \tag{2.8}$$

It is continuous but not linear in v. $\qquad\square$

When f has a Gâteaux semiderivative in all directions v at a point x that is continuous with respect to v, it has a Hadamard semiderivative in all directions

$$d_H f(x; v) = \lim_{w \to v} d_H f(x; w),$$

but again it is not necessarily linear in v.

2.2 Locally Lipschitz Functions

The following theorem gives a sufficient condition for the equivalence of Gâteaux and Hadamard semiderivatives (resp., Gâteaux and Fréchet derivatives).

Theorem 2.1. *Let E be a normed vector space. Given a function $f : U \to \mathbf{R}$ which is uniformly Lipschitzian in a neighborhood U of x in E, that is,*

$$\exists c(x) > 0, \forall y, z \in U, \quad |f(y) - f(z)| \leq c(x) |y - z|_E. \tag{2.9}$$

(i) *If $df(x; v)$ exists, then $d_H f(x; v)$ exists and $df(x; v) = d_H f(x; v)$. Moreover, if $df(x; v)$ exists for all $v \in E$, then*

$$\forall v_2, v_1 \in E, \quad |d_H f(x; v_2) - d_H f(x; v_1)| \leq c |v_2 - v_1|_E.$$

(ii) *If f has a Gâteaux derivative at x, then it has a Fréchet derivative at x and they are equal.*

Proof. (i) There exist $\bar{\varepsilon} > 0$ and a neighborhood W of v in E such that

$$\forall w \in W, \ \forall \varepsilon, \ 0 < \varepsilon \leq \bar{\varepsilon}, \quad x + \varepsilon w \in U \text{ and } x + \varepsilon v \in U.$$

Then

$$\frac{1}{\varepsilon}\big[f(x + \varepsilon w) - f(x)\big] = \frac{1}{\varepsilon}\big[f(x + \varepsilon w) - f(x + \varepsilon v)\big] + \frac{1}{\varepsilon}\big[f(x + \varepsilon v) - f(x)\big]$$

and

$$\left| \frac{1}{\varepsilon}\big[f(x + \varepsilon w) - f(x)\big] - df(x; v) \right|$$

$$\leq \left| \frac{1}{\varepsilon}\big[f(x + \varepsilon v) - f(x)\big] - df(x; v) \right| + c(x)|w - v|_E.$$

So as $\varepsilon \to 0$ and $w \to v$, $d_H f(x; v) = df(x; v)$.

(ii) This part of the proof is obtained from Lemma 3.1 in Chapter 4 extended to a normed vector space. $\qquad\square$

2.3   Chain Rule for Semiderivatives

The main difference between the two semiderivatives is that the composition of two functions that are semidifferentiable in the sense of Hadamard is semidifferentiable in the sense of Hadamard.

Theorem 2.2. *Let E and F be two topological vector spaces and let h be the composition of two mappings f and g:*

$$h(x) = f\big(g(x)\big) \tag{2.10}$$

in a neighborhood U of a point x in E, where

$$g : U \subset E \to F \text{ and } f : g(U) \to \mathbf{R}. \tag{2.11}$$

Assume that

(i) g has a Gâteaux (resp., Hadamard) semiderivative at x in the direction v, and

(ii) $d_H f(g(x); dg(x; v))$ exists.

Then

$$dh(x; v) = d_H f(g(x); dg(x; v))$$
$$\big(resp., \ d_H h(x; v) = d_H f(g(x); d_H g(x; v))\big). \tag{2.12}$$

Proof. (a) For $\varepsilon > 0$ small enough let

$$m(\varepsilon) = \frac{g(x + \varepsilon v) - g(x)}{\varepsilon} - dg(x; v) \text{ in } F.$$

By assumption, $m(\varepsilon) \to 0$ as $\varepsilon \to 0$. By the definition of $dh(x; v)$ we want to find the limit of the differential quotient

$$d(\varepsilon) = \frac{f\big(g(x + \varepsilon v)\big) - f\big(g(x)\big)}{\varepsilon},$$

which can be rewritten as

$$d(\varepsilon) = \frac{f\big(g(x) + \varepsilon\big(dg(x; v) + m(\varepsilon)\big)\big) - f\big(g(x)\big)}{\varepsilon},$$

where

$$dg(x; v) + m(\varepsilon) \to dg(x; v) \text{ as } \varepsilon \to 0.$$

So by the definition of $d_H f$,

$$\lim_{\varepsilon \searrow 0} d(\varepsilon) = d_H f\big(g(x); dg(x; v)\big).$$

(b) When g is Hadamard semidifferentiable we replace $m(\varepsilon)$ and $d(\varepsilon)$ by

$$m(\varepsilon, w) = \frac{g(x + \varepsilon w) - g(x)}{\varepsilon} - dg(x; v)$$

and

$$d(\varepsilon, w) = \frac{1}{\varepsilon}\Big[f\big(g(x) + \varepsilon\big(dg(x; v) + m(\varepsilon, w)\big)\big) - f\big(g(x)\big)\Big]$$

and proceed as in part (a). $\square$

In general we cannot improve the semiderivative of h by improving the semi-derivative of g when f is not Hadamard semidifferentiable.

Example 2.4.
Consider the composition of the function $f : \mathbf{R}^2 \to \mathbf{R}$ in Example 2.1 and the map

$$g : \mathbf{R} \to \mathbf{R}^2, \quad g(x) = (x, x^2). \tag{2.13}$$

The map f is Gâteaux, but not Hadamard semidifferentiable, and the map g is infinitely differentiable. However, the composition

$$h(x) = f\big(g(x)\big) = \frac{1}{x^2} \tag{2.14}$$

is not even Gâteaux semidifferentiable at 0. $\square$

We reiterate that the Hadamard semidifferentiability is a key property for the "chain rule." In general, the composition $h = f \circ g$ of two maps will fail to have a semiderivative unless f has a Hadamard semiderivative—even if the map $g : E \to F$ is Fréchet differentiable at the point x.

2.4 Semiderivatives of Convex Functions

Finally, the class of functions that are Hadamard semidifferentiable is not restrictive since it contains the classical continuously differentiable functions and the convex continuous functions.

Theorem 2.3. *Let $f : U \subset E \to \mathbf{R}$ be a convex function defined in a convex neighborhood U of a point x of a topological vector space E.*

(i) *There exists a neighborhood V of x such that*

$$\forall y \in V, \forall v \in E, \quad df(y; v) \text{ exists.} \tag{2.15}$$

(ii) *If f is continuous at x, then there exists a neighborhood W of x such that*

$$\forall y \in W, \forall v \in E, \quad d_H f(y; v) \text{ exists.} \tag{2.16}$$

Proof. Part (ii) is a consequence of part (i) and the fact that a continuous convex function at x is locally Lipschitzian in a neighborhood of x (cf. Ekeland and Temam [1, pp. 11–12]) by direct application of Theorem 2.1. We give the proof of part (i) for completeness. Let $\theta \in \,]0, 1]$. Notice that for fixed x and v,

$$\exists \alpha, 0 < \alpha < 1, \text{ such that } x - \alpha v \in U,$$
$$\exists \theta_0, 0 < \theta_0 < 1, \text{ such that } \forall \theta \in \,]0, \theta_0] \quad x + \theta v \in U.$$

For this fixed α, we show that

$$\boxed{\forall \theta \in \,]0, \theta_0[, \quad \frac{f(x) - f(x - \alpha v)}{\alpha} \leq \frac{f(x + \theta v) - f(x)}{\theta}.} \tag{2.17}$$

This follows from the identity

$$x = \frac{\alpha}{\alpha + \theta}(x + \theta v) + \frac{\theta}{\alpha + \theta}(x - \alpha v)$$

and the convexity of f,

$$f(x) \leq \frac{\alpha}{\alpha + \theta} f(x + \theta v) + \frac{\theta}{\alpha + \theta} f(x - \alpha v).$$

This can be rewritten

$$\frac{\theta}{\theta + \alpha}\big[f(x) - f(x - \alpha v)\big] \leq \frac{\alpha}{\theta + \alpha}\big[f(x + \theta v) - f(x)\big]$$

and yields (2.17). Define

$$\varphi(\theta) = \frac{f(x + \theta v) - f(x)}{\theta}, \quad 0 < \theta < \theta_0.$$

Now we show that φ is a *monotone increasing* function of $\theta > 0$. For all θ_1 and θ_2, $0 < \theta_1 < \theta_2 < \theta_0$:

$$f(x + \theta_1 v) - f(x) = f\left(\frac{\theta_1}{\theta_2}(x + \theta_2 v) + \left(1 - \frac{\theta_1}{\theta_2}\right)x\right) - f(x)$$

$$\le \frac{\theta_1}{\theta_2}f(x + \theta_2 v) + \left(1 - \frac{\theta_1}{\theta_2}\right)f(x) - f(x)$$

which implies that $\varphi(\theta_1) \le \varphi(\theta_2)$. As φ is a monotone increasing function which is bounded below, its limit exists as θ goes to 0. By definition it is equal to $df(x; v)$. $\quad\square$

2.5 Conditions for Fréchet Differentiability

It turns out that for finite-dimensional spaces E there is a complete analogy between the Gâteaux derivative and the linearity and continuity of the Gâteaux semiderivative on one hand and the Fréchet derivative and the linearity and continuity of the Hadamard semiderivative on the other hand.

Theorem 2.4. *If a function f has a Fréchet derivative at x, then f has a Hadamard semiderivative at x for all $v \in E$ and the map*

$$v \mapsto d_H f(x; v) \; : \; E \to \mathbf{R}$$

is linear and continuous. The converse is true when E is finite dimensional.

Proof. $(\Rightarrow)$ Consider the quotient

$$q(t, w) \stackrel{\text{def}}{=} \frac{f(x + tw) - f(x)}{t}$$

as $w \to v$ and $t \searrow 0$. Hence $h(t, w) \stackrel{\text{def}}{=} tw \to 0$ as $w \to v$ and $t \searrow 0$. Then

$$q(t, w) = Q(h(t, w))\,|w|_E + \langle \nabla f(x), w \rangle_E,$$

where

$$Q(h) \stackrel{\text{def}}{=} \frac{f(x + h) - f(x) - \langle \nabla f(x), h \rangle_E}{|h|_E}$$

and 0 if $h = 0$. By assumption, $Q(h(t, w)) \to 0$ since $h(t, w) = tw \to 0$, $v = 0$, and $\langle \nabla f(x), w \rangle_E \to \langle \nabla f(x), v \rangle_E$ as $w \to v$ by continuity. Therefore,

$$\lim_{\substack{w \to v \\ t \searrow 0}} q(t, w) = \langle \nabla f(x), v \rangle_E$$

and $d_H f(x; v)$ exists and $d_H f(x; v) = \langle \nabla f(x), v \rangle_E$ is linear and continuous with respect to v.

$(\Leftarrow)$ Define the element $\nabla f(x)$ of E' as

$$\forall v \in E, \quad \langle \nabla f(x), v \rangle_E \stackrel{\text{def}}{=} d_H f(x; v).$$

Denote by $\overline{Q}$ the limsup of $Q(h)$ as $|h| \to 0$, which can possibly be infinite. Notice that for $h \neq 0$

$$Q(h) = Q\left(|h| \frac{h}{|h|}\right).$$

Since E is finite dimensional, there exists a sequence $\{h_n\}$ and $v \in E$ such that

$$|Q(h_n)| \to \overline{Q} \text{ and } w_n \stackrel{\text{def}}{=} \frac{h_n}{|h_n|} \to v \in E.$$

By letting $t_n = |h_n|$,

$$Q(h_n) = \frac{f(x + t_n w_n) - f(x)}{t_n} - \langle \nabla f(x), w_n \rangle_E$$
$$= q(t_n, w_n) - \langle \nabla f(x), w_n \rangle_E \to d_H f(x; v) - \langle \nabla f(x), v \rangle_E = 0.$$

Therefore, $\overline{Q} = 0$ and f is Fréchet differentiable at x. $\qquad\square$

We complete this section with a classical sufficient condition for the Fréchet differentiability.

Theorem 2.5. *Given a normed vector space E and a map $f : E \to \mathbf{R}$, assume that there exists a neighborhood $V(x)$ of $x \in E$ such that*

(i) *for all $y \in V(x)$, f has a Gâteaux derivative $\nabla f(y)$, and*

(ii) *the map $\nabla f : V(x) \to E'$ is continuous at x.*

Then f has a Fréchet derivative at x.

Proof. There exists $\rho > 0$ such that the open ball $B(x, \rho)$ is contained in $V(x)$. For $h \in E$, $0 < |h| < \rho$, consider the quotient

$$q(h) \stackrel{\text{def}}{=} \frac{|f(x + h) - f(x) - \langle \nabla f(x), h \rangle_E|}{|h|_E}.$$

There exists α, $0 < \alpha < 1$, such that

$$f(x + h) - f(x) = \langle \nabla f(x + \alpha h), h \rangle_E$$
$$\Rightarrow q(h) = \left\langle \nabla f(x + \alpha h) - \nabla f(x), \frac{h}{|h|_E} \right\rangle_E$$
$$\Rightarrow |q(h)| \leq |\nabla f(x + \alpha h) - \nabla f(x)|_{E'} \to 0 \text{ as } h \to 0.$$

This shows that f is Fréchet differentiable at x. $\qquad\square$

2.6 Hadamard Semiderivative and Velocity Method

In section 2 of Chapter 7 we have drawn an analogy between Gâteaux and Hadamard semiderivatives on one hand and shape semiderivatives obtained by a perturbation of the identity and the velocity method on the other hand. The next theorem relates the Hadamard semiderivative and the semiderivative obtained by the velocity method for real functions defined on $\mathbf{R}^N$.

Theorem 2.6. *Let $f : N_X \to \mathbf{R}$ be a real function defined in a neighborhood N_X of a point X in $\mathbf{R}^N$. Then f is Hadamard semidifferentiable at (X, v) if and only if there exists $\tau > 0$ such that for all velocity fields $V : [0, \tau] \to \mathbf{R}$ satisfying the assumptions*

(a) (V_1) $\forall x \in \mathbf{R}^N, V(\cdot, x) \in C\big([0, \tau]; \mathbf{R}^N\big)$;

(b) (V_2) $\exists c > 0, \ \forall x, y \in \mathbf{R}^N, \quad \|V(\cdot, y) - V(\cdot, x)\|_{C([0,\tau];\mathbf{R}^N)} \leq c|y - x|$;

(c) *the limit*

$$df(X; V) \stackrel{\text{def}}{=} \lim_{t \searrow 0} \frac{f\big(T_t(V)(X)\big) - f(X)}{t} \tag{2.18}$$

exists and for all V satisfying (a) *and* (b) *and* $V(0) = v$,

$$df(X; V) = df(X; \widetilde{V}), \quad \widetilde{V}(t) \stackrel{\text{def}}{=} v \ \forall t \geq 0, \tag{2.19}$$

where $T_t(V)(X) \stackrel{\text{def}}{=} x(t)$ is the solution of the differential equation

$$\begin{cases} \dfrac{dx}{dt}(t) = V\big(t, x(t)\big), & 0 < t < \tau, \\[2mm] x(0) = X. \end{cases}$$

Proof. ($\Rightarrow$) Let V be a vector field satisfying conditions (a), (b), and (c). Define

$$w(t) \stackrel{\text{def}}{=} \frac{1}{t} \int_0^t V\big(s, x(s)\big)\, ds, \quad 0 < t \leq \tau.$$

It is continuous on $]0, \tau]$ and

$$w(t) - v = \frac{1}{t} \int_0^t \Big[V\big(s, x(s)\big) - V(0, X) \Big]\, ds.$$

Therefore

$$|w(t) - v| \leq c \max_{[0,t]} |x(s) - X| + \max_{[0,t]} |V(s, X) - V(0, X)|$$

$$\Rightarrow \lim_{t \searrow 0} w(t) = v.$$

So

$$\lim_{t \searrow 0} \frac{f\big(x(t;X)\big) - f(X)}{t} = \lim_{\substack{w(t) \to v \\ t \searrow 0}} \frac{f\big(X + tw(t)\big) - f(X)}{t} = d_H f(X;v),$$

since f is Hadamard semidifferentiable at (X, v). The limit only depends on $V(0) = v$.

($\Leftarrow$) Conversely, saying that the $d_H f(x;v)$ exists is equivalent to saying that for the sequence $t_n = 2^{-n}$ and any sequence $w_n \to v$ the limit

$$\frac{f(x + t_n w_n) - f(x)}{t_n}$$

exists and only depends on v. We now show that we can associate with such a sequence $\{w_n\}$ a velocity field V satisfying (a) to (c) such that $V(t_n, X) = w_n$ and $T_{t_n}(V)(X) = X + t_n w_n$. Then from properties (2.18) and (2.19) we conclude that $d_H f(x;v)$ exists. Set $t_0 = 1$, $t_n = 2^{-n}$, $x_n = X + t_n w_n$ and observe that $t_n - t_{n+1} = -2^{-(n+1)}$. Define for $m \geq 1$ the following C^m-interpolation in $(0, 1]$: for t in $[t_{n+1}, t_n]$

$$\boxed{\begin{aligned}
T_t(X) &\stackrel{\text{def}}{=} x_n + p\left(\frac{t - t_n}{t_{n+1} - t_n}\right)(x_{n+1} - x_n) \\
&\quad + q_0\left(\frac{t - t_n}{t_{n+1} - t_n}\right)(t_{n+1} - t_n)w_n \\
&\quad + q_1\left(\frac{t - t_n}{t_{n+1} - t_n}\right)(t_{n+1} - t_n)w_{n+1}, \quad T_0(X) \stackrel{\text{def}}{=} X,
\end{aligned}}$$

where $p, q_0, q_1 \in P^{2m+1}[0, 1]$ are polynomials of order $2m + 1$ on $[0, 1]$ such that $p(0) = 0$ and $p(1) = 1$ and $p^{(\ell)}(0) = 0 = p^{(\ell)}(1)$, $1 \leq \ell \leq m$, $q_0(0) = 0 = q_0(1)$, $q_0'(0) = 1$, $q_0'(1) = 0$ and $q_0^{(\ell)}(0) = 0 = q_0^{(\ell)}(1)$, $2 \leq \ell \leq m$, $q_1(0) = 0 = q_1(1)$, $q_1'(0) = 0$, $q_1'(1) = 1$ and $q_1^{(\ell)}(0) = 0 = q_1^{(\ell)}(1)$, $2 \leq \ell \leq m$. By definition $T(\cdot, X) \in C^m((0, 1]; \mathbf{R}^N)$. Moreover, for $1 \leq \ell \leq m$,

$$T_{t_n}(X) = x_n, \ T_{t_{n+1}}(X) = x_{n+1}, \quad \frac{\partial T}{\partial t}(t_n, X) = w_n, \ \frac{\partial T}{\partial t}(t_{n+1}, X) = w_{n+1},$$

$$\frac{\partial T}{\partial t}(t, X) = p'\left(\frac{t - t_n}{t_{n+1} - t_n}\right)\frac{x_{n+1} - x_n}{t_{n+1} - t_n} + q_0'\left(\frac{t - t_n}{t_{n+1} - t_n}\right)w_n$$
$$+ q_1'\left(\frac{t - t_n}{t_{n+1} - t_n}\right)w_{n+1}.$$

We now show that T satisfies conditions (a) and (b) which are equivalent to condition (4.2) in Chapter 7 on V. First observe that $T_t(X) - X$ and $\partial T/\partial t$ are both independent of X and $(T_t - I)(X) \in C^m((0, 1]; \mathbf{R}^N)$. Define

$$f(t) \stackrel{\text{def}}{=} T_t(X) - X \quad \Rightarrow \quad \frac{\partial T}{\partial t}(t, X) = \frac{\partial f}{\partial t}.$$

Hence T and $\partial T/\partial t$ clearly satisfy condition (T1) in (4.8) in Chapter 7. It satisfies condition (T2) since $T_t^{-1}(Y) = Y - f(t)$, and a fortiori T_t^{-1} satisfies condition (T3). Therefore, the velocity field

$$V(t,x) \overset{\text{def}}{=} \frac{\partial T}{\partial t}(t, T_t^{-1}(x)) = \frac{\partial f}{\partial t}(t)$$

satisfies condition (V) given by (4.2) of Chapter 7. Thus properties (a) and (b) are satisfied. It remains to show that $\partial f/\partial t(t) \to v$ as $t \to 0$. It is easy to check that by construction[1] $-p'(\tau) + q_0'(\tau) + q_1'(\tau) = 1$ in $[0,1]$ and that on each interval $[t_{n+1}, t_n]$

$$\frac{x_{n+1} - x_n}{t_{n+1} - t_n} = w_{n+1} - 2w_n,$$

$$\frac{\partial f}{\partial t}(t) = p'\left(\frac{t - t_n}{t_{n+1} - t_n}\right)(w_{n+1} - 2w_n) + q_0'\left(\frac{t - t_n}{t_{n+1} - t_n}\right)w_n$$
$$+ q_1'\left(\frac{t - t_n}{t_{n+1} - t_n}\right)w_{n+1},$$

$$\frac{\partial f}{\partial t}(t) - v = p'\left(\frac{t - t_n}{t_{n+1} - t_n}\right)[(w_{n+1} - v) - 2(w_n - v)]$$
$$+ q_0'\left(\frac{t - t_n}{t_{n+1} - t_n}\right)(w_n - v) + q_1'\left(\frac{t - t_n}{t_{n+1} - t_n}\right)(w_{n+1} - v).$$

The left-hand side now clearly converges since the three polynomials are bounded by a constant independent of n. This is sufficient to prove the theorem. $\qquad\square$

Remark 2.1.

The equivalence Theorem 2.6 shows how to extend the Hadamard semiderivative from linear spaces to submanifolds of $\mathbf{R}^N$. $\qquad\square$

3 First-Order Semiderivatives and Shape Gradient

Recall for the unconstrained case conditions (V) given by (4.2) in Chapter 7:

$$(\text{V}) \qquad \begin{aligned} &\exists \tau > 0, \ \forall x \in \mathbf{R}^N, \quad V(\cdot, x) \in C([0,\tau]; \mathbf{R}^N), \\ &\exists c > 0, \ \forall x, y \in \mathbf{R}^N, \quad \|V(\cdot, y) - V(\cdot, x)\|_{C([0,\tau];\mathbf{R}^N)} \le c|y - x|. \end{aligned} \qquad (3.1)$$

Similarly, in the constrained case for a closed subset D of $\mathbf{R}^N$ recall the conditions (V1_D) and (V2_D) given by (5.5) in Chapter 7 in their following equivalent form (V_D):

$$(\text{V1}_D) \qquad \begin{aligned} &\exists \tau > 0, \ \forall x \in D, \ V(\cdot, x) \in C([0,\tau]; \mathbf{R}^N), \ \exists c > 0, \\ &\forall x, y \in D, \quad \|V(\cdot, y) - V(\cdot, x)\|_{C([0,\tau];\mathbf{R}^N)} \le c|y - x|, \end{aligned} \qquad (3.2)$$

$$(\text{V2}_D) \qquad \forall x \in \partial D, \ \forall t \in [0,\tau], \quad V(t,x) \in L_D(x),$$

---

[1] By interpolation of the function $g(\tau) = \tau - 1$, $g(\tau) = g(0)p(\tau) + g'(0)q_0(\tau) + g'(1)q_1(\tau) = -p(\tau) + q_0(\tau) + q_1(\tau)$, and necessarily $1 = g'(\tau) = -p'(\tau) + q_0'(\tau) + q_1'(\tau)$.

where $L_D(x) = C_D(x) \cap \{-C_D(x)\}$ is the *linear tangent space* to D at the point $x \in \partial D$. Introduce the following linear subspace of $\mathrm{Lip}\,(D, \mathbf{R}^N)$:

$$\boxed{\mathrm{Lip}\,_L(D, \mathbf{R}^N) \overset{\text{def}}{=} \left\{ \theta \in \mathrm{Lip}\,(D, \mathbf{R}^N) \; : \; \forall X \in \partial D,\; \theta(X) \in L_D(X) \right\}.} \tag{3.3}$$

Under the action of a velocity field V satisfying conditions (3.2), a domain Ω in D is transformed into a new domain,

$$\boxed{\Omega_t(V) \overset{\text{def}}{=} T_t(V)(\Omega) = \{ T_t(V)(X) \; : \; \forall X \in \Omega \},} \tag{3.4}$$

also contained in D.

3.1   Definitions of Semiderivatives and Derivatives

We give the definitions in the general constrained case. Recall from Definition 2.1 of section 2 in Chapter 7 that a function $J(\Omega)$ defined on a family of subsets Ω of $\mathbf{R}^N$ (resp., D) is called a *shape function* if for any transformation T of $\mathbf{R}^N$ (resp., D such that $T(\overline{D}) = \overline{D}$) $T(\Omega) = \Omega$ implies that $J(T(\Omega)) = J(\Omega)$.

Definition 3.1.
Let Θ be a topological vector subspace of $\mathrm{Lip}\,_L(D, \mathbf{R}^N)$ and J a real-valued shape function.

  (i) Given a velocity field V satisfying conditions (3.2), J is said to have a *Eulerian semiderivative* at Ω in the direction V if the following limit exists and is finite:

$$dJ(\Omega; V) \overset{\text{def}}{=} \lim_{t \searrow 0} \frac{J(\Omega_t(V)) - J(\Omega)}{t}. \tag{3.5}$$

  (ii) For $\theta \in \mathrm{Lip}\,_L(D, \mathbf{R}^N)$ and the autonomous velocity field

$$\forall t \in [0, \tau],\; \forall x \in D, \quad \tilde{\theta}(t)(x) \overset{\text{def}}{=} \theta(x) \tag{3.6}$$

we shall use the notation $dJ(\Omega; \tilde{\theta})$ or simply $dJ(\Omega; \theta)$.

  (iii) Given $\theta \in \Theta$, J is said to have a *Hadamard semiderivative at Ω in the direction θ with respect to Θ* if for all V satisfying conditions (3.1), $V(t) \in \Theta$, and $V(0) = \theta$,

$$dJ(\Omega; V) \text{ exists and only depends on } V(0) = \theta. \tag{3.7}$$

In that case the semiderivative will be denoted $d_H J(\Omega; \theta)$ and necessarily

$$d_H J(\Omega; V(0)) = dJ(\Omega; V(0)).$$

  (iv) J is said to be *differentiable* at Ω in Θ' if it has a Eulerian semiderivative at Ω in all directions $\theta \in \Theta$ and the map

$$\theta \mapsto dJ(\Omega; \theta) \; : \; \Theta \to \mathbf{R} \tag{3.8}$$

is linear and continuous. The map (3.8) is denoted $G(\Omega)$ and referred to as the *gradient* of J in the topological dual Θ' of Θ. $\square$

The definition of a Eulerian semiderivative is quite general. For instance, it readily applies to shape functions defined on closed submanifolds D of $\mathbf{R}^N$. It includes cases where $dJ(\Omega; V)$ is dependent not only on $V(0)$ but also on $V(t)$ in a neighborhood of $t = 0$. We shall see that this will not occur under some continuity assumption on the map $V \mapsto dJ(\Omega; V)$. When $dJ(\Omega; V)$ only depends on $V(0)$, the analysis can be specialized to autonomous vector fields V, and the semiderivative can be related to the gradient of J. If J has a *Hadamard semiderivative* at Ω in the direction θ, it has a Eulerian semiderivative at Ω in the direction θ and

$$\boxed{d_H J(\Omega; \theta) = dJ(\Omega; \theta).}$$

Example 3.1.
For any measurable subset Ω of $\mathbf{R}^N$, consider the volume function

$$J(\Omega) = \int_\Omega dx. \tag{3.9}$$

For Ω with finite volume and V in $C([0, \tau]; C_0^1(\mathbf{R}^N, \mathbf{R}^N))$, consider the transformations $T_t(\Omega)$ of Ω:

$$J(T_t(\Omega)) = \int_{T_t(\Omega)} dx = \int_\Omega |\det(DT_t)|\, dx = \int_\Omega \det(DT_t)\, dx$$

for t small since $\det(DT_0) = \det I = 1$. This yields the Eulerian semiderivative

$$dJ(\Omega; V) = \int_\Omega \operatorname{div} V(0)\, dx. \tag{3.10}$$

By definition it only depends on $V(0)$ and hence J has a Hadamard semiderivative

$$d_H J(\Omega; V(0)) = \int_\Omega \operatorname{div} V(0)\, dx \tag{3.11}$$

and even a gradient $G(\Omega)$ for $\Theta = C_0^1(\mathbf{R}^N, \mathbf{R}^N)$. $\qquad\square$

The following simple continuity condition can also be used to obtain the Hadamard semidifferentiability.

Theorem 3.1. *Let Θ be a Banach subspace of $\operatorname{Lip}_L(D, \mathbf{R}^N)$, J a real-valued shape function, and Ω a subset of $\mathbf{R}^N$.*

(i) *Given $\theta \in \Theta$, if*

$$\forall V \in C([0, \tau]; \Theta) \text{ such that } V(0) = \theta, \quad dJ(\Omega; V) \text{ exists,} \tag{3.12}$$

and if the map

$$V \mapsto dJ(\Omega; V) : C([0, \tau]; \Theta) \to \mathbf{R} \tag{3.13}$$

is continuous for the subspace of V's such that $V(0) = \theta$, then J is Hadamard semidifferentiable at Ω in the direction θ with respect to Θ and

$$\forall V \in C([0, \tau]; \Theta),\ V(0) = \theta, \quad dJ(\Omega; V) = dJ\big(\Omega; V(0)\big) = d_H J\big(\Omega; \theta\big).$$
$$\tag{3.14}$$

(ii) *If for all V in $C([0, \tau]; \Theta)$, $dJ(\Omega; V)$ exists and the map*

$$V \mapsto dJ(\Omega; V) : C([0, \tau]; \Theta) \to \mathbf{R} \qquad (3.15)$$

is continuous, then J is Hadamard semidifferentiable at Ω in the direction $V(0)$ with respect to Θ and

$$\forall V \in C([0, \tau]; \Theta), \quad dJ(\Omega; V) = dJ(\Omega; V(0)) = d_H J(\Omega; V(0)). \qquad (3.16)$$

Proof. Given V in $C([0, \tau]; \Theta)$ such that $V(0) = \theta$, construct the sequence

$$V_n(t) \stackrel{\text{def}}{=} \begin{cases} V(t), & 0 \le t \le \frac{\tau}{n} \\ V\left(\frac{\tau}{n}\right), & \frac{\tau}{n} < t \le \tau \end{cases} \quad \text{for all integers } n \ge 1.$$

Note that for each $n \ge 1$, $dJ(\Omega; V_n) = dJ(\Omega; V)$ since for $t < \tau/n$, $T_t(V_n) = T_t(V)$. By continuity of V, $\{V_n\}$ converges in $C([0, \tau]; \Theta)$ to the autonomous field $\widetilde{V}(t) = V(0)$:

$$\forall n, \ \forall V \in C([0, \tau]; \Theta),$$
$$\forall \varepsilon > 0, \ \exists \delta > 0, \ \forall t', \ |t'| < \delta, \quad \|V(t') - V(0)\|_\Theta < \varepsilon.$$

Hence

$$\forall \varepsilon > 0, \ \exists N > 0, \ \forall n \ge N, \ \frac{\tau}{n} < \delta$$
$$\Rightarrow \quad \sup_{t \in [0, \frac{\tau}{n}]} \|V(t) - V(0)\|_\Theta < \varepsilon \quad \Rightarrow V_n \to \widetilde{V}.$$

Hence by continuity of the map (3.15)

$$dJ(\Omega; V_n) \to dJ(\Omega; \widetilde{V}) \quad \Rightarrow dJ(\Omega; V) = dJ(\Omega; \widetilde{V}).$$

But since $dJ(\Omega; \widetilde{V})$ is only dependent on $V(0)$, for all $W \in \Theta$ such that $W(0) = V(0)$ we have the identity $dJ(\Omega; W) = dJ(\Omega; \widetilde{V})$. Therefore, J has a Hadamard semiderivative at Ω in the direction $V(0)$ and

$$d_H J(\Omega; V(0)) = dJ(\Omega; \widetilde{V}) = dJ(\Omega; V).$$

This concludes the proof of the theorem. $\qquad\qquad\square$

For simplicity, the last theorem was <u>only</u> given for a Banach space Θ. This includes the spaces $\mathcal{C}_0^k(\mathbf{R}^N)$, $\mathcal{C}^k(\overline{\mathbf{R}^N})$, $C^{0,1}(\overline{\mathbf{R}^N})$, and $\mathcal{B}^k(\mathbf{R}^N, \mathbf{R}^N)$ considered at the end of Chapter 2. However, its conclusions are not limited to Banach spaces. Other constructions can be used as illustrated below in the unconstrained case. Consider velocity fields in

$$\boxed{\vec{\mathcal{V}}^{m,k} \stackrel{\text{def}}{=} \varinjlim_K \left\{ V_K^{m,k} : \forall K \text{ compact in } \mathbf{R}^N \right\},} \qquad (3.17)$$

where for $m \geq 0$,

$$\boxed{V_K^{m,k} \stackrel{\text{def}}{=} C^m\left([0,\tau]; \mathcal{D}^k(K, \mathbf{R}^{\mathrm{N}}) \cap \mathrm{Lip}(\mathbf{R}^{\mathrm{N}}, \mathbf{R}^{\mathrm{N}})\right)} \qquad (3.18)$$

and $\varinjlim$ denotes the inductive limit set with respect to K endowed with its natural inductive limit topology. For autonomous fields, this construction reduces to

$$\overrightarrow{\mathcal{V}}^k \stackrel{\text{def}}{=} \varinjlim_K \left\{ V_K^k : \forall K \text{ compact in } \mathbf{R}^{\mathrm{N}} \right\}, \qquad (3.19)$$

$$V_K^k \stackrel{\text{def}}{=} \begin{cases} \mathcal{D}^0(K, \mathbf{R}^{\mathrm{N}}) \cap \mathrm{Lip}\,(K, \mathbf{R}^{\mathrm{N}}), & k = 0, \\ \mathcal{D}^k(K, \mathbf{R}^{\mathrm{N}}), & 1 \leq k \leq \infty. \end{cases} \qquad (3.20)$$

For $k \geq 1$, $\overrightarrow{\mathcal{V}}^k = \mathcal{D}^k(\mathbf{R}^{\mathrm{N}}, \mathbf{R}^{\mathrm{N}})$. In all cases conditions (3.1) are satisfied.

Theorem 3.2. *Let J be a real-valued shape function, Ω a subset of $\mathbf{R}^{\mathrm{N}}$, and $k \geq 0$ an integer.*

(i) *Given $\theta \in \overrightarrow{\mathcal{V}}^k$, assume that*

$$\forall V \in \overrightarrow{\mathcal{V}}^{0,k}, \; V(0) = \theta, \quad dJ(\Omega; V) \text{ exists}, \qquad (3.21)$$

and that the map

$$V \mapsto dJ(\Omega; V) : \overrightarrow{\mathcal{V}}^{0,k} \to \mathbf{R} \qquad (3.22)$$

is continuous for all V's such that $V(0) = \theta$. Then J is Hadamard semidifferentiable in Ω in the direction θ with respect to $\mathcal{V}^k$ and

$$\forall V \in \overrightarrow{\mathcal{V}}^{0,k}, \; V(0) = \theta, \quad d_H J(\Omega; \theta) = dJ(\Omega; V) = dJ(\Omega; V(0)). \qquad (3.23)$$

(ii) *Assume that for all $V \in \overrightarrow{\mathcal{V}}^{0,k}$, $dJ(\Omega; V)$ exists and that the map*

$$V \mapsto dJ(\Omega; V) : \overrightarrow{\mathcal{V}}^{0,k} \to \mathbf{R} \qquad (3.24)$$

is continuous. Then J is Hadamard semidifferentiable in Ω in the direction $V(0)$ with respect to $\mathcal{V}^k$ and

$$\forall V \in \overrightarrow{\mathcal{V}}^{0,k}, \quad d_H J(\Omega; V(0)) = dJ(\Omega; V) = dJ(\Omega; V(0)). \qquad (3.25)$$

Proof. It is sufficient to prove the theorem for any compact subset K of $\mathbf{R}^{\mathrm{N}}$ and hence only for velocities in $\mathcal{V}_K^{0,k} = C([0,\tau]; \mathcal{V}_K^k)$, where $\mathcal{V}_K^k$ is a Banach space contained in $C_0^k(\mathbf{R}^{\mathrm{N}}, \mathbf{R}^{\mathrm{N}})$. So the theorem follows from Theorem 3.1. $\qquad \square$

3.2 Perturbations of the Identity and Fréchet Derivative

In the unconstrained case ($D = \mathbf{R}^N$), we have introduced in section 2 of this chapter and section 4 of Chapter 7 a notion of directional semiderivative associated with first- and second-order perturbations of the identity. Even if this approach does not naturally extend to the constrained case, it is interesting to compare the definitions and results with those associated with the velocity method of Definition 3.1.

Go back to the generic metric spaces of Micheletti in section 9 of Chapter 2 and their Courant metric. They all use a Banach subspace Θ of transformations from D to $\mathbf{R}^N$, $\Theta \subset \mathrm{Lip}_L(D, \mathbf{R}^N)$, and the space

$$\mathcal{F}(\Theta) \stackrel{\text{def}}{=} \left\{ F : D \to D : F - I \in \Theta \text{ and } F^{-1} - I \in \Theta \right\}.$$

Associate with $F \in \mathcal{F}(\Theta)$

$$d(I, F) \stackrel{\text{def}}{=} \inf_{(f_1, \ldots, f_n)} \sum_{i=1}^{n} \|f_i\|_\Theta + \inf_{(g_1, \ldots, g_m)} \sum_{i=1}^{m} \|g_i\|_\Theta, \tag{3.26}$$

where the infima are taken over all finite factorizations in $\mathcal{F}(\Theta)$ of the form

$$F = (I + f_n) \circ \cdots \circ (I + f_1) \text{ and } F^{-1} = (I + g_m) \circ \cdots \circ (I + g_1)$$

for f_i, g_i in Θ. Extend this definition to all pairs F and G in $\mathcal{F}(\Theta)$:

$$d(F, G) \stackrel{\text{def}}{=} d(I, G \circ F^{-1}). \tag{3.27}$$

Define for some open or closed subset Ω of D the subgroup

$$\mathcal{G}(\Omega) \stackrel{\text{def}}{=} \{F \in \mathcal{F}(\Theta) : F(\Omega) = \Omega\}$$

and the Courant metric on $\mathcal{F}(\Theta)/\mathcal{G}(\Omega)$,

$$\rho(F, H) \stackrel{\text{def}}{=} \inf_{G, \tilde{G} \in \mathcal{G}(\Omega)} d(F \circ G, H \circ \tilde{G}). \tag{3.28}$$

Assume that $\mathcal{F}(\Theta)$ is complete.

Here we consider the unconstrained case ($D = \mathbf{R}^N$) and assume that there exists a ball B_ε of radius $\varepsilon > 0$ in Θ and a constant $c > 0$ such that

$$\forall f \in B_\varepsilon, \quad \|[I + f]^{-1} - I\|_\Theta \leq c \|f\|_\Theta. \tag{3.29}$$

This is true in all cases considered in Chapter 2. Hence the maps

$$f \mapsto [I + f] \mapsto [I + f] : B_\varepsilon \subset \Theta \to \mathcal{F}(\Theta) \to \mathcal{F}(\Theta)/\mathcal{G}(\Omega)$$

are well defined and continuous in $f = 0$ since

$$\rho(I, I + f) \leq d(I, I + f) \leq \|f\|_\Theta + \|[I + f]^{-1} - I\|_\Theta \leq (1 + c) \|f\|_\Theta.$$

For a shape function J, the map

$$[I + f] \mapsto J_\Omega(f) \stackrel{\text{def}}{=} J([I + f](\Omega)) : \mathcal{F}(\Theta)/\mathcal{G}(\Omega) \to \mathbf{R}$$

is well defined since J is invariant on $\mathcal{G}(\Omega)$ and the map

$$f \mapsto J_\Omega(f) \stackrel{\text{def}}{=} J([I + f](\Omega)) : \Theta \to \mathbf{R}$$

is continuous in $f = 0$ if J is continuous in Ω for the Courant metric on $\mathcal{F}(\Theta)/\mathcal{G}(\Omega)$.

The following definitions are now the standard definitions of section 2 applied to the function $J_\Omega(f)$ defined on the ball B_ε in the topological vector space Θ. They parallel the ones of Definition 3.1.

Definition 3.2.
Let J be a real-valued shape function and Θ a topological vector subspace of $\mathrm{Lip}\,(\mathbf{R}^N, \mathbf{R}^N)$. For $f \in \Theta$ such that $[I + f] \in \mathcal{F}(\Theta)$, denote $[I + f](\Omega)$ by Ω_f.

(i) J_Ω is said to have a *Gâteaux semiderivative* at f in the direction $\theta \in \Theta$ if the following limit exists and is finite:

$$dJ_\Omega(f; \theta) \stackrel{\text{def}}{=} \lim_{t \searrow 0} \frac{J\big([I + f + t\theta](\Omega)\big) - J([I + f](\Omega))}{t}. \tag{3.30}$$

(ii) J_Ω is said to be *Gâteaux differentiable* at f if it has a *Gâteaux semiderivative* at f in all directions $\theta \in \Theta$ and the map

$$\theta \mapsto dJ_\Omega(f; \theta) : \Theta \to \mathbf{R} \tag{3.31}$$

is linear and continuous. The map (3.31) is denoted $\nabla J_\Omega(f)$ and referred to as the gradient of J_Ω in the topological dual Θ' of Θ.

(iii) If, in addition, Θ is a normed vector space, we say that J is *Fréchet differentiable* at f if J is Gâteaux differentiable at f and

$$\lim_{\|\theta\|_\Theta \to 0} \frac{|J([I + f + \theta](\Omega)) - J([I + f](\Omega)) - \langle \nabla J_\Omega(f), \theta \rangle_\Theta|}{\|\theta\|_\Theta} = 0. \tag{3.32}$$

$\square$

The semiderivatives of J and J_Ω are related.

Theorem 3.3. *Let J be a real-valued shape function.*

(i) *Assume that J_Ω has a Gâteaux semiderivative at f in the direction $\theta \in \Theta$; then J has a Eulerian semiderivative at Ω_f in the direction V_θ^f and*

$$\boxed{dJ_\Omega(f; \theta) = dJ(\Omega_f; V_\theta^f), \quad V_\theta^f(t) \stackrel{\text{def}}{=} \theta \circ [I + f + t\theta]^{-1}.} \tag{3.33}$$

(ii) *If J has a Hadamard semiderivative at Ω_f in the direction $\theta \circ [I + f]^{-1}$, then J_Ω has a Gâteaux semiderivative at f in the direction θ and*

$$\boxed{dJ_\Omega(f;\theta) = d_H J(\Omega_f; \theta \circ [I + f]^{-1}).} \tag{3.34}$$

Conversely, if J_Ω has a Gâteaux semiderivative at f in the direction $\theta \circ [I+f]$, then J has a Hadamard semiderivative at Ω_f in the direction θ. If either $dJ_\Omega(f;\theta)$ or $d_H J(\Omega_f;\theta)$ is linear and continuous with respect to all θ in Θ, so is the other and

$$\forall \theta \in \Theta, \quad \boxed{\begin{aligned} \langle \nabla J_\Omega(f), \theta \rangle_\Theta &= \langle G(\Omega_f), \theta \circ [I + f]^{-1} \rangle_\Theta, \\ \langle G(\Omega_f), \theta \rangle_\Theta &= \langle \nabla J_\Omega(f), \theta \circ [I + f] \rangle_\Theta. \end{aligned}} \tag{3.35}$$

Proof. By definition,

$$\begin{aligned} dJ_\Omega(f;\theta) &= \lim_{t \searrow 0} \frac{J\big([I + f + t\theta](\Omega)\big) - J([I + f](\Omega))}{t} \\ &= \lim_{t \searrow 0} \frac{J\big(T_t([I + f](\Omega))\big) - J([I + f](\Omega))}{t} = dJ([I + f](\Omega); V_\theta^f) \end{aligned}$$

for the family of transformations

$$T_t^f \stackrel{\text{def}}{=} [I + f + t\theta] \circ [I + f]^{-1},$$

and from Theorem 4.1 in section 4.1 of Chapter 7, T_t^f corresponds to the velocity field

$$V_\theta^f(t) \stackrel{\text{def}}{=} \frac{\partial T_t^f}{\partial t} \circ (T_t^f)^{-1} = \theta \circ [I + f + t\theta]^{-1}.$$

Identity (3.34) now follows from the fact that J has a Hadamard semiderivative at Ω_f. The other properties readily follow from the definitions. $\square$

We have standard sufficient conditions for the Fréchet differentiability from Theorem 2.5.

Theorem 3.4. *Let J be a real-valued shape function. Let Ω be a subset of $\mathbf{R}^N$ and Θ be $C_0^{k+1}(\mathbf{R}^N)$, $C^{k+1}(\overline{\mathbf{R}^N})$, $C^{k,1}(\overline{\mathbf{R}^N})$, $\mathcal{B}^{k+1}(\mathbf{R}^N, \mathbf{R}^N)$, $k \geq 0$. If J_Ω is Gâteaux differentiable for all f in B_ε and the map*

$$f \mapsto \nabla J_\Omega(f) : B_\varepsilon \to \Theta'$$

is continuous in $f = 0$, then J_Ω is Fréchet differentiable in $f = 0$.

3.3 Shape Gradient and Structure Theorem

In view of the previous discussion we now specialize to autonomous vector fields V to further study the properties and the structure of $dJ(\Omega; V)$. For simplicity we also specialize to the unconstrained case in $\mathbf{R}^N$. The constrained case yields similar result but is technically more involved (cf. Delfour and Zolésio [14] for D open in $\mathbf{R}^N$).

The choice of a *shape gradient* depends on the choice of the topological vector subspace Θ of $\mathrm{Lip}\,(\mathbf{R}^N, \mathbf{R}^N)$. We choose to work in the classical framework of the Theory of Distributions (cf. Schwartz [3]) with $\Theta = \mathcal{D}(\mathbf{R}^N, \mathbf{R}^N)$, the space of all infinitely differentiable transformations θ of $\mathbf{R}^N$ with compact support. For these velocity fields V, conditions (3.1) are satisfied.

Definition 3.3.
Let J be a real-valued shape function. Let Ω be a subset of $\mathbf{R}^N$.

(i) The function J is said to be *shape differentiable* at Ω if it is differentiable at Ω for all θ in $\mathcal{D}(\mathbf{R}^N, \mathbf{R}^N)$.

(ii) The map (3.8) defines a vector distribution $G(\Omega)$ in $\mathcal{D}(\mathbf{R}^N, \mathbf{R}^N)'$, which will be referred to as the *shape gradient* of J at Ω.

(iii) When, for some finite $k \geq 0$, $G(\Omega)$ is continuous for the $\mathcal{D}^k(\mathbf{R}^N, \mathbf{R}^N)$-topology, we say that the shape gradient $G(\Omega)$ is of order k. $\square$

The next theorem gives additional properties of shape differentiable functions.

Notation 3.1.
Associate with a subset A of $\mathbf{R}^N$ and an integer $k \geq 0$ the set

$$L_A^k \stackrel{\mathrm{def}}{=} \left\{ V \in \mathcal{D}^k(\mathbf{R}^N, \mathbf{R}^N) : \forall x \in A, V(x) \in L_A(x) \right\},$$

where $L_A(x) = \{-C_A(x)\} \cap C_A(x)$ and $C_A(x)$ is given by (5.27) in Chapter 7. $\square$

Theorem 3.5 (structure theorem). *Let J be a real-valued shape function. Assume that J has a shape gradient $G(\Omega)$ for some subset Ω of $\mathbf{R}^N$ with boundary Γ.*

(i) *The support of the shape gradient $G(\Omega)$ is contained in Γ.*

(ii) *If Ω is open or closed in $\mathbf{R}^N$ and the shape gradient is of order k for some $k \geq 0$, then there exists $[G(\Omega)]$ in $(\mathcal{D}^k/L_\Omega^k)'$ such that for all V in $\mathcal{D}^k \stackrel{\mathrm{def}}{=} \mathcal{D}^k(\mathbf{R}^N, \mathbf{R}^N)$,*

$$dJ(\Omega; V) = \langle [G(\Omega)], q_L V \rangle_{\mathcal{D}^k/L_\Omega^k}, \tag{3.36}$$

where $q_L : \mathcal{D}^k \to \mathcal{D}^k/L_\Omega^k$ is the canonical quotient surjection. Moreover,

$$G(\Omega) = (q_L)^* [G(\Omega)], \tag{3.37}$$

where $(q_L)^*$ denotes the transpose of the linear map q_L.

Proof. (i) Any V in $\mathcal{D}$ such that $V = 0$ on Γ satisfies assumptions $(V1_\Omega)$ and $(V2_\Omega)$ in (5.5) (with Ω in place of D) and $V \in L_\Omega^\infty$. Then by Theorem 5.1 (i) of Chapter 7, $T_s : \bar{\Omega} \to \bar{\Omega}$ is a homeomorphism and $(\bar{\Omega})_s = T_s(\bar{\Omega}) = \bar{\Omega}$ and $(\operatorname{int}\Omega)_s = T_s(\operatorname{int}\Omega) = \operatorname{int}\Omega$. Thus when Ω is closed ($\Omega = \bar{\Omega}$) or open ($\Omega = \operatorname{int}\Omega$),

$$\forall s \geq 0, \quad T_s(\Omega) = \Omega \quad \Rightarrow \quad J(\Omega_s) = J(\Omega) \quad \Rightarrow \quad dJ(\Omega; V) = 0.$$

(ii) It is sufficient to prove that $dJ(\Omega; V) = 0$ for all V in L_Ω^k. The other statements follow by standard arguments and the fact that L_Ω^k is a closed linear subspace of $\mathcal{D}^k$. From part (i) we know that the result is true for all V in L_Ω^∞ and hence by a density argument for all V in L_Ω^k. $\qquad\square$

Remark 3.1.
When the boundary Γ of Ω is compact and J is shape differentiable at Ω, the distribution $G(\Omega)$ is of finite order. Once this is known, the conclusions of Theorem 3.5 (ii) apply with k equal to the order of $G(\Omega)$. Hence $G(\Omega)$ will belong to a Hilbert space $H^{-s}(\mathbf{R}^N)$ for some $s \geq 0$. $\qquad\square$

The quotient space is very much related to a trace on the boundary Γ, and when the boundary Γ is sufficiently smooth we can indeed make that identification.

Corollary 1. *Assume that the assumptions of Theorem* 3.5 *are satisfied for an open domain* Ω, *that the order of* $G(\Omega)$ *is* $k \geq 0$, *and that the boundary* Γ *of* Ω *is* C^{k+1}. *Then for all* x *in* Γ, $L_\Omega(x)$ *is an* $(N-1)$-*dimensional hyperplane to* Ω *at* x *and there exists a unique outward unit normal* $n(x)$ *which belongs to* $C^k(\Gamma; \mathbf{R}^N)$. *As a result, the kernel of the map*

$$V \mapsto \gamma_\Gamma(V) \cdot n : \mathcal{D}^k(\mathbf{R}^N, \mathbf{R}^N) \to C^k(\Gamma) \tag{3.38}$$

coincides with L_Ω^k, *where* $\gamma_\Gamma : \mathcal{D}^k(\mathbf{R}^N, \mathbf{R}^N) \to C^k(\Gamma, \mathbf{R}^N)$ *is the trace of* V *on* Γ. *Moreover, the map* $p_L(V)$

$$q_L(V) \mapsto p_L\big(q_L(V)\big) \overset{\text{def}}{=} \gamma_\Gamma(V) \cdot n : \mathcal{D}^k/L_\Omega^k \to C^k(\Gamma) \tag{3.39}$$

is a well-defined isomorphism. In particular, there exists a scalar distribution $g(\Gamma)$ *in* $\mathbf{R}^N$ *with support in* Γ *such that* $g(\Gamma) \in C^k(\Gamma)'$ *and for all* V *in* $\mathcal{D}^k(\mathbf{R}^N, \mathbf{R}^N)$

$$\boxed{dJ(\Omega; V) = \langle g(\Gamma), \gamma_\Gamma(V) \cdot n \rangle_{C^k(\Gamma)}} \tag{3.40}$$

and

$$G(\Omega) = {}^*(q_L)[G(\Omega)], \quad [G(\Omega)] = {}^*(p_L)g(\Gamma). \tag{3.41}$$

When $g(\Gamma) \in L^1(\Gamma)$

$$dJ(\Omega; V) = \int_\Gamma g\, V \cdot n \, d\Gamma \ \text{and} \ G = \gamma_\Gamma^*(g\, n), \tag{3.42}$$

where γ_Γ *is the trace operator on* Γ.

Proof. The surjectivity of (3.39) is a consequence of the fact that Γ is compact and that for a C^{k+1} boundary, $k \geq 0$, it is always possible to construct an extension N of the unit normal n on Γ which belongs $\mathcal{D}^k(\mathbf{R}^N, \mathbf{R}^N)$ with support in a neighborhood of Γ. Then for any v in $C^k(\Gamma)$, there exists also an extension $\tilde{v}$ in $\mathcal{D}^k(\mathbf{R}^N)$ with support in a neighborhood of Γ, and the vector $V = \tilde{v}N$ belongs to $\mathcal{D}^k(\mathbf{R}^N, \mathbf{R}^N)$ and coincides with vn on Γ. $\square$

Remark 3.2.

In 1907, Hadamard [1] used displacements along the normal to the boundary Γ of a C^∞-domain (as in section 3.1 of Chapter 7) to compute the derivative of the first eigenvalue of the clamped plate. Theorem 3.5 and its corollary are generalizations to arbitrary shape functions of that property to open or closed domains with an arbitrary boundary. The structure theorem for shape functions on open domains with a C^{k+1}-boundary is due to Zolésio [12] in 1979 and not to Hadamard even if the formula (3.42) is often called the *Hadamard formula*. $\square$

Example 3.2.

For any measurable subset Ω of $\mathbf{R}^N$, consider the volume shape function (3.9) of Example 3.1:

$$J(\Omega) = \int_\Omega dx.$$

For Ω with finite volume and V in $\mathcal{D}^1(\mathbf{R}^N, \mathbf{R}^N)$, we have seen in Example 3.1 that

$$dJ(\Omega; V) = \int_\Omega \operatorname{div} V \, dx = \int_{\mathbf{R}^N} \chi_\Omega \operatorname{div} V \, dx, \tag{3.43}$$

and this is formula (3.40) with $k = 1$. For an open domain Ω with a C^1 compact boundary Γ,

$$dJ(\Omega; V) = \int_\Gamma V \cdot n \ d\Gamma, \tag{3.44}$$

which is also continuous with respect to V in $\mathcal{D}^0(\mathbf{R}^N, \mathbf{R}^N)$. Here the smoothness of the boundary decreases the order of the distribution $G(\Omega)$. This raises the question of the characterization of the family of all subsets Ω of $\mathbf{R}^N$ for which the map

$$V \mapsto \int_\Omega \operatorname{div} V \, dx : \mathcal{D}^1(\mathbf{R}^N, \mathbf{R}^N) \to \mathbf{R} \tag{3.45}$$

can be continuously extended to $\mathcal{D}^0(\mathbf{R}^N, \mathbf{R}^N)$. But this is the family of *locally finite perimeter sets*: sets Ω whose characteristic function belongs to $BV_{\text{loc}}(\mathbf{R}^N)$. $\square$

4 Elements of Shape Calculus

In this section we recall a number of basic formulae from Zolésio [7, 8] for the derivative of *domain and boundary integrals*. The reader is also referred to the *companion book* of Sokołowski and Zolésio [9] for the computation of shape derivatives associated with a wide range of partial differential equations.

A more modern treatment of the derivative of boundary integrals is given that emphasizes a recent and apparently new approach to the *tangential calculus* on a C^2 submanifold of $\mathbf{R}^N$ of codimension one. This development took place in the context of the theory of shells, where a simple and self-contained intrinsic differential calculus has been developed by using the oriented distance function of Chapter 5. It completely avoids the use of local maps and bases and Christoffel's symbols. In this section a new, considerably simplified proof of the formula for the derivative of boundary integrals is presented by using the oriented distance function.

4.1   Basic Formula for Domain Integrals

The simplest examples of domain functions are given by *volume integrals* over a bounded open domain Ω in $\mathbf{R}^N$. They use a basic formula in connection with the family of transformations $\{T_t : 0 \le t \le \tau\}$. Assume that condition (3.1) is satisfied by the velocity field $\{V(t) : 0 \le t \le \tau\}$. Further assume that $V \in C^0([0,\tau]; C^1_{\mathrm{loc}}(\mathbf{R}^N, \mathbf{R}^N))$ and that $\tau > 0$ is such that the *Jacobian* J_t is strictly positive:

$$\forall t \in [0,\tau], \quad J_t(X) \overset{\mathrm{def}}{=} \det DT_t(X) > 0, \quad (DT_t)_{ij} = \partial_j T_i,$$

where $DT_t(X)$ is the *Jacobian matrix* of the transformation $T_t = T_t(V)$ associated with the velocity vector field V. Given a function φ in $W^{1,1}_{\mathrm{loc}}(\mathbf{R}^N)$, consider for $0 \le t \le \tau$ the volume integral

$$J(\Omega_t(V)) \overset{\mathrm{def}}{=} \int_{\Omega_t(V)} \varphi \, dx, \tag{4.1}$$

where $\Omega_t(V) \overset{\mathrm{def}}{=} T_t(V)(\Omega)$. By the change of variable formula,

$$J(\Omega_t(V)) = \int_{\Omega_t(V)} \varphi \, dx = \int_\Omega \varphi \circ T_t \, J_t \, dx, \tag{4.2}$$

and the following formulae and results are easy to check.

Theorem 4.1. *Let φ be a function in $W^{1,1}_{\mathrm{loc}}(\mathbf{R}^N)$. Assume that the vector field $V = \{V(t) : 0 \le t \le \tau\}$ satisfies condition (V).*

(i) *For each $t \in [0,\tau]$ the map*

$$\varphi \mapsto \varphi \circ T_t : W^{1,1}_{\mathrm{loc}}(\mathbf{R}^N) \to W^{1,1}_{\mathrm{loc}}(\mathbf{R}^N)$$

and its inverse are both locally Lipschitzian and

$$\nabla(\varphi \circ T_t) = {}^*DT_t \, \nabla \varphi \circ T_t.$$

(ii) *If $V \in C^0([0,\tau]; C^1_{\mathrm{loc}}(\mathbf{R}^N, \mathbf{R}^N))$, then the map*

$$t \mapsto \varphi \circ T_t : [0,\tau] \to W^{1,1}_{\mathrm{loc}}(\mathbf{R}^N)$$

is well defined and for each t

$$\frac{d}{dt}\varphi \circ T_t = (\nabla\varphi \cdot V(t)) \circ T_t \in L^1_{\mathrm{loc}}(\mathbf{R}^N). \tag{4.3}$$

Hence the function

$$t \mapsto \varphi \circ T_t \ \text{belongs to } C^1([0,\tau]; L^1_{\mathrm{loc}}(\mathbf{R}^N)) \cap C^0([0,\tau]; W^{1,1}_{\mathrm{loc}}(\mathbf{R}^N)).$$

(iii) *If $V \in C^0([0,\tau]; C^1_{\mathrm{loc}}(\mathbf{R}^N, \mathbf{R}^N))$, then the map*

$$t \mapsto J_t \ : \ [0,\tau] \to C^0_{\mathrm{loc}}(\mathbf{R}^N)$$

is differentiable and

$$\frac{dJ_t}{dt} = [\,\mathrm{div}\,V(t)\,] \circ T_t \, J_t \in C^0_{\mathrm{loc}}(\mathbf{R}^N). \tag{4.4}$$

Hence the map $t \mapsto J_t$ belongs to $C^1([0,\tau]; C^0_{\mathrm{loc}}(\mathbf{R}^N))$.

Indeed it is easy to check that

$$\frac{d}{dt}DT_t(X) = DV(t, T_t(X))\, DT_t(X), \quad DT_0(X) = I,$$

$$\frac{d}{dt}\,\det DT_t(X) = \mathrm{tr}\, DV(t, T_t(X))\, \det DT_t(X)$$

$$\Rightarrow \frac{d}{dt}\,\det DT_t(X) = \mathrm{div}\, V(t, T_t(X))\, \det DT_t(X), \quad \det DT_0(X) = 1,$$

and (4.4) follows directly by definition of $J_t(X)$.

From (4.2), (4.3), and (4.4)

$$dJ(\Omega; V) = \frac{d}{dt} J(\Omega_t(V))\Big|_{t=0} = \int_\Omega \nabla\varphi \cdot V(0) + \varphi\,\mathrm{div}\,V(0)\, dx$$

$$\Rightarrow dJ(\Omega; V) = \int_\Omega \mathrm{div}\,(\varphi\, V(0))\, dx.$$

If Ω has a Lipschitzian boundary, then by Stokes's theorem

$$dJ(\Omega; V) = \int_\Gamma \varphi\, V(0) \cdot n\, d\Gamma.$$

Theorem 4.2. *Assume that there exists $\tau > 0$ such that the velocity field $V(t)$ satisfies conditions (V) and $V \in C^0([0,\tau]; C^1_{\mathrm{loc}}(\mathbf{R}^N, \mathbf{R}^N))$. Given a function $\varphi \in C(0,\tau; W^{1,1}_{\mathrm{loc}}(\mathbf{R}^N)) \cap C^1(0,\tau; L^1_{\mathrm{loc}}(\mathbf{R}^N))$ and a bounded measurable domain Ω with boundary Γ, the semiderivative of the function*

$$\boxed{\ J_V(t) \stackrel{\text{def}}{=} \int_{\Omega_t(V)} \varphi(t)\, dx\ } \tag{4.5}$$

at $t = 0$ is given by

$$dJ_V(0) = \int_\Omega \varphi'(0) + \operatorname{div}\left(\varphi(0)\,V(0)\right)\,dx, \tag{4.6}$$

where $\varphi(0)(x) \overset{\text{def}}{=} \varphi(0,x)$ and $\varphi'(0)(x) \overset{\text{def}}{=} \partial\varphi/dt(0,x)$. If, in addition, Ω is an open domain with a Lipschitzian boundary Γ, then

$$dJ_V(0) = \int_\Omega \varphi'(0)\,dx + \int_\Gamma \varphi(0)\,V(0)\cdot n\,dx. \tag{4.7}$$

4.2   Basic Formula for Boundary Integrals

Given ψ in $H^2_{\text{loc}}(\mathbf{R}^N)$, consider for some bounded open Lipschitzian domain Ω in $\mathbf{R}^N$ the shape function

$$J(\Omega) \overset{\text{def}}{=} \int_\Gamma \psi\,d\Gamma. \tag{4.8}$$

This integral is invariant with respect to homeomorphisms which map Ω onto itself (and hence Γ onto itself). Given the velocity field V and $t \geq 0$, consider the expression

$$J(\Omega_t(V)) \overset{\text{def}}{=} \int_{\Gamma_t(V)} \psi\,d\Gamma_t.$$

Using the change of variable $T_t(V)$ and the material introduced in (3.13) of section 3.2 and (5.12) of section 5 in Chapter 2, this integral can be brought back from Γ_t to Γ:

$$J(\Omega_t(V)) \overset{\text{def}}{=} \int_{\Gamma_t} \psi\,d\Gamma_t = \int_\Gamma \psi\circ T_t\,\omega_t\,d\Gamma, \tag{4.9}$$

where the density ω_t is given as

$$\omega_t = |M(DT_t)n|, \tag{4.10}$$

n is the outward normal field on Γ, and $M(DT_t)$ is the cofactor matrix of DT_t, that is,

$$M(DT_t) = J_t\,{}^*(DT_t)^{-1} \quad \Rightarrow \quad \boxed{\omega_t = J_t\,|\,{}^*(DT_t)^{-1}n|.} \tag{4.11}$$

It can easily be checked from (4.10) and (4.11) that $t \mapsto \omega_t$ is differentiable in $C^0(\Gamma)$ and that the limit

$$\omega' = \lim_{t\searrow 0}\frac{1}{t}(\omega_t - \omega) = \operatorname{div} V(0) - DV(0)n\cdot n \tag{4.12}$$

in the $C^0(\Gamma)$-norm is linear and continuous with respect to $V(0)$ in the $C^1_{\text{loc}}(\mathbf{R}^N, \mathbf{R}^N)$ Fréchet topology. Hence

$$dJ(\Omega; V) = \int_\Gamma \nabla\psi \cdot V(0) + \psi \left(\text{div}\, V(0) - DV(0)n \cdot n\right) d\Gamma. \qquad (4.13)$$

From Corollary 1 to the structure theorem (Theorem 3.5), $dJ(\Omega; V)$ depends only on the normal component v_n of the velocity field $V(0)$ on Γ:

$$v_n \overset{\text{def}}{=} v \cdot n, \quad v \overset{\text{def}}{=} V(0)|_\Gamma, \qquad (4.14)$$

through Hadamard's formula (3.40). In view of this property any other velocity field with the same smoothness and normal component on Γ will yield the same limit. Given $k > 0$ consider the *tubular neighborhood*

$$\boxed{S_k(\Gamma) \overset{\text{def}}{=} \left\{x \in \mathbf{R}^N \; : \; |b(x)| < k\right\}} \qquad (4.15)$$

of Γ in $\mathbf{R}^N$ for the oriented distance function $b = b_\Omega$ associated with Ω. Assuming that Γ is compact and of class C^2, there exists $h > 0$ such that $b \in C^2(S_{2h}(\Gamma))$. Let $\varphi \in \mathcal{D}(\mathbf{R}^N)$ be such that $\varphi = 1$ in $S_h(\Gamma)$ and $\varphi = 0$ outside of $S_{2h}(\Gamma)$. Consider the velocity field

$$W(t) \overset{\text{def}}{=} (V(0) \cdot \nabla b)\, \nabla b\, \varphi.$$

Clearly, the normal component of $W(0)$ on Γ coincides with v_n. Moreover, in $S_h(\Gamma)$

$$\nabla\psi \cdot W = \nabla\psi \cdot \nabla b\, V(0) \cdot \nabla b \quad \Rightarrow \quad \boxed{\nabla\psi \cdot W|_\Gamma = \nabla\psi \cdot n\, V(0) \cdot n = \frac{\partial\psi}{\partial n}\, v_n,}$$

$$DW = V(0) \cdot \nabla b\, D^2 b + \nabla b\; {}^*\nabla(V(0) \cdot \nabla b),$$

$$\text{div}\, W = V(0) \cdot \nabla b\, \Delta b + \nabla b \cdot \nabla(V(0) \cdot \nabla b),$$

$$DW \nabla b \cdot \nabla b = V(0) \cdot \nabla b\, D^2 b \nabla b \cdot \nabla b + \nabla(V(0) \cdot \nabla b) \cdot \nabla b$$

$$= \nabla(V(0) \cdot \nabla b) \cdot \nabla b,$$

$$\text{div}\, W - DW \nabla b \cdot \nabla b = V(0) \cdot \nabla b\, \Delta b$$

$$\Rightarrow \quad \boxed{\text{div}\, W - DW \nabla b \cdot \nabla b|_\Gamma = V(0) \cdot n\, H = H\, v_n}$$

since $\nabla b|_\Gamma = n$, $D^2 b \nabla b = 0$, and $H = \Delta b$ is the *additive curvature*, that is, the sum of the $N - 1$ curvatures of Γ or $N - 1$ times the mean curvature $\overline{H}$. Finally,

$$dJ(\Omega; V) = \int_\Gamma \left(\frac{\partial\psi}{\partial n} + \psi H\right) v_n\, d\Gamma. \qquad (4.16)$$

We have proved the following result.

Theorem 4.3. *Let Γ be the boundary of a bounded open subset Ω of $\mathbf{R}^N$ of class C^2 and ψ an element of $C^1([0,\tau]; H^2_{\mathrm{loc}}(\mathbf{R}^N))$. Assume that $V \in C^0([0,\tau]; C^1_{\mathrm{loc}}(\mathbf{R}^N, \mathbf{R}^N))$. Consider the function*

$$J_V(t) \stackrel{\text{def}}{=} \int_{\Gamma_t(V)} \psi(t)\, d\Gamma_t.$$

Then the derivative of $J_V(t)$ with respect to t in $t = 0$ is given by the expression

$$
\begin{aligned}
dJ_V(0) &= \int_\Gamma \psi'(0) + \left(\frac{\partial\psi}{\partial n} + H\psi\right) V(0) \cdot n\, d\Gamma \\
&= \int_\Gamma \psi'(0) + \nabla\psi \cdot V(0) + \psi\,(\operatorname{div} V(0) - DV(0)n \cdot n)\, d\Gamma,
\end{aligned}
\tag{4.17}
$$

where $\psi'(0)(x) \stackrel{\text{def}}{=} \partial\psi/\partial t(0, x)$.

Note that, as in the case of the volume integral, we have two formulae. Hence the following identity:

$$
\begin{aligned}
\int_\Gamma &\left(\frac{\partial\psi}{\partial n} + H\psi\right) V(0) \cdot n\, d\Gamma \\
&= \int_\Gamma \nabla\psi \cdot V(0) + \psi\,(\operatorname{div} V(0) - DV(0)n \cdot n)\, d\Gamma.
\end{aligned}
\tag{4.18}
$$

4.3   Examples of Shape Derivative

4.3.1   Volume of Ω and Area of Γ

Consider the volume shape function

$$J(\Omega) = \int_\Omega dx.$$

This shape function is used as a constraint on the domain in several examples of shape optimization problems. We get

$$dJ(\Omega; V) = \int_\Omega \operatorname{div} V(0)\, dx \tag{4.19}$$

and, if Γ is Lipschitzian,

$$dJ(\Omega; V) = \int_\Gamma V(0) \cdot n\, d\Gamma. \tag{4.20}$$

A sufficient condition on the field $V(0)$ to preserve the volume is $\operatorname{div} V(0) = 0$ in Ω and, if Γ is Lipschitzian, $V(0) \cdot n = 0$ on Γ.

Consider the (shape) area function

$$J(\Omega) = \int_\Gamma d\Gamma.$$

Assuming that Γ is of class C^2 we get from (4.17)

$$dJ(\Omega; V) = \int_\Gamma H\, V(0) \cdot n\, d\Gamma, \tag{4.21}$$

where $H = \Delta b$ is the additive curvature. The condition for the surface of Γ to be preserved is that $V(0) \cdot n$ be orthogonal (in $L^2(\Gamma)$) to H.

4.3.2 $H^1(\Omega)$-Norm

Given ϕ and ψ in $H^2_{\mathrm{loc}}(\mathbf{R}^N)$, consider the shape function

$$J(\Omega) = \int_\Omega \nabla\phi \cdot \nabla\psi\, dx.$$

By using the change of variable $T_t(V)$, $\Omega_t = \Omega_t(V) = T_t(V)(\Omega)$ and

$$\int_{\Omega_t} \nabla\phi \cdot \nabla\psi\, dx = \int_\Omega [A(V)(t)\nabla(\phi \circ T_t)] \cdot \nabla(\psi \circ T_t)\, dx, \tag{4.22}$$

where $A(V)$ is the following matrix associated with the field V:

$$A(V)(t) = J(t)\,(DT_t)^{-1}\,{}^*(DT_t)^{-1} \tag{4.23}$$

and $J(t) = \det(DT_t)$. Expression (4.23) is easily obtained from the identity

$$(\nabla\phi) \circ T_t = {}^*(DT_t)^{-1}\nabla(\phi \circ T_t). \tag{4.24}$$

If $V \in C^0([0,\tau]; C^k_{\mathrm{loc}}(\mathbf{R}^N; \mathbf{R}^N))$, $k \geq 1$, T and T^{-1} belong to $C^1([0,\tau]; C^k_{\mathrm{loc}}(\mathbf{R}^N, \mathbf{R}^N))$ and $A(V)$ to $C^1([0,\tau], C^{k-1}_{\mathrm{loc}}(\mathbf{R}^N; \mathbf{R}^{N^2}))$. Then if ϕ, ψ belongs to $H^2_{\mathrm{loc}}(\mathbf{R}^N)$ we get $t \mapsto \phi \circ T_t$ which is differentiable in $H^1_{\mathrm{loc}}(\mathbf{R}^N)$, with $\partial\phi/\partial t \circ T_t|_{t=0} = \nabla\phi \cdot V(0)$, which belongs to $H^1_{\mathrm{loc}}(\mathbf{R}^N)$. Finally, we obtain

$$\begin{aligned}
dJ(\Omega; V) = &\int_\Omega [A'(V)\nabla\phi] \cdot \nabla\psi\, dx \\
&+ \int_\Omega \{\nabla(\nabla\phi \cdot V(0)) \cdot \nabla\psi + \nabla\phi \cdot \nabla(\nabla\psi \cdot V(0))\}\, dx,
\end{aligned} \tag{4.25}$$

where $A'(V)$ is the derivative in the $C^{k-1}_{\mathrm{loc}}(\mathbf{R}^N, \mathbf{R}^{N^2})$-norm

$$A'(V) \stackrel{\mathrm{def}}{=} \frac{\partial}{\partial t} A(V)(t)|_{t=0} = \operatorname{div} V(0)\, I - 2\varepsilon(V(0)) \tag{4.26}$$

and $\varepsilon(V(0))$ is the symmetrized Jacobian matrix (the strain tensor associated to the field $V(0)$ in elasticity)

$$\varepsilon(V(0)) = \frac{1}{2}[{}^*DV(0) + DV(0)]. \tag{4.27}$$

We finally obtain the *volume expression*

$$\boxed{\begin{aligned} dJ(\Omega; V) = \int_\Omega & [\operatorname{div} V(0)\, I - 2\varepsilon(V(0))]\, \nabla\phi \cdot \nabla\psi \\ & + [\nabla(\nabla\phi \cdot V(0)) \cdot \nabla\psi + \nabla\phi \cdot \nabla(\nabla\psi \cdot V(0))]\, dx. \end{aligned}} \tag{4.28}$$

When Γ is a C^1-submanifold, $\phi, \psi \in H^2_{\mathrm{loc}}(\mathbf{R}^N)$, we obtain the simpler *boundary expression* by directly using formula (4.17):

$$\boxed{dJ(\Omega; V) = \int_\Gamma \nabla\phi \cdot \nabla\psi\, V(0) \cdot n\, d\Gamma.} \tag{4.29}$$

This simple example nicely illustrates the notion of density gradient g. From expression (4.28) it was obvious that the mapping $V \mapsto dJ(\Omega; V)$ was well defined, linear, and continuous on $C^1([0, \tau]; C^1_{\mathrm{loc}}(\mathbf{R}^N; \mathbf{R}^N))$. By the structure theorem and Hadamard's formula, we knew that dJ could be written in the form

$$\int_\Gamma g\, V(0) \cdot n\, d\Gamma.$$

But from (4.29) we know that g, which is an element of $\mathcal{D}^1(\Gamma)'$, is an element of $W^{1/2,1}(\Gamma)$ given by

$$g = \nabla\phi \cdot \nabla\psi \ \text{(traces on } \Gamma). \tag{4.30}$$

The direct calculation of g from expression (4.24) would have been very fastidious.

4.3.3   Normal Derivative

Let Γ be of class C^2 and $\phi \in H^2_{\mathrm{loc}}(\mathbf{R}^N)$ be given. Consider the following shape function:

$$J(\Omega) \overset{\text{def}}{=} \int_\Gamma \left|\frac{\partial\phi}{\partial n}\right|^2 d\Gamma = \int_\Gamma |\nabla\phi \cdot n|^2\, d\Gamma.$$

By the change of variable formula we get, with $\Omega_t = T_t(V)(\Omega)$ and $\Gamma_t = T_t(V)(\Gamma)$,

$$J(\Omega_t) \overset{\text{def}}{=} \int_{\Gamma_t} |\nabla\phi \cdot n_t|^2\, d\Gamma_t = \int_\Gamma \left\{ {}^*(DT_t)^{-1}\nabla(\phi \circ T_t) \cdot (n_t \circ T_t)\right\}^2 \omega_t\, d\Gamma, \tag{4.31}$$

where $n_t \circ T_t$ is the transported normal field n_t from Γ_t onto Γ. The derivative can be obtained by using formula (4.17) of Theorem 4.3 and one of the above two expressions. However, the first expression first requires the construction of an extension N_t of the normal n_t in a neighborhood of Γ. In both cases the following result will be useful.

Theorem 4.4. *Let $k \geq 1$ be an integer. Given a velocity field $V(t)$ satisfying condition (V) such that $V \in C([0, \tau]; C^k_{\mathrm{loc}}(\mathbf{R}^N, \mathbf{R}^N))$, then*

$$\boxed{\;n_t \circ T_t = \frac{{}^*(DT_t)^{-1} n}{|\,{}^*(DT_t)^{-1} n|} = \frac{M(DT_t) n}{|M(DT_t) n|}\,,\;} \qquad (4.32)$$

where n and n_t are the respective outward normals to Ω and Ω_t on Γ and Γ_t and $M(DT_t)$ is the cofactor's matrix of DT_t.

Proof. Go back to Definition 3.1 in section 3.1 of Chapter 2. We have shown in that section that for a domain Ω of class C^k the *unit outward normal* at any point $y \in \Gamma_x = \Gamma \cap U(x)$ is given by expression (3.9)

$$n(y) = -\frac{{}^*(Dh_x)^{-1} e_N}{|\,{}^*(Dh_x)^{-1} e_N|}(h_x^{-1}(y)) \quad \forall y \in \Gamma_x,$$

where h_x is the local diffeomorphism specified by (3.3):

$$g_x \in C^k\big(U(x), B\big), \quad h_x = g_x^{-1} \in C^k\big(B, U(x)\big).$$

For $\Omega_t = \Omega_t(V)$ and $x_t \overset{\mathrm{def}}{=} T_t(x)$, choose the following new neighborhood and local diffeomorphism

$$U_t \overset{\mathrm{def}}{=} T_t(U(x)), \ h_t \overset{\mathrm{def}}{=} T_t \circ h_x : B \to U_t, \ g_t = h_t^{-1} \overset{\mathrm{def}}{=} g_x \circ T_t^{-1} : U_t \to B.$$

On $\Gamma_t(V) \cap U_t$ the normal in given by the same expression but with h_t in place of h_x:

$$n_t = -\frac{{}^*(Dh_t)^{-1} \circ h_t^{-1} e_N}{|\,{}^*(Dh_t)^{-1} \circ h_t^{-1} e_N|}.$$

However,

$$D(T_t \circ h_x) = DT_t \circ h_x \, Dh_x,$$
$$D(T_t \circ h_x) \circ (T_t \circ h_x)^{-1}, = \big[DT_t \, Dh_x \circ h_x^{-1} \big] \circ T_t^{-1},$$
$${}^*(DT_t \circ h_x)^{-1} \circ (T_t \circ h_x)^{-1} e_N = \big[{}^*(DT_t)^{-1} \, {}^*(Dh_x)^{-1} \circ h_x^{-1} e_N \big] \circ T_t^{-1}.$$

Therefore, by using the previous expressions for n and n_t,

$$n_t = \frac{{}^*(DT_t)^{-1} n}{|\,{}^*(DT_t)^{-1} n|} \circ T_t^{-1} \quad \Rightarrow \quad n_t \circ T_t = \frac{{}^*(DT_t)^{-1} n}{|\,{}^*(DT_t)^{-1} n|}. \qquad (4.33)$$

The other expression for $n_t \circ T_t$ in terms of the cofactor matrix $M(DT_t)$ of DT_t readily follows from the identity $M(DT_t) = J(t)\, {}^*(DT_t)^{-1}$. $\square$

Recalling the expression for ω_t,

$$\omega_t = |M(DT_t) n| = J(t)\, |\,{}^*(DT_t)^{-1} n|,$$

our boundary integral becomes

$$\int_{\Gamma_t} |\nabla\phi \cdot n_t|^2 \, d\Gamma_t = \int_{\Gamma} |[A(t)\nabla(\phi \circ T_t)] \cdot n\,|^2 \, \omega_t^{-1} \, d\Gamma. \qquad (4.34)$$

Using expression (4.26) of $A'(V)$ and expression (4.12) of ω' we get

$$dJ(\Omega; V)$$

$$= 2\int_{\Gamma} \frac{\partial\phi}{\partial n} \left[A'(V)\nabla\phi \cdot n + \nabla(\nabla\phi \cdot V(0)) \cdot n\right] - \left|\frac{\partial\phi}{\partial n}\right|^2 \omega' \, d\Gamma$$

$$= 2\int_{\Gamma} \frac{\partial\phi}{\partial n} \left\{ \left[\operatorname{div} V(0) \, I - DV(0) - {}^*DV(0)\right] \nabla\phi \cdot n + \nabla(\nabla\phi \cdot V(0)) \cdot n \right\}$$

$$\qquad - \left|\frac{\partial\phi}{\partial n}\right|^2 (\operatorname{div} V(0) - DV(0)n \cdot n) \, d\Gamma$$

$$= \int_{\Gamma} 2\frac{\partial\phi}{\partial n} \left\{ \operatorname{div} V(0) \frac{\partial\phi}{\partial n} - DV(0)\,\nabla\phi \cdot n + D^2\phi\, V(0) \cdot n \right\}$$

$$\qquad - \left|\frac{\partial\phi}{\partial n}\right|^2 (\operatorname{div} V(0) - DV(0)n \cdot n) \, d\Gamma$$

$$= \int_{\Gamma} \left|\frac{\partial\phi}{\partial n}\right|^2 (\operatorname{div} V(0) - DV(0)n \cdot n)$$

$$\qquad + 2\frac{\partial\phi}{\partial n} \left\{ DV(0)n \cdot n \frac{\partial\phi}{\partial n} - DV(0)\,\nabla\phi \cdot n + D^2\phi\, V(0) \cdot n \right\} d\Gamma.$$

This formula can be somewhat simplified by using identity (4.18) with

$$\psi = |\nabla\phi \cdot \nabla b|^2 \quad \Rightarrow \quad \psi|_\Gamma = \left|\frac{\partial\phi}{\partial n}\right|^2,$$

$$\nabla\psi = 2\,\nabla\phi \cdot \nabla b\, \nabla(\nabla\phi \cdot \nabla b) = 2\,\nabla\phi \cdot \nabla b\, \left[D^2\phi\nabla b + D^2 b\nabla\psi\right]$$

$$\Rightarrow \quad \nabla\psi|_\Gamma = 2\frac{\partial\phi}{\partial n} \left[D^2\phi\, n + D^2 b\, \nabla\phi\right]$$

$$\Rightarrow \quad \nabla\psi \cdot V(0)|_\Gamma = 2\frac{\partial\phi}{\partial n} \left[D^2\phi\, n + D^2 b\, \nabla\phi\right] \cdot V(0)$$

$$\Rightarrow \quad \frac{\partial\psi}{\partial n} = \nabla\psi \cdot \nabla b|_\Gamma = 2\frac{\partial\phi}{\partial n} \left[D^2\phi\, n + D^2 b\, \nabla\phi\right] \cdot \nabla b = 2\frac{\partial\phi}{\partial n} D^2\phi\, n \cdot n.$$

We obtain

$$\int_{\Gamma} \left(2\frac{\partial\phi}{\partial n} D^2\phi\, n \cdot n + H \left|\frac{\partial\phi}{\partial n}\right|^2 \right) V(0) \cdot n \, d\Gamma$$

$$\qquad (4.35)$$

$$= \int_{\Gamma} 2\frac{\partial\phi}{\partial n} \left[D^2\phi\, n + D^2 b\, \nabla\phi\right] \cdot V(0) + \left|\frac{\partial\phi}{\partial n}\right|^2 (\operatorname{div} V(0) - DV(0)n \cdot n) \, d\Gamma,$$

and hence

$$
\begin{aligned}
dJ(\Omega; V) &= \int_\Gamma \left(2\frac{\partial\phi}{\partial n} D^2\phi n \cdot n + H \left|\frac{\partial\phi}{\partial n}\right|^2 \right) V(0) \cdot n \\
&\quad + 2\frac{\partial\phi}{\partial n} \left\{ \frac{\partial\phi}{\partial n} DV(0)n \cdot n - \nabla\phi \cdot \left({}^*DV(0)\,n + D^2b\,V(0) \right) \right\} d\Gamma.
\end{aligned}
\tag{4.36}
$$

This formula can be more readily obtained from the first expression (4.31) and the extension

$$
N_t = \frac{{}^*(DT_t)^{-1}\nabla b}{|\,{}^*(DT_t)^{-1}\nabla b|} \circ T_t^{-1}
\tag{4.37}
$$

of the normal n_t. To compute N', decompose N_t as follows:

$$
N_t = f(t) \circ T_t^{-1}, \quad f(t) = \frac{g(t)}{\sqrt{g(t) \cdot g(t)}}, \quad g(t) = {}^*(DT_t)^{-1}\nabla b,
$$

$$
N' = f' - Df(0)V(0), \quad g' = -\,{}^*DV(0)\nabla b, \quad f(0) = \nabla b,
$$

$$
f' = \frac{g'\,|g(0)| - g(0) \cdot g'\,g(0)/|g(0)|}{|g(0)|^2}
$$

$$
= g' - g' \cdot \nabla b\,\nabla b = DV(0)\nabla b \cdot \nabla b\,\nabla b - {}^*DV(0)\nabla b.
$$

So finally

$$
N'|_\Gamma = (DV(0)n \cdot n)\,n - {}^*DV(0)n - D^2b\,V(0)
\tag{4.38}
$$

and

$$
\frac{\partial}{\partial t}|\nabla\phi \cdot N_t|^2\Big|_{t=0} = 2\frac{\partial\phi}{\partial n}\nabla\phi \cdot N'
$$

$$
= 2\frac{\partial\phi}{\partial n}\nabla\phi \cdot \left\{ (DV(0)n \cdot n)n - {}^*DV(0)\,n - D^2b\,V(0) \right\}.
$$

The first part of the integral (4.36) explicitly depends on the normal component of $V(0)$. Yet we know from the structure theorem that for this function, the shape derivative only depends on the normal component of $V(0)$. To make this explicit, it is necessary to introduce some elements of tangential calculus.

5   Elements of Tangential Calculus

In this section some basic elements of the differential calculus on a C^2-submanifold of codimension 1 are introduced. The approach avoids local bases and coordinates by using the intrinsic tangential derivatives. A more systematic treatment combined with the use of the distance function has been successfully used in the theory of thin and asymptotic shells with a $C^{1,1}$-midsurface.

Let Ω be an open domain of class C^2 in $\mathbf{R}^N$ with compact boundary Γ. Therefore, there exists $h > 0$ such that $b = b_\Omega \in C^2(S_{2h}(\Gamma))$. The *projection* of a point x onto Γ is given by

$$p(x) \stackrel{\text{def}}{=} x - b(x)\, \nabla b(x),$$

and the *orthogonal projection operator* of a vector onto the tangent plane $T_{p(x)}\Gamma$ is given by

$$P(x) \stackrel{\text{def}}{=} I - \nabla b(x)\, {}^*\nabla b(x).$$

Notice that, as a transformation of $T_{p(x)}\Gamma$,

$$P(x) \,:\, T_{p(x)}\Gamma \to T_{p(x)}\Gamma$$

is the identity transformation on $T_{p(x)}\Gamma$. In fact $P(x)$ coincides with the *first fundamental form*. Similarly $D^2 b$ can be considered as a transformation of $T_{p(x)}$ since $D^2 b(x) n(x) = D^2 b(x) \nabla b(x) = 0$. We shall show in section 5.6 that

$$D^2 b(x) \,:\, T_{p(x)} \to T_{p(x)}$$

coincides with the *second fundamental form* of Γ. Similarly $D^2 b(x)^2$ coincides with the *third fundamental form*. Finally,

$$Dp(x) = I - \nabla b(x)\, {}^*\nabla b(x) - b\, D^2 b(x) \quad \text{and} \quad Dp|_\Gamma = P.$$

5.1   Intrinsic Definition of the Tangential Gradient

The classical way to define the tangential gradient of a scalar function $f : \Gamma \to \mathbf{R}$ is through an appropriately smooth extension F of f in a neighborhood of Γ using the fact that the resulting expression on Γ is independent of the choice of the extension F. In this section an equivalent direct intrinsic definition is given in terms of the extension $f \circ p$ of the function f. This is the basis of a simple differential calculus on Γ which uses the Euclidean differential calculus in the ambient neighborhood of Γ.

Given $f \in C^1(\Gamma)$, let $F \in C^1(S_{2h}(\Gamma))$ be a C^1-extension of f. Define

$$g(F) \stackrel{\text{def}}{=} \nabla F|_\Gamma - \frac{\partial F}{\partial n}\, n \quad \text{on } \Gamma.$$

This is the orthogonal projection $P(x)\nabla F(x)$ of $\nabla F(x)$ onto the tangent plane $T_x \Gamma$ to Γ at x. To get something intrinsic, $g(F)$ must be independent of the choice of F. It is sufficient to show that $g(F) = 0$ for $f = 0$. But $F = f = 0$ on Γ and the tangential component of ∇F is 0 on Γ, and

$$\nabla F|_\Gamma = \frac{\partial F}{\partial n}\, n \quad \Rightarrow \quad g(F) = \nabla F|_\Gamma - \frac{\partial F}{\partial n} n = 0.$$

Definition 5.1 (general extension).
Assume that Γ is compact and that there exists $h > 0$ such that $b_\Omega \in C^2(S_{2h}(\Gamma))$. Given an extension $F \in C^1(S_{2h}(\Gamma))$ of $f \in C^1(\Gamma)$, the *tangential gradient* of f in a point of Γ is defined as

$$\nabla_\Gamma f \stackrel{\text{def}}{=} \nabla F|_\Gamma - \frac{\partial F}{\partial n}\, n. \qquad \square$$

The notation is quite natural. The subscript Γ of $\nabla_\Gamma f$ indicates that the gradient is with respect to the variable x in the submanifold Γ.

Theorem 5.1. *Assume that Γ is compact and that there exists $h > 0$ such that $b_\Omega \in C^2(S_{2h}(\Gamma))$ and that $f \in C^1(\Gamma)$. Then*

(i) $\nabla_\Gamma f = (P\,\nabla F)|_\Gamma$ *and* $n \cdot \nabla_\Gamma f = \nabla b \cdot \nabla_\Gamma f = 0$;

(ii) $\nabla(f \circ p) = [I - b\, D^2 b]\, \nabla_\Gamma f \circ p$ *and* $\nabla(f \circ p)|_\Gamma = \nabla_\Gamma f$.

Proof. (i) By definition

$$P\,\nabla F = (I - \nabla b\, {}^*\nabla b)\,\nabla F = \nabla F - \nabla F \cdot \nabla b\, \nabla b,$$

$$(P\,\nabla F)|_\Gamma = (\nabla F)|_\Gamma - \frac{\partial F}{\partial n}\, n = \nabla_\Gamma f.$$

Moreover,

$$\nabla b \cdot P\,\nabla F = (I - \nabla b\, {}^*\nabla b)\,\nabla b \cdot \nabla F = 0$$
$$\Rightarrow (\nabla b)|_\Gamma \cdot (P\,\nabla F)|_\Gamma = 0 \quad \Rightarrow n \cdot \nabla_\Gamma f = 0.$$

(ii) $F = f \circ p$ is a C^1-extension of f and

$$\nabla(f \circ p) = \nabla(f \circ p \circ p) = Dp\,\nabla(f \circ p) \circ p$$
$$= [I - \nabla b\, {}^*\nabla b - b\, D^2 b]\,\nabla(f \circ p)|_\Gamma \circ p.$$

But by definition of $\nabla_\Gamma f$

$$\nabla(f \circ p)|_\Gamma = \nabla_\Gamma f + \frac{\partial(f \circ p)}{\partial n}\, n$$

$$\Rightarrow \nabla(f \circ p) = [I - \nabla b\, {}^*\nabla b - b\, D^2 b]\left(\nabla_\Gamma f + \frac{\partial(f \circ p)}{\partial n}\, n\right) \circ p.$$

Recall, from Theorem 4.3(i) in Chapter 5, that $n \circ p = \nabla b$ so that

$$\nabla(f \circ p) = [I - \nabla b\, {}^*\nabla b - b\, D^2 b]\left(\nabla_\Gamma f \circ p + \frac{\partial(f \circ p)}{\partial n}\, \nabla b\right)$$
$$= [I - \nabla b\, {}^*\nabla b - b\, D^2 b]\,\nabla_\Gamma f \circ p.$$

Again

$$\nabla b \cdot \nabla_\Gamma f \circ p = n \circ p \cdot \nabla_\Gamma f \circ p = (n \cdot \nabla_\Gamma f) \circ p = 0$$

from part (i) and

$$\nabla(f \circ p) = [I - bD^2b]\, \nabla_\Gamma f \circ p \quad \Rightarrow \quad \nabla(f \circ p)|_\Gamma = \nabla_\Gamma f. \qquad \square$$

In view of part (ii) of the theorem, $f \circ p$ plays the role of a *canonical extension* of a map $f : \Gamma \to \mathbf{R}$ to a neighborhood $S_{2h}(\Gamma)$ of Γ, and its gradient is tangent to the level sets of b. This suggests the use of the following definition of tangential gradient which will be the clue to the tangential differential calculus.

Definition 5.2 (canonical extension).
Under the assumptions of Definition 5.1 on b_Ω, associate with $f \in C^1(\Gamma)$

$$\nabla_\Gamma f \stackrel{\text{def}}{=} \nabla(f \circ p)|_\Gamma. \qquad (5.1)$$

$$\square$$

Remark 5.1.
This definition naturally extends to nonempty sets A such that d_A^2 belongs to $C^{1,1}(S_{2h}(A))$, since the projection onto A,

$$p_A(x) = x - \frac{1}{2}\nabla d_A^2(x),$$

is $C^{0,1}$. They are the sets of positive reach introduced by Federer. They include convex sets and submanifolds of codimension larger than or equal to 1. $\qquad \square$

Theorem 5.2. *Under the assumption of Definition 5.1 on b_Ω for $f \in C^1(\Gamma)$,*

(i) $\nabla b \cdot \nabla(f \circ p) = 0$ *in* $S_h(\Gamma)$ *and* $n \cdot \nabla_\Gamma f = 0$ *on* Γ.

(ii) $\nabla F|_\Gamma - \frac{\partial F}{\partial n} n = (P\nabla F)|_\Gamma = \nabla_\Gamma f$ *and* $\nabla(f \circ p) = [I - b\,D^2b]\,\nabla_\Gamma f \circ p$ *in* Γ.

Proof. (i) Consider

$$\begin{aligned}
\nabla(f \circ p) \cdot \nabla b &= \nabla(f \circ p \circ p) \cdot \nabla b = Dp\,\nabla(f \circ p) \circ p \cdot \nabla b \\
&= \nabla(f \circ p) \circ p \cdot Dp\,\nabla b.
\end{aligned}$$

But

$$Dp\,\nabla b = [I - \nabla b\, {}^*\nabla b - b\,D^2b]\,\nabla b = 0$$

and

$$\nabla(f \circ p) \cdot \nabla b = 0, \quad \nabla_\Gamma f \cdot n = \nabla(f \circ p)|_\Gamma \cdot \nabla b|_\Gamma = 0.$$

(ii) By definition

$$P\nabla F = (I - \nabla b\, {}^*\nabla b)\,\nabla F.$$

However, since $F \circ p = f \circ p$ on Γ

$$\nabla(F \circ p) = \nabla(F \circ p \circ p) = Dp\,\nabla(F \circ p) \circ p = Dp\,\nabla(f \circ p) \circ p = Dp\,\nabla_\Gamma f \circ p$$

and

$$\nabla(F \circ p) = Dp\,\nabla F \circ p \quad \Rightarrow \quad \boxed{Dp\,\nabla F \circ p = Dp\,\nabla_\Gamma f \circ p.}$$

By restricting to Γ

$$Dp|_\Gamma\,\nabla F|_\Gamma = Dp|_\Gamma\,\nabla_\Gamma f \quad \Rightarrow \quad P\nabla F|_\Gamma = P\nabla_\Gamma f = \nabla_\Gamma f. \qquad \square$$

5.2 First-Order Derivatives

The *tangential Jacobian matrix* of a vector function $v \in C^1(\Gamma)^M$, $M \geq 1$, is defined in the same way as the gradient

$$D_\Gamma v \stackrel{\text{def}}{=} D(v \circ p)|_\Gamma \quad \text{or} \quad (D_\Gamma v)_{ij} = (\nabla_\Gamma v_i)_j. \tag{5.2}$$

If $^*v = (v_1, \ldots, v_M)$, then

$$^*D_\Gamma v = (\nabla_\Gamma v_1, \ldots, \nabla_\Gamma v_M),$$

where $\nabla_\Gamma v_i$ is a column vector. From the previous theorems about the tangential gradient we can recover the definition from an extension $V \in C^1(S_{2h}(\Gamma))^M$ of v:

$$^*D_\Gamma v = (P\,\nabla V_1, \ldots, P\,\nabla V_M)|_\Gamma = (I - \nabla b\,{}^*\nabla b)\,{}^*DV|_\Gamma$$
$$= {}^*DV|_\Gamma - \nabla b\,{}^*(DV\,\nabla b)|_\Gamma$$

and

$$D_\Gamma v = DV|_\Gamma - DV n\,{}^*n = (DV\,P)|_\Gamma. \tag{5.3}$$

Also,

$$^*D(v \circ p) = [I - b\,D^2 b]\,(\nabla_\Gamma v_1, \ldots, \nabla_\Gamma v_M) \circ p = [I - b\,D^2 b]\,{}^*D_\Gamma v \circ p,$$

and we have for the extension

$$D(v \circ p) = D_\Gamma v \circ p\,[I - b\,D^2 b]. \tag{5.4}$$

Note that

$$D(v \circ p)\,\nabla b = 0 \quad \text{and} \quad D_\Gamma v\,n = 0. \tag{5.5}$$

For a vector function $V \in C^1(\Gamma)^N$ define the *tangential divergence* as

$$\text{div}_\Gamma v \stackrel{\text{def}}{=} \text{div}\,(v \circ p)|_\Gamma, \tag{5.6}$$

and it is easy to show that

$$\text{div}_\Gamma v = \text{div}\,(v \circ p)|_\Gamma = \text{tr}\,D(v \circ p)|_\Gamma = \text{tr}\,D_\Gamma v,$$

$$\text{div}_\Gamma v = \text{tr}\,[DV|_\Gamma - DV\,n\,{}^*n] = \text{div}\,V|_\Gamma - DV\,n \cdot n.$$

The *tangential linear strain tensor* of linear elasticity is given by

$$\varepsilon_\Gamma(v) \stackrel{\text{def}}{=} \frac{1}{2}(D_\Gamma v + {}^*D_\Gamma v), \tag{5.7}$$

$$\varepsilon(v \circ p) = \frac{1}{2}(D(v \circ p) + {}^*D(v \circ p))$$
$$= \varepsilon_\Gamma(v) \circ p - \frac{b}{2}\,[D_\Gamma v \circ p\,D^2 b + D^2 b\,{}^*D_\Gamma v \circ p] \tag{5.8}$$
$$\Rightarrow \varepsilon_\Gamma(v) = \varepsilon(v \circ p)_{|\Gamma}. \tag{5.9}$$

The tangential Jacobian matrix of the normal n is especially interesting since, from Theorem 4.3 (i) in Chapter 5, $n \circ p = \nabla b = \nabla b \circ p$. As a result,

$$\boxed{D_\Gamma(n) = D^2 b|_\Gamma = {}^*D_\Gamma(n)} \quad \Rightarrow \quad \boxed{\varepsilon_\Gamma(n) = D^2 b|_\Gamma.} \tag{5.10}$$

The *tangential vectorial divergence* of a matrix or tensor function A is defined as

$$(\overrightarrow{\operatorname{div}}_\Gamma A)_i \overset{\text{def}}{=} \operatorname{div}_\Gamma A_{i,\cdot}. \tag{5.11}$$

5.3 Second-Order Derivatives

Assume that Ω is of class C^3. The simplest second-order derivative is the *Laplace–Beltrami operator* of a function $f \in C^2(\Gamma)$, which is defined as

$$\boxed{\Delta_\Gamma f \overset{\text{def}}{=} \operatorname{div}_\Gamma(\nabla_\Gamma f).} \tag{5.12}$$

Recall from Theorem 5.2 the identity

$$\nabla(f \circ p) = [I - b\, D^2 b]\, \nabla_\Gamma f \circ p.$$

Then

$$\operatorname{div}(\nabla_\Gamma f \circ p) = \operatorname{div}(\nabla(f \circ p)) + \operatorname{div}(b\, D^2 b\, \nabla_\Gamma f \circ p),$$
$$\operatorname{div}(\nabla_\Gamma f \circ p) = \Delta(f \circ p) + b\operatorname{div}(D^2 b\, \nabla_\Gamma f \circ p) + (D^2 b\, \nabla_\Gamma f \circ p) \cdot \nabla b,$$
$$\operatorname{div}(\nabla_\Gamma f \circ p) = \Delta(f \circ p) + b\operatorname{div}(D^2 b\, \nabla_\Gamma f \circ p),$$

and by taking restrictions to Γ,

$$\boxed{\Delta_\Gamma f = \Delta(f \circ p)|_\Gamma.}$$

The *tangential Hessian matrix* of second-order derivatives is defined as

$$\boxed{D_\Gamma^2 f \overset{\text{def}}{=} D_\Gamma(\nabla_\Gamma f).} \tag{5.13}$$

Here the curvatures of the submanifold begin to appear. The Hessian matrix is not symmetrical and does not coincide with the restriction of the Hessian matrix of the canonical extension. Specifically,

$$D^2(f \circ p) = D(\nabla(f \circ p)) = D([I - b\, D^2 b]\, \nabla_\Gamma f \circ p)$$
$$= D(\nabla_\Gamma f \circ p) - D(b\, D^2 b\, \nabla_\Gamma f \circ p)$$
$$= D(\nabla_\Gamma f \circ p) - b\, D(D^2 b\, \nabla_\Gamma f \circ p) - D^2 b\, \nabla_\Gamma f \circ p\, {}^*\nabla b,$$
$$D^2(f \circ p)|_\Gamma = D_\Gamma(\nabla_\Gamma f) - D^2 b\, \nabla_\Gamma f\, {}^*\nabla b = D_\Gamma^2 f - D^2 b\, \nabla_\Gamma f\, {}^*n.$$

Of course, since $D^2(f \circ p)$ is symmetrical we also have

$$
\begin{aligned}
D^2(f \circ p) &= {}^*D(\nabla(f \circ p)) = {}^*D([I - b\,D^2b]\,\nabla_\Gamma f \circ p) \\
&= {}^*D(\nabla_\Gamma f \circ p) - {}^*D(b\,D^2b\,\nabla_\Gamma f \circ p) \\
&= {}^*D(\nabla_\Gamma f \circ p) - b\,{}^*D(D^2b\,\nabla_\Gamma f \circ p) - \nabla b\,{}^*(D^2b\,\nabla_\Gamma f \circ p),
\end{aligned}
$$
$$
D^2(f \circ p)|_\Gamma = {}^*D_\Gamma(\nabla_\Gamma f) - \nabla b\,{}^*(D^2b\,\nabla_\Gamma f) = {}^*D_\Gamma^2 f - n\,{}^*(D^2b\,\nabla_\Gamma f).
$$

As a final result we have the following identity:

$$
\boxed{\;D_\Gamma^2 f - (D^2b\,\nabla_\Gamma f)\,{}^*n = D^2(f \circ p)|_\Gamma = {}^*D_\Gamma^2 f - n\,{}^*(D^2b\,\nabla_\Gamma f),\;}
\tag{5.14}
$$

since by definition ${}^*D_\Gamma^2 f = {}^*(D_\Gamma^2 f)$. So the Hessian and its transpose differ by terms that contain first-order derivatives, as is well known in differential geometry. Note that $D^2b\,\nabla_\Gamma f = -{}^*D_\Gamma(\nabla_\Gamma f)n = -{}^*D_\Gamma^2 f\,n$ and that we also can write

$$
\boxed{
\begin{aligned}
D_\Gamma^2 f + {}^*D_\Gamma^2 f\,[n\,{}^*n] &= D^2(f \circ p)|_\Gamma = {}^*D_\Gamma^2 f + [n\,{}^*n]\,D_\Gamma^2 f \\
&\Rightarrow PD_\Gamma^2 f = {}^*(PD_\Gamma^2 f).
\end{aligned}
}
\tag{5.15}
$$

5.4 A Few Useful Formulae and the Chain Rule

Associate with $F \in C^1(S_{2h}(\Gamma))$ and $V \in C^1(S_{2h}(\Gamma))^N$

$$
f \overset{\text{def}}{=} F|_\Gamma, \quad v \overset{\text{def}}{=} V|_\Gamma, \quad v_n \overset{\text{def}}{=} v \cdot n, \quad v_\Gamma \overset{\text{def}}{=} v - v_n\,n,
\tag{5.16}
$$

where v_Γ and v_n are the respective tangential part and the normal component of v. In view of the previous definitions the following identities are easy to check:

$$
\nabla F|_\Gamma = \nabla_\Gamma f + \frac{\partial F}{\partial n}\,n,
\tag{5.17}
$$
$$
DV|_\Gamma = D_\Gamma v + DVn\,{}^*n,
\tag{5.18}
$$
$$
\operatorname{div} V|_\Gamma = \operatorname{div}_\Gamma v + DVn \cdot n,
\tag{5.19}
$$
$$
D_\Gamma v\,n = 0, \quad D^2b\,n = 0.
\tag{5.20}
$$

Decomposing v into its tangential part and its normal component,

$$
D_\Gamma v = D_\Gamma v_\Gamma + v_n D^2b + n\,{}^*\nabla_\Gamma v_n,
\tag{5.21}
$$
$$
\operatorname{div}_\Gamma v = \operatorname{div}_\Gamma v_\Gamma + \Delta b\,v_n = \operatorname{div}_\Gamma v_\Gamma + H\,v_n,
\tag{5.22}
$$
$$
\nabla_\Gamma v_n = {}^*D_\Gamma v\,n + D^2b\,v_\Gamma.
\tag{5.23}
$$

Given $f \in C^1(\Gamma)$ and $g \in C^1(\Gamma;\Gamma)$, consider the canonical extensions $f \circ p \in C^1(S_{2h}(\Gamma))$ and $g \circ p \in C^1(S_{2h}(\Gamma);\mathbf{R}^N)$ and the gradient of the composition

$$
\nabla(f \circ p \circ g \circ p) = {}^*D(g \circ p)\,\nabla(f \circ p) \circ g \circ p
$$
$$
\Rightarrow \boxed{\;\nabla_\Gamma(f \circ g) = {}^*D_\Gamma g\,\nabla_\Gamma f \circ g,\;}
\tag{5.24}
$$

and for a vector-valued function $v \in C^1(\Gamma;\mathbf{R}^N)$,

$$
\boxed{\;D_\Gamma(v \circ g) = D_\Gamma v \circ g\,D_\Gamma g.\;}
\tag{5.25}
$$

5.5 The Stokes and Green Formulae

One interesting application of the shape calculus in connection with the tangential calculus is the tangential Stokes formula. Given $v \in C^1(\Gamma)^N$, consider Stokes's formula in $\mathbf{R}^N$ for the vector function $v \circ p$:

$$\int_\Omega \operatorname{div}(v \circ p)\, dx = \int_\Gamma v \cdot n\, d\Gamma.$$

Given an autonomous velocity field V, differentiate both sides of Stokes's formula with respect to t:

$$\int_{\Omega_t(V)} \operatorname{div}(v \circ p)\, dx = \int_{\Gamma_t(V)} (v \circ p) \cdot n_t = \int_{\Gamma_t(V)} (v \circ p) \cdot N_t\, d\Gamma_t,$$

where N_t is the extension (4.37) of n_t. This gives the new identity

$$\int_\Gamma \operatorname{div}(v \circ p)\, V \cdot n\, d\Gamma = \int_\Gamma v \cdot N' + \left\{ \frac{\partial}{\partial n}[(v \circ p) \cdot \nabla b] + H\, v \cdot n \right\} V \cdot n\, d\Gamma.$$

Now choose the velocity field $V = \nabla b\, \psi$, where $\psi \in \mathcal{D}(S_{2h}(\Gamma))$ is chosen in such a way that $\psi = 1$ on $S_h(\Gamma)$. Using expression (4.38) for N',

$$V \cdot n = n \cdot n = 1, \quad \operatorname{div}(v \circ p)|_\Gamma = \operatorname{div}_\Gamma v,$$
$$N'|_\Gamma = DVn \cdot n\, n - {}^*DVn - D^2b\, V$$
$$= D^2b\nabla b \cdot \nabla b\, \nabla b - D^2b\nabla b - D^2b\nabla b = 0,$$
$$\nabla((v \circ p) \cdot \nabla b) = D^2b\, v \circ p + {}^*D(v \circ p)\nabla b,$$
$$\frac{\partial}{\partial n}[(v \circ p) \cdot \nabla b] = \left[D^2b\, v + {}^*D_\Gamma v\, n \right] \cdot n = v \cdot D^2b\nabla b + n \cdot D_\Gamma v\, n = 0.$$

Finally we get the *tangential Stokes formula* with $H = \Delta b$:

$$\boxed{\int_\Gamma \operatorname{div}_\Gamma v\, d\Gamma = \int_\Gamma H\, v \cdot n\, d\Gamma.} \tag{5.26}$$

For a function $f \in C^1(\Gamma)$ and a vector $v \in C^1(\Gamma)^N$, the above formula also yields the *tangential Green's formula*,

$$\boxed{\int_\Gamma f\operatorname{div}_\Gamma v + \nabla_\Gamma f \cdot v\, d\Gamma = \int_\Gamma H\, f\, v \cdot n\, d\Gamma.} \tag{5.27}$$

5.6 Relation between Tangential and Covariant Derivatives

One of the simplest examples of a tangential derivative is when Ω is the half-space

$$H_+ \stackrel{\text{def}}{=} \{\zeta \in \mathbf{R}^N : \zeta \cdot e_N > 0\}$$

for some orthonormal basis $\{e_1, \ldots, e_N\}$ in $\mathbf{R}^N$ and Γ is the boundary of H_+ denoted

$$H \stackrel{\text{def}}{=} \left\{ \zeta \in \mathbf{R}^N : \zeta \cdot e_N = 0 \right\}.$$

If p_H denotes the projection onto H, then $p_H(\zeta) = p_H(\zeta', \zeta_N) = \zeta' \stackrel{\text{def}}{=} (\zeta_1, \ldots, \zeta_{N-1}) \in H$, and for any function $\varphi \in C^1(H)$ the tangential gradient

$$\nabla_H \varphi = \nabla(\varphi \circ p_H)|_H$$

coincides with the gradient of φ in H:

$$\boxed{\nabla_H \varphi \cdot e_\alpha = \frac{\partial \varphi}{\partial \zeta_\alpha}, \quad 1 \le \alpha \le N - 1.}$$

With the notation of section 3.1 of Chapter 2, consider a set Ω locally of class C^2 in $\mathbf{R}^N$. Its boundary $\Gamma = \partial\Omega$ is an $(N-1)$-dimensional submanifold of $\mathbf{R}^N$ of class C^2. At each point $x \in \Gamma$, there is a C^2-diffeomorphism h_x from the open unit ball B onto a neighborhood $U(x)$ of x such that

$$h_x(B_0) = \Gamma_x \stackrel{\text{def}}{=} \Gamma \cap U(x), \quad h_x(B_+) = \Omega \cap U(x),$$

where $B_+ = H_+ \cap B$ and $B_0 = H \cap B$ and H_+ and H are as defined above for some appropriate orthonormal basis.

For a function $f \in C^1(\Gamma)$, $f \circ \Phi \in C^1(B_0)$ with $\Phi \stackrel{\text{def}}{=} h_x|_{B_0} : B_0 \to \Gamma$. Observing that $f \circ \Phi \circ p_H = f \circ p \circ \Phi \circ p_H$, we have

$$\nabla(f \circ \Phi \circ p_H) = \nabla(f \circ p \circ \Phi \circ p_H),$$

$$\nabla(f \circ \Phi \circ p_H) = {}^*D(\Phi \circ p_H)\nabla(f \circ p) \circ \Phi \circ p_H,$$

$$\boxed{\nabla_H(f \circ \Phi) = {}^*D_H\Phi \, \nabla_\Gamma f \circ \Phi.}$$

The covariant basis associated with Φ is defined as

$$\boxed{a_\alpha \stackrel{\text{def}}{=} \frac{\partial \Phi}{\partial \zeta_\alpha}, \quad 1 \le \alpha \le N - 1, \quad a_N \stackrel{\text{def}}{=} \frac{{}^*(Dh_x)^{-1}e_N}{|{}^*(Dh_x)^{-1}e_N|}\bigg|_{B_0},}$$

where from identity (3.9) of section 3.1 in Chapter 2 we know that $a_N \circ \Phi^{-1}$ is the inward unit normal to Ω, that is,

$$\boxed{n = -a_N \circ \Phi^{-1}.} \tag{5.28}$$

The *covariant partial derivatives* of f are defined as

$$\boxed{f_\alpha \stackrel{\text{def}}{=} \frac{\partial(f \circ \Phi)}{\partial \zeta_\alpha}, \quad 1 \le \alpha \le N - 1.}$$

But from the previous observations

$$a_\alpha = \frac{\partial \Phi}{\partial \zeta_\alpha} = D_H \Phi e_\alpha, \quad f_\alpha = \frac{\partial (f \circ \Phi)}{\partial \zeta_\alpha} = e_\alpha \cdot \nabla_H (f \circ \Phi),$$

$f \circ \Phi \circ p_H = f \circ p \circ \Phi \circ p_H$ and

$$\nabla (f \circ \Phi \circ p_H) = \nabla (f \circ p \circ \Phi \circ p_H) = {}^*D(\Phi \circ p_H) \nabla (f \circ p) \circ \Phi \circ p_H,$$

$$\boxed{\nabla_H (f \circ \Phi) = {}^*D_H(\Phi) \nabla_\Gamma f \circ \Phi,}$$

$$f_\alpha = e_\alpha \cdot \nabla_H (f \circ \Phi) = e_\alpha \cdot {}^*D_H \Phi \nabla_\Gamma f \circ \Phi = D_H \Phi e_\alpha \cdot \nabla_\Gamma f \circ \Phi$$
$$= a_\alpha \cdot \nabla_\Gamma f \circ \Phi$$

$$\Rightarrow \boxed{f_\alpha = a_\alpha \cdot \nabla_\Gamma f \circ \Phi = e_\alpha \cdot \nabla_H (f \circ \Phi).}$$

For a vector function $v \in C^1(\Gamma; \mathbf{R}^N)$

$$\boxed{v_{,\alpha} \overset{\text{def}}{=} \frac{\partial (v \circ \Phi)}{\partial \zeta_\alpha}, \quad 1 \le \alpha \le N - 1.}$$

Again, $v \circ \Phi \circ p_H = v \circ p \circ \Phi \circ p_H$ and

$$D(v \circ \Phi \circ p_H) = D(v \circ p \circ \Phi \circ p_H) = D(v \circ p) \circ \Phi \circ p_H \, D(\Phi \circ p_H),$$

$$\boxed{D_H(v \circ \Phi) = D_\Gamma v \circ \Phi \, D_H \Phi} \quad \Rightarrow \quad \boxed{v_{,\alpha} = D_H(v \circ \Phi) e_\alpha = D_\Gamma v \circ \Phi \, a_\alpha.}$$

The *second fundamental form* of Γ is defined from the inward normal a_N,

$$\boxed{b_{\alpha\beta} \overset{\text{def}}{=} -a_\alpha \cdot a_{N,\beta},}$$

where from (5.28) $a_N = -n \circ \Phi$, and from (5.10) $D_\Gamma n = D^2 b|_\Gamma$,

$$D_H a_N = -D_\Gamma n \circ \Phi \, D_H \Phi = -D^2 b \circ \Phi \, D_H \Phi,$$
$$a_{N,\beta} = -D_H a_N \, e_\beta = -D^2 b \circ \Phi \, D_H \Phi \, e_\beta = -D^2 b \circ \Phi \, a_\beta,$$

$$\boxed{b_{\alpha\beta} = a_\alpha \cdot D^2 b \circ \Phi \, a_\beta.}$$

Recall that since $|\nabla b| = 1$, $D^2 b \nabla b = 0$ and ∇b is an eigenvector for the eigenvalue 0. As a result the eigenvalues of $D^2 b$ at a point of Γ are the eigenvalues of the second fundamental form $b_{\alpha\beta}$ (the $N - 1$ *principal curvatures*) plus $\{0\}$. In particular,

$$\Delta b = \operatorname{tr} D^2 b = H = (N - 1)\bar{H},$$

where $\bar{H}$ is the *mean curvature* of Γ and H is the *additive curvature* (cf. section 3.3 of Chapter 2).

5.7 Back to the Example of Section 4.3.3

Coming back to formula (4.36) for $dJ(\Omega; V(0))$ of the boundary integral of the square of the normal derivative in section 4.3.3, the tangential calculus is now used on the term

$$\int_\Gamma 2\frac{\partial \phi}{\partial n}\left\{\frac{\partial \phi}{\partial n} DV(0)n \cdot n - \nabla\phi \cdot \left({}^*DV(0)\,n + D^2 b\, V(0)\right)\right\}\,d\Gamma,$$

$$T \stackrel{\text{def}}{=} \frac{\partial \phi}{\partial n} DV(0)n \cdot n - \nabla\phi \cdot \left({}^*DV(0)\,n + D^2 b\, V(0)\right).$$

From identity (5.18) with $v = V(0)|_\Gamma$ and identity (5.23),

$$^*DV(0)n = {}^*D_\Gamma v\, n + DV(0)n \cdot n\, n,$$

$$\nabla\phi \cdot {}^*DV(0)n = \nabla\phi \cdot {}^*D_\Gamma v\, n + \frac{\partial \phi}{\partial n} DV(0)n \cdot n,$$

$$\Rightarrow T = -\nabla\phi \cdot \left[{}^*D_\Gamma v\, n + D^2 b\, v_\Gamma\right] = -\nabla\phi \cdot \nabla_\Gamma v_n = -\nabla_\Gamma\phi \cdot \nabla_\Gamma v_n.$$

Therefore, by using the tangential Stokes formula (5.26),

$$\int_\Gamma 2\frac{\partial \phi}{\partial n} T\, d\Gamma = -\int_\Gamma 2\frac{\partial \phi}{\partial n} \nabla_\Gamma\phi \cdot \nabla_\Gamma v_n\, d\Gamma$$

$$= \int_\Gamma -2\,\text{div}_\Gamma\left\{\frac{\partial \phi}{\partial n} v_n\, \nabla_\Gamma\phi\right\} + 2\,\text{div}_\Gamma\left\{\frac{\partial \phi}{\partial n} \nabla_\Gamma\phi\right\} v_n\, d\Gamma$$

$$= \int_\Gamma -2H\frac{\partial \phi}{\partial n} v_n\, \nabla_\Gamma\phi \cdot n + 2\,\text{div}_\Gamma\left\{\frac{\partial \phi}{\partial n} \nabla_\Gamma\phi\right\} v_n\, d\Gamma$$

$$= \int_\Gamma 2\,\text{div}_\Gamma\left\{\frac{\partial \phi}{\partial n} \nabla_\Gamma\phi\right\} v_n\, d\Gamma = \int_\Gamma 2\,\text{div}_\Gamma\left\{\frac{\partial \phi}{\partial n} \nabla_\Gamma\phi\right\} V(0) \cdot n\, d\Gamma.$$

Substituting into expression (4.36), we finally get the explicit formula in term of $V(0) \cdot n$ as predicted by the structure theorem

$$\boxed{\begin{aligned} &dJ(\Omega; V) \\ &= \int_\Gamma\left\{2\frac{\partial \phi}{\partial n} D^2\phi\, n \cdot n + H\left|\frac{\partial \phi}{\partial n}\right|^2 + 2\,\text{div}_\Gamma\left(\frac{\partial \phi}{\partial n} \nabla_\Gamma\phi\right)\right\} V(0) \cdot n\, d\Gamma. \end{aligned}} \tag{5.29}$$

6 Second-Order Semiderivative and Shape Hessian

The object of this section is to study second-order derivatives and semiderivatives by the *velocity method* for smooth and nonsmooth domains and discuss its relationship with the *method of perturbations of the identity* in the unconstrained case. The analysis of this section is motivated by first computing the second-order derivative of a domain integral by using the combined strengths of the shape and tangential calculi in section 6.1. A basic formula for the second-order semiderivative of the domain integral is given in section 6.2 which also reveals the general structure of this derivative. Structure theorems for the second-order Eulerian semiderivative $d^2 J(\Omega; V; W)$ of a function $J(\Omega)$ for two vector fields V and W are given in sections 6.3 and 6.4. A first theorem shows that under some natural continuity assumptions,

$$d^2 J(\Omega; V; W) = d^2 J(\Omega; V(0); W(0)) + dJ(\Omega; V'(0)),$$

where $V'(0)$ is the time-partial derivative $\partial_t V(t, x)$ at $t = 0$. As in the study of first-order Eulerian semiderivatives, this first theorem reduces the study of second-order Eulerian semiderivatives to the autonomous case. So we then specialize to fields in $\mathcal{D}^k(D, \mathbf{R}^N)$ and give the equivalent of Hadamard's structure theorem for the term $d^2 J(\Omega; V(0); W(0))$.

This bilinear term is decomposed into a symmetrical term plus the gradient acting on the first half of the Lie bracket $[V(0), W(0)]$ in section 6.5. The symmetrical part is itself decomposed into a symmetrical part that depends only on the normal component of the velocity fields and a symmetrical term made up of the gradient acting on a generic group of terms, which occurs in all examples considered in this section.

6.1 Second-Order Derivative of the Domain Integral

Given $f \in C^2_{\mathrm{loc}}(\mathbf{R}^N)$ and a domain Ω of class C^2, consider the function

$$J(\Omega_{t,s}(V, W)) \stackrel{\text{def}}{=} \int_{\Omega_{t,s}(V,W)} f \, dx, \tag{6.1}$$

where

$$\Omega_{t,s}(V, W) \stackrel{\text{def}}{=} T_s(W)(\Omega_t(V)) = T_s(W)(T_t(V)(\Omega)),$$

$$J_{V,W}(t, s) \stackrel{\text{def}}{=} J(\Omega_{t,s}(V, W))$$

for some pair of autonomous velocity fields V and W satisfying condition (3.1) and additional smoothness conditions as necessary. The objective is to compute

$$\boxed{d^2 J_{V,W} \stackrel{\text{def}}{=} \frac{\partial}{\partial s} \left\{ \frac{\partial}{\partial t} J_{V,W}(t, s) \Big|_{t=0} \right\} \Big|_{s=0}.}$$

From formula (4.7) in Theorem 4.2 of the previous section, we already know that

$$\frac{\partial}{\partial t} J_{V,W}(t, s) \Big|_{t=0} = dJ(\Omega_s(W); V) = \int_{\Gamma_s(W)} f \, V \cdot n_s \, d\Gamma_s = \int_{\Gamma_s(W)} f \, V \cdot N_s \, d\Gamma_s,$$

where $N_s = N_s(W)$ is the extension (4.37) of the normal n_s. So we can readily use formula (4.17) from Theorem 4.3:

$$\boxed{d^2 J_{V,W} = \int_\Gamma f \, V \cdot N'(W) + \left\{ \frac{\partial}{\partial n} (f \, V \cdot \nabla b) + H f \, V \cdot n \right\} W \cdot n \, d\Gamma,} \tag{6.2}$$

where $N'(W)$ is the derivative of the extension $N_s = N_s(W)$ given by expression (4.38). This yields

$$
d^2 J_{V,W}
$$
$$
= \int_\Gamma f\, V \cdot \left\{ (DWn \cdot n)\, n - {}^*DW - D^2 b\, W \right\}
$$
$$
+ \left(\frac{\partial}{\partial n} (f\, V \cdot \nabla b) + H f\, V \cdot n \right) W \cdot n\, d\Gamma
$$
$$
= \int_\Gamma f\, \left\{ V \cdot \left\{ (DWn \cdot n)\, n - {}^*DWn - D^2 b\, W \right\} + \frac{\partial}{\partial n} (V \cdot \nabla b)\, W \cdot n \right\}
$$
$$
+ \left(\frac{\partial f}{\partial n} + H f \right) V \cdot n\, W \cdot n\, d\Gamma.
$$

It remains to untangle the following term in the first part of the integral:

$$
T \overset{\text{def}}{=} V \cdot \left\{ (DWn \cdot n)\, n - {}^*DWn - D^2 b\, W \right\} + \frac{\partial}{\partial n} (V \cdot \nabla b)\, W \cdot n.
$$

Using the notation $v = V|_\Gamma$ and $w = W|_\Gamma$,

$$
\nabla (V \cdot \nabla b) = {}^*DV \nabla b + D^2 b\, V,
$$
$$
\nabla (V \cdot \nabla b) \cdot \nabla b = {}^*DV \nabla b \cdot \nabla b + D^2 b\, V \cdot \nabla b = DV \nabla b \cdot \nabla b
$$
$$
\Rightarrow \frac{\partial}{\partial n} (V \cdot \nabla b) = DVn \cdot n,
$$
$$
V \cdot {}^*DWn = V \cdot \left[{}^*D_\Gamma w + n\, {}^*(DWn) \right] n
$$
$$
= V \cdot \left[\nabla_\Gamma w_n - D^2 b\, w \right] + DWn \cdot n\, V \cdot n,
$$
$$
V \cdot \left[{}^*DWn + D^2 b\, W \right] = \nabla_\Gamma w_n \cdot v_\Gamma + DWn \cdot n\, v_n.
$$

Finally,

$$
T = DWn \cdot n\, v_n - (v_\Gamma \cdot \nabla_\Gamma w_n + DWn \cdot n\, v_n) + DVn \cdot n\, w_n
$$
$$
= DVn \cdot n\, w_n - v_\Gamma \cdot \nabla_\Gamma w_n,
$$

and by using the tangential Stokes formula (5.26)

$$
\int_\Gamma f\, T\, d\Gamma = \int_\Gamma f\, \left\{ DVn \cdot n\, w_n - v_\Gamma \cdot \nabla_\Gamma w_n \right\} d\Gamma
$$
$$
= \int_\Gamma f\, DVn \cdot n\, w_n - \mathrm{div}_\Gamma\, (f w_m v_\Gamma) + \mathrm{div}_\Gamma\, (f v_\Gamma)\, w_n\, d\Gamma
$$
$$
= \int_\Gamma \left\{ f\, DVn \cdot n + \mathrm{div}_\Gamma\, (f v_\Gamma) \right\} w_n - f w_n\, v_\Gamma \cdot n\, d\Gamma
$$
$$
= \int_\Gamma \left\{ f\, (DVn \cdot n + \mathrm{div}_\Gamma v_\Gamma) + \nabla_\Gamma f \cdot v_\Gamma \right\} w_n\, d\Gamma.
$$

Finally, we get two equivalent expressions

$$
\begin{aligned}
d^2 &J_{V,W} \\
&= \int_\Gamma \left(\frac{\partial f}{\partial n} + H\,f \right) v_n\,w_n + f\,(DVn \cdot n\,w_n - v_\Gamma \cdot \nabla_\Gamma w_n)\,d\Gamma \\
&= \int_\Gamma \left\{ \left(\frac{\partial f}{\partial n} + H\,f \right) v_n + f\,(DVn \cdot n + \mathrm{div}_\Gamma v_\Gamma) + \nabla_\Gamma f \cdot v_\Gamma \right\} w_n\,d\Gamma.
\end{aligned}
\tag{6.3}
$$

Since the above expressions involved the composition $T_s(W) \circ T_t(V)$, it is expected that the condition for the symmetry of expression (6.3) will involve the *Lie bracket* $[V, W] = DVW - DWV$. Indeed, by using identity (5.23)

$$
\begin{aligned}
DVW \cdot n &= (D_\Gamma v + DVn\,{}^*n\}\,W \cdot n \\
&= w \cdot {}^*D_\Gamma v\,n + DVn \cdot n\,w_n \\
&= w_\Gamma \cdot (\nabla_\Gamma v_n - D^2 b\,v_\Gamma) + DVn \cdot n\,w_n
\end{aligned}
$$

and substituting in the first expression, we get a symmetrical term plus the first half of the Lie bracket

$$
\begin{aligned}
d^2 &J_{V,W} \\
&= \int_\Gamma \left(\frac{\partial f}{\partial n} + H\,f \right) v_n\,w_n + f\,\left(D^2 b\,v_\Gamma \cdot w_\Gamma - v_\Gamma \cdot \nabla_\Gamma w_n - w_\Gamma \cdot \nabla_\Gamma v_n \right) \\
&\qquad\qquad + f\,DVW \cdot n\,d\Gamma.
\end{aligned}
\tag{6.4}
$$

Thus

$$
d^2_{V,W} = d^2_{W,V} \quad \Longleftrightarrow \quad \int_\Gamma f\,[V,W] \cdot n\,d\Gamma = 0,
\tag{6.5}
$$

from which either $f\,[V,W] \cdot n = 0$ on Γ or $\mathrm{div}\,(f\,[V,W]) = 0$ on Ω can be used as sufficient conditions.

Example 6.1.
Let $\Omega = \{(x,y) : x^2 + y^2 < 1\}$ be the unit ball in $\mathbf{R}^2$ with boundary $\Gamma = \{(x,y) : x^2 + y^2 = 1\}$. Choose

$$
V(x,y) = (1,0), \quad W(x,y) = (x^2/2, 0), \quad n = (x,y)/\sqrt{x^2 + y^2}
$$

$$
\Rightarrow\ DVW - DWV = (x,0) \ \Rightarrow\ [DVW - DWV] \cdot n = x^2/\sqrt{x^2 + y^2} = x^2. \quad \square
$$

6.2 Basic Formula for Domain Integrals

We have proved the following result.

Theorem 6.1. *Let $f \in C^2([0,\tau] \times [0,\tau]; H^2_{\mathrm{loc}}(\mathbf{R}^N))$ and Γ be the boundary of a bounded open subset Ω of $\mathbf{R}^N$ of class C^2. Assume that V belongs to $C^0([0,\tau]; C^2_{\mathrm{loc}}(\mathbf{R}^N, \mathbf{R}^N))$ and W to $C^0([0,\tau]; C^1_{\mathrm{loc}}(\mathbf{R}^N, \mathbf{R}^N))$. Consider the function*

$$\boxed{\, J_{V,W}(t,s) \stackrel{\text{def}}{=} \int_{\Omega_{t,s}(V,W)} f(t,s)\, dx. \,} \tag{6.6}$$

Then the partial derivative of $J_{V,W}(t,s)$ with respect to t in $t=0$ is given by

$$
\begin{aligned}
\frac{\partial}{\partial t} J_{V,W}(t,s)\Big|_{t=0} &= \int_{\Omega_s(W)} \frac{\partial f}{\partial t}(0,s) + \operatorname{div}\left(f(0,s)V(0)\right) dx \\
&= \int_{\Omega_s(W)} \frac{\partial f}{\partial t}(0,s)\, dx + \int_{\Gamma_s(W)} f(0,s)\, V(0) \cdot n_s \, d\Gamma_s
\end{aligned}
\tag{6.7}
$$

and the second-order mixed derivative of $J_{V,W}(t,s)$ in $(t,s)=(0,0)$,

$$d^2 J_{V,W} \stackrel{\text{def}}{=} \frac{\partial}{\partial s}\left\{ \frac{\partial}{\partial t} J_{V,W}(t,s)\Big|_{t=0} \right\}\Big|_{s=0}, \tag{6.8}$$

is given by the expression

$$
\boxed{
\begin{aligned}
d^2 J_{V,W} &= \int_\Omega \frac{\partial}{\partial s}\left(\frac{\partial f}{\partial t}\right) + \operatorname{div}\left(\frac{\partial f}{\partial s}V(0) + \frac{\partial f}{\partial t}W(0)\right) \\
&\qquad\qquad + \operatorname{div}\left[\operatorname{div}(f\,V(0))\,W(0)\right] dx \\
&= \int_\Omega \frac{\partial}{\partial s}\left(\frac{\partial f}{\partial t}\right) dx + \int_\Gamma \left(\frac{\partial f}{\partial s}V(0) + \frac{\partial f}{\partial t}W(0)\right) \cdot n \\
&\qquad\qquad + \operatorname{div}(f\,V(0))\,W(0)\cdot n\, d\Gamma.
\end{aligned}
}
\tag{6.9}
$$

The last term in the second integral can be expressed in terms of $v = V(0)|_\Gamma$ and $w = W(0)|_\Gamma$ as follows:

$$
\boxed{
\begin{aligned}
&\int_\Gamma \operatorname{div}(f\,V(0))\,W(0)\cdot n\, d\Gamma \\
&= \int_\Gamma \left\{ \left(\frac{\partial f}{\partial n} + H f\right) v_n + f\left(DVn\cdot n + \operatorname{div}_\Gamma v_\Gamma\right) + \nabla_\Gamma f \cdot v_\Gamma \right\} w_n\, d\Gamma \\
&= \int_\Gamma \left(\frac{\partial f}{\partial n} + H f\right) v_n\, w_n + f\left(D^2 b\, v_\Gamma \cdot w_\Gamma - v_\Gamma \cdot \nabla_\Gamma w_n - w_\Gamma \cdot \nabla_\Gamma v_n\right) \\
&\qquad\qquad + f\, DVW\cdot n\, d\Gamma.
\end{aligned}
}
\tag{6.10}
$$

6.3 Nonautonomous Case

The framework introduced in sections 4 and 5 of Chapter 7 has reduced the computation of the Eulerian semiderivative of $J(\Omega)$ to the computation of the derivative of the function

$$j(t) \stackrel{\text{def}}{=} J(\Omega_t(V)) \tag{6.11}$$

for a velocity field $V \in C([0,\tau]; C^k_{loc}(\mathbf{R}^N; \mathbf{R}^N))$. In $t \geq 0$

$$j'(t) = dJ\big(\Omega_t(V); V_t\big) \tag{6.12}$$

since $T_{s+t}(V) = T_s(V_t) \circ T_t(V)$, where $V_t(s) \overset{\text{def}}{=} V(t+s)$ and $V_t(0) = V(t)$.
   This suggests the following definition.

Definition 6.1.
Let J be a real-valued shape function. Let V and W satisfy condition (V) and
assume that for all $t \in [0,\tau]$, $dJ\big(\Omega_t(W); V_t\big)$ exists for $\Omega_t(W) = T_t(W)(\Omega)$. The
function J is said to have a *second-order Eulerian semiderivative* at Ω in the directions (V, W) if the following limit exists:

$$\lim_{t \searrow 0} \frac{dJ(\Omega_t(W); V_t) - dJ(\Omega; V)}{t}. \tag{6.13}$$

When it exists, it is denoted $d^2 J(\Omega; V; W)$.   □

   If, for all t, J has a Hadamard semiderivative at $\Omega_t(W)$, recall that

$$dJ(\Omega_t(W); V_t) = d_H J(\Omega_t(W); V_t(0))$$
$$= d_H J(\Omega_t(W); V(t)) = dJ(\Omega_t(W); V(t)),$$

and the above definition reduces to

$$d^2 J(\Omega; V; W) = \lim_{t \searrow 0} \frac{dJ\big(\Omega_t(W); V(t)\big) - dJ\big(\Omega; V(0)\big)}{t}.$$

Remark 6.1.
This last definition is compatible with the second-order expansion of $j(t)$ with respect to t around $t = 0$:

$$j(t) \cong j(0) + tj'(0) + \frac{t^2}{2} j''(0), \tag{6.14}$$

where

$$j(0) = J(\Omega), \quad j'(0) = dJ(\Omega; V), \quad j''(0) = d^2 J(\Omega; V; V). \tag{6.15}$$

□

   The next theorem is the analogue of Theorem 3.2 and provides the canonical structure of the second-order Eulerian semiderivative (cf. (3.17) to (3.19) in
section 3.1 for the definitions of $\vec{\mathcal{V}}^{m,\ell}$ and $\mathcal{V}^\ell$).

Theorem 6.2. *Let J be a real-valued shape function, Ω a subset of $\mathbf{R}^N$, and $m \geq 0$
and $\ell \geq 0$ two integers. Assume that*

   (i) $\forall V \in \vec{\mathcal{V}}^{m+1,\ell}, \forall W \in \vec{\mathcal{V}}^{m,\ell}, \quad d^2 J(\Omega; V; W)$ *exists;*

(ii) $\forall W \in \vec{\mathcal{V}}^{m,\ell}$, $\forall t \in [0,\tau]$, J has a shape gradient of order ℓ and is Hadamard semidifferentiable at $\Omega_t(W)$;

(iii) $\forall U \in \mathcal{V}^\ell$, the map

$$W \mapsto d^2 J(\Omega; U; W) : \vec{\mathcal{V}}^{m,\ell} \to \mathbf{R} \qquad (6.16)$$

is continuous.

Then for all V in $\vec{\mathcal{V}}^{m+1,\ell}$ and all W in $\vec{\mathcal{V}}^{m,\ell}$,

$$d^2 J(\Omega; V; W) = d^2 J(\Omega; V(0); W(0)) + dJ(\Omega; V'(0)), \qquad (6.17)$$

where

$$V'(0)(x) \overset{\text{def}}{=} \lim_{t \searrow 0} \frac{V(t,x) - V(0,x)}{t}. \qquad (6.18)$$

Proof. The differential quotient (6.13) can be split into the sum of two terms

$$\frac{dJ(\Omega_t(W); V(0)) - dJ(\Omega; V(0))}{t} + \frac{dJ(\Omega_t(W); V(t)) - dJ(\Omega_t(W); V(0))}{t}. \qquad (6.19)$$

In view of (i) and (iii), for all U in $\mathcal{V}^\ell$,

$$d^2 J(\Omega; U; W) = d^2 J(\Omega; U; W(0))$$

by the same argument as in the proof of Theorem 3.2 for the gradient. Hence the first term converges to

$$d^2 J(\Omega; V(0); W) = d^2 J(\Omega; V(0); W(0)).$$

For the second term recall that V belongs to $\vec{\mathcal{V}}^{m+1,\ell}$ and observe that the vector field

$$\widehat{V}(t) = \frac{V(t) - V(0)}{t}$$

belongs to $\vec{\mathcal{V}}^{m,\ell}$ and that $\widehat{V}(0) = V'(0)$. Thus by linearity of $dJ(\Omega; V)$, the second term in (6.19) can be written as

$$dJ\left(\Omega_t(W); \frac{V(t) - V(0)}{t}\right) = dJ(\Omega_t(W); \widehat{V}(t)),$$

$$dJ(\Omega_t(W); \widehat{V}(t)) = t\frac{1}{t}\left[dJ(\Omega_t(W); \widehat{V}(t)) - dJ(\Omega; \widehat{V}(0))\right] + dJ(\Omega; \widehat{V}(0)).$$

But for any V in $\vec{\mathcal{V}}^{m+2,\ell}$, $\widehat{V}$ belongs to $\vec{\mathcal{V}}^{m+1,\ell}$. Then by assumption (i),

$$\lim_{t \searrow 0} \frac{dJ(\Omega_t(W); \widehat{V}(t)) - dJ(\Omega; \widehat{V}(0))}{t} = d^2 J(\Omega; \widehat{V}; W),$$

which implies that

$$\lim_{t \searrow 0} dJ\big(\Omega_t(W); \widehat{V}(t)\big) = dJ\big(\Omega; \widehat{V}(0)\big) = dJ\big(\Omega; V'(0)\big).$$

Now by assumption (ii), the map $U \mapsto dJ(\Omega; U)$ is linear and continuous on $\mathcal{D}^\ell(\mathbf{R}^N, \mathbf{R}^N)$, and the map

$$V \mapsto V'(0) \mapsto dJ\big(\Omega; V'(0)\big) : \vec{\mathcal{V}}^{m+2,\ell} \to \mathcal{V}^\ell \to \mathbf{R}$$

is linear and continuous (hence uniformly continuous) for the topology $\vec{\mathcal{V}}^{m+1,\ell}$ for all V in the dense subspace $\vec{\mathcal{V}}^{m+2,\ell}$. Hence it uniquely and continuously extends to all elements of $\vec{\mathcal{V}}^{m+1,\ell}$. This completes the proof of the theorem. $\qquad\square$

This important theorem gives the canonical structure of the second-order Eulerian semiderivative: a first term that depends on $V(0)$ and $W(0)$ and a second term that is equal to $dJ(\Omega; V'(0))$. When V is autonomous the second term disappears and the semiderivative coincides with $d^2 J(\Omega; V; W(0))$ which can be separately studied for autonomous vector fields in $\mathcal{V}^\ell$.

We conclude this section with the explicit computation of the second-order Eulerian semiderivative for a shape function $J(\Omega)$ with respect to two velocity fields V and W satisfying the conditions of Theorem 6.2 and such that the shape gradient at any t is of the form

$$\boxed{\; dJ(\Omega_t(W); V(t)) = \int_{\Gamma_t(W)} g(t)\, V(t) \cdot n_t \, d\Gamma_t \;} \qquad (6.20)$$

for some function $g(t) \in C(\Gamma_t(W))$. Further assume that the family of functions $g(t)$ has an extension $Q \in C^1([0,\tau]; C^k_{\mathrm{loc}}(N(\Gamma); \mathbf{R}^N))$ to an open neighborhood $N(\Gamma)$ of Γ such that $\cup \{\Gamma_t(W) : 0 \le t \le \tau\} \subset N(\Gamma)$. Therefore using the extension $N_t(W)$ of the normal n_t on $\Gamma_t(W)$, it amounts to differentiating the expression

$$j(t) = \int_{\Gamma_t(W)} Q(t)\, V(t) \cdot N_t(W) \, d\Gamma_t.$$

Apply the first formula (4.17) of Theorem 4.3 to get

$$j'(0) = \int_\Gamma (Q'_W(0)\, V(0) + Q(0)\, V'(0)) \cdot n + Q(0)\, V(0) \cdot N'(W)$$

$$+ \left(\frac{\partial}{\partial n} (Q(0)\, V(0) \cdot \nabla b) + H\, Q(0)\, V(0) \cdot n \right) W(0) \cdot n \, d\Gamma,$$

where $Q'_W(0)$ only depends on W. But the last three terms have already been computed in several forms. They constitute expression (6.2) in section 6.1, which yields (6.3) and (6.4) with $f = Q(0)$, $V = V(0)$, and $W = W(0)$. This yields with the notation $v = V(0)|_\Gamma$ and $w = W(0)|_\Gamma$

$$
\begin{aligned}
d^2 J(V;W) &= \int_\Gamma Q'_W(0)\, v_n + Q(0)\, V'(0)\cdot n + \left(\frac{\partial Q(0)}{\partial n} + H\,Q(0)\right) v_n\, w_n \\
&\quad + Q(0)\left(D^2 b\, v_\Gamma \cdot w_\Gamma - v_\Gamma \cdot \nabla_\Gamma w_n - w_\Gamma \cdot \nabla_\Gamma v_n\right) \\
&\quad + Q(0)\, DVW \cdot n\, d\Gamma \\
&= \int_\Gamma Q'_W(0)\, v_n + Q(0)\, V'(0)\cdot n + \left(\frac{\partial Q(0)}{\partial n} + H\,Q(0)\right) v_n\, w_n \\
&\quad + \left\{ Q(0)\left(DVn\cdot n + \operatorname{div}_\Gamma v_\Gamma\right) + \nabla_\Gamma Q(0)\cdot v_\Gamma\right\} w_n\, d\Gamma.
\end{aligned}
\tag{6.21}
$$

Remark 6.2.

When V is autonomous the term in $V'(0)$ disappears. In that case the first half of the Lie bracket can be eliminated by restarting the computation with $\mathcal{V}(t) = V \circ T_t^{-1}(W)$ in place of V since $\mathcal{V}(0) = V$ and

$$
\mathcal{V}'(0) = -DV\,W \quad \Rightarrow \quad \int_\Gamma Q(0)\,\mathcal{V}'(0)\cdot n + Q(0)\,DVW\cdot n\, d\Gamma = 0
$$
$$
\Rightarrow d^2 J(V,W) = d^2 J(\mathcal{V},W) + dJ(\Omega; DVW). \qquad \square
$$

Remark 6.3.

Except for the terms that contain the first half of the Lie bracket DVW and $V'(0)$, the only term that might not be symmetrical in the first expression (6.21) is the one in $Q'_W(0)$. In fact, according to the second expression and our theorem, $Q'_W(0) = Q'_{W(0)}$ only depends on $W(0)$. Now choose autonomous velocity fields V and W. Furthermore, assume that W is of the form $W = w_\Gamma \circ p$. Since $W \cdot n = 0$ on Γ, $dJ(\Omega_t(W); V(0)) = dJ(\Omega; V(0))$ and necessarily $d^2 J(\Omega; V; w_\Gamma \circ p) = 0$. Therefore, $d^2 J(\Omega; V; W)$ only depends on w_n and hence $Q'_W(0) = Q'_{w_n}(0)$ and the integral

$$
\int_\Gamma Q'_{w_n}(0)\, v_n\, d\Gamma
$$

only depends on w_n and v_n. $\qquad \square$

The above expressions give valuable information on the structure of the second-order Eulerian derivative. Other expressions can also be obtained. For instance, if $j(t)$ is transformed into the volume integral

$$
j(t) = dJ(\Omega_t(W); V(t)) = \int_{\Gamma_t(W)} Q(t)\, V(t)\cdot n_t\, d\Gamma_t = \int_{\Omega_t(W)} \operatorname{div}\left(Q(t)\, V(t)\right) dx,
$$

we get from formula (4.6) in Theorem 4.2 the equivalent volume expression

$$
d^2 J(\Omega; V; W)
$$
$$
= \int_\Omega \operatorname{div}\left\{ Q'_{W(0)}(0)\, V(0) + Q(0)\, V'(0) + \operatorname{div}\left(Q(0)\, V(0)\right) W(0)\right\} dx,
$$

which can obviously be transformed into a boundary expression.

6.4 Autonomous Case

Definition 6.2.
Let J be a real-valued shape function. Let Ω be a subset of $\mathbf{R}^N$.

(i) The function $J(\Omega)$ is said to be *twice shape differentiable* at Ω if

$$\forall V, \forall W \in \mathcal{D}(\mathbf{R}^N, \mathbf{R}^N), \quad d^2 J(\Omega; V; W) \text{ exists} \tag{6.22}$$

and the map

$$(V, W) \mapsto d^2 J(\Omega; V; W) : \mathcal{D}(\mathbf{R}^N, \mathbf{R}^N) \times \mathcal{D}(\mathbf{R}^N, \mathbf{R}^N) \to \mathbf{R} \tag{6.23}$$

is bilinear and continuous. We denote by h the map (6.23).

(ii) Denote by $H(\Omega)$ the vector distribution in $(\mathcal{D}(\mathbf{R}^N, \mathbf{R}^N) \otimes \mathcal{D}(\mathbf{R}^N, \mathbf{R}^N))'$ associated with h:

$$d^2 J(\Omega; V; W) = \langle H(\Omega), V \otimes W \rangle = h(V, W), \tag{6.24}$$

where $V \otimes W$ is the tensor product of V and W defined as

$$(V \otimes W)_{ij}(x, y) = V_i(x) W_j(y), \quad 1 \leq i, j \leq N, \tag{6.25}$$

and $V_i(x)$ (resp., $W_j(y)$) is the ith (resp., jth) component of the vector V (resp., W) (cf. Schwartz's [1] kernel theorem and Gelfand and Vilenkin [1]). $H(\Omega)$ will be called the *shape Hessian* of J at Ω.

(iii) When there exists a finite integer $\ell \geq 0$ such that $H(\Omega)$ is continuous for the $\mathcal{D}^\ell(\mathbf{R}^N, \mathbf{R}^N) \otimes \mathcal{D}^\ell(\mathbf{R}^N, \mathbf{R}^N)$-topology, we say that $H(\Omega)$ is of order ℓ.

In what follows, the compact notation $\mathcal{D}^\ell$ will be used in place of $\mathcal{D}^\ell(\mathbf{R}^N, \mathbf{R}^N)$. $\square$

Theorem 6.3. *Let J be a real-valued shape function and Ω a subset of $\mathbf{R}^N$ with boundary Γ. Assume that J is twice shape differentiable.*

(i) *The vector distribution $H(\Omega)$ has support in $\Gamma \times \Gamma$.*

(ii) *If Ω is an open or closed domain in $\mathbf{R}^N$ and $H(\Omega)$ is of order $\ell \geq 0$, then there exists a continuous bilinear form*

$$[h] : \left(\mathcal{D}^\ell / D_\Gamma^\ell\right) \times \left(\mathcal{D}^\ell / L_\Omega^\ell\right) \to \mathbf{R} \tag{6.26}$$

such that for all $[V]$ in $\mathcal{D}^\ell / D_\Gamma^\ell$ and $[W]$ in $\mathcal{D}^\ell / L_\Omega^\ell$,

$$\boxed{d^2 J(\Omega; V; W) = [h]\big(q_D(V), q_L(W)\big),} \tag{6.27}$$

where $q_D : \mathcal{D}^\ell \to \mathcal{D}^\ell / D_\Gamma^\ell$ and $q_L : \mathcal{D}^\ell \to \mathcal{D}^\ell / L_\Omega^\ell$ are the canonical quotient surjections and

$$\boxed{D_\Gamma^\ell = \{V \in \mathcal{D}^\ell(\mathbf{R}^N, \mathbf{R}^N) : \partial^\alpha V = 0 \text{ on } \Gamma, \forall \alpha, |\alpha| \leq \ell\}.} \tag{6.28}$$

Proof. (i) It is sufficient to prove the following two properties:

(a) $\forall V, W \in \mathcal{D}$ such that $W = 0$ in a neighborhood of Γ, $d^2 J(\Omega; V; W) = 0$.

(b) $\forall V, W \in \mathcal{D}$ such that $V = 0$ in a neighborhood of Γ, $d^2 J(\Omega; V; W) = 0$.

In case (a) the proof is similar to the one in Theorem 3.5 for the gradient and we prove the stronger result that for W such that $W = 0$ on Γ,

$$\Omega_t(W) = \Omega, \forall t \geq 0 \implies dJ(\Omega_t(W); V) = dJ(\Omega; V)$$
$$\implies d^2 J(\Omega; V; W) = 0.$$

In case (b) $V = 0$ in a neighborhood N of Γ and in $\complement K$, the complement of the compact support K of V. So $U = \complement K$ is a neighborhood of Γ where $V = 0$. By construction $U \cap K = \varnothing$ and there exists a bounded neighborhood $\mathcal{U}$ of K such that $\overline{\mathcal{U}} \cap \Gamma = \varnothing$. Since $\overline{\mathcal{U}}$ is compact and Γ is closed, the minimum distance d from $\overline{\mathcal{U}}$ to Γ is finite and nonzero. Let

$$N(\Gamma) = \left\{ y \in \mathbf{R}^N : d_\Gamma(y) < d/2 \right\},$$

where

$$d_\Gamma(y) = \inf\{|y - x| : x \in \Gamma\}.$$

For all X in Γ

$$T_t(X) - X = \int_0^t W(T_s(X))\, ds = tW(X) + \int_0^t \left[W(T_s(X)) - W(X) \right] ds,$$

and by condition (3.1) on W,

$$|T_t(X) - X| \leq t|W(X)| + ct \max_{[0,t]} |T_s(X) - X|,$$

and it can easily be shown that for $t < 1/c$

$$\max_{[0,t]} |T_s(X) - X| < \frac{t}{1 - ct} |W(X)|.$$

Thus

$$\sup_{X \in \Gamma} \max_{[0,t]} |T_s(X) - X| \leq \frac{t}{1 - ct} \sup_{X \in \Gamma} |W(X)|.$$

But W is continuous with compact support. Therefore,

$$\sup_{X \in \Gamma} |W(X)| \leq \sup_{X \in \operatorname{supp} W} |W(X)| = \|W\|_{C(\mathbf{R}^N; \mathbf{R}^N)} < \infty,$$

and there exist $\tau > 0$ such that

$$\forall s \in [0, \tau], \quad \frac{s}{1 - cs} \|W\|_C < \frac{d}{2}.$$

By definition and the previous inequalities

$$d_\Gamma(T_s(X)) = \inf_{Y \in \Gamma} |T_s(X) - Y| \leq |T_s(X) - X| < \frac{d}{2}$$

for all s in $[0, \tau]$ and all $X \in \Gamma$. This implies that

$$\forall s \in [0, \tau], \forall X \in \Gamma, \quad \Gamma_s(W) = T_s(W)(\Gamma) \subset N(\Gamma).$$

By construction, $V = 0$ in $N(\Gamma)$ since the distance from K to Γ is greater than or equal to d. Therefore,

$$\forall s \in [0, \tau], \quad V \in L^\infty_{\Omega_s(W)},$$

and as in the proof of Theorem 3.5, $dJ(\Omega_s(W); V) = 0$ and necessarily $d^2 J(\Omega; V; W) = 0$.

(ii) We have already established in (i) that the bilinear form

$$(V, W) \mapsto h(V, W) : \mathcal{D} \times \mathcal{D} \to \mathbf{R}$$

is zero for all $V \in \mathcal{D}$ and $W \in \mathcal{D}$ such that $W = 0$ on Γ and also zero for all $W \in \mathcal{D}$ and $V \in \mathcal{D}$ for which $V = 0$ in a neighborhood of Γ. By density all this is still true in $\mathcal{D}^\ell$, and now by the same argument as in the proof of Theorem 3.5 for all V in $\mathcal{D}^\ell$,

$$[W] \mapsto h(V, W) : \mathcal{D}^\ell / L^\ell \to \mathbf{R}$$

is well defined, linear, and continuous. For the first component it is necessary to show that for all W in

$$D_\Gamma^\ell = \left\{ V \in \mathcal{D}^\ell(\mathbf{R}^N, \mathbf{R}^N) : \partial^\alpha V = 0 \text{ on } \Gamma, \forall \alpha, |\alpha| \le \ell \right\}$$

the bilinear form $h(V, W) = 0$. We first prove the result for the subspace

$$A = \mathcal{D}(\Omega; \mathbf{R}^N) \oplus \mathcal{D}(\complement\bar\Omega; \mathbf{R}^N).$$

Then by density and continuity the result holds for the $\mathcal{D}^\ell(\mathbf{R}^N, \mathbf{R}^N)$-closure $\bar A$ of A. Finally we prove that $\bar A = D_\Gamma^\ell$. For any V in A, there exist $V_1 \in \mathcal{D}(\Omega; \mathbf{R}^N)$ and $V_2 \in \mathcal{D}(\complement\bar\Omega; \mathbf{R}^N)$ such that $V = V_1 + V_2$. Moreover,

$$K_1 = \operatorname{supp} V_1 \subset \Omega \text{ and } K_2 = \operatorname{supp} V_2 \subset \complement\bar\Omega$$

are compact subsets of the open sets Ω and $\complement\bar\Omega$, respectively. Hence $V_1 = 0$ (resp., $V_2 = 0$) in the open neighborhood $\complement K_1$ (resp., $\complement K_2$) of Γ and necessarily $V = V_1 + V_2 = 0$ in the neighborhood $U = \complement(K_1 \cup K_2)$ of Γ. Hence from part (i) $h(V, W) = 0$. By definition of D_Γ^ℓ, $D_\Gamma^\ell \subset \mathcal{D}^\ell(\bar\Omega; \mathbf{R}^N) \oplus \mathcal{D}^\ell(\complement\Omega; \mathbf{R}^N)$. Now $A \subset D_\Gamma^\ell$,

$$\bar A = \overline{\mathcal{D}(\Omega; \mathbf{R}^N)} \oplus \overline{\mathcal{D}(\complement\bar\Omega; \mathbf{R}^N)},$$

and

$$\overline{\mathcal{D}(\Omega; \mathbf{R}^N)} = \mathcal{D}^\ell(\bar\Omega; \mathbf{R}^N), \quad \overline{\mathcal{D}(\complement\bar\Omega; \mathbf{R}^N)} = \mathcal{D}^\ell(\complement\Omega; \mathbf{R}^N).$$

By construction each V in $\bar A$ is of the form $V = V_1 + V_2$ for

$$V_1 \in \mathcal{D}^\ell(\bar\Omega; \mathbf{R}^N), \quad K_1 = \operatorname{supp} V_1 \text{ compact in } \bar\Omega,$$
$$\forall |\alpha| \le \ell \quad \partial^\alpha V_1 = 0 \text{ on } \Gamma,$$
$$V_2 \in \mathcal{D}^\ell(\complement\Omega; \mathbf{R}^N), \quad K_2 = \operatorname{supp} V_2 \text{ compact in } \complement\Omega,$$
$$\forall |\alpha| \le \ell \quad \partial^\alpha V_2 = 0 \text{ on } \Gamma.$$

Hence

$$\operatorname{supp} V = K_1 \cup K_2 \text{ compact in } \left[\bar{\Omega}\right] \cup \left[\mathbf{R}^{\mathrm{N}} \setminus \Omega\right] = \mathbf{R}^{\mathrm{N}}$$

and $V \in \mathcal{D}^{\ell}(\mathbf{R}^{\mathrm{N}}, \mathbf{R}^{\mathrm{N}})$. Moreover,

$$\forall \alpha, |\alpha| \leq \ell, \quad \partial^{\alpha} V = \partial^{\alpha} V_1 + \partial^{\alpha} V_2 = 0 \text{ on } \Gamma.$$

This proves that $\bar{A} \subset D_{\Gamma}^{\ell}$ and thence $\bar{A} = D_{\Gamma}^{\ell}$. To complete the proof notice that by continuity of $V \mapsto h(V, W)$, for all W in $\mathcal{D}^{\ell}$ the map

$$[V] \mapsto h(V, W,) : \mathcal{D}^{\ell}/D_{\Gamma}^{\ell} \to \mathbf{R}$$

is well defined, linear, and continuous. Finally, the map

$$([V], [W]) \mapsto h(V, , W) : (\mathcal{D}^{\ell}/L^{\ell}) \times (\mathcal{D}^{\ell}/D_{\Gamma}^{\ell}) \to \mathbf{R}$$

is well defined, bilinear, and continuous. $\qquad\square$

The next and last result is the extension of the structure theorem, Theorem 3.5, to second-order Eulerian semiderivatives. We need the result established in the corollary to Theorem 3.5. For a domain Ω with a boundary Γ which is $C^{\ell+1}$, $\ell \geq 0$, the map

$$q_L(W) \mapsto p_L\big(q_L(W)\big) = \gamma_{\Gamma}(W) \cdot n : \mathcal{D}_{\Omega}^{\ell}/L_{\Omega}^{\ell} \to C^{\ell}(\Gamma) \tag{6.29}$$

is a well-defined isomorphism. This will be used for the V-component. For the W-component we need the following lemma.

Lemma 6.1. *Assume that the boundary Γ of Ω is $C^{\ell+1}$, $\ell \geq 0$. Then the map*

$$q_D(V) \mapsto p_D\big(q_D(V)\big) = \gamma_{\Gamma}(V) : \mathcal{D}^{\ell}/D_{\Gamma}^{\ell} \to C^{\ell}(\Gamma, \mathbf{R}^{\mathrm{N}}) \tag{6.30}$$

is a well-defined isomorphism, where

$$p_D : \mathcal{D}^{\ell} \to \mathcal{D}^{\ell}/D_{\Gamma}^{\ell} \tag{6.31}$$

is the canonical surjection and D_{Γ}^{ℓ} is given by (6.28).

Proof. The proof follows by standard arguments. $\qquad\square$

Theorem 6.4. *Let J be a real-valued shape function. Assume that the conditions of Theorem 6.3 (ii) are satisfied and that the boundary Γ of the open domain Ω is $C^{\ell+1}$ for $\ell \geq 0$.*

(i) *The map*

$$\begin{cases} (v, w) \mapsto h_{D \times L}(v, w) = [h](p_D^{-1} v, p_L^{-1} w) \\ : C^{\ell}(\Gamma, \mathbf{R}^{\mathrm{N}}) \times C^{\ell}(\Gamma) \to \mathbf{R} \end{cases} \tag{6.32}$$

is bilinear and continuous, and for all V and W in $\mathcal{D}^{\ell}(\mathbf{R}^{\mathrm{N}}, \mathbf{R}^{\mathrm{N}})$

$$d^2 J(\Omega; V; W) = h_{D \times L}\Big(\gamma_{\Gamma} P(V), \big((\gamma_{\Gamma} W) \cdot n\big)\Big), \tag{6.33}$$

where $P(V)$ is a linear combination of derivatives of V up to order ℓ.

(ii) *This induces a vector distribution $h(\Gamma \otimes \Gamma)$ on $C^\ell(\Gamma, \mathbf{R}^{\mathrm{N}}) \otimes C^\ell(\Gamma)$ of order ℓ,*

$$h(\Gamma \otimes \Gamma) : C^\ell(\Gamma, \mathbf{R}^{\mathrm{N}}) \otimes C^\ell(\Gamma), \to \mathbf{R}, \qquad (6.34)$$

such that for all V and W in $\mathcal{D}^\ell(\mathbf{R}^{\mathrm{N}}, \mathbf{R}^{\mathrm{N}})$

$$\langle h(\Gamma \otimes \Gamma), (\gamma_\Gamma V) \otimes ((\gamma_\Gamma W) \cdot n) \rangle = d^2 J(\Omega; V; W), \qquad (6.35)$$

where $(\gamma_\Gamma V) \otimes ((\gamma_\Gamma W) \cdot n)$ is defined as the tensor product

$$\Big((\gamma_\Gamma V) \otimes ((\gamma_\Gamma W) \cdot n)\Big)_i(x, y) = (\gamma_\Gamma V_i)(x)((\gamma_\Gamma W) \cdot n)(y), \quad x, y \in \Gamma, \qquad (6.36)$$

$V_i(x)$ is the ith component of $V(x)$, and

$$\forall y \in \Gamma, \quad \big(\gamma_\Gamma(W) \cdot n\big)(y) = (\gamma_\Gamma W)(y) \cdot n(y). \qquad (6.37)$$

In the regular case the reader is also referred to the early papers of Zolésio [24, p. 434] and Bucur and Zolésio [12].

Remark 6.4.

Finally, under the assumptions of Theorems 6.3 and 6.4

$$\begin{aligned}
&d^2 J(\Omega; V; W) \\
&= \langle h(\Gamma \otimes \Gamma), (\gamma_\Gamma P(V(0))) \otimes (l(\gamma_\Gamma W(0)) \cdot nr) \rangle + \langle (g(\Gamma), (\gamma_\Gamma V'(0)) \cdot n) \rangle
\end{aligned} \qquad (6.38)$$

for all V in $\overrightarrow{V}^{m+1,\ell}$ and W in $\overrightarrow{V}^{m,\ell}$. $\qquad\square$

Example 6.2.

Go back to the example of the domain integral in section 6.1,

$$J(\Omega) \overset{\text{def}}{=} \int_\Omega f \, dx.$$

For $V \in \mathcal{D}^1(\mathbf{R}^{\mathrm{N}}, \mathbf{R}^{\mathrm{N}})$

$$dJ(\Omega; V) = \int_\Gamma f \, V \cdot n \, d\Gamma = \int_\Omega \operatorname{div}(fV) \, dx.$$

Using the domain expression with V in $\mathcal{D}^2(\mathbf{R}^{\mathrm{N}}, \mathbf{R}^{\mathrm{N}})$ and W in $\mathcal{D}^1(\mathbf{R}^{\mathrm{N}}, \mathbf{R}^{\mathrm{N}})$,

$$dJ(\Omega_s(W); V) = \int_{\Omega_s(W)} \operatorname{div}(fV) \, dx,$$

we readily get

$$d^2 J(\Omega; V; W) = \int_\Gamma \operatorname{div}(fV) \, W \cdot n \, d\Gamma = \int_\Omega \operatorname{div}\big[\operatorname{div}(fV) \, W\big] \, dx, \qquad (6.39)$$

and if Γ is C^1,

$$d^2 J(\Omega; V; W) = \int_\Gamma \operatorname{div} V \, W \cdot n \, d\Gamma \qquad (6.40)$$

is continuous for pairs

$$(V, W) \in \mathcal{D}^1(\mathbf{R}^{\mathrm{N}}, \mathbf{R}^{\mathrm{N}}) \times \mathcal{D}^0(\mathbf{R}^{\mathrm{N}}, \mathbf{R}^{\mathrm{N}}) \text{ or } C^1(\Gamma, \mathbf{R}^{\mathrm{N}}) \times C^0(\Gamma, \mathbf{R}^{\mathrm{N}}). \qquad\square$$

6.5 Decomposition of $d^2 J(\Omega; V(0), W(0))$

One important observation in the explicit computation of the second-order Eulerian semiderivative of the domain integral (6.1) in section 6.1 was the lack of symmetry and the appearance of the first half of the Lie bracket in (6.4). The same phenomenon was observed in the final form of the basic formula (6.10) for domain integrals (6.6) in Theorem 6.1, and also in section 6.3 for the derivative of the shape gradient (6.21) when it can be represented in integral form (6.20). In this section perturbations of the identity will be used to show that $d^2 J(\Omega; V(0), W(0))$ can be further decomposed into a symmetric term plus the gradient applied to the velocity $DV(0)W(0)$:

$$d^2 J(\Omega; V(0), W(0)) = \langle d^2 J_\Omega(0)V(0), W(0)\rangle + dJ(\Omega; DV(0)W(0)).$$

Furthermore, the symmetrical term can be obtained by the velocity method

$$d^2 J_\Omega(0; V(0), W(0)) = \frac{d}{dt} J\left(\Omega_t(W(0); V(0) \circ T_t^{-1}(W(0))\right)\Big|_{t=0}.$$

Theorem 6.5. *Let J be a real-valued shape function and Θ a Banach subspace of* $\mathrm{Lip}\,(\mathbf{R}^N; \mathbf{R}^N)$.

(i) *Given f, θ, and ξ in B_ε, assume that there exists $\tau > 0$ such that*

$$\forall t \in [0, \tau], \quad d^2 J_\Omega(f + t\xi; \theta) \text{ exists.}$$

Then

$$\boxed{d^2 J_\Omega(f; \theta; \xi) \text{ exists} \iff d^2 J(\Omega_f; \mathcal{V}; W_\xi) \text{ exists}} \tag{6.41}$$

for $\Omega_f = [I + f](\Omega)$ and the velocity fields

$$\boxed{W_\xi(t) \overset{\text{def}}{=} \xi \circ [I + f + t\xi]^{-1}} \text{ and } \boxed{\mathcal{V}(t) \overset{\text{def}}{=} \theta \circ [I + f + t\xi]^{-1}.} \tag{6.42}$$

(ii) *If f, θ belong to $\mathcal{V}^{\ell+1}$ and ξ to $\mathcal{V}^\ell$, and $\mathcal{V}$ and W_ξ satisfy the conditions of Theorem 6.2, then*

$$\boxed{\begin{aligned} d^2 J_\Omega(f; \theta; \xi) &= d^2 J(\Omega_f; \theta \circ [I + f]^{-1}; \xi \circ [I + f]^{-1}) \\ &\quad - dJ(\Omega_f; D(\theta \circ [I + f]^{-1})\xi \circ [I + f]^{-1}), \end{aligned}} \tag{6.43}$$

$$d^2 J(\Omega_f; \theta; \xi) = d^2 J_\Omega(f; \theta \circ [I + f]; \xi \circ [I + f]) + dJ(\Omega_f; D\theta\,\xi). \tag{6.44}$$

(iii) *If V and W satisfy the conditions of Theorem 6.2 and V belongs to $\overrightarrow{\mathcal{V}}^{m+1,\ell+1}$, then*

$$\begin{aligned} &d^2 J(\Omega; V(0); W(0)) \\ &= d^2 J_\Omega(0; V(0); W(0)) + dJ(\Omega; DV(0)\,W(0)), \end{aligned} \tag{6.45}$$

$$\boxed{\begin{aligned} &d^2 J(\Omega; V; W) \\ &= d^2 J_\Omega(0; V(0); W(0)) + dJ(\Omega; V'(0) + DV(0)\,W(0)), \end{aligned}} \tag{6.46}$$

and

$$\boxed{\begin{aligned}
&d^2 J_\Omega(0; V(0); W(0)) \\
&= \frac{d}{dt} dJ(\Omega_t(W(0)); V(0) \circ T_t^{-1}(W(0)))\Big|_{t=0}.
\end{aligned}}$$

(6.47)

Proof. (i) Assume that $d^2 J_\Omega(f; \theta; \xi)$ exists and consider the differential quotient

$$q(t) \overset{\text{def}}{=} \frac{1}{t} [dJ_\Omega(f + t\xi; \theta) - dJ_\Omega(f; \theta)] \to d^2 J_\Omega(f; \theta; \xi).$$

From Theorem 3.3 (ii),

$$dJ_\Omega(f + t\xi; \theta) = dJ([I + f + t\xi](\Omega); \theta \circ [I + f + t\xi]^{-1}).$$

Define

$$T_t \overset{\text{def}}{=} [I + f + t\xi] \circ [I + f]^{-1}, \quad \mathcal{V}(t) \overset{\text{def}}{=} \theta \circ [I + f + t\xi]^{-1}.$$

From Theorem 4.1 in Chapter 7, $T_t = T_t(W_\xi)$ for the velocity field

$$W_\xi(t) \overset{\text{def}}{=} \frac{\partial T_t}{\partial t} \circ T_t^{-1} = \xi \circ [I + f + t\xi]^{-1}.$$

Therefore

$$dJ_\Omega(f + t\xi; \theta) = dJ(T_t(W_\xi)(\Omega_f); \mathcal{V}(t)),$$

$$q(t) = \frac{1}{t} [dJ(T_t(W_\xi)(\Omega_f); \mathcal{V}(t)) - dJ(\Omega_f; \mathcal{V}(0))] \to d^2(\Omega_f; \mathcal{V}; W_\xi)$$

since $q(t)$ converges as t goes to 0. The converse is obvious.

(ii) is now a direct consequence of Theorem 6.2 and the fact that

$$\mathcal{V}(t) = \theta \circ [I + f]^{-1} \circ T_t^{-1}(W_\xi)$$
$$\Rightarrow \mathcal{V}'(0) = -D(\theta \circ [I + f]^{-1}) \xi \circ [I + f]^{-1}.$$

(iii) From Theorem 6.2 and part (ii) with $f = 0$, $\theta = V(0)$ and $\xi = W(0)$. $\qquad\square$

If $D^2 J_\Omega(f)$ exists in a neighborhood of $f = 0$ and if it is continuous at $f = 0$, then $D^2 J_\Omega(0)$ is symmetrical and this completes the decomposition of the shape gradient into a symmetrical operator and the gradient applied to the first half of the Lie bracket $[V(0), W(0)]$

$$d^2 J(\Omega; V(0); W(0))$$
$$= \langle D^2 J_\Omega(0) V(0), W(0) \rangle + \langle G(\Omega), DV(0) W(0) \rangle.$$

But this is not the end of the story. From the computation of the Hessian of the domain integral (6.4) in section 6.1,

$$d^2 J_{V,W}$$

$$= \int_\Gamma \left(\frac{\partial f}{\partial n} + H\,f \right) v_n\,w_n + \boxed{f\left(D^2 b\,v_\Gamma \cdot w_\Gamma - v_\Gamma \cdot \nabla_\Gamma w_n - w_\Gamma \cdot \nabla_\Gamma v_n\right)}$$

$$+ f\,DVW \cdot n\,d\Gamma,$$

the result of the computation (6.10) in section 6.2,

$$\int_\Gamma \mathrm{div}\,(f\,V(0))\,W(0) \cdot n\,d\Gamma$$

$$= \int_\Gamma \left(\frac{\partial f}{\partial n} + H\,f \right) v_n\,w_n + \boxed{f\left(D^2 b v_\Gamma \cdot w_\Gamma - v_\Gamma \cdot \nabla_\Gamma w_n - w_\Gamma \cdot \nabla_\Gamma v_n\right)}$$

$$+ f\,DVW \cdot n\,d\Gamma,$$

and expression (6.21) of the derivative of (6.20) in section 6.3,

$$d^2 J(V;W) = \int_\Gamma Q'_{w_n}(0)\,v_n + Q(0)\,V'(0) \cdot n + \left(\frac{\partial Q(0)}{\partial n} + H\,Q(0) \right) v_n\,w_n$$

$$+ \boxed{Q(0)\left(D^2 b\,v_\Gamma \cdot w_\Gamma - v_\Gamma \cdot \nabla_\Gamma w_n - w_\Gamma \cdot \nabla_\Gamma v_n\right)}$$

$$+ Q(0)\,DVW \cdot n\,d\Gamma,$$

it is readily seen that the symmetrical term $\langle D^2 J_\Omega(0)\,V(0), W(0) \rangle$ further decomposes into a symmetrical term which only depends on the normal components v_n and w_n of $V(0)$ and $W(0)$ and another symmetrical term, which is boxed in the above expressions. This last term is the same in all expressions and only depends on the trace of $G(0)$ (here of f and $Q(0)$) on Γ and the group of terms

$$\boxed{D^2 b\,v_\Gamma \cdot w_\Gamma - v_\Gamma \cdot \nabla_\Gamma w_n - w_\Gamma \cdot \nabla_\Gamma v_n} \tag{6.48}$$

involving v_Γ, w_Γ, and tangential derivatives of v_n and w_n on Γ.

It would be interesting to further investigate this structure. In the context of a domain optimization problem, if the shape gradient is zero, both the term (6.48) and the first half of the Lie bracket will be multiplied by zero. Thus they won't contribute to the Hessian which will reduce to the symmetrical part that depends only on v_n and w_n; that is, for autonomous velocity fields V and W,

$$\int_\Gamma \left(\frac{\partial f}{\partial n} + H\,f \right) v_n\,w_n\,d\Gamma,$$

$$\int_\Gamma Q'_{w_n}(0)\,v_n + \left(\frac{\partial Q(0)}{\partial n} + H\,Q(0) \right) v_n\,w_n\,d\Gamma,$$

and the term

$$\int_\Gamma Q'_{w_n}(0)\,v_n\,d\Gamma = \int_\Gamma Q'_{v_n}(0)\,w_n\,d\Gamma$$

is symmetrical. For earlier results on the structure of the second-order shape derivative the reader is referred to Zolésio [24, p. 434] and Bucur and Zolésio [12]. Of course, the task of establishing similar results in the constrained case ($D \neq \mathbf{R}^N$) remains to be done.

Chapter 9
Shape Gradients under a State Equation Constraint

1 Introduction

When a shape functional depends on the solution of a boundary value problem defined on the underlying domain, it is said to be *constrained*. This special type of constraint is to be distinguished from constraints on the geometry such as the volume, perimeter, curvatures, etc. Using the generic terminology of *control theory* the solution of the boundary value problem will be called the *state* and its corresponding equation or inequality the *state equation* or *inequality*. The domains will be identified with the *controls*, and the constraints on the domains with the *control constraints*. Additional constraints on the solution of the state equation are usually called *state constraints*. Such problems have received a considerable amount of attention over the last half-century, and a rich and abundant literature can be found in mechanics, control theory, and optimization. Treating the state equation as an equality constraint and using *Lagrange multipliers* naturally yields a *dual variable* which is solution of the *adjoint equation* of the linearized state equation. This dual variable is called the *adjoint state*.

Of course, under appropriate differentiability assumptions with respect to the state and the control variables, necessary conditions can be obtained within the classical framework of the *calculus of variations*. Probably one of the most influential contributions of the last half of the 20th century to *optimal control theory* was made by Pontryagin et al. [1] in the 1960s, when they showed that the differentiability with respect to the control could be relaxed and replaced by the pointwise maximization with respect to the control variable of the *Hamiltonian* constructed from the objective function and the coupled state and adjoint state equations, the so-called *maximum principle*.

One of the important technical advantages of the control theory approach is to avoid the differentiation of the state with respect to the control. This is not only true for the characterization of optimal controls, but also to obtain explicit expressions of the derivative of a *state equation constrained* objective function with respect to the control. Using a Lagrangian or Hamiltonian formulation, the derivative of the constrained objective function with respect to the control is typically equal to the "partial" derivative of the Lagrangian or the Hamiltonian with respect to the control, where the adjoint variable is a solution of an appropriate adjoint state equation that is coupled with the state equation.

This chapter will concentrate on two generic examples often encountered in shape optimization. The first one is associated with the so-called *compliance problems*, where the shape functional is equal to the minimum of a domain-dependent energy functional. The special feature of such functionals is that the adjoint state coincides with the state. This obviously leads to considerable simplifications in the analysis. In that case it will be shown that theorems on the differentiability of the minimum of a functional with respect to a real parameter readily give explicit expressions of the Eulerian semiderivative even when the minimizer is not unique. The second one will deal with a shape functional which can be expressed as the saddle point of some appropriate Lagrangian. As in the first example, theorems on the differentiability of the saddle point of a functional with respect to a real parameter readily give explicit expressions of the Eulerian semiderivative even when the solution of the saddle point equations is not unique.

Avoiding the differentiation of the state equation with respect to the domain is particularly advantageous in shape problems. Here the state variable (the solution of the boundary value problem) lives in a function space (Banach, Hilbert, Sobolev, etc.) which depends on the control (*underlying variable domain*)! Thus the notion of derivative of the state with respect to the domain is more delicate. In this chapter two techniques will be presented to get around this difficulty: *function space parametrization* and *function space embedding*.

Function space parametrization consists in transporting the functions on variable domains back onto the initial domain, where they can be compared. It is related to the notion of *material derivative* in continuum and structural mechanics. This will be illustrated by computing the volume and boundary integral expressions of the Eulerian semiderivative of a simple shape functional for the homogeneous Dirichlet and Neumann boundary value problems by the theorems on the differentiation of a minimum and a saddle point.

Function space embedding consists in constructing extensions of the domain-dependent boundary value problems to a larger fixed domain. For homogeneous Dirichlet boundary conditions, it is sufficient to consider extensions by zero to $\mathbf{R}^N$. The nonhomogeneous case requires the introduction of a multiplier to take into account the boundary condition. This technique will be illustrated by computing the boundary integral expression of the Eulerian semiderivative of a simple shape functional for the nonhomogeneous Dirichlet boundary value problem by the theorems on the differentiation of a saddle point.

In addition to the generic example, the theorem on the differentiation of an infimum with respect to a parameter will be applied to the example of the buckling of columns considered in section 4 of Chapter 3. In section 3 an explicit expression of the semiderivative of *Euler's buckling load* with respect to the cross-sectional area will be given. A necessary and sufficient condition will also be given to characterize the maximum Euler's buckling load with respect to a family of cross-sectional areas. The theory is further illustrated in section 4 by providing the semiderivative of the first eigenvalue of several boundary problems over a bounded open domain: Laplace equation, bi-Laplace equation, linear elasticity. In general the first eigenvalue is not simple over an arbitrary bounded open domain and the eigenvalue is not differentiable; yet the main theorem provides explicit domain expressions for a bounded open domain and boundary expressions of the semiderivatives for a sufficiently smooth bounded open domain.

2 Min Formulation

2.1 An Illustrative Example and a Shape Variational Principle

Let Ω be a bounded open domain in $\mathbf{R}^N$ with a smooth boundary Γ. Let $y = y(\Omega)$ be the solution of the Dirichlet problem

$$-\triangle y = f \text{ in } \Omega, \quad y = 0 \text{ on } \Gamma, \tag{2.1}$$

where f is a fixed function in $H^1(\mathbf{R}^N)$. The solution of (2.1) is the minimizing element in $H_0^1(\Omega)$ of the energy functional

$$E(\Omega, \varphi) = \int_\Omega \left[\frac{1}{2}|\nabla\varphi|^2 - f\varphi \right] dx. \tag{2.2}$$

Introduce the shape function

$$J(\Omega) \overset{\text{def}}{=} \inf_{\varphi \in H_0^1(\Omega)} E(\Omega, \varphi) = -\int_\Omega \frac{1}{2}|\nabla y|^2 \, dx. \tag{2.3}$$

We want to show that

$$dJ(\Omega; V) = -\frac{1}{2} \int_\Gamma \left| \frac{\partial y}{\partial n} \right|^2 V \cdot n \, d\Gamma. \tag{2.4}$$

This example is the prototype of a *free boundary* problem[1] which can be obtained from the following *shape variational* principle:

$$dJ(\Omega; V) = 0 \quad \forall V. \tag{2.5}$$

It yields the extra boundary condition[2]

$$\frac{\partial y}{\partial n} = 0 \text{ on } \Gamma. \tag{2.6}$$

Equations (2.1) and (2.6) characterize a free boundary problem. It is the simplest example of a large family of problems where the shape function is an extremal of a natural internal energy (cf. section 6 of Chapter 3). They correspond to the so-called "compliance problems" in elasticity theory. They also occur in fracture theory and image segmentation. The first-order variation of this shape functional yields the extra boundary condition, which is characteristic of a free boundary problem.

---

[1] Usually with other constraints, such as a volume constraint on Ω and/or an area constraint on Γ.

[2] With a volume equality constraint we would get that $|\partial y/\partial n|^2$ is equal to a constant on Γ.

2.2 Function Space Parametrization

To compute the first-order derivative of $J(\Omega)$ we perturb the bounded open domain Ω by a velocity field V, which generates the family of transformations $\{T_t : 0 \leq t \leq \tau\}$ of $\mathbf{R}^N$ and the family of domains $\{\Omega_t = T_t(\Omega) : 0 \leq t \leq \tau\}$. At t,

$$J(\Omega_t) = \inf_{\varphi \in H_0^1(\Omega_t)} E(\Omega_t, \varphi) \tag{2.7}$$

and the minimizing element $y_t = y(\Omega_t)$ is the solution of the Dirichlet problem

$$-\triangle y_t = f \text{ in } \Omega_t, \quad y_t = 0 \text{ on } \Gamma_t, \tag{2.8}$$

where Γ_t is the boundary of Ω_t. We want to compute the derivative

$$dj(0) = \lim_{t \searrow 0} \frac{j(t) - j(0)}{t} \tag{2.9}$$

of the function

$$j(t) \stackrel{\text{def}}{=} J(\Omega_t). \tag{2.10}$$

We need a theorem that would give the derivative of a Min with respect to a parameter $t \geq 0$ at $t = 0$. The difficulty here is that the function space $H^1(\Omega_t)$ depends on the parameter t. To get around this and obtain a Min with respect to a function space that is independent of t, introduce the following parametrization:

$$H_0^1(\Omega_t) = \left\{ \varphi \circ T_t^{-1} : \varphi \in H_0^1(\Omega) \right\}. \tag{2.11}$$

Notice that since T_t is a homeomorphism, it transforms the open domain Ω into the open domain Ω_t and sends the boundary Γ of Ω onto the boundary Γ_t of Ω_t. In particular, when V is sufficiently smooth for all φ in $H_0^1(\Omega)$, $\varphi \circ T_t^{-1} \in H_0^1(\Omega_t)$, and conversely, for all ψ in $H_0^1(\Omega_t)$, $\psi \circ T_t \in H_0^1(\Omega)$. This parametrization does not affect the value of the minimum $J(\Omega_t)$ but changes the functional E:

$$J(\Omega_t) = \inf_{\varphi \in H_0^1(\Omega)} E(T_t(\Omega), \varphi \circ T_t^{-1}). \tag{2.12}$$

This parametrization is typical in *shape analysis*. It amounts to introducing the new energy functional

$$\widetilde{E}(t, \varphi) = E(T_t(\Omega), \varphi \circ T_t^{-1}), \quad \varphi \in H_0^1(\Omega). \tag{2.13}$$

Our objective in the next section is to compute the limit (2.9) for

$$j(t) = \inf_{\varphi \in H_0^1(\Omega)} \widetilde{E}(t, \varphi). \tag{2.14}$$

This will be done. Before closing, it is interesting to characterize the minimizing element y^t in $H_0^1(\Omega)$ of

$$\widetilde{E}(t, \varphi) = \int_{\Omega_t} \left[\frac{1}{2} |\nabla(\varphi \circ T_t^{-1})|^2 - f(\varphi \circ T_t^{-1}) \right] dx \tag{2.15}$$

which is the solution of the following variational equation: to find $y^t \in H_0^1(\Omega)$ such that for all $\varphi \in H_0^1(\Omega)$

$$\int_{\Omega_t} \left\{ \nabla(y^t \circ T_t^{-1}) \cdot \nabla(\varphi \circ T_t^{-1}) - f(\varphi \circ T_t^{-1}) \right\} dx = 0. \qquad (2.16)$$

Compare this expression with the characterization of the minimizing element y_t of $E(\Omega_t, \varphi)$ on $H_0^1(\Omega_t)$: to find $y_t \in H_0^1(\Omega_t)$ such that for all $\varphi \in H_0^1(\Omega_t)$

$$\int_{\Omega_t} \left\{ \nabla y_t \cdot \nabla \varphi - f\varphi \right\} dx = 0. \qquad (2.17)$$

It is easy to verify that

$$y_t = y^t \circ T_t^{-1} \text{ and } y^t = y_t \circ T_t. \qquad (2.18)$$

So y^t is the solution y_t of (2.8) transported back onto the fixed domain Ω by the change of variable induced by T_t.

In view of this expression (2.15) can be rewritten on the fixed domain Ω

$$\widetilde{E}(t, \varphi) = \int_\Omega \frac{1}{2} \left(A(t)\nabla\varphi \right) \cdot \nabla\varphi - (f \circ T_t)\, \varphi\, J_t\, dx, \qquad (2.19)$$

where for t in $[0, \tau]$ small

$$DT_t = \text{Jacobian matrix of } T_t \qquad (2.20)$$
$$J_t = |\det DT_t| = \det DT_t \text{ for } t \geq 0 \text{ small} \qquad (2.21)$$
$$A(t) = J_t [DT_t]^{-1} \, {}^*[DT_t]^{-1}. \qquad (2.22)$$

With this change of variable y^t is now characterized by the variational equation

$$\begin{cases} y^t \in H_0^1(\Omega) \text{ and } \forall \varphi \in H_0^1(\Omega) \\ \int_\Omega A(t)\nabla y^t \cdot \nabla\varphi - J_t(f \circ T_t)\varphi\, dx = 0. \end{cases} \qquad (2.23)$$

2.3 Differentiability of a Minimum with Respect to a Parameter

Consider a functional

$$G \colon [0, \tau] \times X \to \mathbf{R} \qquad (2.24)$$

for some $\tau > 0$ and some set X. For each t in $[0, \tau]$ define

$$g(t) \overset{\text{def}}{=} \inf\{G(t, x) : x \in X\}, \qquad (2.25)$$
$$X(t) \overset{\text{def}}{=} \{x \in X : G(t, x) = g(t)\}. \qquad (2.26)$$

The objective is to characterize the limit

$$dg(0) \overset{\text{def}}{=} \lim_{t \searrow 0} \frac{g(t) - g(0)}{t} \qquad (2.27)$$

when $X(t)$ is not empty for $0 \le t \le \tau$.

When $X(t) = \{x^t\}$ is a singleton, $0 \le t \le \tau$, and the derivative

$$\dot{x} = \lim_{t \searrow 0} \frac{x^t - x^0}{t} \tag{2.28}$$

of x is known, then it is easy to obtain $dg(0)$ under appropriate differentiability of the functional G with respect to t and x. When $\dot{x}$ is not readily available or when the sets $X(t)$ are not singletons, this direct approach fails or becomes very intricate. In this section we present a theorem that gives an explicit expression for $dg(0)$, the derivative of the Min of the functional G with respect to t at $t = 0$. Its originality is that the differentiability of x^t is replaced by a continuity assumption on the set-valued function and the existence of the partial derivative of the functional G with respect to the parameter t. In other words this technique does not require a priori knowledge of the derivative $\dot{x}$ of the minimizing elements x^t with respect to t.

Theorem 2.1. *Let X be an arbitrary set, $\tau > 0$, and $G \colon [0,\tau] \times X \to \mathbf{R}$ a well-defined functional. Assume that the following conditions are satisfied:*

(H1) *For all $t \in [0,\tau]$, $X(t) \ne \varnothing$;*

(H2) *for all x in $\bigcup_{t \in [0,\tau]} X(t)$, $\partial_t G(t,x)$ exists everywhere in $[0,\tau]$;*

(H3) *there exists a topology $\mathcal{T}_X$ on X such that for any sequence $\{t_n\} \subset]0,\tau]$, $t_n \to t_0 = 0$, $\exists x_0 \in X(0)$, $\exists$ a subsequence $\{t_{n_k}\}$ of $\{t_n\}$, and for each $k \ge 1$, $\exists x_{n_k} \in X(t_{n_k})$ such that*

    (i) *$x_{n_k} \to x^0$ in the $\mathcal{T}_X$-topology*

    (ii) *and*

$$\liminf_{\substack{k \to \infty \\ t \searrow 0}} \partial_t G(t, x_{n_k}) \ge \partial_t G(0, x^0);$$

(H4) *for all x in $X(0)$, the map $t \mapsto \partial_t G(t,x)$ is upper semicontinuous at $t = 0$.*

Then there exists $x^0 \in X(0)$ such that

$$dg(0) = \lim_{t \searrow 0} \frac{g(t) - g(0)}{t} = \inf_{x \in X(0)} \partial_t G(0,x) = \partial_t G(0, x^0). \tag{2.29}$$

Remark 2.1.
In the literature condition (H3)(i) is known as *sequential semicontinuity for set-valued functions*. When $X(0)$ is a singleton $\{x_0\}$ we readily get $dg(0) = \partial_t G(0, x_0)$.
$\qquad\square$

Remark 2.2.
This theorem, and in particular the last part of property (2.29), extends a former result by Lemaire [1, Thm. 2.1, p. 38], where sequential compactness of the set X was assumed. It also completes and extends Theorem 1 in Zolésio [7] and Delfour and Zolésio [3].
$\qquad\square$

Proof of Theorem 2.1. (i) We first establish upper and lower bounds to the differential quotient

$$\frac{\Delta(t)}{t}, \quad \Delta(t) \stackrel{\text{def}}{=} g(t) - g(0).$$

Choose arbitrary x_0 in $X(0)$ and x_t in $X(t)$. Then, by definition,

$$G(t, x_t) = g(t) \le G(t, x_0),$$
$$-G(0, x_t) \le -g(0) = -G(0, x_0).$$

Add up the above two inequalities to obtain

$$G(t, x_t) - G(0, x_t) \le \Delta(t) \le G(t, x_0) - G(0, x_0).$$

By assumption (H2), there exist θ_t, $0 < \theta_t < 1$, and α_t, $0 < \alpha_t < 1$, such that

$$G(t, x_t) - G(0, x_t) = t\partial_t G(\theta_t t, x_t),$$
$$G(t, x_0) - G(0, x_0) = t\partial_t G(\alpha_t t, x_0),$$

and by dividing by $t > 0$

$$\partial_t G(\theta_t t, x_t) \le \frac{\Delta(t)}{t} \le \partial_t G(\alpha_t t, x_0). \tag{2.30}$$

(ii) Define

$$\underline{dg}(0) = \liminf_{t \searrow 0} \frac{\Delta(t)}{t}, \quad \overline{dg}(0) = \limsup_{t \searrow 0} \frac{\Delta(t)}{t}.$$

There exists a sequence $\{t_n : 0 < t_n \le \tau\}$, $t_n \to 0$, such that

$$\lim_{n \to \infty} \frac{\Delta(t_n)}{t_n} = \underline{dg}(0).$$

By assumption (H3), $\exists x^0 \in X(0)$, $\exists$ a subsequence $\{t_{n_k}\}$ of $\{t_n\}$, and for each $k \ge 1$, $\exists x_{n_k} \in X(t_{n_k})$ such that $x_{n_k} \to x^0$ in $\mathcal{T}_X$ and

$$\liminf_{\substack{t \searrow 0 \\ k \to \infty}} \partial_t G(t, x_{n_k}) \ge \partial_t G(0, x^0).$$

So, from the first part of the estimate (2.30) for $t = t_{n_k}$,

$$\partial_t G(\theta_{t_{n_k}} t_{n_k}, x_{n_k}) \le \frac{\Delta(t_{n_k})}{t_{n_k}}$$

and

$$\partial_t G(0, x^0) \le \liminf_{k \to \infty} \partial_t G(\theta_{t_{n_k}} t_{n_k}, x_{n_k}) \le \lim_{k \to \infty} \frac{\Delta(t_{n_k})}{t_{n_k}} = \underline{dg}(0).$$

Therefore,

$$\exists x^0 \in X(0), \quad \partial_t G(0, x^0) \leq \underline{dg}(0),$$

and

$$\inf_{x \in X(0)} \partial_t G(0, x) \leq \partial_t G(0, x^0) \leq \underline{dg}(0).$$

From the second part of (2.30) and assumption (H4) we also obtain

$$\forall x \in X(0), \quad \partial_t G(0, x) \geq \overline{dg}(0),$$

$$\overline{dg}(0) \leq \inf_{x \in X(0)} \partial_t G(0, x), \tag{2.31}$$

and necessarily

$$\inf_{x \in X(0)} \partial_t G(0, x) = \underline{dg}(0) = \overline{dg}(0) = \inf_{x \in X(0)} \partial_t G(0, x).$$

In particular, from (2.3) and (2.31)

$$\partial_t G(0, x^0) = dg(0) = \inf_{x \in X(0)} \partial_t G(0, x)$$

and x^0 is a minimizing point of $\partial_t G(0, \cdot)$. $\square$

2.4   Application of the Theorem

Our example has a unique minimizing point y^t for $t \geq 0$ small. Here $X = H_0^1(\Omega)$, $X(t) = \{y^t\}$, and it is sufficient to establish the continuity of the map $t \mapsto y^t$ at $t = 0$ for an appropriate topology on $H_0^1(\Omega)$.

We now check assumptions (H1) to (H4). Assume that V belongs to $C^0\big([0, \tau];$ $\mathcal{D}^2(\mathbf{R}^N, \mathbf{R}^N)\big)$ and that $f \in H^1(\mathbf{R}^N)$. Choose $\tau > 0$ small enough such that

$$J_t = |J_t|, \quad 0 \leq t \leq \tau, \tag{2.32}$$

and that there exist constants $0 < \alpha < \beta$ such that

$$\forall \xi \in \mathbf{R}^N, \quad \alpha |\xi|^2 \leq A(t)\, \xi \cdot \xi \leq \beta |\xi|^2, \text{ and } \alpha \leq J_t \leq \beta. \tag{2.33}$$

Since the bilinear form associated with (2.23) is coercive, there exists a unique solution y^t to (2.23) and

$$\forall t \in [0, \tau], \quad X(t) = \{y^t\} \neq \varnothing. \tag{2.34}$$

So assumption (H1) is satisfied. To check (H2) use expression (2.19) and compute

$$\partial_t \widetilde{E}(t, \varphi) = \int_\Omega \left\{ \frac{1}{2} [A'(t)\nabla\varphi] \cdot \nabla\varphi - [\operatorname{div} V_t \, (f \circ T_t) + J_t \nabla f \cdot V_t] \, \varphi \right\} dx, \tag{2.35}$$

where

$$V_t(X) = V(t, T_t(X)), \quad DV_t(X) = DV(t, T_t(X)), \tag{2.36}$$

$$A'(t) = (\operatorname{div} V_t)I - (DV_t)^* - DV_t. \tag{2.37}$$

By assumptions on V and f, $\partial_t \widetilde{E}(t, \varphi)$ exists everywhere in $[0, \tau]$ for all φ in $H_0^1(\Omega)$ and assumption (H2) is satisfied.

To check assumption (H3)(i) we first show that $\{y^t\}$ is bounded in $H_0^1(\Omega)$. From (2.33)

$$\alpha \|\nabla y^t\|_{L^2(\Omega)}^2 \leq \int_\Omega A(t) \nabla y^t \cdot \nabla y^t \, dx$$
$$= \int_\Omega J_t(f \circ T_t) \, y^t \, dx \leq \|J_t \, f \circ T_t\|_{L^2(\Omega)} \|y^t\|_{L^2(\Omega)}. \tag{2.38}$$

By using the norm

$$\|\varphi\|_{H_0^1(\Omega)} = \|\nabla \varphi\|_{L^2(\Omega)}$$

and the continuous injection of $H_0^1(\Omega)$ into $L^2(\Omega)$,

$$\exists c > 0, \quad \|\varphi\|_{L^2(\Omega)} \leq c \|\varphi\|_{H_0^1(\Omega)}.$$

So from (2.38)

$$\|y^t\|_{H_0^1(\Omega)} \leq \frac{c}{\alpha} \|J_t f \circ T_t\|_{L^2(\Omega)}. \tag{2.39}$$

But $J_t \to 1$ as $t \to 0$ and $f \circ T_t \to f$ in $L^2(\Omega)$ by the following lemma.

Lemma 2.1. *Assume that $V \in C^0\big([0, \tau]; \mathcal{D}^1(\mathbf{R}^N, \mathbf{R}^N)\big)$ satisfies assumption (V) of Chapter 7 and that $f \in L^2(\mathbf{R}^N)$. Then*

$$\lim_{t \searrow 0} f \circ T_t = f \quad \text{and} \quad \lim_{t \searrow 0} f \circ T_t^{-1} = f \quad \text{in } L^2(\mathbf{R}^N). \tag{2.40}$$

So by (2.39) y^t is bounded:

$$\exists c > 0, \quad \sup_{t \in [0, \tau]} \|y^t\|_{H_0^1(\Omega)} \leq c. \tag{2.41}$$

The next step is to prove the continuity by subtracting (2.23) at $t > 0$ from (2.23) at $t = 0$:

$$\int_\Omega \nabla y^t \cdot \nabla \varphi \, dx + \int_\Omega (A(t) - I) \nabla y^t \cdot \nabla \varphi \, dx$$
$$= \int_\Omega f \varphi \, dx + \int_\Omega [J_t(f \circ T_t) - f] \varphi \, dx,$$
$$\int_\Omega \nabla y \cdot \nabla \varphi \, dx = \int_\Omega f \varphi \, dx.$$

Subtract and set $\varphi = y^t - y$:

$$\int_\Omega |\nabla(y^t - y)|^2 \, dx$$

$$= - \int_\Omega (A(t) - I) \nabla y^t \cdot \nabla(y^t - y) + [J_t(f \circ T_t) - f](y^t - y) \, dx$$

$$\leq |A(t) - I| \|\nabla y^t\|_{L^2(\Omega)} \|\nabla(y^t - y)\|_{L^2(\Omega)}$$
$$+ \|J_t \, f \circ T_t - f\|_{L^2(\Omega)} \|y^t - y\|_{L^2(\Omega)}$$

and

$$\|y^t - y\|_{H_0^1(\Omega)} \le c\,\{|A(t) - I| + \|J_t\, f \circ T_t - f\|_{L^2(\Omega)}\}.$$

But $A(t) - I \to 0$ and $J_t\, f \circ T_t \to f$ in $L^2(\Omega)$, and finally $y^t \to y$ in $H_0^1(\Omega)$. So assumption (H3)(i) is satisfied for $H_0^1(\Omega)$-strong.

For assumption (H3)(ii) we have

$$\partial_t G(t, \varphi) = \int_\Omega \left\{ \frac{1}{2}\,[A'(t)\nabla\varphi] \cdot \nabla\varphi - [\operatorname{div} V_t(f \circ T_t) + J_t \nabla f \circ V_t]\,\varphi \right\}\, dx$$

and for φ in $H_0^1(\Omega)$, f in $H^1(\mathbf{R}^N)$, and V in $C^0([0, \tau]; \mathcal{D}^2(\mathbf{R}^N, \mathbf{R}^N))$

$$\partial_t G(t, \varphi) - \partial_t G(0, y)$$
$$= \int_\Omega \left\{ \frac{1}{2}(A'(t) - A'(0))\nabla\varphi \cdot \nabla\varphi \right.$$
$$\left. - [\operatorname{div} V_t(f \circ T_t) - J_t \nabla f \cdot V_t - \operatorname{div} V(0)f + \nabla f \cdot V(0)]\,\varphi \right\}\, dx$$
$$+ \int_\Omega \frac{1}{2}\,\{A'(0)\nabla\varphi \cdot \nabla\varphi - A'(0)\nabla y \cdot \nabla y\}\, dx.$$

As φ goes to y in $H_0^1(\Omega)$ and $t \to 0$, the first term converges to zero since φ is bounded and

$$A'(t) \to A'(0),$$
$$\operatorname{div} V_t(f \circ T_t) \to \operatorname{div} V(0)f \text{ in } L^2(\Omega),$$
$$J_t \nabla f \cdot V_t \to \nabla f \cdot V(0) \text{ in } L^2(\Omega).$$

The second term is continuous with respect to φ in $H_0^1(\Omega)$ and goes to zero as $\varphi \to y$ in $H_0^1(\Omega)$. So assumption (H3)(ii) is satisfied.

Finally, (H4) is satisfied since

$$t \mapsto \partial_t \widetilde{E}(t, y) = \int_\Omega \left\{ \frac{1}{2} A'(t)\nabla y \cdot \nabla y - [\operatorname{div} V_t f + \nabla f \cdot V_t]\, y \right\}\, dx \qquad (2.42)$$

is continuous in $[0, \tau]$. So all the assumptions of Theorem 2.1 are satisfied and for V in $C^0([0, \tau]; \mathcal{D}^2(\mathbf{R}^N, \mathbf{R}^N))$ and f in $H^1(\mathbf{R}^N)$,

$$dJ(\Omega; V) = \int_\Omega \left\{ \frac{1}{2} A'(0)\nabla y \cdot \nabla y - [\operatorname{div} V(0)f + \nabla f \cdot V(0)]\, y \right\}\, dx. \qquad (2.43)$$

For V autonomous, expression (2.43) is continuous with respect to the space $\mathcal{D}^1(\mathbf{R}^N, \mathbf{R}^N)$, and the shape gradient is of order 1. We know by the structure theorem, Theorem 3.5 of Chapter 8, that for Ω open and a C^2-boundary Γ,

$$\exists g(\Gamma) \in \mathcal{D}^1(\Gamma)', \quad dJ(\Omega; V) = \langle g(\Gamma), V \rangle_{\mathcal{D}^1(\Gamma)}.$$

The characterization of $g(\Gamma)$ will be given in the next section.

We complete this section with the proof of Lemma 2.1.

Proof of Lemma 2.1. (i) By density of $\mathcal{D}^1(\mathbf{R}^N)$ in $L^2(\mathbf{R}^N)$ for all $\varepsilon > 0$, there exists f_ε such that

$$\|f - f_\varepsilon\|_{L^2} < \frac{\varepsilon}{\max\{J_t^{-1} : 0 \le t \le \tau\}} \qquad (\le \varepsilon\alpha \le \varepsilon).$$

Hence

$$\|f \circ T_t - f\| \le \|f_\varepsilon \circ T_t - f_\varepsilon\| + \|f \circ T_t - f_\varepsilon \circ T_t\| + \|f - f_\varepsilon\|.$$

The last term is less than ε and the middle term can be rewritten after a change of variable

$$\int_{\mathbf{R}^N} |f \circ T_t - f_\varepsilon \circ T_t|^2 dx = \int_{\mathbf{R}^N} |f - f_\varepsilon|^2 J_t^{-1}\, dx \le \varepsilon^2.$$

A function f_ε in $\mathcal{D}^1(\mathbf{R}^N)$ has a compact support and is uniformly Lipschitz continuous, that is,

$$\exists c > 0, \forall x, y \in \mathbf{R}^N, \quad |f_\varepsilon(y) - f_\varepsilon(x)| \le c|y - x|.$$

Thus for all X in $\mathbf{R}^N$

$$|f_\varepsilon(T_t(X)) - f_\varepsilon(X)| \le c|T_t(X) - X|. \tag{2.44}$$

Now

$$T_t(X) = X + \int_0^t V(s, X)\, ds + \int_0^t [V(s, T_s(X)) - V(s, X)]\, ds.$$

Since $V \in C^0([0, \tau]; \mathcal{D}^1(\mathbf{R}^N, \mathbf{R}^N))$ is also uniformly Lipschitz continuous by assumption (V)

$$\exists c > 0, \forall (X, Y), \forall s \in [0, \tau], \quad |V(s, Y) - V(s, X)| \le c\,|Y - X|,$$

and for all t in $[0, \tau]$

$$|T_t(X) - X| \le t\|V(\cdot, X)\|_{C^0([0,\tau];\mathbf{R}^N)} + \int_0^t c|T_s(X) - X|\, ds.$$

It is now easy to verify that there exists a constant $c > 0$ such that

$$\max_{s \in [0,t]} |T_s(X) - X| \le ct\|V(\cdot, X)\|_{C^0([0,\tau];\mathbf{R}^N)}. \tag{2.45}$$

Finally, in view of (2.44) and (2.45) the term

$$\int_{\mathbf{R}^N} |f_\varepsilon(T_t(X)) - f_\varepsilon(X)|^2 dX \le ct^2 \int_{\mathbf{R}^N} \|V(\cdot, X)\|_{C^0}^2\, dX,$$

$$\int_{\mathbf{R}^N} |f_\varepsilon(T_t(X)) - f_\varepsilon(X)|^2 dX = \int_{K_\varepsilon} |f_\varepsilon(T_t(X)) - f_\varepsilon(X)|^2\, dX$$

$$\le c\,t^2 \int_{K_\varepsilon} \|V(\cdot, X)\|_{C^0}^2\, dX \le c't^2.$$

So for t small enough the right-hand side of (2.4) is less than 3ε and this completes the proof of the first part of (2.41).

(ii) The second part of (2.41) can be obtained by the following change of variable:

$$\int_{\mathbf{R}^N} |f \circ T_t^{-1} - f|^2 \, dx = \int_{\mathbf{R}^N} |f - f \circ T_t|^2 J_t^{-1} \, dX$$

and the fact that $\beta^{-1} < J_t^{-1} < \alpha^{-1}$. This completes the proof of the lemma. $\square$

2.5 Domain and Boundary Integral Expressions of the Shape Gradient

Expression (2.43) for the shape gradient is a volume (or domain) integral, and it is easy to check that the map

$$V \mapsto dJ(\Omega; V) : \vec{\mathcal{V}}^{0,1} \to \mathbf{R} \tag{2.46}$$

is linear and continuous (cf. section 3.3 in Chapter 8). So by Corollary 1 to the structure theorem, Theorem 3.5 in Chapter 8, we know that for a domain Ω with a C^2-boundary Γ there exists a scalar distribution $g(\Gamma)$ in $\mathcal{D}(\Gamma)'$ such that

$$dJ(\Omega; V) = \langle g(\Gamma), V(0) \cdot n \rangle_{\mathcal{D}^1(\Gamma)}. \tag{2.47}$$

The next objective is to further characterize the boundary expression. Recall that we have assumed that $f \in H^1(\mathbf{R}^N)$. So for a C^2-boundary Γ the solution $y = y(\Omega)$ of the problem

$$-\triangle y = f \text{ in } \Omega, \quad y = 0 \text{ on } \Gamma$$

belongs to $H^2(\Omega)$. For velocity fields V in $C^0([0,\tau]; \mathcal{D}^1(\mathbf{R}^N, \mathbf{R}^N))$ satisfying assumption (V) the transported solution y^t in $H^2(\Omega)$ is the solution of the system

$$-\text{div}\,(A(t)\nabla y^t) = J_t f \circ T_t \text{ in } \Omega, \quad y^t = 0 \text{ on } \Gamma. \tag{2.48}$$

Knowing that for all t in $[0, \tau]$, $y^t \in H^2(\Omega) \cap H_0^1(\Omega)$, we can repeat the computation of $\partial_t \widetilde{E}(t, \varphi)$ for φ in $H^2(\Omega) \cap H_0^1(\Omega)$ instead of $H_0^1(\Omega)$. With this extra smoothness we can use the formula

$$\left. \frac{d}{dt} \int_{\Omega_t} F(t, x) \, dx \right|_{t=0} = \int_\Gamma F(0, x) V(0) \cdot n \, d\Gamma + \int_\Omega \frac{\partial F}{\partial t}(0, x) \, dx \tag{2.49}$$

for a sufficiently smooth function $F : [0, \tau] \times \mathbf{R}^N \to \mathbf{R}$. Prior to applying this formula to expression (2.15) in section 2.2, notice that for φ in $H^2(\Omega)$,

$$\dot{\varphi} = \left. \frac{d}{dt} \varphi \circ T_t^{-1} \right|_{t=0} = -\nabla \varphi \cdot V(0) \in H^1(\Omega), \tag{2.50}$$

but it generally does not belong to $H_0^1(\Omega)$. Then

$$\partial_t \widetilde{E}(t,\varphi)\big|_{t=0} = \int_\Gamma \left\{ \frac{1}{2}|\nabla\varphi|^2 - f\varphi \right\} V(0) \cdot n \, d\Gamma + \int_\Omega \{\nabla\varphi \cdot \nabla\dot\varphi - f\dot\varphi\} \, dx. \quad (2.51)$$

Substitute $\varphi = y$ in (2.51):

$$\partial_t \widetilde{E}(t,y)\big|_{t=0} = \int_\Gamma \left\{ \frac{1}{2}|\nabla y|^2 - fy \right\} V(0) \cdot n \, d\Gamma \\ - \int_\Omega \{\nabla y \cdot \nabla(\nabla y \cdot V(0)) - f(\nabla y \cdot V(0))\} \, dx. \quad (2.52)$$

But $y \in H^2(\Omega) \cap H_0^1(\Omega)$ and

$$\int_\Omega \nabla y \cdot \nabla(\nabla y \cdot V(0)) \, dx = - \int_\Omega \triangle y \, \nabla y \cdot V(0) \, dx + \int_\Gamma \frac{\partial y}{\partial n} \nabla y \cdot V(0) \, d\Gamma$$

and

$$\partial_t \widetilde{E}(t,y)\big|_{t=0} = \int_\Gamma \left\{ \left[\frac{1}{2}|\nabla y|^2 - fy \right] V(0) \cdot n - \frac{\partial y}{\partial n} \nabla y \cdot V(0) \right\} d\Gamma. \quad (2.53)$$

But

$$y = 0 \text{ on } \Gamma \;\Rightarrow\; \nabla y = \frac{\partial y}{\partial n} n \text{ on } \Gamma,$$

and finally

$$\partial_t \widetilde{E}(t,y)\big|_{t=0} = - \int_\Gamma \frac{1}{2} \left| \frac{\partial y}{\partial n} \right|^2 V(0) \cdot n \, d\Gamma. \quad (2.54)$$

This is the boundary expression that is continuous for $V(0)$ in the space $\mathcal{D}^0(\mathbf{R}^N, \mathbf{R}^N)$. It has been obtained via a parametrization of the function space appearing in the Min formulation. Thus

$$dJ(\Omega; V) = - \int_\Omega \frac{1}{2} \left| \frac{\partial y}{\partial n} \right|^2 V(0) \cdot n \, d\Gamma, \quad (2.55)$$

as predicted in section 2.1.

Remark 2.3.
An original application of the computations made for the generic example can be found in Delfour, Payre, and Zolésio [3]. In that paper the derivative of the energy function with respect to the nodes in a P^1 finite element approximation is obtained from the volume expression of the shape semiderivative of the continuous problem. This technique also applies to a broad class of boundary value problems and to mixed hybrid finite element approximations. As observed by several authors, the boundary integral expression is not suitable since the finite element solution does not have the appropriate smoothness under which the boundary integral formula is obtained. This technique is used to obtain a triangularization which minimizes

the approximation error of the solution. Other formulae of the same type have been obtained for mixed finite element approximations in Delfour, Mghzali, and Zolésio [1] and several other boundary value problems. The reader is also referred to Delfour, Payre, and Zolésio [1, 2, 4, 5] for application of those techniques to thermal problems such as diffusers and space radiators. The last paper combines the above techniques with the systematic handling of parametrized geometries. $\square$

3 Buckling of Columns

Auchmuty's dual principle was used for the functional associated with the optimal design of a column against buckling in section 4 of Chapter 3. In this section we first use this construction to compute the directional semiderivative with respect to the cross-sectional area and then use it to give a necessary and sufficient condition that characterizes the maximizers. Indeed, recall the identity

$$\mu(A) = -\frac{1}{2\lambda(A)}$$

and notice that A is a maximizer of λ over $\mathcal{A}$ if and only if A is a maximizer of μ over $\mathcal{A}$.

We have seen in Theorem 4.3 of section 4 of Chapter 3 that the concave upper semicontinuous functional

$$\mu(A) \overset{\text{def}}{=} \inf \left\{ L(A,v) \,:\, v \in H_0^2(0,1) \right\}, \tag{3.1}$$

$$L(A,v) \overset{\text{def}}{=} \frac{1}{2} \int_0^1 A\,|v''|^2\,dx - \left[\int_0^1 |v'|^2\,dx \right]^{1/2} \tag{3.2}$$

has maximizers over the weakly compact convex set $\mathcal{A}$. Therefore, if $\mu(A)$ has a directional semiderivative $d\mu(A; B)$, a maximizer is completely characterized by

$$\exists A \in \mathcal{A}, \ \forall B \in \mathcal{A}, \quad d\mu(A; B - A) \le 0.$$

This directional semiderivative always exists for concave continuous functions, and Theorem 2.1 can be used to get an explicit expression of $d\mu(A; B)$.

Given $A \in \mathcal{A}$, $B \in L^\infty(0,1)$, and $t \ge 0$, let

$$A_t \overset{\text{def}}{=} A + tB.$$

In view of the fact that $A \ge A_0 > 0$, there exists $\tau > 0$ such that for all $0 \le t \le \tau$

$$\forall t, \ 0 \le t \le \tau, \quad A_t \ge A_0/2.$$

Define

$$\tilde{L}(t,v) \overset{\text{def}}{=} \frac{1}{2} \int_0^1 A_t\,|v''|^2\,dx - \left[\int_0^1 |v'|^2\,dx \right]^{1/2}$$

$$\Rightarrow \mu(A_t) = \inf \left\{ \tilde{L}(t,v) \,:\, v \in H_0^2(0,1) \right\}, \quad X(t) \overset{\text{def}}{=} E(A_t).$$

Theorem 3.1. *Let B be an arbitrary function in $L^\infty(0,1)$ and A an element of $\mathcal{A}$.*

(i) *The directional semiderivative at A in the direction B is given by the following expression:*

$$d\mu(A; B) = \inf_{v \in E(A)} \frac{1}{2} \int_0^1 B\,|v''|^2\,dx, \tag{3.3}$$

and since μ is concave and continuous, $d_H\mu(A; B)$ also exists.

(ii) *The maximizing elements of $\mu(A)$ or $\lambda(A)$ in $\mathcal{A}$ are completely characterized by the variational inequality*

$$\exists A \in \mathcal{A},\ \forall B \in \mathcal{A}, \quad \inf_{v \in E(A)} \frac{1}{2} \int_0^1 (B - A)\,|v''|^2\,dx \le 0 \tag{3.4}$$

or equivalently

$$\exists A \in \mathcal{A},\ \forall B \in \mathcal{A}, \quad \inf_{v \in E(A)} \frac{1}{2} \int_0^1 B\,|v''|^2\,dx \le \frac{1}{2\lambda(A)} = -\mu(A). \tag{3.5}$$

If we replace $E(A)$ by

$$\tilde{E}(A) \stackrel{\text{def}}{=} \left\{ u \in H_0^2(0,1) : \begin{array}{l} \displaystyle\int_0^1 |u'|^2\,dx = 1 \ \text{and}\ \forall v \in H_0^2(0,1) \\[2mm] \displaystyle\int_0^1 A\,u''\,v''\,dx = \lambda(A) \int_0^1 u'\,v'\,dx \end{array} \right\},$$

then the maximizing elements of $\mu(A)$ or $\lambda(A)$ in $\mathcal{A}$ are completely characterized by the variational inequality

$$\exists A \in \mathcal{A},\ \forall B \in \mathcal{A}, \quad \inf_{v \in \tilde{E}(A)} \int_0^1 B\,|v''|^2\,dx \le \lambda(A). \tag{3.6}$$

Proof. (i) From Theorem 4.2 in Chapter 3, $X(t) \ne \varnothing$ and assumption (H1) is satisfied. The partial derivative of $\tilde{L}(t, v)$ with respect to t is given by the expression

$$\partial_t \tilde{L}(t, v) = \frac{1}{2} \int_0^1 B\,|v''|^2\,dx, \tag{3.7}$$

which is independent of t and exists for all $B \in L^\infty(0,1)$. Hence assumptions (H2) and (H4) are trivially satisfied. From (4.7) in Theorem 4.1 of Chapter 3

$$\exists u_t \in H_0^2(0,1),\ \forall v \in H_0^2(0,1), \quad \int_0^1 A_t\,u_t''\,v''\,dx = \lambda(A_t) \int_0^1 u_t'\,v'\,dx \tag{3.8}$$

$$\Rightarrow \frac{A_0}{2} \int_0^1 |u_t''|^2\,dx \le \int_0^1 A_t\,|u_t''|^2\,dx = \lambda(A_t) \int_0^1 |u_t'|^2\,dx = \frac{1}{\lambda(A_t)}. \tag{3.9}$$

But

$$0 < \frac{A_0}{2} \le A_t \le A_B \stackrel{\text{def}}{=} A_1 + \tau\|B\|_{L^\infty}$$

$$\Rightarrow \frac{1}{2}\mu(A_0) \le \mu(A_t) \le \mu(A_B) \text{ and } 2\lambda(A_0) \le \lambda(A_t) \le \lambda(A_B).$$

Therefore for any sequence $t_n \searrow 0$, there exists a subsequence $\{t_{n_k}\}$, μ, and $u \in H_0^2(0,1)$ such that

$$\mu_k \stackrel{\text{def}}{=} \mu(A_{t_{n_k}}) \to \mu < 0, \quad \lambda_k \stackrel{\text{def}}{=} \lambda(A_{t_{n_k}}) = -\frac{1}{2\mu_k} \to \lambda = -\frac{1}{2\mu} > 0,$$

$$u_k \rightharpoonup u \text{ in } H_0^2(0,1)\text{-weak} \quad \Rightarrow u_k \to u \text{ in } H_0^1(0,1).$$

Going to the limit in (3.8)

$$\forall v \in H_0^2(0,1), \quad \int_0^1 A_{t_{n_k}} u_k'' v'' \, dx = \lambda(A_{t_{n_k}}) \int_0^1 u_k' v' \, dx, \quad \|u_k'\|_{L^2} = \frac{1}{\lambda_k},$$

we get the following variational equation for $u \in H_0^2(0,1)$:

$$\forall v \in H_0^2(0,1), \quad \int_0^1 A u'' v'' \, dx = \lambda \int_0^1 u' v' \, dx,$$

$$\|u'\|_{L^2} = \lim_{k\to\infty} \|u_k'\|_{L^2} = \lim_{k\to\infty} \frac{1}{\lambda_k} = \frac{1}{\lambda} > 0.$$

To show that $u \in X(0) = E(A)$, let u_0 be any element of $E(A)$. By definition,

$$\frac{1}{2} \int_0^1 A_{t_{n_k}} |u_k''|^2 \, dx - \|u_k'\|_{L^2} \le \frac{1}{2} \int_0^1 A_{t_{n_k}} |u_0''|^2 \, dx - \|u_0'\|_{L^2}. \tag{3.10}$$

But $A_{t_{n_k}} \to A$ in $L^\infty(0,1)$-strong and

$$\liminf_{k\to\infty} \int_0^1 A_{t_{n_k}} |u_k''|^2 \, dx = \liminf_{k\to\infty} \int_0^1 (A_{t_{n_k}} - A) |u_k''|^2 \, dx + \int_0^1 A |u_k''|^2 \, dx$$

$$\ge \liminf_{k\to\infty} -\|A_{t_{n_k}} - A\|_{L^\infty} \int_0^1 |u_k''|^2 \, dx + \int_0^1 A |u_k''|^2 \, dx$$

$$\ge \int_0^1 A |u''|^2 \, dx$$

since $\|u_k''\|_{L^2}$ is bounded. Since $\|u_k'\|_{L^2} \to \|u'\|_{L^2}$, by going to the limit in (3.10), we get

$$L(A, u) = \frac{1}{2} \int_0^1 A |u''|^2 \, dx - \|u'\|_{L^2}$$

$$\le \liminf_{k\to\infty} \frac{1}{2} \int_0^1 A_{t_{n_k}} |u_k''|^2 \, dx - \|u_k'\|_{L^2}$$

$$\le \liminf_{k\to\infty} \frac{1}{2} \int_0^1 A_{t_{n_k}} |u_0''|^2 \, dx - \|u_0'\|_{L^2} = \frac{1}{2} \int_0^1 A |u_0''|^2 \, dx - \|u_0'\|_{L^2}$$

$$= \mu(A) = \inf_{v \in V} L(A, v).$$

Therefore, by definition of the minimum, $u \in E(A) = X(0)$ and assumption (H3)(i) is satisfied.

To check the second part of (H3) we first show that $\{u_k\}$ converges not only in $H_0^2(0,1)$-weak, but also in $H_0^2(0,1)$-strong. Then assumption (H4) directly follows from that property:

$$\frac{1}{2}\partial\tilde{L}(t, u_k) = \int_0^1 B\,|u_k''|^2\,dx \to \int_0^1 B\,|u''|^2\,dx = \partial\tilde{L}(t, u).$$

The strong convergence follows from the following chain of inequalities:

$$
\begin{aligned}
\frac{A_0}{2}\int_0^1 |u_k'' - u''|^2\,dx &\leq \int_0^1 \frac{A_0}{2}|u_k''|^2\,dx + \int_0^1 \frac{A_0}{2}|u''|^2\,dx - \int_0^1 A_0 u_k''\,u''\,dx \\
&\leq \int_0^1 A_k|u_k''|^2\,dx + \int_0^1 A_0|u''|^2\,dx - 2\int_0^1 A_0 u_k''\,u''\,dx \\
&= \lambda_k \int_0^1 |u_k'|^2\,dx + \int_0^1 A_0|u''|^2\,dx - 2\int_0^1 A_0 u_k''\,u''\,dx \\
&\to \lambda \int_0^1 |u'|^2\,dx - \int_0^1 A_0|u''|^2\,dx = 0
\end{aligned}
$$

since $\lambda = \lambda(A)$. This complete the proof of part (i).

(ii) The first characterization now follows directly from the fact that μ is concave and continuous and $\mathcal{A}$ is closed and convex. The second characterization uses the following identity, which comes from (3.8) for the eigenvectors and their normalization (4.11) in Theorem 4.2 of Chapter 3 as elements of $E(A)$: for all $v \in E(A)$

$$\frac{1}{2}\int_0^1 A\,|v''|^2\,dx = \lambda(A)\,\frac{1}{2}\int_0^1 |v'|^2\,dx = \frac{1}{2\lambda(A)} = -\mu(A). \qquad \square$$

4 Eigenvalue Problems

The first eigenvalue $\lambda(\Omega)$ of a linear boundary value problem defined on a bounded open subset Ω of $\mathbf{R}^N$ is a classical example of shape functional which comes in the form of an infimum. It is generally not differentiable since the eigenvalue can be repeated, but Theorem 2.1 of section 2.3 can be used to compute its Eulerian semiderivative.

In this section we give the *volume expression* of the Eulerian semiderivative for the Laplacian and the bi-Laplacian with a homogeneous Dirichlet boundary condition for a general bounded open domain and its *boundary expression* when the domain is sufficiently smooth. In the case of the Laplacian over a smooth domain, the first eigenvalue is simple and differentiable. As in section 3 it is technically advantageous to work with Auchmuty's [1] dual principle rather than with the Rayleigh quotient. It is also technically advantageous to embed the problem of the computation of the Eulerian semiderivative for the domain Ω into a larger and sufficiently smooth hold-all D, which will contain all the perturbations $\Omega_t = T_t(V)(\Omega)$ of

Ω for t sufficiently small. The smoothness conditions that would normally occur on Ω will then occur on D. Once the expression for the semiderivative is obtained, D can be thrown away, leaving a final expression that does not require any smoothness assumption on Ω.

4.1 Transport of $H_0^k(\Omega)$ by $W^{k,\infty}$-Transformations of $\mathbf{R}^N$

Recall Theorem 2.1 of section 2 in Chapter 6 on the embedding of $H_0^k(\Omega)$ into $H_0^k(D)$. As a consequence, the homogeneous Dirichlet boundary value problem in Ω: find $y \in H_0^1(\Omega)$ such that

$$\forall \varphi \in H_0^1(\Omega), \quad \int_\Omega \nabla y \cdot \nabla \varphi \, dx = \int_\Omega f \, \varphi \, dx,$$

is completely equivalent to the variational problem, find $Y \in H_0^1(\Omega; D)$ such that

$$\forall \Phi \in H_0^1(\Omega; D), \quad \int_D \nabla Y \cdot \nabla \Phi \, dx = \int_D f \, \Phi \, dx$$

and $Y|_\Omega = y$.

Let $k \geq 1$ and let T be a transformation of $\mathbf{R}^N$ such that

$$T, \, T^{-1} \in W^{k,\infty}(\mathbf{R}^N, \mathbf{R}^N).$$

Associate with T the map

$$\varphi \mapsto \mathcal{T}(\varphi) \overset{\text{def}}{=} \varphi \circ T^{-1} \, : \, \mathcal{D}(\mathbf{R}^N) \to H_0^k(\mathbf{R}^N).$$

By assumption on T, the norms $\|\varphi\|_{H^k}$ and $\|\mathcal{T}(\varphi)\|_{H^k} = \|\varphi \circ T^{-1}\|_{H^k}$ are equivalent and by density $\mathcal{T}$ extends to a linear bijection

$$\varphi \mapsto \mathcal{T}(\varphi) \overset{\text{def}}{=} \varphi \circ T^{-1} \, : \, H_0^k(\mathbf{R}^N) \to H_0^k(\mathbf{R}^N)$$

such that both $\mathcal{T}$ and $\mathcal{T}^{-1}$ are uniformly Lipschitzian. Given any bounded open subset Ω of $\mathbf{R}^N$, the map

$$\varphi \mapsto \mathcal{T}_\Omega(\varphi) \overset{\text{def}}{=} \mathcal{T}(\varphi)|_\Omega \, : \, H_0^k(\Omega) \to H_0^k(T(\Omega))$$

is again a linear bijection such that both $\mathcal{T}_\Omega$ and $\mathcal{T}_\Omega^{-1}$ are uniformly Lipschitzian. Given a bounded open subset D of $\mathbf{R}^N$, further assume that T is such that

$$T(D) = D.$$

As a result $T(\mathbf{R}^N \setminus D) = \mathbf{R}^N \setminus D$ and $T(\partial D) = \partial D$ and

$$\mathcal{T}_D \, : \, H_0^k(D) \to H_0^k(D).$$

Let Ω be an open subset of $\mathbf{R}^N$ such that $\Omega \subset D$. Then $T(\Omega) \subset D$, $T^{-1}(\Omega) \subset D$, and $H_0^1(\Omega)$ and $H_0^1(T(\Omega))$ can be identified with the subspaces $H_0^1(\Omega; D)$ and $H_0^1(T(\Omega); D)$ of $H_0^1(D)$. Moreover,

$$\forall \varphi \in \mathcal{D}(\Omega), \quad e_0(\varphi) \circ T^{-1} = e_0(\varphi \circ T^{-1})$$

and the following diagram commutes:

$$
\begin{array}{ccc}
H_0^k(\Omega) & \overset{e_\Omega}{\to} & H_0^k(D) \\[2mm]
\downarrow T_\Omega & & \downarrow T_D \\[2mm]
H_0^k(T(\Omega)) & \overset{e_\Omega}{\to} & H_0^k(D)
\end{array}
$$

$$
\Rightarrow \boxed{T_D(H_0^k(\Omega; D)) = H_0^k(T(\Omega); D).}
$$

4.2 Laplacian and Bi-Laplacian

For an arbitrary bounded open subset Ω of $\mathbf{R}^N$ the bounded open hold-all D can be arbitrarily chosen such that $\overline{\Omega} \subset D$. For instance, D can be a large enough open ball. This will make it possible to throw on D any smoothness assumption that would normally occur on Ω and work with an arbitrary Ω. In this set-up, D is used as an intermediate step and can be thrown away once the semiderivative has been computed.

Given a velocity field

$$
V \in C([0, \tau]; W_0^{k,\infty}(D; \mathbf{R}^N)), \tag{4.1}
$$

the flow mapping $T_t = T_t(V)$ maps D onto itself and ∂D onto itself and is equal to the identity on $\mathbf{R}^N \setminus D$. So it transports $H_0^1(D)$ onto itself.

Theorem 4.1. *For $k = 1, 2$ and $0 \leq t < \tau$,*

$$
\varphi \in H_0^k(D) \iff \varphi \circ T_t(V) \in H_0^k(D),
$$
$$
\varphi \in H_0^k(\Omega; D) \iff \varphi \circ T_t(V) \in H_0^k(\Omega; D).
$$

Hence for $\Omega_t = T_t(\Omega)$,

$$
H_0^k(\Omega_t; D) = \left\{ \varphi \circ T_t^{-1}(V) \; : \; \forall \varphi \in H_0^k(\Omega; D) \right\}. \tag{4.2}
$$

Proof. The second statement follows from the fact that T_t and its inverse T_t^{-1} are both Lipschitz continuous and that, by the previous considerations or Lemma 7.3 in Chapter 6, they transport sets of zero capacity onto sets of zero capacity: $T_t(D \setminus \Omega) = D \setminus T_t(\Omega) = D \setminus \Omega_t$. $\qquad \square$

As $\overline{\Omega}$ is compact and $\overline{\Omega} \subset D$, there exists $\tau_V > 0$ such that for all t, $0 \leq t < \tau_V$, $\overline{\Omega_t(V)} \subset D$. In both cases the first eigenvalue is given by the Rayleigh quotient:

$$
\lambda(\Omega_t(V)) = \inf \left\{ \frac{a_{\Omega_t}^k(\varphi, \varphi)}{\int_{\Omega_t} \varphi^2 \, dx} \; : \; \forall \varphi \in H_0^k(\Omega_t), \; \varphi \neq 0 \right\}, \quad k = 1, 2,
$$

where

$$
a_{\Omega_t}^1(\varphi, \psi) \overset{\text{def}}{=} \int_{\Omega_t} \nabla \varphi \cdot \nabla \psi \, dx \quad \text{and} \quad a_{\Omega_t}^2(\varphi, \psi) \overset{\text{def}}{=} \int_{\Omega_t} \Delta \varphi \, \Delta \psi \, dx.
$$

In view of the previous constructions $H_0^k(\Omega_t)$ can be replaced by $H_0^k(\Omega_t; D)$:

$$\lambda(\Omega_t(V)) = \inf\left\{\frac{a_D^k(\varphi, \varphi)}{\int_D \varphi^2\, dx} : \forall \varphi \in H_0^k(\Omega_t; D),\ \varphi \neq 0\right\}, \quad k = 1, 2. \tag{4.3}$$

Note that if $\varphi \neq 0$ is a minimizer, then for any real number $\alpha \neq 0$, $\alpha\varphi$ is also a minimizer.

Theorem 4.2. *Given $k = 1, 2$, let Ω be a bounded open subset of $\mathbf{R}^N$. Assume that*

$$V \in C([0, \tau]; W_0^{k,\infty}(D; \mathbf{R}^N))$$

for some bounded open domain D in $\mathbf{R}^N$ such that $\Omega \subset D$. There exists at least one nonzero solution $\varphi \in H_0^k(\Omega_t; D)$ to the minimization problem (4.3), $\lambda(\Omega_t(V)) \geq \lambda(D) > 0$, and

$$\forall \varphi \in H_0^1(D), \quad \int_D \varphi^2\, dx \leq \lambda(D)^{-1} \int_D |\nabla\varphi|^2\, dx,$$

$$\forall \varphi \in H_0^2(D), \quad \int_D \varphi^2\, dx \leq \lambda(D)^{-1} \int_D |\Delta\varphi|^2\, dx.$$

The solutions are completely characterized by the following variational equation: there exists $\varphi \in H_0^k(\Omega_t; D)$ such that

$$\forall \psi \in H_0^k(\Omega_t; D), \quad a_D^k(\varphi, \psi) = \lambda(\Omega_t(V)) \int_D \varphi\psi\, dx \tag{4.4}$$

or equivalently

$$\forall \psi \in H_0^k(\Omega_t), \quad a_{\Omega_t}^k(\varphi, \psi) = \lambda(\Omega_t(V)) \int_{\Omega_t} \varphi\psi\, dx. \tag{4.5}$$

Proof. Same technique as in the proof of Theorem 2.2 in Chapter 6. $\qquad\square$

The minimization problem can be rewritten on the unit sphere in $L^2(D)$ by normalization:

$$\lambda(\Omega_t(V)) = \inf\left\{\int_D |\nabla\varphi|^2\, dx : \forall \varphi \in H_0^1(\Omega_t; D),\ \|\varphi\|_{L^2(D)} = 1\right\}. \tag{4.6}$$

The minimizing elements of (4.6) cannot be zero since the injection of $H_0^1(D)$ into $L^2(D)$ is compact (cf. Theorem 2.1 in Chapter 6).

Remark 4.1.
One of the consequences of the use of the embedding of $H_0^1(\Omega)$ into $H_0^1(D)$ is that the characterization of the first eigenvalue on Ω only requires that Ω be a bounded open subset of $\mathbf{R}^N$. It is not necessary to assume that Ω is Lipschitzian in order to use Rellich's theorem, since D can be chosen sufficiently large ($\overline{\Omega} \subset D$) and smooth in order to contain all variations $\overline{\Omega}_t \subset D$ as t goes to zero. This technique will be exploited in the computation of the Eulerian semiderivative of $\lambda(\Omega)$. $\qquad\square$

Auchmuty's [1] dual variational principle for this eigenvalue problem can be chosen as

$$\mu(\Omega_t) \stackrel{\text{def}}{=} \inf \left\{ L^k(\Omega_t, \varphi) \; : \; \varphi \in H_0^k(\Omega_t) \right\}, \tag{4.7}$$

$$L^k(\Omega_t, \varphi) \stackrel{\text{def}}{=} \frac{1}{2} a_{\Omega_t}^k(\varphi, \varphi) - \left[\int_{\Omega_t} \varphi^2 \, dx \right]^{1/2}. \tag{4.8}$$

By using the embedding of $H_0^1(\Omega_t)$ into $H_0^1(D)$, this problem can be rewritten as

$$\mu(\Omega_t) \stackrel{\text{def}}{=} \inf \left\{ L^k(D, \varphi) \; : \; \varphi \in H_0^k(\Omega_t; D) \right\}, \tag{4.9}$$

$$L^k(D, \varphi) \stackrel{\text{def}}{=} \frac{1}{2} a_D^k(\varphi, \varphi) - \left[\int_D \varphi^2 \, dx \right]^{1/2}. \tag{4.10}$$

Theorem 4.3. *Given $k = 1, 2$, let Ω be a bounded open subset of $\mathbf{R}^N$. Assume that*

$$V \in C^0([0, \tau[\, ; W_0^{k,\infty}(D; \mathbf{R}^N))$$

for some bounded open domain D in $\mathbf{R}^N$ such that $\Omega \subset D$. Then for $0 \le t < \tau$,

$$\mu(\Omega_t) = -\frac{1}{2\lambda(\Omega_t)} \tag{4.11}$$

and the set of minimizers of (4.7) is given by

$$E^k(\Omega_t) \stackrel{\text{def}}{=} \left\{ \varphi \in H_0^k(\Omega_t; D) \; : \; \begin{array}{c} \varphi \text{ is solution of (4.4) and} \\ \left\{ \int_D |\varphi|^2 \, dx \right\}^{1/2} = 1/\lambda(\Omega_t) \end{array} \right\}. \tag{4.12}$$

Proof. From the previous theorem the set $E(\Omega_t)$ is not empty, and for any $\varphi \in E(\Omega_t)$

$$\mu(\Omega_t) \le L(t, \varphi) = -\frac{1}{2\lambda(\Omega_t)} < 0.$$

Therefore, the minimizers of (4.8) are different from the zero functions. For $\varphi \ne 0$, the functional $L(t, \varphi)$ is differentiable and its directional derivative is given by

$$dL(t, \varphi; \psi) = a_D^k(\varphi, \psi) - \frac{1}{\|\varphi\|_{L^2(D)}} \int_D \varphi \psi \, dx, \tag{4.13}$$

and any minimizer of $L(t, \varphi)$ is a stationary point of $dL(t, \varphi; \psi)$; that is,

$$\forall \psi \in H_0^k(\Omega_t; D), \quad a_D^k(\varphi, \psi) - \frac{1}{\|\varphi\|_{L^2(D)}} \int_D \varphi \psi \, dx = 0. \tag{4.14}$$

Therefore φ is a solution of the eigenvalue problem with

$$\lambda' = \frac{1}{\|\varphi\|_{L^2(D)}} \quad \Rightarrow \quad -\frac{1}{2\lambda'} = \mu(\Omega_t) \le -\frac{1}{2\lambda(\Omega_t)}.$$

By minimality of $\lambda(\Omega_t)$, we necessarily have $\lambda' = \lambda(\Omega_t)$ and this concludes the proof of the theorem. $\qquad\square$

We now turn to the computation of the Eulerian semiderivative $d\lambda(\Omega; V)$ via $d\mu(\Omega; V)$ from the identity

$$\lambda(\Omega_t) = -\frac{1}{2\mu(\Omega_t)},$$

since whenever $d\mu(\Omega; V)$ exists

$$d\lambda(\Omega; V) = \frac{1}{2\mu(\Omega)^2}\, d\mu(\Omega; V) = 2\,\lambda(\Omega)^2\, d\mu(\Omega; V).$$

Assuming that V satisfies assumption (4.1), we use the function space parametrization of section 2.4 in conjunction with Theorem 2.1. From the characterization (4.2) of $H_0^k(\Omega_t; D)$, define the following new functional: for each $\varphi \in H_0^1(D)$,

$$\tilde{L}(t, \varphi) \stackrel{\text{def}}{=} L(D, \varphi \circ T_t^{-1}(V))$$

$$= \frac{1}{2} a_D^k(\varphi \circ T_t^{-1}(V), \varphi \circ T_t^{-1}(V)) - \left\{ \int_D |\varphi \circ T_t^{-1}(V)|^2\, dx \right\}^{1/2}.$$

After a change of variable for $k = 1$ and $\varphi \in H_0^1(D)$,

$$\tilde{L}(t, \varphi) = \frac{1}{2} \int_D A(t)\nabla\varphi \cdot \nabla\varphi\, dx - \left\{ \int_D J_t\, |\varphi|^2\, dx \right\}^{1/2},$$

$$A(t) = J_t\, DT_t^{-1}\, {}^*DT_t^{-1}, \quad J_t = \det(DT_t),$$

and for $k = 2$ and $\varphi \in H_0^2(D)$

$$\tilde{L}(t, \varphi) = \frac{1}{2} \int_D |\mathrm{div}\,(B(t)\nabla\varphi)|^2\, J_t\, dx - \left\{ \int_D J_t\, |\varphi|^2\, dx \right\}^{1/2},$$

$$B(t) = DT_t^{-1}\, {}^*DT_t^{-1}.$$

To apply Theorem 2.1, choose

$$X(t) \stackrel{\text{def}}{=} \left\{ \varphi^t \stackrel{\text{def}}{=} \varphi_t \circ T_t\; : \; \forall \varphi_t \in E^k(\Omega_t) \right\},$$

endowed with the weak topology of $H_0^1(D)$. Assumption (H1) is clearly satisfied. For all $\varphi \neq 0$, then $\tilde{L}(t, \varphi)$ is differentiable. For $k = 1$ and $0 \neq \varphi \in H_0^1(D)$

$$\partial_t \tilde{L}(t, \varphi) = \frac{1}{2} \int_D A'(t)\nabla\varphi \cdot \nabla\varphi\, dx - \frac{1}{\left\{ \int_D J_t\, |\varphi|^2\, dx \right\}^{1/2}} \int_D |\varphi|^2\, J_t'\, dx,$$

and for $k = 2$ and $0 \neq \varphi \in H_0^2(D)$

$$\partial_t \tilde{L}(t, \varphi) = \int_D \mathrm{div}\,(B'(t)\nabla\varphi)\, \mathrm{div}\,(B(t)\nabla\varphi)\, J_t\, dx$$

$$+ \frac{1}{2} \int_D |\mathrm{div}\,(B(t)\nabla\varphi)|^2\, J_t'\, dx$$

$$- \frac{1}{\left\{ \int_D J_t\, |\varphi|^2\, dx \right\}^{1/2}} \int_D |\varphi|^2\, J_t'\, dx.$$

Hence assumptions (H2) and (H4) are satisfied. In $t = 0$ the above expressions simplify. For $k = 1$ and $\varphi \in E^1(\Omega)$,

$$\partial_t \tilde{L}(0, \varphi) = \frac{1}{2} \int_D A'(0) \nabla \varphi \cdot \nabla \varphi \, dx - \lambda(\Omega) \int_D |\varphi|^2 \operatorname{div} V(0) \, dx$$

$$= - \int_D \varepsilon(V(0)) \nabla \varphi \cdot \nabla \varphi \, dx + \int_D \left[\frac{1}{2} |\nabla \varphi|^2 - \lambda(\Omega)|\varphi|^2 \right] \operatorname{div} V(0) \, dx,$$

$$A'(0) = \operatorname{div} V(0) I - DV(0) - {}^*DV(0), \quad \varepsilon(U) \stackrel{\text{def}}{=} \frac{1}{2} [DU + {}^*DU],$$

and for $k = 2$ and $\varphi \in E^2(\omega)$,

$$\partial_t \tilde{L}(0, \varphi) = 2 \int_D \operatorname{div}\left(\varepsilon(V(0)) \nabla \varphi\right) \Delta \varphi \, dx$$

$$+ \int_D \left[\frac{1}{2} |\Delta \varphi|^2 - \lambda(\Omega)|\varphi|^2 \right] \operatorname{div} V(0) \, dx,$$

$$B'(0) = -DV(0) - {}^*DV(0) = -2\varepsilon(V(0)).$$

For volume-preserving velocities, $\operatorname{div} V(0) = 0$, and the above expressions reduce to the first integral term.

In order to apply Theorem 2.1 it remains to check assumption (H3). This is the first example where the set of minimizers is not necessarily unique and for which we have a complete description of the Eulerian semiderivative.

Theorem 4.4. *Given $k = 1, 2$, let Ω be a bounded open subset of $\mathbf{R}^N$. Assume that*

$$V \in C^0([0, \tau[\, ; W_0^{k, \infty}(D; \mathbf{R}^N))$$

for some bounded open domain D in $\mathbf{R}^N$ such that $\overline{\Omega} \subset D$. Then for $k = 1$

$$\frac{1}{2} d\lambda(\Omega; V) = \inf_{\varphi \in \tilde{E}^1(\Omega)} - \int_\Omega \varepsilon(V(0)) \nabla \varphi \cdot \nabla \varphi \, dx$$

$$+ \int_\Omega \left[\frac{1}{2} |\nabla \varphi|^2 - \lambda(\Omega)|\varphi|^2 \right] \operatorname{div} V(0) \, dx,$$

$$\varepsilon(U) \stackrel{\text{def}}{=} \frac{1}{2} [DU + {}^*DU],$$

$$\tilde{E}^1(\Omega) \stackrel{\text{def}}{=} \left\{ \varphi \in H_0^1(\Omega) : \begin{array}{l} -\Delta \varphi = \lambda(\Omega)\varphi \ \text{in } \mathcal{D}(\Omega)' \\ \text{and } \int_\Omega |\varphi|^2 \, dx = 1 \end{array} \right\};$$

for $k = 2$

$$\frac{1}{2} d\lambda(\Omega; V) = \inf_{\varphi \in \tilde{E}^2(\Omega)} \int_\Omega 2 \operatorname{div}\left(\varepsilon(V(0)) \nabla \varphi\right) \Delta \varphi \, dx$$

$$+ \int_\Omega \left[\frac{1}{2} |\Delta \varphi|^2 - \lambda(\Omega)|\varphi|^2 \right] \operatorname{div} V(0) \, dx,$$

$$\tilde{E}^2(\Omega_t) \stackrel{\text{def}}{=} \left\{ \varphi \in H_0^2(\Omega) : \begin{array}{l} \Delta(\Delta \varphi) = \lambda(\Omega)\varphi \ \text{in } \mathcal{D}(\Omega)' \\ \text{and } \int_\Omega |\varphi|^2 \, dx = 1 \end{array} \right\}.$$

In both cases $\lambda(\Omega)$ has a Hadamard semiderivative in the sense of Definition 3.1 (iii) in Chapter 8, which is continuous with respect to $V(0) \in W_0^{k,\infty}(D; \mathbf{R}^N)$

$$d\lambda(\Omega; V) = d_H \lambda(\Omega; V(0)).$$

Proof. As $\overline{\Omega}$ is compact and $\overline{\Omega} \subset D$, there exists $\tau_V > 0$ such that for all t, $0 \le t < \tau_V$, $\Omega_t(V) \subset D$. Observe that

$$\mu(D) = \inf_{\varphi \in H_0^1(D)} L(\varphi) \le \inf_{\varphi \in H_0^1(\Omega_t; D)} L(\varphi) = \mu(\Omega_t).$$

By continuity of $t \mapsto \tilde{L}(t, \varphi)$, for any $\varphi^t \in X(t)$ and $\varphi^0 \in X(0)$,

$$\tilde{L}(t, \varphi^t) \le \tilde{L}(t, \varphi^0) \;\Rightarrow\; \limsup_{t \searrow 0} \tilde{L}(t, \varphi^t) \le \tilde{L}(0, \varphi^0) = \mu(\Omega)$$

$$\Rightarrow\; \mu(D) \le \limsup_{t \searrow 0} \mu(\Omega_t) \le \mu(\Omega).$$

Therefore there exists τ', $0 < \tau' \le \tau_V$, such that for all t, $0 \le t \le \tau'$,

$$\mu(D) \le \mu(\Omega_t) \le \frac{1}{2}\mu(\Omega) < 0 \;\;\Rightarrow\; 0 < \lambda(D) \le \lambda(\Omega_t) \le 2\,\lambda(\Omega).$$

Since $A(t) \to I$ and $J_t \to 1$ as t goes to zero, there exist $c > 0$ and $0 < \tau'' \le \tau$ such that for all $0 \le t \le \tau''$,

$$\tilde{L}(t, \varphi) = \frac{1}{2}\int_D A(t)\nabla\varphi \cdot \nabla\varphi \, dx - \left\{ \int_D J_t \, |\varphi|^2 \, dx \right\}^{1/2}$$

$$\ge c \left\{ \|\nabla\varphi\|_{L^2(D)}^2 - \|\varphi\|_{L^2(D)} \right\} \ge c\|\nabla\varphi\|_{L^2(D)}^2 - c'\|\nabla\varphi\|_{L^2(D)}$$

for some $c' > 0$ by Poincaré's inequality. Therefore for all $\varphi^t \in X(t)$,

$$c\|\nabla\varphi^t\|_{L^2(D)}^2 - c'\|\nabla\varphi^t\|_{L^2(D)} \le \tilde{L}(t, \varphi^t) \le 0 \;\;\Rightarrow\; \|\nabla\varphi^t\|_{L^2(D)} \le c'/c.$$

For any sequence $t_n \searrow 0$, there exists subsequences μ and $\varphi_0 \in H_0^1(\Omega; D)$ such that

$$\mu_n = \mu(\Omega_{t_n}) \to \mu \text{ and } \varphi_n = \varphi^{t_n} \rightharpoonup \varphi_0 \text{ in } H_0^1(\Omega; D)\text{-weak,}$$

$$\lambda_n = \lambda(\Omega_{t_n}) \to \lambda = -\frac{1}{2\mu} > 0.$$

Therefore, since $A(t_n) \to I$, $J_{t_n} \to 1$, and $\{\varphi_n\}$ converges in H^1-weak,

$$\forall\psi \in H_0^1(\Omega; D), \quad \frac{1}{2}\int_D A(t_n)\nabla\varphi_n \cdot \nabla\psi \, dx = \lambda_n \int_D \varphi_n \, \psi \, dx,$$

$$\Rightarrow \forall\psi \in H_0^1(\Omega; D), \quad \frac{1}{2}\int_D A(0)\nabla\varphi_0 \cdot \nabla\psi \, dx = \lambda \int_D \varphi_0 \, \psi \, dx,$$

$$\lambda_n = \int_D J_{t_n} \, |\varphi_n|^2 \, dx \to \int_D |\varphi_0|^2 \, dx \;\;\Rightarrow\; \int_D |\varphi_0|^2 \, dx = \lambda > 0,$$

and λ is an eigenvalue for the problem on Ω. But we know that

$$\mu(D) \le \limsup_{n\to\infty} \mu(\Omega_{t_n}) = -\frac{1}{2\lambda}, \quad \mu(D) \le -\frac{1}{2\lambda} \le -\frac{1}{2\lambda(\Omega)},$$

$$\lambda(D) \le \lambda \le \lambda(\Omega) \quad \Rightarrow \quad \lambda = \lambda(\Omega),$$

since $\lambda(\Omega)$ is minimal and hence $\varphi_0 \in X(0)$. This proves condition (H3)(i). To prove condition (H3)(ii) and complete the proof we first prove that the weakly convergent subsequence strongly converges in $H_0^1(\Omega; D)$. By compactness of the injection of $H_0^1(D)$ into $L^2(D)$ (cf. Theorem 2.1 in Chapter 6)

$$\varphi_n \rightharpoonup \varphi_0 \text{ in } H_0^1(D)\text{-weak} \quad \Rightarrow \quad \varphi_n \to \varphi_0 \text{ in } L^2(D)\text{-strong}, \tag{4.15}$$

and for some $\alpha > 0$ independent of n,

$$\alpha \int_D |\nabla(\varphi_n - \varphi_0)|^2 \, dx \le \int_D A(t_n)\nabla(\varphi_n - \varphi_0) \cdot \nabla(\varphi_n - \varphi_0) \, dx$$

$$= \int_D A(t_n)\nabla\varphi_n \cdot \nabla\varphi_n - 2A(t_n)\nabla\varphi_0 \cdot \nabla\varphi_n$$

$$+ A(t_n)\nabla\varphi_0 \cdot \nabla\varphi_0 \, dx$$

$$= \int_D \lambda_n J_{t_n} \varphi_n \varphi_n - 2A(t_n)\nabla\varphi_0 \cdot \nabla\varphi_n$$

$$+ A(t_n)\nabla\varphi_0 \cdot \nabla\varphi_0 \, dx$$

$$\to \int_D \lambda \varphi_0 \varphi_0 - 2\nabla\varphi_0 \cdot \nabla\varphi_0$$

$$+ \nabla\varphi_0 \cdot \nabla\varphi_0 \, dx = 0.$$

By the same technique as in section 2.4 when $n \to \infty$ and $t \searrow 0$

$$\partial_t \tilde{L}(t, \varphi_n) = \frac{1}{2} \int_D A'(t)\nabla\varphi_n \cdot \nabla\varphi_n \, dx$$

$$- \frac{1}{\left\{ \int_D J_t \, |\varphi_n|^2 \, dx \right\}^{1/2}} \int_D |\varphi_n|^2 \, J'_t \, dx$$

$$\to \frac{1}{2} \int_D A'(0)\nabla\varphi_0 \cdot \nabla\varphi_0 \, dx$$

$$- \frac{1}{\left\{ \int_D |\varphi_0|^2 \, dx \right\}^{1/2}} \int_D |\varphi_0|^2 \, \mathrm{div}\, V(0) \, dx = \partial_t \tilde{L}(0, \varphi_0).$$

The case $k = 2$ is analogous. $\qquad\square$

If Ω is assumed to be of class C^{2k}, the eigenvector functions belong to $H_0^k(\Omega) \cap H^{2k}(\Omega)$ and the volume expressions of the previous theorem can be expressed as boundary integrals as in section 2.5. In that case we can use formulae (2.49) and (2.50) to compute the partial derivative of $L_t = L(\Omega_t, \varphi \circ T_t^{-1})$. For $k = 1$,

$$L_t = \frac{1}{2} \int_{\Omega_t} |\nabla(\varphi \circ T_t^{-1})|^2 \, dx - \left\{ \int_{\Omega_t} |\varphi \circ T_t^{-1}|^2 \, dx \right\}^{1/2}.$$

From formula (2.49)

$$L' \stackrel{\text{def}}{=} \partial_t L_t|_{t=0} = \frac{1}{2} \int_\Gamma |\nabla\varphi|^2 \, V(0) \cdot n \, d\Gamma - \frac{1}{2} \frac{\int_\Gamma |\varphi|^2 \, V(0) \cdot n \, d\Gamma}{\left\{ \int_\Omega |\varphi|^2 \, dx \right\}^{1/2}}$$
$$+ \int_\Omega \nabla\varphi \cdot \nabla\dot\varphi \, dx - \frac{\int_\Omega \varphi\dot\varphi \, dx}{\left\{ \int_\Omega |\varphi|^2 \, dx \right\}^{1/2}}.$$

For $\varphi \in E^1(\Omega)$, $\varphi = 0$ on Ω, $-\Delta\varphi = \lambda(\Omega)\varphi$ in Ω, and $\lambda(\Omega)\|\varphi\|_{L^2(\Omega)} = 1$,

$$L' = \int_\Gamma \frac{1}{2}|\nabla\varphi|^2 \, V(0) \cdot n \, d\Gamma + \int_\Omega \nabla\varphi \cdot \nabla\dot\varphi \, dx - \lambda(\Omega) \int_\Omega \varphi\dot\varphi \, dx$$
$$= \int_\Gamma \frac{1}{2}|\nabla\varphi|^2 \, V(0) \cdot n \, d\Gamma + \int_\Gamma \frac{\partial\varphi}{\partial n} \dot\varphi \, d\Gamma.$$

But $\varphi = 0$ on Γ implies that $\nabla\varphi = \partial\varphi/\partial n \, n$ on Γ, and from identity (2.50)

$$\dot\varphi = -\nabla\varphi \cdot V(0) \quad \Rightarrow \quad \dot\varphi = -\frac{\partial\varphi}{\partial n} V(0) \cdot n \text{ on } \Gamma$$
$$\Rightarrow L' = -\int_\Gamma \frac{1}{2}\left|\frac{\partial\varphi}{\partial n}\right|^2 V(0) \cdot n \, d\Gamma.$$

For $k = 2$ the computation of the partial derivative of $L_t = L(\Omega_t, \varphi \circ T_t^{-1})$ is similar, with obvious changes:

$$L_t = \frac{1}{2} \int_{\Omega_t} |\Delta(\varphi \circ T_t^{-1})|^2 \, dx - \left\{ \int_{\Omega_t} |\varphi \circ T_t^{-1}|^2 \, dx \right\}^{1/2}.$$

From formula (2.49)

$$L' \stackrel{\text{def}}{=} \partial_t L_t|_{t=0} = \frac{1}{2} \int_\Gamma |\Delta\varphi|^2 \, V(0) \cdot n \, d\Gamma - \frac{1}{2} \frac{\int_\Gamma |\varphi|^2 \, V(0) \cdot n \, d\Gamma}{\left\{ \int_\Omega |\varphi|^2 \, dx \right\}^{1/2}}$$
$$+ \int_\Omega \Delta\varphi \, \Delta\dot\varphi \, dx - \frac{\int_\Omega \varphi\dot\varphi \, dx}{\left\{ \int_\Omega |\varphi|^2 \, dx \right\}^{1/2}}.$$

For $\varphi \in E^2(\Omega)$, $\varphi = 0$ on Ω, $\Delta(\Delta\varphi) = \lambda(\Omega)\varphi$ in Ω, and $\lambda(\Omega)\|\varphi\|_{L^2(\Omega)} = 1$,

$$L' = \int_\Gamma \frac{1}{2}|\Delta\varphi|^2 \, V(0) \cdot n \, d\Gamma + \int_\Omega \Delta\varphi \, \Delta\dot\varphi \, dx - \lambda(\Omega) \int_\Omega \varphi\dot\varphi \, dx$$
$$= \int_\Gamma \frac{1}{2}|\Delta\varphi|^2 \, V(0) \cdot n \, d\Gamma + \int_\Gamma \frac{\partial\Delta\varphi}{\partial n} \dot\varphi - \Delta\varphi \frac{\partial\dot\varphi}{\partial n} \, d\Gamma.$$

But $\varphi = 0$ and $\partial\varphi/\partial n = 0$ on Γ imply that $\nabla\varphi = 0$ on Γ and

$$D^2\varphi = (D^2\varphi \, n) \, {}^*n \text{ on } \Gamma \quad \Rightarrow \quad \Delta\varphi = (D^2\varphi \, n) \cdot n \text{ on } \Gamma.$$

From identity (2.50) we have on Γ

$$\dot\varphi = -\nabla\varphi \cdot V(0) \quad \Rightarrow \dot\varphi = 0 \text{ on } \Gamma$$

$$\Rightarrow \nabla\dot\varphi = -\nabla(\nabla\varphi \cdot V(0)) = -D^2\varphi\, V(0) - {}^*DV(0)\nabla\varphi = -D^2\varphi\, V(0)$$

$$\Rightarrow \frac{\partial\dot\varphi}{\partial n} = -D^2\varphi\, V(0)\cdot n = -D^2\varphi\, n\cdot n\, V(0)\cdot n = -\Delta\varphi\, V(0)\cdot n,$$

$$L' = -\int_\Gamma \frac{1}{2}|\Delta\varphi|^2\, V(0)\cdot n\, d\Gamma.$$

The quantity $D^2\varphi\, n\cdot n$ is equal to $\partial^2\varphi/\partial n^2$. Let b_Ω be the oriented function associated with the domain Ω of class C^4. It is C^4 in a neighborhood of Γ. Therefore, in that neighborhood

$$\psi \overset{\text{def}}{=} \nabla\varphi\cdot\nabla b_\Omega, \quad \psi|_\Gamma = \frac{\partial\varphi}{\partial n}, \quad \frac{\partial}{\partial n}\left(\frac{\partial\varphi}{\partial n}\right) = \frac{\partial\psi}{\partial n} = \nabla\psi\cdot\nabla b_\Omega|_\Gamma,$$

$$\nabla\psi = \nabla(\nabla\varphi\cdot\nabla b_\Omega) = D^2 b_\Omega\nabla\varphi + D^2\varphi\nabla b_\Omega,$$

$$\nabla\psi\cdot\nabla b_\Omega = D^2 b_\Omega\nabla\varphi\cdot\nabla b_\Omega + D^2\varphi\nabla b_\Omega\cdot\nabla b_\Omega = D^2\varphi\nabla b_\Omega\cdot\nabla b_\Omega,$$

$$\boxed{\frac{\partial^2\varphi}{\partial n^2} = \frac{\partial}{\partial n}\left(\frac{\partial\varphi}{\partial n}\right) = D^2\varphi\nabla b_\Omega\cdot\nabla b_\Omega|_\Gamma = D^2\varphi|_\Gamma\, n\cdot n = \Delta\varphi|_\Gamma.}$$

We summarize the results in the next theorem.

Theorem 4.5. *Given $k = 1, 2$, let Ω be a bounded open subset of $\mathbf{R}^N$ of class C^{2k}. Assume that*

$$V \in C^0([0,\tau[\,;W_0^{k,\infty}(D;\mathbf{R}^N))$$

for some bounded open domain D in $\mathbf{R}^N$ such that $\overline\Omega \subset D$. Then for $k = 1$

$$d\lambda(\Omega;V) = \inf_{\varphi\in\tilde E^1(\Omega)} -\int_\Gamma \left|\frac{\partial\varphi}{\partial n}\right|^2 V(0)\cdot n\, d\Gamma,$$

$$\tilde E^1(\Omega) \overset{\text{def}}{=} \left\{\varphi \in H_0^1(\Omega)\cap H^2(\Omega) : \begin{array}{l} -\Delta\varphi = \lambda(\Omega)\varphi \ \text{in } \Omega \\ \text{and} \int_\Omega |\varphi|^2\, dx = 1 \end{array}\right\};$$

for $k = 2$

$$d\lambda(\Omega;V) = \inf_{\varphi\in\tilde E^2(\Omega)} -\int_\Gamma \left|\frac{\partial^2\varphi}{\partial n^2}\right|^2 V(0)\cdot n\, d\Gamma$$

$$\tilde E^2(\Omega_t) \overset{\text{def}}{=} \left\{\varphi \in H_0^2(\Omega)\cap H^4(\Omega) : \begin{array}{l} \Delta(\Delta\varphi) = \lambda(\Omega)\varphi \ \text{in } \Omega \\ \text{and} \int_\Omega |\varphi|^2\, dx = 1 \end{array}\right\}.$$

In both cases $\lambda(\Omega)$ has a Hadamard semiderivative in the sense of Definition 3.1 (iii) in Chapter 8, which is continuous with respect to $V(0) \in C_0(D;\mathbf{R}^N)$:

$$d\lambda(\Omega;V) = d_H\lambda(\Omega;V(0)).$$

4.3   Linear Elasticity

The constructions and results of the previous section readily extend to the vectorial case of linear elasticity: to find $U \in H_0^1(\Omega)^3$ such that

$$\forall W \in H_0^1(\Omega)^3, \quad \int_\Omega C\varepsilon(U) \cdots \varepsilon(W)\, dx = \int_\Omega F \cdot W\, dx \qquad (4.16)$$

for some distributed loading $F \in L^2(\Omega)^3$ and a *constitutive law* C which is a bilinear symmetric transformation of

$$\mathrm{Sym} \overset{\text{def}}{=} \left\{ \tau \in \mathcal{L}(\mathbf{R}^3; \mathbf{R}^3) : {}^*\tau = \tau \right\}$$

($\mathcal{L}(\mathbf{R}^3; \mathbf{R}^3)$ is the space of all linear transformations of $\mathbf{R}^3$ or 3×3-matrices) under the following assumption.

Assumption 4.1.
The *constitutive law* is a linear bijective and symmetric transformation $C \colon \mathrm{Sym} \to \mathrm{Sym}$ for which there exists a constant $\alpha > 0$ such that $C\tau \cdots \tau \geq \alpha\, \tau \cdots \tau$ for all $\tau \in \mathrm{Sym}$. $\qquad \square$

For instance, for the Lamé constants $\mu > 0$ and $\lambda \geq 0$, the special constitutive law $C\tau = 2\mu\, \tau + \lambda \operatorname{tr} \tau\, I$ satisfies Assumption 4.1 with $\alpha = 2\mu$.

The associated bilinear form is

$$a_\Omega(U, W) \overset{\text{def}}{=} \int_\Omega C\varepsilon(U) \cdots \varepsilon(W)\, dx,$$

where the unknown is now a vector function. To make sense of (4.16) we shall use Korn's inequality on a larger bounded open Lipschitzian domain D, $\Omega \subset D$,

$$\exists c_D > 0, \ \forall W \in H_0^1(D)^3, \quad \int_D |W|^2\, dx \leq c_D \int_D |\varepsilon(W)|^2\, dx.$$

In view of the considerations of the previous section, for a bounded open domain Ω and the velocity fields $V \in C^0([0, \tau[\, ; W_0^{1,\infty}(D; \mathbf{R}^3))$ for all $0 \leq t < \tau$

$$\exists c_D > 0, \ \forall W \in H_0^1(\Omega_t(V))^3, \quad \int_{\Omega_t(V)} |W|^2\, dx \leq c_D \int_{\Omega_t(V)} |\varepsilon(W)|^2\, dx.$$

So for any $F \in L^2(D)^3$, the variation equation (4.16) with Ω_t in place of Ω has a unique solution U_t in $H_0^1(\Omega_t(V))^3$.

With the same assumptions, the first eigenvalue is given by the Rayleigh quotient:

$$\lambda(\Omega_t(V)) = \inf \left\{ \frac{a_{\Omega_t}(U, U)}{\int_{\Omega_t} |U|^2\, dx} : \forall U \in H_0^1(\Omega_t)^3, \ U \neq 0 \right\}.$$

In view of the previous constructions, $H_0^1(\Omega_t)$ can be replaced by $H_0^1(\Omega_t; D)$:

$$\lambda(\Omega_t(V)) = \inf \left\{ \frac{a_{\Omega_t}(U, U)}{\int_{\Omega_t} |U|^2\, dx} : \forall U \in H_0^1(\Omega_t; D)^3, \ U \neq 0 \right\}. \qquad (4.17)$$

Theorem 4.6. *Let Ω be a bounded open subset of $\mathbf{R}^3$. Assume that*

$$V \in C^0([0, \tau[\,; W_0^{1,\infty}(D; \mathbf{R}^3))$$

for some bounded open Lipschitzian domain D in $\mathbf{R}^3$ such that $\Omega \subset D$. There exists at least one nonzero solution $U \in H_0^1(\Omega_t; D)^3$ to the minimization problem (4.17), $\lambda(\Omega_t(V)) \geq \lambda(D) > 0$, and

$$\forall W \in H_0^1(D)^3, \quad \int_D |W|^2\, dx \leq \lambda(D)^{-1} \int_D \|\varepsilon(W)\|^2\, dx.$$

The solutions are completely characterized by the following variational equation: there exists $U \in H_0^1(\Omega_t; D)^3$ such that

$$\forall W \in H_0^1(\Omega_t; D)^3, \quad a_D(U, W) = \lambda(\Omega_t(V)) \int_D U \cdot W\, dx, \tag{4.18}$$

or equivalently

$$\forall W \in H_0^1(\Omega_t)^3, \quad a_{\Omega_t}(U, W) = \lambda(\Omega_t(V)) \int_{\Omega_t} U \cdot W\, dx. \tag{4.19}$$

Auchmuty's [1] dual variational principle for this eigenvalue problem can be chosen as

$$\mu(\Omega_t) \stackrel{\text{def}}{=} \inf \left\{ L(\Omega_t, U) \; : \; U \in H_0^1(\Omega_t)^3 \right\}, \tag{4.20}$$

$$L(\Omega_t, U) \stackrel{\text{def}}{=} \frac{1}{2} a_{\Omega_t}(U, U) - \left[\int_{\Omega_t} |U|^2\, dx \right]^{1/2}. \tag{4.21}$$

By using the embedding of $H_0^1(\Omega_t)$ into $H_0^1(D)$, this problem can be rewritten as

$$\mu(\Omega_t) \stackrel{\text{def}}{=} \inf \left\{ L(D, U) \; : \; U \in H_0^1(\Omega_t; D)^3 \right\}, \tag{4.22}$$

$$L(D, U) \stackrel{\text{def}}{=} \frac{1}{2} a_D(U, U) - \left[\int_D |U|^2\, dx \right]^{1/2}. \tag{4.23}$$

Theorem 4.7. *Let Ω be a bounded open subset of $\mathbf{R}^3$. Assume that*

$$V \in C^0([0, \tau[\,; W_0^{1,\infty}(D; \mathbf{R}^3))$$

for some bounded open Lipschitzian domain D in $\mathbf{R}^3$ such that $\Omega \subset D$. Then for $0 \leq t < \tau$,

$$\mu(\Omega_t) = -\frac{1}{2\lambda(\Omega_t)} \tag{4.24}$$

and the set of minimizers of (4.20) is given by

$$E(\Omega_t) \stackrel{\text{def}}{=} \left\{ U \in H_0^1(\Omega_t; D)^3 \; : \; \begin{array}{c} U \text{ is solution of (4.18) and} \\ \left\{ \int_D |U|^2\, dx \right\}^{1/2} = 1/\lambda(\Omega_t) \end{array} \right\}. \tag{4.25}$$

From the identity

$$\lambda(\Omega_t) = -\frac{1}{2\mu(\Omega_t)},$$

if $d\mu(\Omega; V)$ exists, then $d\lambda(\Omega; V)$ exists and

$$d\lambda(\Omega; V) = \frac{1}{2\mu(\Omega)^2}\, d\mu(\Omega; V) = 2\,\lambda(\Omega)^2\, d\mu(\Omega; V).$$

We again use the function space parametrization of section 2.4 in conjunction with Theorem 2.1. From the characterization (4.2) of $H_0^1(\Omega_t; D)$, define the new functional: for each $U \in H_0^1(D)^3$

$$\tilde{L}(t, U) \stackrel{\text{def}}{=} L(D, U \circ T_t^{-1}(V))$$

$$= \frac{1}{2} a_D(U \circ T_t^{-1}(V), U \circ T_t^{-1}(V)) - \left[\int_D |U \circ T_t^{-1}(V)|^2 \, dx \right]^{1/2}.$$

After a change of variables for $U \in H_0^1(D)^3$,

$$\tilde{L}(t, U) = \frac{1}{2} \int_D C\varepsilon(U \circ T_t^{-1}) \circ T_t \cdot\cdot\, \varepsilon(U \circ T_t^{-1}) \circ T_t\, J_t\, dx - \left[\int_D J_t\, |U|^2\, dx \right]^{1/2},$$

where $J_t = \det(DT_t)$ and

$$D(U \circ T_t^{-1}) \circ T_t = D(U)(DT_t)^{-1},$$
$$2\,\varepsilon(U \circ T_t^{-1}) \circ T_t = D(U)(DT_t)^{-1} + {}^*(DT_t)^{-1}\, {}^*D(U).$$

Defining the transformation

$$4\,\widetilde{C}(t)\, \tau \cdot\cdot\, \sigma \stackrel{\text{def}}{=} C\left\{ \tau(DT_t)^{-1} + {}^*(DT_t)^{-1}\, {}^*\tau \right\} \cdot\cdot\, \left\{ \sigma(DT_t)^{-1} + {}^*(DT_t)^{-1}\, {}^*\sigma \right\},$$

the previous expression can be written in the following more compact form:

$$\tilde{L}(t, U) = \frac{1}{2} \int_D \widetilde{C}(t)D(U) \cdot\cdot\, D(U)\, J_t\, dx - \left[\int_D J_t\, |U|^2\, dx \right]^{1/2}.$$

To apply Theorem 2.1, choose

$$X(t) \stackrel{\text{def}}{=} \left\{ U^t \stackrel{\text{def}}{=} U_t \circ T_t \ : \ \forall U_t \in E(\Omega_t) \right\}$$

endowed with the weak topology of $H_0^1(D)^3$. Assumption (H1) is clearly satisfied. For all $U \neq 0$, $\tilde{L}(t, U)$ is differentiable. For $0 \neq U \in H_0^1(D)^3$

$$\partial_t \tilde{L}(t, U) = \frac{1}{2} \int_D \widetilde{C}'(t)D(U) \cdot\cdot\, D(U)\, J_t + \widetilde{C}(t)D(U) \cdot\cdot\, D(U)\, J_t'\, dx$$

$$- \frac{1}{\left\{ \int_D J_t\, |U|^2\, dx \right\}^{1/2}} \int_D |U|^2\, J_t'\, dx,$$

where

$$4\,\widetilde{C}'(t)\,\tau \cdots \sigma = C\left\{\tau T'(t) + {}^*T'(t)\,{}^*\tau\right\}\cdots\left\{\sigma(DT_t)^{-1} + {}^*(DT_t)^{-1}\,{}^*\sigma\right\}$$
$$+ C\left\{\tau(DT_t)^{-1} + {}^*(DT_t)^{-1}\,{}^*\tau\right\}\cdots\left\{\sigma T'(t) + {}^*T'(t)\,{}^*\sigma\right\},$$
$$T(t) \stackrel{\text{def}}{=} (DT_t)^{-1}, \quad T'(t) = -(DT_t)^{-1}DV(t)\circ T_t.$$

Hence assumptions (H2) and (H4) are satisfied. In $t = 0$ the above expressions simplify. For $U \in E(\Omega)$

$$\partial_t \tilde{L}(0,U) = \frac{1}{2}\int_D \widetilde{C}'(0)D(U)\cdots D(U) + C\,\varepsilon(U)\cdots\varepsilon(U)\,\mathrm{div}\,V(0)\,dx$$
$$- \frac{1}{\left\{\int_D |U|^2\,dx\right\}^{1/2}}\int_D |U|^2\,\mathrm{div}\,V(0)\,dx,$$

and for all U and W

$$2\,\widetilde{C}'(0)\,D(U)\cdots D(W) = -\,C\left\{D(U)\,DV(0) + {}^*DV(0)\,{}^*D(U)\right\}\cdots\varepsilon(W)$$
$$-\,C\,\varepsilon(U)\cdots\left\{D(W)\,DV(0) + {}^*DV(0)\,{}^*D(W)\right\}.$$

This is another example of the application of Theorem 2.1 to a case where the set of minimizers is not a singleton and for which we have a complete description of the Eulerian semiderivative.

Theorem 4.8. *Let Ω be a bounded open subset of $\mathbf{R}^3$. Assume that*

$$V \in C^0([0,\tau[\,;W_0^{1,\infty}(D;\mathbf{R}^3))$$

for some bounded open Lipschitzian domain D in $\mathbf{R}^3$ such that $\overline{\Omega} \subset D$. Then

$$\frac{1}{2}d\lambda(\Omega;V) = \inf_{U\in\tilde{E}(\Omega)} -\int_\Omega \widetilde{C}'(0)D(U)\cdots D(U)\,dx$$
$$+\int_\Omega\left[\frac{1}{2}C\,\varepsilon(U)\cdots\varepsilon(U) - \lambda(\Omega)|U|^2\right]\mathrm{div}\,V(0)\,dx$$
$$= \inf_{U\in\tilde{E}(\Omega)} -\int_\Omega C\,\varepsilon(U)\cdots\left\{D(U)\,DV(0) + {}^*DV(0)\,{}^*D(U)\right\}\,dx$$
$$+\int_\Omega\left[\frac{1}{2}C\,\varepsilon(U)\cdots\varepsilon(U) - \lambda(\Omega)|U|^2\right]\mathrm{div}\,V(0)\,dx,$$
$$\tilde{E}(\Omega) \stackrel{\text{def}}{=} \left\{U \in H_0^1(\Omega)^3 : \begin{array}{c} -\,\overrightarrow{\mathrm{div}}\,(\varepsilon(U)) = \lambda(\Omega)U \ \ in\ \mathcal{D}(\Omega)' \\ and \int_\Omega |U|^2\,dx = 1 \end{array}\right\}.$$

$\lambda(\Omega)$ has a Hadamard semiderivative in the sense of Definition 3.1 (iii) in Chapter 8, which is continuous with respect to $V(0) \in W_0^{1,\infty}(D;\mathbf{R}^3)$:

$$d\lambda(\Omega;V) = d_H\lambda(\Omega;V(0)).$$

If Ω is assumed to be of class C^2, the eigenvector functions belong to $H_0^1(\Omega)^3 \cap H^2(\Omega)^3$ and the volume expressions of the previous theorem can be expressed as boundary integrals as in section 2.5. In that case we can use formulae (2.49) and (2.50) to compute the partial derivative of $L_t = L(\Omega_t, U \circ T_t^{-1})$:

$$L_t = \frac{1}{2} \int_{\Omega_t} C \, \varepsilon(U \circ T_t^{-1}) \cdot\cdot \, \varepsilon(U \circ T_t^{-1}) \, dx - \left\{ \int_{\Omega_t} |U \circ T_t^{-1}|^2 \, dx \right\}^{1/2}.$$

From formula (2.49)

$$L' \stackrel{\text{def}}{=} \partial_t L_t|_{t=0} = \frac{1}{2} \int_\Gamma C \, \varepsilon(U) \cdot\cdot \, \varepsilon(U) \, V(0) \cdot n \, d\Gamma - \frac{1}{2} \frac{\int_\Gamma |U|^2 \, V(0) \cdot n \, d\Gamma}{\left\{ \int_\Omega |U|^2 \, dx \right\}^{1/2}}$$

$$+ \int_\Omega C \, \varepsilon(U) \cdot\cdot \, \varepsilon(\dot{U}) \, dx - \frac{\int_\Omega U \cdot \dot{U} \, dx}{\left\{ \int_\Omega |U|^2 \, dx \right\}^{1/2}}.$$

For $U \in E(\Omega)$, $U = 0$ on Γ, $\lambda(\Omega)\|U\|_{L^2(\Omega)} = 1$, and $-\overrightarrow{\text{div}}\,(\varepsilon(U)) = \lambda(\Omega)U$ in Ω,

$$L' = \frac{1}{2} \int_\Gamma C \, \varepsilon(U) \cdot\cdot \, \varepsilon(U) \, V(0) \cdot n \, d\Gamma + \int_\Omega C \, \varepsilon(U) \cdot\cdot \, \varepsilon(\dot{U}) - \lambda(\Omega) \, U \cdot \dot{U} \, dx$$

$$= \int_\Gamma \frac{1}{2} C \, \varepsilon(U) \cdot\cdot \, \varepsilon(U) \, V(0) \cdot n + [C \, \varepsilon(U)]n \cdot \dot{U} \, d\Gamma.$$

But $U = 0$ on Γ implies that $D_\Gamma(U) = 0$, $D(U) = [D(U)]n \, {}^*n$ on Γ, and from identity (2.50),

$$\dot{U} = -D(U)V(0) \in H_0^1(\Omega)^3 \quad \Rightarrow \quad \dot{U} = -[D(U)]n \, V(0) \cdot n \text{ on } \Gamma$$

$$\Rightarrow L' = - \int_\Gamma [C \, \varepsilon(U)]n \cdot [D(U)]n \, V(0) \cdot n \, d\Gamma.$$

This expression can be rewritten only in terms of $\varepsilon(U)$ as follows:

$$2 \, \varepsilon(U) = D(U)n \, {}^*n + n \, {}^*(D(U)n), \quad 2 \, \varepsilon(U)n = D(U)n + (D(U)n) \cdot n \, n,$$

$$2 \, \varepsilon(U)n \cdot n = 2 \, D(U)n \cdot n,$$

$$[C \, \varepsilon(U)]n \cdot D(U)n = 2 \, [C \, \varepsilon(U)]n \cdot \varepsilon(U)n - [C \, \varepsilon(U)]n \cdot n \, (\varepsilon(U)n) \cdot n.$$

We summarize the results in the next theorem.

Theorem 4.9. *Let Ω be a bounded open subset of $\mathbf{R}^3$ of class C^2. Assume that*

$$V \in C^0([0, \tau[\, ; W_0^{1,\infty}(D; \mathbf{R}^3))$$

for some bounded open Lipschitzian domain D in $\mathbf{R}^3$ such that $\overline{\Omega} \subset D$. Then

$$d\lambda(\Omega; V) = \inf_{U \in \tilde{E}(\Omega)} - \int_{\Gamma} [C\,\varepsilon(U)]n \cdot [D(U)]n\, V(0) \cdot n\, d\Gamma,$$

$$d\lambda(\Omega; V) = \inf_{U \in \tilde{E}(\Omega)} - \int_{\Gamma} 2\,[C\,\varepsilon(U)]n \cdot \varepsilon(U)n$$

$$- [C\,\varepsilon(U)]n \cdot n\,(\varepsilon(U)n) \cdot n\, V(0) \cdot n\, d\Gamma,$$

$$\tilde{E}(\Omega) \stackrel{\text{def}}{=} \left\{ U \in H_0^1(\Omega)^3 \cap H^2(\Omega)^3 \; : \; \begin{array}{c} -\overrightarrow{\text{div}}\,(\varepsilon(U)) = \lambda(\Omega)\,U \;\; in \;\; \Omega \\[2mm] and \; \int_{\Omega} |U|^2\, dx = 1 \end{array} \right\}.$$

For the special constitutive law $C\tau = 2\mu\tau + \lambda \operatorname{tr}\tau\, I$,

$$d\lambda(\Omega; V) = \inf_{U \in \tilde{E}(\Omega)} - \int_{\Gamma} \left\{ 4\,\mu |[\varepsilon(U)]n|^2 + (\lambda - 2\mu)\,|\operatorname{tr}\varepsilon(U)|^2 \right\} V(0) \cdot n\, d\Gamma.$$

$\lambda(\Omega)$ has a Hadamard semiderivative in the sense of Definition 3.1 (iii) in Chapter 8, which is continuous with respect to $V(0) \in C_0(D; \mathbf{R}^3)$:

$$d\lambda(\Omega; V) = d_H\lambda(\Omega; V(0)).$$

5 Saddle Point Formulation and Function Space Parametrization

5.1 An Illustrative Example

Let Ω be a bounded open domain in $\mathbf{R}^N$ with a smooth boundary Γ. Let $y = y(\Omega)$ be the solution of the Neumann problem

$$-\triangle y + y = f \;\; \text{in} \;\; \Omega, \qquad \frac{\partial y}{\partial n} = 0 \;\; \text{on} \;\; \Gamma, \tag{5.1}$$

where f is a fixed function in $H^1(\mathbf{R}^N)$. Associate with $y(\Omega)$ the objective function

$$J(\Omega) = \frac{1}{2} \int_{\Omega} |y(\Omega) - y_d|^2\, dx, \tag{5.2}$$

where y_d is a fixed function in $H^1(\mathbf{R}^N)$.

The solution of (5.1) coincides with the minimizing element of the following variational problem:

$$\inf \left\{ E(\Omega, \varphi) : \varphi \in H^1(\Omega) \right\}, \tag{5.3}$$

$$E(\Omega, \varphi) \stackrel{\text{def}}{=} \frac{1}{2} \int_{\Omega} \left\{ |\nabla \varphi|^2 + \varphi^2 - 2\,f\varphi \right\} dx. \tag{5.4}$$

The minimizing element y of (5.4) is the solution in $H^1(\Omega)$ of Euler's equation:

$$dE(\Omega, y; \varphi) = 0, \qquad \forall \varphi \in H^1(\Omega), \tag{5.5}$$

$$dE(\Omega, y; \varphi) = \int_{\Omega} [\nabla y \cdot \nabla \varphi + y\varphi - f\varphi]\, dx \tag{5.6}$$

which is the *variational equation* for y.

The *objective function* $J(\Omega)$ is a shape functional, and the solution of (5.1) will be called the *state*. It is convenient to introduce the *objective function*

$$F(\Omega, \varphi) \stackrel{\text{def}}{=} \frac{1}{2} \int_\Omega |\varphi - y_d|^2 \, dx \tag{5.7}$$

which clearly expresses the dependence on Ω and φ. To sum up, we consider the objective function

$$J(\Omega) = F\big(\Omega, y(\Omega)\big), \tag{5.8}$$

where $y = y(\Omega)$ is the solution of

$$y \in H^1(\Omega), \quad \forall \varphi \in H^1(\Omega), \quad dE(\Omega, y; \varphi) = 0. \tag{5.9}$$

We wish to find an expression for the shape derivative $dJ(\Omega; V)$.

5.2   Saddle Point Formulation

The basic approach is the one of control theory. Equation (5.1) (or, in its variational form, (5.9)) is considered as a state constraint in the minimization problem. We construct a Lagrangian functional by introducing a *Lagrange multiplier* function or the so-called *adjoint state* ψ:

$$G(\Omega, \varphi, \psi) = F(\Omega, \varphi) + dE(\Omega, \varphi; \psi). \tag{5.10}$$

Then the objective function is given by

$$J(\Omega) = \min_{\varphi \in H^1(\Omega)} \sup_{\psi \in H^1(\Omega)} G(\Omega, \varphi, \psi) \tag{5.11}$$

since

$$\sup_{\psi \in H^1(\Omega)} G(\Omega, \varphi, \psi) = \begin{cases} F(\Omega, y(\Omega)) & \text{if } \varphi = y(\Omega), \\ +\infty & \text{if } \varphi \neq y(\Omega). \end{cases} \tag{5.12}$$

In our example the Lagrangian G is convex and continuous with respect to the variable φ and concave and continuous with respect to the variable ψ. Moreover, the space $H^1(\Omega)$ is convex and closed. So the functional G has a saddle point if and only if the saddle point equations have a solution (y, p) (cf. Ekeland and Temam [1]):

$$p \in H^1(\Omega), \quad dG(\Omega, y, p; 0, \psi) = 0, \ \forall \psi \in H^1(\Omega), \tag{5.13}$$

$$y \in H^1(\Omega), \quad dG(\Omega, y, p; \varphi, 0) = 0, \ \forall \varphi \in H^1(\Omega). \tag{5.14}$$

They are completely equivalent to

$$y \in H^1(\Omega), \quad dE(\Omega, y; \psi) = 0, \ \forall \psi \in H^1(\Omega), \tag{5.15}$$

$$p \in H^1(\Omega), \quad dF(\Omega, y; \varphi) + d^2 E(\Omega, y; p; \varphi) = 0, \ \forall \varphi \in H^1(\Omega), \tag{5.16}$$

or

$$-\triangle y + y = f \text{ in } \Omega, \quad \frac{\partial y}{\partial n} = 0 \text{ on } \Gamma, \tag{5.17}$$

$$-\triangle p + p + y - y_d = 0 \text{ in } \Omega, \quad \frac{\partial p}{\partial n} = 0 \text{ on } \Gamma. \tag{5.18}$$

System (5.15)–(5.16) has a unique solution in $H^1(\Omega) \times H^1(\Omega)$ which coincides with the unique saddle point of $G(\Omega, \varphi, \psi)$ in $H^1(\Omega) \times H^1(\Omega)$.

5.3 Function Space Parametrization

We have shown that the objective function $J(\Omega)$ can be expressed as a MinMax of a functional G with a unique saddle point (y, p) which is completely characterized by the variational equations (5.15)–(5.16). The same result holds when Ω is transformed into a domain $\Omega_t = T_t(\Omega)$ under the action of the velocity field V for $t \geq 0$:

$$J(\Omega_t) = \min_{\varphi \in H^1(\Omega_t)} \sup_{\psi \in H^1(\Omega_t)} G(\Omega_t, \varphi, \psi), \tag{5.19}$$

where the saddle point (y_t, p_t) is completely characterized by

$$y_t \in H^1(\Omega_t), \quad dE(\Omega_t, y_t; \psi) = 0, \quad \forall \psi \in H^1(\Omega_t), \tag{5.20}$$

$$p_t \in H^1(\Omega_t), \quad dF(\Omega_t, y_t; \varphi) + d^2 E(\Omega_t, y_t; p_t, \varphi) = 0, \quad \forall \varphi \in H^1(\Omega_t). \tag{5.21}$$

We are looking for a theorem that will give an expression for the derivative of a MinSup with respect to a parameter $t \geq 0$. However, in (5.19) the space $H^1(\Omega_t)$ depends on the parameter t. To get around this difficulty and obtain a MinSup expression for $J(\Omega_t)$ over spaces that are independent of $t \geq 0$, we introduce the following parametrization:

$$H^1(\Omega_t) = \left\{ \varphi \circ T_t^{-1} : \varphi \in H^1(\Omega) \right\} \tag{5.22}$$

since T_t and T_t^{-1} are diffeomorphisms. This parametrization does not affect the value of the saddle point $J(\Omega_t)$ but will change the parametrization of the functional G:

$$J(\Omega_t) = \inf_{\varphi \in H^1(\Omega)} \sup_{\psi \in H^1(\Omega)} G(\Omega_t, \varphi \circ T_t^{-1}, \psi \circ T_t^{-1}). \tag{5.23}$$

This parametrization is apparently unique to *shape analysis*. It amounts to introducing the new Lagrangian functional for all φ and ψ in $H^1(\Omega)$:

$$\widetilde{G}(t, \varphi, \psi) = G\left(T_t(\Omega), \varphi \circ T_t^{-1}, \psi \circ T_t^{-1} \right). \tag{5.24}$$

Our next objective is to find an expression for the limit

$$dg(0) = \lim \frac{g(t) - g(0)}{t}, \tag{5.25}$$

where

$$g(t) = J(\Omega_t) = \inf_{\varphi \in H^1(\Omega)} \sup_{\psi \in H^1(\Omega)} \widetilde{G}(t, \varphi, \psi). \tag{5.26}$$

This will be done in the next section.

Before closing, it is useful to look at the expression for $\widetilde{G}$ and the resulting saddle point (y^t, p^t) in $H^1(\Omega) \times H^1(\Omega)$. By definition, $\widetilde{G}$ is given by the expression

$$\begin{aligned}
\widetilde{G}(t, \varphi, \psi) = &\frac{1}{2} \int_{\Omega_t} |\varphi \circ T_t^{-1} - y_d|^2 \, dx \\
&+ \int_{\Omega_t} \left[\nabla(\varphi \circ T_t^{-1}) \cdot \nabla(\psi \circ T_t^{-1}) \right. \\
&\left. \qquad + (\varphi \circ T_t^{-1}) - f\,(\psi \circ T_t^{-1}) \right] \, dx,
\end{aligned} \tag{5.27}$$

and its saddle point is the solution of the variational equations

$$y^t \in H^1(\Omega), \text{ and for all } \psi \text{ in } H^1(\Omega),$$

$$\begin{aligned}
\int_{\Omega_t} &\left[\nabla(y^t \circ T_t^{-1}) \cdot \nabla(\psi \circ T_t^{-1}) \right. \\
&\left. + (y^t \circ T_t^{-1})(\psi \circ T_t^{-1}) - f(\psi \circ T_t^{-1}) \right] dx = 0,
\end{aligned} \tag{5.28}$$

$$p^t \in H^1(\Omega), \text{ and for all } \phi \text{ in } H^1(\Omega),$$

$$\begin{aligned}
\int_{\Omega_t} &[(y^t \circ T_t^{-1} - y_d)(\varphi \circ T_t^{-1}) + \nabla(y^t \circ T_t^{-1}) \cdot \nabla(\varphi \circ T_t^{-1}) \\
&+ (y^t \circ T_t^{-1})(\varphi \circ T_t^{-1})] \, dx = 0.
\end{aligned} \tag{5.29}$$

It is readily seen that $(y^t \circ T_t^{-1}, p^t \circ T_t^{-1})$ coincide with the saddle point (y_t, p_t) in $H^1(\Omega_t) \times H^1(\Omega_t)$:

$$y_t = y^t \circ T_t^{-1}, \quad p_t = p^t \circ T_t^{-1},$$

or equivalently

$$y^t = y_t \circ T_t, \quad p^t = p_t \circ T_t. \tag{5.30}$$

The solutions (y^t, p^t) can easily be interpreted as the solutions (y_t, p_t) on Ω_t transported back to the fixed domain Ω by the transformation T_t.

In view of this observation, we can rewrite expressions (5.27) to (5.29) on the fixed domain Ω by using the coordinate transformation T_t. Expression (5.27) becomes

$$\begin{aligned}
\widetilde{G}(t, \varphi, \psi) = &\frac{1}{2} \int_{\Omega} |\varphi - y_d \circ T_t|^2 J_t dx \\
&+ \int_{\Omega} A(t) \nabla\varphi \cdot \nabla\psi + J_t \left[\varphi\psi - (f \circ T_t)\psi \right] dx,
\end{aligned} \tag{5.31}$$

where for $t > 0$ small,

$$DT_t = \text{Jacobian matrix of } T_t, \tag{5.32}$$

$$J_t = \det DT_t \quad (\text{since } \det DT_t = |\det DT_t| \text{ for } t \geq 0 \text{ small}), \tag{5.33}$$

$$A(t) = J_t [DT_t]^{-1}\, {}^*[DT_t]^{-1}. \tag{5.34}$$

Similarly the variational equations (5.28)–(5.29) reduce to

$$y_t \in H^1(\Omega) \text{ and } \forall \psi \in H^1(\Omega)$$
$$\int_\Omega \left\{ A(t)\nabla y^t \cdot \nabla \psi + J_t \left[y^t \psi - (f \circ T_t), \psi \right] \right\} dx = 0, \tag{5.35}$$

$$p^t \in H^1(\Omega) \text{ and } \forall \varphi \in H^1(\Omega)$$
$$\int_\Omega \left\{ A(t)\nabla p^t \cdot \nabla \varphi + J_t \left[p^t \varphi + (y^t - y_d \circ T_t)\varphi \right] \right\} dx = 0. \tag{5.36}$$

5.4 Differentiability of a Saddle Point with Respect to a Parameter

Consider a functional

$$G \colon [0, \tau] \times X \times Y \to \mathbf{R} \tag{5.37}$$

for some $\tau > 0$ and sets X and Y. For each t in $[0, \tau]$ define

$$g(t) = \inf_{x \in X} \sup_{y \in Y} G(t, x, y) \tag{5.38}$$

and the sets

$$X(t) = \left\{ x^t \in X : \sup_{y \in Y} G(t, x^t, y) = g(t) \right\}, \tag{5.39}$$

$$Y(t, x) = \left\{ y^t \in Y : G(t, x, y^t) = \sup_{y \in Y} G(t, x, y) \right\}. \tag{5.40}$$

Similarly define

$$h(t) = \sup_{y \in Y} \inf_{x \in X} G(t, x, y) \tag{5.41}$$

and the sets

$$Y(t) = \left\{ y^t \in Y : \inf_{x \in X} G(t, x, y^t) = h(t) \right\}, \tag{5.42}$$

$$X(t, y) = \left\{ x^t \in X : G(t, x^t, y) = \inf_{x \in X} G(t, x, y) \right\}. \tag{5.43}$$

In general we always have the inequality

$$h(t) \leq g(t). \tag{5.44}$$

To complete the set of notations, we introduce the set of *saddle points*

$$S(t) = \{(x, y) \in X \times Y : g(t) = G(t, x, y) = h(t)\} \tag{5.45}$$

which may be empty.

Our objective is to find realistic conditions under which the limit

$$dg(0) = \lim_{t \searrow 0} \frac{g(t) - g(0)}{t} \tag{5.46}$$

exists. A case of special interest is when G has a saddle point for all t in $[0, \tau]$. It can be viewed as an extension of Theorem 2.1 in section 2.3 on the differentiability of a Min with respect to a parameter. It is used when the functional to be minimized is a function of the state, which is itself a function of the domain through the boundary value problem. In that case the saddle point equations coincide with the "state equation" and the "adjoint state equation" as illustrated in the previous section. The main advantage of this approach is to avoid the problem of the existence and characterization of the derivative of the state x^t with respect to t. In a control problem this would be the directional derivative of the state with respect to the control variable. In particular, it is not necessary to invoke any implicit function theorem with possibly restrictive differentiability conditions. It will be sufficient to check two continuity conditions for the set-valued maps $X(\cdot)$ and $Y(\cdot)$. To complete this discussion we recall the following.

Lemma 5.1. *Fix t in $[0, \tau]$. Then*

$$\forall (x^t, y^t) \in X(t) \times Y(t), \quad h(t) \le G(t, x^t, y^t) \le g(t), \tag{5.47}$$

and if $h(t) = g(t)$,

$$X(t) \times Y(t) = S(t). \tag{5.48}$$

Proof. (i) If $X(t) \times Y(t) = \varnothing$ there is nothing to prove. If there exist $x^t \in X(t)$ and $y^t \in Y(t)$, then by definition

$$h(t) = \inf_{x \in X} G(t, x, y^t) \le G(t, x^t, y^t) \le \sup_{y \in Y} G(t, x^t, y^t) = g(t). \tag{5.49}$$

(ii) If $h(t) = g(t)$, then in view of (5.49), $X(t) \times Y(t) \subset S(t)$. Conversely, if there exists $(x^t, y^t) \in S(t)$, then $h(t) = G(t, x^t, y^t) = g(t)$, and by the definition of $X(t)$ and $Y(t)$, $(x^t, y^t) \in X(t) \times Y(t)$. $\qquad\square$

It is important to keep in mind that identity (5.48) is always true when

$$h(t) = \sup_{y \in Y} \inf_{x \in X} G(t, x, y) = \inf_{x \in X} \sup_{y \in Y} G(t, x, y) = g(t)$$

but that $S(t)$ may be empty.

Theorem 5.1 (Correa and Seeger [1]). *Let the sets X and Y, the real number $\tau > 0$, and the functional*

$$G \colon [0, \tau] \times X \times Y \to \mathbf{R}$$

be given. Assume that the following assumptions hold:

(H1) $S(t) \neq \varnothing$, $0 \leq t \leq \tau$;

(H2) *for all (x, y) in $[\cup\{X(t) : 0 \leq t \leq \tau\} \times Y(0)] \cup [X(0) \times \cup\{Y(t) : 0 \leq t \leq \tau\}]$ the partial derivative $\partial_t G(t, x, y)$ exists everywhere in $[0, \tau]$;*

(H3) *there exists a topology $\mathcal{T}_X$ on X such that for any sequence $\{t_n : 0 < t_n \leq \tau\}$, $t_n \to t_0 = 0$, $\exists x^0 \in X(0)$, $\exists$ a subsequence $\{t_{n_k}\}$ of $\{t_n\}$, and for each $k \geq 1$, $\exists x_{n_k} \in X(t_{n_k})$ such that*

> (i) $x_{n_k} \to x^0$ *in the $\mathcal{T}_X$-topology, and*
>
> (ii) *for all y in $Y(0)$,*

$$\liminf_{\substack{t \searrow 0 \\ k \to \infty}} \partial_t G(t, x_{n_k}, y) \geq \partial_t G(0, x^0, y); \tag{5.50}$$

(H4) *there exists a topology $\mathcal{T}_Y$ on Y such that for any sequence $\{t_n : 0 < t_n \leq \tau\}$, $t_n \to t_0 = 0$, $\exists y^0 \in Y(0)$, $\exists$ a subsequence $\{t_{n_k}\}$ of $\{t_n\}$, and for each $k \geq 1$, $\exists y_{n_k} \in Y(t_{n_k})$ such that*

> (i) $y_{n_k} \to y^0$ *in the $\mathcal{T}_Y$-topology, and*
>
> (ii) *for all x in $X(0)$,*

$$\limsup_{\substack{t \searrow 0 \\ k \to \infty}} \partial_t G(t, x, y_{n_k}) \leq \partial_t G(0, x, y^0). \tag{5.51}$$

Then there exists $(x^0, y^0) \in X(0) \times Y(0)$ such that

$$dg(0) = \inf_{x \in X(0)} \sup_{y \in Y(0)} \partial_t G(0, x, y) = \partial_t G(0, x^0, y^0)$$

$$= \sup_{y \in Y(0)} \inf_{x \in X(0)} \partial_t G(0, x, y). \tag{5.52}$$

Thus (x^0, y^0) is a saddle point of $\partial_t G(0, x, y)$ on $X(0) \times Y(0)$.

Proof. (i) We first establish upper and lower bounds to the differential quotient

$$\frac{\Delta(t)}{t}, \quad \Delta(t) \overset{\text{def}}{=} g(t) - g(0).$$

Choose arbitrary x_0 in $X(0)$, x_t in $X(t)$, y_0 in $Y(0)$, and y_t in $Y(t)$. Then by definition,

$$G(t, x_t, y_0) \leq G(t, x_t, y_t) \leq G(t, x_0, y_t),$$
$$-G(0, x_t, y_0) \leq -G(0, x_0, y_0) \leq -G(0, x_0, y_t).$$

Add up the above two chains of inequalities to obtain

$$G(t, x_t, y_0) - G(0, x_t, y_0) \le \Delta(t) \le G(t, x_0, y_t) - G(0, x_0, y_t).$$

By assumption (H2), there exist θ_t, $0 < \theta_t < 1$, and α_t, $0 < \alpha_t < 1$, such that

$$G(t, x_t, y_0) - G(0, x_t, y_0) = t\partial_t G(\theta_t t, x_t, y_0),$$
$$G(t, x_0, y_t) - G(0, x_0, y_t) = t\partial_t G(\alpha_t t, x_0, y_t),$$

and by dividing by $t > 0$,

$$\partial_t G(\theta_t t, x_t, y_0) \le \frac{\Delta(t)}{t} \le \partial_t G(\alpha_t t, x_0, y_t). \tag{5.53}$$

(ii) Define

$$\underline{dg}(0) = \liminf_{t \searrow 0} \frac{\Delta(t)}{t}, \quad \overline{dg}(0) = \limsup_{t \searrow 0} \frac{\Delta(t)}{t}.$$

There exists a sequence $\{t_n : 0 < t_n \le \tau\}$, $t_n \to 0$, such that

$$\lim_{n \to \infty} \frac{\Delta(t_n)}{t_n} = \underline{dg}(0).$$

By assumption (H3), $\exists x^0 \in X(0)$, $\exists$ a subsequence $\{t_{n_k}\}$ of $\{t_n\}$ for each $k \ge 1$, $\exists x_{n_k} \in X(t_{n_k})$ such that $x_{n_k} \to x^0$ in $\mathcal{T}_X$, and

$$\forall y \in Y(0), \quad \liminf_{\substack{t \searrow 0 \\ k \to \infty}} \partial_t G(t, x_{n_k}, y) \ge \partial_t G(0, x^0, y).$$

Thus from the first part of the estimate (5.53) for any $y \in Y(0)$ and $t = t_{n_k}$,

$$\partial_t G(\theta_{t_{n_k}} t_{n_k}, x_{n_k}, y) \le \frac{\Delta(t_{n_k})}{t_{n_k}}$$

and

$$\partial_t G(0, x^0, y) \le \liminf_{k \to \infty} \partial_t G(\theta_{t_{n_k}} t_{n_k}, x_{n_k}, y) \le \lim_{k \to \infty} \frac{\Delta(t_{n_k})}{t_{n_k}} = \underline{dg}(0).$$

Therefore

$$\exists x^0 \in X(0), \forall y \in Y(0), \quad \partial_t G(0, x^0, y) \le \underline{dg}(0)$$

and

$$\inf_{x \in X(0)} \sup_{y \in Y(0)} \partial_t G(0, x, y) \le \sup_{y \in Y(0)} \partial_t G(0, x^0, y) \le \underline{dg}(0). \tag{5.54}$$

By a dual argument and assumption (H4) we also obtain

$$\exists y^0 \in Y(0), \forall x \in X(0), \quad \partial_t G(0, x, y^0) \ge \overline{dg}(0),$$
$$\overline{dg}(0) \le \inf_{x \in X(0)} \partial_t G(0, x, y^0) \le \sup_{y \in Y(0)} \inf_{x \in X(0)} \partial_t G(0, x, y), \tag{5.55}$$

and necessarily

$$\inf x \in X(0) \sup_{y \in Y(0)} \partial_t G(0, x, y)$$

$$= \underline{dg}(0) = \overline{dg}(0) = \sup_{y \in Y(0)} \inf_{x \in X(0)} \partial_t G(0, x, y).$$

In particular, from (5.54) and (5.55),

$$\sup_{y \in Y(0)} \partial_t G(0, x^0, y) = dg(0) = \inf_{x \in X(0)} \inf \partial_t G(0, x, y^0)$$

and (x^0, y^0) is a saddle point of $\partial_t G(0, \cdot, \cdot)$. $\square$

Remark 5.1.
In the applications this formulation of the theorem presents some definite technical
advantages over its original version.

(i) From identity (5.52), $\partial_t G(0, \cdot, \cdot)$ has a saddle point with respect to $X(0) \times Y(0)$.

(ii) Another important feature is the use of subsequences in assumptions (H3)
and (H4). This makes it possible to work with weak topologies in reflexive
Banach spaces and use the eventual boundedness of the sets of saddle points.

(iii) Finally, assumption (H2) and conditions (5.50) and (5.51) in (H3) and (H4)
need only be checked on the family of saddle points at $t = 0$. For instance, the
first part of assumptions (H3) and (H4) could be satisfied in $H^1(\Omega) \times H^1(\Omega)$.
Yet, if the saddle points are smoother, say in $H^2(\Omega) \times H^2(\Omega)$, this extra
smoothness can be used to satisfy (H2) and (5.50) and (5.51) in (H3) and
(H4).

5.5 Application of the Theorem

Our example has a unique saddle point (y^t, p^t) for $t \geq 0$ small, and we can use
the corollary to Theorem 5.1. The set-valued maps X and Y reduce to ordinary
functions

$$t \mapsto X(t) = y^t, \quad t \mapsto Y(t) = p^t, \tag{5.56}$$

and it is sufficient to show their continuity at $t = 0$ in $H^1(\Omega)$. So we now check
assumptions (H1) to (H4).

Assume that V belongs to $\mathcal{V}^1$, that is, $\mathcal{D}^1(\mathbf{R}^N, \mathbf{R}^N)$, and that f and y belong
to $H^1(\mathbf{R}^N)$. Choose $\tau > 0$ small enough such that

$$J_t = \det DT_t = |\det DT_t| = |J_t|, \quad 0 \leq t \leq \tau, \tag{5.57}$$

and that there exist constants $0 < \alpha < \beta$ such that

$$\forall \xi \in \mathbf{R}^N, \quad \alpha|\xi|^2 \leq A(t)\,\xi \cdot \xi \leq \beta|\xi|^2 \text{ and } \alpha \leq J_t \leq \beta. \tag{5.58}$$

Since the bilinear forms associated with (5.35) and (3.34) are coercive, there exists a unique pair (y^t, p^t) solution of the system (5.35)–(5.36). Hence

$$\forall t \in [0, \tau], \quad X(t) = \{y^t\} \neq \varnothing, \ Y(t) = \{p^t\} \neq \varnothing. \tag{5.59}$$

So assumption (H1) is satisfied. To check (H2) we use expression (5.31) and compute for φ and ψ in $H^1(\Omega)$:

$$\partial_t \widetilde{G}(t, \varphi, \psi)$$
$$= \int_\Omega \left\{ \frac{1}{2}(\varphi - y_d \circ T_t)^2 \operatorname{div} V_t - (\varphi - y_d \circ T_t)\nabla y_d \cdot V_t J_t \right\} dx \tag{5.60}$$
$$+ \int_\Omega \left\{ A'(t)\nabla\varphi \cdot \nabla\psi + \operatorname{div} V_t(\varphi\psi - f \circ T_t \psi) - J_t \nabla f \cdot V_t \psi \right\} dx,$$

where

$$V_t(X) \stackrel{\text{def}}{=} V(T_t(X)), \ A'(t) = (\operatorname{div} V_t)I - {}^*DV_t - DV_t, \tag{5.61}$$

I is the identity matrix on $\mathbf{R}^N$, and DV_t is the Jacobian matrix of V_t. By the choice of V in $\mathcal{D}^1(\mathbf{R}^N, \mathbf{R}^N)$, $t \mapsto V_t$ and $t \mapsto DV_t$ are continuous on $[0, \tau]$. Moreover, f and y_d belong to $H^1(\mathbf{R}^N)$. As a result, expression (5.60) is well defined and $\partial_t \widetilde{G}(t, \varphi, \psi)$ exists everywhere in $[0, \tau]$ for all φ and ψ in $H^1(\Omega)$. This can be proved in many ways. For instance, we establish (5.60) for f and y_d in $\mathcal{D}(\mathbf{R}^N)$. Then we show that the affine map $(f, y_d) \mapsto \partial_t \widetilde{G}(\cdot, \varphi, \psi)$ is continuous from $H^1(\mathbf{R}^N) \times H^1(\mathbf{R}^N)$ to $C^1([0, \tau])$. So it extends by uniform continuity to all (f, y_d) in $H^1(\mathbf{R}^N) \times H^1(\mathbf{R}^N)$ and density of $\mathcal{D}(\mathbf{R}^N)$ in $H^1(\mathbf{R}^N)$. Assumption (H2) is satisfied.

To check assumptions (H3)(i) and (H4)(i), we first show that for any sequence $\{t_n\} \subset [0, \tau]$, $t_n \to 0$, there exists a subsequence of $\{y^{t_n}\}$, still denoted $\{y^{t_n}\}$, such that

$$y^{t_n} \rightharpoonup y^0 = y \text{ in } H^1(\Omega)\text{-weak},$$
$$p^{t_n} \rightharpoonup p^0 = p \text{ in } H^1(\Omega)\text{-weak},$$

where (y, p) is the solution of system (5.15)–(5.16) or (5.17)–(5.18). By the choice of τ satisfying condition (5.58), there exists a constant $c > 0$ such that

$$\alpha \|y^t\|_{H^1(\Omega)} \leq \beta c \|f\|_{L^2(\mathbf{R}^N)},$$
$$\alpha \|p^t\|_{H^1(\Omega)} \leq \beta c \|y^t - y_d\|_{L^2(\Omega)}.$$

So the pair $\{y^t, p^t\}$ is bounded in $H^1(\Omega) \times H^1(\Omega)$ and there exists a subsequence $\{y^{t_n}, p^{t_n}\}$ and a pair (z, q) in $H^1(\Omega) \times H^1(\Omega)$ such that

$$y^{t_n} \rightharpoonup z \text{ in } H^1(\Omega)\text{-weak, and } p^{t_n} \rightharpoonup q \text{ in } H^1(\Omega)\text{-weak}.$$

The pair (z, q) can be characterized by going to the limit in the variational equations (5.35)–(5.36):

$$\int_\Omega \left\{ A(t_n)\nabla y^{t_n} \cdot \nabla\psi + J_{t_n}\left[y^{t_n}\psi - (f \circ T_{t_n})\psi\right] \right\} dx = 0,$$
$$\int_\Omega \left\{ A(t_n)\nabla p^{t_n} \cdot \nabla\varphi + J_{t_n}\left[p^{t_n}\varphi + (y^{t_n} - y_d \circ T_{t_n})\varphi\right] \right\} dx = 0.$$

So we proceed as in section 2.4 and use Lemma 2.1 to obtain

$$\forall \psi, \quad \int_\Omega \{\nabla z \cdot \nabla \psi + z\psi - f\psi\}\, dx = 0,$$

$$\forall \varphi, \quad \int_\Omega \{\nabla q \cdot \nabla \varphi + q\varphi + (z - y_d)\varphi\}\, dx = 0.$$

By uniqueness $(z, q) = (y, p)$. We now proceed as in section 2.4 and prove that

$$y^{t_n} \to y \text{ in } H^1(\Omega)\text{-strong}, \quad p^{t_n} \to p \text{ in } H^1(\Omega)\text{-strong}$$

by the same argument. So assumptions (H3)(i) and (H4)(i) are satisfied for the strong topology of $H^1(\Omega)$. Finally, assumptions (H3)(ii) and (H4)(ii) are readily satisfied in view of the strong continuity of $(t, \varphi) \mapsto \partial_t \widetilde{E}(t, \varphi, \psi)$ and $(t, \psi) \mapsto \partial_t \widetilde{E}(t, \varphi, \psi)$. In fact it would have been sufficient to check assumption (H3) with $H^1(\Omega)$-strong and (H4) with $H^1(\Omega)$-weak.

So all assumptions of Theorem 5.1 are satisfied and

$$dJ(\Omega; V) = \int_\Omega \left\{ \frac{1}{2}(y - y_d)^2 \mathrm{div}\, V - (y - y_d)\nabla y_d \cdot V \right\} dx$$

$$+ \int_\Omega \{A'(0)\nabla y \cdot \nabla p + \mathrm{div}\, V(0), (yp - fp) - \nabla f \cdot V(0)p\}\, dx, \quad (5.62)$$

where (y, p) is the solution of equations (5.17)–(5.18) or, in variational form,

$$y \in H^1(\Omega), \quad \forall \psi \in H^1(\Omega), \quad \int_\Omega \{\nabla y \cdot \nabla \psi + y\psi - f\psi\}\, dx = 0, \quad (5.63)$$

$$p \in H^1(\Omega), \quad \forall \varphi \in H^1(\Omega), \quad \int_\Omega \{\nabla p \cdot \nabla \varphi + p\varphi + (y - y_d)\varphi\}\, dx = 0. \quad (5.64)$$

5.6 Domain and Boundary Expressions for the Shape Gradient

Expression (5.62) for the shape gradient is a volume or domain integral. For y_d and f in $H^1(\mathbf{R}^N)$ it is readily seen that the map

$$V \mapsto dJ(\Omega; V) : \mathcal{D}^1(\mathbf{R}^N, \mathbf{R}^N) \to \mathbf{R} \quad (5.65)$$

is linear and continuous. So by Corollary 1 to the structure theorem (Theorem 3.5 in section 3.3 of Chapter 8), we know that for a domain Ω with a C^2-boundary Γ, there exists a scalar distribution $g(\Gamma)$ in $\mathcal{D}^1(\Gamma)'$ such that

$$dJ(\Omega; V) = \langle g(\Gamma), V \cdot n \rangle. \quad (5.66)$$

We now further characterize this boundary expression. In view of the assumptions on f, y_d, and Ω, the pair (y, p) is the solution in $H^2(\Omega) \times H^2(\Omega)$ of the system

$$-\triangle y + y = f \text{ in } \Omega, \quad \frac{\partial y}{\partial n} = 0 \text{ on } \Gamma, \quad (5.67)$$

$$-\triangle p + p + (y - y_d) = 0 \text{ in } \Omega, \quad \frac{\partial p}{\partial n} = 0 \text{ on } \Gamma. \quad (5.68)$$

Similarly, for V in $\mathcal{D}^1(\mathbf{R}^N, \mathbf{R}^N)$ the system

$$-\operatorname{div}\left[A(t)\nabla y^t\right] + J_t y^t = J_t f \circ T_t \text{ in } \Omega, \quad \frac{\partial y^t}{\partial n} = 0 \text{ on } \Gamma, \qquad (5.69)$$

$$-\operatorname{div}\left[A(t)\nabla p^t\right] + J_t p^t + (y^t - y_d \circ T_t)J_t = 0 \text{ in } \Omega, \quad \frac{\partial p^t}{\partial n} = 0 \text{ on } \Gamma \qquad (5.70)$$

has a unique solution in $H^2(\Omega) \times H^2(\Omega)$ instead of $H^1(\Omega) \times H^1(\Omega)$. With this extra smoothness we can use the formula

$$\frac{d}{dt}\int_{\Omega_t} F(t,x)\,dx\bigg|_{t=0} = \int_\Gamma F(0,x)V(0)\cdot n\,d\Gamma + \int_\Omega \frac{\partial F}{\partial t}(0,x)\,dx \qquad (5.71)$$

for a sufficiently smooth function $F\colon [0,\tau]\times\mathbf{R}^N \to \mathbf{R}$. We easily obtain

$$\partial_t\widetilde{G}(0,\varphi,\psi) = \int_\Gamma \left\{\frac{1}{2}(\varphi - y_d)^2 + \nabla\varphi\cdot\nabla\psi + \varphi\psi - f\psi\right\}\,d\Gamma$$

$$+ \int_\Omega \left\{(\varphi - y_d)\dot\varphi + \nabla\psi\cdot\nabla\dot\varphi + \psi\dot\varphi\right\}\,dx$$

$$+ \int_\Omega \left\{\nabla\varphi\cdot\nabla\dot\psi + \varphi\dot\psi - f\dot\psi\right\}\,dx, \qquad (5.72)$$

where

$$\dot\varphi = \frac{d}{dt}\varphi\circ T_t^{-1}\bigg|_{t=0} = -\nabla\varphi\cdot V(0) \qquad (5.73)$$

and

$$\dot\psi = \frac{d}{dt}\psi\circ T_t^{-1}\bigg|_{t=0} = -\nabla\psi\cdot V(0). \qquad (5.74)$$

Now substitute for (φ, ψ) the solution (y, p) of (5.63)–(5.64):

$$\partial_t\widetilde{G}(0,y,p)$$
$$= \int_\Gamma \left\{\frac{1}{2}(y - y_d)^2 + \nabla y\cdot\nabla p + yp - fp\right\}V\cdot n\,d\Gamma$$

$$+ \int_\Omega \left\{(y - y_d)(-\nabla y\cdot V) + \nabla p\cdot\nabla(-\nabla y\cdot V) + p(-\nabla p\cdot V)\right\}\,dx$$

$$+ \int_\Omega \left\{\nabla y\cdot\nabla(-\nabla p\cdot V) + y(-\nabla p\cdot V) - f(-\nabla p\cdot V)\right\}\,dx. \qquad (5.75)$$

We recognize that the second term is (5.64) with $\varphi = -\nabla y\cdot V$ and that the third term is (3.63) with $\psi = -\nabla p\cdot V$. So they are both zero, and finally

$$dJ(\Omega;V) = \int_\Gamma \left\{\frac{1}{2}(y - y_d)^2 + \nabla y\cdot\nabla p + yp - fp\right\}V\cdot n\,d\Gamma. \qquad (5.76)$$

It must be emphasized that this last expression has been obtained under the assumption that both y and p belong to $H^2(\Omega)$. We shall see later that shape gradient can also be obtained by our technique for the finite element approximations of y and p. However, for piecewise linear elements, formula (5.76) fails since the finite element solutions y_h and p_h belong to $H^1(\Omega_h)$ but not to $H^2(\Omega_h)$. However, the domain formula (5.62) will remain true. The crucial point is that for the continuous problem, a smooth boundary plus f and y_d in $H^1(\mathbf{R}^N)$ put the solution (y,p) in $H^2(\Omega) \times H^2(\Omega)$. However, the smoothness of the finite element solution (y_h, p_h) cannot be improved.

6 Multipliers and Function Space Embedding

6.1 The Nonhomogeneous Dirichlet Problem

Let Ω be a bounded open domain in $\mathbf{R}^N$ with a sufficiently smooth boundary Γ. Let $y = y(\Omega)$ be the solution of the nonhomogeneous Dirichlet problem

$$-\Delta y = f \text{ in } \Omega, \quad y = g \text{ on } \Gamma, \tag{6.1}$$

where f and g are fixed functions in $H^{1/2+\varepsilon}(\mathbf{R}^N)$ and $H^{2+\varepsilon}(\mathbf{R}^N)$, respectively, for some arbitrary fixed $\varepsilon > 0$. Associate with the solution of (6.1) the objective function

$$J(\Omega) = \frac{1}{2} \int_{\Omega} |y(\Omega) - y_d|^2 \, dx \tag{6.2}$$

for some fixed function y_d in $H^{1/2+\varepsilon}(\mathbf{R}^N)$ and some arbitrary fixed $\varepsilon > 0$. We want to compute the derivative of $J(\Omega)$ with respect to Ω subject to the state equation system (6.1). Our objective is to transform this problem into finding the saddle point of a volume Lagrangian functional. This technique can be applied to other boundary value problems with Dirichlet conditions.

6.2 A Saddle Point Formulation of the State Equation

When $g = 0$ problem (6.1) is equivalent to a variational problem on $H_0^1(\Omega)$. When $g \neq 0$ the extra constraint $\phi = g$ makes the Sobolev space dependent on g. To get around this difficulty, we introduce a Lagrange multiplier and the new functional

$$L(\phi, \psi, \mu) = \int_{\Omega} (\Delta\phi + f)\psi \, dx + \int_{\Gamma} (\phi - g)\mu \, d\Gamma \tag{6.3}$$

for all $\psi \in H^2(\Omega)$ and $\mu \in H^{1/2}(\Gamma)$. This is a convex-concave functional with a unique saddle point $(\hat\phi, \hat\psi, \hat\mu)$ which is completely characterized by the equations

$$\Delta\hat\phi + f = 0 \text{ in } \Omega, \quad \hat\phi - g = 0 \text{ in } \Gamma, \tag{6.4}$$

$$\forall\phi \in H^2(\Omega), \quad \int_{\Omega} \Delta\phi\,\hat\psi \, dx + \int_{\Gamma} \phi\hat\mu \, d\Gamma = 0. \tag{6.5}$$

The last equation characterizes $\hat{\psi}$ and $\hat{\mu}$:

$$\Delta\hat{\psi} = 0 \text{ in } \Omega, \quad \hat{\psi} = 0 \text{ on } \Gamma, \tag{6.6}$$

$$\hat{\mu} = \frac{\partial\hat{\psi}}{\partial n} \text{ on } \Gamma \tag{6.7}$$

(cf., for instance, Ekeland and Temam [1, Prop. 1.6]). Of course, this implies that the saddle point is unique and given by

$$(\hat{\phi}, \hat{\psi}, \hat{\mu}) = (y, 0, 0). \tag{6.8}$$

The purpose of the above computation was to find out the form of the multiplier $\hat{\mu}$:

$$\hat{\mu} = \frac{\partial\hat{\psi}}{\partial n} \text{ on } \Gamma \tag{6.9}$$

in order to rewrite the previous functional as a function of two variables instead of three:

$$L(\Omega, \phi, \psi) = \int_\Omega (\Delta\phi + f)\psi \, dx + \int_\Gamma (\phi - g)\frac{\partial\psi}{\partial n} \, d\Gamma \tag{6.10}$$

for (ϕ, ψ) in $H^2(\Omega) \times H^2(\Omega)$. It is also advantageous for shape problems to get rid of boundary integrals whenever possible. So noting that

$$\int_\Gamma (\phi - g)\frac{\partial\psi}{\partial n} \, d\Gamma = \int_\Omega \text{div} \left[(\phi - g)\nabla\psi\right] dx, \tag{6.11}$$

we finally use the functional

$$L(\Omega, \phi, \psi) = \int_\Omega \left\{(\Delta\phi + f)\psi + (\phi - g)\Delta\psi + \nabla(\phi - g) \cdot \nabla\psi\right\} dx \tag{6.12}$$

on $H^2(\Omega) \times H^2(\Omega)$. It is readily seen that it has a unique saddle point $(\hat{\phi}, \hat{\psi})$ in $H^2(\Omega) \times H^2(\Omega)$, which is completely characterized by the saddle point equations:

$$\Delta\hat{\phi} + f = 0 \text{ in } \Omega, \quad \hat{\phi} = g \text{ on } \Gamma, \tag{6.13}$$

$$\Delta\hat{\psi} = 0 \text{ in } \Omega, \quad \hat{\psi} = 0 \text{ on } \Gamma. \tag{6.14}$$

6.3 Saddle Point Expression of the Objective Function

Now repeat the above constructions taking into account the objective function. First introduce the objective function

$$F(\Omega, \phi) = \frac{1}{2}\int_\Omega |\phi - y_d|^2 \, dx \tag{6.15}$$

and the new Lagrangian functional

$$G(\Omega, \phi, \psi) = F(\Omega, \phi) + L(\Omega, \phi, \psi).$$

Then it is easy to verify that

$$J(\Omega) = \mathrm{Min}_{\phi \in H^2(\Omega)} \mathrm{Max}_{\psi \in H^2(\Omega)} G(\Omega, \phi, \psi). \tag{6.16}$$

The Lagrangian $G(\Omega, \phi, \psi)$ is given by the expression

$$G(\Omega, \phi, \psi) = \frac{1}{2} \int_\Omega |\phi - y_d|^2 \, dx$$
$$+ \int_\Omega \{(\Delta\phi + f)\psi + (\phi - g)\Delta\psi + \nabla(\phi - g) \cdot \nabla\psi\} \, dx \tag{6.17}$$

on $H^2(\Omega) \times H^2(\Omega)$. It is readily seen that it has a unique saddle point $(\hat{\phi}, \hat{\psi})$ which is completely characterized by the following saddle point equations:

$$\Delta\hat{\phi} + f = 0 \text{ in } \Omega, \quad \hat{\phi} = g \text{ on } \Gamma, \tag{6.18}$$

$$\forall \phi \in H^2(\Omega), \quad \int_\Omega \{(\hat{\phi} - y_d)\phi + \Delta\phi\hat{\psi} + \phi\Delta\hat{\psi} + \nabla\phi \cdot \nabla\hat{\psi}\} \, dx = 0. \tag{6.19}$$

But the last equation is equivalent to

$$\forall \phi \in H^2(\Omega), \quad \int_\Omega [(\hat{\phi} - y_d) + \Delta\hat{\psi}] \, \phi \, dx + \int_\Gamma \frac{\partial\phi}{\partial n} \, \hat{\psi} \, d\Gamma = 0 \tag{6.20}$$

or

$$\Delta\hat{\psi} + (\hat{\phi} - y_d) = 0 \text{ in } \Omega, \quad \hat{\psi} = 0 \text{ on } \Gamma \tag{6.21}$$

by using the theorem on the surjectivity of the trace. In what follows, we shall use the notation (y, p) for the saddle point $(\hat{\phi}, \hat{\psi})$. As a result, we have

$$J(\Omega) = \mathrm{Min}_{\phi \in H^2(\Omega)} \mathrm{Max}_{\psi \in H^2(\Omega)} G(\Omega, \phi, \psi). \tag{6.22}$$

We shall now use the above Lagrangian formulation combined with the velocity method to compute the shape gradient of $J(\Omega)$. Given a velocity field V in $\mathcal{D}^1(\mathbf{R}^N, \mathbf{R}^N)$ and the parametrized domains $\Omega_t = T_t(\Omega)$,

$$J(\Omega_t) = \mathrm{Min}_{\phi \in H^2(\Omega)} \mathrm{Max}_{\psi \in H^2(\Omega)} G(\Omega_t, \phi, \psi). \tag{6.23}$$

There are two methods to get rid of the time dependence in the underlying function spaces:

- the *function space parametrization* and

- the *function space embedding.*

In the first case, we parametrize the functions in $H^2(\Omega_t)$ by elements of $H^2(\Omega)$ through the transformation

$$\phi \mapsto \phi \circ T_t^{-1} = H^2(\Omega) \to H^2(\Omega_t), \tag{6.24}$$

where "∘" denotes the composition of the two maps and we introduce the *parametrized Lagrangian*,

$$\widetilde{G}(t,\phi,\psi) = G(T_t(\Omega),\phi \circ T_t^{-1},\psi \circ T_t^{-1}) \qquad (6.25)$$

on $H^2(\Omega) \times H^2(\Omega)$. In the function space embedding method, we introduce a large enough domain, D, which contains all the transformations $\{\Omega_t : 0 \le t \le \bar{t}\,\}$ of Ω for some small $\bar{t} > 0$.

In this section, we use the function space embedding method with $D = \mathbf{R}^N$ and

$$J(\Omega_t) = \min_{\Phi \in H^2(\mathbf{R}^N)} \max_{\Psi \in H^2(\mathbf{R}^N)} G(\Omega_t,\Phi,\Psi). \qquad (6.26)$$

As can be expected, the price to pay for the use of this method is the fact that the set of saddle points

$$S(t) = X(t) \times Y(t) \subset H^2(\mathbf{R}^N) \times H^2(\mathbf{R}^N) \qquad (6.27)$$

is not a singleton anymore since

$$X(t) = \{\Phi \in H^2(\mathbf{R}^N) : \Phi|_{\Omega_t} = y_t\}, \qquad (6.28)$$

$$Y(t) = \{\Psi \in H^2(\mathbf{R}^N) : \Psi|_{\Omega_t} = p_t\}, \qquad (6.29)$$

where (y_t,p_t) is the unique solution in $H^2(\Omega_t) \times H^2(\Omega_t)$ to the previous saddle point equations on Ω_t:

$$\Delta y_t + f = 0 \text{ in } \Omega_t, \quad y_t = g \text{ on } \Gamma_t, \qquad (6.30)$$

$$\Delta p_t + (y_t - y_d) = 0 \text{ in } \Omega_t, \quad p_t = 0 \text{ on } \Gamma_t. \qquad (6.31)$$

We can now apply the theorem of Correa and Seeger [1], which says that under appropriate assumptions (to be checked in the next section)

$$dJ(\Omega; V) = \min_{\Phi \in X(0)} \max_{\Psi \in Y(0)} \partial_t G(\Omega_t,\Phi,\Psi)\big|_{t=0}. \qquad (6.32)$$

Since we have already characterized $X(0)$ and $Y(0)$, we only need to compute the partial derivative of

$$G(\Omega_t,\Phi,\Psi) = \int_{\Omega_t} \left\{ \frac{1}{2}|\Phi - y_d|^2 + (\Delta\Phi + f)\Psi + (\Phi - g)\Delta\Psi \right.$$
$$\left. + \nabla(\Phi - g) \cdot \nabla\Psi \right\} dx. \qquad (6.33)$$

If we assume that Ω_t is sufficiently smooth, then

$$f, y_d \in H^{1/2+\varepsilon}(\mathbf{R}^N) \text{ and } g \in H^{2+\varepsilon}(\mathbf{R}^N) \ \Rightarrow\ p \in H^{5/2+\varepsilon}(\Omega), \qquad (6.34)$$

and we can choose to consider our saddle points $S(t)$ in $H^{5/2+\varepsilon}(\mathbf{R}^N) \times H^{5/2+\varepsilon}(\mathbf{R}^N)$ rather than $H^2(\mathbf{R}^N) \times H^2(\mathbf{R}^N)$. If the functions Φ and Ψ belong to $H^{5/2+\varepsilon}(\mathbf{R}^N)$, then

$$\partial_t G(\Omega_t, \Phi, \Psi) = \int_{\Gamma_t} \left\{ \frac{1}{2}|\Phi - y_d|^2 + (\Delta\Phi + f)\Psi + (\Phi - g)\Delta\Psi \right.$$
$$\left. + \nabla(\Phi - g)\cdot\nabla\Psi \right\} V\cdot n_t\, d\Gamma_t. \tag{6.35}$$

This expression is an integral over the boundary Γ_t which will not depend on Φ and Ψ outside of $\overline{\Omega}_t$. As a result, the Min and the Max can be dropped in expression (6.32), which reduces to

$$dJ(\Omega; V) = \int_{\Gamma} \left\{ \frac{1}{2}(y - y_d)^2 + (\Delta y + f)p + (y - g)\Delta p \right.$$
$$\left. + \nabla(y - g)\cdot\nabla p \right\} V\cdot n\, d\Gamma. \tag{6.36}$$

However, $p = 0$ and $y - g = 0$ imply

$$\nabla p = \frac{\partial p}{\partial n}n \text{ and } \nabla(y - g) = \frac{\partial}{\partial n}(y - g)n \text{ on } \Gamma \tag{6.37}$$

and, finally,

$$dJ(\Omega; V) = \int_{\Gamma} \left\{ \frac{1}{2}|g - y_d|^2 + \frac{\partial}{\partial n}(y - g)\frac{\partial p}{\partial n} \right\} V\cdot n\, d\Gamma. \tag{6.38}$$

6.4   Verification of the Assumptions of Theorem 5.1

As we have seen, the computation of the shape gradient is both quick and easy. We now turn to the step by step verification of the assumptions of Theorem 5.1. Many of the constructions given below are "canonical" and can be repeated for different problems in different contexts.

Let y_d and $f \in H^1(\mathbf{R}^N)$ and $g \in H^{5/2}(\mathbf{R}^N)$ so that

$$X = Y = H^3(\mathbf{R}^N). \tag{6.39}$$

The saddle points $S(t) = X(t) \times Y(t)$ are given by

$$X(t) = \{\Phi \in X : \Phi|_{\Omega_t} = y_t\}, \tag{6.40}$$
$$Y(t) = \{\Psi \in Y : \Psi|_{\Omega_t} = p_t\}. \tag{6.41}$$

The sets $X(t)$ and $Y(t)$ are not empty since it is always possible to construct a continuous linear extension

$$\Pi^m : H^m(\Omega) \to H^m(\mathbf{R}^N) \tag{6.42}$$

for each $m \geq 1$. For instance, with $m = 1$ and a boundary Γ which is $W^{1,\infty}$, see Agmon, Douglis, and Nirenberg [1, 2], and for $m > 1$, Babič [1] (cf. also Nečas [1]). Using this Π^m, we define the following extension:

$$\Pi_t^m : H^m(\Omega_t) \to H^m(\mathbf{R}^N), \tag{6.43}$$

$$\Pi_t^m(\phi) = [\Pi^m(\phi \circ T_t)] \circ T_t^{-1}. \tag{6.44}$$

In what follows m is fixed and equal to 3, so we shall drop the superscript m and define the extensions

$$Y_t = \Pi_t y_t, \quad P_t = \Pi_t p_t \tag{6.45}$$

of y_t and p_t, respectively. Hence,

$$Y_t \in X(t) \text{ and } P_t \in Y(t) \Rightarrow S(t) \neq \varnothing. \tag{6.46}$$

So condition (H1) is satisfied. Condition (H2) follows from the assumptions on f, y_d, and g. To check conditions (H3) and (H4), we need two general theorems, which can be used in various contexts and problems.

Theorem 6.1. *For $V \in \mathcal{D}^1(\mathbf{R}^N, \mathbf{R}^N)$ and $\Phi \in L^2(\mathbf{R}^N)$,*

$$\lim_{t \searrow 0} \Phi \circ T_t = \Phi \text{ and } \lim_{t \searrow 0} \Phi \circ T_t^{-1} = \Phi \text{ in } L^2(\mathbf{R}^N). \tag{6.47}$$

Proof. (i) The space $\mathcal{D}(\mathbf{R}^N)$ of continuous functions with compact support in $\mathbf{R}^N$ is dense in $L^2(\mathbf{R}^N)$. So given $\varepsilon > 0$, there exists Φ_ε in $\mathcal{D}(\mathbf{R}^N)$ such that

$$\|\Phi - \Phi_\varepsilon\|_{L^2}^2 < \frac{\varepsilon^2}{\max}\{J_t^{-1} : 0 \leq t \leq \tau\}.$$

Hence,

$$\|\Phi \circ T_t - \Phi\| \leq \|\Phi_\varepsilon \circ T_t - \Phi_\varepsilon\| + \|\Phi \circ T_t - \Phi_\varepsilon \circ T_t\| + \|\Phi - \Phi_\varepsilon\|. \tag{6.48}$$

But

$$\forall t \in [0, \tau], \quad \int_{\mathbf{R}^N} |\Phi \circ T_t - \Phi_\varepsilon \circ T_t|^2 \, dx = \int_{\mathbf{R}^N} |\Phi - \Phi_\varepsilon|^2 J_t^{-1} \, dx \leq \varepsilon^2.$$

So the last two terms in (6.48) are less than 2ε. It remains to evaluate the first term for a fixed function Φ_ε with compact support K in $\mathbf{R}^N$. Recall that, since $\Phi_\varepsilon = 0$ on the boundary ∂K of K, $T_t(K) = K$ for all t in $[0, \tau]$ (use Nagumo's [1] theorem twice as in the proof of Theorem 5.1 (i) in Chapter 7). Moreover, by the compactness of K, Φ_ε is uniformly continuous on $\mathbf{R}^N$ and

$$\exists \delta > 0, \forall x, y \in \mathbf{R}^N, \quad |x - y| < \delta \implies |\Phi_\varepsilon(y) - \Phi_\varepsilon(x)| < \frac{\varepsilon}{m(K)^{1/2}}.$$

However, T_t is also uniformly continuous on K and

$$\exists \eta > 0, \forall t, 0 \leq t < \eta, \forall x \in K, \quad |T_t x - x| < \delta.$$

By construction,
$$\operatorname{supp}(\Phi_\varepsilon \circ T_t) = T_t(\operatorname{supp}\Phi_\varepsilon) \subset K,$$

and
$$\Phi_\varepsilon = 0 \text{ and } \Phi_\varepsilon \circ T_t = 0 \text{ outside of } K.$$

Finally,
$$\int_{\mathbf{R}^N} |\Phi_\varepsilon(T_t x) - \Phi_\varepsilon(x)|^2 \, dx = \int_K |\Phi_\varepsilon(T_t x) - \Phi_\varepsilon(x)|^2 \, dx \le \varepsilon^2,$$

and this implies that
$$\forall \varepsilon > 0, \exists \eta > 0, \forall 0 \le t \le \eta, \quad \|\Phi \circ T_t - \Phi\|_{L^2(\mathbf{R}^N)} \le 3\varepsilon.$$

(ii) For the second part of (6.47), we make a change of variable and use the result of part (i):
$$\int_{\mathbf{R}^N} |\Phi \circ T_t^{-1} - \Phi|^2 \, dx = \int_{\mathbf{R}^N} |\Phi - \Phi \circ T_t|^2 J_t \, dx \le \varepsilon^2.$$

This completes the proof. $\qquad\qquad\square$

Corollary 1. *Under the assumptions of Theorem* 6.1 *for* $m \ge 1$, V *in* $\mathcal{D}^m(\mathbf{R}^N, \mathbf{R}^N)$, *and* $\Phi \in H^m(\mathbf{R}^N)$,
$$\lim_{t \searrow 0} \Phi \circ T_t = \Phi \text{ and } \lim_{t \searrow 0} \Phi \circ T_t^{-1} = \Phi \text{ in } H^m(\mathbf{R}^N). \tag{6.49}$$

Theorem 6.2. *Under the assumptions of Corollary* 1 *to Theorem* 6.1,
$$y^t \to y^0 \text{ in } H^m(\Omega)\text{-strong (resp., weak)} \tag{6.50}$$

implies that
$$Y_t \to Y_0 \text{ in } H^m(\mathbf{R}^N)\text{-strong (resp., weak)}.$$

Proof. The strong case is obvious. We prove the weak case for $m = 0$. By definition,
$$Y_t = (\Pi y^t) \circ T_t^{-1},$$

and for all Φ in $L^2(\mathbf{R}^N)$, we consider
$$\int_{\mathbf{R}^N} Y_t \Phi \, dx = \int_{\mathbf{R}^N} (\Pi y^t) \circ T_t^{-1} \Phi \, dx = \int_{\mathbf{R}^N} \Pi y^t \Phi \circ T_t J_t \, dx.$$

We have shown in Theorem 6.1 that
$$\Phi \circ T_t \to \Phi \text{ in } L^2(\mathbf{R}^N)\text{-strong.}$$

In addition, $J_t \to 1$, and by linearity and continuity of Π,
$$\Pi y^t \to \Pi y \text{ in } L^2(\mathbf{R}^N)\text{-weak.}$$

Hence,
$$\forall \Phi \in L^2(\mathbf{R}^N), \int_{\mathbf{R}^N} Y_t \Phi \, dx \to \int_{\mathbf{R}^N} \Pi y \Phi \, dx =\to \int_{\mathbf{R}^N} Y_0 \Phi \, dx.$$

This proves the weak convergence. $\qquad\qquad\square$

To satisfy condition (H3), we transform (y_t, p_t) on Ω_t to $(y^t, p^t) = (y_t \circ T_t, p_t \circ T_t)$ on Ω. The pair (y^t, p^t) is the transported pair of solutions from Ω_t to Ω. It is the unique solution in $H^1(\Omega) \times H^1(\Omega)$ of the system

$$-\text{div}\,[A(t)\nabla y^t] = J_t f \circ T_t \text{ in } \Omega, \quad y^t = g \circ T_t \text{ on } \Gamma, \tag{6.51}$$

$$-\text{div}\,[A(t)\nabla p^t] = J_t(y^t - y_d \circ T_t) \text{ in } \Omega, \quad p^t = 0 \text{ on } \Gamma, \tag{6.52}$$

where

$$A(t) = J_t[DT_t]^{-1*}[DT_t]^{-1}, \quad J_t = |\det DT_t|, \tag{6.53}$$

DT_t is the Jacobian matrix of T_t, and $^*[DT_t]^{-1}$ is the transpose of $[DT_t]^{-1}$.

For sufficiently smooth domains Ω and vector fields V, the pair $\{y^t, p^t\}$ is bounded in $H^1(\Omega) \times H^1(\Omega)$ as t goes to zero. Since $H^1(\Omega)$ is a Hilbert space, we can extract weakly convergent subsequences to some $(\bar{y}, \bar{p})$ in $H^1(\Omega) \times H^1(\Omega)$. However, by linearity of the equation with respect to (y^t, p^t) and continuity of the coefficients with respect to t, the limit point $(\bar{y}, \bar{p})$ will coincide with (y^0, p^0), since the system has a unique solution at $t = 0$. Then we go back to the equation for y^t and y and show that the convergence is strong in $H^1(\Omega)$. Finally, by using the regularity of the data and the classical regularity theorems, we show that $(y^t, p^t) \to (y,p)$ in $H^3(\Omega) \times H^3(\Omega)$.

For the verification of condition (H4), we go back to expression (6.35), which can be rewritten as a volume integral:

$$\partial_t G(\Omega_t, \Phi, \Psi) = \int_{\Omega_t} \text{div} \left\{ \left[\frac{1}{2}(\Phi - y_d)^2 + (\Delta\Phi + f)\Psi \right. \right.$$
$$\left. \left. + (\Phi - g)\Delta\Psi + \nabla(\Phi - g) \cdot \nabla\Psi \right] V \right\} dx \tag{6.54}$$

for $(\Phi, \Psi) \in H^3(\mathbf{R}^N) \times H^3(\mathbf{R}^N)$. Now introduce the map

$$(\Phi, \Psi) \mapsto F(\Phi, \Psi)$$
$$= \left[\frac{1}{2}(\Phi - y_d)^2 + (\Delta\Phi + f)\Psi + (\Phi - g)\Delta\Psi + \nabla(\Phi - g) \cdot \nabla\Psi \right] V$$
$$: H^3(\mathbf{R}^N) \times H^3(\mathbf{R}^N) \to (H^1(\mathbf{R}^N))^N.$$

It is bilinear and continuous. Finally, the map

$$(t, F) \mapsto \int_{\Gamma_t} F \circ n_t \, d\Gamma = \int_{\Omega_t} F \, dx = \int_{\Omega} (\text{div}\, F) \circ T_t J_t^{-1} \, dx \tag{6.55}$$

from $[0, \tau] \times H^1(\mathbf{R}^N)$ to $\mathbf{R}$ is continuous. Then

$$(t, \Phi, \Psi) \mapsto \partial_t G(\Omega_t, \Phi, \Psi) = \int_{\Gamma_t} F(\Psi, \Psi) \cdot n_t \, d\Gamma_t \tag{6.56}$$

is continuous and condition (H4) is satisfied. This completes the verification of the four conditions of Theorem 5.1.

Elements of Bibliography

R.A. ADAMS

[1] *Sobolev spaces*, Academic Press, New York, London, 1975.

R. ADAMS, N. ARONSZAJN, AND K.T. SMITH

[1] *Theory of Bessel potentials. Part II.* Ann. Inst. Fourier (Grenoble) **17** (1967) fasc. 2, 1–133 (1968).

S. AGMON

[1] *Lectures on elliptic boundary value problems*, Van Nostrand-Reinhold, New York, Toronto, London, 1965.

S. AGMON, A. DOUGLIS, AND L. NIRENBERG

[1] *Estimates near the boundary for solutions of elliptic partial differential equations satisfying general boundary conditions* I, Comm. Pure Appl. Math. **12** (1959), 623–727.

[2] *Estimates near the boundary for solutions of elliptic partial differential equations satisfying general boundary conditions* II, Comm. Pure Appl. Math. **17** (1964), 35–92.

J.C. AGUILAR AND J.-P. ZOLÉSIO

[1] *Shape control of a hydrodynamic wake*, in "Partial differential equations methods in control and shape analysis" (Pisa), G. Da Prato and J.P. Zolésio, eds., 1–10, Lecture Notes in Pure and Appl. Math., 188, Dekker, New York, 1997.

[2] *Coque fluide intrinsèque sans approximation géométrique* (French) [Intrinsic fluid shell without geometrical approximation], C. R. Acad. Sci. Paris Sér. I Math. **326** (1998), 1341–1346.

G. ALLAIRE AND R.V. KOHN

[1] *Optimal design for minimum weight and compliance in plane stress using extremal microstructures*, European J. Mech. A Solids **12** (1993), 839–878.

[2] *Topology optimization and optimal shape design using homogenization*, in "Topology design of structures," 207–218, NATO Adv. Sci. Inst. Ser. E Appl. Sci., 227, Kluwer, Dordrecht, 1993.

[3] *Optimal bounds on the effective behavior of a mixture of two well-ordered materials*, Quart. Appl. Math. **51** (1993), 643–674.

W.K. ALLARD

[1] *On the first variation of a varifold*, Ann. of Math. (2) **95** (1972), 417–491.

[2] *On the first variation of area and generalized mean curvature*, in Geometric Measure Theory and Minimal Surfaces, E. Bombieri, coord., 1–30, C.I.M.E., Edizioni Cremonese, Rome, 1973.

W.K. ALLARD AND F.J. ALMGREN, JR.

[1] *Geometric measure theory and the calculus of variations*, Proc. Sympos. Pure Math., 44, AMS, Providence, RI, 1986.

F.J. ALMGREN, JR.

[1] *The theory of varifolds—a variational calculus in the large for the k-dimensional area integrand*, mimeographed notes, Princeton University, Princeton, NJ, 1965.

[2] *Plateau's problem*, Benjamin, New York, Amsterdam, 1966.

[3] *Existence and regularity almost everywhere of solutions to elliptic variational problems among surfaces of varying topological type and singularity structure*, Ann. of Math. (2) **87** (1968), 321–391.

H.W. ALT

[1] *Strömungen durch inhomogene poröse Medien mit freien Rand*, J. Reine Angew. Math. **305** (1979), 89–115.

H.W. ALT AND L.A. CAFFARELLI

[1] *Existence and regularity for a minimum problem with free boundary*, J. Reine Angew. Math. **325** (1981), 105–144.

H.W. ALT AND G. GILARDI

[1] *The behavior of the free boundary for the dam problem*, Ann. Scuola Norm. Sup. Pisa Cl. Sci. **9** (1982), 571–626.

L. AMBROSIO AND G. BUTTAZZO

[1] *An optimal design problem with perimeter penalization*, Calc. Var. Partial Differential Equations **1** (1993), 55–69.

Y. AMIRAT AND J. SIMON

[1] *Riblets and drag minimization*, in "Optimization methods in partial differential equations," 9–17, Contemp. Math., 209, AMS, Providence, RI, 1997.

G. ARUMUGAM AND O. PIRONNEAU

[1] *Sur le problème des "riblets,"* Rapport de recherche R87027, Publications du Laboratoire d'analyse numérique, Université Pierre et Marie Curie, Paris, 1987.

[2] *On the problems of riblets as a drag reduction device*, Optimal Control Appl. Methods **12** (1989), 93–112.

J.-P. AUBIN

[1] *L'analyse non linéaire et ses motivations économiques*, Masson, Paris, New York, Barcelona, Milano, Mexico, Sao Paulo, 1984.

[2] *Exercices d'analyse non linéaire*, Masson, Paris, New York, Barcelona, Milano, Mexico, Sao Paulo, 1987.

[**3**] *Mutational and morphological analysis, tools for shape evolution and morphogenesis*, Birkäuser, Boston, Basel, Berlin 1999.

J.-P. Aubin and A. Cellina

[**1**] *Differential inclusions*, Springer-Verlag, Berlin, 1984.

J.-P. Aubin and H. Frankowska

[**1**] *Set-valued analysis*, Birkhäuser, Boston, Basel, Berlin, 1990.

G. Auchmuty

[**1**] *Dual variational principles for eigenvalue problems*, Nonlinear Functional Analysis and Its Applications, F. Browder, ed., 55–71, AMS, Providence, RI, 1986.

V.M. Babič

[**1**] *On the extension of functions*, Uspekhi Mat. Nauk **8** (1953), 111–113 (in Russian).

C. Baiocchi and A. Capelo

[**1**] *Variational and quasivariational inequalities*, Wiley, Chichester, New York, 1984.

C. Baiocchi, V. Comincioli, E. Magenes, and G.A. Pozzi

[**1**] *Free boundary problems in the theory of fluid flow through porous media: Existence and uniqueness theorem*, Ann. Mat. Pura Appl. **97** (1973), 1–82.

V. Barbu, G. Da Prato, and J.-P. Zolésio

[**1**] *Feedback controllability of the free boundary of the one phase Stefan problem*, Differential Integral Equations **4** (1991), 225–239.

J.A. Bello, E. Fernández-Cara, J. Lemoine, and J. Simon

[**1**] *On drag differentiability for Lipschitz domains*, in "Control of partial differential equations and applications," 11–22, Lecture Notes in Pure and Appl. Math., 174, Dekker, New York, 1996.

[**2**] *The differentiability of the drag with respect to the variations of a Lipschitz domain in a Navier–Stokes flow*, SIAM J. Control Optim. **35** (1997), 626–640.

J.A. Bello, E. Fernández Cara, and J. Simon

[**1**] *Optimal shape design for Navier-Stokes flow*, in "System modelling and optimization," 481–489, Lecture Notes in Control and Inform. Sci., 180, Springer-Verlag, Berlin, 1992.

[**2**] *The variation of the drag with respect to the domain in Navier-Stokes flow*, in "Optimization, optimal control and partial differential equations," 287–296, Internat. Ser. Numer. Math., 107, Birkhäuser, Basel, 1992.

A. BENDALI

[1] *Existence et régularisation dans un problème d'identification de domaine. Application à l'analyse numérique d'un cas modèle*, Thèse de 3e cycle, Université d'Alger, Alger 1975.

M. BENDSØE

[1] *Methods for optimization of structural topology, shape and material*, Springer-Verlag, Berlin, 1995.

M. BENDSØE AND N. KIKUCHI

[1] *Generating optimal topologies in structural design using a homogenization method*, Comput. Methods Appl. Mech. Engrg. **71** (1988), 197–224.

[2] *Optimal shape design as a material distribution problem*, Struct. Optim. **1** (1989), 193–202.

M. BENDSØE AND C.A. MOTA SOARES

[1] *Topology design of structures*, NATO Adv. Sci. Inst. Ser. E: Appl. Sci., 227, Kluwer, Dordrecht, 1993.

M. BENDSØE, N. OLHOFF, AND J. SOKOŁOWSKI

[1] *Sensitivity analysis of problems of elasticity with unilateral constraints*, J. Structural Mech. **13** (1985), 201–222.

M. BENDSØE AND J. SOKOŁOWSKI

[1] *Sensitivity analysis and optimal design of elastic plates with unilateral point supports*, Mech. Structures Mach. **15** (1987), 383–393.

[2] *Design sensitivity analysis of elastic-plastic analysis problems*, Mech. Structures Mach. **16** (1988), 81–102.

[3] *Shape sensitivity analysis of optimal compliance functionals*, Mech. Structures Mach. (1) **23** (1995), 35–58.

[4] *Shape sensitivity analysis of nonsmooth shape functionals*, in "Recent developments in optimization" (Dijon, 1994), 36–45, Lecture Notes in Econom. and Math. Systems, 429, Springer-Verlag, Berlin, 1995.

[5] *Shape sensitivity analysis of optimal compliance functionals*, Mech. Structures Mach. **23** (1995), 35–58.

B. BENEDICT, J. SOKOŁOWSKI, AND J.-P. ZOLÉSIO

[1] *Shape optimization for contact problems*, in "System modelling and optimization" (Copenhagen, 1983), 790–799, Lecture Notes in Control and Inform. Sci., 59, Springer-Verlag, Berlin, New York, 1984.

M. BERGER

[1] *Geometry* I, II, Springer-Verlag, Berlin, Heidelberg, New York, 1987 (transl. from the French *Géométrie*, Vols. 1, 2, 3, 4, 5, 6, CEDIC/Fernand Nathan, Paris, 1977).

M. BERGER AND B. GOSTIAUX

[1] *Differential geometry: Manifolds, curves and surfaces*, Springer-Verlag, New York, 1988 (transl. from the French 1987 edition of *Géométrie différentielle: Variétés, courbes et surfaces*, 2nd ed., Presses Universitaire de France, Paris, 1992).

A. BERN

[1] *Contribution à la modélisation des surfaces libres en régime permanent*, thèse de l'École Nationale Supérieure des Mines de Paris, CEMEF, Sophia-Antipolis, France, 1987.

J. BLUM

[1] *Numerical simulation and optimal control in plasma physics with applications to tokamacs*, Wiley, New York, 1989.

S. BOISGERAULT AND N. GOMEZ

[1] *Personal communication*, ISIA, Ecole des Mines, Sophia-Antipolis, France.

G. BOUCHITTÉ, G. BUTTAZZO, AND P. SEPPECHER

[1] *Shape optimization solutions via Monge-Kantorovich equation*, C. R. Acad. Sci. Paris Sér. I Math. **324** (1997), 1185–1191.

D. BRESCH AND J. SIMON

[1] *Sur les variations normales d'un domaine* (French) [On the normal variations of a domain], ESAIM Control Optim. Calc. Var. **3** (1998), 251–261 (electronic).

D. BUCUR

[1] *Capacitary extension and two dimensional shape optimization*, Preprint I.N.L.N. 93.24, Institute Non Linéaire de Nice, Sophia-Antipolis, France, 1994.

[2] *Un théorème de caractérisation pour la γ-convergence* [A characterization theorem for γ-convergence], C. R. Acad. Sci. Paris Sér. I Math. **323** (1996), 883–888.

[3] *Shape analysis for nonsmooth elliptic operators*, Appl. Math. Lett. **9** (1996), 11–16.

[4] *Concentration-compacité et γ-convergence* [Concentration-compactness principle and γ-convergence], C. R. Acad. Sci. Paris Sér. I Math. **327** (1998), 255–258.

[5] *Characterization for the Kuratowski limits of a sequence of Sobolev spaces*, J. Differential Equations **151** (1999), 1–19.

D. BUCUR, G. BUTTAZZO, AND I. FIGUEIREDO

[1] *On the attainable eigenvalues of the Laplace operator*, SIAM J. Math. Anal. **30** (1999), 527–536.

D. BUCUR, G. BUTTAZZO, AND A. HENROT

[1] *Existence results for some optimal partition problems*, Adv. Math. Sci. Appl. **8** (1998), 571–579.

D. BUCUR, G. BUTTAZZO, AND P. TREBESCHI

[1] *An existence result for optimal obstacles*, J. Funct. Anal. **162** (1999), 96–119.

D. Bucur and P. Trebeschi

[1] *Shape optimisation problems governed by nonlinear state equations*, Proc. Roy. Soc. Edinburgh Sect. A **128** (1998), 945–963.

D. Bucur and N. Varchon

[1] *Stabilité de la solution d'un problème de Neumann pour des variations de frontière*, C. R. Acad. Sci. Paris Sér. I Math. **331** (2000), 371.

D. Bucur and J.-P. Zolésio

[1] *Flat cone condition and shape analysis*, in "Control of partial differential equations" (Trento, 1993), 37–49, Lecture Notes in Pure and Appl. Math., 165, Dekker, New York, 1994.

[2] *Continuité par rapport au domaine dans le problème de Neumann* (French) [Continuity with respect to shape for the Neumann problem], C. R. Acad. Sci. Paris Sér. I Math. **319** (1994), 57–60.

[3] *Optimisation de forme sous contrainte capacitaire* (French) [Shape optimization under the uniform capacity condition], C. R. Acad. Sci. Paris Sér. I Math. **318** (1994), 795–800.

[4] *Shape optimization for elliptic problems under Neumann boundary conditions*, in "Calculus of variations, homogenization and continuum mechanics" (Marseilles, 1993), 117–129, Ser. Adv. Math. Appl. Sci., 18, World Scientific, River Edge, NJ, 1994.

[5] *N-dimensional shape optimization under capacitary constraint*, J. Differential Equations **123** (1995), 504–522.

[6] *Pseudo-courbure dans l'optimisation de forme* (French) [Pseudo-curvature in shape optimization], C. R. Acad. Sci. Paris Sér. I Math. **321** (1995), 387–390.

[7] *Shape analysis of eigenvalues on manifolds*, in "Homogenization and applications to material sciences" (Nice, 1995), 67–79, GAKUTO Internat. Ser. Math. Sci. Appl., 9, Gakkōtosho, Tokyo, 1995.

[8] *Free boundary problems and density perimeter*, J. Differential Equations **126** (1996), 224–243.

[9] *Shape continuity for Dirichlet-Neumann problems*, in "Progress in partial differential equations: The Metz surveys 4," 53–65, Pitman Res. Notes Math. Ser., 345, Longman, Harlow, UK, 1996.

[10] *Boundary optimization under pseudo curvature constraint*, Ann. Scuola Norm. Sup. Pisa Cl. Sci. (4) **23** (1996), 681–699 (1997).

[11] *Wiener's criterion and shape continuity for the Dirichlet problem*, Boll. Un. Mat. Ital. B (7) **11** (1997), 757–771.

[12] *Anatomy of the shape Hessian via Lie brackets*, Ann. Mat. Pura Appl. (4) **173** (1997), 127–143.

[13] *Stabilité du spectre d'un opérateur elliptique par rapport au domaine* (French) [Stability of the spectrum of an elliptic operator under domain perturbations], C. R. Acad. Sci. Paris Sér. I Math. **324** (1997), 191–194.

[14] *Spectrum stability of an elliptic operator to domain perturbations*, J. Convex Anal. **5** (1998), 19–30.

G. BUTTAZZO AND G. DAL MASO

[1] *An existence result for a class of shape optimization problems*, Arch. Rational Mech. Anal. **122** (1993), 183–195.

G. BUTTAZZO, G. DAL MASO, A. GARRONI, AND A. MALUSA

[1] *On the relaxed formulation of some shape optimization problems*, Adv. Math. Sci. Appl. **7** (1997), 1–24.

G. BUTTAZZO AND A. VISINTIN

[1] *Motion by mean curvature and related topics*, Proceedings of the International Conference held in Trento, July 20–24, 1992, G. Buttazzo and A. Visintin, eds., de Gruyter, Berlin, 1994.

R. CACCIOPPOLI

[1] *Misura e integrazione sugli insiemi dimensionalmente orientati* I *and* II, Atti Accad. Naz. Lincei Rend. Cl. Sci. Fis. Mat. Natur. (8), **12** (1952), 3–11; 137–146.

J. CAGNOL AND J.-P. ZOLÉSIO

[1] *Hidden shape derivative in the wave equation with Dirichlet boundary condition*, C. R. Acad. Sci. Paris Sér. I Math. **326** (1998), 1079–1084.

[2] *Shape derivative in the wave equation with Dirichlet boundary conditions*, J. Differential Equations **158** (1999), 175–210.

[3] *Hidden shape derivative in the wave equation*, in "Systems modelling and optimization" (Detroit, MI, 1997), 42–52, Chapman & Hall/CRC Res. Notes Math., 396, Chapman & Hall/CRC, Boca Raton, FL, 1999.

A.P. CALDERON

[1] *Lebesgue spaces of differentiable functions and distributions*, Proc. Sympos. Pure Math., 4, 33–49, AMS, Providence, RI, 1961.

G. CANADAS, F. CHAPEL, M. CUER, AND J.-P. ZOLÉSIO

[1] *Shape interfaces in an inverse problem related to the wave equation*, in "Inverse problems: An interdisciplinary study" (Montpellier, 1986), 533–551, Adv. Electron. Electron Phys., Suppl. 19, Academic Press, London, 1987.

P. CANNARSA, G. DA PRATO, AND J.-P. ZOLÉSIO

[1] *Dynamical shape control of the heat equation*, Systems Control Lett. **12** (1989), 103–109.

[2] *Evolution equations in noncylindrical domains*, Atti Accad. Naz. Lincei Rend. Cl. Sci. Fis. Mat. Natur. (8) **83** (1989), 73–77 (1990).

[3] *Riccati equations in noncylindrical domains*, in "Stabilization of flexible structures" (Montpellier, 1989), 148–155, Lecture Notes in Control and Inform. Sci., 147, Springer-Verlag, Berlin, 1990.

[4] *The damped wave equation in a moving domain*, J. Differential Equations **85** (1990), 1–16.

P. CANNARSA AND H.M. SONER

[1] *On the singularities of the viscosity solutions to Hamilton-Jacobi-Bellman equations*, Indiana Univ. Math. J. **36** (1987), 501–524.

J. CÉA

[1] *Quelques méthodes sur la recherche d'un domaine optimal*, in "Applied Inverse Problems," 135–146, Lecture Notes in Phys., 85, Springer-Verlag, Berlin, New York, 1978.

[2] *Problems of shape optimal design*, in "Optimization of Distributed Parameter Structures," II, E. J. Haug and J. Céa, eds., Sijhoff and Noordhoff, Alphen aan den Rijn, the Netherlands, 1981, 1005–1048.

[3] *Numerical methods of shape optimal design*, in "Optimization of Distributed Parameter Structures," II, E. J. Haug and J. Céa, eds., Sijhoff and Noordhoff, Alphen aan den Rijn, the Netherlands, 1981, 1049–1087.

[4] *Conception optimale ou identification de forme: Calcul rapide de la dérivée directionnelle de la fonction coût*, RAIRO Modél. Math. Anal. Numér. **20** (1986), 371–402.

J. CÉA, A.J. GIOAN, AND J. MICHEL

[1] *Quelques résultats sur l'identification de domaines*, CALCOLO **10** (1974), 207–232.

J. CÉA AND K. MALANOWSKI

[1] *An example of a max-min problem in partial differential equations*, SIAM J. Control **8** (1970), 305–316.

D. CHENAIS

[1] *Un résultat d'existence dans un problème d'identification de domaine*, C. R. Acad. Sci. Paris Sér. A **276** (1973), 547–550.

[2] *Un résultat de compacité d'un ensemble de parties de R^n*, C. R. Acad. Sci. Paris Sér. A-B **277** (1973), A905–A907.

[3] *Un ensemble de variétés à bord lipschitziennes dans R^n: compacité, théorème de prolongement dans certains espaces de Sobolev*, C. R. Acad. Sci. Paris Sér. A-B **280** (1975), A1145–A1147.

[4] *On the existence of a solution in a domain identification problem*, J. Math. Anal. Appl. **52** (1975), 189–219.

[5] *Homéomorphismes entre ouverts voisins uniformément lipschitziens*, C. R. Acad. Sci. Paris Sér. A-B **283** (1976), A461–A464.

[6] *Sur une famille de variétés à bord lipschitziennes, application à un problème d'identification de domaines*, Ann. Inst. Fourier (Grenoble) **27** (1977), 201–231.

[7] *Champ de vecteurs lipschitziens jouant le rôle de normales pour des ouverts lipschitziens de R^n*, C. R. Acad. Sci. Paris Sér. A-B **284** (1977), A1369–A1372.

[8] *Homéomorphisme entre ouverts lipschitziens*, Ann. Mat. Pura Appl. (4) **118** (1978), 343–398.

D. CHENAIS AND B. ROUSSELET

[1] *Différentiation du champ de déplacements dans une arche par rapport à la forme de la surface moyenne en élasticité linéaire* [Differentiation of displacement in an arch in relation to the form of the mid-surface in linear elasticity], C. R. Acad. Sci. Paris Sér. I Math. **298** (1984), 533–536.

[2] *Dependence of the buckling load of a nonshallow arch with respect to the shape of its midcurve*, RAIRO Modél. Math. Anal. Numér. **24** (1990), 307–341.

D. CHENAIS, B. ROUSSELET, AND R. BENEDICT

[1] *Design sensitivity for arch structures with respect to midsurface shape under static loading*, J. Optim. Theory Appl. **58** (1988), 225–239.

K.-T. CHENG AND N. OLHOFF

[1] *An investigation concerning optimal design of solid elastic plates*, Internat. J. Solids Structures **17** (1981), 305–323.

[2] *Regularized formulation for optimal design of axisymmetric plates*, Internat. J. Solids Structures **18** (1982), 153–169.

A. CHERKAEV AND R. KOHN

[1] *Topics in the mathematical modelling of composite materials*, Progress in Nonlinear Differential Equations Appl., 31, Birkhäuser, Boston, 1997.

G. CHOQUET

[1] *Theory of capacities*, Ann. Inst. Fourier **5** (1953–1954), 131–295.

M. CHOULLI AND A. HENROT

[1] *Use of the domain derivative to prove symmetry results in partial differential equations*, Math. Nachr. **192** (1998), 91–103.

M. CHOULLI, A. HENROT, AND M. MOKHTAR-KHARROUBI

[1] *Domain derivative of the leading eigenvalue of a model transport operator*, Transport Theory Statist. Phys. **28** (1999), 403–418.

D. CIORANESCU AND F. MURAT

[1] *Un terme étrange venu d'ailleurs*, Nonlinear partial differential equations, Collège de France Seminar, II (H. Brézis and J.-L. Lions, eds.), 98–138, Pitman, Boston, London, Melbourne, 1982.

[2] *Un terme étrange venu d'ailleurs* II, Nonlinear partial differential equations, Collège de France Seminar, III (H. Brézis and J.-L. Lions, eds.), 154–178, Pitman, Boston, London, Melbourne, 1982.

N. CLARIOND AND J.-P. ZOLÉSIO

[1] *Hydroelastic behavior of an inextensible membrane*, in "Boundary control and variation" (Sophia Antipolis, 1992), 73–92, Lecture Notes in Pure and Appl. Math., 163, Dekker, New York, 1994.

F.H. CLARKE

[1] *Optimization and Nonsmooth Analysis*, Wiley, New York, Chichester, Brisbane, Toronto, Singapore, 1983.

R. CORREA AND A. SEEGER

[1] *Directional derivative of a mimimax function*, Nonlinear Anal. **9** (1985), 13–22.

S.J. COX

[1] *The shape of the ideal column*, Math. Intelligencer **14** (1992), 16–24.

[2] *The generalized gradient at a multiple eigenvalue*, J. Funct. Anal. **133** (1995), 30–40.

S.J. COX, B. KAWOHL, AND P.X. UHLIG

[1] *On the optimal insulation of conductors*, J. Optim. Theory Appl. **100** (1999), 253–263.

S.J. COX AND C.M. MCCARTHY

[1] *The shape of the tallest column*, SIAM J. Math. Anal. **29** (1998), 547–554.

S.J. COX AND M.L. OVERTON

[1] *On the optimal design of columns against buckling*, SIAM J. Math. Anal. **23** (1992), 287–325.

M. CUER AND J.-P. ZOLÉSIO

[1] *Control of singular problem via differentiation of a min-max*, Systems Control Lett. **11** (1988), 151–158.

E.N. DANCER

[1] *The effect of domain shape on the number of positive solutions of certain non linear equations*, J. Differential Equations **74** (1988), 120–156.

D. DANERS

[1] *Domain perturbation for linear and non linear parabolic equations*, J. Differential Equations **129** (1996), 358–402.

G. DA PRATO AND J.-P. ZOLÉSIO

[1] *An optimal control problem for a parabolic equation in noncylindrical domains*, Systems Control Lett. **11** (1988), 73–77.

[2] *Existence and optimal control for wave equation in moving domain*, in "Stabilization of flexible structures" (Montpellier, 1989), 167–190, Lecture Notes in Control and Inform. Sci., 147, Springer-Verlag, Berlin, 1990.

[3] *Boundary control for inverse free boundary problems*, in "Boundary control and boundary variation" (Sophia-Antipolis, 1990), 163–173, Lecture Notes in Control and Inform. Sci., 178, Springer-Verlag, Berlin, 1992.

R. DAUTRAY AND J.-L. LIONS

[1] *Analyse mathématique et calcul numérique*, 4, Masson, Paris, Milano, Barcelona, Mexico, 1988.

E. DE GIORGI

[1] *Su una teoria generale della misura*, Ann. Mat. Pura Appl. (4) **36** (1954), 191–213

[2] *Nuovi teoremi relativi alle misure $(r - 1)$-dimensionali in uno spazio ad r dimensioni*, Ricerche Mat. **4** (1955), 95–113.

[3] *Frontiere orientate di misura minima*, Sem. Mat. Scuola Norm. Sup. Pisa (1960–1961), Editrice Tecnico Scientifica, Pisa, 1961.

[4] *Complementi alla teoria della misura $(n - 1)$-dimensionale in uno spazio n-dimensionale*, Sem. Mat. Scuola Norm. Sup. Pisa (1960–1961), Editrice Tecnico Scientifica, Pisa, 1961.

E. DE GIORGI, F. COLOMBINI, AND L.C. PICCININI

[1] *Frontiere orientate di misura minima e questioni collegate*, Editrice Tecnico Scientifica, Pisa, 1972.

M.C. DELFOUR

[1] *Shape derivatives and differentiability of min max*, in "Shape optimization and free boundaries," 35–111, NATO Adv. Sci. Inst. Ser. C Math. Phys. Sci., 380, Kluwer, Dordrecht, 1992.

[2] *Boundaries, interfaces, and transitions*, M.C. Delfour, ed., CRM Proc. Lecture Notes, 13, AMS, Providence, RI, 1998.

[3] *Intrinsic differential geometric methods in the asymptotic analysis of linear thin shells*, in "Boundaries, interfaces, and transitions" (Banff, AB, 1995), 19–90, CRM Proc. Lecture Notes, 13, AMS, Providence, RI, 1998.

[4] *Intrinsic $P(2, 1)$ thin shell model and Naghdi's models without a priori assumption on the stress tensor*, in "Proc. International Conference on Optimal Control of Partial Differential Equations," K.H. Hoffmann, G. Leugering, and F. Tröltzsch, eds., 99–113, Internat. Ser. Numer. Math., 133, Birkhäuser, Basel, 1999.

[5] *Membrane shell equation: Characterization of the space of solutions*, in "Control of Distributed Parameter and Stochastic Systems," S. Chen, X. Li, J. Yong, and X.Y. Zhou, eds., 21–29, Chapman & Hall, New York, 1999.

[6] *Characterization of the space of the membrane shell equation for arbitrary $C^{1,1}$ midsurfaces*, Control and Cybernetics **28** (1999), 481–501.

[7] *Tangential differential calculus and functional analysis on a $C^{1,1}$ submanifold*, in "Differential-geometric methods in the control of partial differential equations," Contemp. Math. 268, R. Gulliver, W. Littman, and R. Triggiani, eds., 83–115, AMS, Providence, RI, 2000.

M.C. DELFOUR, Z. MGHZALI, AND J.-P. ZOLÉSIO

[1] *Computation of shape gradients for mixed finite element formulation*, in "Partial differential equation methods in control and shape analysis," 77–93, G. Da Prato and J.-P. Zolésio, eds., Lecture Notes in Pure and Appl. Math., 188, Dekker, New York, 1997.

M.C. DELFOUR AND J. MORGAN

[1] *Differentiability of min max and saddle points under relaxed assumptions*, in "Boundary control and boundary variations," J.-P. Zolésio, ed., 174–185, Lecture Notes in Control and Inform. Sci., 178, Springer-Verlag, Berlin, 1992.

[2] *A complement to the differentiability of saddle points and Min Max*, Optimal Control Appl. Methods **13** (1992), 87–94.

[3] *One-sided derivative of minmax and saddle points with respect to a parameter*, Optimization **31** (1994), 343–358.

M.C. DELFOUR, G. PAYRE, AND J.-P. ZOLÉSIO

[1] *Optimal design of minimum weight thermal diffuser with constraint on the output thermal power flux*, Appl. Math. Optim. **9** (1982/1983), 225–262.

[2] *Shape optimal design of a radiating fin*, in "System modelling and optimization" (Copenhagen, 1983), P. Thoft-Christensen, ed., 810–818, Lecture Notes in Control and Inform. Sci., 59, Springer-Verlag, Berlin, 1984.

[3] *An optimal triangulation for second order elliptic problems*, Comput. Methods Appl. Mech. Engrg. **50** (1985), 231–261.

[4] *Some problems in shape optimal design for communications satellites*, in "Distributed parameter systems" (Vorau, 1984), 129–144, Lecture Notes in Control and Inform. Sci., 75, Springer-Verlag, Berlin, 1985.

[5] *Optimal parametrized design of thermal diffusers*, Optimal Control Appl. Methods **7** (1986), 163–184.

M.C. DELFOUR AND G. SABIDUSSI

[1] *Shape optimization and free boundaries*, M.C. Delfour and G. Sabisussi, eds., NATO Adv. Sci. Inst. Ser. C Math. Phys. Sci., 380, Kluwer, Dordrecht, Boston, London, 1992.

M.C. DELFOUR AND J.-P. ZOLÉSIO

[1] *Dérivation d'un min max et application à la dérivation par rapport au contrôle d'une observation non-différentiable de l'état* (French) [Minimax differentiability and application to a nondifferentiable cost function in optimal control], C. R. Acad. Sci. Paris Sér. I Math. **302** (1986), 571–574.

[2] *Differentiability of a min max and application to optimal control and design problems*, in "Control Problems for Systems Described as Partial Differential Equations and Applications" (Gainesville, FL, 1986), I. Lasiecka and R. Triggiani, eds., part I, 204–219, part II, 220–229, Lecture Notes in Control and Inform. Sci., 97, Springer-Verlag, Berlin, New York, 1987.

[3] *Shape sensitivity analysis via min max differentiability*, SIAM J. Control Optim. **26** (1988), 834–862.

[4] *Further developments in shape sensitivity analysis via a penalization method*, in "Boundary Control and Boundary Variations," J.-P. Zolésio, ed., 153–191, Lecture Notes in Comput. Sci., 100, Springer-Verlag, Berlin, Heidelberg, New York, Tokyo, 1988.

[5] *Shape sensitivity analysis via a penalization method*, Ann. Mat. Pura Appl. (4) **151** (1988), 179–212.

[6] *Analyse des problèmes de forme par la dérivation des minimax* (French) [Analysis of design problems by differentiation of the minimax], Analyse non linéaire (Perpignan, 1987), Ann. Inst. H. Poincaré Anal. Non Linéaire **6** (1989), suppl., 211–227.

[7] *Shape analysis and optimization: Lagrangian methods and applications*, in "Control of boundaries and stabilization," 95–108, Lecture Notes in Control and Inform. Sci., 125, Springer-Verlag, Berlin, New York, 1989.

[8] *Further developments in the application of min max differentiability to shape sensitivity analysis*, in "Control of partial differential equations," 108–119, Lecture Notes in Control and Inform. Sci., 114, Springer-Verlag, Berlin, New York, 1989.

[9] *Computation of the shape Hessian by a Lagrangian method*, in "Fifth Symp. on Control of Distributed Parameter Systems," A. El Jai and M. Amouroux, eds., 85–90, Pergamon Press, Oxford, New York, 1989.

[10] *Shape Hessian by the velocity method: A Lagrangian approach*, in "Stabilization of flexible structures" (Montpellier, 1989), J.-P. Zolésio, ed., 255–279, Lecture Notes in Control and Inform. Sci., 147, Springer-Verlag, Berlin, 1990.

[11] *Shape derivatives for non-smooth domains*, in "Optimal control of partial differential equations," K.-H. Hoffmann and W. Krabs, eds., 38–55, Lecture Notes in Control and Inform. Sci., 149, Springer-Verlag, Berlin, Heidelberg, New York, 1991.

[12] *Velocity method and Lagrangian formulation for the computation of the shape Hessian*, SIAM J. Control Optim. **29** (1991), 1414–1442.

[13] *Anatomy of the shape Hessian*, Ann. Mat. Pura Appl. (4) **159** (1991), 315–339.

[14] *Structure of shape derivatives for nonsmooth domains*, J. Funct. Anal. **104** (1992), 1–33.

[15] *Adjoint state in the control of variational inequalities*, in "Emerging applications in free boundary problems," Pitman Res. Notes Math. Ser., 280, J.M. Chadam and H. Rasmussen, eds., 215–223, Longman Sci. Tech., Harlow, UK, 1993.

[16] *Functional analytic methods in shape analysis*, in "Boundary control and variation," J.P. Zolésio, ed., 105–139, Lecture Notes in Pure and Appl. Math., 163, Dekker, New York 1994.

[17] *Shape analysis via oriented distance functions*, J. Funct. Anal. **123** (1994), 129–201.

[18] *Oriented distance function in shape analysis and optimization*, in "Control and optimal design of distributed parameter systems," J. Lagnese, D.L. Russell, L. White, eds., 39–72, IMA Vol. Math. Appl., 70, Springer-Verlag, New York, 1995.

[19] *On a variational equation for thin shells*, in "Control and optimal design of distributed parameter systems," J. Lagnese, D.L. Russell, L. White, eds., 25–37, IMA Vol. Math. Appl., 70, Springer-Verlag, New York, 1995.

[20] *A boundary differential equation for thin shells*, J. Differential Equations **119** (1995), 426–449.

[21] *Modelling and control of thin/shallow shells*, in "Proc. IFIP Conference on Control of Partial Differential Equations," E. Casas and L.A. Fernández, eds., Dekker, New York 1995.

[22] *The optimal design problem of Céa and Malanowski revisited*, in "Optimal design and control," J. Borggaard, J. Burkardt, M. Gunzburger, and J. Peterson, eds., 133–150, Progr. Systems Control Theory, 19, Birkhäuser, Boston, Basel, Berlin, 1995.

[23] *Distance functions, curvatures and application to the theory of shells*, in "Curvature flows and related topics," 47–66, GAKUTO Internat. Ser. Math. Sci. Appl., 5, Gakkōtosho, Tokyo, 1995.

[24] *On the design and control of systems governed by differential equations on submanifolds*, in "Distributed parameter systems: Modelling and control" (Warsaw, 1995), Control Cybernet. **25** (1996), 497–514.

[25] *Tangential differential equations for dynamical thin/shallow shells*, J. Differential Equations **128** (1996), 125–167.

[26] *Intrinsic modelling of shells*, in "Modelling and optimization of distributed parameter systems" (Warsaw, 1995), 16–30, Chapman & Hall, New York, 1996.

[27] *Modelling and control of thin/shallow shells*, in "Control of partial differential equations and applications" (Laredo, 1994), 51–62, Lecture Notes in Pure and Appl. Math., 174, Dekker, New York, 1996.

[28] *Differential equations for linear shells: Comparison between intrinsic and classical models*, in "Advances in mathematical sciences: CRM's 25 years" (Montreal, PQ, 1994), 41–124, CRM Proc. Lecture Notes, 11, AMS, Providence, RI, 1997.

[29] *On a geometrical bang-bang principle for some compliance problems*, in "Partial differential equations methods in control and shape analysis," G. Da Prato and J.P. Zolésio, eds., 95–109, Lecture Notes in Pure and Appl. Math., 188, Dekker, New York 1997.

[30] *Convergence to the asymptotic model for linear thin shells*, in "Optimization methods in partial differential equations" (South Hadley, MA, 1996), 75–93, Contemp. Math., 209, AMS, Providence, RI, 1997.

[31] *Hidden boundary smoothness for some classes of differential equations on submanifolds*, in "Optimization methods in partial differential equations" (South Hadley, MA, 1996), 59–73, Contemp. Math., 209, AMS, Providence, RI, 1997.

[32] *Shape analysis via distance functions: Local theory*, in "Boundaries, interfaces, and transitions," 91–123, CRM Proc. Lecture Notes, 13, AMS, Providence, RI, 1998.

[33] *Convergence of the linear $P(1,1)$ and $P(2,1)$ thin shells to asymptotic shells*, in "Plates and shells" (Québec, QC, 1996), 125–158, CRM Proc. Lecture Notes, 21, AMS, Providence, RI, 1999.

[34] *Hidden boundary smoothness in hyperbolic tangential problems on nonsmooth domains*, in "Systems modelling and optimization" (Detroit, MI, 1997), 53–61, CRC Res. Notes Math., 396, Chapman & Hall/CRC, Boca Raton, FL, 1999.

[35] *Tangential calculus and shape derivatives*, in "Shape optimization and optimal design: Proc. of the IFIP Conference," J. Cagnol, M.P. Polis, and J.P. Zolésio, eds., 37–60, Dekker, New York, Basel, 2001.

[36] *Velocity method and Courant metric topologies in shape analysis of partial differential equations*, in "Control of nonlinear distributed parameter systems," G. Chen, I. Lasiecka, and J. Zhou, eds., 45–68, Marcel Dekker, New York, Basel, 2000.

C. DELLACHERIE

[1] *Ensembles analytiques, capacités, mesures de Hausdorff*, Lecture Notes in Applied Mathematics, 295, Springer-Verlag, Berlin, Heidelberg, New York, 1972.

C. DERVIEUX AND B. PALMÉRIO

[1] *Une formule de Hadamard dans les problèmes d'identification de domaines*, C. R. Acad. Sci. Paris **280** (1975), 1697–1700, 1761–1764.

F.R. DESAINT AND J.-P. ZOLÉSIO

[1] *Sensitivity analysis of all eigenvalues of a symmetrical elliptic operator*, in "Boundary control and variation" (Sophia Antipolis, 1992), 141–160, Lecture Notes in Pure and Appl. Math., 163, Dekker, New York, 1994.

[2] *Dérivées par rapport au domaine des valeurs propres du laplacien* (French) [Shape derivatives of the eigenvalues of the Laplacian], C. R. Acad. Sci. Paris Sér. I Math. **321** (1995), 1337–1340.

[3] *Dérivation de forme dans les problèmes elliptiques tangentiels* (French) [Shape derivative for tangential elliptic problems], C. R. Acad. Sci. Paris Sér. I Math. **323** (1996), 781–786.

[4] *Manifold derivative in the Laplace-Beltrami equation*, J. Funct. Anal. **151** (1997), 234–269.

[5] *Shape derivative for the Laplace-Beltrami equation*, in "Partial differential equation methods in control and shape analysis" (Pisa), 111–132, Lecture Notes in Pure and Appl. Math., 188, Dekker, New York, 1997.

[6] *Shape boundary derivative for an elastic membrane*, in "Advances in mathematical sciences: CRM's 25 years" (Montreal, PQ, 1994), 481–491, CRM Proc. Lecture Notes, 11, AMS, Providence, RI, 1997.

PH. DESTUYNDER

[1] *Étude d'un algorithme numérique en optimisation de forme*, Ph.D. thesis, Université de Paris VI, 1976.

J. DETEIX

[1] *Conception optimale d'une digue à deux matériaux*, Ph.D. thesis, Université de Montréal, 1997.

J. DIEUDONNÉ

[1] *Treatrise on analysis*, Vols. II, III, Academic Press, New York, London, 1970, 1972.

[2] *Foundations of modern analysis*, Academic Press Inc., New York, London, 1960 (French transl. *Éléments d'analyse 1. Fondement de l'analyse moderne*, Gauthier-Villars, Paris, 1969; Bordas, 1979.

A. DJADANE

[1] *Régularité d'une fonctionnelle de domaines et application à l'analyse numérique d'un problème modèle*, Thèse de 3e cycle, Université d'Alger, Alger, 1975.

J. DUGUNDJI

[1] *Topology*, Allyn and Bacon, Boston, 1966.

H.B. DWIGHT

[1] *Tables of integrals and other mathematical data*, McMillan, New York, 1961.

R. DZIRI AND J.-P. ZOLÉSIO

[1] *Shape derivative for the heat-conduction equation with Lipschitz continuous co-efficients*, Boll. Un. Mat. Ital. B (7) **10** (1996), 569–594.

[2] *Shape-sensitivity analysis for nonlinear heat convection*, Appl. Math. Optim. **35** (1997), 1–20.

[3] *An energy principle for a free boundary problem for Navier-Stokes equations*, in "Partial differential equation methods in control and shape analysis" (Pisa), 133–151, Lecture Notes in Pure and Appl. Math., 188, Dekker, New York, 1997.

[4] *Interface minimisant l'énergie dans un écoulement stationnaire de Navier-Stokes* (French) [Interface minimizing the energy of a stationary Navier-Stokes flow], C. R. Acad. Sci. Paris Sér. I Math. **324** (1997), 1439–1444.

[5] *Shape existence in Navier-Stokes flow with heat convection*, Ann. Scuola Norm. Sup. Pisa Cl. Sci. (4) **24** (1997), 165–192.

[6] *Dynamical shape control in non-cylindrical Navier-Stokes equation*, J. Convex Anal. **6** (1999), 1–26.

[7] *Dynamical shape control in non-cylindrical hydrodynamics*, in "Conference on Inverse Problems, Control and Shape Optimization" (Carthage, 1998), Inverse Problems **15** (1999), 113–122.

I. EKELAND AND R. TEMAM

[1] *Analyse convexe et problèmes variationnels*, Dunod, Gauthier-Villars, Bordas, Paris, Bruxelles, Montréal, 1974.

S. EL YACOUBI AND J. SOKOŁOWSKI

[1] *Domain optimization problems for parabolic control systems*, in "Shape optimization and scientific computations," Appl. Math. Comput. Sci. **6** (1996), 277–289.

L.C. EVANS AND R.F. GARIEPY

[1] *Measure theory and the properties of functions*, CRC Press, Boca Raton, FL, 1992.

M. FARSHAD

[1] *Variations in eigenvalues and eigenvectors in continuum mechanics*, AIAA J. **12** (1974), 560–561.

H. FEDERER

[1] *Hausdorff measure and Lebesgue area*, Proc. Nat. Acad. Sci. U.S.A. **37** (1951), 90–94.

[2] *Curvature measures*, Trans. Amer. Math. Soc. **93** (1959), 418–419.

[3] *Geometric measure theory*, Springer-Verlag, Berlin, Heidelberg, New York, 1969.

[4] *Some properties of distributions whose partial derivatives are representable by integration*, Bull. Amer. Math. Soc. **74** (1968), 183–186.

[5] *A note on the Gauss-Green theorem*, Proc. Amer. Math. Soc. **9** (1958), 447–451.

H. FEDERER AND W.H. FLEMING

[1] *Normal and integral currents*, Ann. of Math. **72** (1960), 458–520.

W.H. FLEMING

[1] *Functions whose partial derivatives are measures*, Illinois J. Math. **4** (1960), 452–478.

W.H. FLEMING AND R. RISHEL

[1] *An integral formula for total gradient variation*, Arch. Math. **11** (1960), 218–222.

M. FORTIN AND R. GLOWINSKI

[1] *Augmented Lagrangian methods: Applications to the numerical solution of boundary-value problems*, Stud. Math. Appl., 15, North-Holland, Amsterdam, 1983.

J.H.G. FU

[1] *Tubular neighborhoods in Euclidean spaces*, Duke Math. J. **52** (1985), 1025–1046.

N. FUJII

[1] *Domain optimization problems with a boundary value problem as a constraint*, Control of Distributed Parameter Systems, H.E. Rauch, ed., 5–9, Pergamon Press, Oxford, New York, 1986.

[2] *Second variation and its application in a domain optimization problem*, Control of Distributed Parameter Systems, H.E. Rauch, ed., 431–436, Pergamon Press, Oxford, New York, 1986.

P.R. GARABEDIAN AND M. SCHIFFER

[1] *Identities in the theory of conformal mapping*, Trans. Amer. Math. Soc. **65** (1949), 187–238.

[2] *On existence theorems of potential theory and conformal mapping*, Ann. of Math. **52** (1950), 164–187.

[3] *Variational problems in the theory of elliptic partial differential equations*, J. Rational Mech. Anal. **2** (1953), 137–171.

[4] *Convexity of domain functionals*, J. Anal. Math. **2** (1953), 281–368.

[5] *The local maximum theorem for the coefficients of univalent functions*, Arch. Rational Mech. Anal. **26** (1967), 1–32.

B.R. GELBAUM AND J.M.H. OLMSTED

[1] *Counterexamples in analysis*, Holden-Day, San Francisco, 1964.

M. GELFAND AND N.Y. VILENKIN

[1] *Les distributions, applications de l'analyse harmonique*, 4, Dunod, Paris, 1967 (translated from Russian by G. Rideau).

T. GHEMIRES

[1] *Optimisation de forme et son application à des problèmes thermiques*, Ph.D. thesis, Université de Montréal, 1988.

D. GILBARG AND N.S. TRUDINGER

[1] *Elliptic partial differential equations of second order*, Springer-Verlag, Berlin, Heidelberg, New York, Tokyo, 1983.

E. GIUSTI

[1] *Minimal surfaces and functions of bounded variation*, Birkhäuser, Boston, Basel, Stuttgart, 1984.

A. GORRE

[1] *Évolution de tubes opérables gouvernée par des équations mutationnelles*, Ph.D. thesis, Université Paris-Dauphine, Paris 1996.

D. GATAREK AND J. SOKOŁOWSKI

[1] *Shape sensitivity analysis for stochastic evolution equations*, in "Boundary control and boundary variation," 215–220, Lecture Notes in Control and Inform. Sci., 178, Springer-Verlag, Berlin, 1992.

[2] *Shape sensitivity analysis of optimal control problems for stochastic parabolic equations*, in "Control of distributed systems," Span.-Fr. Days, 103–114, Grupo Anal. Mat. Apl. Univ. Málaga, Málaga, Spain 1990.

[3] *Shape sensitivity analysis for stochastic PDEs*, in "Modelling uncertain data," 12–21, Math. Res., 68, Akademie-Verlag, Berlin, 1992.

H. GRAD, P.N. NU, AND D.C. STEVENS

[1] *Adiabatic evolution of plasma equilibrium*, Bull. Phys. Soc. **19** (1974), 865.

[2] *Adiabatic evolution of plasma equilibrium*, Proc. Nat. Acad Sci. U.S.A. (10) **72** (1975), 3789–3793.

Y. GUIDO AND J.-P. ZOLÉSIO

[1] *Shape Hessian for a nondifferentiable variational free boundary problem*, in "Partial differential equation methods in control and shape analysis" (Pisa), G. Da Prato and J.-P. Zolésio, eds., 205–213, Lecture Notes in Pure and Appl. Math., 188, Dekker, New York, 1997.

J. HADAMARD

[1] *Mémoire sur le problème d'analyse relatif à l'équilibre des plaques élastiques encastrées*, Mém. Sav. Étrang., 33, 1907; in "Œuvres de Jacques Hadamard," 515–641, Editions du C.N.R.S., Paris, 1968.

Y. HAMDOUCH

[1] *Optimisation des plaques*, M.S. thesis, Université de Montréal, 1996.

E.J. HAUG AND J.S. ARORA

[1] *Applied optimal design*, Wiley, New York, 1979.

E.J. HAUG AND J. CÉA

[1] *Optimization of distributed parameter structures*, I and II, NATO Adv. Sci. Inst. Ser. E Appl. Sci., 50, Sijhoff and Noordhoff, Alphen aan den Rijn, the Netherlands, 1981.

E.J. HAUG, K.K. CHOI, AND V. KOMKOV

[1] *Design sensitivity analysis of structural systems*, Academic Press, New York, 1986.

E.J. HAUG AND B. ROUSSELET

[1] *Design sensitivity in structural analysis. I. Static response variations*, J. Struct. Mech. **8** (1980), 17–41; II. *Eigenvalue variations*, J. Struct. Mech. **8** (1980), 161–186.

[2] *Design sensitivity analysis of eigenvalue variations*, in "Optimization of distributed parameter structures," II, 1371–1396, NATO Adv. Sci. Inst. Ser. E Appl. Sci., 50, Nijhoff, the Hague, 1981.

[3] *Design sensitivity analysis of static response variations*, in "Optimization of distributed parameter structures," II, 1345–1370, NATO Adv. Sci. Inst. Ser. E Appl. Sci., 50, Nijhoff, the Hague, 1981.

F. HAUSDORFF

[1] *Dimension und äusseres Mass*, Math. Ann. **79** (1919), 157–179.

L.I. HEDBERG

[1] *Spectral synthesis in Sobolev spaces and uniqueness of solutions of dirichlet problems*, Acta Math. **147** (1982), 237–263.

J. HEINONEN, T. KILPELAINEN, AND O. MARTIO

[1] *Nonlinear potential theory of degenerate elliptic equations*, Clarendon Press, Oxford, 1993.

A. HENROT

[1] *Continuity with respect to the domain for the Laplacian: A survey*, in "Shape design and optimization," Control Cybernet. **23** (1994), 427–443.

[2] *How to prove symmetry in shape optimization problems?*, in "Shape optimization and scientific computations," Control Cybernet. **25** (1996), 1001–1013.

A. HENROT, J.-P. BRANCHER, AND M. PIERRE

[1] *Existence of equilibria in electromagnetic casting*, in "Proceedings of the fifth international symposium on numerical methods in engineering 1, 2," 221–228, Comput. Mech., Southampton, 1989.

A. HENROT, W. HORN, AND J. SOKOŁOWSKI

[1] *Domain optimization problem for stationary heat equation*, in "Shape optimization and scientific computations," Appl. Math. Comput. Sci. **6** (1996), 353–374.

A. HENROT AND M. PIERRE

[1] *About existence of a free boundary in electromagnetic shaping*, in "Recent advances in nonlinear elliptic and parabolic problems," 283–293, Pitman Res. Notes Math. Ser., 208, Longman Sci. Tech., Harlow, UK, 1989.

[2] *Un problème inverse en formage des métaux liquides* [An inverse problem in the shaping of molten metals], RAIRO Modél. Math. Anal. Numér. **23** (1989), 155–177.

[3] *About existence of equilibria in electromagnetic casting*, Quart. Appl. Math. **49** (1991), 563–575.

[4] *About critical points of the energy in an electromagnetic shaping problem*, in "Boundary control and boundary variation," 238–252, Lecture Notes in Control and Inform. Sci., 178, Springer-Verlag, Berlin, 1992.

A. HENROT AND J. SOKOŁOWSKI

[1] *A shape optimization problem for the heat equation*, in "Optimal control," 204–223, Appl. Optim., 15, Kluwer, Dordrecht, 1998.

K.-H. HOFFMANN AND J. SOKOŁOWSKI

[1] *Interface optimization problems for parabolic equations*, in "Shape design and optimization," Control Cybernet. **23** (1994), 445–451.

K.-H. HOFFMANN AND D. TIBA

[1] *Fixed domain methods in variable domain problems*, in "Free boundary problems: Theory and applications" (Toledo, 1993), 123–146, Pitman Res. Notes Math. Ser., 323, Longman Sci. Tech., Harlow, UK, 1995.

[2] *Control of a plate with nonlinear shape memory alloy reinforcements*, Adv. Math. Sci. Appl. **7** (1997), 427–436.

P. HOLNICKI, J. SOKOŁOWSKI, AND A. ŻOCHOWSKI

[1] *Sensitivity analysis of an optimal control problem arising from air quality control in urban area*, in "System modelling and optimization," 331–339, Lecture Notes in Control and Inform. Sci., 84, Springer-Verlag, Berlin, 1986.

J. HORVÁTH

[1] *Topological vector spaces and distributions*, I, Addison-Wesley, Reading, MA, 1966.

D. JERISON

[1] *The direct method in the calculus of variations for convex bodies*, Adv. Math. **122** (1996), 262–279.

J.L. KELLEY

[1] *General topology*, Van Nostrand, Toronto, New York, London, 1955.

A. KHLUDNEV AND J. SOKOŁOWSKI

[1] *Modelling and control in solid mechanics*, Internat. Ser. Numer. Math., 122, Birkhäuser, Basel, 1997.

V.L. KLEE

[1] *Convexity of Chebyshev sets*, Math. Ann. **142** (1961), 292–304.

R.V. KOHN

[1] *Numerical structural optimization via a relaxed formulation*, in "Shape optimization and free boundaries," 173–210, NATO Adv. Sci. Inst. Ser. C Math. Phys. Sci., 380, Kluwer, Dordrecht, 1992.

R.V. KOHN AND G. STRANG

[1] *Explicit relaxation of a variational problem in optimal design*, Bull. Amer. Math. Soc. (N.S.) **9** (1983), 211–214.

[2] *Optimal design and relaxation of variational problems.* I, Comm. Pure Appl. Math. **39** (1986), 113–137; II, 139–182; III, 353–377.

S.R. KULKARNI, S. MITTER, AND T.J. RICHARDSON

[1] *An existence theorem and lattice approximations for a variational problem arising in computer vision,* in "Signal processing, Part I. Signal processing theory," L. Auslander, T. Kailath, and S. Mitter, eds., 189–210, IMA Vol. Math. Appl., 22, Springer-Verlag, Heidelberg, Berlin, New York, 1990.

S. LANG

[1] *Differential and Riemannian manifolds,* Springer-Verlag, New York, 1995.

P. LAURENCE

[1] *Une nouvelle formule pour dériver par rapport aux lignes de niveau et applications,* C. R. Acad. Sci. Paris Sér. I Math. **306** (1988), 455–460.

H. LEBESGUE

[1] *Leçons sur l'intégration et la recherche des fonctions primitives,* Gauthier-Villars, Paris, 1904.

[2] *Sur l'intégration des fonctions discontinues,* Ann. Ecole Norm. **27** (1910), 361–450.

B. LEMAIRE

[1] *Problèmes min-max et applications au contrôle optimal de systèmes gouvernés par des équations aux dérivées partielles linéaires,* Ph.D. thesis, Université de Montpellier, Montpellier, 1970.

W. LIU AND J. E. RUBIO

[1] *Local convergence and optimal shape design,* SIAM J. Control Optim. **30** (1992), 49–62.

W. LIU, P. NEITTAANMÄKI, AND D. TIBA

[1] *Sur les problèmes d'optimisation structurelle,* C. R. Acad. Sci. Paris Sér. I Math. **331** (2000), 101–106.

P. LUR'E AND A.V. CHERKAEV

[1] *Prager theorem application to optimal design of thin plates,* MTT Mechanics of Solids **11** (1976), 157–159 (in Russian).

J.-P. MARMORAT, G. PAYRE, AND J.-P. ZOLÉSIO

[1] *A convergent finite element scheme for a wave equation with a moving boundary,* in "Boundary control and boundary variation" (Sophia-Antipolis, 1990), 297–308, Lecture Notes in Control and Inform. Sci., 178, Springer-Verlag, Berlin, 1992.

J.-P. MARMORAT AND J.-P. ZOLÉSIO

[1] *Static and dynamic behavior of a fluid-shell system,* in "Partial differential equation methods in control and shape analysis" (Pisa), G. Da Prato and J.-P. Zolésio, eds., 245–257, Lecture Notes in Pure and Appl. Math., 188, Dekker, New York, 1997.

G. MATHERON

[1] *Examples of topological properties of skeletons*, in "Image analysis and mathematical morphology," J. Serra, ed., 217–238, Academic Press, London, 1988.

V.G. MAZ'JA

[1] *Sobolev spaces*, Springer-Verlag, Berlin, Heidelberg, New York, Tokyo, 1980.

A.M. MICHELETTI

[1] *Metrica per famiglie di domini limitati e proprietà generiche degli autovalori*, Ann. Scuola Norm. Sup. Pisa Ser. (3) **26** (1972), 683–694.

M. MIRANDA

[1] *Sul minimo dell'integrale del gradienti di una funzione*, Ann. Scuola Norm. Sup. Pisa (3) **19** (1965), 627–665.

F. MORGAN

[1] *Geometric measure theory, a beginner's guide*, Academic Press, Boston, San Diego, 1988.

C.B. MORREY, JR.

[1] *Multiple integrals in the calculus of variations*, Springer-Verlag, New York, 1966.

J. MOSSINO AND J.-P. ZOLÉSIO

[1] *Formulation variationnelle de problèmes issus de la physique des plasmas*, C. R. Acad. Sci. Paris Sér. A **285** (1977), 1003–1007.

F. MURAT

[1] *Un contre-exemple pour le problème de contrôle dans les coefficients*, C. R. Acad. Sci. Paris **273** (1971), 708–711.

[2] *Contre-exemples pour divers problèmes où le contrôle intervient dans les coefficients*, Ann. Mat. Pura Appl. **112** (1977), 49–68.

F. MURAT AND J. SIMON

[1] *Sur le contrôle par un domaine géométrique*, Rapport 76015, Université Pierre et Marie Curie, Paris, 1976.

F. MURAT AND L. TARTAR

[1] *Calcul des variations et homogénéisation* (French) [Calculus of variations and homogenization], in "Homogenization methods: Theory and applications in physics" (Bréau-sans-Nappe, 1983), 319–369, Collect. Dir. Études Rech. Elec. France, 57, Eyrolles, Paris, 1985.

[2] *Optimality conditions and homogenization*, in "Nonlinear variational problems" (Isola d'Elba, 1983), 1–8, Pitman Res. Notes Math. Ser., 127, Pitman, Boston, London, 1985.

[3] *Calculus of variations and homogenization*, in "Topics in the mathematical modelling of composite materials," A. Cherkaev and R. Kohn, eds., 139–173, Progr. Nonlinear Differential Equations Appl., 31, Birkhäuser, Boston, 1997.

[4] *H-convergence*, in "Topics in the mathematical modelling of composite materials," A. Cherkaev and R. Kohn, eds., 21–43, Progr. Nonlinear Differential Equations Appl., 31, Birkhäuser, Boston, 1997.

[5] *On the control of coefficients in partial differential equations*, in "Topics in the mathematical modelling of composite materials," A. Cherkaev and R. Kohn, eds., 1–8, Progr. Nonlinear Differential Equations Appl., 31, Birkhäuser, Boston, 1997.

A. Myśliński, J. Sokołowski, and J.-P. Zolésio

[1] *On nondifferentiable plate optimal design problems*, in "System modelling and optimization" (Copenhagen, 1983), 829–838, Lecture Notes in Control and Inform. Sci., 59, Springer-Verlag, Berlin, 1984.

M. Nagumo

[1] *Über die Loge der Integralkurven gewöhnlicher Differentialgleichungen*, Proc. Phys. Math. Soc. Japan **24** (1942), 551–559.

J. Nečas

[1] *Les méthodes directes en théorie des équations elliptiques*, Masson, Paris, and Academia, Prague, 1967.

P. Neittaanmäki and D. Tiba

[1] *An embedding of domains approach in free boundary problems and optimal design*, SIAM J. Control Optim. **33** (1995), 1587–1602.

P. Neittaanmäki, J. Sokołowski, and J.-P. Zolésio

[1] *Optimization of the domain in elliptic variational inequalities*, Appl. Math. Optim. **18** (1988), 85–98.

A. Novruzi and J.-R. Roche

[1] *Second order derivatives, Newton method, application to shape optimization*, Rapport no. 95/8, Institut Élie Cartan Nancy, France.

N. Olhoff

[1] *On singularities, local optima and formation of stiffeners in optimal design of plates*, Proc. IUTAM Symp. on Optimization in Structural Design, A. Sawczuk and Z. Mróz, eds., 82–103 Springer-Verlag, Berlin, 1975.

K. Piekarski and J. Sokołowski

[1] *Shape control of elastic plates*, in "Systems modelling and optimization" (Detroit, MI, 1997), 80–88, CRC Res. Notes Math., 396, Chapman & Hall/CRC, Boca Raton, FL, 1999.

M. Pierre and J. Sokołowski

[1] *Differentiability of projection and applications*, in "Control of partial differential equations and applications (Laredo, 1994), 231–240, Lecture Notes in Pure and Appl. Math., 174, Dekker, New York, 1996.

O. Pironneau

[1] *Optimal design for elliptic systems*, Springer-Verlag, New York, 1984.

J. A. F. PLATEAU

[1] *Statique expérimentale et théorique des liquides soumis aux seules forces moléculaires*, Gauthier-Villars, Paris, 1873.

L. PONTRYAGIN, V. BOLTYANSKII, R. GAMKRELIDZE, AND E. MISHCHENKO

[1] *The mathematical theory of optimal processes*, transl. from the Russian by K.N. Trirogoff; L.W. Neustadt, ed., Interscience, Wiley, New York, London, 1962.

J.M. RAKOTOSON AND R. TEMAM

[1] *Relative rearrangement in quasilinear elliptic variational inequalities*, Indiana Univ. Math. J. **36** (1987), 757–810.

M. RAO AND J. SOKOŁOWSKI

[1] *Shape sensitivity analysis of state constrained optimal control problems for distributed parameter systems*, in "Control of partial differential equations" (Santiago de Compostela, 1987), 236–245, Lecture Notes in Control and Inform. Sci., 114, Springer-Verlag, Berlin, New York, 1989.

[2] *Differential stability of perturbed optimization with applications to parameter estimation*, in "Stabilization of flexible structures" (Montpellier, 1989), 298–311, Lecture Notes in Control and Inform. Sci., 147, Springer-Verlag, Berlin, 1990.

[3] *Sensitivity analysis of shallow shell with obstacle*, in "Modelling and inverse problems of control for distributed parameter systems" (Laxenburg, 1989), 135–144, Lecture Notes in Control and Inform. Sci., 154, Springer-Verlag, Berlin, 1991.

[4] *Polyhedricity of convex sets in Sobolev space $H_0^2(\Omega)$*, Nagoya Math. J. **130** (1993), 101–110.

[5] *Sensitivity analysis of unilateral problems in $H_0^2(\Omega)$ and applications*, Numer. Funct. Anal. Optim. **14** (1993), 125–143.

[6] *Stability of solutions to unilateral problems*, "Emerging applications in free boundary problems" (Montreal, PQ, 1990), 235–240, Pitman Res. Notes Math. Ser., 280, Longman Sci. Tech., Harlow, UK, 1993.

[7] *Optimal shape design of plates and shells with obstacles*, in "World Congress of Nonlinear Analysts '92," I–IV (Tampa, FL, 1992), 2595–2604, de Gruyter, Berlin, 1996.

J. RAUCH AND M. TAYLOR

[1] *Potential and scattering theory on wildly perturbed domains*, J. Funct. Anal. **18** (1975), 27–59.

R. REISS

[1] *Optimal compliance criterion for axisymmetric solid plates*, Internat. J. Solids Structures **4** (1976), 319–329.

T.J. RICHARDSON

[1] *Scale independent piecewise smooth segmentation of images via variational methods*, Report CICS-TH-194, Center for Intelligent Control Systems, Massachusetts Institute of Technology, Cambridge, MA, 1990.

A. Rivière

[1] *Classification de points d'un ouvert d'un espace euclidien relativement à la distance au bord, étude topologique et quantitative des classes obtenues*, thèse de doctorat, Université de Paris-Sud, Centre d'Orsay, 1987.

T.L. Rockafellar

[1] *Convex analysis*, Princeton University Press, Princeton, NJ, 1972.

C.A. Rogers

[1] *Hausdorff measure*, Cambridge University Press, Cambridge, UK, 1970.

B. Rousselet

[1] *Étude de la régularité de valeurs propres par rapport à des déformations bilipschitziennes du domaine géométrique*, C. R. Acad. Sci. Paris Sér. A-B **283** (1976), A507–A509.

[2] *Identification de domaines et problèmes de valeurs propres* (French) [Identification of domains, and eigenvalue problems], Thèse de 3e cycle, Université de Nice, 1977.

[3] *Optimal design and eigenvalue problems*, in "Optimization techniques," Part 1, 343–352. Lecture Notes in Control and Information Sci., 6, Springer-Verlag, Berlin, 1978.

[4] *Design sensitivity analysis in structural mechanics*, in "Nonlinear analysis," 245–254, Abh. Akad. Wiss. DDR, Abt. Math. Naturwiss. Tech., 1981, 2, Akademie-Verlag, Berlin, 1981.

[5] *Singular dependence of repeated eigenvalues*, in "Optimization of distributed parameter structures," II, 1443–1456, NATO Adv. Sci. Inst. Ser. E Appl. Sci., 50, Nijhoff, the Hague, 1981.

[6] *Quelques résultats en optimisation de domaines*, Ph.D. thesis, Université de Nice, 1982.

[7] *Note on the design differentiability of the static response of elastic structures*, J. Structural Mech. **10** (1982/1983), 353–358.

[8] *Shape design sensitivity of a membrane*, J. Optim. Theory Appl. **40** (1983), 595–623.

[9] *Static and dynamical loads, pointwise constraint in structural optimization*, in "Optimization: Theory and algorithms," 225–238, Lecture Notes in Pure and Appl. Math., 86, Dekker, New York, 1983.

[10] *Shape design sensitivity of a membrane*, J. Optim. Theory Appl. **40** (1983), 595–623.

[11] *Static and dynamical loads, pointwise constraint in structural optimization*, in "Optimization: Theory and algorithms," 225–238, Lecture Notes in Pure and Appl. Math., 86, Dekker, New York, 1983.

[12] *Optimisation d'un système surfacique (équation de la chaleur)* (French) [Optimization of a surface system (the heat equation)], in "Control of boundaries and stabilization," 231–240, Lecture Notes in Control and Inform. Sci., 125, Springer-Verlag, Berlin, New York, 1989.

[13] *Shape optimization of structures with pointwise state constraints*, in "Control of partial differential equations," 255–264, Lecture Notes in Control and Inform. Sci., 114, Springer-Verlag, Berlin, New York, 1989.

[14] *A finite strain rod model and its design sensitivity*, Mech. Structures Mach. **20** (1992), 415–432.

[15] *Introduction to shape sensitivity—three-dimensional and surface systems*, in "Shape and layout optimization of structural systems and optimality criteria methods," 243–278, CISM Courses and Lectures, 325, Springer-Verlag, Vienna, 1992.

[16] *A finite strain rod model and its design sensitivity*, in "Optimization of large structural systems," I, II, 433–453, NATO Adv. Sci. Inst. Ser. E Appl. Sci., 231, Kluwer, Dordrecht, 1993.

[17] *Introduction to shape sensitivity: Three-dimensional and surface systems*, in "Optimization of large structural systems," I, II, 397–432, NATO Adv. Sci. Inst. Ser. E Appl. Sci., 231, Kluwer, Dordrecht, 1993.

[18] *Introduction to shape sensitivity: Three-dimensional and surface systems*, in "Shape optimization and scientific computations," Control Cybernet. **25** (1996), 937–967.

B. ROUSSELET AND D. CHENAIS

[1] *Continuité et différentiabilité d'éléments propres: Application à l'optimisation de structures* (French) [Continuity and differentiability of eigenelements: Application to the optimization of structures], Appl. Math. Optim. **22** (1990), 27–59.

B. ROUSSELET AND E.J. HAUG

[1] *Design sensitivity analysis of shape variation*, in "Optimization of distributed parameter structures," II, 1397–1442, NATO Adv. Sci. Inst. Ser. E Appl. Sci., 50, Nijhoff, the Hague, 1981.

[2] *Design sensitivity analysis in structural mechanics. III. Effects of shape variation*, J. Structural Mech. **10** (1982/83), 273–310.

B. ROUSSELET, J. PIEKARSKI, AND A. MYŚLIŃSKI

[1] *Shape optimization of hyperelastic rod*, in "System modelling and optimization," 218–225, Chapman & Hall, London, 1996.

[2] *Design sensitivity for a hyperelastic rod in large displacement with respect to its midcurve shape*, J. Optim. Theory Appl. **96** (1998), 683–708.

G. ROZVANY

[1] *Structural design via optimality criteria*, Kluwer, Dordrecht 1989.

W. RUDIN

[1] *Principles of mathematical analysis*, McGraw-Hill, New York, 1967.

L. SCHWARTZ

[1] *Théorie des noyaux*, in "Proceedings of the International Congress of Mathematicians," Cambridge, MA, Vol. I, 1950, 220–230, AMS, Providence, RI, 1952.

[2] *Théorie des distributions*, Hermann, Paris, 1966.

[3] *Analyse mathématique* I, Hermann, Paris, 1967.

J.A. SETHIAN

[1] *Level set methods and fast marching methods. Evolving interfaces in computational geometry, fluid mechanics, computer vision, and materials science*, Cambridge Monogr. Appl. Comput. Math., 3, 2nd ed., Cambridge University Press, Cambridge, 1999.

J. SIMON

[1] *Control of boundaries and stabilization*, in "Proceedings of the IFIP-WG 7.2 Conference held in Clermont-Ferrand," J. Simon, ed., Lecture Notes in Control and Inform. Sci., 125, Springer-Verlag, Berlin, New York, 1989.

[2] *Differentiation on a Lipschitz manifold*, in "Control of partial differential equations," 277–283, Lecture Notes in Control and Inform. Sci., 114, Springer-Verlag, Berlin, New York, 1989.

[3] *Second variations for domain optimization problems*, in "Control and estimation of distributed parameter systems," 361–378, Internat. Ser. Numer. Math., 91, Birkhäuser, Basel, 1989.

[4] *Domain variation for drag in Stokes flow*, in "Control theory of distributed parameter systems and applications," 28–42, Lecture Notes in Control and Inform. Sci., 159, Springer-Verlag, Berlin, 1991.

J. SOKOŁOWSKI

[1] *Sensitivity analysis for a class of variational inequalities*, in "Optimization of distributed parameter structures," II (Iowa City, IA, 1980), 1600–1609, NATO Adv. Sci. Inst. Ser. E Appl. Sci., 50, Nijhoff, the Hague, 1981.

[2] *Differential stability of projection in Hilbert space onto convex set. Applications to sensitivity analysis of optimal control problems*, in "Analysis and algorithms of optimization problems," 1–37, Lecture Notes in Control and Inform. Sci., 82, Springer-Verlag, Berlin, New York, 1986.

[3] *Shape sensitivity analysis of boundary optimal control problems for parabolic systems*, SIAM J. Control Optim. **26** (1988), 763–787.

[4] *Shape sensitivity analysis of nonsmooth variational problems*, in "Boundary control and boundary variations" (Nice, 1986), 265–285, Lecture Notes in Comput. Sci., 100, Springer-Verlag, Berlin, New York, 1988.

[5] *Optimal design and control of structures*, in "Proceedings of the Second Symposium held in Jabłonna," Z. Mróz and J. Sokołowski, eds., Control Cybernet. **19** (1990), (1991), 1–341.

[6] *Sensitivity analysis of shape optimization problems*, in "Optimal design and control of structures" (Jabłonna, 1990), Control Cybernet. **19** (1990), 271–286 (1991).

[7] *Shape sensitivity analysis of variational inequalities*, in "Shape optimization and free boundaries" (Montreal, PQ, 1990), 287–319, NATO Adv. Sci. Inst. Ser. C Math. Phys. Sci., 380, Kluwer, Dordrecht, 1992.

[8] *Nonsmooth optimal design problems for the Kirchhoff plate with unilateral conditions*, Kybernetika (Prague) **29** (1993), 284–290.

[9] *Applications of material derivative method*, in "Topology design of structures" (Sesimbra, 1992), 307–311, NATO Adv. Sci. Inst. Ser. E Appl. Sci., 227, Kluwer, Dordrecht, 1993.

[10] *Stability of solutions to shape estimation problems*, Mech. Structures Mach. **21** (1993), 67–94.

[11] *Modelling, identification, sensitivity analysis and control of structures*, J. Sokołowski, ed., Control Cybernet. **23** (1994), 583–812.

[12] *Shape design and optimization*, J. Sokołowski, ed., Control Cybernet. **23** (1994), 303–581.

[13] *Shape optimization and scientific computations*, papers from the Joint CIMPA/GDR/TEMPUS-JEP Intensive School on Shape Optimal Design, Theory, Application and Software held in Warsaw July 3–14, 1995, J. Sokołowski, ed., Control Cybernet. **25** (1996), special issue.

[14] *Shape optimization and scientific computations*, papers from the First French-Polish Summer School on Shape Optimization and Scientific Computing held in Warsaw, July 10–14, 1994, J. Sokołowski, ed., Appl. Math. Comput. Sci. **6** (1996).

[15] *Displacement derivatives in shape optimization of thin shells*, in "Optimization methods in partial differential equations" (South Hadley, MA, 1996), 247–266, Contemp. Math., 209, AMS, Providence, RI, 1997.

[16] *Recent advances in structural modelling and optimization*, J. Sokołowski and J. Taylor, eds., Control Cybernet. **27** (1998), 159–330.

J. SOKOŁOWSKI AND J. SPREKELS

[1] *Dynamical shape control of nonlinear thin rods*, in "Optimal control of partial differential equations" (Irsee, 1990), 202–208, Lecture Notes in Control and Inform. Sci., 149, Springer-Verlag, Berlin, 1991.

[2] *Dynamical shape control and the stabilization of nonlinear thin rods*, Math. Methods Appl. Sci. **14** (1991), 63–78.

J. SOKOŁOWSKI AND A. ZOCHOWSKI

[1] *On the topological derivative in shape optimization*, SIAM J. Control Optim. **37** (1999), 1251–1272.

[2] *Topological derivatives for elliptic problems*, in "Conference on Inverse Problems, Control and Shape Optimization" (Carthage, 1998), Inverse Problems **15** (1999), 123–134.

J. SOKOŁOWSKI AND J.-P. ZOLÉSIO

[1] *Dérivée par rapport au domaine de la solution d'un problème unilatéral* (French) [Shape derivative for the solutions of variational inequalities], C. R. Acad. Sci. Paris Sér. I Math. **301** (1985), 103–106.

[2] *Differential stability of solutions to unilateral problems*, in "Free boundary problems: Application and theory, IV (Maubuisson, 1984), 537–547, Pitman Res. Notes Math. Ser., 121, Pitman, Boston, London, 1985.

[3] *Sensitivity analysis of elastic-plastic torsion problem*, in "System modelling and optimization" (Budapest, 1985), 845–853, Lecture Notes in Control and Inform. Sci., 84, Springer-Verlag, Berlin, 1986.

[4] *Shape design sensitivity analysis of plates and plane elastic solids under unilateral constraints*, J. Optim. Theory Appl. **54** (1987), 361–382.

[5] *Shape sensitivity analysis of unilateral problems*, SIAM J. Math. Anal. **18** (1987), 1416–1437.

[6] *Shape sensitivity analysis in the presence of singularities*, in "Some topics on inverse problems" (Montpellier, 1987), 294–303, World Scientific, Singapore, 1988.

[7] *Shape sensitivity analysis of contact problem with prescribed friction*, Nonlinear Anal. **12** (1988), 1399–1411.

[8] *Shape sensitivity analysis of hyperbolic problems*, in "Stabilization of flexible structures" (Montpellier, 1989), 280–297, Lecture Notes in Control and Inform. Sci., 147, Springer-Verlag, Berlin, 1990.

[9] *Introduction to shape optimization. Shape sensitivity analysis*, Springer Ser. Comput. Math., 16, Springer-Verlag, Berlin, 1992.

M. SOULI AND J.-P. ZOLÉSIO

[1] *Semidiscrete and discrete gradient for nonlinear water wave problems* in "Boundary control and boundary variations" (Nice, 1986), 297–310, Lecture Notes in Comput. Sci., 100, Springer-Verlag, Berlin, New York, 1988.

[2] *Le contrôle de domaine dans le problème de résistance de vague non linéaire en hydrodynamique* [Control of the domain in the problem of resistance of nonlinear waves in hydrodynamics], Ann. Sci. Math. Québec **15** (1991), 203–214.

[3] *Shape derivative of discretized problems*, in "Optimal control of partial differential equations" (Irsee, 1990), 209–228, Lecture Notes in Control and Inform. Sci., 149, Springer-Verlag, Berlin, 1991.

[4] *Shape derivative of discretized problems*, Comput. Methods Appl. Mech. Engrg. **108** (1993), 187–199.

[5] *Finite element method for free surface flow problems*, Comput. Methods Appl. Mech. Engrg. **129** (1996), 43–51.

J. SPREKELS AND D. TIBA

[1] *Propriétés de bang-bang généralisées dans l'optimisation des plaques* (French) [Generalized bang-bang properties in the optimization of plates], C. R. Acad. Sci. Paris Sér. I Math. **327** (1998), 705–710.

[2] *A duality-type method for the design of beams*, Adv. Math. Sci. Appl. **9** (1999), 89–102.

[3] *A duality approach in the optimization of beams and plates*, SIAM J. Control Optim. **37** (1999), 486–501.

[4] *Sur les arches lipschitziennes*, C. R. Acad. Sci. Paris Sér. I Math. **331** (2000), 179.

G. STAMPACCHIA

[1] *Équations elliptiques du second ordre à coefficients discontinus*, Sém. Math. Sup., Presses Univ. Montréal, Montréal, 1966.

G. STRANG AND R.V. KOHN

[1] *Optimal design in elasticity and plasticity*, Internat. J. Numer. Methods Engrg. **22** (1986), 183–188.

[2] *Fibered structures in optimal design*, in "Ordinary and partial differential equations," 205–216, Pitman Res. Notes Math. Ser., 157, Longman Sci. Tech., Harlow, UK, 1987.

V. ŠVERÁK

[1] *On shape optimal design*, C. R. Acad. Sci. Paris Sér. I Math. **315** (1992), 545–549.

[2] *On optimal shape design*, J. Math. Pures Appl. **72** (1993), 537–551.

R. TEMAM

[1] *Monotone rearrangement of a function and the Grad-Mercier equation of plasmas physics*, in "Proceedings of the international meeting on recent methods in nonlinear analysis," Rome, May 8–12, 1978, E. De Giorgi, E. Magenes, and U. Mosco, eds., Pitagora Editrice, Bologna, 1979.

[2] *Problèmes mathématiques en plasticité*, Gauthier-Villars, Paris, 1983.

D. TIBA

[1] *Propriétés de controlabilité pour les systèmes elliptiques, la méthode des domaines fictifs et problèmes de design optimal* [Controllability properties for elliptic systems, the fictitious domain method and optimal shape design problems], in "Optimization, optimal control and partial differential equations" (Iaşi, 1992), 251–261, Internat. Ser. Numer. Math., 107, Birkhäuser, Basel, 1992.

[2] *A property of Sobolev spaces and existence in optimal design*, Preprint 5, Institute of Mathematics, Romanian Academy, Bucharest, Romania, 1999.

T. TIIHONEN AND J.-P. ZOLÉSIO

[1] *Gradient with respect to nodes for nonisoparametric finite elements*, in "Boundary control and boundary variations" (Nice, 1986), 311–316, Lecture Notes in Comput. Sci., 100, Springer-Verlag, Berlin, 1988.

I. TODHUNTER AND K. PEARSON

[1] *A history of the theory of elasticity and of the strength of materials*, Cambridge University Press, Cambridge, 1886.

F.A. VALENTINE

[1] *Convex sets*, McGraw-Hill, New York, 1964.

P. VILLAGGIO AND J.-P. ZOLÉSIO

[1] *New results in shape optimization*, in "Boundary control and boundary variation" (Sophia-Antipolis, 1990), 356–361, Lecture Notes in Control and Inform. Sci., 178, Springer-Verlag, Berlin, 1992.

[2] *Suboptimal shape of a plate stretched by planar forces*, in "Partial differential equation methods in control and shape analysis" (Pisa), 321–331, G. Da Prato and J.-P. Zolésio, eds., Lecture Notes in Pure and Appl. Math., 188, Dekker, New York, 1997.

H. WHITNEY

[1] *Geometric integration theory*, Princeton Math. Ser., 21, Princeton University Press, Princeton, 1957.

L.C. YOUNG

[1] *Partial area* I, Riv. Mat. Univ. Parma **10** (1959), 103–113.

[2] *Lectures on the calculus of variations and optimal control theory*, Saunders, Philadelphia, 1969.

E.H. ZARANTONELLO

[1] *Projections on convex sets in Hilbert spaces and spectral theory*, Parts I and II, in "Contributions to nonlinear functional analysis," E.H. Zarantonello, ed., 237–424, Academic Press, New York, London, 1971.

W.P. ZIEMER

[1] *Weakly differentiable functions. Sobolev spaces and functions of bounded variation*, Grad. Texts in Math., 120, Springer-Verlag, New York, 1989.

J.-P. ZOLÉSIO

[1] *Localisation d'un objet de forme convexe donnée* (French), C. R. Acad. Sci. Paris Sér. A-B **274** (1972), A850–A852.

[2] *Sur la localisation d'un domaine*, Ph.D. thesis, Université de Nice, 1973.

[3] *Un résultat d'existence de vitesse convergente dans des problèmes d'identification de domaine*, C. R. Acad. Sci. Paris Sér. A-B **283** (1976), A855–A858.

[4] *An optimal design procedure for optimal control support*, in "Convex analysis and its applications" (Proc. Conf., Muret-le-Quaire, 1976), 207–219, Lecture Notes in Econom. and Math. Systems, 144, Springer-Verlag, Berlin, 1977.

[5] *Localisation du support d'un contrôle optimal* (French), C. R. Acad. Sci. Paris Sér. A-B **284** (1977), A191–A194.

[6] *Solution variationnelle d'un problème de valeur propre non linéaire et frontière libre en physique des plasmas* (French), C. R. Acad. Sci. Paris Sér. A-B **288** (1979), A911–A913.

[7] *Identification de domaines par déformation*, Ph.D. thesis, Université de Nice, 1979.

[8] *The material derivative (or speed) method for shape optimization*, in "Optimization of distributed parameter structures," II, Iowa City, IA, 1980, E.J. Haug and J. Céa, eds., 1089–1151, NATO Adv. Sci. Inst. Ser. E Appl. Sci., 50, Sijhofff and Nordhoff, Alphen aan den Rijn, 1981 (Nijhoff, the Hague).

[9] *Domain variational formulation for free boundary problems*, in "Optimization of distributed parameter structures," II, Iowa City, IA, 1980, E.J. Haug and J. Céa, eds., 1152–1194, NATO Adv. Sci. Inst. Ser. E Appl. Sci., 50, Sijhofff and Nordhoff, Alphen aan den Rijn, 1981 (Nijhoff, the Hague).

[10] *Semiderivatives of repeated eigenvalues*, in "Optimization of distributed parameter structures," II, Iowa City, IA, 1980, E.J. Haug and J. Céa, eds., 1457–1473, NATO Adv. Sci. Inst. Ser. E Appl. Sci., 50, Sijhofff and Nordhoff, Alphen aan den Rijn, 1981 (Nijhoff, the Hague).

[11] *Shape controllability for free boundaries*, in "System modelling and optimization" (Copenhagen, 1983), 354–361, Lecture Notes in Control and Inform. Sci., 59, Springer-Verlag, Berlin, New York, 1984.

[12] *Numerical algorithms and existence result for a Bernoulli-like steady free boundary problem*, Large Scale Systems **6** (1984), 263–278.

[13] *Les derivées par rapport aux nœuds des triangulations en identification de domaines* [Derivatives with respect to triangularization nodes and their uses in domain identification], Ann. Sci. Math. Québec **8** (1984), 97–120.

[14] *Shape stabilization of flexible structure*, in "Distributed parameter systems" (Vorau, 1984), 446–460, Lecture Notes in Control and Inform. Sci., 75, Springer-Verlag, Berlin, 1985.

[15] *Shape variational solution for a Bernoulli like steady free boundary problem*, Distributed Parameter Systems, Lecture Notes Comput. Sci., 102, F. Kappel, K. Kunish, and W. Schappacher, eds., 333–343, Springer-Verlag, Berlin, Heidelberg, New York, 1987.

[16] *Boundary control and boundary variations*, in "Proceedings of the IFIP WG 7.2 Conference held in Nice," June 10–13, 1986, J.-P. Zolésio, ed., Lecture Notes in Comput. Sci., 100, Springer-Verlag, Berlin, New York, 1988.

[17] *Shape derivatives and shape acceleration*, in "Control of partial differential equations" (Santiago de Compostela, 1987), 309–318, Lecture Notes in Control and Inform. Sci., 114, Springer-Verlag, Berlin, New York, 1989.

[18] *Stabilization of flexible structures*, papers from the Third Working Conference held in Montpellier, January 1989, J.-P. Zolésio, ed., Lecture Notes in Control and Inform. Sci., 147, Springer-Verlag, Berlin, 1990.

[19] *Identification of coefficients with bounded variation in the wave equation*, in "Stabilization of flexible structures" (Montpellier, 1989), 248–254, Lecture Notes in Control and Inform. Sci., 147, Springer-Verlag, Berlin, 1990.

[20] *Galerkin approximation for wave equation in moving domain*, in "Stabilization of flexible structures" (Montpellier, 1989), 191–225, Lecture Notes in Control and Inform. Sci., 147, Springer-Verlag, Berlin, 1990.

[21] *Eulerian Galerkin approximation for wave equation in moving domain*, in "Progress in partial differential equations: The Metz surveys," 112–127, Pitman Res. Notes Math. Ser., 249, Longman Sci. Tech., Harlow, UK, 1991.

[22] *Shape formulation for free boundary problems with nonlinearised Bernoulli condition*, Lecture Notes in Control and Inform. Sci., 178, 362–392, Springer-Verlag, Berlin, Heidelberg, New York, 1991.

[23] *Some results concerning free boundary problems solved by domain (or shape) optimization of energy*, in "Modelling and inverse problems of control for distributed parameter systems" (Laxenburg, 1989), 161–170, Lecture Notes in Control and Inform. Sci., 154, Springer-Verlag, Berlin, 1991.

[24] *Shape optimization problems and free boundary problems*, in "Shape optimization and free boundaries," M.C. Delfour, ed., 397–457, Kluwer, Dordrecht, Boston, London, 1992.

[25] *Introduction to shape optimization problems and free boundary problems*, in "Shape optimization and free boundaries" (Montreal, PQ, 1990), 397–457, NATO Adv. Sci. Inst. Ser. C Math. Phys. Sci., 380, Kluwer, Dordrecht, 1992.

[26] *Boundary control and boundary variation*, in "Proceedings of the IFIP WG 7.2 Conference held in Sophia-Antipolis," October 15–17, 1990, J.-P. Zolésio, ed., Lecture Notes in Control and Inform. Sci., 178, Springer-Verlag, Berlin, 1992.

[**27**] *Shape formulation of free boundary problems with nonlinearized Bernoulli condition*, in "Boundary control and boundary variation" (Sophia-Antipolis, 1990), 362–392, Lecture Notes in Control and Inform. Sci., 178, Springer-Verlag, Berlin, 1992.

[**28**] *Boundary control and variation*, "Proceedings of the IFIP WG 7.2 Conference held in Sophia Antipolis," June 1992, J.-P. Zolésio, ed., Lecture Notes in Pure and Appl. Math., 163, Dekker, New York, 1994.

[**29**] *Weak shape formulation of free boundary problems*, Ann. Scuola Norm. Sup. Pisa Cl. Sci. **21** (1994), 11–44.

[**30**] *Approximation for the wave equation in a moving domain*, in "Control of partial differential equations" (Trento, 1993), 271–279, Lecture Notes in Pure and Appl. Math., 165, Dekker, New York, 1994.

[**31**] *Partial differential equation methods in control and shape analysis*, papers from the IFIP Working Group 7.2 Conference on Control and Shape Optimization held at the Scuola Normale Superiore di Pisa, Pisa, G. Da Prato and J.-P. Zolésio, eds., Lecture Notes in Pure and Appl. Math., 188, Dekker, New York, 1997.

[**32**] *Shape differential equation with a non-smooth field*, in "Computational methods for optimal design and control" (Arlington, VA, 1997), 427–460, Progr. Systems Control Theory, 24, Birkhäuser, Boston, 1998.

J.-P. Zolésio and C. Truchi

[**1**] *Shape stabilization of wave equation*, in "Boundary control and boundary variations" (Nice, 1986), 372–398, Lecture Notes in Comput. Sci., 100, Springer-Verlag, Berlin, New York, 1988.

Index of Notation

Index